Lectures on Electromagnetic Field Theory: A Comprehensive Overview

Lectures on Electromagnetic Field Theory: A Comprehensive Overview

WENG CHO CHEW

Purdue University Press, West Lafayette, Indiana

Copyright © 2025 by Weng Cho Chew. All rights reserved.

Cataloging-in-Publication Data on file at the Library of Congress.

978-1-62671-096-2 (paperback)

978-1-62671-097-9 (epdf)

DEDICATED TO THE MEMORY OF

ANDREW M. WEINER

Contents

[1]This section is mathematically complicated. It can be skipped on first reading.

[2]The derivation in this section is complex, but worth the pain, since this knowledge was not discovered until the 1960s. It is not discussed in many textbooks.

[3]May be skipped on first reading.

[4]This lecture has enough subtleties and depth that one should read it with care. It will not be until the end of the lecture book that such subtleties are encountered.

[5]This term was used by me in [1]. It seems to be used by other authors later to mean other things, but I will continue to use this term in this course.

[6]This section can be skipped on first reading.

[7]These polarizations are also variously know as TE_z, or the s and p polarizations, a descendent from the notations for acoustic waves where s and p stand for shear and pressure waves, respectively.

[8]Also known as TM_z polarization.

[9]May be skipped on first reading. This section is too complicated for a first course in graduate electromagnetics. But if you have the will, you can learn it.

Preface

This set of lecture notes is from my teaching of ECE 604 Electromagnetic Field Theory, since Fall of 2018, at Purdue University in West Lafayette, Indiana. It is intended for entry-level graduate students. Because different universities have different undergraduate requirements in electromagnetic field theory, this is a course intended to "level the playing field". From this point onward, hopefully, all students will have the fundamental background in electromagnetic field theory needed to take advanced-level courses and do research at Purdue.

In developing this course, I have drawn heavily upon knowledge of our predecessors in this area. I have listed many of the textbooks and papers used in the reference list. Being a practitioner in this field for over forty years, I have seen electromagnetic theory impacting modern technology development unabated. Despite its age, the set of Maxwell's equations has had an enduring legacy and continues to be important, from statics to optics to extreme-ultra-violet (EUV) light, from classical to quantum, and from sub-nanometer (subatomic) length scales to galactic length scales. The applications of electromagnetic technologies have also been tremendously wide-ranging, including geophysical exploration, remote sensing, electrical power systems, renewable and clean energy, biomedical engineering, optics, photonics integrated circuits, computer circuit design, computer systems, quantum computers, quantum communications, and many more. Electromagnetic field theory is not everything, but it remains an indispensable component of modern technology developments.

The challenge in teaching this course is on how to teach over 150 years of knowledge in one semester; this is, of course, mission impossible! To do this, we use the traditional wisdom of engineering education: distill the knowledge, make it as simple as possible, and teach the fundamental big ideas in one short semester. Because of this, you may find the flow of the lectures erratic and rapid. Sometimes, I feel the need to touch on certain big ideas before moving on, resulting in the choppiness of the curriculum.

Also, in this course, I exploit mathematical "homomorphism" as much as possible to simplify the teaching. After years of practicing in this area, I find that many complex and advanced concepts become simpler if mathematical homomorphism is exploited between the advanced concepts and simpler ones. An example of this is on waves in layered media. The problem is homomorphic to the transmission line problem: hence, using transmission line theory, one can simplify the derivations of some complicated formulas.

A large part of modern applied electromagnetic technologies is based on heuristics: this is difficult to teach, as it relies on physical insight and experience. Modern commercial software has reshaped this landscape: the field of math-physics modeling through numerical simulations, known as computational electromagnetics (CEM), has made rapid advances in recent years. Many cut-and-try laboratory experiments, based on heuristics, have been replaced by cut-and-try computer experiments, which are a lot cheaper.

An exciting modern development is the role of electromagnetics and Maxwell's equations in quantum technologies. We will connect Maxwell's equations to quantum electromagnetics toward the end of this course. This is a challenge, as it has never been done before in an entry-level course to my knowledge.[10]

The first vacuum tube computer, ENIAC, was built around 1945. After that, in the 1950s, a series of vacuum tube plus transistor computers were built, including the ILLIAC series at the University of Illinois. Those computers can fill a whole room. After some 70 years, with the exponential growth of nanotechnologies, we can now carry a pocket-size cell phone packed with billions of transistors. A change in *modus operandi* is that engineering designs are now done increasingly more with software to reduce cost rather than cut-and-try experiments. Thus, an important field of CEM has emerged in recent years. Virtual prototyping and digital twins rely greatly on the fidelity of simulation software: engineering designs can be done with software rather than hardware. In fact, 95 percent of a computer chip design is now done with software simulation, greatly reducing the design cost. Unfortunately, we can only spend two lectures on CEM to convey some of the big ideas to the students of electromagnetics. The devil is in the details in the implementation of these big ideas, which can be pursued in other courses.

Weng Cho Chew

January 13, 2026, Purdue University

[10]Thomas Roth is now teaching a quantum technology course to advanced undergraduates at Purdue, and the link to his lecture notes can be found at "https://nanohub.org/resources/38552".

Acknowledgements

I would like to thank Dan Jiao and the late Andy Weiner for sharing their lecture notes and their experiences for teaching this course. At the very outset, Andy Weiner was the person in charge of this course at Purdue University, but he generously and freely allowed me to teach this course unfettered. Mark Lundstrom gave me useful feedback on Lectures 38 and 39 on the quantum theory of electromagnetics. I also thank Luis Gomez for his enthusiasm in EM/CEM that we discuss almost every week on related topics. He and his students also share with me their knowledge on bioelectromagnetics, and generously devotes his time to help grow our industry consortium. Thomas Roth is avidly interested in quantum technologies, and I thank him and his students for teaching me a lot about superconducting qubit technologies. I thank many colleagues, such as the late Andy Weiner, Dan Jiao, Kevin Webb, Zubin Jacob, Peter Bermel, Minghao Qi, Shreya Sen, Peter Bermel, Luis Gomez, and Thomas Roth, who have shared their research and invited me to sit in the doctoral committee of their students from whom I learn a great deal.

Mitch Daniels and Mung Chiang, and the university leadership and colleagues, have sustained 14 years of tuition freeze. Because of that, many of my former students/postdocs are sending their children to study at Purdue. This gives us the rare mini-reunion we have at Purdue. Thanks are also due Frances Cordova for leaving in her legacy excellent gym facilities for us to exercise regularly.

At Illinois, I would like to thank Erhan Kudeki, who always takes an active interest in my writing on this subject matter. He suggested that I teach an ECE course related to quantum technologies. Farzad Kamalabadi there also showed me what the state-variable approach is in control theory. Jose Schutt-Aine shared with me his latest knowledge on high-speed interconnects with me. Wen-Mei Hwu co-supervised Mert Hidayetoglu to use advanced GPUs to solve large-scale inverse problems. Thanks also to the late Narayana Rao, while being the associate head then, let me teach almost every undergraduate ECE course at Illinois: signal processing, electric machinery, and microelectronics, in addition to electromagnetics. I also thank the late Y.T. Lo, S.W. (Andy) Lee, the late S.L. (Shun-Lien) Chuang[11] who introduced us to the Chinese community in Urbana-Champaign. Y.T. and Andy also mentored me throughout my career. The Chinese community provided us the cultural support to grow our young family, with our daughter Huibin Amelia, and son Shinen Ethan. Other colleagues in the Electromagnetics Laboratory, late Paul Mayes, Paul Klock, and Raj Mittra also taught me lots. I had interesting collaborations with Eric Michielssen, Jianming

[11]We knew Shun-Lien and his wife, Shu-Jung (Lolita) since we were graduate students at MIT. My wife also got her PhD from MIT, but due to the boundary conditions imposed by the immigration law, she decided to stay home to raise our lovely children.

Jin and many more on novel fast algorithms. The Pauls helped me set up an experimental lab at Illinois after I received some equipment donations from Schlumberger. Paul Mayes showed us, (me and my former PhD student, Bill Weedon) on how to make broadband Vivaldi antennas. With them, we could demonstrate, with Fu-Chiarng Chen, the first evidence of sub-wavelength resolution with experimental measurements. Raj Mittra showed us that with resourcefulness, one can do tons with electromagnetics.

I am thankful to people I have interacted with: Dong-Yeop Na, an immaculate thinker, for helping teach part of this course. Collaborating with Dong-Yeop allowed us to make important forays and advances in quantum electromagnetics. Thanks are also due to Purdue U for a startup package that allowed me to explore this new exciting area. Other members of our research team, Boyuan Zhang, Jie Zhu (co-supervised with Peter Bermel), Ivan Okhmatovskii (now studying with Gerhard Klimeck), Akila Murugesan, Sina Vaezi, and Masood Nekoei, in addition to their research, supported (support) this course. Thanks are to Robert H.Y. Chao, who read the lecture notes and gave me some very useful feedback; also, to Muhammad Arshad who helped with the typesetting and artwork, Rebecca Emanuelsen for her sharp-eyed copy-editing of these notes, and careful proof-reading by Ryan Wans. I had many students at Illinois who have done stellar research: I learned much about quantum technologies from Aiyin Liu, Thomas Roth, and Chris Ryu.

Thomas Roth, Dong-Yeop Na, and I, in recent years, taught short courses on quantum electromagnetics at AP-S/URSI in Singapore (2021), Denver (2022), Portland (2023), Florence, Italy (2024), and Ottawa, Canada (2025). Many researchers such as Zhen Peng, Paolo Rocca, Gabriele Gradoni, Nicola Anselmi, and Emanuel Colella also contributed to these short courses.I enjoyed learning from them at these short courses. Some of the materials for the short courses are factored into the last two lectures of these notes. I would like to thank Thomas Roth for his dogged persistence in organizing these short courses. Thanks to URSI EMTS 2025 for inviting me for a plenary lecture at the historical city of Bologna, Italy, giving me the urgency to document these lecture notes.

I am thankful to many who have invited me to give keynote talks at many international conferences such as Giuliano Manara for EMTS, Erping Li for APEMC, many ICEAA meetings by George Uslenghi, Roberto Graglia, and Don Wilton plus a few more in China, Taiwan, and Japan. I was exposed to many IMA, SIAM, and Physics meetings where I enriched my experiences. We are thankful for many companies who shared their knowledge and supported our research, now and before, such as Cadence, Siemens, ASML, ASUSTEK, Synopsys, Ansys, Intel, and Schlumberger. I thank the NSF for their support and AFOSR, ONR, and ARO for their past support.

Recently, Dezhi Wang, Zekui Jia, Luis Gomez, Chris Ryu, Masood Nekoie, and Thomas Roth helped proofread these lecture notes, especially on Lectures 38 and 39, giving feedback to make them easier to read. I am thankful to Justin Race and Katherine Purple of Purdue U Press for accommodating my schedule, and also to Chris Brannan for the cover design. Cristina Farmus and her team, especially Ashlee Messersmith for helping us put the finishing touches to this book.

When I first came to Purdue in 2017, I was tasked to teach a microelectronic course, a course that I had not taught for decades. I am thankful to Mark Lundstrom, Zhihong Chen, and Peide Ye for sharing their lecture slides on microelectronics with me. Zhihong Chen also offered her help when we co-taught the microelectronic course together.

I spent four years at Schlumberger-Doll Research after MIT, at the high point of the oil industry.

I am grateful for the knowledge I have imbibed there, meeting many top researchers and scholars from different fields, as well as working with many young researchers from top schools in New England. I was exposed to many top scholars and am thankful for their mentorship, especially from Pabitra Sen and the late Mike Ekstrom, who believed in me.

My four years as the Dean of Engineering at Hong Kong U (HKU) was rewarding. While there, I learnt a great deal about physics, chemistry, and different aspects of engineering science from Fuchun Zhang, Jian Wang, Guanhua Chen, and Hong Guo (McGill U) from the HKU science faculty. My visiting appointments earlier at NUS and NTU in Singapore, Superlec in Paris, and later at HKU, Tsinghua, and National Taiwan U, Brown U, exposed me to myriads of ideas.

We were funded by the HK RGC on "Theory, Modeling, and Simulation of Emerging Electronics" to encourage collaboration between Science and Engineering. While there, I learned a lot from my engineering colleagues, many of whom are now world-class researchers. While at HKU, I had a chance to collaborate with Lijun Jiang, SHA Wei, SUN Sheng, and a budding young scientist, Yongpin Chen, plus many more too numerous to mention. My four years of immersion into the diverse and dynamic HK culture was an unforgettable experience. I am thankful for the mentorship and support of Richard Wong and Lap-Chee Tsui.

I am indebted to my wife, Chew-Chin Phua, for her life-long support, love, patience, and understanding, especially when I was busy putting together these lecture notes. We also have to adjust by moving to Purdue U from the U of Illinois, where I have taught for 32 years.

Thanks also to Brian Johnston, a peace corps who served in Kuantan, Malaysia, who encouraged me to study at MIT. I thank my electromagnetics colleagues worldwide, especially those from AP-S/URSI, who form the community of scholars that I see almost every year. They give me inspiration and support to grow in this important area. I owe Ajay Poddar, Anisha Apte, Meisong Tong, Yahya Antar, and many more for helping with the COPE program.

I am indebted to the late Prof. Jin-Au Kong: he introduced me to the beauty of electromagnetic theory at MIT, including relativistic electromagnetic theory. He was a very spirited teacher of the undergraduate electromagnetic course at MIT as well: he was my academic advisor since I was a freshman at MIT. We (including my wife) were often invited to their home, hosted by Mrs. Kong (Wen Yuan) and his mother. Kong started the PIERS meetings and PIER Journals, which have been kept alive up to this day under the leadership of my academic brother, Prof. Leung Tsang, and also by Prof. Sailing He of Zhejiang U. I thank Sailing He for his organizational genius and resourcefulness that help make PIERS a leading meeting in our area. I am also thankful to Dr. Jiang Qin for her tireless support of the journals and the conferences. Thanks are due to Sailing He and Jiang Qin, who nurture a group of young researchers to carry on the torch for Prof. Kong. Also, Prof. Kazuya Kobayashi of Japan and Prof. Saibun Tjuatja, U Texas Arlington, have steadfastly supported PIERS all these years. Kazuya Kobayashi worked untiringly to have the emperor of Japan grace PIERS Chiba recently.

I thank God for letting us live through the age of transformation, where we witness the compounding effect of nanotechnology, the miracle of wireless communication, and many more. With a click of a button, we can now communicate daily with our relatives and friends living 10,000 miles away! They help to bring us closer to our relatives and friends in Singapore and Malaysia. I thank the pastors at WUMC and TUMC for sharing the teaching of Christ with us.

Part I

Fundamentals, Complex Media, Theorems and Principles

Lecture 1

Introduction, Maxwell's Equations

In the beginning, this field was known as either the field of electricity and magnetism or the field of optics. But later, as we shall discuss, these two fields were found to be based on the same set equations, Maxwell's equations. Maxwell's equations unified these two fields, and it is common to call the study of electromagnetic theory based on Maxwell's equations *electromagnetics*. Electromagnetics has wide-ranging applications, from statics to ultraviolet (UV) light, in the present world, with impacts on many different technologies.

1.1 Importance of Electromagnetics

In this section, we will explain why electromagnetics is so important and its impact on many different areas. Then we will give a brief history of electromagnetics and how it has evolved in the modern world. Next, we will go briefly over Maxwell's equations in their full glory. But we will begin the study of electromagnetics by focusing on static problems, which are valid in the long-wavelength limit or at zero frequency.

Electromagnetics has been based on Maxwell's equations, which are the result of the seminal work of James Clerk Maxwell and were completed in 1865, after his presentation to the British Royal Society in 1864. He was very much inspired by the experimentally motivated Coulomb's law and Gauss's law (1785), Ampère's law (1823), and Faraday's law (1835). Maxwell's equations were completed over 150 years ago now, and this is a long time compared to the leaps and bounds of progress we have made in technological advancements. Nevertheless, research in electromagnetics has continued unabated despite the field's age. The reason for this is that electromagnetics is extremely useful, as it is pervasive and has impacted a large sector of modern technologies.

To understand why electromagnetics is so useful, we have to understand first a few points about Maxwell's equations.

- Maxwell's equations are valid over a vast length scale, from subatomic dimensions to galactic dimensions. Hence, these equations are valid over a vast range of wavelengths: from static to UV wavelengths.[1]

- Maxwell's equations are relativistic invariant in the parlance of special relativity [2]. In fact, Einstein was motivated with the theory of special relativity in 1905 by Maxwell's equations [3]. These equations look the same, irrespective of what inertial reference frame[2] one is in.

- Maxwell's equations are valid in the quantum regime, as it was demonstrated by Paul Dirac in 1927 [4]. Hence, many methods of calculating the response of a medium to a classical field can be applied in the quantum regime as well. When electromagnetic theory is combined with quantum theory, the field is called quantum optics came into existance. Roy Glauber won a Nobel Prize in 2005 because of his work in this area [5]. Alain Aspect, John F. Clausser, and Anton Zeilinger won the Nobel Prize in 2022 for their works in quantum optics.

- Maxwell's equations and the pertinent gauge theory has inspired The Yang-Mills theory (1954) [6], which is also known as a generalized electromagnetic theory. Yang-Mills theory is motivated by differential forms in differential geometry [7]. To quote from Misner, Thorne, and Wheeler, "Differential forms illuminate electromagnetic theory, and electromagnetic theory illuminates differential forms" [8, 9].

- Maxwell's equations are some of the most accurate physical equations that have been validated by experiments. In 1985, Richard Feynman wrote that electromagnetic theory had been validated to one part in a billion.[3] Now, it has been validated to one part in a trillion (Aoyama et al 2012).[4]

- As a consequence, electromagnetics has permeated many technologies and has had a tremendous impact on science and technology. This is manifested in electrical engineering, optics, wireless and optical communications, computers, remote sensing, subsurface sensing, biomedical engineering, etc. It is expected that quantum electromagnetics (the quantum extension of electromagnetics) will grow in importance as quantum technologies develop.

---

[1] Current lithography process is working with using deep UV light with a wavelength of 193 nm and, subsequently, 13.5 nm.

[2] An inertial reference frame is a coordinate frame that is traveling at a velocity v.

[3] This means that if a jet is to fly from New York to Los Angeles, an error of one part in a billion means an error of a few millmeters.

[4] This means an error of a hairline, if one were to shoot a light beam from the earth to the moon.

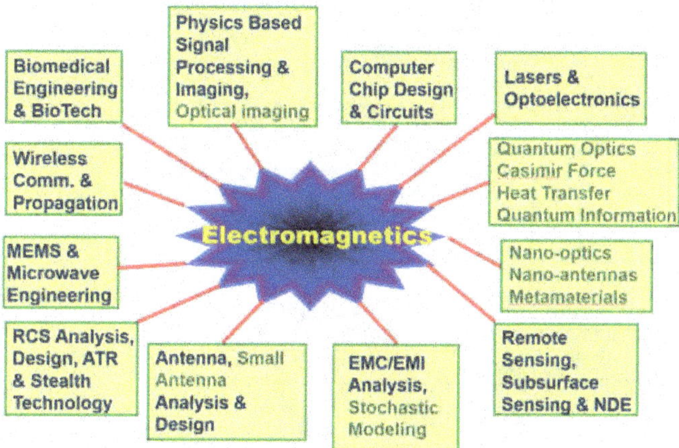

Figure 1.1: The impact of electromagnetics in many technologies. The areas in blue are prevalent areas impacted by electromagnetics some 20 years ago [10], and the areas in light brown are modern emerging areas impacted by electromagnetics.

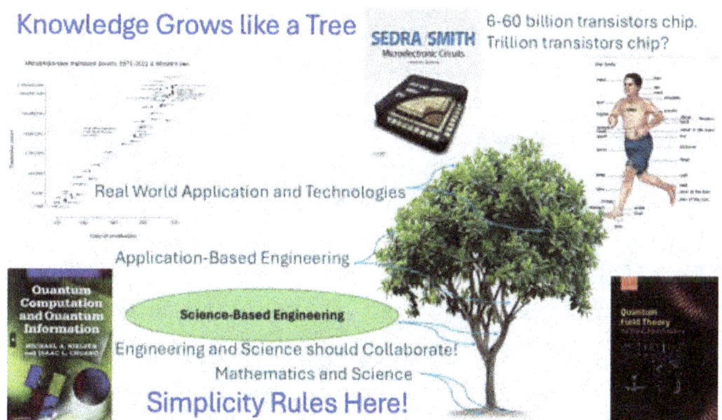

Figure 1.2: Knowledge grows like a tree. Engineering knowledge and real-world applications are driven by fundamental knowledge from math and the sciences. At a university, we do science-based engineering research that can impact wide-ranging real-world applications. But everyone is equally important in transforming our society. Just like the parts of the human body: no one can claim that one is more important than the others (1 Corinthians 12).

Figure 1.2 shows how knowledge is driven by basic math and science. Its growth is like a tree. Due to the vast ocean of knowledge that we are immersed in, it is important that we collaborate, especially between engineering and science, to develop technologies that can transform this world.

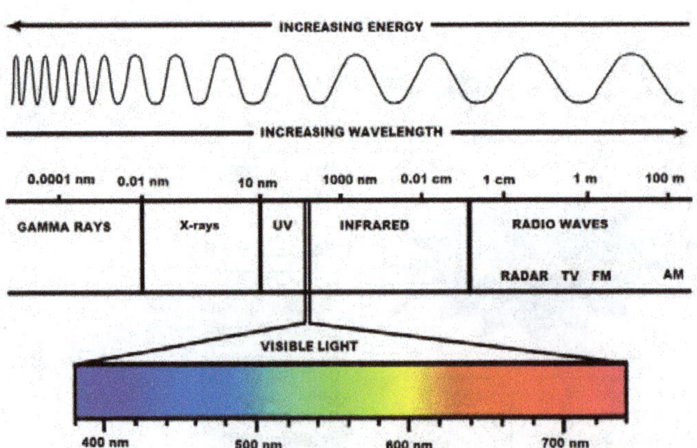

Figure 1.3: The electromagnetic spectrum goes from statics to UV. Deep UV and extreme UV (EUV) are now used in nanolithography, while statics is used in circuit theory and geophysical exploration, (Figure by Elert, Glenn, "The Electromagnetic Spectrum").

1.1.1 The Electromagnetic Spectrum

The electromagnetic field has been used from very low frequencies to very high frequencies. Electromagnetics incldes very low frequencies, ultralow frequency (ULF) <3, Hz, extremely low frequency (ELF) 3-,3000 Hz ; and very low frequency (VLF) 3 KHz-30 KHz- to probe the earth surface's, and for submarine communications because of their deeper penetration depths. The AM radio stations operating in the several hundred kilohertz range have wavelengths of several hundred meters. FM radio stations are in the range of several hundred megahertz, while TV stations operate in the several 100 MHz. Microwaves have wavelengths in the order of centimeters, and infrared light ranges from 1,000 μm to 1 μm. The visible spectrum ranges from 700 nm to 400 nm. UV light ranges from 400 nm to 100 nm, while X-ray is generally from 100 nm to 1 nm. Gamma ray is generally below 1 nm.

Historically, UV light of 193 nm was used for nanolithography while extreme UV (EUV) of 13.5 nm is now also used. X-ray is important for imaging, while gamma ray is used for some medical applications. The lights above UV are harmful to the human body.

1.1.2 A Brief History of Electromagnetics

Electricity and magnetism have been known to humankind for a long time. Also, the physical properties of light have been known. But the field of electricity and magnetism, now termed electromagnetics in the modern world, has been thought to be governed by different physical laws as opposed to those for optics. This is understandable, as the physics of electricity and magnetism is quite different from the physics of optics as they were known to humans in the past.

For example, lodestone was known to the ancient Greek and Chinese around 600 BC to 400 BC. Compasses were used in China begining around 200 BC. Static electricity was reported by the

Greeks as early as 400 BC. But these curiosities did not make an impact until the age of telegraphy. The coming about of telegraphy was due to the invention of the voltaic cell or the galvanic cell in the late 1700s, by Luigi Galvani and Alessandro Volta [11]. It was soon discovered that two pieces of wire, connected to a voltaic cell, can transmit information at a distance.

By the early 1800s, this possibility had spurred the development of telegraphy. Both Andr'e-Marie Amp'ere in 1823 [12, 13] and Michael Faraday in 1838 [14] did experiments to better understand the properties of electricity and magnetism. And hence, Ampère's law and Faraday's law were named after them. Kirchhoff's voltage and current laws were also developed in 1845 to help better understand telegraphy [15, 16]. Despite these laws, the technology of telegraphy was poorly understood at the time. For instance, it was not known why the telegraphy signal was distorted. Ideally, the signal should be a digital signal switching between ones and zeros, but the digital signal lost its shape rapidly along a telegraphy line.[5]

It was not until 1865 that James Clerk Maxwell [18] put in the missing term in Ampère's law, the displacement current term, and only then was the mathematical theory for electricity and magnetism completed. Ampère's law is now known as generalized Ampère's law. The complete set of equations are now named Maxwell's equations in honor of James Clerk Maxwell.[6]

The rousing success of Maxwell's theory was that it predicted wave phenomena as they have been observed along telegraphy lines. But it was not until 23 years later, in 1888, that Heinrich Hertz [20] did experiments to prove that an electromagnetic field can propagate through space across a room. This illustrates the historical difficulty of slow knowledge dissemination when new knowledge was discovered. Moreover, from experimental measurement of the permittivity and permeability of matter, it was decided that electromagnetic waves move at a tremendous speed. But the velocity of light has been known for a long while from astronomical observations (Roemer 1676) [21]. The interference phenomenon in light was observed in Newton's ring (1704) [22]. Newton did not publish his analysis until after Hooke's death, as part of his treatise "Opticks" published in 1704 When these pieces of information were combined together, it was decided that electricity and magnetism, and optics, were actually governed by the same physical law. or Maxwell's equations. Thus, optics and electromagnetics were unified into one field!

In Figure 1.4, a brief history of electromagnetics and optics is depicted. In the beginning, it was thought that electricity and magnetism, on the one hand and optics on the other were governed by different physical laws. Low-frequency electromagnetics was governed by the understanding of fields and their interaction with media. Optical phenomena were governed by ray optics, reflection and refraction of light. But the advent of Maxwell's equations in 1865 revealed that these fields could be unified under electromagnetic theory. Solving Maxwell's equations then became a rewarding mathematical endeavor.

---

[5]As a side note, in 1837, Samuel Morse invented the Morse code for telegraphy [17]. There was cross-pollination of ideas across the Atlantic Ocean despite the distance. In fact, Benjamin Franklin associated lightning with electricity in the latter part of the eighteenth century. Also, notice that electrical machinery was invented in 1832 even though electromagnetic theory was not fully understood.

[6]However, it was Oliver Heaviside (1850–1925) who distilled Maxwell's equations into the four equations that are found in electromagnetics textbooks today [19].

- Lode stone 400BC, Compass 200BC
- Static electricity, Greek, 400 BC
- Ampere's Law 1823;
- Faraday Law 1838;
- KCL/KVL 1845
- Telegraphy (Morse) 1837;
- Electrical machinery (Sturgeon) 1832;

- Pinhole camera, 400BC, Mozi,
- Ibn Sahl, refraction 984;
- Snell, 1621;
- Huygens/Newton 1660;
- Fresnel 1814;
- Kirchhoff 1883;

- Maxwell's equations 1864/1865;
- Heaviside, Hertz, Rayleigh, Sommerfeld, Debye, Mie, Kirchhoff, Love, Lorentz (plus many unsung heroes);
- Quantum electrodynamics 1927 (Dirac, Feynman, Schwinger, Tomonaga);
- Electromagnetic technology;

- Nano-fabrication technology;
- Single-photon measurement;
- Quantum optics/Nano-optics 1980s;
- Quantum information/Bell's theorem 1980s;
- Quantum electromagnetics/optics (coming).

Figure 1.4: A brief history of electromagnetics and optics is depicted. In the early days, it was thought that optics was a different discipline from electricity and magnetism. After 1865, the two fields were unified and governed by Maxwell's equations.

The photoelectric effect [23, 24] and Planck's radiation law [25] point to the fact that electromagnetic energy is manifested in terms of packets of energy, indicating the corpuscular nature of light. Each unit of this energy is now known as the photon. A photon carries an energy packet equal to $\hbar\omega$, where ω is the angular frequency of the photon and the Planck constant $\hbar = 6.626 \times 10^{-34}$ Js, which is a very small constant. Hence, the higher the frequency, the easier it is to detect this packet of energy, or feel the graininess of electromagnetic energy. Eventually, in 1927 [4], quantum theory was incorporated into electromagnetics, and the quantum nature of light gave rise to the field of quantum optics. Recently, even microwave photons have been measured [26, 27]. They are difficult to detect because of the low frequency of microwaves (10^9 Hz) compared to optics (10^{15} Hz); a microwave photon carries a packet of energy about a million times smaller than that of an optical photon.

The progress in nanofabrication [28] allows one to make optical components that are of a sub-optical wavelength, as the wavelength of blue light is about 450 nm.[7] As a result, interaction of light with nanoscale optical components requires the solution of Maxwell's equations in its full glory, whereas traditionally ray optics were used to describe many optical phenomena.

In the early days of quantum theory, there were two prevailing theories of quantum interpretation. Quantum measurements were found to be random. In order to explain the probabilistic nature of quantum measurements, Albert Einstein posited that a random hidden variable caused the

[7]The size of the smallest transistor now is about 5 nm, while the size of the coronavirus is about 50 to 140 nm.

random outcome of an experiment. On the other hand, the Copenhagen school of interpretation led by Niels Bohr asserted that the outcome of a quantum measurement is not known until after a measurement [29].

In the 1960s, Bell's theorem (by John Stewart Bell) [30] said that an inequality should be satisfied if Albert Einstein's hidden variable theory was correct. Otherwise, the violation of the inequality implies that the Copenhagen school of interpretation should prevail. Experimental measurement, showed that the inequality was violated, favoring the Copenhagen school of quantum interpretation [29]. This interpretation says that a quantum state is in a linear superposition of states before a measurement. but after a measurement, a quantum state "collapses" to the state that is measured. This implies that quantum information can be hidden *incognito* in a quantum state. Hence, for a quantum particle, such as a photon, its quantum state is unknown until after its measurement. In other words, quantum theory is "spooky" or "weird". This also has the profound and beautiful implication that "our karma is not written on our forehead when we are born; our future is in our own hands!" This leads to growing interest in quantum information and quantum communication using photons. Quantum technology with the use of photons, an electromagnetic quantum particle, is a subject of growing interest.

1.2 Maxwell's Equations in Integral Form

Even though experimentally motivated, Maxwell's equations can be presented as fundamental postulates.[8] We will present them in their integral forms but will not belabor them until later.

$$\oint_C \mathbf{E} \cdot d\boldsymbol{l} = -\frac{d}{dt} \iint_S \mathbf{B} \cdot d\boldsymbol{S} \qquad \text{Faraday's law (1838)} \tag{1.2.1}$$

$$\oint_C \mathbf{H} \cdot d\boldsymbol{l} = \frac{d}{dt} \iint_S \mathbf{D} \cdot d\boldsymbol{S} + I \qquad \text{Ampère's law (1823)} \tag{1.2.2}$$

$$\oiint_S \mathbf{D} \cdot d\boldsymbol{S} = Q \qquad \text{Gauss's or Coulomb's law (1785)} \tag{1.2.3}$$

$$\oiint_S \mathbf{B} \cdot d\boldsymbol{S} = 0 \qquad \text{Gauss's law (1835)} \tag{1.2.4}$$

The units of the basic quantities above are given as:

E: V/m	**H**: A/m
D: C/m^2	**B**: W/m^2
I: A	Q: C

where V = volts, A = amperes, C = coulombs, and W = webers.

[8]Postulates in physics are similar to axioms in mathematics. They are assumptions that need not be proved.

In this course, we use a boldface to denote a vector, and a hat to denote a unit vector. Hence, a vector can be written as $\mathbf{E} = \hat{x}E_x + \hat{y}E_y + \hat{z}E_z$. where $\hat{x}$, $\hat{y}$, and $\hat{z}$ are unit vectors in Cartesian coordinates. In some books, alternatively, a vector is written as $\mathbf{E} = (E_x, E_y, E_z)$.

Before we close this section, it is to be noted that (1.2.3) and (1.2.4) are derivable (1.2.1) and (1.2.2) for time-varying problems. Therefore, for time-varying problems, not all four of Maxwell's equations are independent of each other. They are only independent of each other for electrostatic problems. We will verify this fact later.

1.3 Static Electromagnetics

In statics, the field is assumed to be non-time-varying. Hence all the time dependence terms can be removed from Maxwell's equations, and we have

$$\oint_C \mathbf{E} \cdot d\boldsymbol{l} = 0 \qquad \text{Faraday's law} \tag{1.3.1}$$

$$\oint_C \mathbf{H} \cdot d\boldsymbol{l} = I \qquad \text{Ampère's law} \tag{1.3.2}$$

$$\oiint_S \mathbf{D} \cdot d\boldsymbol{S} = Q \qquad \text{Gauss's or Coulomb's law} \tag{1.3.3}$$

$$\oiint_S \mathbf{B} \cdot d\boldsymbol{S} = 0 \qquad \text{Gauss's law} \tag{1.3.4}$$

The first equation above, which is the static form of Faraday's law, also gives rise to Kirchhoff's voltage law. The second equation is the original form of Ampère's law where displacement current was ignored. The third and the fourth equations remain unchanged compared to the time-varying (dynamic) form of Maxwell's equations.

1.3.1 Coulomb's Law (Statics)

This law, developed in 1785 [31], expresses the force between two charges q_1 and q_2. If these charges are positive, the force is repulsive, and it is given by

$$\mathbf{f}_{1\to 2} = \frac{q_1 q_2}{4\pi\varepsilon r^2}\hat{\mathbf{r}}_{12} \tag{1.3.5}$$

where the units are:

$\mathbf{f}$ (force): newton

q (charge): coulomb

ε (permittivity): farad/meter

r (distance between q_1 and q_2): meter

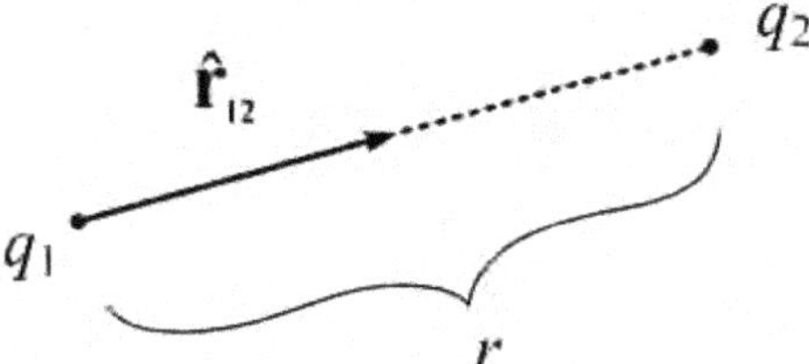

Figure 1.5: The force between two charges q_1 and q_2. The force is repulsive if the two charges have the same sign.

$\hat{\mathbf{r}}_{12}$ = unit vector pointing from charge 1 to charge 2

$$\hat{\mathbf{r}}_{12} = \frac{\mathbf{r}_2 - \mathbf{r}_1}{|\mathbf{r}_2 - \mathbf{r}_1|}, \qquad r = |\mathbf{r}_2 - \mathbf{r}_1| \tag{1.3.6}$$

Using the definition for unit vector, the force between two charges can also be rewritten as

$$\mathbf{f}_{1\to 2} = \frac{q_1 q_2 (\mathbf{r}_2 - \mathbf{r}_1)}{4\pi\varepsilon|\mathbf{r}_2 - \mathbf{r}_1|^3}, \qquad (\mathbf{r}_1, \mathbf{r}_2 \text{ are position vectors}) \tag{1.3.7}$$

1.3.2 Electric Field (Statics)

The electric field $\mathbf{E}$ is defined as the force per unit charge [32]. For two charges, one of charge q and the other one of incremental charge Δq, the force between the two charges, according to Coulomb's law (1.3.5), is

$$\mathbf{f} = \frac{q\Delta q}{4\pi\varepsilon r^2}\hat{\mathbf{r}} \tag{1.3.8}$$

where $\hat{\mathbf{r}}$ is a unit vector pointing from charge q to the incremental charge Δq. Then the electric field $\mathbf{E}$, which is the force per unit charge, is given by

$$\mathbf{E} = \frac{\mathbf{f}}{\Delta q} = \frac{q}{4\pi\varepsilon r^2}\hat{\mathbf{r}}, \qquad (\text{V/m}) \tag{1.3.9}$$

Therefore, in general, the electric field $\mathbf{E}(\mathbf{r})$ at location $\mathbf{r}$ from a point charge q at $\mathbf{r}'$, using the definition for $\hat{\mathbf{r}}$, is given by

$$\mathbf{E}(\mathbf{r}) = \frac{q(\mathbf{r} - \mathbf{r}')}{4\pi\varepsilon|\mathbf{r} - \mathbf{r}'|^3} \tag{1.3.10}$$

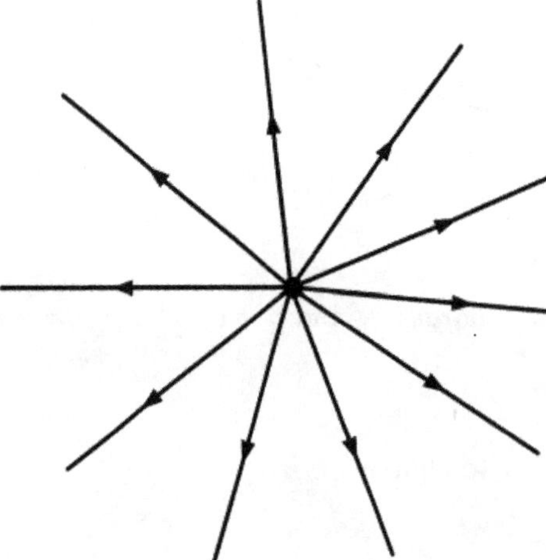

Figure 1.6: Emanating **E** field from an electric point charge as depicted by (1.3.9) and (1.3.10). The physical meaning of the figure is that, a positive charge q is placed at the center.

If one knows **E** due to a point charge , one will know **E** due to any charge distribution because any charge distribution can be decomposed into the sum of point charges. For instance, if there are N point charges each with amplitude q_i, then by the principle of linear superposition, assuming that linearity holds, the total field produced by these N charges is

$$\mathbf{E}(\mathbf{r}) = \sum_{i=1}^{N} \frac{q_i(\mathbf{r} - \mathbf{r}_i)}{4\pi\varepsilon|\mathbf{r} - \mathbf{r}_i|^3} \tag{1.3.11}$$

where $q_i = \varrho(\mathbf{r}_i)\Delta V_i$ is the incremental charge at $\mathbf{r}_i$ enclosed in the volume ΔV_i. In the continuum limit, the above becomes

$$\mathbf{E}(\mathbf{r}) = \int_V \frac{\varrho(\mathbf{r}')(\mathbf{r} - \mathbf{r}')}{4\pi\varepsilon|\mathbf{r} - \mathbf{r}'|^3} dV \tag{1.3.12}$$

In other words, the total field, by the principle of linear superposition, is the integral summation of the contributions from the distributed charge density $\varrho(\mathbf{r})$.

1.3.3 Gauss's Law for Electric Flux (Statics)

This law is also known as Coulomb's law, as the two are closely related to each other. Apparently, this simple law was first expressed by Joseph-Louis Lagrange in 1773 [33] and later reexpressed by Gauss in 1813 (Wikipedia) This law can be expressed as

$$\oiint_S \mathbf{D} \cdot d\mathbf{S} = Q \tag{1.3.13}$$

where $\mathbf{D}$ is electric flux density with unit C/m^2 and $\mathbf{D} = \varepsilon \mathbf{E}$, $d\mathbf{S}$ is an incremental surface at the point on S given by $dS\hat{n}$ where $\hat{n}$ is the unit normal pointing outward away from the surface, and Q is the total charge enclosed by the surface S.

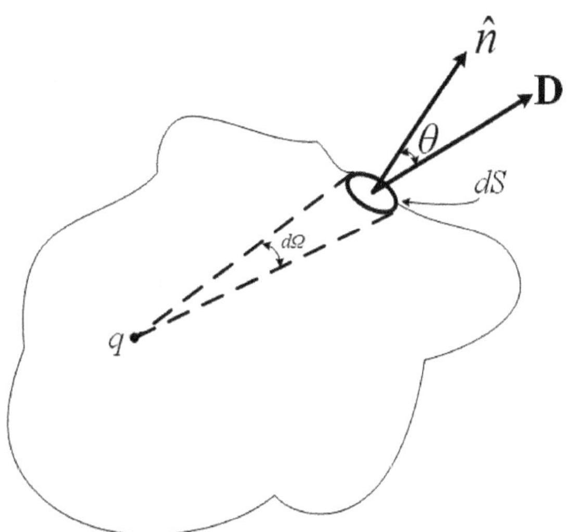

Figure 1.7: Electric flux through an incremental surface dS where $\hat{n}$ is the unit normal and $\mathbf{D}$ is the electric flux density passing through the incremental surface.

The left-hand side of (1.3.13) represents a surface integral over a closed surface S. To understand it, one can break the surface into a sum of incremental surfaces ΔS_i, with a local unit normal $\hat{n}_i$ associated with it. The surface integral can then be approximated by a summation

$$\oiint_S \mathbf{D} \cdot d\mathbf{S} \approx \sum_i \mathbf{D}_i \cdot \hat{n}_i \Delta S_i = \sum_i \mathbf{D}_i \cdot \Delta \mathbf{S}_i \tag{1.3.14}$$

where one has defined the incremental surface $\Delta \mathbf{S}_i = \hat{n}_i \Delta S_i$. In the limit when ΔS_i becomes infinitesimally small, the summation becomes a surface integral.

1.3.4 Gauss's Law from Coulomb's Law (Statics)

From Coulomb's law, the ensuing electric field due to a point charge and the electric flux is

$$\mathbf{D} = \varepsilon \mathbf{E} = \frac{q}{4\pi r^2}\hat{\mathbf{r}} \tag{1.3.15}$$

When a closed spherical surface S is drawn around the point charge q, by symmetry, the electric flux though every point of the surface is the same. Moreover, the normal vector $\hat{n}$ on the surface is just $\hat{\mathbf{r}}$. Consequently, $\mathbf{D} \cdot \hat{n} = \mathbf{D} \cdot \hat{\mathbf{r}} = q/(4\pi r^2)$, which is a constant on a spherical of radius r. Hence, we conclude that for a point charge q and the pertinent electric flux $\mathbf{D}$ that it produces on a spherical surface satisfies,

$$\oiint_S \mathbf{D} \cdot d\mathbf{S} = 4\pi r^2 \mathbf{D} \cdot \hat{n} = 4\pi r^2 D_r = q \tag{1.3.16}$$

Therefore, Gauss's law is satisfied by a point charge.

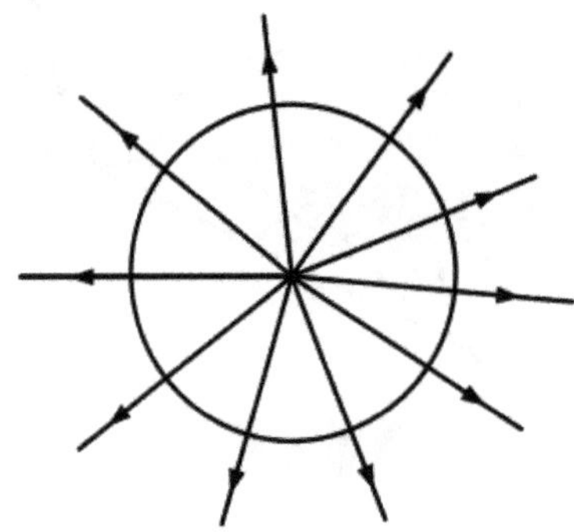

Figure 1.8: Electric flux from a point charge satisfies Gauss's law.

Even when the shape of the spherical surface S is distorted from a sphere to an arbitrary shape surface S, it can be shown that the total flux through S is still q. In other words, the total flux through sufaces S_1 and S_2 in Figure 1.9 are the same.

This can be appreciated by taking a sliver of the angular sector as shown in Figure 1.10. Here, ΔS_1 and ΔS_2 are two incremental surfaces intercepted by this sliver of angular sector. The amount of flux passing through this incremental surface is given by $d\mathbf{S} \cdot \mathbf{D} = \hat{n} \cdot \mathbf{D}\Delta S = \hat{n} \cdot \hat{\mathbf{r}} D_r \Delta S$. Here, $\mathbf{D} = \hat{\mathbf{r}} D_r$ is pointing in the $\hat{\mathbf{r}}$ direction. In ΔS_1, $\hat{n}$ is pointing in the $\hat{\mathbf{r}}$ direction. But in ΔS_2, the incremental area has been enlarged by that $\hat{n}$ not aligned with $\mathbf{D}$. But this enlargement is compensated by $\hat{n} \cdot \hat{\mathbf{r}}$. Also, ΔS_2 has grown bigger, but the flux at ΔS_2 has grown smaller by the ratio of $(r_2/r_1)^2$. Finally, the two fluxes are equal in the limit that the sliver of angular sector becomes infinitesimally small. This proves the assertion that the total fluxes through S_1 and S_2 are equal. Since the total flux from a point charge q through a closed surface is independent of its shape but always equal to q, if we have a total charge Q, it can be expressed as the sum of point

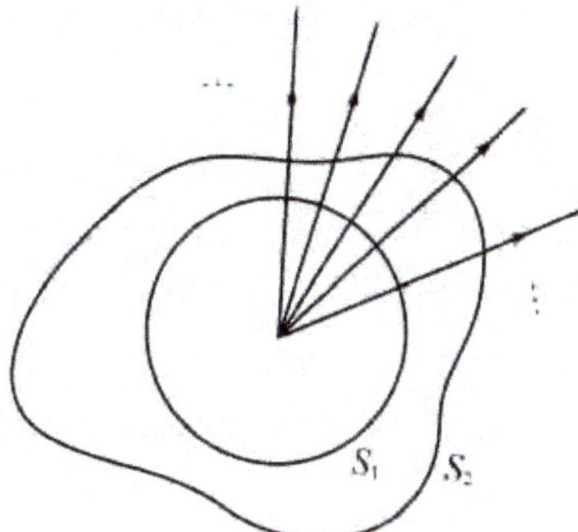

Figure 1.9: Same amount of electric flux from a point charge passes through two surfaces S_1 and S_2. This allows Gauss's law for electric flux to be derivable from Coulomb's law for statics.

charges, namely:

$$Q = \sum_i q_i \tag{1.3.17}$$

Then the total flux through a closed surface equals the total charge enclosed by it, which is the statement of Gauss's law or Coulomb's law.

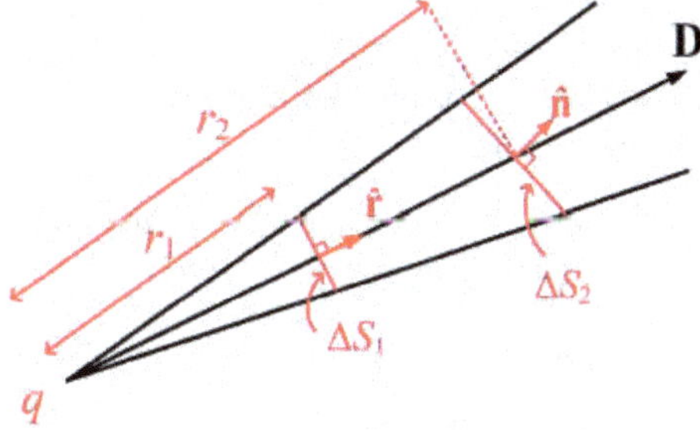

Figure 1.10: When a sliver of angular sector is taken, the same amount of electric flux from a point charge passes through two incremental surfaces ΔS_1 and ΔS_2 at different distances from the point charge.

Exercises for Lecture 1

Problem 1-1:

(i) Explain why the electric flux going through ΔS_1 and ΔS_2 are the same in Figure 1.10. To make it simpler, you can explain for the 2D case.

(ii) Find the field due to a ring of charges with line charge density ϱ C/m as shown in the figure. **Hint:** Use symmetry.

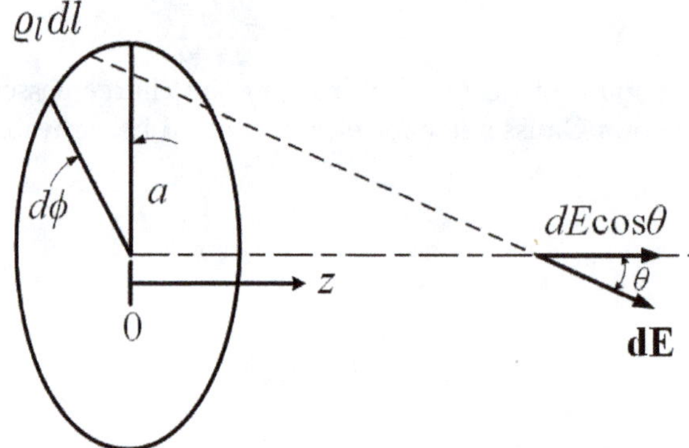

Figure 1.11: Electric field of a ring of charge (Figure by Ramo, Whinnery, and Van Duzer [34]).

(iii) What is the electric field between coaxial cylinders with surface charge densities of ϱ_s and $-\varrho_s$ of unit length in a coaxial cable?

Hint: Use symmetry and cylindrical coordinates to express $\mathbf{E} = \hat{\rho}E_\rho$ and apply Gauss's law.

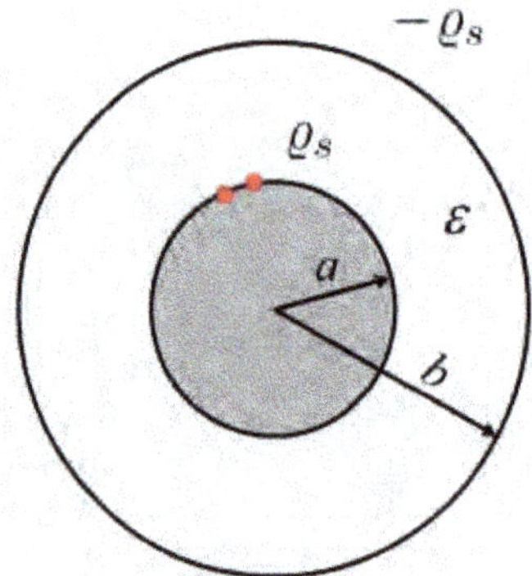

Figure 1.12: Figure for Problem 1-1 for a coaxial cylinder.

(iv) The figure shows a sphere of uniform volume charge density ϱ. Find the electric field $\mathbf{E}$ as a function of distance r from the center of the sphere. **Hint:** Again, use symmetry and spherical coordinates to express $\mathbf{E} = \hat{r}E_r$ and apply Gauss's law.

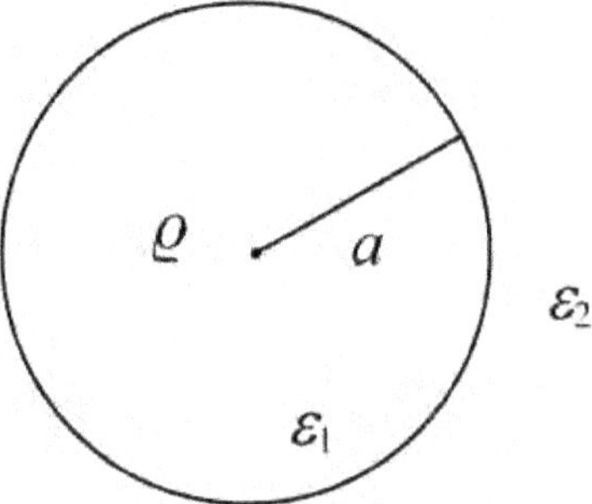

Figure 1.13: Figure for Problem 1-1 for a sphere with uniform volume charge density ϱ.

(v) Given an infinitely long cylindrical circular wire carrying a DC current I, find the magnetic field around the wire using symmetry argument and Ampère's law.

Lecture 2

Maxwell's Equations, Derivative Form

Maxwell's equations were originally written in integral form, as was shown in the previous lecture. Integral forms have nice physical meaning and can be easily related to experimental measurements. However, the differential-operator form[1] can be easily converted to differential equations or partial differential equations where a whole sleuth of mathematical methods and numerical methods can be deployed. Therefore, it is prudent to derive the differential-operator form of Maxwell's equations.

2.1 Gauss's Divergence Theorem

We will first prove Gauss's divergence theorem.[2] The divergence theorem is one of the most important theorems in vector calculus [34, 35, 36, 37]. It says that:

$$\iiint_V dV \nabla \cdot \mathbf{D} = \oiint_S \mathbf{D} \cdot d\mathbf{S} \tag{2.1.1}$$

The right-hand side of the above is the total electric flux $\mathbf{D}$ that comes out of the surface S. In the above, assuming that $V \to \Delta V$, an infinitesimal volume, then $\nabla \cdot \mathbf{D}$ is defined as

$$\nabla \cdot \mathbf{D} = \lim_{\Delta V \to 0} \frac{\oiint_{\Delta S} \mathbf{D} \cdot d\mathbf{S}}{\Delta V} \tag{2.1.2}$$

The above can be used to derive the explicit form of $\nabla \cdot \mathbf{D}$, as shall be shown later. The above implies that the divergence of the electric flux $\mathbf{D}$, or $\nabla \cdot \mathbf{D}$, is given by first computing the flux

[1]We caution ourselves not to use the term "differential forms" which has a different meaning used in differential geometry for another form of Maxwell's equations.

[2]Named after Carl Friedrich Gauss, a precocious genius who lived between 1777 and 1855.

coming (or oozing) out of a small volume ΔV surrounded by a small surface ΔS and taking their ratio as shown on the right-hand side of the above. As shall be shown, the ratio has a finite limit and, eventually, we will find a simplified expression for it. We know that if $\Delta V \approx 0$ or small, then the above implies that

$$\Delta V \nabla \cdot \mathbf{D} \approx \oiint_{\Delta S} \mathbf{D} \cdot d\mathbf{S} \tag{2.1.3}$$

First, we assume that a volume V has been discretized[3] into a sum of small cuboids, where the i-th cuboid has a volume of ΔV_i as shown in Figure 2.1. Then

$$V \approx \sum_{i=1}^{N} \Delta V_i \tag{2.1.4}$$

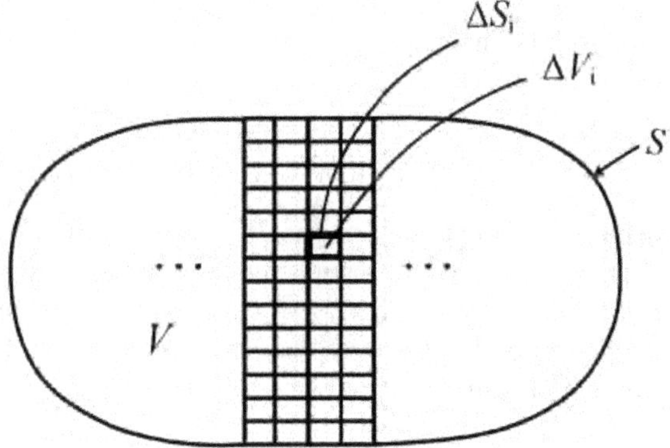

Figure 2.1: The discretization of a volume V into a sum of small volumes ΔV_i, each of which is a small cuboid. a staircasing error occurs near the boundary of the volume V, but the error diminishes as $\Delta V_i \to 0$.

[3]Other terms used are "tesselated", "meshed", or "gridded".

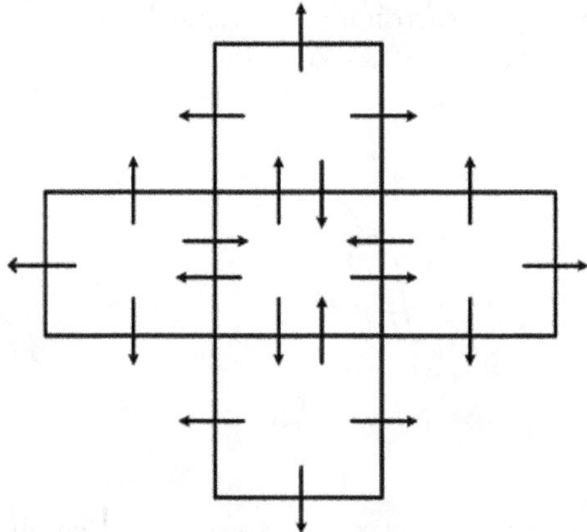

Figure 2.2: Fluxes from adjacent cuboids cancel each other, leaving only the fluxes at the boundary, that remain uncanceled. Please imagine that there is a third dimension of the cuboids in this picture where it comes out of the paper.

Then from (2.1.2) and (2.1.3), for the i-th cuboid,

$$\Delta V_i \nabla \cdot \mathbf{D}_i \approx \oiint_{\Delta S_i} \mathbf{D}_i \cdot d\mathbf{S}_i \tag{2.1.5}$$

By summing the above over all the cuboids, or over i, one gets

$$\sum_i \Delta V_i \nabla \cdot \mathbf{D}_i \approx \sum_i \oiint_{\Delta S_i} \mathbf{D}_i \cdot d\mathbf{S}_i \approx \oiint_S \mathbf{D} \cdot d\mathbf{S} \tag{2.1.6}$$

The last approximation follows, because it is easily seen that the fluxes out of the inner surfaces of the cuboids cancel each other, leaving only fluxes flowing out of the cuboids at the edge of the volume V, as explained in Figure 2.2. The right-hand side of the above equation (2.1.6) becomes a surface integral over the surface S except for the staircasing approximation (see Figure 2.1). However, this approximation becomes increasingly accurate as $\Delta V_i \to 0$. Moreover, the left-hand side of (2.1.6) becomes a volume integral, and we have

$$\iiint_V dV \nabla \cdot \mathbf{D} = \oiint_S \mathbf{D} \cdot d\mathbf{S} \tag{2.1.7}$$

The above is the well-known Gauss's divergence theorem.

2.1.1 Some Details

Next, we will derive the details of the definition embodied in (2.1.2). To this end, we evaluate the numerator of the right-hand side carefully, in accordance to Figure 2.3.

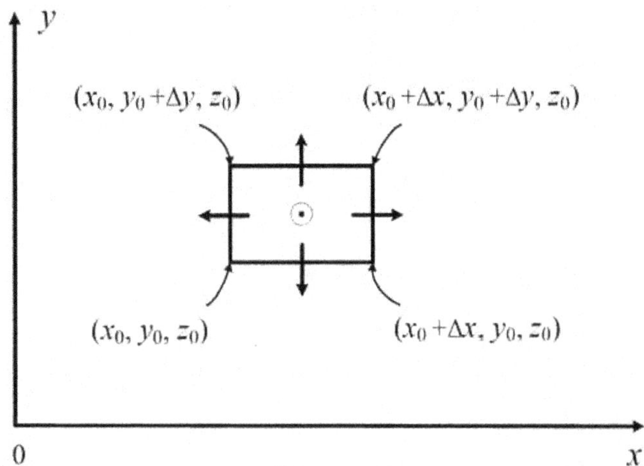

Figure 2.3: Illustration of the calculation of fluxes from a small cuboid where a corner of the cuboid is located at (x_0, y_0, z_0). There is a third z dimension of the cuboid not shown and coming out of the paper. Hence, this cuboid, unlike that shown in the figure, has six faces.

Accounting for the fluxes going through all of the six faces and assigning the appropriate signs in accordance with the fluxes leaving and entering the cuboid, one arrives at the following six terms.

$$\begin{aligned}\oiint_{\Delta S} \mathbf{D}\cdot d\mathbf{S} \approx & -\mathrm{D}_x(x_0,y_0,z_0)\Delta y\Delta z + \mathrm{D}_x(x_0+\Delta x, y_0, z_0)\Delta y \Delta z \\ & -\mathrm{D}_y(x_0,y_0,z_0)\Delta x\Delta z + \mathrm{D}_y(x_0, y_0+\Delta y, z_0)\Delta x\Delta z \\ & -\mathrm{D}_z(x_0,y_0,z_0)\Delta x\Delta y + \mathrm{D}_z(x_0,y_0,z_0+\Delta z)\Delta x\Delta y \end{aligned} \tag{2.1.8}$$

Factoring out the volume of the cuboid $\Delta V = \Delta x \Delta y \Delta z$ in the above, one gets

$$\begin{aligned}\oiint_{\Delta S} \mathbf{D}\cdot d\mathbf{S} \approx \Delta V \{ & \left[D_x(x_0+\Delta x, \ldots) - D_x(x_0,\ldots)\right]/\Delta x \\ & + \left[D_y(\ldots, y_0+\Delta y, \ldots) - D_y(\ldots, y_0, \ldots)\right]/\Delta y \\ & + \left[D_z(\ldots, z_0+\Delta z) - D_z(\ldots, z_0)\right]/\Delta z\} \end{aligned} \tag{2.1.9}$$

Or that

$$\frac{\oiint \mathbf{D}\cdot d\mathbf{S}}{\Delta V} \approx \frac{\partial D_x}{\partial x} + \frac{\partial D_y}{\partial y} + \frac{\partial D_z}{\partial z} \tag{2.1.10}$$

In the limit when $\Delta V \to 0$, then

$$\lim_{\Delta V\to 0} \frac{\oiint \mathbf{D}\cdot d\mathbf{S}}{\Delta V} = \frac{\partial D_x}{\partial x} + \frac{\partial D_y}{\partial y} + \frac{\partial D_z}{\partial z} = \nabla\cdot\mathbf{D} \tag{2.1.11}$$

where (2.1.2) has been used for the above definition of $\nabla \cdot \mathbf{D}$. Furthermore,

$$\nabla = \hat{x}\frac{\partial}{\partial x} + \hat{y}\frac{\partial}{\partial y} + \hat{z}\frac{\partial}{\partial z} \tag{2.1.12}$$

$$\mathbf{D} = \hat{x}D_x + \hat{y}D_y + \hat{z}D_z \tag{2.1.13}$$

The above is the definition of the divergence operator in Cartesian coordinates. The divergence operator $\nabla\cdot$ has its complicated representations in cylindrical and spherical coordinates, a subject that we would not delve into in this course. But they can be derived, and are best looked up at the back of some textbooks on electromagnetics [37].

Consequently, one obtains the Gauss's divergence theorem given by

$$\iiint_V dV \nabla \cdot \mathbf{D} = \oiint_S \mathbf{D} \cdot d\mathbf{S} \tag{2.1.14}$$

2.1.2 Physical Meaning of Divergence Operator

The physical meaning of divergence is that if $\nabla \cdot \mathbf{D} \neq 0$ at a point in space, it is implied that there are fluxes oozing or exuding from that point in space [38]. On the other hand, if $\nabla \cdot \mathbf{D} = 0$, it implies no net flux oozing out from that point in space. In other words, whatever flux goes into the point must come out of it. The flux is then termed divergence free. Thus, $\nabla \cdot \mathbf{D}$ is a measure of how many sources or sinks exist for the flux at a point. The sum of these sources or sinks gives the amount of flux leaving or entering the surface that surrounds the sources or sinks.

Moreover, if one integrates a divergence-free flux over a volume V and invokes Gauss's divergence theorem, one gets

$$\oiint_S \mathbf{D} \cdot d\mathbf{S} = 0 \tag{2.1.15}$$

In such a scenerio, whatever flux enters the surface S must leave it. In other words, what comes in must go out of the volume V, or that flux is conserved. This is true of incompressible fluid flow [39], electric flux flow in a source free region, and magnetic flux flow, where the flux is conserved.

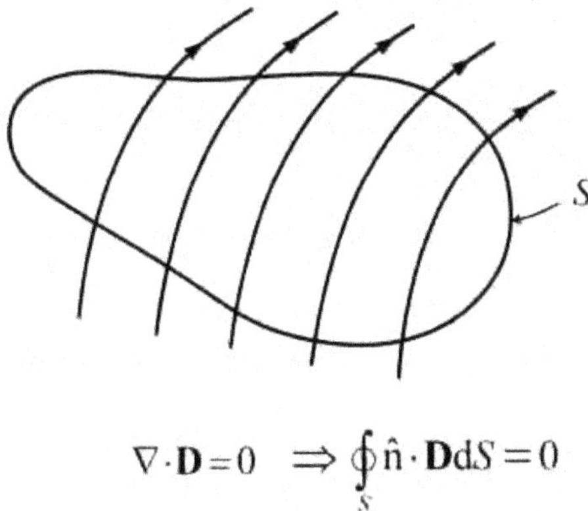

Figure 2.4: In an incompressible flux flow, flux is conserved: whatever flux enters a volume V must leave the volume V.

2.1.3 Gauss's Law in Derivative Form

By further using Gauss's or Coulomb's law, it is implied that

$$\oiint_S \mathbf{D} \cdot d\mathbf{S} = Q = \iiint dV \varrho \tag{2.1.16}$$

We can replace the left-hand side of the above by (2.1.14) to arrive at

$$\iiint_V dV \nabla \cdot \mathbf{D} = \iiint_V dV \varrho \tag{2.1.17}$$

When $V \to 0$, we arrive at the pointwise relationship, a relationship at an arbitrary point in space. Therefore,

$$\nabla \cdot \mathbf{D} = \varrho \tag{2.1.18}$$

2.2 Stokes's Theorem

The mathematical description of fluid flow was well established before the establishment of electromagnetic theory [40]. Hence, much mathematical description of electromagnetic theory uses the language of fluid. In mathematical notations, Stokes's theorem is[4].

$$\oint_C \mathbf{E} \cdot d\mathbf{l} = \iint_S \nabla \times \mathbf{E} \cdot d\mathbf{S} \tag{2.2.1}$$

[4]Named after George Gabriel Stokes, who lived between 1819 and 1903

In the above, the contour C is a closed contour, whereas the surface S is not closed.[5]

From the above definition, we can derive an explicit form for $\nabla \times \mathbf{E}$. First, applying Stokes's theorem to a small surface ΔS, we define a curl operator[6] $\nabla\times$ at a point to be measured as

$$(\nabla \times \mathbf{E}) \cdot \hat{n} = \lim_{\Delta S \to 0} \frac{\oint_{\Delta C} \mathbf{E} \cdot d\mathbf{l}}{\Delta S} \tag{2.2.2}$$

In the above, $\mathbf{E}$ is a force per unit charge, and $\nabla \times \mathbf{E}$ is a vector. The above can be viewed as the definition of $\nabla \times \mathbf{E}$. Taking $\oint_{\Delta C} \mathbf{E} \cdot d\mathbf{l}$ as a measure of the torque or rotation of the field $\mathbf{E}$ around a small loop ΔC, the ratio of this rotation to the area of the loop ΔS has a limit when ΔS becomes infinitesimally small. This ratio is related to $(\nabla \times \mathbf{E}) \cdot \hat{n}$ where $\hat{n}$ is a unit normal to the surface ΔS. As in angular momentum, the direction of the torque is along the rotation axis of the force.

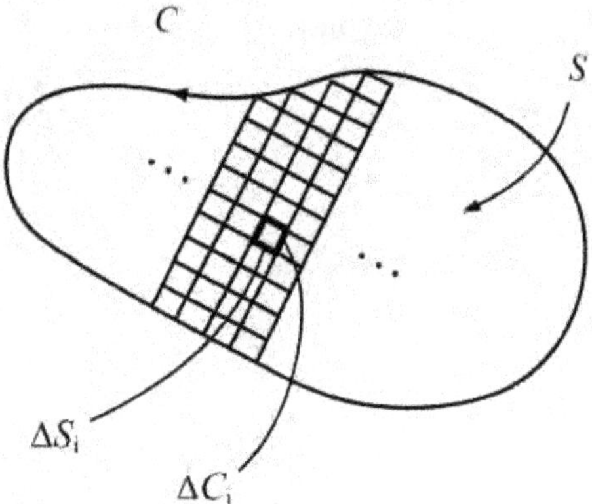

Figure 2.5: In proving Stokes's theorem, a closed contour C is assumed to enclose an open flat surface S. Then the surface S is tessellated into the sum of small rects (rectangles) as shown. The staircasing error at the boundary C vanishes in the limit when the rects are made vanishingly small.

First, the flat surface S enclosed by C is tessellated (also referred to as meshed, gridded, or discretized) into the sum of small rects (rectangles), as shown in Figure 2.5. Then (2.2.2) is applied to one of these small rects to arrive at

$$\oint_{\Delta C_i} \mathbf{E}_i \cdot d\mathbf{l}_i = (\nabla \times \mathbf{E}_i) \cdot \Delta \mathbf{S}_i \tag{2.2.3}$$

where one defines $\Delta \mathbf{S}_i = \hat{n}\Delta S$. Next, we sum the above equation over i or over all the small rects

[5]In other words, C has no boundary whereas S has boundary. A closed surface S has no boundary like when we were proving Gauss's divergence theorem previously.

[6]Sometimes called a rotation operator.

to arrive at

$$\sum_i \oint_{\Delta C_i} \mathbf{E}_i \cdot d\mathbf{l}_i = \sum_i \nabla \times \mathbf{E}_i \cdot \Delta \mathbf{S}_i \tag{2.2.4}$$

Again, on the left-hand side of the above, all the contour integrals over the small rects cancel each other internal to S, save for those on the boundary. In the limit when $\Delta S_i \to 0$, the left-hand side becomes a contour integral over the larger contour C, and the right-hand side becomes a surface integral over S. One arrives at Stokes's theorem, which is

$$\oint_C \mathbf{E} \cdot d\mathbf{l} = \iint_S (\nabla \times \mathbf{E}) \cdot d\mathbf{S} \tag{2.2.5}$$

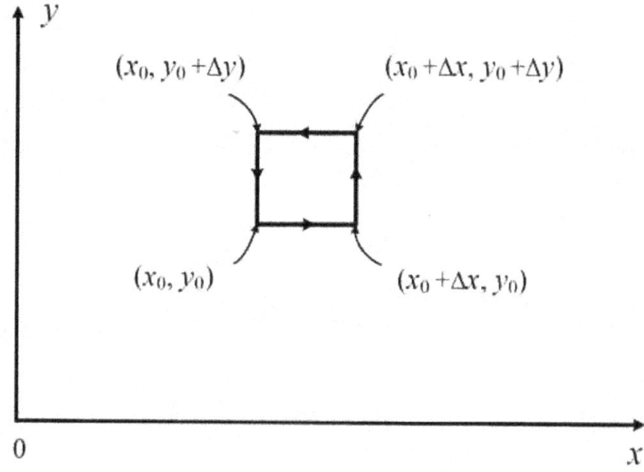

Figure 2.6: An approximation of the integration over a small rect. There are four edges to this small rect.

As of this point, $\nabla \times \mathbf{E}$ is not defined explicitly. Hence, we next need to prove the details of definition (2.2.2) using Figure 2.6. Performing the integral over the small rect, one gets

$$\begin{aligned}\oint_{\Delta C} \mathbf{E} \cdot d\mathbf{l} &= E_x(x_0, y_0, z_0)\Delta x + E_y(x_0 + \Delta x, y_0, z_0)\Delta y \\ &\quad - E_x(x_0, y_0 + \Delta y, z_0)\Delta x - E_y(x_0, y_0, z_0)\Delta y \\ &= \Delta x \Delta y \left(\frac{E_x(x_0, y_0, z_0)}{\Delta y} - \frac{E_x(x_0, y_0 + \Delta y, z_0)}{\Delta y} \right. \\ &\quad \left. - \frac{E_y(x_0, y_0, z_0)}{\Delta x} + \frac{E_y(x_0 + \Delta x, y_0, z_0)}{\Delta x} \right)\end{aligned} \tag{2.2.6}$$

We have picked the normal to the incremental surface ΔS to be $\hat{z}$ in the above example, and hence, the above gives rise to the identity that

$$\lim_{\Delta S \to 0} \frac{\oint_{\Delta S} \mathbf{E} \cdot d\mathbf{l}}{\Delta S} = \frac{\partial}{\partial x} E_y - \frac{\partial}{\partial y} E_x = \hat{z} \cdot \nabla \times \mathbf{E} \tag{2.2.7}$$

Picking different ΔS with different orientations and normals $\hat{n}$ where $\hat{n} = \hat{x}$ or $\hat{n} = \hat{y}$, one gets

$$\frac{\partial}{\partial y} E_z - \frac{\partial}{\partial z} E_y = \hat{x} \cdot \nabla \times \mathbf{E} \tag{2.2.8}$$

$$\frac{\partial}{\partial z} E_x - \frac{\partial}{\partial x} E_z = \hat{y} \cdot \nabla \times \mathbf{E} \tag{2.2.9}$$

The above gives the x, y, and z components of $\nabla \times \mathbf{E}$. It is to be noted that $\nabla \times \mathbf{E}$ is a vector. In other words, one gets

$$\begin{aligned} \nabla \times \mathbf{E} = \hat{x} \left(\frac{\partial}{\partial y} E_z - \frac{\partial}{\partial z} E_y \right) + \hat{y} \left(\frac{\partial}{\partial z} E_x - \frac{\partial}{\partial x} E_z \right) \\ + \hat{z} \left(\frac{\partial}{\partial x} E_y - \frac{\partial}{\partial y} E_x \right) \end{aligned} \tag{2.2.10}$$

where

$$\nabla = \hat{x} \frac{\partial}{\partial x} + \hat{y} \frac{\partial}{\partial y} + \hat{z} \frac{\partial}{\partial z} \tag{2.2.11}$$

The above gives an explicit formula for $\nabla \times \mathbf{E}$ in Cartesian coordinates.

2.2.1 Physical Meaning of the Curl Operator

The curl operator $\nabla \times$ is a measure of the rotation, the torque, or the circulation of a field at a point in space.[7] On the other hand, $\oint_{\Delta C} \mathbf{E} \cdot d\mathbf{l}$ is a measure of the circulation of the field $\mathbf{E}$ around the loop formed by C. To see if a field has a nonzero curl, one can imagine a paddle wheel placed in such a field. If the field is uniform, the paddle wheel will not rotate implying that the curl of a uniform field is zero. However, if the field is varying in the z direction, then the paddle wheel will rotate, implying that a nonuniform field has nonzero curl.

[7] In many old textbooks, the notation "rot" is still used for the curl or $\nabla \times$ operator.

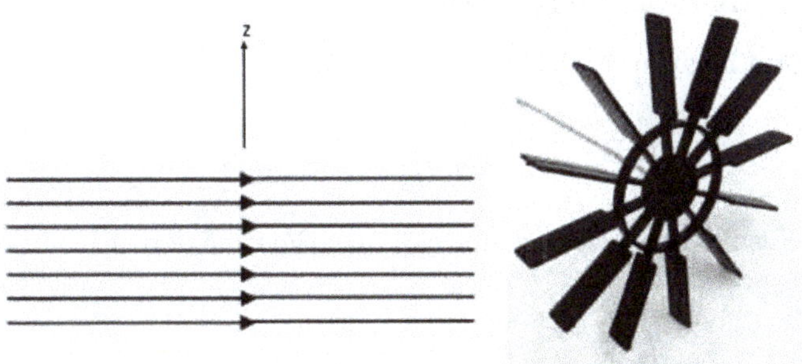

Figure 2.7: A paddle wheel can be used to test if a field has zero or nonzero curl, as explained in the text.

Again, the curl operator has its complicated representations in other coordinate systems such as cylindrical or spherical coordinates, a subject that will not be discussed in detail here [37].

It is to be noted that our proof of the Stokes's theorem is for a flat open surface S and not for a general curved open surface. Since all curved surfaces can be tessellated into a union of flat triangular surfaces according to the tiling theory of simplices,[8] the generalization of the above proof to curved surfaces are straightforward. An example of such a triangulation of a curved surface into a union of flat triangular surfaces is shown in Figure 2.8.

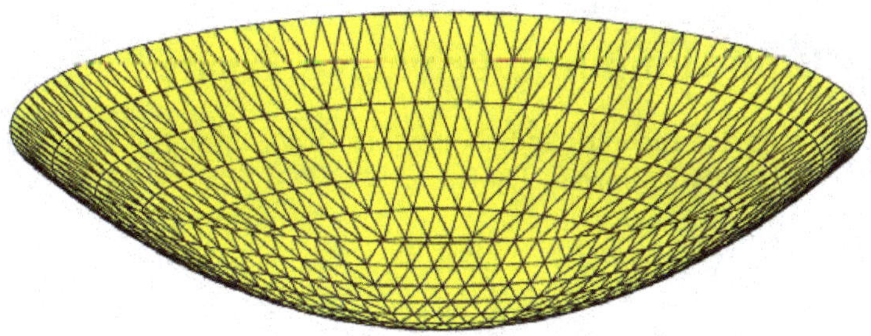

Figure 2.8: An arbitrary curved surface can be triangulated with flat triangular patches, called simplices. The triangulation can be made arbitrarily accurate by making the patches arbitrarily small.

[8]It says that any curve in 1D can be approximated by a union of line segments; a 2D surface can be approximated by a union of triangles; and a 3D volume can be approximated by a union of tetrahedrons. Line segments, triangles, and tetrahedrons are simplices in 1D, 2D, and 3D, respectively.

2.2.2 Faraday's Law in Derivative Form

Faraday's law in integral form is given by[9]

$$\oint_C \mathbf{E} \cdot d\mathbf{l} = -\frac{d}{dt} \iint_S \mathbf{B} \cdot d\mathbf{S} \tag{2.2.12}$$

Assuming that the surface S is not time varying, one can take the time derivative into the integrand and rewrite the above as

$$\oint_C \mathbf{E} \cdot d\mathbf{l} = -\iint_S \frac{\partial}{\partial t} \mathbf{B} \cdot d\mathbf{S} \tag{2.2.13}$$

In the above, d/dt becomes $\partial/\partial t$ inside the integrand since $\mathbf{B} = \mathbf{B}(\mathbf{r}, t)$ is a multivariable function. One can replace the left-hand side with the use of Stokes's theorem to arrive at

$$\iint_S \nabla \times \mathbf{E} \cdot d\mathbf{S} = -\iint_S \frac{\partial}{\partial t} \mathbf{B} \cdot d\mathbf{S} \tag{2.2.14}$$

In the above, $d\mathbf{S}$ is an arbitrary elemental surface that can be made very small. Then the integral can be removed, and one has

$$\nabla \times \mathbf{E}(\mathbf{r}, t) = -\frac{\partial}{\partial t} \mathbf{B}(\mathbf{r}, t) \tag{2.2.15}$$

The above is Faraday's law in differential-operator form.

In the static limit, $\frac{\partial \mathbf{B}}{\partial t} = 0$, giving

$$\nabla \times \mathbf{E} = 0 \tag{2.2.16}$$

2.3 Maxwell's Equations in Derivative Form

With the use of Gauss's divergence theorem and Stokes's theorem, Maxwell's equations can now be written more elegantly in Derivative forms. They are:

$$\nabla \times \mathbf{E}(\mathbf{r}, t) = -\frac{\partial \mathbf{B}}{\partial t}(\mathbf{r}, t) \tag{2.3.1}$$

$$\nabla \times \mathbf{H}(\mathbf{r}, t) = \frac{\partial \mathbf{D}}{\partial t}(\mathbf{r}, t) + \mathbf{J}(\mathbf{r}, t) \tag{2.3.2}$$

$$\nabla \cdot \mathbf{D}(\mathbf{r}, t) = \varrho(\mathbf{r}, t) \tag{2.3.3}$$

$$\nabla \cdot \mathbf{B}(\mathbf{r}, t) = 0 \tag{2.3.4}$$

These equations are pointwise relations, as they relate the left-hand side and right-hand side field values at a given point in space. Moreover, they are not independent of each other. For instance,

[9]Faraday's law is experimentally motivated. Michael Faraday (1791–1867) was an extraordinary experimentalist who documented this law with meticulous care. It was only decades later that a mathematical description of this law was arrived at.

one can take the divergence of the first equation (2.3.1) and making use of the vector identity that $\nabla \cdot (\nabla \times \mathbf{E}) = 0$, one gets

$$-\frac{\partial \nabla \cdot \mathbf{B}}{\partial t} = 0 \rightarrow \nabla \cdot \mathbf{B} = \text{constant} \tag{2.3.5}$$

This constant corresponds to magnetic charges, and since they have not been experimentally observed, one can set the constant to zero. Thus the fourth of Maxwell's equations, (2.3.4), follows from the first (2.3.1).

Similarly, by taking the divergence of the second equation (2.3.2) and making use of the current continuity equation that

$$\nabla \cdot \mathbf{J} + \frac{\partial \varrho}{\partial t} = 0 \tag{2.3.6}$$

one can obtain the second to last equation (2.3.3). Notice that in (2.3.3), the charge density ϱ can be time varying, whereas in the previous lecture, we "derived" this equation from Coulomb's law using electrostatic theory.

The above logic follows if $\partial/\partial t \neq 0$ and is not valid for static case. In other words, for statics, the third and the fourth equations are not derivable from the first two. Hence, all four of Maxwell's equations are needed for static problems. For electrodynamic problems, only solving the first two equations suffices.

Something is amiss in the above. If $\mathbf{J}$ is known, then solving the first two equations implies solving for four vector unknowns, $\mathbf{E}, \mathbf{H}, \mathbf{B}$, and $\mathbf{D}$, which have 12 scalar unknowns. But there are only two vector equations or six scalar equations in the first two equations. Thus, one needs more equations. These are provided by the constitutive relations that we shall discuss next.

2.4 Historical Notes

There are several interesting historical notes about Maxwell.

- It is to be noted that when James Clerk Maxwell first wrote his equations down, they were in many equations and very difficult to digest [18, 41, 42]. It was an eccentric English genius, Oliver Heaviside, an electrical engineer by training, together with the Maxwellians [43], who distilled those equations into the present form found in textbooks. Putatively, most cannot read Maxwell's treatise [41] beyond the first 50 pages [19].
- Maxwell wrote many poems in his short lifespan (1831-1879) and they can be found at [44].
- An ancestor of James Clerk Maxwell married from the Clerk family into the Maxwell family. One of the conditions of marriage was that all of the descendants of the Clerk family should adopt the family name Clerk Maxwell. That was why Maxwell was addressed as Professor Clerk Maxwell in his papers.

Exercises for Lecture 2

Problem 2-1:

(i) By going through proper flux counting, show that (2.1.9) is valid.

(ii) By going through the math carefully, starting from (2.2.6), show that (2.2.10) is correct.

(iii) Explain why Stokes's theorem can be generalized to curved surfaces.

(iv) In Section 2.3 of lecture 2, show that for the four Maxwell's equations, equations (2.3.3) and (2.3.4) are derivable from the first two Maxwell's equations.

(v) Explain why this derivation is not valid for static electromagnetic fields.

(vi) By converting the current continuity equation into integral form, explain why it is the same as charge conservation.

Lecture 3

Constitutive Relations, Wave Equation, the Static Green's Function

Constitutive relations are important for defining the electromagnetic material properties of the media involved. Also, wave phenomenon is a major triumph of Maxwell's equations. Hence, we will study the derivation of this important phenomenon here. Moreover, to make matters simple, we will reduce the problem to electrostatics to simplify the math to introduce the concept of the Green's function. This is an important concept in electromagnetics, from statics to dynamics.

As mentioned previously, for time-varying problems, only the first two of the four Maxwell's equations suffice because the latter two are derivable from the first two. In other words, the latter two of Maxwell's equations are redundant, and electromagnetics solutions are embedded in the first two. But the first two equations have four unknowns, **E**, **H**, **D**, and **B**. Hence, two more equations are needed to solve for these unknowns. These extra equations come from the constitutive relations.

3.1 Simple Constitutive Relations

The constitution relation between electric flux density **D** and the electric field **E** in free space (or vacuum) is

$$\mathbf{D} = \varepsilon_0 \mathbf{E} \tag{3.1.1}$$

where **D** has the unit coulombs per meter squared (C/m2), **E** has the unit or volts per meter (V/m), and ε has the unit of farads per meter (F/m). The **E** can be thought of as an applied field, giving rise to flux density **D** or flux flow.

It is to be noted that $\frac{\partial \mathbf{D}}{\partial t}$ has the physical meaning of displacement current like those flowing through a capacitor. When a dielectric medium is present between two parallel plates of the capacitor, this displacement current is enhanced. Thus, when material medium is present, one has to add the contribution to **D** by the polarization density **P** in the medium, which is a dipole density.[1] Then [34, 35, 45]

$$\mathbf{D} = \varepsilon_0 \mathbf{E} + \mathbf{P} \tag{3.1.2}$$

The second term **P** above is due to material property, and it is the contribution to the electric flux due to the polarization density of the medium. It is due to the little dipole contribution because of the polar nature of the atoms or molecules that make up a medium.

By the same token, the first term $\varepsilon_0 \mathbf{E}$ can be thought of as the polarization density contribution of vacuum. Vacuum, though it represents nothingness, has electrons and positrons, or electron-positron pairs, lurking in it [46]. Electrons are matter, whereas positrons are antimatter. In the quiescent state, they represent nothingness, but they can be polarized by an electric field **E**. This also explains why electromagnetic waves can propagate through vacuum.

Many media can be approximated by linear media. Then **P** is linearly proportional to the applied field **E**, or $\mathbf{P} = \varepsilon_0 \chi_0 \mathbf{E}$, or[2]

$$\begin{aligned} \mathbf{D} &= \varepsilon_0 \mathbf{E} + \varepsilon_0 \chi_0 \mathbf{E} \\ &= \varepsilon_0 (1 + \chi_0) \mathbf{E} = \varepsilon \mathbf{E}, \quad \varepsilon = \varepsilon_0 (1 + \chi_0) = \varepsilon_0 \varepsilon_r \end{aligned} \tag{3.1.3}$$

where χ_0 is the electric susceptibility. In other words, for linear material media, one can replace the vacuum permittivity ε_0 with an effective permittivity $\varepsilon = \varepsilon_0 \varepsilon_r$ where ε_r is the relative permittivity. Thus, **D** is linearly proportional to **E**. In free space,[3] the permittivity ε is (experimentally determined)

$$\varepsilon = \varepsilon_0 = 8.854 \times 10^{-12} \approx \frac{10^{-8}}{36\pi} \text{ F/m} \tag{3.1.4}$$

The constitutive relation between magnetic flux **B** and magnetic field **H** is given as[4]

$$\mathbf{B} = \mu \mathbf{H}, \qquad \mu \text{ is the permeability in H/m (Henries per meter)} \tag{3.1.5}$$

In free space or vacuum,

$$\mu = \mu_0 = 4\pi \times 10^{-7} \text{ H/m} \tag{3.1.6}$$

In the above, **B** is the magnetic flux density in units of webers per meter (Wb/m) or tesla (T), and **H** has the unit of amperes per meter (A/m), while μ has the unit of henries per meter (H/m). As shall be explained later, this μ_0 is an assigned value, giving it a precise value as shown above. In other materials, the permeability can be written as

$$\mu = \mu_0 \mu_r \tag{3.1.7}$$

[1] Note that a dipole moment is given by $Q\ell$ where Q is its charge in coulombs and ℓ is its length in meters. Hence, dipole density, or polarization density, as a dimension of C/m^2, which is the same unit as that of electric flux **D**.

[2] This is not the most general linear relation between **P** and **E**, but it is the simplest one we can begin with.

[3] It is to be noted that we will use MKS or SI (systéme internationale) units in this course. Another possible unit is the CGS unit used in many physics texts [47]

[4] Again, this is not the most general linear relation but the simplest to begin with.

The above can be derived using a similar argument as that for relative permittivity, where the different permeability is due to the presence of magnetic dipole density in a material medium. In the above, μ_r is termed the relative permeability.

3.2 Emergence of Wave Phenomenon, Triumph of Maxwell's Equations

One of the major triumphs of Maxwell's equations is the prediction of the wave phenomenon. This was experimentally verified by Heinrich Hertz in 1888 [20], some 23 years after the completion of Maxwell's theory in 1865 [18]. To see this, we consider the first two Maxwell's equations for time-varying fields in vacuum or a source-free medium.[5] They are

$$\nabla \times \mathbf{E} = -\mu_0 \frac{\partial \mathbf{H}}{\partial t} \tag{3.2.1}$$

$$\nabla \times \mathbf{H} = \varepsilon_0 \frac{\partial \mathbf{E}}{\partial t} \tag{3.2.2}$$

Taking the curl of (3.2.1), we have

$$\nabla \times \nabla \times \mathbf{E} = -\mu_0 \frac{\partial}{\partial t} \nabla \times \mathbf{H} \tag{3.2.3}$$

It is understood that in the above, the double curl operator implies $\nabla \times (\nabla \times \mathbf{E})$. Substituting (3.2.2) into (3.2.3), we have

$$\nabla \times \nabla \times \mathbf{E} = -\mu_0 \varepsilon_0 \frac{\partial^2}{\partial t^2} \mathbf{E} \tag{3.2.4}$$

The above is the vector wave equation. In the above, the left-hand side can be simplified by using the identity that $\mathbf{a} \times (\mathbf{b} \times \mathbf{c}) = \mathbf{b}(\mathbf{a} \cdot \mathbf{c}) - \mathbf{c}(\mathbf{a} \cdot \mathbf{b})$,[6] but be mindful that the operator ∇ has to operate on a function to its right. Therefore, we arrive at the identity that

$$\nabla \times \nabla \times \mathbf{E} = \nabla \nabla \cdot \mathbf{E} - \nabla^2 \mathbf{E} \tag{3.2.5}$$

Since $\nabla \cdot \mathbf{E} = 0$ in a source-free medium, (3.2.4) becomes

$$\nabla^2 \mathbf{E} - \mu_0 \varepsilon_0 \frac{\partial^2}{\partial t^2} \mathbf{E} = 0 \tag{3.2.6}$$

where

$$\nabla^2 = \nabla \cdot \nabla = \frac{\partial^2}{\partial x^2} + \frac{\partial^2}{\partial y^2} + \frac{\partial^2}{\partial z^2}$$

The above is known as the Laplacian operator. Here, (3.2.6) is the wave equation in three space dimensions [35, 48].

[5] This is because the third and the fourth Maxwell's equations are derivable from the first two when $\partial/\partial_t \neq 0$.

[6] For mnemonics, this formula is also known as the "back-of-the-cab" or "BAC-CAB" formula. It can be proved easily by expanding the vectors in their Cartesian components and going through the algebraic manipulation.

To see the simplest form of wave emerging in the above, we can let $\mathbf{E} = \hat{x}E_x(z,t)$ so that $\nabla \cdot \mathbf{E} = 0$, which is required in the source-free medium. Then (3.2.6) becomes

$$\frac{\partial^2}{\partial z^2}E_x(z,t) - \mu_0\varepsilon_0\frac{\partial^2}{\partial t^2}E_x(z,t) = 0 \tag{3.2.7}$$

Eq. (3.2.7) is known mathematically as the wave equation in one-dimensional space. It can also be written as

$$\frac{\partial^2}{\partial z^2}f(z,t) - \frac{1}{c_0^2}\frac{\partial^2}{\partial t^2}f(z,t) = 0 \tag{3.2.8}$$

where $c_0^2 = (\mu_0\varepsilon_0)^{-1}$. Eq. (3.2.8) can also be factorized as

$$\left(\frac{\partial}{\partial z} - \frac{1}{c_0}\frac{\partial}{\partial t}\right)\left(\frac{\partial}{\partial z} + \frac{1}{c_0}\frac{\partial}{\partial t}\right)f(z,t) = 0 \tag{3.2.9}$$

or

$$\left(\frac{\partial}{\partial z} + \frac{1}{c_0}\frac{\partial}{\partial t}\right)\left(\frac{\partial}{\partial z} - \frac{1}{c_0}\frac{\partial}{\partial t}\right)f(z,t) = 0 \tag{3.2.10}$$

The above can be verified easily by direct expansion, and using the fact that

$$\frac{\partial}{\partial t}\frac{\partial}{\partial z} = \frac{\partial}{\partial z}\frac{\partial}{\partial t} \tag{3.2.11}$$

The above implies that we have either

$$\left(\frac{\partial}{\partial z} + \frac{1}{c_0}\frac{\partial}{\partial t}\right)f_+(z,t) = 0 \tag{3.2.12}$$

or

$$\left(\frac{\partial}{\partial z} - \frac{1}{c_0}\frac{\partial}{\partial t}\right)f_-(z,t) = 0 \tag{3.2.13}$$

Eq. (3.2.12) and (3.2.13) are known as the one-way wave equations or the advective equations [49]. From the above factorization, it is seen that the solutions to these one-way wave equations are also the solutions of the original wave equation given by (3.2.8). Their general solutions are then of the form

$$f_+(z,t) = F_+(z - c_0t) \tag{3.2.14}$$
$$f_-(z,t) = F_-(z + c_0t) \tag{3.2.15}$$

We can verify the above by back substitution into (3.2.12) and (3.2.13). Eq. (3.2.14) constitutes a right-traveling wave function of any shape while (3.2.15) constitutes a left-traveling wave function of any shape. Because (3.2.14) and (3.2.15) are also solutions to (3.2.8), we can write the general solution to the wave equation as

$$f(z,t) = F_+(z - c_0t) + F_-(z + c_0t) \tag{3.2.16}$$

This is a wonderful result since F_+ and F_- are arbitrary functions[7] of any shape (see Figure 3.1); they can be used to encode information for communication, as has happened in wireless communication, which has transformed the modern world!

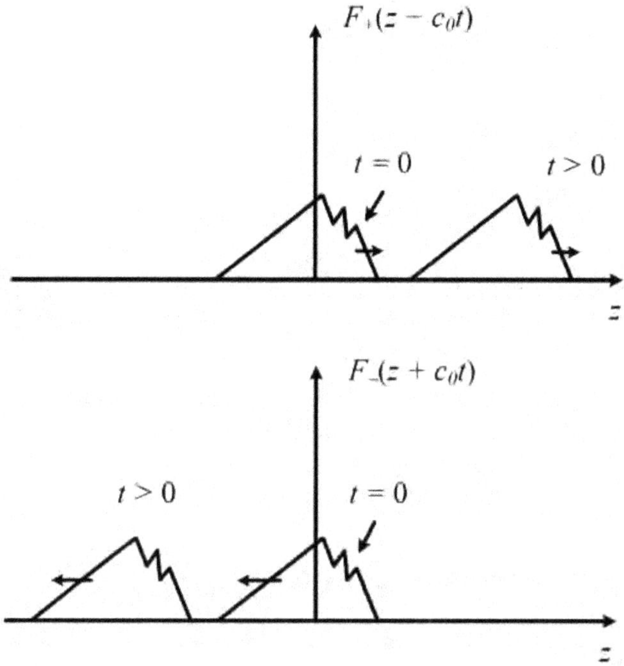

Figure 3.1: Solutions of the wave equation can be single-value functions of any shape. In the above, F_+ travels in the positive z direction, while F_- travels in the negative z direction as t increases. The pulses are delayed version of each other, and this is the retardation physics.

Furthermore, one can calculate the velocity of this wave to be

$$c_0 = 299{,}792{,}458\text{m/s} \simeq 3 \times 10^8\text{m/s} \tag{3.2.17}$$

where $c_0 = \sqrt{1/\mu_0\varepsilon_0}$. It is to be noted that the value of ε and μ can be changed by working in different units, but the velocity of light cannot be changed [50][p. 781].

Maxwell's equations (3.2.1) and (3.2.2) imply that $\mathbf{E}$ and $\mathbf{H}$ are linearly proportional to each other. One can define a new magnetic field variable $\mathbf{H} = \alpha\mathbf{H}'$ and see that the constants in these two

[7]Single-value functions where their derivative exists.

equations will change. The resulting equations are

$$\nabla \times \mathbf{E} = -\mu_0 \alpha \frac{\partial \mathbf{H}'}{\partial t} \tag{3.2.18}$$

$$\nabla \times \mathbf{H}' = \frac{\varepsilon_0}{\alpha} \frac{\partial \mathbf{E}}{\partial t} \tag{3.2.19}$$

One can think of the above Maxwell's equations as having a new $\mu_0' = \mu_0 \alpha$, and $\varepsilon_0' = \frac{\varepsilon_0}{\alpha}$. However, upon eliminating the magnetic field from these two equations, the resulting equation is still (3.2.4) with a universal constant proportional to $\mu_0 \varepsilon_0 = 1/c_0^2$, which is the velocity of light. Thus, there is only one independent constant to be determined in the wave equation, which is c_0. The value of μ_0 is defined neatly to be $4\pi \times 10^{-7}$ H/m, while the value of ε_0 has been measured to be about 8.854×10^{-12} farad m^{-1}. Now it has been decided that the velocity of light is used as a standard and is defined to be the integer given in (3.2.17). A meter is now defined to be the distance traveled by light in 1/(299792458) seconds. Hence, the more accurate that unit of time, or seconds that can be calibrated, the more accurate we can calibrate the unit of length, or meters. Thus, the design of an accurate clock like an atomic clock is an important research area.

The value of ε_0 was measured in the laboratory quite early. Then it was realized that electromagnetic waves propagate at a tremendous velocity, the velocity of light.[8] This was also the defining moment that revealed that the field of electricity and magnetism and the field of optics were both described by Maxwell's equations or electromagnetic theory!

3.3 Static Electromagnetics–Revisted

We have seen static electromagnetics previously in integral form. Now we look at them in differential operator form. When the fields and sources are not time varying, namely that $\partial/\partial t = 0$, we arrive at the static Maxwell's equations for electrostatics and magnetostatics, namely [35, 34, 51]

$$\nabla \times \mathbf{E} = 0 \tag{3.3.1}$$

$$\nabla \times \mathbf{H} = \mathbf{J} \tag{3.3.2}$$

$$\nabla \cdot \mathbf{D} = \varrho \tag{3.3.3}$$

$$\nabla \cdot \mathbf{B} = 0 \tag{3.3.4}$$

Notice that the electrostatic field system is decoupled from the magnetostatic field system. However, in a resistive system where

$$\mathbf{J} = \sigma \mathbf{E} \tag{3.3.5}$$

the two systems are coupled again. This is known as resistive coupling between them. But if $\sigma \to \infty$ in the case of a perfect conductor , or superconductor , then for a finite $\mathbf{J}$, $\mathbf{E}$ has to be zero. The two systems are decoupled again.

Also, one can arrive at the equations above by letting $\mu_0 \to 0$ and $\epsilon_0 \to 0$. In this case, the velocity of light becomes infinite, or retardation (or delay) effect is negligible. In other words, it takes no time (it is instantaneous) for signal propagation through the system in the static approximation.

[8]The velocity of light was known long ago in astronomy by Roemer (1676) [21].

Finally, it is important to note that in statics, the latter two Maxwell's equations are not derivable from the first two. Hence, all four equations have to be considered when one seeks solutions in the static regime or the long-wavelength regime.

3.3.1 Electrostatics

Faraday's law in the static limit is

$$\nabla \times \mathbf{E} = 0 \tag{3.3.6}$$

One way to satisfy the above is to let $\mathbf{E} = -\nabla\Phi$ because of the identity $\nabla \times \nabla = 0$.[9] Alternatively, one can assume that $\mathbf{E}$ is a constant to satisfy (3.3.6). But we usually are interested in solutions that vanish at infinity, and hence, the latter is not a viable solution. Therefore, we let

$$\mathbf{E} = -\nabla\Phi \tag{3.3.7}$$

where Φ is the scalar potential. Furthermore, (3.3.1) has no unique solution. To obtain a unique solution for Φ, (3.3.1) has to be solved in tandem with (3.3.3), which we shall discuss next.

3.3.2 Electrostatics and Kirchoff's Voltage Law

Kirchhoff's voltage law (KVL) is related to electrostatics. By applying Stokes's theorem to (3.3.6), one obtains

$$\oint_C \mathbf{E} \cdot d\mathbf{l} = 0 \tag{3.3.8}$$

The above integration around a loop can be written as a sum of integral over line segments. Since $\mathbf{E} = -\nabla\Phi$, the integral over a line segment can be performed in closed form, yielding

$$\int_a^b \mathbf{E} \cdot d\mathbf{l} = -\Phi_b - \Phi_a \tag{3.3.9}$$

Hence, (3.3.8) implies that the potential drop around a loop is zero, a statement of KVL.

3.3.3 Poisson's Equation

As a consequence of the above discussion,

$$\nabla \cdot \mathbf{D} = \varrho \Rightarrow \nabla \cdot \varepsilon\mathbf{E} = \varrho \Rightarrow -\nabla \cdot \varepsilon\nabla\Phi = \varrho \tag{3.3.10}$$

In the last equation above, if ε is a constant of space, or independent of $\mathbf{r}$, then ε and $\nabla\cdot$ commute, or ε can be moved to the left of $\nabla\cdot$, and one arrives at the simple Poisson's equation, which is a partial differential equation

$$\nabla^2\Phi = -\frac{\varrho}{\varepsilon} \tag{3.3.11}$$

[9]One can easily go through the algebra in Cartesian coordinates to convince oneself of this.

Here, the Laplacian operator

$$\nabla^2 = \nabla \cdot \nabla = \frac{\partial^2}{\partial x^2} + \frac{\partial^2}{\partial y^2} + \frac{\partial^2}{\partial z^2}$$

Solving a partial differential equation is quite involved, but we can find a simple solution to the above partial differential equation in the following simple case. For a point source, we know from Coulomb's law that

$$\mathbf{E} = \frac{q}{4\pi\varepsilon r^2}\hat{r} = -\nabla\Phi \tag{3.3.12}$$

From the above, we deduce that[10]

$$\Phi = \frac{q}{4\pi\varepsilon r} \tag{3.3.13}$$

Therefore, we know the solution to Poisson's equation (3.3.11) is given by the above when the source ϱ represents a point source in the above.

Since this is a *linear equation*, we can use the principle of linear superposition to find the solution when the charge density $\varrho(\mathbf{r})$ is arbitrary. To this end, a point source located at $\mathbf{r}'$ is described by a charge density as

$$\varrho(\mathbf{r}) = q\delta(\mathbf{r} - \mathbf{r}') \tag{3.3.14}$$

where $\delta(\mathbf{r} - \mathbf{r}')$ is a shorthand notation for $\delta(x - x')\delta(y - y')\delta(z - z')$. Therefore, from (3.3.11), the corresponding partial differential equation for a point source is

$$\nabla^2\Phi(\mathbf{r}) = -\frac{q\delta(\mathbf{r} - \mathbf{r}')}{\varepsilon} \tag{3.3.15}$$

The solution to the above equation, from Coulomb's law in accordance to (3.3.13), has to be

$$\Phi(\mathbf{r}) = \frac{q}{4\pi\varepsilon|\mathbf{r} - \mathbf{r}'|} \tag{3.3.16}$$

where (3.3.13) is for a point source at the origin but (3.3.16) is for a point source located at $\mathbf{r} = \mathbf{r}'$. The above is a coordinate independent form of the solution. Here, we adopt the notation that $\mathbf{r} = \hat{x}x + \hat{y}y + \hat{z}z$ and $\mathbf{r}' = \hat{x}x' + \hat{y}y' + \hat{z}z'$, and $|\mathbf{r} - \mathbf{r}'| = \sqrt{(x - x')^2 + (y - y')^2 + (z - z')^2}$.

3.3.4 Static Green's Function

The response to a linear time-invariant system driven by an impulse function is known as the impulse response in the time domain. In the space domain, an analogous solution when the source is a point source is called the Green's function.

Let us define a partial differential equation given by

$$\nabla^2 g(\mathbf{r} - \mathbf{r}') = -\delta(\mathbf{r} - \mathbf{r}') \tag{3.3.17}$$

[10]One can always take the gradient or ∇ of Φ to verify this. Mind you, this is best done in spherical coordinates.

The above is similar to Poisson's equation with a point source on the right-hand side as in (3.3.15). Such a solution, which is a response to a point source, is called the Green's function.[11] By comparing equations (3.3.15) and (3.3.17), then making use of (3.3.16), we deduce that the static Green's function is

$$g(\mathbf{r}-\mathbf{r}') = \frac{1}{4\pi|\mathbf{r}-\mathbf{r}'|} \tag{3.3.18}$$

An arbitrary source, using the sifting property of a delta function, it can be expressed as

$$\varrho(\mathbf{r}) = \iiint_V dV'\varrho(\mathbf{r}')\delta(\mathbf{r}-\mathbf{r}') \tag{3.3.19}$$

where V is the volume over which the charge density $\varrho(\mathbf{r}')$ is nonzero. It is also called the support of the charge density. The above is just the statement that an arbitrary charge distribution $\varrho(\mathbf{r})$ can be expressed as a linear superposition of point sources $\delta(\mathbf{r}-\mathbf{r}')$. Using the above in (3.3.11), we have

$$\nabla^2\Phi(\mathbf{r}) = -\frac{1}{\varepsilon}\iiint_V dV'\varrho(r')\delta(\mathbf{r}-\mathbf{r}') \tag{3.3.20}$$

Hence, we can let

$$\Phi(\mathbf{r}) = \frac{1}{\varepsilon}\iiint_V dV' g(\mathbf{r}-\mathbf{r}')\varrho(\mathbf{r}') \tag{3.3.21}$$

By substituting the above into the left-hand side of (3.3.20), exchanging the order of integration and differentiation, and then making use of equation (3.3.17), it can be shown that (3.3.21) indeed satisfies (3.3.11). The above is just a 3D convolutional integral. Hence, the potential $\Phi(\mathbf{r})$ due to an arbitrary source distribution $\varrho(\mathbf{r})$ can be found by using 3D convolution-namely,

$$\Phi(\mathbf{r}) = \frac{1}{4\pi\varepsilon}\iiint_V \frac{\varrho(\mathbf{r}')}{|\mathbf{r}-\mathbf{r}'|}dV' \tag{3.3.22}$$

In a nutshell, the solution of Poisson's equation when it is driven by an arbitrary source ϱ, is the 3D convolution of the source $\varrho(\mathbf{r})$ with the static Green's function $g(\mathbf{r})$, which is a point source response. In the above, $dV' = dx'dy'dz'$, which is a 3D integration. It is also variously written as $d\mathbf{r}'$ or $d^3\mathbf{r}'$.

[11] George Green (1793–1841), the son of a Nottingham miller, was self-taught, but his work has had a profound impact in our world.

3.3.5 Laplace's Equation

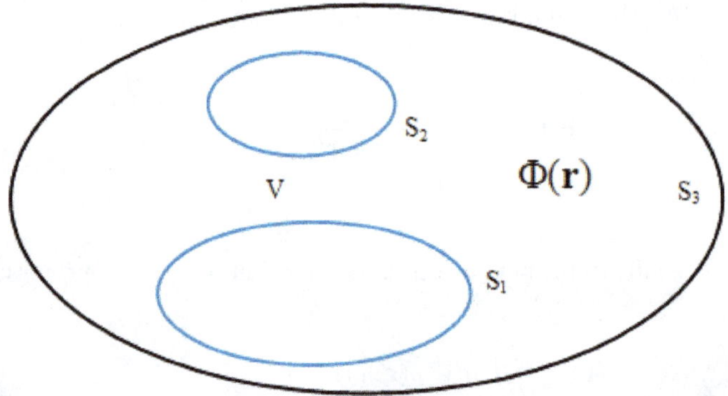

Figure 3.2: A boundary value problem (BVP) for a partial differential equation is usually solved in a region V bounded by a surface S. In this case, the volume V is bounded by three surfaces S_1, S_2, and S_3. The field $\Phi(\mathbf{r})$ needs to satisfy the partial differential equation inside the volume V. Boundary conditions are specified for the field $\Phi(\mathbf{r})$ to be satisfied on disconnected surfaces S_1, S_2, and S_3.

If $\varrho = 0$, or if we are in a source-free region, then for electrostatics,

$$\nabla^2\Phi = 0 \tag{3.3.23}$$

which is the Laplace's equation with $\Phi = 0$ in the trivial case. For the nontrivial solution, Laplace's equation is usually solved as a boundary value problem (BVP). In such a problem, the potential $\Phi(\mathbf{r})$ needs to be found inside a volume V bounded by a surface S. The value of $\Phi(\mathbf{r})$ is stipulated on the boundary of a region with a certain boundary condition, and then the solution is sought in the volume V so as to match the boundary condition on the surface S. Examples of such BVPs are given at the end of the lecture.

Exercises for Lecture 3

Problem 3-1:

(i) Show that (3.2.14) and (3.2.15) are solutions to (3.2.12) and (3.2.13), respectively.

(ii) For static electromagnetics, explain why when a resistive medium exists, the electrostatic system is not decoupled from the magnetostatic system.

(iii) Explain why (3.3.16) is the solution to (3.3.15).

Problem 3-2: Excellent examples of boundary value problems are found in Chapter 6 of Sadiku [37], and in Chapter 7 of Ramo, Whinnery, and Vanduzer [34]. Show that equation (6) of Example 7.12 of [34] is a solution of the boundary value problem defined in Fig. 7.12a.

Lecture 4

Magnetostatics, Boundary Conditions, Jump Conditions

In the previous lecture, Maxwell's equations become greatly simplified in the static limit. We have looked at how the electrostatic problems are solved. We now look at the magnetostatic case. In addition, we will study boundary conditions and jump conditions at an interface and how they can be derived from Maxwell's equations. Maxwell's equations can be first solved in different physical regions. Then the solutions are pieced (or sewn) together by imposing boundary conditions at the boundaries or interfaces of the regions. Such problems are BVPs, and some excellent examples are found in [34] and Sadiku [37].

4.1 Magnetostatics

From Maxwell's equations, we can deduce that the magnetostatic equations for the magnetic field and flux when $\partial/\partial t = 0$ are [34, 35, 51]

$$\nabla \times \mathbf{H} = \mathbf{J} \tag{4.1.1}$$

$$\nabla \cdot \mathbf{B} = 0 \tag{4.1.2}$$

Neither of the equations above has a unique solution. Hence, to solve the above, we have to invoke the constitutive relation that $\mathbf{B} = \mu \mathbf{H}$. These two equations are greatly simplified and, hence, are easier to solve compared to the time-varying case. One way to satisfy the second equation is to let

$$\mathbf{B} = \nabla \times \mathbf{A} \tag{4.1.3}$$

In the above, (4.1.2) does not have a unique solution. By letting $\mathbf{B}$, as in (4.1.3), because of the vector identity

$$\nabla \cdot (\nabla \times \mathbf{A}) = 0 \tag{4.1.4}$$

(4.1.2) is always satisfied. Eq. (4.1.4) is zero for the same algebraic reason that $\mathbf{a}\cdot(\mathbf{a}\times\mathbf{b})=0$. Here, $\mathbf{A}$ is also known as the magnetic vector potential for magnetic field, analogous to Φ being the scalar potential for electric field. In this manner, Gauss's law in (4.1.2) is automatically satisfied.

From (4.1.1), we have

$$\nabla\times\left(\frac{\mathbf{B}}{\mu}\right)=\mathbf{J} \tag{4.1.5}$$

Then using (4.1.3) with the above, we get

$$\nabla\times\left(\frac{1}{\mu}\nabla\times\mathbf{A}\right)=\mathbf{J} \tag{4.1.6}$$

In a homogeneous medium,[1] μ or $1/\mu$ is a constant and it commutes with the differential ∇ operator, or it can be taken outside the differential operator. As such, one arrives at

$$\nabla\times(\nabla\times\mathbf{A})=\mu\mathbf{J} \tag{4.1.7}$$

We use the vector identity that (see the BAC-CAB formula in the previous lecture)

$$\begin{aligned}\nabla\times(\nabla\times\mathbf{A})&=\nabla(\nabla\cdot\mathbf{A})-(\nabla\cdot\nabla)\mathbf{A}\\&=\nabla(\nabla\cdot\mathbf{A})-\nabla^2\mathbf{A}\end{aligned} \tag{4.1.8}$$

where ∇^2 is a shorthand notation for $\nabla\cdot\nabla$. The above is best understood in the Cartesian coordinates. As a result, we arrive at [52]

$$\nabla(\nabla\cdot\mathbf{A})-\nabla^2\mathbf{A}=\mu\mathbf{J} \tag{4.1.9}$$

By imposing the Coulomb gauge that $\nabla\cdot\mathbf{A}=0$ (which will be elaborated in the next section), we arrive at the simplified equation

$$\nabla^2\mathbf{A}=-\mu\mathbf{J} \tag{4.1.10}$$

The above is also known as the vector Poisson's equation [53], which is a partial differential equation. In Cartesian coordinates, the above can be viewed as three scalar Poisson's equations. Each of the Poisson's equations can be solved using the Green's function method previously described. Consequently, in free space

$$\mathbf{A}(\mathbf{r})=\frac{\mu}{4\pi}\iiint_V\frac{\mathbf{J}(\mathbf{r}')}{R}dV' \tag{4.1.11}$$

where

$$R=|\mathbf{r}-\mathbf{r}'| \tag{4.1.12}$$

is the distance between the source point $\mathbf{r}'$ and the observation point $\mathbf{r}$.

[1]It's prudent to warn the reader of the use of the word "homogeneous". In the math community, it usually refers to something to be set to zero. But in the electromagnetics community, it refers to something "nonheterogeneous".

4.1.1 More on the Coulomb Gauge

Gauge is a very important concept in physics [50], and we will further elaborate on it here. First, notice that $\mathbf{A}$ in (4.1.3) is not unique because one can always define

$$\mathbf{A}' = \mathbf{A} - \nabla\Psi \tag{4.1.13}$$

Then

$$\nabla \times \mathbf{A}' = \nabla \times (\mathbf{A} - \nabla\Psi) = \nabla \times \mathbf{A} = \mathbf{B} \tag{4.1.14}$$

where we have made use of $\nabla \times \nabla\Psi = 0$. Hence, the $\nabla\times$ of both $\mathbf{A}$ and $\mathbf{A}'$ produce the same $\mathbf{B}$, implying that $\mathbf{A}$ is nonunique.

To find $\mathbf{A}$ uniquely, we have to define or set the divergence of $\mathbf{A}$ or provide a gauge condition or gauge fixing. One way is to set the divergence of $\mathbf{A}$ to be zero, namely that

$$\nabla \cdot \mathbf{A} = 0 \tag{4.1.15}$$

This will pin the value of $\mathbf{A}$. Then from (4.1.13),

$$\nabla \cdot \mathbf{A}' = \nabla \cdot \mathbf{A} - \nabla^2\Psi \neq \nabla \cdot \mathbf{A} \tag{4.1.16}$$

The last nonequal sign follows if $\nabla^2\Psi \neq 0$. However, if we always impose the Coulomb gauge when seeking the solutions or we further stipulate that $\nabla \cdot \mathbf{A}' = \nabla \cdot \mathbf{A} = 0$, then $-\nabla^2\Psi = 0$. (This does not necessary imply that $\nabla\Psi = 0$, but if we impose the condition that $\Psi \to$ constant when $\mathbf{r} \to \infty$, then $\Psi =$ constant everywhere.[2]) By so doing, $\mathbf{A}$ and $\mathbf{A}'$ are equal to each other implying uniqueness, and we obtain (4.1.10) and (4.1.11).

To see this more simply, the above is akin to the idea that given a vector $\mathbf{a}$, just stipulating that $\mathbf{b}\times\mathbf{a} = \mathbf{c}$ is not enough to determine $\mathbf{a}$. We also need to stipulate what $\mathbf{b}\cdot\mathbf{a}$ is. Here, $\mathbf{a}$, $\mathbf{b}$, and $\mathbf{c}$ are independent vectors. Another way of saying this is that the vector $\mathbf{a}$ can be written as $\mathbf{a} = \mathbf{a}_{\|} + \mathbf{a}_{\perp}$ where $\mathbf{a}_{\|}$ is parallel to the vector $\mathbf{b}$, while $\mathbf{a}_{\perp}$ is perpendicular to $\mathbf{b}$. Here, $\mathbf{a}_{\|}$ is indeterminate now, since $\mathbf{b} \times \mathbf{a}_{\|} = 0$. But by letting $\mathbf{b} \cdot \mathbf{a} = 0$, it will force $\mathbf{a}_{\|} = 0$ and define $\mathbf{a}$ uniquely.

4.1.2 Magnetostatics and Kirchhoff's Current Law

Kirchhoff's current law (KCL) is intimately related to magnetostatics. If one takes the divergence of (4.1.1), one gets $\nabla \cdot \mathbf{J} = 0$. This says that the net current flowing into a point is zero, which is that statement of KCL. One can also integrate this equation about a small volume and use Gauss's divergence theorem to relate it to a surface integral. Then KCL is obviated by such a formula.

4.2 Boundary Conditions—1D Poisson's Equation

To simplify the solutions of Maxwell's equations, they are usually solved in a homogeneous medium. As mentioned before, a complex problem can be divided into piecewise homogeneous regions first,

[2]This is the same statement that a monopole-only line source is absent. All sources are finite or dipolar so that current can flow between the two poles of the source. Then it is a property of the Laplace BVP that if $\Psi = 0$ on a closed surface S, then $\Psi = 0$ everywhere inside S. Earnshaw's theorem [34] is useful for proving this assertion.

and then the solution in each region can be sought separately. Then the total solution must satisfy boundary conditions at the interface between the piecewise homogeneous regions.

What are these boundary conditions? Boundary conditions are actually embedded in the partial differential equations (of which Maxwell's equations and Poisson's equation are) that the potential or the field satisfy. Two important concepts to keep in mind are:

- Differentiation of a function with discontinuous slope will give rise to step discontinuity.
- Differentiation of a function with step discontinuity will give rise to a Dirac delta function. This is also called the jump condition, a term often used by the mathematics community [54].

Take, for example, a one-dimensional Poisson's equation that

$$\frac{d}{dx}\varepsilon(x)\frac{d}{dx}\Phi(x) = -\varrho(x) \tag{4.2.1}$$

where $\varepsilon(x)$ represents material property that has the functional form given in Figure 4.1. One can actually say something about $\Phi(x)$ given $\varrho(x)$ on the right-hand side. If $\varrho(x)$ has a delta function singularity, it implies that $\varepsilon(x)\frac{d}{dx}\Phi(x)$ has a step discontinuity since the derivative of a step function is a delta function. If $\varrho(x)$ is finite everywhere, then $\varepsilon(x)\frac{d}{dx}\Phi(x)$ must be continuous everywhere.

Furthermore, if $\varepsilon(x)\frac{d}{dx}\Phi(x)$ is finite everywhere, it implies that $\Phi(x)$ must be continuous everywhere.

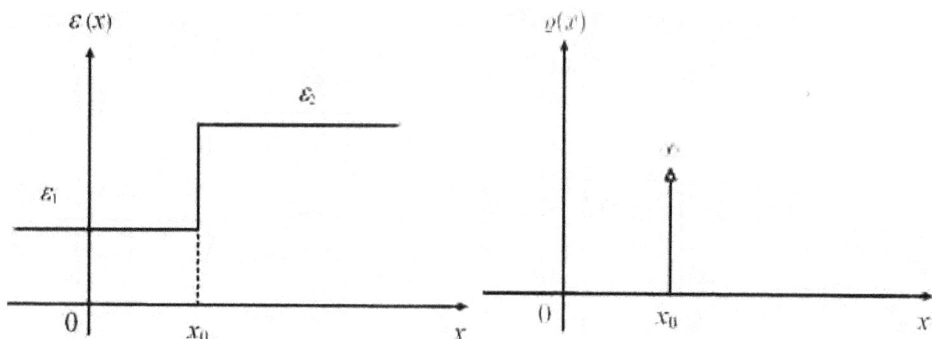

Figure 4.1: A figure showing a charge sheet $\varrho(x) = \varrho_s\delta(x - x_0)$ at the interface between two dielectric media. Because it is a surface charge sheet, the volume charge density $\varrho(x)$ is infinite at the sheet location x_0.

To see this in greater detail, we illustrate it with the following example. In the above, $\varrho(x)$ represents a singular charge distribution given by $\varrho(x) = \varrho_s\delta(x - x_0)$. In this case, the charge distribution is zero everywhere except at the location of the surface charge sheet, where the charge density is infinite; it is represented mathematically by a delta function[3] in space.

[3]This function has been attributed to Dirac, who used it pervasively, but Cauchy was aware of such a function.

To find the boundary condition of the potential $\Phi(x)$ at x_0, we integrate (4.2.1) over an infinitesimal width around x_0, the location of the charge sheet, namely

$$\int_{x_0-\Delta}^{x_0+\Delta} dx \left[\frac{d}{dx}\varepsilon(x)\frac{d}{dx}\Phi(x)\right] = -\int_{x_0-\Delta}^{x_0+\Delta} dx\varrho(x) = -\int_{x_0-\Delta}^{x_0+\Delta} dx\varrho_s\delta(x-x_0) \tag{4.2.2}$$

Since the integrand of the left-hand side is an exact derivative, we get

$$\varepsilon(x)\frac{d}{dx}\Phi(x)\Big|_{x_0-\Delta}^{x_0+\Delta} = -\varrho_s \tag{4.2.3}$$

whereas on the right-hand side, we pick up the contribution from the delta function. Evaluating the left-hand side at its limits, one arrives at

$$\varepsilon(x_0^+)\frac{d}{dx}\Phi(x_0^+) - \varepsilon(x_0^-)\frac{d}{dx}\Phi(x_0^-) \cong -\varrho_s, \tag{4.2.4}$$

where $x_0^\pm = \lim_{\Delta\to 0} x_0 \pm \Delta$. In other words, the jump discontinuity is in $\varepsilon(x)\frac{d}{dx}\Phi(x)$ and the amplitude of the jump discontinuity is proportional to the amplitude of the delta function, ϱ_s.

Since $\mathbf{E} = -\nabla\Phi$, then

$$E_x(x) = -\frac{d}{dx}\Phi(x), \tag{4.2.5}$$

The above, together with (4.2.4), implies the boundary condition that

$$\varepsilon(x_0^+)E_x(x_0^+) - \varepsilon(x_0^-)E_x(x_0^-) = \varrho_s \tag{4.2.6}$$

or

$$D_x(x_0^+) - D_x(x_0^-) = \varrho_s \tag{4.2.7}$$

where

$$D_x(x) = \varepsilon(x)E_x(x) \tag{4.2.8}$$

If $\varrho_s = 0$, then the boundary condition becomes $D_x(x_0^+) = D_x(x_0^-)$.

The lesson learned from above is that the boundary condition is obtained by integrating the pertinent differential equation over an infinitesimal small segment. In this mathematical way of looking at the boundary condition, one can also eyeball the differential equation and ascertain the terms that will have the jump discontinuity and whose derivatives will yield the delta function on the right-hand side.

4.3 Boundary Conditions—Maxwell's Equations

As seen previously, boundary conditions for a field are embedded in the differential equation or, in general, the partial differential equation, that the field satisfies. Hence, boundary conditions can be derived from the differential operator forms of Maxwell's equations. In most textbooks, boundary conditions are obtained by integrating Maxwell's equations over a small pill box [34, 35, 52]. To derive these boundary conditions, we will take an unconventional view: namely, to see what sources can induce jump conditions (or jump discontinuities) on the pertinent fields. Boundary conditions are needed at media interfaces, as well as across current or charge sheets. As shall be shown, each of the Maxwell's equations induces a boundary condition at the interface between two media or two regions separated by surface sources that are infinitely thin. Hopefully, this will give you more physical insight into the reasons for these boundary conditions.

4.3.1 Faraday's Law

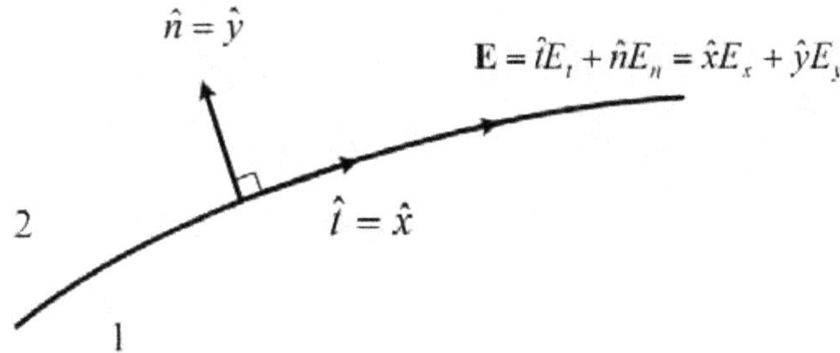

Figure 4.2: This figure is for the derivation of the boundary condition induced by Faraday's law. A local coordinate system can be used to see the boundary condition more lucidly. Here, the normal $\hat{n} = \hat{y}$ and the tangential component $\hat{t} = \hat{x}$.

For this, we start with Faraday's law, which implies that

$$\nabla \times \mathbf{E} = -\frac{\partial \mathbf{B}}{\partial t} \tag{4.3.1}$$

The right-hand side of this equation is a derivative of a time-varying magnetic flux, which is a finite quantity. One quick answer we could ask is that if the right-hand side of the above equation is finite everywhere, could there be any jump discontinuity on the field $\mathbf{E}$ on the left-hand side? The answer is a resounding no!

To see this quickly, one can project the tangential field component and normal field component to a local coordinate system. In other words, one can think of $\hat{t}$ and $\hat{n}$ as the local $\hat{x}$ and $\hat{y}$ coordinates. Then writing the curl operator in these local coordinates, one gets

$$\begin{aligned} \nabla \times \mathbf{E} &= \left(\hat{x}\frac{\partial}{\partial x} + \hat{y}\frac{\partial}{\partial y}\right) \times (\hat{x}E_x + \hat{y}E_y) \\ &= \hat{z}\frac{\partial}{\partial x}E_y - \hat{z}\frac{\partial}{\partial y}E_x \end{aligned} \tag{4.3.2}$$

In simplifying the above, we have used the distributive property of cross product and evaluated the cross product in Cartesian coordinates. The cross product gives four terms, but only two of the four terms are nonzero as shown above.

Since the right-hand side of (4.3.1) is finite, the above implies that $\frac{\partial}{\partial x}E_y$ and $\frac{\partial}{\partial y}E_x$ have to be finite. In other words, E_x is continuous in the y direction and E_y is continuous in the x direction. Since in the local coordinate system $E_x = E_t$, then E_t is continuous across the boundary. The above implies that

$$E_{1t} = E_{2t} \tag{4.3.3}$$

or the tangential components of the electric field are continuous at the interface. To express this in a compact coordinate independent manner, we have

$$\hat{n} \times \mathbf{E}_1 = \hat{n} \times \mathbf{E}_2 \tag{4.3.4}$$

where $\hat{n}$ is the unit normal at the interface, and $\hat{n} \times \mathbf{E}$ always extracts the tangential component of a vector $\mathbf{E}$ (convince yourself).

4.3.2 Gauss's Law for Electric Flux

From Gauss's law, we have

$$\nabla \cdot \mathbf{D} = \varrho \tag{4.3.5}$$

where ϱ is the volume charge density. We would like to express this equation at an interface in terms of a local self-coordinate system.

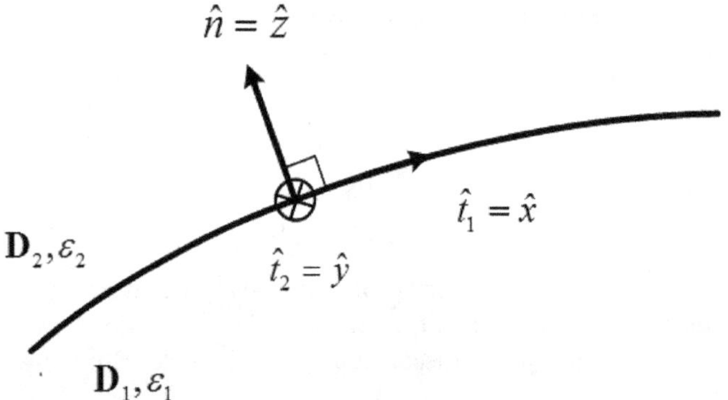

Figure 4.3: A figure showing the derivation of boundary condition for Gauss's law. Again, a local coordinate system can be introduced for simplicity.

Expressing the above in local coordinates x, y, and z as shown in Figure 4.3), then

$$\nabla \cdot \mathbf{D} = \frac{\partial}{\partial x}D_x + \frac{\partial}{\partial y}D_y + \frac{\partial}{\partial z}D_z = \varrho \tag{4.3.6}$$

The boundary condition for the electric flux can be found by *singularity matching.* If there is a surface layer charge at the interface, then the volume charge density must be infinitely large or singular; hence, it can be expressed in terms of a delta function, or $\varrho = \varrho_s \delta(z)$ in local coordinates. Here, ϱ_s is the surface charge density. By looking at the above expression in local coordinates, the only term that can possibly produce a $\delta(z)$ is from $\frac{\partial}{\partial z} D_z$. In other words, D_z must have a jump discontinuity at $z = 0$; the other terms do not. Then

$$\frac{\partial}{\partial z} D_z = \varrho_s \delta(z) \tag{4.3.7}$$

Integrating the above over an infinitesimal gap from $0 - \Delta$ to $0 + \Delta$, we get

$$D_z(z) \Big|_{0-\Delta}^{0+\Delta} = \varrho_s \tag{4.3.8}$$

or in the limit when $\Delta \to 0$,

$$D_z(0^+) - D_z(0^-) = \varrho_s \tag{4.3.9}$$

where $0^+ = \lim_{\Delta \to 0} 0 + \Delta$, and $0^- = \lim_{\Delta \to 0} 0 - \Delta$. Since $D_z(0^+) = D_{2n}$, $D_z(0^-) = D_{1n}$ (where D_{in} is the normal field at the i-th interface), the above becomes

$$D_{2n} - D_{1n} = \varrho_s \tag{4.3.10}$$

In other words, physically, a charge sheet ϱ_s can give rise to a jump discontinuity in the normal component of the electric flux $\mathbf{D}$. Expressed in a compact, coordinate independent form, the boundary condition is

$$\hat{n} \cdot (\mathbf{D}_2 - \mathbf{D}_1) = \varrho_s \tag{4.3.11}$$

Using the physical notion that an electric charge has electric flux $\mathbf{D}$ exuding from it, Figure 4.4 shows an intuitive sketch as to why a charge sheet gives rise to a discontinuous normal component of the electric flux $\mathbf{D}$.

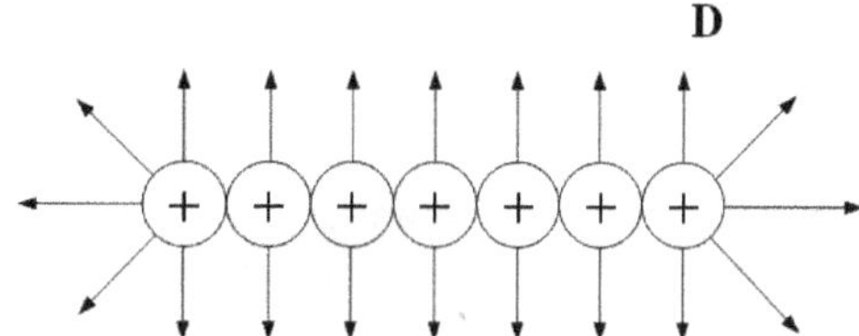

Figure 4.4: A figure intuitively showing why a sheet of electric charge gives rise to a jump discontinuity in the normal component of the electric flux **D**.

4.3.3 Ampère's Law

Ampère's law, or the generalized one, stipulates that

$$\nabla \times \mathbf{H} = \mathbf{J} + \frac{\partial \mathbf{D}}{\partial t} \tag{4.3.12}$$

Again, if the right-hand side is finite everywhere, then $\mathbf{H}$ is a continuous field everywhere. However, if the right-hand side has a delta function singularity, due to a current sheet $\mathbf{J}$ in 3D space, and $\frac{\partial \mathbf{D}}{\partial t}$ is regular or finite everywhere, then the only place where the singularity can be matched on the left-hand side is from the derivative of the magnetic field $\mathbf{H}$ or $\nabla \times \mathbf{H}$. In a word, $\mathbf{H}$ is not continuous. For instance, we can project the above equation onto a local coordinate system just as we did for Faraday's law.

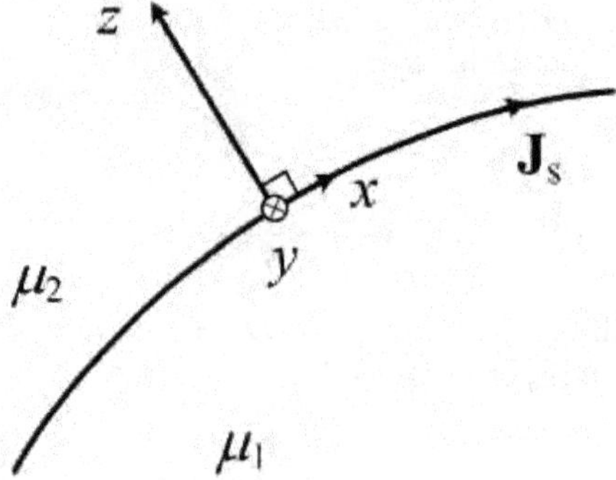

Figure 4.5: A figure showing the derivation of boundary condition for Ampère's law. A local coordinate system is used for simplicity.

To be general, we also include the presence of a current sheet at the interface. We expect the current sheet to induce a jump discontinuity in the magnetic field. This is illustrated intuitively in Figure 4.6.

Figure 4.6: A figure intuitively showing that with the understanding of how a single line current source generates a magnetic field (right), a cluster of them forming a current sheet will generate a jump discontinuity in the tangential component of the magnetic field $\mathbf{H}$ (left).

A current sheet, or a surface current density, becomes a delta function singularity when expressed as a volume current density. Thus, rewriting (4.3.12) in a local coordinate system, assuming that $\mathbf{J} = \hat{x} J_{sx} \delta(z)$,[4] then singularity matching in local coordinates,

$$\nabla \times \mathbf{H} = \hat{x} \left(\frac{\partial}{\partial y} H_z - \frac{\partial}{\partial z} H_y \right) = \hat{x} J_{sx} \delta(z) \tag{4.3.13}$$

The displacement current term on the right-hand side is ignored since it is regular or finite and will not induce a jump discontinuity on the field; hence, we have the form of the right-hand side of the above equation. From the above, the only term that can produce a $\delta(z)$ singularity on the left-hand side is the $-\frac{\partial}{\partial z} H_y$ term. Therefore, by singularity matching, we conclude that

$$-\frac{\partial}{\partial z} H_y = J_{sx} \delta(z) \tag{4.3.14}$$

In other words, physically, H_y has to have a jump discontinuity at the interface where the current sheet resides; or

$$H_y(z = 0^+) - H_y(z = 0^-) = -J_{sx} \tag{4.3.15}$$

The above implies that

$$H_{2y} - H_{1y} = -J_{sx} \tag{4.3.16}$$

But H_y is just the tangential component of the $\mathbf{H}$ field. In a word, physically, the current sheet J_{sx} induces a jump discontinuity on the y component of the magnetic field. Now if we repeat the same exercise with a current with a y component, or $\mathbf{J} = \hat{y} J_{sy} \delta(z)$, at the interface, we have

$$H_{2x} - H_{1x} = J_{sy} \tag{4.3.17}$$

Now, (4.3.16) and (4.3.17) can be rewritten using a cross product as

$$\hat{z} \times (\hat{y} H_{2y} - \hat{y} H_{1y}) = \hat{x} J_{sx} \tag{4.3.18}$$

$$\hat{z} \times (\hat{x} H_{2x} - \hat{x} H_{1x}) = \hat{y} J_{sy} \tag{4.3.19}$$

The above two equations can be consolidated, and written in a coordinate independent form to give

$$\hat{n} \times (\mathbf{H}_2 - \mathbf{H}_1) = \mathbf{J}_s \tag{4.3.20}$$

where, in this case, $\hat{n} = \hat{z}$. In other words, a current sheet $\mathbf{J}_s$ can give rise to a jump discontinuity in the tangential components of the magnetic field, $\hat{n} \times \mathbf{H}$.

[4]The form of this equation can be checked by dimensional analysis. Here, $\mathbf{J}$ has the unit of A m^{-2}, $\delta(z)$ has unit of m^{-1}, and J_{sx}, a current sheet density, has the unit A m^{-1}.

4.3.4 Gauss's Law for Magnetic Flux

Similarly, from Gauss's law for magnetic flux, or

$$\nabla \cdot \mathbf{B} = 0 \tag{4.3.21}$$

one deduces that

$$\hat{n} \cdot (\mathbf{B}_2 - \mathbf{B}_1) = 0 \tag{4.3.22}$$

or that the normal magnetic fluxes are continuous at an interface. In other words, since magnetic charges do not exist, the normal component of the magnetic flux has to be continuous.

The take-home message here is that the boundary conditions are buried in the differential operators and source singularities. If there are singular terms such as sheet sources in Maxwell's equations, then via the differential operators, the boundary conditions can be deduced. These boundary conditions at an interface are also known as jump conditions if a current or a source sheet is present.

4.3.5 Locally Flat Surfaces

It is to be noted that the above boundary conditions are derived for locally flat surfaces, but all surfaces can be formed by the union of curved surfaces. When an observation point is very close to a curved surface compared to the radius of curvature of the surface, it can be assumed to be locally flat. As we shall learn later, an electromagnetic field cannot distinguish between a sharp surface and a curved surface when the radius of curvature of the curved surface is small compared to the wavelength.

Exercises for Lecture 4

Problem 4-1:

Earnshaw's theorem says that a minimum or maximum point cannot appear in the potential of the solution to Laplace's equation. The detail of the proof is quite long, but we can motivate this theorem in the following way:

(i) Laplace's equation can be solved by the separation of variables, namely that in 2D, its solution can be written as $\Phi(x,y) = A\cos(ax)\exp(-ay)$. Show that this is a solution to Laplace's equation and that this function does not have a maximum or a minimum point except at the boundary.

(ii) However, if we write $\Phi(x,y) = A\cos(ax)\cos(by)$, show that this function does have a maximum or a minimum but is not a solution of Laplace's equation. Earnshaw's theorem means that if Φ is a solution to Laplace's equation in a region V, Φ can only have maximum or minimum value at the boundary of V.

(iii) Use this to explain that if a region V is bounded by a surface S, and if Φ is constant on S, then it is the same constant everywhere in V.

(iv) Use this fact to explain how the Faraday's cage works.

(v) Explain why the Coulomb gauge can be used to guarantee a unique vector potential $\mathbf{A}$.

Problem 4-2:

(i) By back substitution, show that (4.1.11) in fact satisfies (4.1.10).

(ii) Coulomb's law give the scalar potential for a monopole charge to be

$$\Phi = \frac{q}{4\pi\varepsilon r}$$

Assuming that two monopoles are aligned on the z-axis and are spaced infinitesimally apart, show that by differentiating this expression with respect to z, one can get the scalar potential for a dipole to be

$$\Phi_d = \frac{\ell q\cos(\theta)}{4\pi\varepsilon r^2}$$

where ℓ is the length of the dipole.

(iii) Give the physical meaning of this mathematical procedure.

(iv) When the above Φ_d is back substituted into Poission's equation, what do you expect the charge density to be on the right-hand side of Poisson's equation?

(v) Derive the jump condition or the boundary condition induced by Ampère's law when there is a current sheet at the interface between two media.

Lecture 5

The Biot-Savart Law, Conductive Media Interface, Instantaneous Poynting's Theorem

The Biot-Savart law, like Ampère's law, was experimentally determined in around 1820 and is discussed in a number of textbooks [50, 35, 34]. This is the cumulative work of Ampère, Oersted, Biot, and Savart. At this stage of the course, we have learned enough mathematical tools to derive this law from Ampère's law and Gauss's law for magnetostatics, so it is appropriate at this point to show the power of mathematical logic in deriving a law inferred experimentally eons ago. In addition, we will study the boundary conditions at conductive media interfaces and introduce the instantaneous Poynting's theorem.

5.1 Derivation of the Biot-Savart Law

The Bio-Savart law allows us to derive the magnetic field due to the electric current flowing in a filamental wire. We assume that the wire is long, or that it forms a closed loop so that the current is uniform on the wire. To this end, we break the wire into union of tiny segments, and calculate the magnetic field from each of these tiny segments. From Gauss's law and Ampère's law in the static limit, and using the definition of the Green's function, we have derived that

$$\mathbf{A}(\mathbf{r}) = \frac{\mu}{4\pi} \iiint_V \frac{\mathbf{J}(\mathbf{r}')}{R} dV' \tag{5.1.1}$$

where $R = |\mathbf{r} - \mathbf{r}'|$, $\mathbf{r}$ is the field point (observation point), and $\mathbf{r}'$ is the source point. The above is just a three-dimensional convolutional integral between the Green's function and the source $\mathbf{J}(\mathbf{r}')$.

When the current element is small, the current is uniform across its cross section, and the current is carried by a wire of cross-sectional area Δa as shown in Figure 5.1.

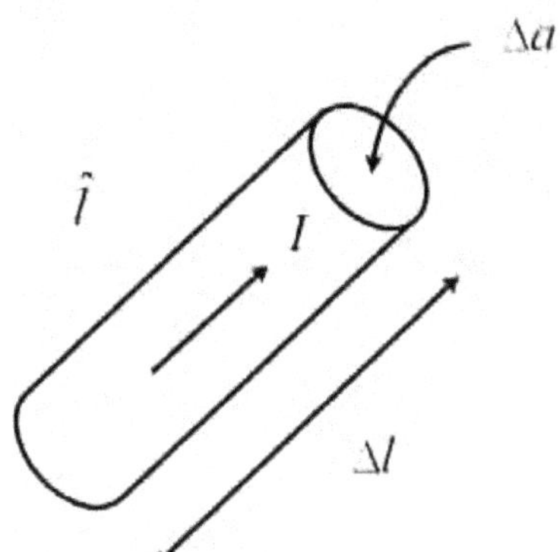

Figure 5.1: A current element used to illustrate the derivation of the Biot-Savart law. The current element generates a magnetic field due to Ampère's law in the static limit. This law was established experimentally, but here, we will derive this law based on our mathematical knowledge so far.

We can approximate the integrand as

$$\mathbf{J}(\mathbf{r}')dV' \approx \mathbf{J}(\mathbf{r}')\Delta V' = \underbrace{\hat{l}I/\Delta a}_{\mathbf{J}(\mathbf{r}')}\underbrace{(\Delta a)\Delta l'}_{\Delta V'} = \hat{l}I\Delta l' \tag{5.1.2}$$

In the above, $\hat{l}I/\Delta a = \mathbf{J}(\mathbf{r}')$ is the current density, and $\Delta V = (\Delta a)\Delta l$ since $\mathbf{J}$ has the unit of $\mathrm{A/m^2}$. Here, $\hat{l}$ is a unit vector pointing in the direction of the current flow or the axis of the wire. Hence, we can let the current element be

$$\mathbf{J}(\mathbf{r}')\Delta V' = I\Delta\mathbf{l}' \tag{5.1.3}$$

where the vector $\Delta\mathbf{l}' = \Delta l\hat{l}$, and $'$ indicates that it is located at $\mathbf{r}'$. Therefore, the incremental vector potential due to an incremental current element $\mathbf{J}(\mathbf{r}')\Delta V'$, according to (5.1.1), is

$$\Delta\mathbf{A}(\mathbf{r}) = \frac{\mu}{4\pi}\left(\frac{\mathbf{J}(\mathbf{r}')\Delta V'}{R}\right) = \frac{\mu}{4\pi}\frac{I\Delta\mathbf{l}'}{R} \tag{5.1.4}$$

Since $\mathbf{B} = \nabla\times\mathbf{A}$, we derive that this incremental $\mathbf{B}$ flux, $\Delta\mathbf{B}$ due to the incremental current $I\Delta\mathbf{l}'$, is

$$\Delta\mathbf{B} = \nabla\times\Delta\mathbf{A}(\mathbf{r}) = \frac{\mu I}{4\pi}\nabla\times\frac{\Delta\mathbf{l}'}{R} = \frac{-\mu I}{4\pi}\Delta\mathbf{l}'\times\nabla\frac{1}{R} \tag{5.1.5}$$

where ∇ operates on the field point, or the unprimed coordinates, and

$$R = \sqrt{(x-x')^2 + (y-y')^2 + (z-z')^2}$$

And we have made use of the fact that $\nabla \times \mathbf{a}f(\mathbf{r}) = -\mathbf{a} \times \nabla f(\mathbf{r})$ when $\mathbf{a}$ is a constant vector. The above can be simplified further by making use of the fact that[1]

$$\nabla \frac{1}{R} = -\frac{1}{R^2}\hat{R} \tag{5.1.6}$$

where $\hat{R}$ is a unit vector pointing in the $\mathbf{r} - \mathbf{r}'$ direction. Consequently, assuming that the incremental length is infinitesimally small, or $\Delta \mathbf{l} \to d\mathbf{l}$, we have, after using (5.1.6) in (5.1.5), that the incremental magnetic flux density $d\mathbf{B}$ is

$$\begin{aligned} d\mathbf{B} &= \frac{\mu I}{4\pi} d\mathbf{l}' \times \frac{1}{R^2}\hat{R} \\ &= \frac{\mu I d\mathbf{l}' \times \hat{R}}{4\pi R^2} \end{aligned} \tag{5.1.7}$$

Since $\mathbf{B} = \mu \mathbf{H}$, the incremental magnetic field is

$$d\mathbf{H} = \frac{I d\mathbf{l}' \times \hat{R}}{4\pi R^2} \tag{5.1.8}$$

or for contribution from the wire, assuming uniform current on the wire, is

$$\mathbf{H}(\mathbf{r}) = \int \frac{I(\mathbf{r}') d\mathbf{l}' \times \hat{R}}{4\pi R^2} \tag{5.1.9}$$

which is the Biot-Savart law, first determined experimentally, but now derived using the rudiments of electromagnetic field theory.

5.2 Shielding by Conductive Media

5.2.1 Boundary Conditions—Conductive Media Case

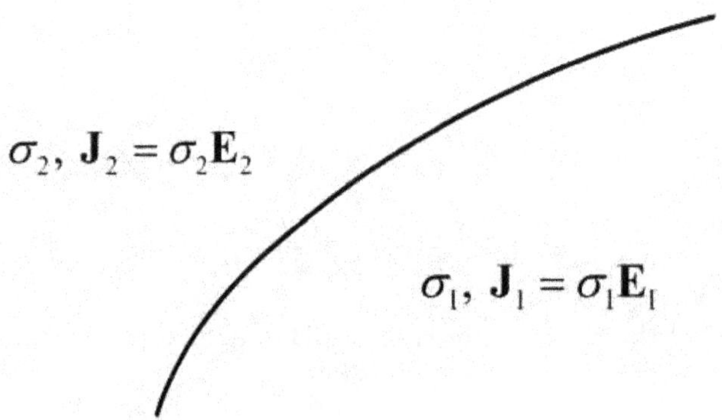

Figure 5.2: The schematics for deriving the boundary condition for the current density $\mathbf{J}$ at the interface of two conductive media.

[1]This is best done by expressing the ∇ operator and R in Cartesian coordinates.

In a conductive medium, $\mathbf{J} = \sigma\mathbf{E}$, which is just a statement of Ohm's law or $I = \frac{V}{R}$. From the current continuity equation, which is derivable from Ampère's law and Gauss's law for electric flux, one gets

$$\nabla \cdot \mathbf{J} = -\frac{\partial \varrho}{\partial t} \tag{5.2.1}$$

The above follows from the charge conservation law, which cannot be violated. If the right-hand side is finite everywhere, it will not induce a jump discontinuity in the current. Moreover, it is zero for static limit. Hence, just like the Gauss's law case, the above implies that the normal component of the current J_n is continuous, or that $J_{1n} = J_{2n}$ in the static limit. In other words, in compact notation, the finiteness of $\nabla \cdot \mathbf{J}$ implies that

$$\hat{n} \cdot (\mathbf{J}_2 - \mathbf{J}_1) = 0 \tag{5.2.2}$$

Hence, using $\mathbf{J} = \sigma\mathbf{E}$, we have

$$\sigma_2 E_{2n} - \sigma_1 E_{1n} = 0 \tag{5.2.3}$$

Note that the above always has to be true in the static limit. But Gauss's law implies the boundary condition that

$$\varepsilon_2 E_{2n} - \varepsilon_1 E_{1n} = \varrho_s \tag{5.2.4}$$

The above equation (5.2.4) is incompatible with (5.2.3) unless $\varrho_s \neq 0$. Hence, surface charge density or charge accumulation is necessary at this interface, unless $\sigma_2/\sigma_1 = \varepsilon_2/\varepsilon_1$. This is found in semiconductor materials that both are conductive and have a permitivity: interfacial charges appear at the interface of two semiconductor materials.

5.2.2 Electric Field Inside a Conductor

Dynamic Case:

For this case, the electric field inside a perfect electric conductor (PEC) has to be zero by the following argument. If medium 1 is a perfect electric conductor, then $\sigma \to \infty$ but $\mathbf{J}_1 = \sigma\mathbf{E}_1$. An infinitesimal $\mathbf{E}_1$ will give rise to an infinite current $\mathbf{J}_1$. To avoid this ludicrous situation, $\mathbf{E}_1$ has to be 0. This implies that $\mathbf{D}_1 = 0$ as well.

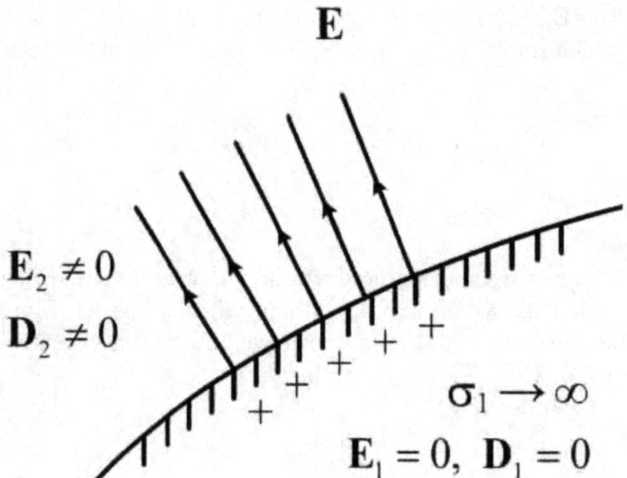

Figure 5.3: The behavior of the electric field and electric flux at the interface of a PEC and free space (or air).

Since tangential **E** is continuous, from Faraday's law, it is still true that

$$E_{2t} = E_{1t} = 0 \tag{5.2.5}$$

or $\hat{n} \times \mathbf{E} = 0$. But since

$$\hat{n} \cdot (\mathbf{D}_2 - \mathbf{D}_1) = \varrho_s \tag{5.2.6}$$

and $\mathbf{D}_1 = 0$,

$$\hat{n} \cdot \mathbf{D}_2 = \varrho_s \tag{5.2.7}$$

So surface charge has to exist at a PEC/air interface. Moreover, normal $\mathbf{D}_2 \neq 0$, but tangential $\mathbf{E}_2 = 0$: Thus the **E** and **D** have to be normal to the PEC surface. The sketch of the electric field in the vicinity of a PEC surface is shown in Figure 5.3.

Static Case:

The above argument for zero electric field inside a perfect conductor is true for electrodynamic problems. However, one does not need the above argument regarding the shielding of the static electric field from a conducting region or an imperfect conductor. In the situation of the two conducting objects example below, as long as the electric fields are nonzero in the objects, currents will keep flowing. They will flow until the charges in the two objects orient themselves so that electric current cannot flow anymore. This happens when the charges produce internal fields that cancel each other, giving rise to zero field inside the two objects. Faraday's law still applies which means that the tangental **E** field has to be continuous but it is zero inside an imperfect conductor. Therefore, the boundary condition that the fields have to be normal to the conducting object surface is still true for electrostatics even if the conductor is imperfect. A sketch of the electric field between two conducting spheres is shown in Figure 5.4. This is also the reason why electric charge can shield off electromagnetic radiation when the frequency is low as in a Faraday cage (see Figure 5.5).

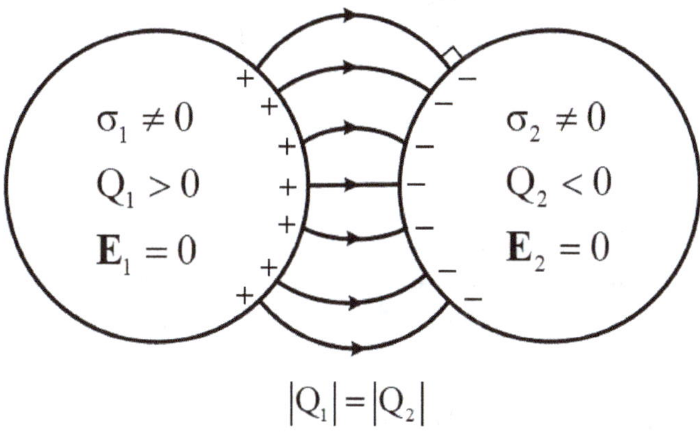

Figure 5.4: The behavior of the electric field and flux outside two conductors in the static limit. The two conductors need not be PEC, and yet the fields are normal to the interface, and the fields are zero inside them.

Figure 5.5: (Left) The Faraday cage can be used to shield out low-frequency electromagnetic fields using charges on the surface of the cage, e.g., it is immune to lightning strikes (Figure by defendershield.com). (Right) It can be used to establish an electromagnetically quiet zone free from RF signals (Source: Wikipedia).

5.2.3 Magnetic Field Inside a Conductor

We have seen that for a finite conductor, as long as $\sigma \neq 0$, the charges will reorient themselves until the electric field is expelled from the conductor; otherwise, the current will keep flowing until $\mathbf{E} = 0$. In a word, static $\mathbf{E}$ is zero inside a conductor.

But for the shielding of the magnetic field, the physics is different. There are no magnetic charges or magnetic conductors in this world. Thus, this physical phenomenon for electric field does not happen for magnetic field; in other words, static magnetic field cannot be expelled from an imperfect electric conductor. However, a magnetic field can be expelled from a perfect conductor or a superconductor. You can only fully understand this physical phenomenon if we study the time-varying form of Maxwell's equations.

In a perfect conductor where $\sigma \to \infty$, it is unstable for the magnetic field $\mathbf{B}$ to be nonzero. As a time-varying magnetic field gives rise to an electric field by the time-varying form of Faraday's law, a small time variation of the $\mathbf{B}$ field will give rise to infinite current flow in a perfect conductor. Therefore, to avoid this ludicrous situation, and to be stable, $\mathbf{B} = 0$ in a perfect conductor or a superconductor.

So if medium 1 is a PEC, then $\mathbf{B}_1 = \mathbf{H}_1 = 0$. The boundary condition then for the magnetic field from Ampère's law is

$$\hat{n} \times (\mathbf{H}_2 - \mathbf{H}_1) = \hat{n} \times \mathbf{H}_2 = \mathbf{J}_s \tag{5.2.8}$$

which is the jump condition for the magnetic field. In other words, a surface current $\mathbf{J}_s$ has to flow at the surface of a PEC in order to support the jump discontinuity in the tangential component of the magnetic field.

From Gauss's law, $\nabla \cdot \mathbf{B} = 0$ implies that $\hat{n} \cdot \mathbf{B}$ is always continuous, or $\hat{n} \cdot (\mathbf{B}_2 - \mathbf{B}_1) = 0$, at an interface because of the absence of magnetic charges. But the magnetic flux $\mathbf{B}_1$ is expelled from the perfect conductor, making $\hat{n} \cdot \mathbf{B}_1 = 0$. Therefore, $\hat{n} \cdot \mathbf{B}_2 = 0$ as well. Hence, there is no normal component of the $\mathbf{B}$ field at the interface of a PEC. Consequently, the boundary condition for $\mathbf{B}_2$ for a PEC becomes

$$\hat{n} \cdot \mathbf{B}_2 = 0 \tag{5.2.9}$$

The $\mathbf{B}$ field in the vicinity of a perfect conductor surface is as shown in Figure 5.6.

When a superconductor cube is placed next to a static magnetic field near a permanent magnet, the $\mathbf{B}$ has to be zero inside a superconductor due to the stability issue. Thus, eddy current will be induced on the superconductor to expel the magnetic field from the permanent magnet, or, in a word, it will produce a magnetic dipole on the superconducting cube that repels the static magnetic field.

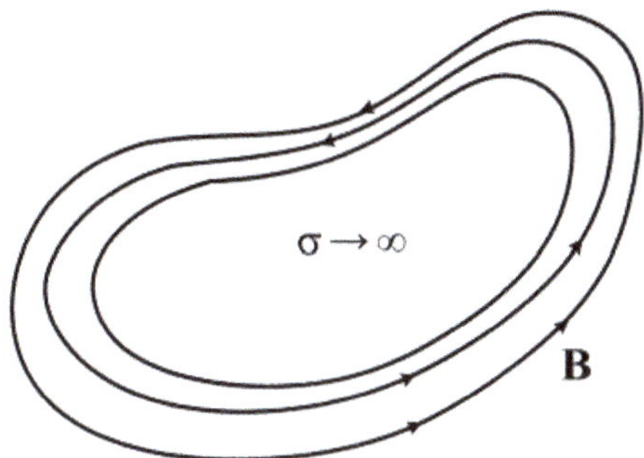

Figure 5.6: Sketch of the magnetic flux **B** around a PEC. As explained in the text, it is seen that $\hat{n} \cdot \mathbf{B} = 0$ at the surface of the perfect electric conductor.

Since these two magnetic dipoles are of opposite polarity, they repel each other. The repulsive force between the magnetic dipoles causes the superconducting cube to levitate on the static magnetic field as shown in Figure 5.7.[2].

Figure 5.7: (Left) Levitation of a superconducting disk on top of a static magnetic field due to expulsion of the magnetic field from the superconductor. This is also known as the Meissner effect (Source: Wikimedia). (Right) Levitation by conjurors in a street in Prague, Czech Republic. This effect is fake.

5.3 Instantaneous Poynting's Theorem

It is habitual to add fictitious magnetic current **M** and fictitious magnetic charge ϱ_m to Maxwell's equations to make them symmetric mathematically.[3] To this end, in general we have, Maxwell's

[2]You may see this demo in a local science museum or your university's physics lab. I saw one in the Boston Museum of Science in 2018

[3]Even though magnetic current does not exist, electric current can be engineered to look like magnetic current, as shall be learned. James Clerk Maxwell also added fictitious magnetic current in his mathematical treatise. Maxwell's equations are almost symmetric and it is tempting to contemplate the addition of the fictitious magnetic current to symmetrize them.

equations as

$$\nabla \times \mathbf{E} = -\frac{\partial \mathbf{B}}{\partial t} - \mathbf{M} \tag{5.3.1}$$

$$\nabla \times \mathbf{H} = \frac{\partial \mathbf{D}}{\partial t} + \mathbf{J} \tag{5.3.2}$$

$$\nabla \cdot \mathbf{D} = \varrho \tag{5.3.3}$$

$$\nabla \cdot \mathbf{B} = \varrho_m \tag{5.3.4}$$

Before we proceed further with studying energy and power, consider the first two of Maxwell's equations where fictitious magnetic current is included and there is a simple isotropic medium such that $\mathbf{B} = \mu_0 \mathbf{H}$ and $\mathbf{D} = \varepsilon_0 \mathbf{E}$. Next, we need to consider only the first two equations (since in electrodynamics, by invoking charge conservation, the third and the fourth equations are derivable from the first two). They are

$$\nabla \times \mathbf{E} = -\frac{\partial \mathbf{B}}{\partial t} - \mathbf{M}_i = -\mu_0 \frac{\partial \mathbf{H}}{\partial t} - \mathbf{M}_i \tag{5.3.5}$$

$$\nabla \times \mathbf{H} = \frac{\partial \mathbf{D}}{\partial t} + \mathbf{J} = \varepsilon_0 \frac{\partial \mathbf{E}}{\partial t} + \mathbf{J}_i + \sigma \mathbf{E} \tag{5.3.6}$$

where $\mathbf{M}_i$ and $\mathbf{J}_i$ are impressed current sources. They are sources that are impressed into the system, and they cannot be changed by their interaction with the environment [55]. They are like current or voltage sources in circuit theory that are immutable by the environment they are inserted into.

Also, for a conductive medium, there is a conduction current or induced current flows in addition to an impressed current. Here, $\mathbf{J} = \sigma \mathbf{E}$ is the induced current source in the conductor. Moreover, $\mathbf{J} = \sigma \mathbf{E}$ is similar to Ohm's law. Maxwell's equations are mathematically elegant, and, hence, they are amenable to manipulations by invoking mathematical logic, as we shall see.

By dot multiplying (5.3.5) with $\mathbf{H}$, and also dot multiplying (5.3.6) with $\mathbf{E}$, we can show that

$$\mathbf{H} \cdot \nabla \times \mathbf{E} = -\mu_0 \mathbf{H} \cdot \frac{\partial \mathbf{H}}{\partial t} - \mathbf{H} \cdot \mathbf{M}_i \tag{5.3.7}$$

$$\mathbf{E} \cdot \nabla \times \mathbf{H} = \varepsilon_0 \mathbf{E} \cdot \frac{\partial \mathbf{E}}{\partial t} + \mathbf{E} \cdot \mathbf{J}_i + \sigma \mathbf{E} \cdot \mathbf{E} \tag{5.3.8}$$

Using the identity, which is the same as the product rule for derivatives, we have[4]

$$\nabla \cdot (\mathbf{E} \times \mathbf{H}) = \mathbf{H} \cdot (\nabla \times \mathbf{E}) - \mathbf{E} \cdot (\nabla \times \mathbf{H}) \tag{5.3.9}$$

Therefore, from (5.3.7), (5.3.8), and (5.3.9), we have

$$\nabla \cdot (\mathbf{E} \times \mathbf{H}) = -\left(\mu_0 \mathbf{H} \cdot \frac{\partial \mathbf{H}}{\partial t} + \varepsilon_0 \mathbf{E} \cdot \frac{\partial \mathbf{E}}{\partial t} + \sigma \mathbf{E} \cdot \mathbf{E} + \mathbf{H} \cdot \mathbf{M}_i + \mathbf{E} \cdot \mathbf{J}_i \right) \tag{5.3.10}$$

[4]The cyclical identity, or the cyclical triple product rule, that $\mathbf{a} \cdot (\mathbf{b} \times \mathbf{c}) = \mathbf{c} \cdot (\mathbf{a} \times \mathbf{b}) = \mathbf{b} \cdot (\mathbf{c} \times \mathbf{a})$ is useful for the derivation.

To elucidate the physical meaning of the above, we first consider $\sigma = 0$ and $\mathbf{M}_i = \mathbf{J}_i = 0$, and the absence of conductive loss and the impressed current sources. Then the above becomes

$$\nabla \cdot (\mathbf{E} \times \mathbf{H}) = -\left(\mu_0 \mathbf{H} \cdot \frac{\partial \mathbf{H}}{\partial t} + \varepsilon_0 \mathbf{E} \cdot \frac{\partial \mathbf{E}}{\partial t}\right) \tag{5.3.11}$$

Rewriting each term on the right-hand side of the above, we have[5]

$$\mu_0 \mathbf{H} \cdot \frac{\partial \mathbf{H}}{\partial t} = \frac{1}{2}\mu_0 \frac{\partial}{\partial t}(\mathbf{H} \cdot \mathbf{H}) = \frac{\partial}{\partial t}\left(\frac{1}{2}\mu_0 |\mathbf{H}|^2\right) = \frac{\partial}{\partial t} W_m \tag{5.3.12}$$

$$\varepsilon_0 \mathbf{E} \cdot \frac{\partial \mathbf{E}}{\partial t} = \frac{1}{2}\varepsilon_0 \frac{\partial}{\partial t}(\mathbf{E} \cdot \mathbf{E}) = \frac{\partial}{\partial t}\left(\frac{1}{2}\varepsilon_0 |\mathbf{E}|^2\right) = \frac{\partial}{\partial t} W_e \tag{5.3.13}$$

where $|\mathbf{H}(\mathbf{r},t)|^2 = \mathbf{H}(\mathbf{r},t) \cdot \mathbf{H}(\mathbf{r},t)$, and $|\mathbf{E}(\mathbf{r},t)|^2 = \mathbf{E}(\mathbf{r},t) \cdot \mathbf{E}(\mathbf{r},t)$ are positive definite. Then (5.3.11) becomes

$$\nabla \cdot (\mathbf{E} \times \mathbf{H}) = -\frac{\partial}{\partial t}(W_m + W_e) \tag{5.3.14}$$

where

$$W_m = \frac{1}{2}\mu_0 |\mathbf{H}|^2, \qquad W_e = \frac{1}{2}\varepsilon_0 |\mathbf{E}|^2 \tag{5.3.15}$$

The above are energy densities in the magnetic field and electric field, respectively.

The vector quantity

$$\mathbf{S}_p = \mathbf{E} \times \mathbf{H} \tag{5.3.16}$$

is called the Poynting vector, and (5.3.14) becomes

$$\nabla \cdot \mathbf{S}_p = -\frac{\partial}{\partial t} W_t \tag{5.3.17}$$

where $W_t = W_e + W_m$ is the total energy density stored in the electric and magnetic fields while $\mathbf{S}_p$ is the power density. It is easy to show that $\mathbf{S}_p$, the power density, has a dimension of watts per square meter (W/m2), and that W_t, the energy density, has a dimension of joules per cubic meter (J/m3).

The above is similar in physical interpretation to the current continuity equation,

$$\nabla \cdot \mathbf{J}(\mathbf{r},t) = -\partial_t \varrho(\mathbf{r},t) \tag{5.3.18}$$

In the current continuity equation, the current density is charge density flow. Hence, power density is energy density flow. We can think of a cube of energy density W_t moving at velocity v, giving rise to power density S_p, and their relationship is

$$S_p = W_t v \tag{5.3.19}$$

[5]The following equality can be established by the product rule of differentiation that $\frac{\partial}{\partial t}(\mathbf{H} \cdot \mathbf{H}) = \mathbf{H} \cdot \frac{\partial \mathbf{H}}{\partial t} + \frac{\partial \mathbf{H}}{\partial t} \cdot \mathbf{H}$.

The right-hand side represents energy density flow while the left-hand side represents power density. One can check the sanity of the above equation using dimensional analysis.

Now, if we let $\sigma \neq 0$, then the term to be included is $\sigma \mathbf{E} \cdot \mathbf{E} = \sigma |\mathbf{E}|^2$ which has the unit of (S m^{-1}) times (V^2 m^{-2}), or (W m^{-3}) where S is siemens. (We arrive at this unit by noticing that $\frac{1}{2}\frac{V^2}{R}$ is the power dissipated in a resistor of R ohms with a unit of watts.) The reciprocal unit of ohms, which used to be called mhos, is now called siemens (S). With $\sigma \neq 0$, (5.3.17) becomes

$$\nabla \cdot \mathbf{S}_p = -\frac{\partial}{\partial t} W_t - \sigma |\mathbf{E}|^2 = -\frac{\partial}{\partial t} W_t - P_d \tag{5.3.20}$$

Here, $\nabla \cdot \mathbf{S}_p$ has the physical meaning of power density oozing out from a point, and $-P_d = -\sigma |\mathbf{E}|^2$ has the physical meaning of power density dissipated (siphoned) at a point by the conductive loss in the medium, which is proportional to $-\sigma |\mathbf{E}|^2$.

Now if we set $\mathbf{J}_i$ and $\mathbf{M}_i$ to be nonzero, (5.3.20) is augmented by the last two terms in (5.3.10), or

$$\nabla \cdot \mathbf{S}_p = -\frac{\partial}{\partial t} W_t - P_d - \mathbf{H} \cdot \mathbf{M}_i - \mathbf{E} \cdot \mathbf{J}_i \tag{5.3.21}$$

The last two terms can be interpreted as the power density supplied by the impressed currents $\mathbf{M}_i$ and $\mathbf{J}_i$ or in short power source P_s. Therefore, (5.3.21) becomes

$$\nabla \cdot \mathbf{S}_p = -\frac{\partial}{\partial t} W_t - P_d + P_s \tag{5.3.22}$$

where

$$P_s = -\mathbf{H} \cdot \mathbf{M}_i - \mathbf{E} \cdot \mathbf{J}_i \tag{5.3.23}$$

and P_s is the power supplied by the impressed current sources $\mathbf{M}_i$ and $\mathbf{J}_i$. These terms are positive if $\mathbf{H}$ and $\mathbf{M}_i$ have opposite signs and if $\mathbf{E}$ and $\mathbf{J}_i$ have opposite signs. The last terms remind us of what happens in a negative resistance device or in a battery.[6] In a battery, positive charges move from a region of lower potential to a region of higher potential (see Figure 5.8), as oppposed to those in a resistor. The positive charges move from one end of a battery to the other end of the battery. Hence, they are doing an "uphill climb" driven by chemical processes within the battery.

[6] A negative resistance has been made by Leo Esaki [56] in an electronic tunnel diode, which won him a share in the Nobel Prize.

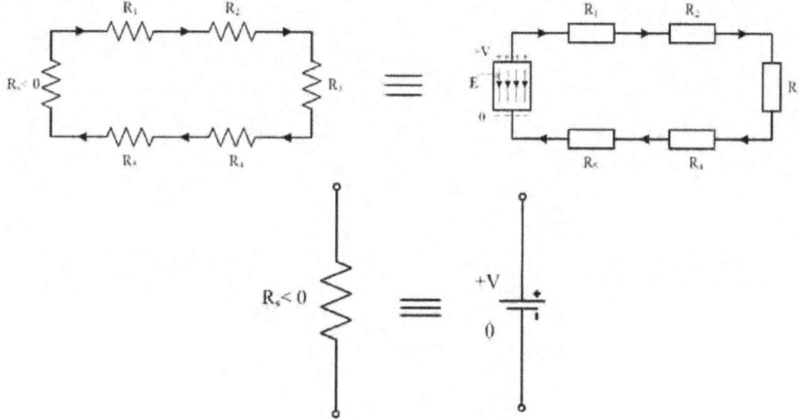

Figure 5.8: The dissipation of energy as the current flows around a loop. A battery can be viewed as having a negative resistance.

In the above, one can easily work out that P_s has the unit of W m^{-3} which is power density supplied. One can also choose to rewrite (5.3.22) in integral form by integrating it over a volume V and invoking the divergence theorem, yielding

$$\int_S d\mathbf{S} \cdot \mathbf{S}_p = -\frac{d}{dt} \int_V W_t dV - \int_V P_d dV + \int_V P_s dV \tag{5.3.24}$$

The left-hand side is

$$\int_S d\mathbf{S} \cdot \mathbf{S}_p = \int_S d\mathbf{S} \cdot (\mathbf{E} \times \mathbf{H}) \tag{5.3.25}$$

which represents the power flowing out of the surface S.

Exercises for Lecture 5

Problem 5-1:

(i) Derive (5.1.5) and (5.1.6).

(ii) Explain why, for electrostatics, a perfect conductor is not needed to shield out the electric field.

(iii) Explain the Meissner effect in a superconductor and why a small piece of superconductor can levitate on a pole of a permanent magnet.

(iv) Provide a physical interpretation to (5.3.20) and the meaning of each of the terms in the equation.

Lecture 6

Time-Harmonic Fields, Complex Power

The analysis of Maxwell's equations can be greatly simplified by assuming the fields to be time harmonic, or sinusoidal (cosinusoidal).[1] Electrical engineers use a method called phasor technique [35, 57] to simplify equations involving time-harmonic signals (these are variously known as monochromatic or constant wave (CW) signals). This is also a poor-man's Fourier transform [58]. That is, one begets the benefits of Fourier transform technique without the full knowledge of Fourier transform. Since only one time-harmonic frequency is involved, this is also called frequency-domain analysis, In other words only one frequency component of the Fourier transform of the signal is involved. Phasors are represented in complex numbers. Therefore, the fields become complex in the frequency domain. From this, we will also discuss the concept of complex power.

6.1 Time-Harmonic Fields—Linear Systems

To learn phasor technique, one makes use a formula by Leonhard Euler (1707–1783),[2]

$$e^{j\alpha} = \cos\alpha + j\sin\alpha \tag{6.1.1}$$

where $j = \sqrt{-1}$ is an imaginary number.[3]

From Euler's formula, one gets

$$\cos\alpha = \Re e\left(e^{j\alpha}\right) \tag{6.1.2}$$

[1]It is simple only for linear systems; for nonlinear systems, such analysis can be quite unwieldy. But rest assured, as we will not discuss nonlinear systems in this course.

[2]As the stamp shows, Euler was blind in one eye.

[3]But lo and behold, in other disciplines such as physics, mathematics, and optics, $\sqrt{-1}$ is denoted by "i", but "i" is too close to the symbol for current. So the preferred symbol for electrical engineering for an imaginary number is j:–a quirkiness of convention, just as positive charges do not carry current in a wire.

where $\Re e$ stands for "the real part of". Hence, all time-harmonic quantities can be written as

$$V(x,y,z,t) = V'(x,y,z)\cos(\omega t + \alpha) = V'(\mathbf{r})\Re e(e^{j(\omega t+\alpha)}) \tag{6.1.3}$$

$$= \Re e\left(V'(\mathbf{r})e^{j\alpha}e^{j\omega t}\right) = \Re e\left(\underset{\sim}{V}(\mathbf{r})e^{j\omega t}\right) \tag{6.1.4}$$

Figure 6.1: A commemorative stamp showing the contribution of Leonhard Euler (Figure by Wikipedia and Pinterest).

Now $\underset{\sim}{V}(\mathbf{r}) = V'(\mathbf{r})e^{j\alpha}$ is a complex number called the phasor representation or phasor of $V(\mathbf{r},t)$, a time-harmonic quantity.[4] Here, the phase $\alpha = \alpha(\mathbf{r})$ can also be a function of position $\mathbf{r}$, or x, y, and z. Consequently, any component of a field can be expressed as[5]

$$E_x(x,y,z,t) = E_x(\mathbf{r},t) = \Re e\left[\underset{\sim}{E}_x(\mathbf{r})e^{j\omega t}\right] \tag{6.1.5}$$

The above can be repeated for y and z components. To express this compactly, for the x, y, and z components together, one can write

$$\mathbf{E}(\mathbf{r},t) = \Re e\left[\underset{\sim}{\mathbf{E}}(\mathbf{r})e^{j\omega t}\right] \tag{6.1.6}$$

[4]We will use a tilde underneath to denote a complex number or a phasor here, but this notation will be dropped later. Whether a variable is complex or real is clear from the context.

[5]In some areas, this is often written as $E_x(\mathbf{r},t) = \frac{1}{2}\underset{\sim}{E}_x(\mathbf{r})e^{j\omega t} + c.c.$ where "c.c." stands for "complex conjugate" or "hermitian conjugate".

$$\mathbf{H}(\mathbf{r},t) = \Re e\left[\underset{\sim}{\mathbf{H}}(\mathbf{r})e^{j\omega t}\right] \tag{6.1.7}$$

where $\underset{\sim}{\mathbf{E}}$ and $\underset{\sim}{\mathbf{H}}$ are complex vector fields. Note that the phasor representation of a field is a complex number but has the same dimension (or unit) as the real field.

Such phasor representations of time-harmonic fields simplify Maxwell's equations. For instance, if one writes

$$\mathbf{B}(\mathbf{r},t) = \Re e\left[\underset{\sim}{\mathbf{B}}(\mathbf{r})e^{j\omega t}\right] \tag{6.1.8}$$

then

$$\frac{\partial}{\partial t}\mathbf{B}(\mathbf{r},t) = \frac{\partial}{\partial t}\Re e\left[\underset{\sim}{\mathbf{B}}(\mathbf{r})e^{j\omega t}\right] = \Re e\left[\frac{\partial}{\partial t}\underset{\sim}{\mathbf{B}}(\mathbf{r})e^{j\omega t}\right] = \Re e\left[\underset{\sim}{\mathbf{B}}(\mathbf{r})j\omega e^{j\omega t}\right] \tag{6.1.9}$$

Therefore, a time derivative can be effected very simply for a time-harmonic field: one just needs to multiply $j\omega$ with the phasor representation of a field or a signal. Hence, given Faraday's law that

$$\nabla \times \mathbf{E} = -\frac{\partial \mathbf{B}}{\partial t} - \mathbf{M} \tag{6.1.10}$$

and assuming that all quantities are time harmonic, then with (6.1.8) and what follows,

$$\mathbf{E}(\mathbf{r},t) = \Re e\left[\underset{\sim}{\mathbf{E}}(\mathbf{r})e^{j\omega t}\right] \tag{6.1.11}$$

$$\mathbf{M}(\mathbf{r},t) = \Re e\left[\underset{\sim}{\mathbf{M}}(\mathbf{r})e^{j\omega t}\right] \tag{6.1.12}$$

Using (6.1.9) and the inserting above into (6.1.10), one gets first

$$\nabla \times \mathbf{E}(\mathbf{r},t) = \Re e\left[\nabla \times \underset{\sim}{\mathbf{E}}(\mathbf{r})e^{j\omega t}\right] \tag{6.1.13}$$

and then equating to the right-hand side of (6.1.10), we have

$$\Re e\left[\nabla \times \underset{\sim}{\mathbf{E}}(\mathbf{r})e^{j\omega t}\right] = -\Re e\left[\underset{\sim}{\mathbf{B}}(\mathbf{r})j\omega e^{j\omega t}\right] - \Re e\left[\underset{\sim}{\mathbf{M}}(\mathbf{r})e^{j\omega t}\right] \tag{6.1.14}$$

Since if

$$\Re e\left\lfloor A(\mathbf{r})e^{j\omega t}\right\rfloor = \Re e\left\lfloor B(\mathbf{r})e^{j\omega t}\right\rfloor, \quad \forall t \tag{6.1.15}$$

then $A(\mathbf{r}) = B(\mathbf{r})$; it must be true from (6.1.14) that[6]

$$\nabla \times \underset{\sim}{\mathbf{E}}(\mathbf{r}) = -j\omega\underset{\sim}{\mathbf{B}}(\mathbf{r}) - \underset{\sim}{\mathbf{M}}(\mathbf{r}) \tag{6.1.16}$$

Therefore, finding the phasor representation of an equation in the frequency domain is clear: whenever we have $\frac{\partial}{\partial t}$, we replace it by $j\omega$. Applying this methodically to the other Maxwell's equations, we have

$$\nabla \times \underset{\sim}{\mathbf{H}}(\mathbf{r}) = j\omega\underset{\sim}{\mathbf{D}}(\mathbf{r}) + \underset{\sim}{\mathbf{J}}(\mathbf{r}) \tag{6.1.17}$$

$$\nabla \cdot \underset{\sim}{\mathbf{D}}(\mathbf{r}) = \underset{\sim}{\varrho_e}(\mathbf{r}) \tag{6.1.18}$$

$$\nabla \cdot \underset{\sim}{\mathbf{B}}(\mathbf{r}) = \underset{\sim}{\varrho_m}(\mathbf{r}) \tag{6.1.19}$$

In the above, the phasors are functions of frequency. For instance, $\underset{\sim}{\mathbf{H}}(\mathbf{r})$ should rightly be written as $\underset{\sim}{\mathbf{H}}(\mathbf{r},\omega)$, but the ω dependence is implied.

[6]The converse is definitely true. If the forward were false, and the converse were true, that would be absurd, and hence, the forward would be true by *reducatio ad absurdum.*

6.2 Fourier Transform Technique

An advantage of phasor technique is that the phasor representations of the fields have the same unit as the original fields in the time domain, as can be seen (6.1.6) and (6.1.7). This is not the case for Fourier transform technique.

In the phasor representation, Maxwell's equations has no time derivatives; hence, the equations are simplified. We can also arrive at the above simplified Maxwell's equations using Fourier transform technique. To this end, we use Faraday's law as an example. By letting

$$\mathbf{E}(\mathbf{r},t) = \frac{1}{2\pi}\int_{-\infty}^{\infty} \mathbf{E}(\mathbf{r},\omega)e^{j\omega t}d\omega \tag{6.2.1}$$

$$\mathbf{B}(\mathbf{r},t) = \frac{1}{2\pi}\int_{-\infty}^{\infty} \mathbf{B}(\mathbf{r},\omega)e^{j\omega t}d\omega \tag{6.2.2}$$

$$\mathbf{M}(\mathbf{r},t) = \frac{1}{2\pi}\int_{-\infty}^{\infty} \mathbf{M}(\mathbf{r},\omega)e^{j\omega t}d\omega \tag{6.2.3}$$

Substituting the above into Faraday's law given by (6.1.10), we get

$$\nabla\times\int_{-\infty}^{\infty} d\omega e^{j\omega t}\mathbf{E}(\mathbf{r},\omega) = -\frac{\partial}{\partial t}\int_{-\infty}^{\infty} d\omega e^{j\omega t}\mathbf{B}(\mathbf{r},\omega) - \int_{-\infty}^{\infty} d\omega e^{j\omega t}\mathbf{M}(\mathbf{r},\omega) \tag{6.2.4}$$

Using the fact that

$$\frac{\partial}{\partial t}\int_{-\infty}^{\infty} d\omega e^{j\omega t}\mathbf{B}(\mathbf{r},\omega) = \int_{-\infty}^{\infty} d\omega\frac{\partial}{\partial t}e^{j\omega t}\mathbf{B}(\mathbf{r},\omega) = \int_{-\infty}^{\infty} d\omega e^{j\omega t}j\omega\mathbf{B}(\mathbf{r},\omega) \tag{6.2.5}$$

and by exchanging the order of differentiation and integration, that

$$\nabla\times\int_{-\infty}^{\infty} d\omega e^{j\omega t}\mathbf{E}(\mathbf{r},\omega) = \int_{-\infty}^{\infty} d\omega e^{j\omega t}\nabla\times\mathbf{E}(\mathbf{r},\omega) \tag{6.2.6}$$

Furthermore, using the fact that

$$\int_{-\infty}^{\infty} d\omega e^{j\omega t}A(\omega) = \int_{-\infty}^{\infty} d\omega e^{j\omega t}B(\omega), \quad \forall t \tag{6.2.7}$$

implies that $A(\omega) = B(\omega)$, and using (6.2.5) and (6.2.6) in (6.2.4), and the property (6.2.7), one gets

$$\nabla\times\mathbf{E}(\mathbf{r},\omega) = -j\omega\mathbf{B}(\mathbf{r},\omega) - \mathbf{M}(\mathbf{r},\omega) \tag{6.2.8}$$

These equations look exactly like the phasor equations we have derived previously, save that the field $\mathbf{E}(\mathbf{r},\omega)$, $\mathbf{B}(\mathbf{r},\omega)$, and $\mathbf{M}(\mathbf{r},\omega)$ are now the Fourier transforms of the field $\mathbf{E}(\mathbf{r},t)$, $\mathbf{B}(\mathbf{r},t)$, and $\mathbf{M}(\mathbf{r},t)$. Moreover, the Fourier transform variables can be complex just like phasors. Repeating the exercise above for the other Maxwell's equations, we obtain equations that look similar to those for their phasor representations. Hence, Maxwell's equations can be simplified either by using phasor technique or Fourier transform technique. However, the dimensions (or units) of the phasors are different from the dimensions of the Fourier-transformed fields: $\underset{\sim}{\mathbf{E}}(\mathbf{r})$, a phasor, and $\mathbf{E}(\mathbf{r},\omega)$, a Fourier transform, do not have the same dimension upon closer examination. The advantage of phasor technique is that the units of the fields are not changed. Also, phasor quantities are related to Fourier transform quantities just by a multiplicative constant.

6.3 Complex Power

Consider now that in the phasor representations, $\underset{\sim}{\mathbf{E}}(\mathbf{r})$ and $\underset{\sim}{\mathbf{H}}(\mathbf{r})$ are complex vectors, and their cross product, $\underset{\sim}{\mathbf{E}}(\mathbf{r}) \times \underset{\sim}{\mathbf{H}}^*(\mathbf{r})$, which still has the unit of power density, has a different physical meaning. First, consider the instantaneous Poynting's vector

$$\mathbf{S}(\mathbf{r},t) = \mathbf{E}(\mathbf{r},t) \times \mathbf{H}(\mathbf{r},t) \tag{6.3.1}$$

where all the quantities are real valued but $\mathbf{E}$ and $\mathbf{H}$ are time-harmonic. Now, we can use phasor technique to analyze the above. Since they are time-harmonic fields, the above can be rewritten as

$$\begin{aligned}\mathbf{S}(\mathbf{r},t) &= \Re e\left[\underset{\sim}{\mathbf{E}}(\mathbf{r})e^{j\omega t}\right] \times \Re e\left[\underset{\sim}{\mathbf{H}}(\mathbf{r})e^{j\omega t}\right] \\ &= \frac{1}{2}\left[\underset{\sim}{\mathbf{E}}e^{j\omega t} + (\underset{\sim}{\mathbf{E}}e^{j\omega t})^*\right] \times \frac{1}{2}\left[\underset{\sim}{\mathbf{H}}e^{j\omega t} + (\underset{\sim}{\mathbf{H}}e^{j\omega t})^*\right]\end{aligned} \tag{6.3.2}$$

where we have made use of the formula that

$$\Re e(Z) = \frac{1}{2}(Z + Z^*) \tag{6.3.3}$$

Then more elaborately, on expanding (6.3.2), we get

$$\mathbf{S}(\mathbf{r},t) = \frac{1}{4}\underset{\sim}{\mathbf{E}} \times \underset{\sim}{\mathbf{H}}e^{2j\omega t} + \frac{1}{4}\underset{\sim}{\mathbf{E}} \times \underset{\sim}{\mathbf{H}}^* + \frac{1}{4}\underset{\sim}{\mathbf{E}}^* \times \underset{\sim}{\mathbf{H}} + \frac{1}{4}\underset{\sim}{\mathbf{E}}^* \times \underset{\sim}{\mathbf{H}}^* e^{-2j\omega t} \tag{6.3.4}$$

On rearranging terms and using (6.3.3) yield

$$\mathbf{S}(\mathbf{r},t) = \frac{1}{2}\Re e\left[\underset{\sim}{\mathbf{E}} \times \underset{\sim}{\mathbf{H}}^*\right] + \frac{1}{2}\Re e\left[\underset{\sim}{\mathbf{E}} \times \underset{\sim}{\mathbf{H}}e^{2j\omega t}\right] \tag{6.3.5}$$

where the first term is independent of time, while the second term is sinusoidal in time. If we define a time-average quantity such that

$$\mathbf{S}_{\text{av}} = \langle \mathbf{S}(\mathbf{r},t) \rangle = \lim_{T\to\infty} \frac{1}{T} \int_0^T \mathbf{S}(\mathbf{r},t)dt \tag{6.3.6}$$

it is quite clear now that the second term of (6.3.5) time-averages to zero since it is sinusoidal, and

$$\mathbf{S}_{\text{av}} = \langle \mathbf{S}(\mathbf{r},t) \rangle = \frac{1}{2}\Re e\left[\underset{\sim}{\mathbf{E}} \times \underset{\sim}{\mathbf{H}}^*\right] \tag{6.3.7}$$

Therefore, in the phasor representation, the quantity

$$\underline{\mathbf{S}} = \underline{\mathbf{E}} \times \underline{\mathbf{H}}^* \tag{6.3.8}$$

is termed the complex Poynting's vector which has the dimension of watts per square meter. The complex power density $\underline{\mathbf{S}}$, is energy density flow associated with it, and is associated with complex power.[7] You should now understand what the real part of this complex power is, and next, we will elucidate the physical meaning of its imaginary part.

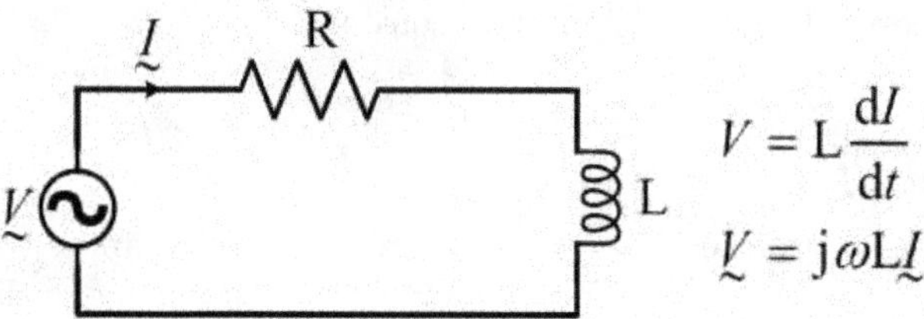

Figure 6.2: A simple circuit to illustrate the concept of complex power in circuit theory. The voltage and current are out of phase, which is a frequency-domain concept.

To understand what complex power is, it is fruitful if we revisit complex power [55, 59] in our circuit theory course. The circuit in Figure 6.2 can be easily solved by using phasor technique. The impedance of the circuit is $Z = R + j\omega L$. Hence,

$$\underline{V} = (R + j\omega L)\underline{I} \tag{6.3.9}$$

where $\underline{V}$ and $\underline{I}$ are the phasors of the voltage and current for time-harmonic signals. Just as in the electromagnetic case, the complex power in watts is taken to be

$$\underline{P} = \underline{V}\underline{I}^* \tag{6.3.10}$$

But the instantaneous power is given by

$$P_{\text{inst}}(t) = V(t)I(t) \tag{6.3.11}$$

where $V(t) = \Re e\{\underline{V}e^{j\omega t}\}$ and $I(t) = \Re e\{\underline{I}e^{j\omega t}\}$. As shall be shown below,

$$P_{\text{av}} = \langle P_{\text{inst}}(t)\rangle = \frac{1}{2}\Re e\left[\underline{P}\right] \tag{6.3.12}$$

It is clear that if $V(t)$ is sinusoidal, it can be written as

$$V(t) = V_0\cos(\omega t) = \Re e\left[\underline{V}e^{j\omega t}\right] \tag{6.3.13}$$

[7]Please note that $\underline{\mathbf{S}}$ is not the phasor representation of $\mathbf{S}(\mathbf{r}, t)$ since the latter is not a time-harmonic signal anymore and has no phasor representation.

where, without loss of generality, we assume that $\underset{\sim}{V} = V_0$. Then from (6.3.9), it is clear that $V(t)$ and $I(t)$ are not in phase. Namely that

$$I(t) = I_0 \cos(\omega t + \alpha) = \Re e\left[\underset{\sim}{I} e^{j\omega t}\right] \tag{6.3.14}$$

where $\underset{\sim}{I} = I_0 e^{j\alpha}$. Then

$$\begin{aligned} P_{\text{inst}}(t) &= V_0 I_0 \cos(\omega t)\cos(\omega t + \alpha) \\ &= V_0 I_0 \cos(\omega t)\left[\cos(\omega t)\cos(\alpha) - \sin(\omega t)\sin\alpha\right] \\ &= V_0 I_0 \cos^2(\omega t)\cos\alpha - V_0 I_0 \cos(\omega t)\sin(\omega t)\sin\alpha \end{aligned} \tag{6.3.15}$$

It can be seen that the first term does not time average to zero, but the second term does! (To see this, we let $\cos(\omega t)\sin(\omega t) = 0.5\sin(2\omega t)$, which time average to zero.) Now taking the time average of (6.3.15), the time average of the first term involves the time average of $\cos^2(\omega t)$, which is 0.5, and we get

$$P_{\text{av}} = \langle P_{\text{inst}} \rangle = \frac{1}{2} V_0 I_0 \cos\alpha = \frac{1}{2}\Re e\left[\underset{\sim}{V}\underset{\sim}{I}^*\right] = \frac{1}{2}\Re e\left[\underset{\sim}{P}\right] \tag{6.3.16}$$

On the other hand, since the power is now complex, the imaginary part of P is called the reactive power. It has a physical meaning that we shall elucidate next.

$$P_{\text{reactive}} = \frac{1}{2}\Im m\left[\underset{\sim}{P}\right] = \frac{1}{2}\Im m\left[\underset{\sim}{V}\underset{\sim}{I}^*\right] = \frac{1}{2}\Im m\left[V_0 I_0 e^{-j\alpha}\right] = -\frac{1}{2} V_0 I_0 \sin\alpha \tag{6.3.17}$$

One sees that amplitude of the time-varying term in (6.3.15) is precisely proportional to $\Im m\left[\underset{\sim}{P}\right]$.[8]

The reason for the existence of the imaginary part of $\underset{\sim}{P}$ is because $V(t)$ and $I(t)$ are out of phase, or $\underset{\sim}{V} = V_0$ but $\underset{\sim}{I} = I_0 e^{j\alpha}$. They are out of phase is because the impedance of the circuit shown with an inductor has a reactive part to it. Hence, the imaginary part of complex power is also called the reactive power [38, 55, 59]. The plots of the instantaneous power in a reactive circuit are shown in Figure 6.3. It is seen that when $\alpha \neq 0$, the instantaneous power can be negative at a certain instant. This means at this instant, the power is flowing from the load to the source instead of flowing from the source to the load. This happens only when the reactive power is nonzero or when a reactive component like an inductor or capacitor exists in the circuit.

When a power company delivers power to our homes, the power is complex because the current and voltage are not in phase due to the presence of rotating machineries, computers, transformers, etc, in our home appliances. Even though the reactive power time averages to zero, the power company still needs to deliver it to our home to run our washing machine, dish washer, fans, air conditioner, etc. And hence, they charge us for it. To be fair, part of this power will be dissipated in the transmission lines that deliver power to our homes. In other words, we have to pay for the use of imaginary power! One advantage of DC power sources is that there is no reactive power associated with them. The con against them is that they cannot be stepped up or down in voltages easily.

[8]Because that complex power is proportional to $\underset{\sim}{V}\underset{\sim}{I}^*$, it is the relative phase between $\underset{\sim}{V}$ and $\underset{\sim}{I}$ that matters. Therefore, α above is the relative phase between the phasor current and phasor voltage.

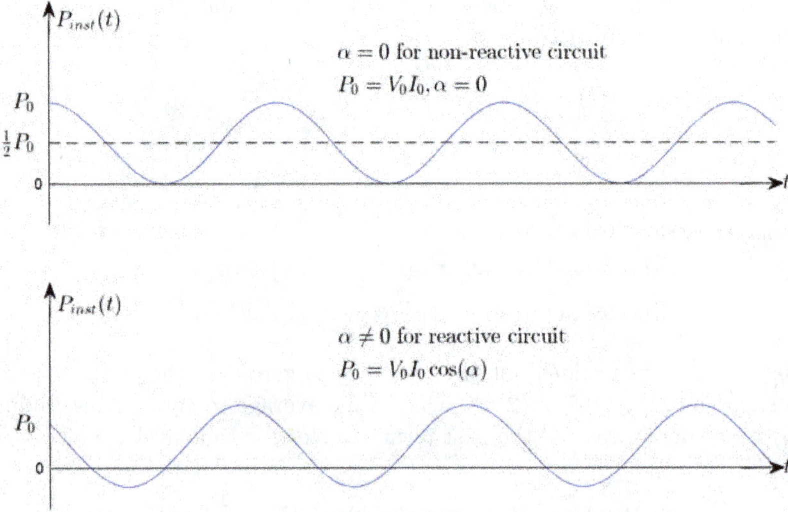

Figure 6.3: (Top) Plots of instantaneous power for when the voltage and the current is in phase ($\alpha = 0$), and (Bottom) when they are out of phase ($\alpha \neq 0$). In the out-of-phase case, there is an additional time-varying term that does not contribute to time-average power as shown in (6.3.15). But the instantaneous power can be negative as shown.

Exercises for Lecture 6

Problem 6-1:

(i) Explain why (6.1.15) and the statement after it is true.

(ii) Explain why (6.2.7) and the statement after it is true.

(iii) Is there a difference in the field quantities obtained from phasor technique and the field quantities obtained from Fourier transform technique?

(iv) Explain the physical meaning of the imaginary part of complex power.

Lecture 7

More on Constitutive Relations, Uniform Plane Wave

As mentioned before, for time-varying problems, the last two of Maxwell's equations are derivable from the first two. But constitutive relations are important for solving only the first two of the four Maxwell's equations. Assuming that **J** is known or zero, the first two vector equations have four vector unknowns:- **E**, **H**, **D**, and **B** which cannot be determined by solving only two equations. The additional two equations come from the constitutive relations. Constitutive relations are useful because they allow us to incorporate material properties into the solutions of Maxwell's equations. The material properties can be frequency dispersive (functions of frequency), anisotropic, bia nisotropic, inhomogeneous, lossy, conductive, nonlinear or spatially dispersive. The use of phasors or frequency domain method will further simplify the characterization of different media. To begin, we will also study uniform plane wave in such media, including lossy conductive media.

7.1 More on Constitutive Relations

As we have discussed, Maxwell's equations are not solvable until the constitutive relations are included. Here, we will look more deeply into various kinds of constitutive relations. Now that we have learned phasor technique, which is a powerful tool for frequency domain analysis, we can study a more general constitutive relationship compared to what we saw earlier.

7.1.1 Isotropic Frequency Dispersive Media

First, let us look at the simple linear constitutive relation previously discussed for dielectric media where [34], [35][p. 82], [50]

$$\mathbf{D} = \varepsilon_0 \mathbf{E} + \mathbf{P} \tag{7.1.1}$$

We begin with a simple linear model where

$$\mathbf{P} = \varepsilon_0 \chi_e \mathbf{E} \tag{7.1.2}$$

where χ_e is the electric susceptibility. Since the displacement current is $\frac{\partial \mathbf{D}}{\partial t}$, when used in the generalized Ampère's law, $\mathbf{P}$, the polarization density, plays an important role for enhancing the flow of the displacement current through space. The polarization density is due to the presence of polar atoms or molecules that behave like little dipoles in the presence of an electric field. For instance, water, whose molecule is H_2O, is a polar molecule that becomes a small dipole when an electric field is applied.

We can think of displacement current flow in polar molecules as capacitive coupling between the dipoles yielding polarization current that flows through space. Namely, for a source-free medium,

$$\nabla \times \mathbf{H} = \frac{\partial \mathbf{D}}{\partial t} = \varepsilon_0 \frac{\partial \mathbf{E}}{\partial t} + \frac{\partial \mathbf{P}}{\partial t} \tag{7.1.3}$$

Figure 7.1: As a series of dipoles lined up end to end, one can see a current, through capacitive coupling, flowing through the line of dipoles as they oscillate back and forth in their polarity. This is similar to how the displacement current flows through a series of capacitors.

For example, for a sinusoidal oscillating field, the dipoles will flip back and forth, giving rise to the flow of the displacement current, just as a time-harmonic electric current can flow through a capacitor as shown in Figure 7.1.

The linear relationship above can be generalized to that of a linear time-invariant system [57], or at any given space point $\mathbf{r}$ [38][p. 212], [50][p. 330],

$$\mathbf{P}(\mathbf{r}, t) = \varepsilon_0 \chi_e(\mathbf{r}, t) \circledast \mathbf{E}(\mathbf{r}, t) \tag{7.1.4}$$

where $\circledast$ implies a convolution. In the frequency domain or the Fourier space, the above linear relationship becomes simpler:

$$\mathbf{P}(\mathbf{r}, \omega) = \varepsilon_0 \chi_e(\mathbf{r}, \omega) \mathbf{E}(\mathbf{r}, \omega), \tag{7.1.5}$$

or

$$\mathbf{D}(\mathbf{r}, \omega) = \varepsilon_0 \left[1 + \chi_e(\mathbf{r}, \omega)\right] \mathbf{E}(\mathbf{r}, \omega) = \varepsilon(\mathbf{r}, \omega) \mathbf{E}(\mathbf{r}, \omega) \tag{7.1.6}$$

where $\varepsilon(\mathbf{r}, \omega) = \varepsilon_0 \left[1 + \chi_e(\mathbf{r}, \omega)\right]$ at any point $\mathbf{r}$ in space. There is a rich variety of ways in which $\chi_e(\omega)$ can manifest itself. Such a permittivity $\varepsilon(\mathbf{r}, \omega)$ is often called the effective permittivity.

Media in which the effective permittivity is a function of frequency are termed dispersive media or frequency dispersive media.

The above concept of simple relation between flux and field can be adapted for magnetic flux and field. By a quirk of history, the magnetic flux density $\mathbf{B}$ is related to the magnetic field $\mathbf{H}$ and magnetization $\mathbf{M}$ as

$$\mathbf{B} = \mu_0(\mathbf{H} + \mathbf{M}) \tag{7.1.7}$$

Defining a magnetic susceptibility χ_m such that $\mathbf{M} = \chi_m \mathbf{H}$, one gets the relationship that

$$\mathbf{B} = \mu_0(1 + \chi_m)\mathbf{H} \tag{7.1.8}$$

which is analogous to the relationship between electric flux $\mathbf{D}$ and electric field $\mathbf{E}$.

7.1.2 Anisotropic Media

For anisotropic media [35][p. 83]

$$\begin{aligned}\mathbf{D} &= \varepsilon_0 \mathbf{E} + \varepsilon_0 \overline{\chi}_e(\omega) \cdot \mathbf{E} \\ &= \varepsilon_0 \left[\overline{\mathbf{I}} + \overline{\chi}_e(\omega)\right] \cdot \mathbf{E} = \overline{\varepsilon}(\omega) \cdot \mathbf{E}\end{aligned} \tag{7.1.9}$$

In the above, $\overline{\varepsilon}$ is a 3×3 matrix also known as a tensor in electromagnetics. The above implies that $\mathbf{D}$ and $\mathbf{E}$ do not necessary point in the same direction —the meaning of anisotropy. (A tensor is a special kind of matrix that is often associated with a physical notion like the relation between two physical fields, whereas a matrix is not.)

Previously, we have assumed that χ_e is frequency independent. This is not usually the case, as all materials have χ_e's that are frequency dependent. (This will become clear later.) Also, since $\overline{\varepsilon}(\omega)$ is frequency dependent, we should view it as a transfer function where the input is $\mathbf{E}$ and the output is $\mathbf{D}$. This implies that in the time domain, the above relation becomes a time-convolution relation as in (7.1.4).

Similarly, for conductive media, from Ohm's law,

$$\mathbf{J} = \sigma \mathbf{E}, \tag{7.1.10}$$

This can be used in Maxwell's equations in the frequency domain to yield the definition of complex permittivity. Using the above in Ampère's law in the frequency domain, we have

$$\nabla \times \mathbf{H}(\mathbf{r}) = j\omega\varepsilon\mathbf{E}(\mathbf{r}) + \sigma\mathbf{E}(\mathbf{r}) = j\omega\underset{\sim}{\varepsilon}(\omega)\mathbf{E}(\mathbf{r}) \tag{7.1.11}$$

where the complex permittivity $\underset{\sim}{\varepsilon}(\omega) = \varepsilon - j\sigma/\omega$. Notice that Ampère's law in the frequency domain with complex permittivity in (7.1.11) is no more complicated than Ampère's law for nonconductive media. The algebra for complex numbers is no more difficult than the algebra for real numbers.[1] This is one of the strengths of phasor technique.

[1] Computer scientists call two systems having the same algebraic structure homomorphic. We will use the term "homomorphism" to denote such systems, even though the term's precise mathematical meaning is quite abstract. We will use the term "homomorphism" to imply similar structures.

For anisotropic conductive media, one has

$$\mathbf{J}(\omega) = \overline{\boldsymbol{\sigma}}(\omega) \cdot \mathbf{E}(\omega), \tag{7.1.12}$$

Here, again, due to the tensorial nature of the conductivity $\overline{\boldsymbol{\sigma}}$, the electric current $\mathbf{J}$ and electric field $\mathbf{E}$ do not necessarily point in the same direction.

The above assumes a local or pointwise linear time-invariant (LTI) relationship between the input and the output of a linear system. This need not be so. In fact, the most general LTI relationship between $\mathbf{P}(\mathbf{r}, t)$ and $\mathbf{E}(\mathbf{r}, t)$ is

$$\mathbf{P}(\mathbf{r}, t) = \varepsilon_0 \int_{-\infty}^{\infty} \iiint_{-\infty}^{\infty} \overline{\boldsymbol{\chi}}(\mathbf{r} - \mathbf{r}', t - t') \cdot \mathbf{E}(\mathbf{r}', t') d\mathbf{r}' dt' \tag{7.1.13}$$

The above is a general convolutional relationship in both space and time. The most notable fact is the nonlocality of $\mathbf{P}(\mathbf{r}, t)$, which depends on the electric field $\mathbf{E}(\mathbf{r}', t')$ at $\mathbf{r}' \neq \mathbf{r}$. In the Fourier transform space, by taking Fourier transform in both space and time, the above becomes

$$\mathbf{P}(\mathbf{k}, \omega) = \varepsilon_0 \overline{\boldsymbol{\chi}}(\mathbf{k}, \omega) \cdot \mathbf{E}(\mathbf{k}, \omega) \tag{7.1.14}$$

where

$$\overline{\boldsymbol{\chi}}(\mathbf{k}, \omega) = \int_{-\infty}^{\infty} \overline{\boldsymbol{\chi}}(\mathbf{r}, t) \exp(j\mathbf{k} \cdot \mathbf{r} - j\omega t) d\mathbf{r} dt \tag{7.1.15}$$

$$\mathbf{P}(\mathbf{k}, \omega) = \int_{-\infty}^{\infty} \mathbf{P}(\mathbf{r}, t) \exp(j\mathbf{k} \cdot \mathbf{r} - j\omega t) d\mathbf{r} dt \tag{7.1.16}$$

$$\mathbf{E}(\mathbf{k}, \omega) = \int_{-\infty}^{\infty} \mathbf{E}(\mathbf{r}, t) \exp(j\mathbf{k} \cdot \mathbf{r} - j\omega t) d\mathbf{r} dt \tag{7.1.17}$$

(The $d\mathbf{r}$ integral above is actually a threefold integral with $d\mathbf{r} = dxdydz$.) Such a medium is termed spatially dispersive as well as frequency dispersive [38][p. 6], [60]. In general,[2]

$$\overline{\boldsymbol{\varepsilon}}(\mathbf{k}, \omega) = \varepsilon_0 (1 + \overline{\boldsymbol{\chi}}(\mathbf{k}, \omega)) \tag{7.1.18}$$

where

$$\mathbf{D}(\mathbf{k}, \omega) = \overline{\boldsymbol{\varepsilon}}(\mathbf{k}, \omega) \cdot \mathbf{E}(\mathbf{k}, \omega) \tag{7.1.19}$$

The above can be extended to magnetic field and magnetic flux, yielding

$$\mathbf{B}(\mathbf{k}, \omega) = \overline{\boldsymbol{\mu}}(\mathbf{k}, \omega) \cdot \mathbf{H}(\mathbf{k}, \omega) \tag{7.1.20}$$

for a general spatial and frequency dispersive magnetic material. In optics, most materials are nonmagnetic, and hence $\mu = \mu_0$, whereas it is quite easy to make anisotropic magnetic materials in radio and microwave frequencies, such as ferrites.

[2]In the following, to be precise, one should replace the 1 with an identity operator, but it is generally implied.

7.1.3 Bianisotropic Media

In the previous section, the electric flux **D** depended only on the electric field **E**; and the magnetic flux **B**, only on the magnetic field **H**. The concept of constitutive relations can be extended to where **D** and **B** depend on both **E** and **H**. In general, one can write a general linear relationship as

$$\mathbf{D} = \overline{\varepsilon}(\omega) \cdot \mathbf{E} + \overline{\xi}(\omega) \cdot \mathbf{H} \tag{7.1.21}$$

$$\mathbf{B} = \overline{\zeta}(\omega) \cdot \mathbf{E} + \overline{\mu}(\omega) \cdot \mathbf{H} \tag{7.1.22}$$

An LTI medium for which the electric or magnetic flux is dependent on both **E** and **H** is known as a bianisotropic medium [35][p. 81].

7.1.4 Inhomogeneous Media

If $\overline{\varepsilon}$, $\overline{\xi}$, $\overline{\zeta}$, or $\overline{\mu}$ is a function of position **r**, the medium is termed an inhomogeneous medium or a heterogeneous medium. There are usually no simple solutions to problems associated with such media [38].

7.1.5 Uniaxial and Biaxial Media

Anisotropic optical materials are often encountered in optics. Examples of them are the biaxial and uniaxial media, and discussions of them are often found in optics books [61, 62, 63]. They are optical materials for which the permittivity tensor can be written as

$$\overline{\varepsilon} = \begin{pmatrix} \varepsilon_1 & 0 & 0 \\ 0 & \varepsilon_2 & 0 \\ 0 & 0 & \varepsilon_3 \end{pmatrix} \tag{7.1.23}$$

When $\varepsilon_1 \neq \varepsilon_2 \neq \varepsilon_3$, the medium is known as a biaxial medium. But when $\varepsilon_1 = \varepsilon_2 \neq \varepsilon_3$, the medium is a uniaxial medium.

In the biaxial medium case, all three components of the electric field "feel" different permittivity constants. But in the uniaxial medium, the electric field in the xy plane feels the same permittivity constant, and the electric field in the z direction feels a different permittivity constant. As shall be shown later, different light polarization will propagate with different behaviors through such a medium. This gives rise to the birefringence phenomenon, which is an interesting optical phenomenon.

7.1.6 Nonlinear Media

In the previous cases, we have assumed that $\overline{\chi}_e$ is independent of the field **E**. The relationships between **P** and **E** can be written more generally as

$$\mathbf{P} = \varepsilon_0 \overline{\chi}_e(\mathbf{E}) \cdot \mathbf{E} \tag{7.1.24}$$

where the relationship can appear in many different forms. For nonlinear media, the relationship can be nonlinear as indicated in the above. It can be easily shown that the principle of linear superposition does not hold for the above equation, a root test of linearity. The nonlinear permittivity effect is important in optics. Here, the wavelength is short, and a small change in the permittivity or refractive index can give rise to cumulative phase delay as the wave has to propagate many wavelengths through a nonlinear optical medium [64, 65, 66]. Kerr optical nonlinearity, discovered in 1875, was one of the earliest nonlinear phenomena observed [35, 61, 64].

Figure 7.2: A uniaxial medium causes different light polarizations to have different velocities, and, hence, to refract differently. (Figure by: David L. Chandler, MIT News Office, January 25, 2011)

For magnetic materials, nonlinearity can occur in the effective permeability of the medium. In other words,

$$\mathbf{B} = \overline{\mu}(\mathbf{H}) \cdot \mathbf{H} \tag{7.1.25}$$

This nonlinearity is important even at low frequencies, as in electric machinery designs [67, 68], and magnetic resonance imaging (MRI) systems [69]. The large permeability in magnetic materials is usually due to the formation of magnetic domains, which can only happen at low frequencies. The **B**-**H** relation in the metal of an electric machine is shown in Figure 7.3. The **B**-**H** is clearly nonlinear. The arrow indicates the time evolution of this relationship, and, clearly, it is not time reversible. The loss of the system is related to the area of the hysteresis loop.

7.2 Wave Phenomenon in the Frequency Domain

We have seen the emergence of wave phenomenon in the time domain previously. Given the simplicity of the frequency domain method, it will be interesting to ask how this phenomenon presents itself for time-harmonic fields. In the frequency domain, the source-free Maxwell's equations are [35][p. 429], [70][p. 107]

$$\nabla \times \mathbf{E}(\mathbf{r}) = -j\omega\mu\mathbf{H}(\mathbf{r}) \tag{7.2.1}$$

$$\nabla \times \mathbf{H}(\mathbf{r}) = j\omega\varepsilon\mathbf{E}(\mathbf{r}) \tag{7.2.2}$$

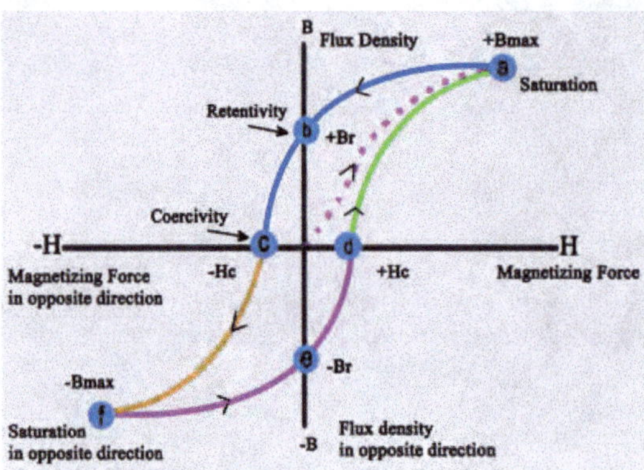

Figure 7.3: In an electric machine, the relation between the **B** flux and the magnetic **H** is usually nonlinear, as shown in the picture. Moreover, the nonlinear system is not time reversible, yielding a **B-H** relation as shown (Figure by: https://electricalacademia.com/electromagnetism/hysteresis-loop-magnetization-curve/).

Taking the curl of (7.2.1) and then substituting (7.2.2) into its right-hand side, one obtains

$$\nabla \times \nabla \times \mathbf{E}(\mathbf{r}) = -j\omega\mu\nabla \times \mathbf{H}(\mathbf{r}) = \omega^2\mu\varepsilon\mathbf{E}(\mathbf{r}) \tag{7.2.3}$$

The above is the vector Helmholtz equation. Again, using the BAC-CAB identity that

$$\nabla \times \nabla \times \mathbf{E} = \nabla(\nabla \cdot \mathbf{E}) - \nabla \cdot \nabla\mathbf{E} = \nabla(\nabla \cdot \mathbf{E}) - \nabla^2\mathbf{E} \tag{7.2.4}$$

and that $\nabla \cdot \mathbf{E} = 0$ in a source-free medium, (7.2.3) becomes

$$(\nabla^2 + \omega^2\mu\varepsilon)\mathbf{E}(\mathbf{r}) = 0 \tag{7.2.5}$$

This is known as the Helmholtz wave equation or just the Helmholtz equation.[3]

To see the wave phenomenon lucidly, we simply let $\mathbf{E} = \hat{x}E_x(z)$, a field pointing in the x direction, but varying only in the z direction. Evidently, $\nabla \cdot \mathbf{E}(\mathbf{r}) = \partial E_x(z)/\partial x = 0$. Then with $\partial/\partial x = 0$ and $\partial/\partial y = 0$, (7.2.5) simplifies to

$$\left(\frac{d^2}{dz^2} + k^2\right) E_x(z) = 0 \tag{7.2.6}$$

[3]For a comprehensive review of this topic, one may read the lecture notes [48].

where $k^2 = \omega^2\mu\varepsilon = \omega^2/c^2$ where $c = 1/\sqrt{\mu\varepsilon}$ is the velocity of light. The general solution to (7.2.6) is of the form

$$E_x(z) = E_{0+}e^{-jkz} + E_{0-}e^{jkz} \tag{7.2.7}$$

Since, time-harmonic field is assumed, one can convert the above back to the time domain using phasor technique, or by using $E_x(z,t) = \Re e[E_x(z,\omega)e^{j\omega t}]$, yielding

$$E_x(z,t) = |E_{0+}|\cos(\omega t - kz + \alpha_+) + |E_{0-}|\cos(\omega t + kz + \alpha_-) \tag{7.2.8}$$

where we have assumed that $E_{0\pm}$ are complex numbers such that

$$E_{0\pm} = |E_{0\pm}|e^{j\alpha_\pm} \tag{7.2.9}$$

The physical picture of the above expressions can be appreciated by rewriting

$$\cos(\omega t \mp kz + \alpha_\pm) = \cos\left[\frac{\omega}{c}(ct \mp z) + \alpha_\pm\right] \tag{7.2.10}$$

where we have used the fact that $k = \frac{\omega}{c}$. The above functions are of the form $F(ct \mp x)$. As mentioned before in (3.2.14) and (3.2.15), these are traveling waves. One can see that the first term on the right-hand side of (7.2.8) is a sinusoidal plane wave traveling to the right, while the second term is a sinusoidal plane wave traveling to the left, both with velocity c.[4] The above plane wave is uniform and a constant in the xy plane but propagating in the z direction. Hence, it is also called a uniform plane wave in 1D.

Moreover, for a fixed t or at $t = 0$, the sinusoidal functions are proportional to $\cos(\mp kz + \alpha_\pm)$. This is a periodic function in z with period $2\pi/k$ which is the wavelength λ, or that

$$k = \frac{2\pi}{\lambda} = \frac{\omega}{c} = \frac{2\pi f}{c} \tag{7.2.11}$$

One can see that because c is a humongous number in free-space electromagnetics, λ can be very large. You can plug in the frequency of your local AM 920 station, operating at 920 KHz, to see that λ is approximately 320 m, the size of several football fields.

The above analysis still holds true even if ε and μ are dispersive, but are real numbers. In this case, the velocity c of the wave is the velocity of its phase, or the phase velocity of the mono-chromatic, time-harmonic, or CW wave. Since ε and μ can be functions of frequency, the velocity $c = 1/\sqrt{\mu\varepsilon}$ can be different for different frequencies. As shall be shown, this gives rise to pulse distortion.

7.3 Uniform Plane Waves in 3D

By repeating the previous derivation for a homogeneous, lossless, dispersive medium, the vector Helmholtz equation for a source-free medium is given by [48]

$$\nabla \times \nabla \times \mathbf{E} - \omega^2\mu\varepsilon\mathbf{E} = 0 \tag{7.3.1}$$

[4]We shall learn later that this is the phase velocity of the wave.

By the same derivation as before for the free-space case, since $\nabla \cdot \mathbf{E} = 0$ due to source-free medium, one has

$$\nabla^2 \mathbf{E} + \omega^2 \mu\varepsilon \mathbf{E} = 0 \tag{7.3.2}$$

The general solution to (7.3.2) is hence

$$\mathbf{E} = \mathbf{E}_0 e^{-jk_x x - jk_y y - jk_z z} = \mathbf{E}_0 e^{-j\mathbf{k}\cdot\mathbf{r}} \tag{7.3.3}$$

where $\mathbf{k} = \hat{x}k_x + \hat{y}k_y + \hat{z}k_z$, $\mathbf{r} = \hat{x}x + \hat{y}y + \hat{z}z$ and $\mathbf{E}_0$ is a constant vector. Here, $\mathbf{k}$ is the k vector or the wave vector, while $\mathbf{r}$ is the position vector. And upon substituting (7.3.3) into (7.3.2), doing the derivatives, it is seen that

$$k_x^2 + k_y^2 + k_z^2 = \omega^2 \mu\varepsilon = \mathbf{k} \cdot \mathbf{k} \tag{7.3.4}$$

This is called the dispersion relation for a plane wave. The above is also the equation for a sphere in a 3D $\mathbf{k}$ space, which is also called the Ewald sphere. Notice that in (7.3.4) is also satisfied if $\mathbf{k} = -\mathbf{k}$, implying that (7.3.3) will admit solution propagating in the $-\mathbf{k}$ direction as well.

In general, k_x, k_y, and k_z can be arbitrary and even complex numbers as long as this relation (7.3.4) is satisfied. To simplify the discussion, we will focus on the case where k_x, k_y, and k_z are all real numbers. When this is the case, the vector function in (7.3.3) represents a uniform plane wave propagating in the $\mathbf{k}$ direction. As can be seen, when $\mathbf{k} \cdot \mathbf{r} =$ constant, it is represented by all points of $\mathbf{r}$ that represents a flat plane (see Figure 7.4). This flat plane represents the constant phase wave front. By increasing the constant, we obtain different planes for progressively changing phase fronts.[5]

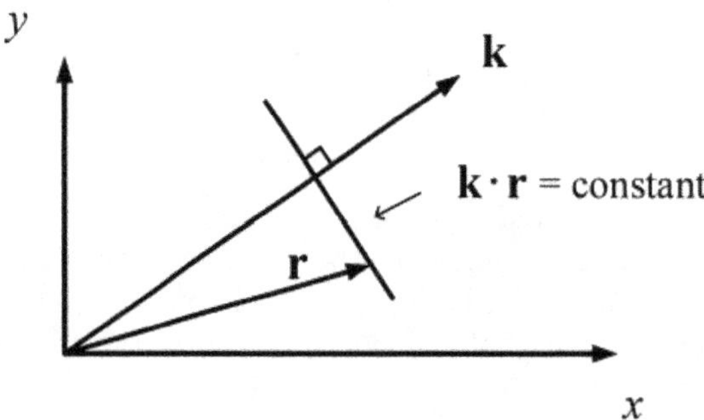

Figure 7.4: A figure showing the geometrical meaning of $\mathbf{k} \cdot \mathbf{r}$ equal to a constant. It is a flat plane that defines the wavefront of a plane wave.

[5]In the $\exp(j\omega t)$ time convention, this phase front is decreasing, whereas in the $\exp(-i\omega t)$ time convention, this phase front is increasing. The $\exp(j\omega t)$ time convention is often used in electrical engineering, while the $\exp(-i\omega t)$ time convention is used in optics and physics.

Further, since $\nabla \cdot \mathbf{E} = 0$, we can show that

$$\begin{aligned}\nabla \cdot \mathbf{E} &= \nabla \cdot \mathbf{E}_0 e^{-jk_x x - jk_y y - jk_z z} = \nabla \cdot \mathbf{E}_0 e^{-j\mathbf{k}\cdot\mathbf{r}} \\ &= (-\hat{x}jk_x - \hat{y}jk_y - \hat{z}jk_z) \cdot \mathbf{E}_0 e^{-j\mathbf{k}\cdot\mathbf{r}} \\ &= -j(\hat{x}k_x + \hat{y}k_y + \hat{z}k_z) \cdot \mathbf{E} = 0 \end{aligned} \tag{7.3.5}$$

or that

$$\mathbf{k} \cdot \mathbf{E}_0 = \mathbf{k} \cdot \mathbf{E} = 0 \tag{7.3.6}$$

Thus, both $\mathbf{E}$ and $\mathbf{E}_0$ are orthogonal to $\mathbf{k}$ for a uniform plane wave.

The above exercise shows that whenever $\mathbf{E}$ is a plane wave, and when the ∇ operator operates on such a vector function, one can simply let $\nabla \rightarrow -j\mathbf{k}$. Hence, in a source-free homogenous medium, Faraday's law becomes

$$\nabla \times \mathbf{E} = -j\omega\mu\mathbf{H} \tag{7.3.7}$$

The above equation simplifies to

$$-j\mathbf{k} \times \mathbf{E} = -j\omega\mu\mathbf{H} \tag{7.3.8}$$

or that

$$\mathbf{H} = \frac{\mathbf{k} \times \mathbf{E}}{\omega\mu} \tag{7.3.9}$$

Similar to (7.3.3), we can define

$$\mathbf{H} = \mathbf{H}_0 e^{-jk_x x - jk_y y - jk_z z} = \mathbf{H}_0 e^{-j\mathbf{k}\cdot\mathbf{r}} \tag{7.3.10}$$

Then using (7.3.3) in (7.3.9), it is clear that

$$\mathbf{H}_0 = \frac{\mathbf{k} \times \mathbf{E}_0}{\omega\mu} \tag{7.3.11}$$

We can assume that $\mathbf{E}_0$ and $\mathbf{H}_0$ are real vectors for easier visualization. Then $\mathbf{E}_0$, $\mathbf{H}_0$ and $\mathbf{k}$ form a right-handed orthogonal system, or that $\mathbf{E}_0 \times \mathbf{H}_0$ points in the direction of $\mathbf{k}$. (This also implies that $\mathbf{E}$, $\mathbf{H}$ and $\mathbf{k}$ form a right-handed orthogonal system as well.) Such a wave, where the electric field and magnetic field are transverse to the direction of propagation, is called a transverse electromagnetic (TEM) wave. Figure 7.5 shows that $\mathbf{k} \cdot \mathbf{E} = 0$, and that $\mathbf{k} \times \mathbf{E}$ points in the direction of $\mathbf{H}$ as shown in (7.3.9). Figure 7.5 also shows that $\mathbf{k}$, $\mathbf{E}$, and $\mathbf{H}$ are orthogonal to each other.

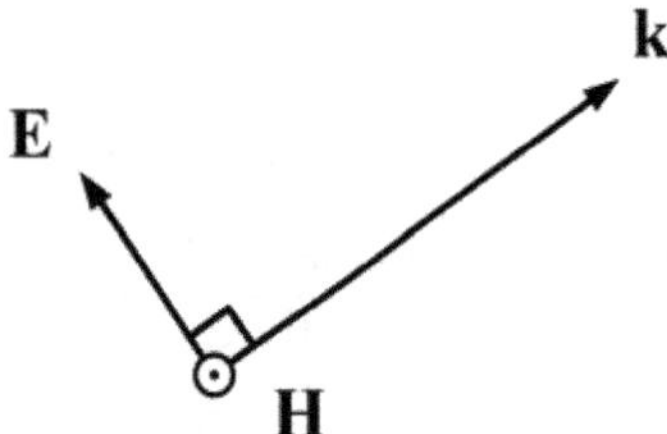

Figure 7.5: The **E**, **H**, and **k** together form a right-hand coordinate system, obeying the right-hand rule. Namely, $\mathbf{E} \times \mathbf{H}$ points in the direction of **k**.

Since in general, $\mathbf{E}_0$ and $\mathbf{H}_0$ can be complex vectors, because they are phasors, we need to show the above relationship for the more general case. From (7.3.9), one can show, using the BAC-CAB formula, assuming **k** is real, that

$$\mathbf{E} \times \mathbf{H}^* = \mathbf{E} \cdot \mathbf{E}^* \frac{\mathbf{k}}{\omega\mu} = |\mathbf{E}|^2 \frac{\mathbf{k}}{\omega\mu} \tag{7.3.12}$$

(It is important to note that the magnitude square of a complex vector $|\mathbf{E}|^2$ is $\mathbf{E} \cdot \mathbf{E}^*$, whereas for a real vector, it is $\mathbf{E} \cdot \mathbf{E}$. The latter definition does not guarantee positive definiteness for complex vectors.) But $\mathbf{E} \times \mathbf{H}^*$ is the direction of power flow, and it is in fact a real vector pointing in the **k** direction. This is also required by the Poynting's theorem.

Furthermore, we can show in general that[6]

$$|\mathbf{H}|^2 = \frac{|\mathbf{k} \times \mathbf{E}|^2}{(\omega\mu)^2} = \frac{\varepsilon}{\mu}|\mathbf{E}|^2 \tag{7.3.13}$$

or that

$$|\mathbf{H}| = \sqrt{\frac{\varepsilon}{\mu}}|\mathbf{E}| = \frac{1}{\eta}|\mathbf{E}| \tag{7.3.14}$$

where the quantity

$$\eta = \sqrt{\frac{\mu}{\varepsilon}} \tag{7.3.15}$$

is call the intrinsic impedance. For vacuum or free space, it is about 377 $\Omega \approx 120\pi\ \Omega$.

Notice that the above analysis holds true as long as ε and μ are real, but they can be frequency dispersive, since we are considering a monochromatic or time-harmonic field. Besides, for a monochromatic signal, the analysis in section 7.2 still applies except that the velocity of light is now given by $c = 1/\sqrt{\mu\varepsilon}$. As we shall see, this velocity is the phase velocity of the monochromatic wave. In the above, when k_x, k_y, and k_z are not all real, the wave is known as inhomogeneous wave.[7]

[6]Since $|\mathbf{k} \times \mathbf{E}|^2 = (\mathbf{k} \times \mathbf{E}) \cdot (\mathbf{k} \times \mathbf{E}^*)$, we can use the cyclic formula $\mathbf{a} \cdot (\mathbf{b} \times \mathbf{c}) = \mathbf{c} \cdot (\mathbf{a} \times \mathbf{b}) = \mathbf{b} \cdot (\mathbf{c} \times \mathbf{a})$ to simplify it.

[7]The term "inhomogeneous plane wave" is used sometimes, but it is a misnomer since there is no more a planar wave front in this case.

Exercises for Lecture 7

Problem 7-1: For uniaxial medium, the permittivity tensor is given by:

$$\overline{\varepsilon} = \begin{pmatrix} \varepsilon & 0 & 0 \\ 0 & \varepsilon & 0 \\ 0 & 0 & \varepsilon_z \end{pmatrix} \tag{E7.1}$$

Assume a plane wave propagating as

$$\mathbf{E} = \mathbf{E}_0 e^{-j\mathbf{k}\cdot\mathbf{r}} \tag{E7.2}$$

(i) From Maxwell's equations, show that the following equation must be satisfied:

$$\mathbf{k} \times \mathbf{k} \times \mathbf{E} = -\omega^2 \mu \overline{\varepsilon} \cdot \mathbf{E} \tag{E7.3}$$

(ii) When the electric field is polarized in the xy plane, ε_z is not felt by the wave. This is called the ordinary wave. Assuming that $\mathbf{k}$ is in the xz plane, show that the dispersion relation from the above equation simplifies to:

$$k_x^2 + k_z^2 = \omega^2 \mu \varepsilon \tag{E7.4}$$

(iii) When the electric field is polarized in the xz plane, ε_z is now felt by the wave. The wave is now called the extraordinary wave. Assuming that $\mathbf{k}$ is in the xz plane, or $k_y = 0$, show that the electric flux has to be of the form:

$$\mathbf{D} = \left(\hat{x} - \hat{z}\frac{k_x}{k_z} \right) \varepsilon E_0 e^{-j\mathbf{k}\cdot\mathbf{r}} \tag{E7.5}$$

And the corresponding electric field is:

$$\mathbf{E} = \left(\hat{x} - \hat{z}\frac{k_x \varepsilon}{k_z \varepsilon_z} \right) E_0 e^{-j\mathbf{k}\cdot\mathbf{r}} \tag{E7.6}$$

Explain your reasoning.

(iv) From (E7.3), for the extraordinary wave, show that the dispersion relation can be reduced to:

$$\frac{k_x^2}{\omega^2 \mu \varepsilon_z} + \frac{k_z^2}{\omega^2 \mu \varepsilon} = 1 \tag{E7.7}$$

(v) The equations (E7.4) and (E7.7) are equations of surfaces known as k-surfaces. Please draw these two surfaces on the same graph (in 2D, it will just be a contour), and explain the physical meanings of the two surfaces.

Lecture 8

Lossy Media, Lorentz Force Law, Drude-Lorentz-Sommerfeld Model

In the previous lecture, we realized the power of phasor technique or the frequency domain analysis. The analysis of a frequency dispersive medium where ε is frequency dependent is similar to that of free space or vacuum. The two problems are mathematically "homomorphic" to each other. In this lecture, we will generalize to the case where ε becomes a complex number, called the complex permittivity. Using phasor technique, this way of solving Maxwell's equations is still homomorphic to solving Maxwell's equations in free space. The analysis is greatly simplified as a result!

8.1 Plane Waves in Lossy Conductive Media

Previously, we have derived the plane wave solution for a lossless homogeneous medium. Since the algebra of complex numbers is similar to that of real numbers, the derivation can be generalized to a conductive medium by invoking mathematical "homomorphism", "since the algebra of real number is similar to the algebra of complex number". In a word, in a conductive medium, one only needs to replace the permittivity with a complex permittivity, as repeated here. When conductive loss is present, $\sigma \neq 0$, and $\mathbf{J} = \sigma\mathbf{E}$. Then generalized Ampère's law becomes

$$\nabla \times \mathbf{H} = j\omega\varepsilon\mathbf{E} + \sigma\mathbf{E} = j\omega\left(\varepsilon + \frac{\sigma}{j\omega}\right)\mathbf{E} \tag{8.1.1}$$

A complex permittivity can be defined as $\underset{\sim}{\varepsilon} = \varepsilon - j\frac{\sigma}{\omega}$. Eq. (8.1.1) can be rewritten as

$$\nabla \times \mathbf{H} = j\omega\underset{\sim}{\varepsilon}\mathbf{E} \tag{8.1.2}$$

This equation is of the same form as the source-free Ampère's law in the frequency domain for a lossless medium where ε is completely real. In a conductive medium, the corresponding Helmholtz equation is

$$\left(\nabla^2 + \omega^2\mu\underset{\sim}{\varepsilon}\right)\mathbf{E} = 0 \tag{8.1.3}$$

Using the same method as before, a wave solution is[1]

$$\mathbf{E} = \mathbf{E}_0 e^{-j\mathbf{k}\cdot\mathbf{r}} \tag{8.1.4}$$

with the dispersion relation which is now given by

$$\mathbf{k}\cdot\mathbf{k} = k_x^2 + k_y^2 + k_z^2 = \omega^2\mu\underset{\sim}{\varepsilon} \tag{8.1.5}$$

Since $\underset{\sim}{\varepsilon}$ is complex now, k_x, k_y, and k_z cannot be all real. Equation (8.1.5) has been derived previously by assuming that $\mathbf{k}$ is a real vector. When $\mathbf{k} = \mathbf{k}' - j\mathbf{k}''$ is a complex vector, some of the previous derivations for a real $\mathbf{k}$ vector may not be correct here for a complex $\mathbf{k}$ vector. It is also difficult to visualize mentally a complex $\mathbf{k}$ vector that is supposed to indicate the direction in which the wave is propagating. Here, $\mathbf{k}'$ and $\mathbf{k}''$ can be vectors pointing in different directions, and the wave can decay and oscillate in different directions.

So again, for simplicity and further physical insight, we look at the simplified case where

$$\mathbf{E} = \hat{x}E_x(z) \tag{8.1.6}$$

so that $\nabla\cdot\mathbf{E} = \partial_x E_x(z) = 0$,[2] and let $\mathbf{k} = \hat{z}k = \hat{z}\omega\sqrt{\mu\underset{\sim}{\varepsilon}} = \hat{z}(k' - jk'')$. This wave is constant in the xy plane, and hence, is a plane wave. Furthermore, in this manner, we are requiring that the wave decays and propagates (or oscillates) only in the z direction. For such a simple plane wave,

$$\mathbf{E} = \hat{x}\mathbf{E}_x(z) = \hat{x}E_0 e^{-jkz} \tag{8.1.7}$$

where $k = \omega\sqrt{\mu\underset{\sim}{\varepsilon}}$ (since $\mathbf{k}\cdot\mathbf{k} = k^2 = \omega^2\mu\underset{\sim}{\varepsilon}$ is still true).

Faraday's law, by letting $\nabla \to j\mathbf{k}$, gives rise to

$$\mathbf{H} = \frac{\mathbf{k}\times\mathbf{E}}{\omega\mu} = \hat{y}\frac{kE_x(z)}{\omega\mu} = \hat{y}\sqrt{\frac{\underset{\sim}{\varepsilon}}{\mu}}E_x(z) \tag{8.1.8}$$

where the $\mathbf{k}$ vector is defined shortly after (8.1.6) above, and $k = \omega\sqrt{\mu\underset{\sim}{\varepsilon}}$, a complex number. It is seen that $\mathbf{H} = \hat{y}H_y$, and that

$$E_x/H_y = \sqrt{\frac{\mu}{\underset{\sim}{\varepsilon}}} \tag{8.1.9}$$

The above is the generalization of the intrinsic impedance defined in (7.3.15) to a lossy conductive medium.

8.1.1 High-Conductivity Case

When the medium is highly conductive, $\sigma \to \infty$, and $\underset{\sim}{\varepsilon} = \varepsilon - j\frac{\sigma}{\omega} \approx -j\frac{\sigma}{\omega}$. In other words, when $|\frac{\sigma}{\omega}| \gg \varepsilon$, the conduction current dominates over the displacement current. Thus, the following

[1] With the assumption of the wave solution below, a derivation operator $\nabla \to -j\mathbf{k}$.

[2] This condition is necessary to arrive at the Helmholtz equation (8.1.3).

approximation can be made, namely,

$$k = \omega\sqrt{\mu \underset{\sim}{\varepsilon}} \simeq \omega\sqrt{-\mu\frac{j\sigma}{\omega}} = \sqrt{-j\omega\mu\sigma} \tag{8.1.10}$$

Taking $\sqrt{-j} = \frac{1}{\sqrt{2}}(1-j)$, we find for a highly conductive medium that[3]

$$k \simeq (1-j)\sqrt{\frac{\omega\mu\sigma}{2}} = k' - jk'' \tag{8.1.11}$$

For a plane wave, e^{-jkz} then becomes

$$e^{-jkz} = e^{-jk'z-k''z} \tag{8.1.12}$$

By converting the above phasor back to the time domain, in the z direction, this plane wave decays exponentially as well as oscillates. The reason being that a conductive medium is lossy, and it absorbs energy from the plane wave. This is similar to the resistive loss we see in the resistive circuit. The penetration depth of this wave is then

$$\delta = \frac{1}{k''} = \sqrt{\frac{2}{\omega\mu\sigma}} \tag{8.1.13}$$

This distance δ, the penetration depth, is called the skin depth of a plane wave propagating in a highly lossy conductive medium where conduction current dominates over displacement current, or $\sigma \gg \omega\varepsilon$. This happens for a radio wave propagating in the saline solution of the ocean, the good Earth, or a wave propagating in highly conductive metal, like your induction cooker.

8.1.2 Low-Conductivity Case

When the conductivity is low, namely, when the displacement current is larger than the conduction current, then $\frac{\sigma}{\omega\varepsilon} \ll 1$, and we have[4]

$$\begin{aligned} k &= \omega\sqrt{\mu\left(\varepsilon - j\frac{\sigma}{\omega}\right)} = \omega\sqrt{\mu\varepsilon\left(1 - \frac{j\sigma}{\omega\varepsilon}\right)} \\ &\approx \omega\sqrt{\mu\varepsilon}\left(1 - j\frac{1}{2}\frac{\sigma}{\omega\varepsilon}\right) = k' - jk'' \end{aligned} \tag{8.1.14}$$

The above is the approximation to $k = k' - jk''$ for a low conductivity medium where conduction current is much smaller than displacement current.

The term $\frac{\sigma}{\omega\varepsilon}$ is called the loss tangent of a lossy medium. It is the ratio of the conduction current to the displacement current in a lossy conductive medium. In general, in a lossy medium $\varepsilon = \varepsilon' - j\varepsilon''$, and $\varepsilon''/\varepsilon'$ is called the loss tangent of the medium. It is to be noted that in the optics and physics community, by a quirk of history, $e^{-i\omega t}$ time convention is preferred. In that case, we need to do the switch $j \rightarrow -i$, and a lossy medium is denoted by $\varepsilon = \varepsilon' + i\varepsilon''$.

[3] A function $z^{1/2}$ is known as a multi-value function. For simplicity, we assume it is a single-value function.

[4] In the following equation, we have made use of the approximation that $(1+x)^n \approx 1+nx$ when x is small, which can also be justified by Taylor series expansion.

8.2 Lorentz Force Law

The Lorentz force law is the generalization of Coulomb's law for forces between two charges. The Lorentz force law includes the presence of a magnetic field. It is given by

$$\mathbf{F} = q\mathbf{E} + q\mathbf{v} \times \mathbf{B} \tag{8.2.1}$$

The first term on the right-hand side is the electric force similar to the statement of Coulomb's law, while the second term is the magnetic force called the $\mathbf{v} \times \mathbf{B}$ force. This law can also be written in terms of the force density $\mathbf{f}$ which is the force on the charge density, instead of force on a single charge. By so doing, we arrive at

$$\mathbf{f} = \varrho\mathbf{E} + \varrho\mathbf{v} \times \mathbf{B} = \varrho\mathbf{E} + \mathbf{J} \times \mathbf{B} \tag{8.2.2}$$

where ϱ is the charge density, and one can identify the current $\mathbf{J} = \varrho\mathbf{v}$. We shall see that the permittivity tensor of a gyrotropic medium is very much governed by this law.

The Lorentz force law can also be derived from the integral form of Faraday's law, if one assumes that the law is applied to a moving loop intercepting a magnetic flux [71]. In other words, the Lorentz force law and Faraday's law are commensurate with each other.

8.3 Drude-Lorentz-Sommerfeld (DLS) Model

In the previous lecture, we have seen how loss can be introduced by having a conduction current flowing in a medium. Now that we have learned the versatility of the frequency domain method and phasor technique, other loss mechanisms can be easily incorporated in the formulation.

First, let us look at the simple constitutive relation where

$$\mathbf{D} = \varepsilon_0\mathbf{E} + \mathbf{P} \tag{8.3.1}$$

We begin with a simple model where

$$\mathbf{P} = \varepsilon_0\chi\mathbf{E} \tag{8.3.2}$$

where χ is the electric susceptibility. To see how $\chi(\omega)$ can be derived, we will study the Drude-Lorentz-Sommerfeld model for a simplified view. This is usually just known as the Drude model or the Lorentz model in many textbooks, although Sommerfeld also contributed to it. These models, the Drude, Debye, and Lorentz models, can be unified in one equation as shall be shown.

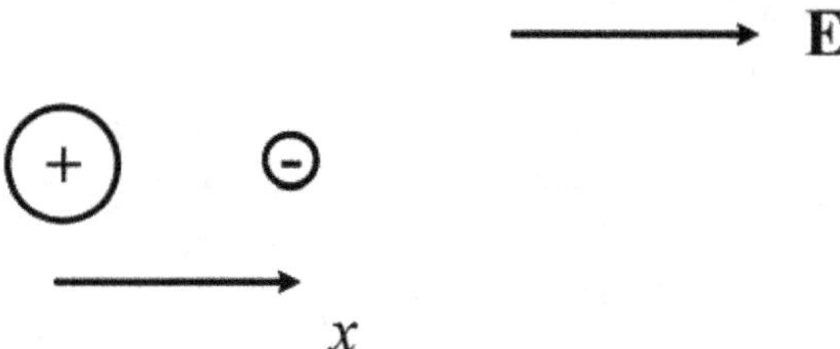

Figure 8.1: Polarization of an atom in the presence of an electric field. Here, it is assumed that the electron is weakly bound or unbound to the nucleus of the atom.

8.3.1 Cold Collisionless Plasma Medium

A plasma medium is one where the electrons of an atom have broken free from the nuclear force of the atom so that the electrons are free to roam about. This happens in a gas that is highly heated, or a gas in the atmosphere that is rarefied, where the atoms are in an environment with very low pressure. Assuming that the free electrons are in abundance, and the atoms are far apart, we can first study, for an isolated atom, how a simple electron is driven by an electric field $\mathbf{E}$ in the absence of a magnetic field $\mathbf{B}$.[5] If the electron is free to move, then the force acting on it, from the Lorentz force law, is just $-e\mathbf{E}$ where $q = -e$ is the charge of the electron (see Figure 8.1). Using a classical model, then from Newton's law, assuming a one-dimensional case, it follows that

$$m_e \frac{d^2x}{dt^2} = -eE \tag{8.3.3}$$

where the left-hand side is due to the inertial force of the mass of the electron, and the right-hand side is the electric force acting on a charge of $-e$ coulomb. Here, we assume that $\mathbf{E}$ points in the x-direction, and we neglect the vector nature of the electric field or that we assume that both x and $\mathbf{E}$ are in the same direction. Writing the above in the frequency domain for time-harmonic fields, and using phasor technique, assuming that x and E are now phasors, one gets

$$-\omega^2 m_e x = -eE \tag{8.3.4}$$

The above implies that the inertial force of the electron, given by $-\omega^2 m_e x$, is of the same polarity as the electric field force on the electron which is $-eE$. From this, we have

$$x = \frac{e}{\omega^2 m_e} E \tag{8.3.5}$$

implying that the displacement x is linearly proportional to the electric field amplitude E, or they are in phase. This, for instance, can happen in a plasma medium where the atoms are ionized, and the electrons are free to roam [73]. Hence, we assume that the positive ions are more massive, sluggish, and move very little compared to the agile electrons when an electric field is applied.

[5]Even if $\mathbf{B} \neq 0$, the $\mathbf{v} \times \mathbf{B}$ force is small if the velocity of the electron is much smaller than the speed of light (see Feynman [72, II-13-6]).

The dipole moment formed by the displaced electron away from the ion due to the electric field is then

$$p = -ex = -\frac{e^2}{\omega^2 m_e} E \tag{8.3.6}$$

for one electron. When there are N electrons per unit volume, the dipole moment density is then given by

$$P = Np = -\frac{Ne^2}{\omega^2 m_e} E \tag{8.3.7}$$

In general, $\mathbf{P}$ and $\mathbf{E}$ point in opposite directions for this medium, and we can write

$$\mathbf{P} = -\frac{Ne^2}{\omega^2 m_e}\mathbf{E} = -\frac{{\omega_p}^2}{\omega^2}\varepsilon_0\mathbf{E} \tag{8.3.8}$$

where we have defined ${\omega_p}^2 = Ne^2/(m_e\varepsilon_0)$ where ω_p is the plasma frequency of the medium. Then,

$$\mathbf{D} = \varepsilon_0\mathbf{E} + \mathbf{P} = \varepsilon_0\left(1 - \frac{{\omega_p}^2}{\omega^2}\right)\mathbf{E} \tag{8.3.9}$$

In this manner, we see that the effective permittivity of the plasma medium is

$$\varepsilon(\omega) = \varepsilon_0\left(1 - \frac{{\omega_p}^2}{\omega^2}\right) \tag{8.3.10}$$

What the above math is saying is that the electric field $\mathbf{E}$ induces a dipole moment density $\mathbf{P}$ that is negative to $\varepsilon_0\mathbf{E}$, or the vacuum part of the contribution to $\mathbf{D}$. This negative dipole density cancels the contribution to the electric flux from the vacuum $\varepsilon_0\mathbf{E}$. For low frequency, the effective permittivity can be negative, disallowing the propagation of a wave as we shall see. Hence, $\varepsilon < 0$ if

$$\omega < \omega_p = \sqrt{N/(m_e\varepsilon_0)}e$$

Since $k = \omega\sqrt{\mu\varepsilon}$, if ε is negative, $k = -j\alpha$ becomes pure imaginary, and a wave such as e^{-jkz} decays exponentially as $e^{-\alpha z}$. This is also known as an evanescent wave. In other words, the wave cannot propagate through such a medium. Our ionosphere is such a medium. The ionospheric plasma shields out electromagnetic waves that are below the plasma frequency ω_p.

Therefore, it was extremely fortuitous that Guglielmo Marconi, in 1901, was able to send a radio signal from Cornwall, England, to Newfoundland, Canada. Naysayers thought his experiment would not succeed, as the radio signal would propagate to outer space and never return. Fortunately, it is the presence of the ionosphere that bounces the radio wave back to Earth, making his experiment a resounding success and a very historic one! Serendipity has often occurred in science and technology development. This historic experiment heralded the age of wireless communications.

This experiment also triggered interests into research on the ionosphere. It was an area again where Oliver Heaviside made contributions; as a result, a layer of the ionosphere is named Heaviside layer or Kennelly-Heaviside layer [74]. If you listen carefully to the Broadway musical "Cats" by Andrew Lloyd Weber, there is a mention of the Heaviside layer in one of the songs!

8.3.2 Bound Electron Case—Heuristics

Before we proceed further, we introduce a heuristic picture of how an electron would move about in a solid. A deeper understanding of this requires understanding the quantum field theory of solids [75], but an approximate picture can be obtained by studying Figure 8.2.

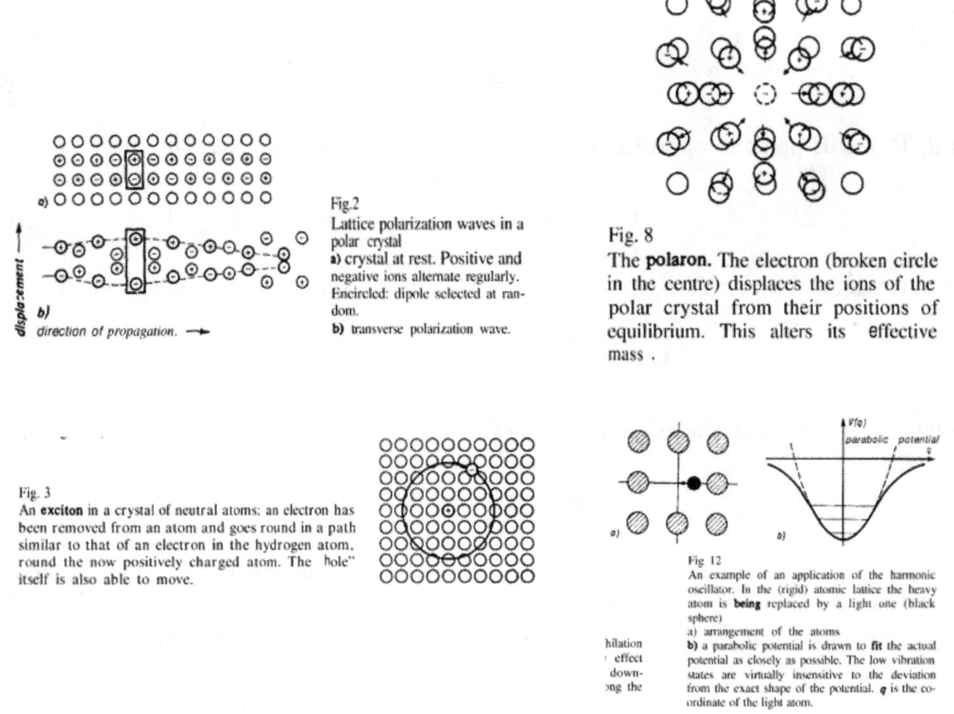

Figure 8.2: These figures are from Quantum Field Theory of Solids by H. Haken. They provide a heuristic explanation of the electromagnetic property of solids (Figure by: Hermann Haken [75]).

In Figure 8.2, the Sub-Fig. 2 illustrates the propagation of an electromagnetic wave through a polarizable medium. Sub-Fig. 3 indicates that in a semiconductor material, the electron is unbound from the nucleus forming a electron-hole pair. The electron is attracted to the hole similar to an electron around the nucleus of the hydrogen atom. Such an electron-hole pair is called an exciton. Sub-Fig. 8 shows that an electron in a medium behaves like a polaron. This term is used to describe an electron with a collection of atoms, and not a free electron, because it polarizes the molecules around it. Due to its coupling to the environment, it moves about with an effective mass. Sub-Fig. 12 shows the trapping potential of an electron in the lattice. When the displacement of the electron is small from the equilibrium point, it behaves like a simple harmonic oscillator. But when its displacement is large, it becomes an anharmonic oscillator. These figures provide us with a heuristic physical understanding of the electromagnetic properties of solids.

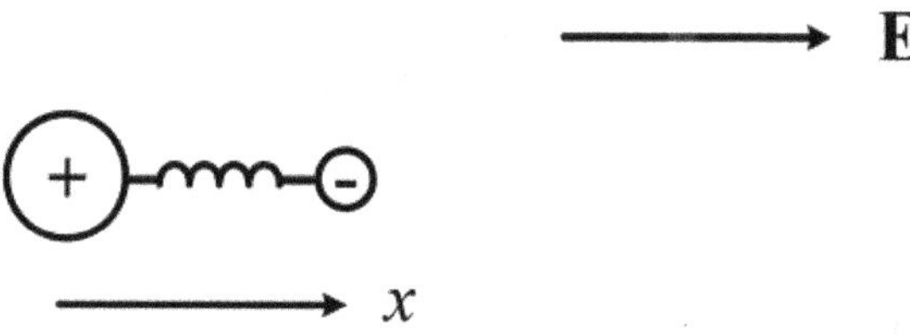

Figure 8.3: The electron with an effective mass is bound to the ion by an attractive force. This can be approximately modeled by a spring providing a restoring force to the electron.

8.3.3 Bound Electron Case—Simple Math Model

The above model of a cold collisionless plasma in (8.3.3) in subsection 8.3.1 can be generalized to the case where the electron is bound to the ion, but the ion now provides a restoring force similar to that of a classical model or a spring (see Figure 8.3), namely,

$$m_e \frac{d^2 x}{dt^2} + \kappa x = -eE \tag{8.3.11}$$

We assume that the ion provides a restoring force just like Hooke's law. Again, for a time-harmonic field, (8.3.11) can be solved easily using phasor technique in the frequency domain to yield

$$x = \frac{e}{(\omega^2 m_e - \kappa)} E = \frac{e}{(\omega^2 - {\omega_0}^2) m_e} E \tag{8.3.12}$$

where we have defined ${\omega_0}^2 m_e = \kappa$. The above is the typical solution of a lossless harmonic oscillator (pendulum) driven by an external force, in this case the electric field. The dipole moment due to an electric field is then

$$p = -ex = -\frac{e^2}{(\omega^2 - {\omega_0}^2) m_e} E, \qquad P = -Np = -\frac{\omega_p^2}{(\omega^2 - {\omega_0}^2)} \varepsilon E \tag{8.3.13}$$

Therefore, when the frequency is low or $\omega = 0$, polarization density P is of the same polarity as the applied electric field E, contributing to a positive dipole moment. It contributes positively to the displacement flux $\mathbf{D}$ via $\mathbf{P}$. However, when $\omega > \omega_0$, P can be out of phase with the applied field E as in the plasma medium.

8.3.4 Damping or Dissipative Case

Using a classical model, equation (8.3.11) can be generalized to the case when frictional, damping, or dissipative forces are present, or that

$$m_e \frac{d^2 x}{dt^2} + m_e \Gamma \frac{dx}{dt} + \kappa x = -eE \tag{8.3.14}$$

The second term on the left-hand side is a force that is proportional to the velocity $v = dx/dt$ of the electron. This is the hallmark of frictional force. Frictional force is due to the collision of the electrons with the background ions or lattice. Hence, it is proportional to the destruction (or change) of momentum ($m_e \frac{dx}{dt} = m_e v$) of an electron. In the average sense, the destruction of the momentum $m_e v$ is given by the product of the collision frequency Γ and the momentum. In the above, Γ has the unit of frequency, and for plasma, and conductor, it can be regarded as a collision frequency. A sanity check shows that the second term above on the left-hand side has the same unit as the first term.

Solving the above in the frequency domain, one gets

$$x = \frac{e}{(\omega^2 - j\omega\Gamma - \omega_0{}^2)m_e} E \tag{8.3.15}$$

Following the same procedure in arriving at (8.3.7), we get

$$P = \frac{\omega_p^2}{(\omega^2 - j\omega\Gamma - \omega_0{}^2)} \varepsilon E \tag{8.3.16}$$

In this, one can identify that

$$\begin{aligned}\chi(\omega) &= \frac{-Ne^2}{(\omega^2 - j\omega\Gamma - \omega_0{}^2)m_e\varepsilon_0} \\ &= -\frac{\omega_p{}^2}{\omega^2 - j\omega\Gamma - \omega_0{}^2}\end{aligned} \tag{8.3.17}$$

where ω_p is as defined before. A function with the above frequency dependence is also called a Lorentzian function. It is the hallmark of a damped harmonic oscillator.

If $\Gamma = 0$, then when $\omega = \omega_0$, one sees an infinite resonance peak exhibited by the detachment location sensitivity (DLS) model. But in the real world, $\Gamma \neq 0$, and when Γ is small, but $\omega \approx \omega_0$, then the peak value of χ is

$$\chi \approx +\frac{\omega_p{}^2}{j\omega\Gamma} = -j\frac{\omega_p{}^2}{\omega\Gamma} \tag{8.3.18}$$

Here, χ exhibits a large negative imaginary part, the character of a dissipative medium, as in the conducting medium we have previously studied. In other words, when $\omega = \omega_0$, the Drude-Lorentz-Sommerfeld (DLS) model is dominated by the dissipation in the medium.

8.3.5 Broad Applicability of Drude-Lorentz-Sommerfeld Model

The DLS model is a wonderful model because it can capture phenomenologically the salient feature of the physics of many electromagnetic media, even though it is purely a classical model.[6] It captures the resonance behavior of an atom absorbing energy from light excitation. When the light wave comes in at the correct frequency, it will excite electronic transition within an atom which can

[6]What we mean here is that only Newton's law has been used, and no quantum theory as yet.

be approximately modeled as a resonator with behavior similar to that of a pendulum oscillator. This electronic resonances will be radiationally damped [76],[7] and the damped oscillation can be modeled by $\Gamma \neq 0$. By picking a mixture of multispecies DLS oscillators, almost any shape of absorption spectra can be curve-fitted [77] (see Figure 8.4).

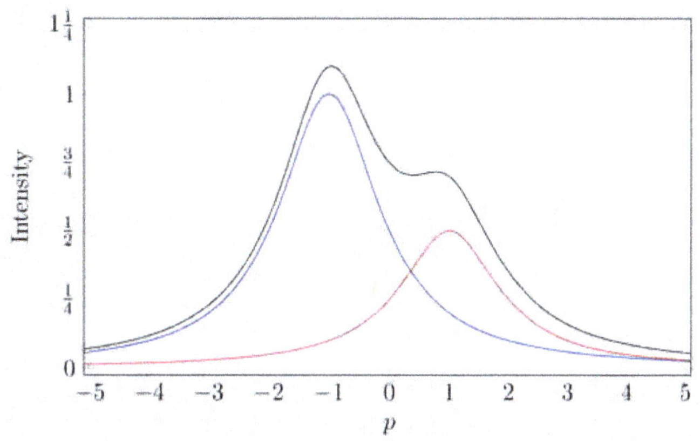

Figure 8.4: A Lorentzian function has almost a bell-shaped curve. By assuming multi-species of DLS oscillators in a medium, one can fit absorption spectra of almost any shape. (Figure by: https://commons.wikimedia.org) ([77]).

Moreover, the above model can also be used to model molecular vibrations. In this case, the mass of the electron will be replaced by the mass of the atom involved. The damping of the molecular vibration is caused by the hindered vibration of the molecule due to interaction with other molecules [78]. The hindered rotation or vibration of water molecules when excited by a microwave is the source of heat in microwave heating.

In the case of plasma, $\Gamma \neq 0$ represents the collision frequency between the free electrons and the ions, giving rise to loss. In the case of a conductor, Γ represents the collision frequency between the conduction electrons in the conduction band with the lattice of the material.[8] Also, if there is no restoring force, then $\omega_0 = 0$. This is true for the sea of electrons moving in the conduction band of a medium. Besides, for sufficiently low frequency, the inertial force can be ignored. Thus, from (8.3.17), when both ω and ω_0 tend to zero, again we have[9]

$$\chi \approx -j\frac{{\omega_p}^2}{\omega\Gamma} \tag{8.3.19}$$

[7]The oscillator radiates as it oscillates, and hence, loses energy to its environment. This causes the decay of the oscillation, just as a damped LC tank circuit losing energy to the resistor.

[8]It is to be noted that electron has a different effective mass in a crystal lattice [79, 80], and hence, the electron mass has to be changed accordingly in the DLS model.

[9]This equation is similar to (8.3.18). In both cases, collision force dominates in the equation of motion (8.3.14).

and

$$\varepsilon = \varepsilon_0(1+\chi) = \varepsilon_0\left(1 - j\frac{{\omega_p}^2}{\omega\Gamma}\right) \tag{8.3.20}$$

We recall that for a conductive medium, we define a complex permittivity to be

$$\varepsilon = \varepsilon_0\left(1 - j\frac{\sigma}{\omega\varepsilon_0}\right) \tag{8.3.21}$$

Comparing (8.3.20) and (8.3.21), we see a relation between σ, ω_p, and Γ, or that

$$\sigma = \varepsilon_0\frac{{\omega_p}^2}{\Gamma} \tag{8.3.22}$$

The above formula for conductivity can be arrived at using collision frequency argument as is done in some textbooks such as in Streetman and Banerjee [81, p. 123].

As such, the DLS model is quite powerful: it can be used to explain a wide variety of phenomena from very low frequency to optical frequency. The electrons in many conductive materials can be modeled as a sea of free electrons moving about quite freely with an effective mass. As such, they behave like a plasma medium, as shall be seen.

8.3.6 Frequency Dispersive Media—A General Discussion

The DLS model shows that, except for vacuum, all media are frequency dispersive. It is prudent to digress and discuss more on the physical meaning of a frequency dispersive medium. The relationship between electric flux and electric field, in the frequency domain, still follows the formula

$$\mathbf{D}(\omega) = \varepsilon(\omega)\mathbf{E}(\omega) \tag{8.3.23}$$

When the effective permittivity, $\varepsilon(\omega)$, is a function of frequency, it implies that the above relationship in the time domain is via convolution, namely,

$$\mathbf{D}(t) = \varepsilon(t) \circledast \mathbf{E}(t) \tag{8.3.24}$$

Since the above represents a linear time-invariant (LTI) system [57], it implies that an input is not followed by an instantaneous output. This can be seen in equation (8.3.3)—that when the electric field applies force to the electron, the electron does not follow the field instantaneously due to the inertia in its mass. Or there is a delay between the input and the output. The reason is because an electron has mass, and it cannot respond immediately to an applied force, or it has inertia. (In other words, the system has a memory of where it was previously when you try to move it.)

Even though the effective permittivity ε is a function of frequency, the frequency domain analysis we have done for a plane wave propagating in a dispersive medium still applies. For a monochromatic signal, it will have a velocity, called the *phase velocity*, given by $v = 1/\sqrt{\mu_0\varepsilon}$. Here, it also implies that different frequency components will propagate with different phase velocities through such a medium. Hence, a narrow pulse will spread in its width because different frequency components are not in phase after a short distance of travel. We see this in a pulse propagating in an optical fiber: the pulse loses its shape after some distance of travel.

Also, the Lorentzian function is great for data fitting, as many experimentally observed resonances have finite Q and a line width. The Lorentzian function models that well. If multiple resonances occur in a medium or an atom, then multispecies DLS model can be used. It is now clear that all media have to be frequency dispersive because of the finite mass of the electron and the inertia it has. Or there is no instantaneous response in a dielectric medium due to the finiteness of the electron mass.

Even at optical frequency, many metals, which have seas of freely moving electrons in the conduction band, can be modeled approximately as a plasma. A metal consists of a sea of electrons in the conduction bands that are not tightly bound to the ions or the lattice. Also, in optics, the inertial force due to the finiteness of the electron mass (in this case effective mass, see Figure 8.5) can be sizeable compared to other forces. Then, $\omega_0 \ll \omega$ or that the restoring force is much smaller than the inertial force, in (8.3.17), and if Γ is small, $\chi(\omega)$ resembles that of a plasma, and ε of a metal can be negative.

Metal	Ag	Au	Bi	Cu	K	Li	Na	Ni	Pt	Zn
$m_e{}^*/m_e$	0.99	1.10	0.047	1.01	1.12	1.28	1.2	28	13	0.85

Figure 8.5: Effective masses of electrons in different metals

8.3.7 Plasmonic Nanoparticles

When a plasmonic nanoparticle made of gold is excited by light, its response is given by (see homework assignment 8-1)

$$\Phi_R = E_0 \frac{a^3 \cos\theta}{r^2} \frac{\varepsilon_s - \varepsilon_0}{\varepsilon_s + 2\varepsilon_0} \tag{8.3.25}$$

In a plasma, ε_s can be negative, and thus, at certain frequency, if $\varepsilon_s = -2\varepsilon_0$, then $\Phi_R \to \infty$. Gold or silver, with a sea of electrons, behaves like a plasma at optical frequencies, since the inertial force in the DLS model is quite large.[10] Therefore, when light interacts with such a particle, it can sparkle brighter than normal. This reminds us of the saying "All that glitters is not gold!" even though this saying has a different intended meaning.

[10]In this case, $\omega^2 \gg \omega_0^2$, and $\omega^2 \gg \omega\Gamma$; the binding force and the collision force can be ignored similar to a cold plasma.

Figure 8.6: Ancient Roman goblets whose laquer coating glisten better under lighting due to the presence of gold nanoparticles. Gold or silver at optical frequencies behaves like plasma (Figure by: https://www.smithsonianmag.com/history/this-1600-year-old-goblet-shows-that-the-romans-were-nanotechnology-pioneers-787224/).

Figure 8.7: Nanoparticles immersed in different solutions will reflect light of different colors (Figure by: nanocomposix.com).

Ancient Romans apparently knew about the potent effect of using gold and silver nanoparticles to enhance the glistening of light in their drinking ware. These nanoparticles were impregnated in the glass or lacquer ware. By impregnating these nanoparticles in different media, the color of light will sparkle at different frequencies, and hence, the color of the glass emulsion can be changed (see website [82]).

Exercises for Lecture 8

Problem 8-1: This solution here, to a boundary value problem for Laplace's equation, can be used to explain why plasmonic particles, when embedded in glass or lacquer, glitter in light. When a dielectric sphere is immersed in a static electric field as shown in the Figure 8.8, the electric field does not satisfy the boundary condition. Hence, the sphere responds by producing a dipolar potential in order to satisfy the boundary condition.

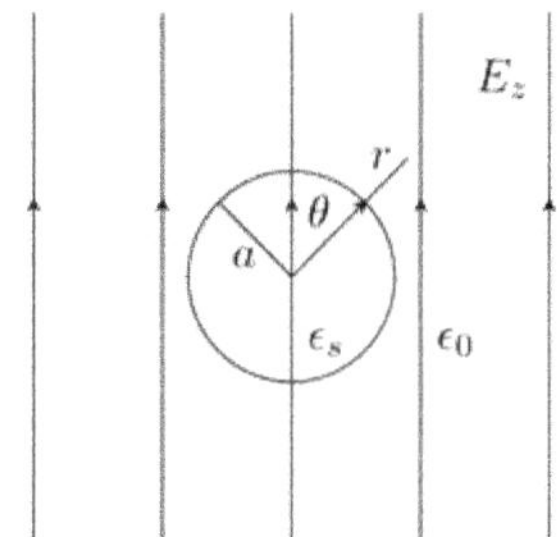

Figure 8.8: Geometry for solving this problem.

(i) Show that the potential outside the sphere can be written as

$$\Phi_{out}(\mathbf{r}) = -E_0 z + \frac{A}{r^2}\cos\theta$$

Explain the physical meanings of the first and second terms on the right-hand side of the above expression.

(ii) The potential inside the sphere can be written as

$$\Phi_{in}(\mathbf{r}) = Bz$$

where B is another unknown coefficient. What kind of electric field corresponds to the above potential?

(iii) Now, assume that the sphere has radius a. Decide on the boundary conditions at the dielectric interface at $r = a$.

(iv) From the boundary conditions, derive the expressions for A and B and show that the second term in (i) is of the form

$$\Phi_R = E_0 \frac{a^3\cos\theta}{r^2}\frac{\varepsilon_s - \varepsilon_0}{\varepsilon_s + 2\varepsilon_0}$$

(v) Now, explain why gold plasmonic nanoparticles can glitter in light.

Problem 8-2:

(i) Estimate the skin depth of the signal in your induction cooker pan. Assume that it operates around 50 KHz, and that the relative permeability μ_r is 100, and that the conductivity is about 10^7 siemens/m.

(ii) Estimate the electron density of the plasma layer in the ionosphere if it is known that radio frequencies below 10 MHz cannot penetrate the ionosphere.

(iii) The conductivity of a conductive medium has been estimated to be

$$\sigma = \varepsilon_0 \frac{\omega_p^2}{\Gamma}$$

using the Drude-Lorentz-Sommerfeld model. Arrive at the same formula using collision frequency argument. **Hint**: You may find the answer in some textbooks like Streetman and Banerjee on Semiconductor Devices [81].

Lecture 9

Waves in Gyrotropic Media, Polarization

We studied TEM uniform plane wave in lecture 7. When the $\mathbf{k}$ vector is pointing in the z direction, for instance, the electric field is polarized in the xy plane. As we shall see, assuming that the electric field is polarized in the x-axis, when such a wave propagates through a gyrotropic medium, its electric field rotates as it propagates. After a propagation distance, it can be polarized in other directions, such as the y direction, after emergence from this medium. Therefore, gyrotropy is an important concept in electromagnetics. In general, when a wave propagates through a gyrotropic medium, the electric field rotates, changing the polarization of the wave. Our ionosphere is such a medium, and it affects radio and microwave communications between the Earth and satellites by affecting the polarization of the wave. We will study this important phenomenon in this lecture, and the general polarization of waves.

9.1 Gyrotropic Media and Faraday Rotation

A gyrotropic medium is anisotropic, and its permittivity can only be described by a tensor. Here, we will go through the beautiful mathematics and derive the effective permittivity tensor of a gyrotropic medium, as encountered in many real-world situations such as the ionosphere or a baised ferrite medium. These media are often biased by a static magnetic field. For instance, our ionosphere is biased by a static magnetic field arising from the Earth's magnetic field [83]. But in this derivation, in order to capture the essence of the physics with a simple model, we assume that the ionosphere has a static magnetic field polarized in the z direction, namely that $\mathbf{B} = \hat{z}B_0$. Now, the equation of motion from the Lorentz force law for an electron with $q = -e$ (in accordance with Newton's second law that $F = ma$ or force equals mass times accelration) becomes

$$m_e \frac{d\mathbf{v}}{dt} = -e(\mathbf{E} + \mathbf{v} \times \mathbf{B}), \qquad \text{or} \qquad m_e \frac{d^2\mathbf{r}}{dt^2} = -e\left(\mathbf{E} + \frac{d\mathbf{r}}{dt} \times \mathbf{B}\right) \tag{9.1.1}$$

The first term of the force on the right-hand side is similar to Coulomb force , while the second term is usually termed the $\mathbf{v} \times \mathbf{B}$ force or the magnetic force.[1]

Next, let us assume that the electric field is polarized in the xy plane. The derivative of $\mathbf{v}$ or $d\mathbf{v}/dt$ is the acceleration of the electron, and also, $\mathbf{v} = d\mathbf{r}/dt$ where $\mathbf{r} = \hat{x}x + \hat{y}y + \hat{z}z$. Again, assuming linearity, we use phasor technique for the analysis. And in the frequency domain, the above equation in the Cartesian coordinates becomes

$$m_e\omega^2 x = e(E_x + j\omega B_0 y) \tag{9.1.2}$$

$$m_e\omega^2 y = e(E_y - j\omega B_0 x) \tag{9.1.3}$$

The above constitutes two equations with two unknowns x and y. They cannot be solved easily for x and y in terms of the electric field because they correspond to a two-by-two matrix system with cross coupling between the unknowns x and y. But they can be simplified and decoupled as follows: We can multiply (9.1.3) by $\pm j$ and add it to (9.1.2) to get two decoupled equations [84]:

$$m_e\omega^2(x + jy) = e[(E_x + jE_y) + \omega B_0(x + jy)] \tag{9.1.4}$$

$$m_e\omega^2(x - jy) = e[(E_x - jE_y) - \omega B_0(x - jy)] \tag{9.1.5}$$

In the above, if we take the new unknowns to be $x \pm jy$, the two equations are decoupled with respect to these two unknowns. Now, defining new variables such that

$$s_\pm = x \pm jy \tag{9.1.6}$$

$$E_\pm = E_x \pm jE_y \tag{9.1.7}$$

then (9.1.4) and (9.1.5) become

$$m_e\omega^2 s_\pm = e(E_\pm \pm \omega B_0 s_\pm) \tag{9.1.8}$$

Thus, solving the above yields

$$s_\pm = \frac{e}{m_e\omega^2 \mp eB_0\omega} E_\pm = C_\pm E_\pm \tag{9.1.9}$$

where[2]

$$C_\pm = \frac{e}{m_e\omega^2 \mp eB_0\omega} \tag{9.1.10}$$

Next, one can define $P_x = -Nex$, $P_y = -Ney$, and that $P_\pm = P_x \pm jP_y = -Nes_\pm$. Then it can be shown that

$$P_\pm = \varepsilon_0 \chi_\pm E_\pm \tag{9.1.11}$$

---

[1] For a plane wave, it can be shown that the $\mathbf{v} \times \mathbf{B}$ force is of order v/c smaller than the Coulomb force, which is termed relativistically small [72, II-13-6]. But it can be made artificially large by a permanent magnet.

[2] By this manipulation, the above equations (9.1.2) and (9.1.3) transform into new equations where there is no cross coupling between $s_\pm$ and $E_\pm$. The mathematical parlance for this is the diagonalization of a matrix equation [85]. Thus, the new equation can be solved easily.

which is a scalar equation! The expression for $\chi_\pm$ can be derived, and they are given as

$$\chi_\pm = -\frac{NeC_\pm}{\varepsilon_0} = -\frac{Ne}{\varepsilon_0}\frac{e}{m_e\omega^2 \mp eB_o\omega} = -\frac{{\omega_p}^2}{\omega^2 \mp \Omega\omega} \tag{9.1.12}$$

where Ω and ω_p are the cyclotron frequency [3] and plasma frequency, respectively, namely,

$$\Omega = \frac{eB_0}{m_e}, \quad {\omega_p}^2 = \frac{Ne^2}{m_e\varepsilon_0} \tag{9.1.13}$$

Notice that at the cyclotron frequency, $|\chi_\pm| \to \infty$, a hallmark of resonance behavior. In other words, $P_\pm$ is finite even when $E_\pm = 0$, or a solution exists to the equation of motion (9.1.1) without a forcing term, which in this case is the electric field. Thus, at this frequency, the solution goes to infinity if the forcing term, $E_\pm$, is not zero. This is like what happens at the resonance of an LC tank circuit whose current or voltage tends to infinity when the forcing term, like the voltage or current, is nonzero. In a word, the solution exists even when the forcing term is zero, similar to the homogeneous solution of an ordinary differential equation. In order to derive the permittivity tensor in the Cartesian coordinates, we first need to express the original variables $P_x,\ P_y,\ E_x, and\ E_y$ in terms of $P_\pm$ and $E_\pm$. With the help of (9.1.11), we arrive at

$$\begin{aligned} P_x = \frac{P_+ + P_-}{2} &= \frac{\varepsilon_0}{2}(\chi_+E_+ + \chi_-E_-) = \frac{\varepsilon_0}{2}[\chi_+(E_x + jE_y) + \chi_-(E_x - jE_y)] \\ &= \frac{\varepsilon_0}{2}[(\chi_+ + \chi_-)E_x + j(\chi_+ - \chi_-)E_y] \end{aligned} \tag{9.1.14}$$

$$\begin{aligned} P_y = \frac{P_+ - P_-}{2j} &= \frac{\varepsilon_0}{2j}(\chi_+E_+ - \chi_-E_-) = \frac{\varepsilon_0}{2j}[\chi_+(E_x + jE_y) - \chi_-(E_x - jE_y)] \\ &= \frac{\varepsilon_0}{2j}[(\chi_+ - \chi_-)E_x + j(\chi_+ + \chi_-)E_y] \end{aligned} \tag{9.1.15}$$

The above relationship in cartesian coordinates can be expressed using a tensor where

$$\mathbf{P} = \varepsilon_0\overline{\chi}\cdot\mathbf{E} \tag{9.1.16}$$

where $\mathbf{P} = [P_x, P_y]$, and $\mathbf{E} = [E_x, E_y]$. From (9.1.14) and (9.1.15) above, $\overline{\chi}$ is a tensor of the form

$$\overline{\chi} = \frac{1}{2}\begin{pmatrix} (\chi_+ + \chi_-) & j(\chi_+ - \chi_-) \\ -j(\chi_+ - \chi_-) & (\chi_+ + \chi_-) \end{pmatrix} = \begin{pmatrix} -\frac{{\omega_p}^2}{\omega^2-\Omega^2} & -j\frac{{\omega_p}^2\Omega}{\omega(\omega^2-\Omega^2)} \\ j\frac{{\omega_p}^2\Omega}{\omega(\omega^2-\Omega^2)} & -\frac{{\omega_p}^2}{\omega^2-\Omega^2} \end{pmatrix} \tag{9.1.17}$$

Notice that in the above, when the $\mathbf{B}$ field is turned off or $\Omega = 0$, then $\overline{\chi}$ above is diagonalized, and it resembles an isotropic medium of a collisionless, cold plasma again. Consequently, for the $\mathbf{B} \neq 0$ case, the above can be generalized to 3D to give

$$\overline{\chi} = \begin{bmatrix} \chi_0 & j\chi_1 & 0 \\ -j\chi_1 & \chi_0 & 0 \\ 0 & 0 & \chi_p \end{bmatrix} \tag{9.1.18}$$

[3] This is also called the gyrofrequency.

where $\chi_p = -\omega_p^2/\omega^2$. Notice that since we assume that $\mathbf{B} = \hat{z}B_0$, the z component of (9.1.1) is unaffected by the $\mathbf{v} \times \mathbf{B}$ force. Hence, the electron moving in the z direction is similar to that in a cold collisionless plasma.

Using the fact that $\mathbf{D} = \varepsilon_0 \mathbf{E} + \mathbf{P} = \varepsilon_0(\overline{\mathbf{I}} + \overline{\chi}) \cdot \mathbf{E} = \overline{\varepsilon} \cdot \mathbf{E}$, the above implies that

$$\overline{\varepsilon} = \varepsilon_0 \begin{bmatrix} 1+\chi_0 & j\chi_1 & 0 \\ -j\chi_1 & 1+\chi_0 & 0 \\ 0 & 0 & 1+\chi_p \end{bmatrix} \tag{9.1.19}$$

Now, $\overline{\varepsilon}$ is in an anisotropic medium, a category that includes gyrotropic media. Please notice that the above tensor is a Hermitian tensor. We shall learn later that this is the property of a lossless medium.

Another characteristic of a gyrotropic medium is that a linearly polarized wave will rotate when passing through it. This is the Faraday rotation effect [84], and we will learn more about this later. This phenomenon poses a severe problem for Earth-to-satellite communications using linear polarized waves, as it requires the alignment of the Earth-to-satellite antennas. Thank goodness, this can be avoided using a rotatingly polarized wave, called a circularly polarized wave, which we will learn about in the next section.

As we have learned, the ionosphere affects our communication systems in two ways: It acts as a mirror for low-frequency electromagnetic or radio waves (making the experiment of Marconi a rousing success). It also affects the polarization of the wave. But the ionosphere of the Earth and the density of electrons that are ionized is highly dependent on temperature, and the effect of the Sun. The fluctuation of particles in the ionosphere gives rise to scintillation effects due to electron motion and collision that affect radio wave communication systems [86].

9.2 Wave Polarization

Studying wave polarization is very important for communication purposes [35]. A wave whose electric field is pointing in the x direction while propagating in the z direction is called a linearly polarized (LP) wave. The same can be said of one with an electric field polarized in the y direction. It turns out that a linearly polarized wave experiences Faraday rotation when it propagates through the ionosphere. For instance, an x polarized wave can become a y polarized wave due to Faraday rotation, so its polarization becomes ambiguous as the wave propagates through the ionosphere. To overcome this, Earth-to-satellite communication is done with circularly polarized (CP) waves [87]. So even if the electric field vector is rotated by Faraday's rotation, it continues to be a CP wave. We will study these polarized waves next. Later, we will study how to make antennas that radiate linearly or circularly polarized waves.

9.2.1 General Polarizations—Elliptical and Circular Polarizations

We can write a general uniform plane wave propagating in the z direction in the time domain for simplicity as

$$\mathbf{E} = \hat{x}E_x(z,t) + \hat{y}E_y(z,t) \tag{9.2.1}$$

Clearly, $\nabla \cdot \mathbf{E} = 0$, and $E_x(z,t)$ and $E_y(z,t)$, by the principle of linear superposition, are solutions to the one-dimensional wave equation as we have discussed in sect:3.2, (3.2.5). For a time-harmonic field, the two components may not be in phase, and in general, we have for time domain that

$$E_x(z,t) = E_1 \cos(\omega t - \beta z) \tag{9.2.2}$$
$$E_y(z,t) = E_2 \cos(\omega t - \beta z + \alpha) \tag{9.2.3}$$

where α denotes the phase difference between these two wave components. We shall study how the linear superposition of these two components behaves for different instances of α. To start, we observe this field at $z = 0$. Then

$$\mathbf{E} = \hat{x} E_1 \cos(\omega t) + \hat{y} E_2 \cos(\omega t + \alpha) \tag{9.2.4}$$

For $\alpha = \frac{\pi}{2}$

$$E_x = E_1 \cos(\omega t),\ E_y = E_2 \cos(\omega t + \pi/2) \tag{9.2.5}$$

Next, we evaluate the above for different instances of ωt

$$\omega t = 0, \qquad E_x = E_1, \qquad E_y = 0 \tag{9.2.6}$$
$$\omega t = \pi/4, \qquad E_x = E_1/\sqrt{2}, \qquad E_y = -E_2/\sqrt{2} \tag{9.2.7}$$
$$\omega t = \pi/2, \qquad E_x = 0, \qquad E_y = -E_2 \tag{9.2.8}$$
$$\omega t = 3\pi/4, \qquad E_x = -E_1/\sqrt{2}, \qquad E_y = -E_2/\sqrt{2} \tag{9.2.9}$$
$$\omega t = \pi, \qquad E_x = -E_1, \qquad E_y = 0 \tag{9.2.10}$$

The tip of the vector field $\mathbf{E}$ traces out an ellipse as show in Figure 9.1. With the left-hand thumb pointing in the z direction, the direction of propagation, and the wave rotating in the direction of the fingers, such a wave is called left-hand elliptically polarized (LHEP) wave.

When $E_1 = E_2$, the ellipse becomes a circle, and we have a left-hand circularly polarized (LHCP) wave. When $\alpha = -\pi/2$, the wave rotates in the counter-clockwise direction, and the wave is either a right-hand elliptically polarized (RHEP) or right-hand circularly polarized (RHCP) wave depending on the ratio of E_1/E_2. Figure 9.2 shows the different polarizations of the wave for different phase differences and amplitude ratios. Figure 9.3 shows an illustration of a CP wave propagating through space.

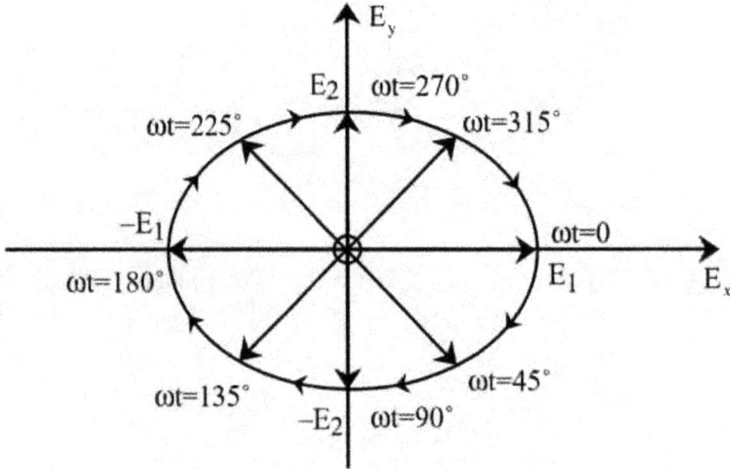

Figure 9.1: If one follows the tip of the electric field vector, it traces out an ellipse as a function of time t.

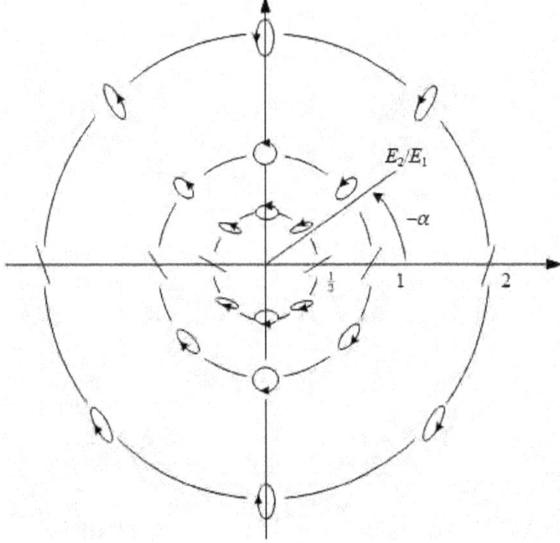

Figure 9.2: Due to phase differences between the E_x and E_y components of the field, and their relative amplitudes E_2/E_1, different polarizations will ensue. The arrow indicates the direction of rotation of the field vector.

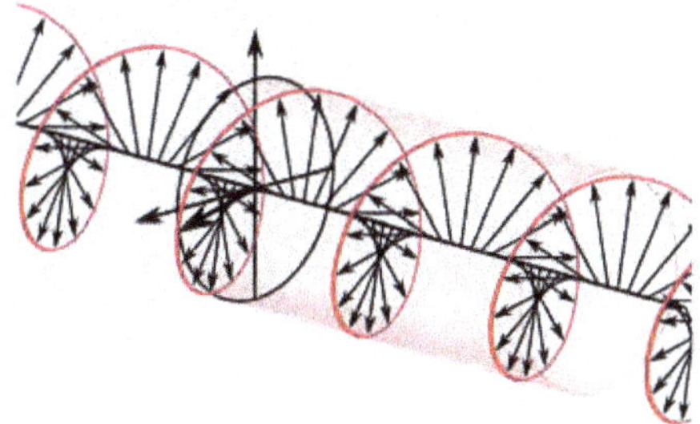

Figure 9.3: The rotation of the field vector of a RHCP wave as it propagates in the right direction [88]. (Figure by Wikipedia.)

9.2.2 Arbitrary Polarization Case and Axial Ratio[4]

As seen before, the tip of the field vector traces out an ellipse in space as it propagates. The axial ratio (AR) is the ratio of the major axis to the minor axis of this ellipse. It is an important figure of merit for designing CP (circularly polarized) antennas (antennas that will radiate circularly polarized waves). The closer this ratio is to 1, the better the antenna design is. We will discuss the general polarization and the AR of a wave.

For the general case for arbitrary α, we let

$$E_x = E_1 \cos\omega t, \qquad E_y = E_2 \cos(\omega t + \alpha) = E_2(\cos\omega t \cos\alpha - \sin\omega t \sin\alpha) \tag{9.2.11}$$

Then from the above, expressing E_y in terms of E_x, one gets

$$E_y = \frac{E_2}{E_1} E_x \cos\alpha - E_2 \left[1 - \left(\frac{E_x}{E_1}\right)^2\right]^{1/2} \sin\alpha \tag{9.2.12}$$

Rearranging and squaring, we get

$$a{E_x}^2 - bE_xE_y + c{E_y}^2 = 1 \tag{9.2.13}$$

where

$$a = \frac{1}{{E_1}^2 \sin^2\alpha}, \qquad b = \frac{2\cos\alpha}{E_1E_2 \sin^2\alpha}, \qquad c = \frac{1}{{E_2}^2 \sin^2\alpha} \tag{9.2.14}$$

After letting $E_x \to x$, and $E_y \to y$, equation (9.2.13) is of the form,

$$ax^2 - bxy + cy^2 = 1 \tag{9.2.15}$$

[4]This section is mathematically complicated. It can be skipped on first reading.

which is the equation of an ellipse. The equation of an ellipse with original coordinates is,

$$\left(\frac{x'}{A}\right)^2 + \left(\frac{y'}{B}\right)^2 = 1 \tag{9.2.16}$$

where A and B are axes of the ellipse as shown in Figure 9.4. We can transform the above back to the (x, y) coordinates to get (9.2.15). To this end, we let

$$x' = x\cos\theta - y\sin\theta \tag{9.2.17}$$

$$y' = x\sin\theta + y\cos\theta \tag{9.2.18}$$

to get

$$x^2\left(\frac{\cos^2\theta}{A^2} + \frac{\sin^2\theta}{B^2}\right) - xy\sin 2\theta\left(\frac{1}{A^2} - \frac{1}{B^2}\right) + y^2\left(\frac{\sin^2\theta}{A^2} + \frac{\cos^2\theta}{B^2}\right) = 1 \tag{9.2.19}$$

Comparing (9.2.13) and (9.2.19), one gets

$$\theta = \frac{1}{2}\tan^{-1}\left(\frac{2\cos\alpha E_1 E_2}{{E_2}^2 - {E_1}^2}\right) \tag{9.2.20}$$

$$\mathrm{AR} = \left(\frac{1+\Delta}{1-\Delta}\right)^{1/2} > 1 \tag{9.2.21}$$

where AR is the axial ratio and

$$\Delta = \left(1 - \frac{4{E_1}^2{E_2}^2\sin^2\alpha}{\left({E_1}^2 + {E_2}^2\right)^2}\right)^{1/2} \tag{9.2.22}$$

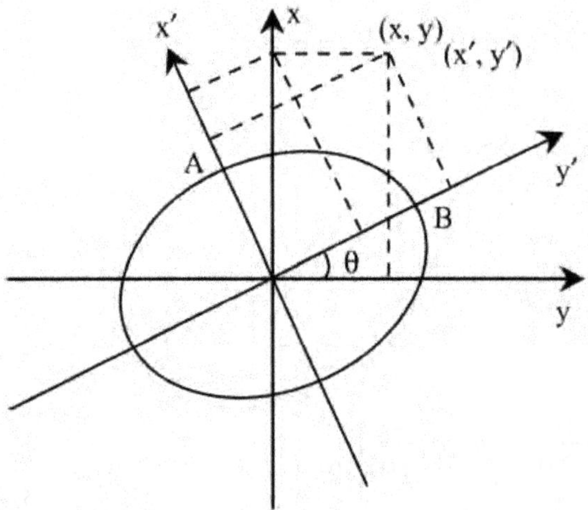

Figure 9.4: This figure shows the parameters used to derive the axial ratio (AR) of an elliptically polarized wave.

9.3 Polarization and Power Flow

For a linearly polarized wave in the time domain,[5]

$$\mathbf{E} = \hat{x}E_0 \cos(\omega t - \beta z), \quad \mathbf{H} = \hat{y}\frac{E_0}{\eta}\cos(\omega t - \beta z) \tag{9.3.1}$$

Hence, the instantaneous power we have learned previously in section 5.3 becomes

$$\mathbf{S}(t) = \mathbf{E}(t) \times \mathbf{H}(t) = \hat{z}\frac{{E_0}^2}{\eta}\cos^2(\omega t - \beta z) \tag{9.3.2}$$

indicating that for a linearly polarized wave, the instantaneous power is a function of both time and space. It travels as lumps of energy through space. In the above, E_0 is the amplitude of the linearly polarized wave. Moreover, taking the time average of the above, we have

$$\langle \mathbf{S}(t) \rangle = \hat{z}\frac{{E_0}^2}{2\eta} \tag{9.3.3}$$

Next, we look at power flow for for elliptically and circularly polarized waves. It is to be noted that in the phasor domain or frequency domain, (9.2.1) now becomes

$$\mathbf{E}(z,\omega) = \hat{x}E_1 e^{-j\beta z} + \hat{y}E_2 e^{-j\beta z + j\alpha} \tag{9.3.4}$$

For an LHEP wave, $E_1 \neq E_2$ and $\alpha = \pi/2$,

$$\mathbf{E}(z,\omega) = e^{-j\beta z}(\hat{x}E_1 + j\hat{y}E_2) \tag{9.3.5}$$

whereas for an LHCP wave, $E_1 = E_2$ and $\alpha = \pi/2$

$$\mathbf{E}(z,\omega) = e^{-j\beta z}E_1(\hat{x} + j\hat{y}) \tag{9.3.6}$$

For an RHEP wave, the above becomes

$$\mathbf{E}(z,\omega) = e^{-j\beta z}(\hat{x}E_1 - j\hat{y}E_2) \tag{9.3.7}$$

whereas for an RHCP wave, it is

$$\mathbf{E}(z,\omega) = e^{-j\beta z}E_1(\hat{x} - j\hat{y}) \tag{9.3.8}$$

Focusing on the circularly polarized wave,

$$\mathbf{E} = (\hat{x} \pm j\hat{y})E_0 e^{-j\beta z} \tag{9.3.9}$$

Using that $\boldsymbol{\beta} = \hat{z}\beta$, and letting $\nabla \rightarrow -j\boldsymbol{\beta}$, Faraday's law becomes

$$\mathbf{H} = \frac{\boldsymbol{\beta} \times \mathbf{E}}{\omega\mu} \tag{9.3.10}$$

[5] In the course, β and k are used interchangeably. The former is preferred in microwaves and the latter in optics.

And then

$$\mathbf{H} = (\mp\hat{x} - j\hat{y})j\frac{E_0}{\eta}e^{-j\beta z} \tag{9.3.11}$$

where $\eta = \sqrt{\mu/\varepsilon}$ is the intrinsic impedance of the medium. Therefore,

$$\mathbf{E}(z,t) = \hat{x}E_0\cos(\omega t - \beta z) \pm \hat{y}E_0\sin(\omega t - \beta z) \tag{9.3.12}$$

$$\mathbf{H}(z,t) = \mp\hat{x}\frac{E_0}{\eta}\sin(\omega t - \beta z) + \hat{y}\frac{E_0}{\eta}\cos(\omega t - \beta z) \tag{9.3.13}$$

Then the instantaneous power becomes

$$\mathbf{S}(z,t) = \mathbf{E}(z,t)\times\mathbf{H}(z,t) = \hat{z}\frac{{E_0}^2}{\eta}\cos^2(\omega t - \beta z) + \hat{z}\frac{{E_0}^2}{\eta}\sin^2(\omega t - \beta z) = \hat{z}\frac{{E_0}^2}{\eta} \tag{9.3.14}$$

In other words, a CP wave delivers constant instantaneous power independent of space and time, as opposed to a linearly polarized wave which delivers a nonconstant instantaneous power as shown in (9.3.2). Moreover, taking the time average of the above, we have

$$\langle\mathbf{S}(z,t)\rangle = \hat{z}\frac{{E_0}^2}{\eta} \tag{9.3.15}$$

which is independent of z, t.

It is to be noted that the complex Poynting's vector for a lossless medium

$$\underset{\sim}{\mathbf{S}} = \mathbf{E}\times\mathbf{H}^* \tag{9.3.16}$$

is real and constant independent of space for linearly, circularly, and elliptically polarized waves. That is, if we were to go through the exercise to obtain $\underset{\sim}{\mathbf{S}}$ for the general case, we will let

$$\mathbf{E} = (\hat{x}E_1 \pm j\hat{y}E_2)e^{-j\beta z} \tag{9.3.17}$$

The corresponding magnetic field can be found as

$$\mathbf{H} = \frac{\boldsymbol{\beta}\times\mathbf{E}}{\omega\mu} = \frac{\beta}{\omega\mu}(\hat{y}E_1 \mp j\hat{x}E_2)e^{-j\beta z} \tag{9.3.18}$$

Using the above, we find that the complex Poynting's vector as

$$\underset{\sim}{\mathbf{S}} = \mathbf{E}\times\mathbf{H}^* = \frac{\beta}{\omega\mu}\hat{z}(|E_1|^2 + |E_2|^2) \tag{9.3.19}$$

Then the time-average power density is

$$\langle\mathbf{S}\rangle = \frac{1}{2}\Re e\left\{\underset{\sim}{\mathbf{S}}\right\} = \frac{1}{2\eta}\hat{z}(|E_1|^2 + |E_2|^2) \tag{9.3.20}$$

When $E_1 = E_2 = E_0$, the above becomes

$$\langle\mathbf{S}\rangle = \frac{1}{2}\Re e\left\{\underset{\sim}{\mathbf{S}}\right\} = \frac{1}{\eta}\hat{z}|E_0|^2 \tag{9.3.21}$$

which is the same as in (9.3.14).

When $E_2 = 0$ for a linearly polarized wave, and $E_1 = E_0$, we have

$$\langle \mathbf{S} \rangle = \frac{1}{2} \Re e \left\{ \underset{\sim}{\mathbf{S}} \right\} = \frac{1}{2\eta} \hat{z} |E_0|^2 \tag{9.3.22}$$

This is the same as what we have found before in (9.3.3). Notice that the Poynting's vector is a constant independent of z. This is because there is no reactive power in a plane wave of any polarization: the stored energy in the plane wave cannot be returned to the source!

Exercises for Lecture 9

Problem 9-1: Show that (9.1.17) becomes an isotropic medium of cold collisionless plasma when the **B** field is turned off.

Problem 9-2: This problem will show the Faraday rotation of a linearly polarized wave as it propagates through the ionosphere. This happens because the ionosphere is a gyrotropic medium. Its permittivity tensor, as shown in this lecture, can be expressed as:

$$\overline{\varepsilon} = \begin{bmatrix} \varepsilon_x & j\varepsilon_g & 0 \\ -j\varepsilon_g & \varepsilon_y & 0 \\ 0 & 0 & \varepsilon_z \end{bmatrix} \tag{E9.1}$$

(i) From the lecture notes, surmise the forms of $\varepsilon_x, \varepsilon_y, \varepsilon_z$, and ε_g. Assuming all these quantities are real, how would you classify this matrix that corresponds to this medium? You will learn later that such a matrix corresponds to a lossless medium.

(ii) Assume that the wave is propagating in the z direction, given as

$$\mathbf{E} = \hat{x}E_0 e^{-jkz}$$

Show that this is the solution to Maxwell's equations as well as the vector Helmholtz equation.

(iii) Show that a linearly polarized wave can be split into a sum of a RHCP wave and a LHCP wave. In other words, show that

$$\mathbf{E} = \hat{x}E_0 e^{-jkz} = \mathbf{E}_{LHCP} + \mathbf{E}_{RHCP}$$

And find the forms of $\mathbf{E}_{LHCP}$ and $\mathbf{E}_{RHCP}$.

(iv) Show that these two waves RHCP and LHCP will propagate with different wave numbers k_+ and k_- or phase velocity in a gyrotropic medium. Find these wave numbers.

(v) Show that the wave rotates (known as Faraday rotation) as it propagates in the z direction, as shown above. After a distance d in the gyrotropic medium, it becomes rotated and the wave acquires a y component as well. Therefore, the field vector is tilted with respect to the x-axis.

(vi) Show that the tilt angle, or the angle of Faraday rotation is:

$$\theta_F = \frac{k_+ - k_-}{2} d$$

Lecture 10

Momentum, Complex Poynting's Theorem, Lossless Condition, Energy Density

In the last lecture, we study circularly polarized waves as well as linearly polarized waves. In addition, these waves can carry power giving rise to power flow. But in addition to carrying power, a traveling wave also carries momentum: for a linearly polarized wave, it carries linear momentum in the direction of the propagation of the traveling wave. But for a circularly polarized wave, it carries angular momentum as well.[1]

Previously, we have studied complex power and the complex Poynting's theorem in the frequency domain with phasors. Here, we will derive the lossless conditions for the permittivity and permeability tensors under which a medium is lossless. As we have shown in the instantaneous Poynting's theorem, energy density is well defined for a lossless dispersionless medium, but we will assume a different formula when the medium is dispersive. From the Drude-Lorentz-Sommerfeld model, it is seen that a medium is never dispersionless save for a vacuum. In other words, it $\mathbf{D}(\omega) = \varepsilon(\omega)\mathbf{E}(\omega)$, if ε is a function of frequency, the change in $\mathbf{D}$ cannot be instantaneous if there is a sudden change in $\mathbf{E}$.

10.1 Spin Angular Momentum and Cylindrical Vector Beam

In this section, we will study the spin angular momentum of a circularly polarized (CP) wave. It is to be noted that in cylindrical coordinates, as shown in Figure 10.1, $\hat{x} = \hat{\rho}\cos\phi - \hat{\phi}\sin\phi$,

[1]At this juncture, you will have to recalibrate your electrical engineering training and learn to think that energy flow and momentum are not carried by electron flow except for some special cases. Most often than not, they are carried by photons or the electromagnetic fields.

$\hat{y} = \hat{\rho}\sin\phi + \hat{\phi}\cos\phi$, then a CP field is proportional to

$$(\hat{x} \pm j\hat{y}) = \hat{\rho}e^{\pm j\phi} \pm j\hat{\phi}e^{\pm j\phi} = e^{\pm j\phi}(\hat{\rho} \pm \hat{\phi}) \tag{10.1.1}$$

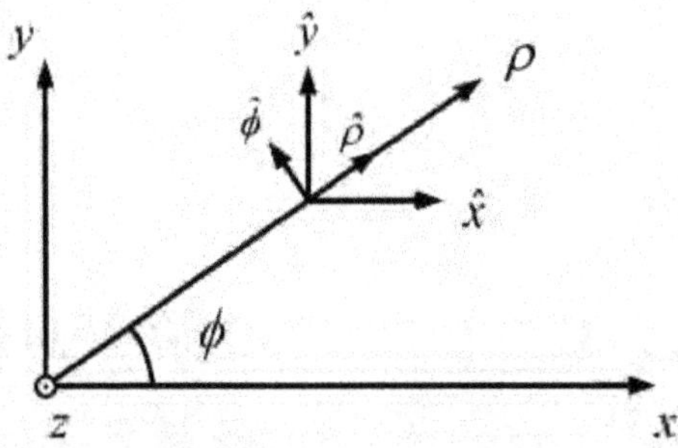

Figure 10.1: The local coordinates used to describe a circularly polarized wave: In Cartesian and polar coordinates.

Therefore, with the $e^{\pm j\phi}$ dependence, the $\hat{\rho}$ and $\hat{\phi}$ of a CP wave is also an azimuthal traveling wave in the $\hat{\phi}$ direction in addition to being a traveling wave $e^{-j\beta z}$ in the $\hat{z}$ direction. This is obviated by rewriting

$$e^{-j\phi} = e^{-jk_\phi \rho\phi} \tag{10.1.2}$$

where $k_\phi = 1/\rho$ is the azimuthal wave number, and $\rho\phi$ is the arc length traversed by the azimuthal wave. Notice that the wavenumber k_ϕ is dependent on p: the larger ρ is, the smaller k_ϕ becomes, and hence, the larger the azimuthal wavelength. Thus, the wave possesses angular momentum called the spin angular momentum (SAM), just as a traveling wave, $e^{-j\beta z}$ possesses linear angular momentum in the $\hat{z}$ direction.

In optics research, the generation of cylindrical vector beam is in vogue. Figure 10.2 shows a method to generate such a beam. A CP light passes through a radial analyzer that will only allow the radial component of (10.1.1) to be transmitted. Then a spiral phase elementspiral phase element (SPE) compensates for the $\exp(\pm j\phi)$ phase shift in the azimuthal direction. Finally, the light is a cylindrical vector beam which is radially polarized without spin angular momentum. Such a beam has been found to have nice focusing property, and hence, has aroused the interest of the researchers in the optics community [89].

10.2 Momentum Density of Electromagnetic Field

We have seen that a traveling wave carries power and has energy density associated with it. In other words, the moving or traveling energy density gives rise to power flow. It turns out that a traveling wave also carries a momentum with it. The momentum density of electromagnetic field is given by

$$\mathbf{G} = \mathbf{D} \times \mathbf{B} \tag{10.2.1}$$

also called the momentum density vector. With it, one can derive the momentum conservation law [35, p. 59] [50]. The derivation is rather long, but we will justify the above formula and simplify the derivation using the particle or the corpuscular nature of light of an electromagnetic field. The following derivation is only valid for plane waves.

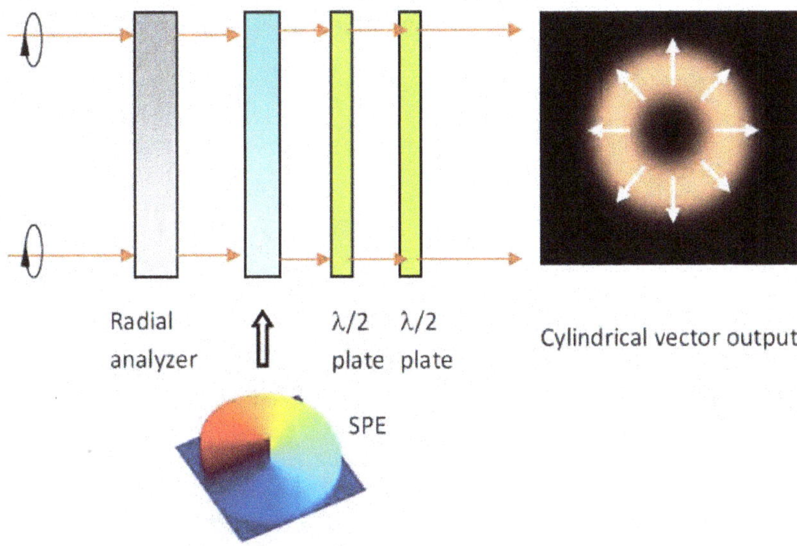

Figure 10.2: A cylindrical vector beam can be generated experimentally. The spiral phase element (SPE) compensates for the $\exp(\pm j\phi)$ phase shift (Figure by: https://www.mdpi.com/2076-3417/10/10/3427 (Zhan Q.) [89]). The half-wave plate rotates the polarization of a wave by 90 degrees.

It has been long known that electromagnetic energy is carried by photons, each of which is associated with a packet of energy given by $\hbar\omega$ (first discovered by Planck [25]). It is also well known that a photon has momentum given by $\hbar k$.[2] Assuming that there are photons, with density of N photons per unit volume streaming through space at the velocity of light c, then the power flow associated with these streaming photons is given by[3]

$$\mathbf{E} \times \mathbf{H} = \hbar\omega N c\hat{z} \tag{10.2.2}$$

[2]It was Louis de Broglie who posited that an electron is both a wave and a particle, the famous particle-wave duality picture, and its momentum is related to $\hbar k$ [90], where k is the wavenumber of the pertinent wave associated with the electron.

[3]Again, we have to recalibrate our thinking. Usually, in classical thinking, we think of a moving particle having energy given by $\frac{1}{2}mv^2$, but photons are massless particles zipping around at the speed of light.

Assuming that the plane wave is propagating in the z direction. Using $k = \omega/c$, we can rewrite the above more suggestively as

$$\mathbf{E} \times \mathbf{H} = \hbar k N c^2 \hat{z} \tag{10.2.3}$$

where $k = \omega/c$. Assuming that each photon has a momentum given by $\hbar k$,[4] we can relate the momentum density vector to be

$$\mathbf{G} = \hbar k N \hat{z} \tag{10.2.4}$$

which has a unit of momentum per unit volume. Then from the above, we deduce that

$$\mathbf{E} \times \mathbf{H} = \mathbf{G} c^2 = \frac{1}{\mu\varepsilon}\mathbf{G} \tag{10.2.5}$$

Or the above can be rewritten as[5]

$$\mathbf{G} = \mathbf{D} \times \mathbf{B} \tag{10.2.6}$$

where $\mathbf{D} = \varepsilon\mathbf{E}$, and $\mathbf{B} = \mu\mathbf{H}$.[6]

10.3 Complex Poynting's Theorem and Lossless Conditions

10.3.1 Complex Poynting's Theorem

It has been previously shown that the vector $\mathbf{E}(\mathbf{r}, t) \times \mathbf{H}(\mathbf{r}, t)$ has a dimension of watts/m^2 which is that of power density. Therefore, it is associated with the direction of power flow [35, 50]. As has been shown for a time-harmonic field, a time average of this vector can be defined as

$$\langle \mathbf{E}(\mathbf{r}, t) \times \mathbf{H}(\mathbf{r}, t) \rangle = \lim_{T\to\infty} \frac{1}{T} \int_0^T \mathbf{E}(\mathbf{r}, t) \times \mathbf{H}(\mathbf{r}, t)\, dt. \tag{10.3.1}$$

Given time-harmonic fields $\mathbf{E}(\mathbf{r}, t)$ and $\mathbf{H}(\mathbf{r}, t)$, whose phasors are $\mathbf{E}(\mathbf{r}, \omega)$ and $\mathbf{H}(\mathbf{r}, \omega)$, respectively, we can show that for a time-harmonic field,

$$\langle \mathbf{E}(\mathbf{r}, t) \times \mathbf{H}(\mathbf{r}, t) \rangle = \frac{1}{2} \Re e\{\mathbf{E}(\mathbf{r}, \omega) \times \mathbf{H}^*(\mathbf{r}, \omega)\}. \tag{10.3.2}$$

Here, the vector $\mathbf{E}(\mathbf{r}, \omega) \times \mathbf{H}^*(\mathbf{r}, \omega)$, as previously discussed, is also known as the complex Poynting's vector. We define the instantaneous Poynting's vector to be

$$\mathbf{S}(\mathbf{r}, t) = \mathbf{E}(\mathbf{r}, t) \times \mathbf{H}(\mathbf{r}, t) \tag{10.3.3}$$

---

[4]Again, we have to reset our thinking as a photon is a massless particle behaving like a wave.

[5]It will be interesting to show that the units on both sides of the equation are the same. The unit of Planck constant is joule-second.

[6]The author is indebted to Wei SHA for this simple derivation.

and the complex Poynting's vector to be

$$\underset{\sim}{\mathbf{S}}(\mathbf{r},\omega) = \mathbf{E}(\mathbf{r},\omega) \times \mathbf{H}^*(\mathbf{r},\omega) \tag{10.3.4}$$

Then for time-harmonic fields,

$$\langle \mathbf{S}(\mathbf{r},t)\rangle = \frac{1}{2}\Re e\left\{\underset{\sim}{\mathbf{S}}(\mathbf{r},\omega)\right\} \tag{10.3.5}$$

The above is often the source of confusion in the definition of Poynting's vector, because $\underset{\sim}{\mathbf{S}}$ is not the phasor representation of $\mathbf{S}$. Also, $\mathbf{S}$ is not time-harmonic and hence, does not have a phasor representation.

Since the above definition of the complex Poynting's vector has the dimension of power density, we will study its conservative property. To do so, we take its divergence and use the appropriate vector identity to obtain[7]

$$\nabla \cdot (\mathbf{E} \times \mathbf{H}^*) = \mathbf{H}^* \cdot \nabla \times \mathbf{E} - \mathbf{E} \cdot \nabla \times \mathbf{H}^*. \tag{10.3.6}$$

Next, using Maxwell's equations for $\nabla \times \mathbf{E}$ and $\nabla \times \mathbf{H}^*$, namely

$$\nabla \times \mathbf{E} = -j\omega \mathbf{B} \tag{10.3.7}$$

$$\nabla \times \mathbf{H}^* = -j\omega \mathbf{D}^* + \mathbf{J}^* \tag{10.3.8}$$

and the constitutive relations for anisotropic media that

$$\mathbf{B} = \overline{\mu} \cdot \mathbf{H}, \quad \mathbf{D}^* = \overline{\varepsilon}^* \cdot \mathbf{E}^* \tag{10.3.9}$$

we have

$$\nabla \cdot (\mathbf{E} \times \mathbf{H}^*) = -j\omega\, \mathbf{H}^* \cdot \mathbf{B} + j\omega\, \mathbf{E} \cdot \mathbf{D}^* - \mathbf{E} \cdot \mathbf{J}^* \tag{10.3.10}$$

$$= -j\omega\, \mathbf{H}^* \cdot \overline{\mu} \cdot \mathbf{H} + j\omega\, \mathbf{E} \cdot \overline{\varepsilon}^* \cdot \mathbf{E}^* - \mathbf{E} \cdot \mathbf{J}^*. \tag{10.3.11}$$

The above is also known as the complex Poynting's theorem. It can also be written in an integral form using Gauss's divergence theorem, namely,

$$\int_S d\mathbf{S} \cdot (\mathbf{E} \times \mathbf{H}^*) = -j\omega \int\limits_V dV (\mathbf{H}^* \cdot \overline{\mu} \cdot \mathbf{H} - \mathbf{E} \cdot \overline{\varepsilon}^* \cdot \mathbf{E}^*) - \int\limits_V dV \mathbf{E} \cdot \mathbf{J}^*. \tag{10.3.12}$$

where S is the surface bounding the volume V.

10.3.2 Lossless Conditions

For a region V that is lossless and source-free, $\mathbf{J} = 0$. There should be no net time-averaged power-flow out of or into this region V. Therefore,

$$\Re e \int\limits_S d\mathbf{S} \cdot (\mathbf{E} \times \mathbf{H}^*) = 0, \tag{10.3.13}$$

[7]The product rule for derivative will be used, and we will drop the argument $\mathbf{r}, \omega$ for the phasors in our discussion next as they will be implied.

Because of the above, and energy conservation, the real part of the right-hand side of (10.3.11), without the $\mathbf{E}\cdot\mathbf{J}^*$ term, must also be zero. In other words, the right-hand side of (10.3.11) should be purely imaginary with no real part. Thus

$$\int_V dV(\mathbf{H}^*\cdot\overline{\mu}\cdot\mathbf{H}-\mathbf{E}\cdot\overline{\varepsilon}^*\cdot\mathbf{E}^*) \tag{10.3.14}$$

must be a purely real quantity. Other than the possibility that the above is zero, the general requirement for (10.3.14) to be real for arbitrary $\mathbf{E}$ and $\mathbf{H}$ is that $\mathbf{H}^*\cdot\overline{\mu}\cdot\mathbf{H}$ and $\mathbf{E}\cdot\overline{\varepsilon}^*\cdot\mathbf{E}^*$ are real quantities. This is only possible if $\overline{\mu}$ is Hermitian.[8] Therefore, the conditions for anisotropic media to be lossless are

$$\overline{\mu}=\overline{\mu}^\dagger, \qquad \overline{\varepsilon}=\overline{\varepsilon}^\dagger, \tag{10.3.15}$$

requiring the permittivity and permeability tensors to be Hermitian. If this is the case, (10.3.14) is always real for arbitrary $\mathbf{E}$ and $\mathbf{H}$, and (10.3.13) is true, implying a lossless region V. Notice that for an isotropic medium, $\overline{\mu}\rightarrow\mu$ and $\overline{\varepsilon}\rightarrow\varepsilon$, these lossless conditions reduce simply to that $\Im m(\mu)=0$ and $\Im m(\varepsilon)=0$, or that μ and ε are pure real quantities. Looking back, many of the effective permittivities or dielectric constants that we have derived using the Drude-Lorentz-Sommerfeld model cannot be lossless when the friction term is nonzero. Looking at the formula for χ as given by (8.3.17), it cannot be real, and hence, it corresponds to a lossy medium. The friction can be due to the collision of the electrons with the lattice.

10.3.3 Anisotropic Medium Case

For a lossy medium, which is conductive and anisotropic, we may define $\mathbf{J}=\overline{\sigma}\cdot\mathbf{E}$ where $\overline{\sigma}$ is a general conductivity tensor. In this case, equation (10.3.12), after combining the last two terms, may be written as

$$\int_S d\mathbf{S}\cdot(\mathbf{E}\times\mathbf{H}^*)=-j\omega\int_V dV\left[\mathbf{H}^*\cdot\overline{\mu}\cdot\mathbf{H}-\mathbf{E}\cdot\left(\overline{\varepsilon}^*+\frac{j\overline{\sigma}^*}{\omega}\right)\cdot\mathbf{E}^*\right] \tag{10.3.16}$$

$$=-j\omega\int dV[\mathbf{H}^*\cdot\overline{\mu}\cdot\mathbf{H}-\mathbf{E}\cdot\tilde{\overline{\varepsilon}}^*\cdot\mathbf{E}^*], \tag{10.3.17}$$

where $\tilde{\overline{\varepsilon}}=\overline{\varepsilon}-\frac{j\overline{\sigma}}{\omega}$ which is the general complex permittivity tensor. In this manner, (10.3.17) has the same structure as the source-free Poynting's theorem. Notice here that the complex permittivity tensor $\tilde{\overline{\varepsilon}}$ is clearly non-Hermitian, corresponding to a lossy medium.

For a lossless medium without the source term, by taking the imaginary part of (10.3.12), we arrive at

$$\Im m\int_S d\mathbf{S}\cdot(\mathbf{E}\times\mathbf{H}^*)=-\omega\int_V dV(\mathbf{H}^*\cdot\overline{\mu}\cdot\mathbf{H}-\mathbf{E}\cdot\overline{\varepsilon}^*\cdot\mathbf{E}^*), \tag{10.3.18}$$

[8] $\mathbf{H}^*\cdot\overline{\mu}\cdot\mathbf{H}$ is real only if its complex conjugate, or conjugate transpose is itself. Using some details from matrix algebra that $(\overline{\mathbf{A}}\cdot\overline{\mathbf{B}}\cdot\overline{\mathbf{C}})^t=\overline{\mathbf{C}}^t\cdot\overline{\mathbf{B}}^t\cdot\overline{\mathbf{A}}^t$, implies that (in physics notation, the transpose of a vector is implied in a dot product) $(\mathbf{H}^*\cdot\overline{\mu}\cdot\mathbf{H})^\dagger=(\mathbf{H}\cdot\overline{\mu}^*\cdot\mathbf{H}^*)^t=\mathbf{H}^*\cdot\overline{\mu}^\dagger\cdot\mathbf{H}=\mathbf{H}^*\cdot\overline{\mu}\cdot\mathbf{H}$. The last equality in the above is possible only if $\overline{\mu}=\overline{\mu}^\dagger$ or that $\overline{\mu}$ is hermitian.

The left-hand side of the above is the reactive power coming out of the volume V, and hence, the right-hand side can be interpreted approximately as reactive power as well. It is to be noted that $\mathbf{H}^* \cdot \overline{\mu} \cdot \mathbf{H}$ and $\mathbf{E} \cdot \overline{\varepsilon}^* \cdot \mathbf{E}^*$ are not to be interpreted as stored energy density when the medium is dispersive. The correct expressions for stored energy density in dispersive media will be derived later in the next section.

But the quantity $\mathbf{H}^* \cdot \overline{\mu} \cdot \mathbf{H}$ for lossless, dispersionless media is associated with the time-averaged energy density stored in the magnetic field, while the quantity $\mathbf{E} \cdot \overline{\varepsilon}^* \cdot \mathbf{E}^*$ for lossless, dispersionless media is associated with the time-averaged energy density stored in the electric field. Then, for lossless, dispersionless, source-free media, the right-hand side of the above can be interpreted as stored energy density. Therefore, the reactive power is roughly proportional to the time rate of change of the difference of the time-averaged energy stored in the magnetic field and the electric field. For example, in a resonant cavity, these two stored energies are equal to each other, and the need for external reactive power is zero, just as in the case of an LC tank circuit. In this case, energies just exchange between each other without external sources.

10.4 Energy Density in Dispersive Media[9]

This section is going to give us some new formulas previously unknown to most of us. The derivation is going to be laborious, but the punchline is that we need to update the energy storage formulas in electromagnetics. It turns out that the energy storage formula should be

$$\langle W_T \rangle = \frac{1}{4}\left[\frac{\partial \omega' \mu}{\partial \omega'}|\mathbf{H}|^2 + \frac{\partial \omega' \varepsilon}{\partial \omega'}|\mathbf{E}|^2\right] \tag{10.4.1}$$

For a nondispersive medium, μ and ε are independent of frequency, the above reverts back to a familiar expression,

$$\langle W_T \rangle = \frac{1}{4}\left[\mu|\mathbf{H}|^2 + \varepsilon|\mathbf{E}|^2\right] \tag{10.4.2}$$

which is what we have derived before.

A dispersive medium alters our concept of what the formula energy density should be.[10] In order to derive the new formula, we assume that the field has complex ω in $e^{j\omega t}$, where $\omega = \omega' - j\omega''$, rather than real ω dependence. In other words, the field is not time harmonic anymore, but quasi-time-harmonic when ω'' is very small.

We take the divergence of the complex power for fields with such a time dependence, and let $e^{j\omega t}$ be attached to the field. So $\mathbf{E}(t)$ and $\mathbf{H}(t)$ are complex field but not exactly like phasors we have studied before since they are not truly time harmonic. In other words, we let

$$\mathbf{E}(\mathbf{r},t) = \underset{\sim}{\mathbf{E}}(\mathbf{r},\omega)e^{j\omega t}, \quad \mathbf{H}(\mathbf{r},t) = \underset{\sim}{\mathbf{H}}(\mathbf{r},\omega)e^{j\omega t} \tag{10.4.3}$$

[9]The derivation in this section is complex, but worth the pain, since this knowledge was not discovered until the 1960s. It is not discussed in many textbooks.

[10]The derivation here is inspired by Hermann Anton Haus, Electromagnetic Noise and Quantum Optical Measurements [91]. Generalization to anisotropic media is given by W.C. Chew, *Lectures on Theory of Microwave and Optical Waveguides* [92].

The above, just like phasors, can be made to satisfy Maxwell's equations where a time derivative becomes $j\omega$ but this time with complex ω. We can study the quantity $\mathbf{E}(\mathbf{r},t)\times\mathbf{H}^*(\mathbf{r},t)$ which still has the unit of power density. In the real ω case, their time dependence will exactly cancel each other and this quantity becomes complex power again. But now, ω is complex, and the field is quasi-time-harmonic, and their time dependences do not cancel because of the complex ω. In other words, $e^{j\omega t}e^{-j\omega^* t} = e^{2\omega'' t} \neq 1$. Therefore, as before,

$$\nabla\cdot[\mathbf{E}(t)\times\mathbf{H}^*(t)] = \mathbf{H}^*(t)\cdot\nabla\times\mathbf{E}(t) - \mathbf{E}(t)\cdot\nabla\times\mathbf{H}^*(t) \tag{10.4.4}$$

where the quantities are still time dependent, whereas in the lossless case, the right-hand side would be time independent. Maxwell's equations for this quasi-time-harmonic fields, when ω is complex, become

$$\nabla\times\mathbf{E} = -j\omega\mathbf{B} \tag{10.4.5}$$
$$\nabla\times\mathbf{H}^* = -j\omega^*\mathbf{D}^* + \mathbf{J}^* \tag{10.4.6}$$

Using the above, in (10.4.4), after using $\mathbf{B}=\mu\mathbf{H}$, $\mathbf{D}=\varepsilon\mathbf{E}$, we arrive at

$$\nabla\cdot[\mathbf{E}(t)\times\mathbf{H}^*(t)] = -\mathbf{H}^*(t)\cdot j\omega\mu\mathbf{H}(t) + \mathbf{E}(t)\cdot j\omega^*\varepsilon^*\mathbf{E}^*(t) \tag{10.4.7}$$

where Maxwell's equations have been used to substitute for $\nabla\times\mathbf{E}(t)$ and $\nabla\times\mathbf{H}^*(t)$. The space dependence of the field is implied, and we assume a source-free medium so that $\mathbf{J}=0$.

If $\mathbf{E}(t)\sim e^{j\omega t}$, then due to ω being complex, $\mathbf{H}^*(t)\sim e^{-j\omega^* t}$. Then the term like $\mathbf{E}(t)\times\mathbf{H}^*(t)$ is not truly time independent but becomes

$$\mathbf{E}(t)\times\mathbf{H}^*(t) \sim e^{j(\omega-\omega^*)t} = e^{2\omega'' t} \tag{10.4.8}$$

And each of the terms above will have similar time dependence. Writing (10.4.4) more explicitly, by letting $\omega = \omega' - j\omega''$, we have

$$\nabla\cdot[\mathbf{E}(t)\times\mathbf{H}^*(t)] = -j(\omega'-j\omega'')\mu(\omega)|\mathbf{H}(t)|^2 + j(\omega'+j\omega'')\varepsilon^*(\omega)|\mathbf{E}(t)|^2 \tag{10.4.9}$$

So far, everything is exact, and no approximation has been made. To simplify the complicated math here, we make some approximations!

Assuming that $\omega'' \ll \omega'$, or that the field is quasi-time-harmonic, after using Taylor series approximation, we have

$$\mu(\omega'-j\omega'') \cong \mu(\omega') - j\omega''\frac{\partial\mu(\omega')}{\partial\omega'}, \quad \varepsilon(\omega'-j\omega'') \cong \varepsilon(\omega') - j\omega''\frac{\partial\varepsilon(\omega')}{\partial\omega'} \tag{10.4.10}$$

Using (10.4.10) in (10.4.9), collecting terms of the same order, and ignoring $(\omega'')^2$ terms yields[11]

$$\begin{aligned}\nabla\cdot[\mathbf{E}(t)\times\mathbf{H}^*(t)] \cong & -j\omega'\mu(\omega')|\mathbf{H}(t)|^2 + j\omega'\varepsilon^*(\omega')|\mathbf{E}(t)|^2 \\ & - \omega''\mu(\omega')|\mathbf{H}(t)|^2 - \omega'\omega''\frac{\partial\mu}{\partial\omega'}|\mathbf{H}(t)|^2 \\ & - \omega''\varepsilon^*(\omega')|\mathbf{E}(t)|^2 - \omega'\omega''\frac{\partial\varepsilon^*}{\partial\omega'}|\mathbf{E}(t)|^2\end{aligned} \tag{10.4.11}$$

[11]This is the general technique of perturbation expansion [49].

The above can be rewritten as

$$\nabla \cdot [\mathbf{E}(t) \times \mathbf{H}^*(t)] \cong -j\omega' \left[\mu(\omega')|\mathbf{H}(t)|^2 - \varepsilon^*(\omega')|\mathbf{E}(t)|^2\right] - \omega'' \left[\frac{\partial \omega'\mu(\omega')}{\partial \omega'}|\mathbf{H}(t)|^2 + \frac{\partial \omega'\varepsilon^*(\omega')}{\partial \omega'}|\mathbf{E}(t)|^2\right] \tag{10.4.12}$$

The above approximation is extremely good when $\omega'' \ll \omega'$. For a lossless medium, $\varepsilon(\omega')$ and $\mu(\omega')$ are purely real, and the first term of the right-hand side is purely imaginary while the second term is purely real. For better physical insight, in the limit when $\omega'' \to 0$, for better physical insight, we take half the imaginary part of the above equation to get

$$\nabla \cdot \frac{1}{2}\Im m\,[\mathbf{E} \times \mathbf{H}^*] \cong -\omega' \left[\frac{1}{2}\mu|\mathbf{H}|^2 - \frac{1}{2}\varepsilon|\mathbf{E}|^2\right] \tag{10.4.13}$$

Now, the left-hand side and right-hand side of the above now can be interpreted as reactive power, something we have learned before in the complex Poynting's theorem.

When half the real part of (10.4.12) is taken, we obtain some new terms,

$$\nabla \cdot \frac{1}{2}\Re e[\mathbf{E} \times \mathbf{H}^*] = -\frac{\omega''}{2}\left[\frac{\partial \omega'\mu}{\partial \omega'}|\mathbf{H}|^2 + \frac{\partial \omega'\varepsilon}{\partial \omega'}|\mathbf{E}|^2\right] \tag{10.4.14}$$

The left-hand side of the above has the physical meaning of time-average power density when $\omega'' \to 0$, or in the time-harmonic limit. Since the right-hand side has time dependence of $e^{2\omega'' t}$, when $\omega'' \neq 0$, it can be written as

$$\nabla \cdot \frac{1}{2}\Re e[\mathbf{E} \times \mathbf{H}^*] \cong -\frac{\partial}{\partial t}\frac{1}{4}\left[\frac{\partial \omega'\mu}{\partial \omega'}|\mathbf{H}|^2 + \frac{\partial \omega'\varepsilon}{\partial \omega'}|\mathbf{E}|^2\right] = -\frac{\partial}{\partial t}\langle W_T\rangle \tag{10.4.15}$$

The above is a restatement of that for a weakly time-harmonic system, the divergence of the time-average power density on the left-hand side is proportional to the time derivative of the stored energy on the right-hand side.[12] This has the same physical meaning as the current continuity equation which is a statement of charge conservation. Therefore, we reproduce the formulas at the beginning of this section: the time-average stored energy density can be identified as

$$\langle W_T\rangle = \frac{1}{4}\left[\frac{\partial \omega'\mu}{\partial \omega'}|\mathbf{H}|^2 + \frac{\partial \omega'\varepsilon}{\partial \omega'}|\mathbf{E}|^2\right] \tag{10.4.16}$$

Again, for a nondispersive medium, μ and ε are independent of frequency, the above reverts back to a familiar expression,

$$\langle W_T\rangle = \frac{1}{4}\left[\mu|\mathbf{H}|^2 + \varepsilon|\mathbf{E}|^2\right] \tag{10.4.17}$$

which is what we have derived before.

[12]Even though we say that this formula is approximately correct, to use the parlance of perturbation theory, it is correct up to leading order. Viz., it is exact when $\omega'' = 0$.

In the above analysis, we have used a quasi-time-harmonic signal with $\exp(j\omega t)$ dependence. In the limit when $\omega'' \to 0$, this signal reverts back to a time-harmonic signal, and to our usual interpretation of complex power. However, by assuming the frequency ω to have a very small imaginary part ω'', it forces the stored energy to grow very slightly, and hence, power has to be supplied to maintain the growth of this stored energy. By so doing, and use of energy conservation, it allows us to identify the expression for energy density for a dispersive medium. These expressions for energy density were not discovered until 1960 by Léon Brillouin [93], as energy density times group velocity should be power flow. More discussion on this topic can be found in Jackson [50].

It is to be noted that if the same analysis is used to study the energy storage in a capacitor or an inductor, the energy storage formulas have to be altered accordingly if the capacitor or inductor is frequency dependent.

Exercises for Lecture 10

Problem 10-1:

(i) Derive (10.2.6) of the lecture notes.

(ii) Explain why the time average of the instantaneous Poynting's vector is half the real part of the complex Poynting's vector.

(iii) What are the conditions of the permittivity and permeability tensors for them to describe a lossless medium?

(iv) Derive equation (10.4.12) of the lecture notes.

Problem 10-2: Explain how the quarter-wave plate and the half-wave plate work. What kind of medium is needed to make it work? (This knowledge can be found in many textbooks or the Internet. A heuristic explanation suffices.)

Problem 10-3: Show that the units on the left-hand side and the right-hand side of (10.2.6) are equal to each other.

Lecture 11

Uniqueness Theorem

The uniqueness of a solution to a linear system of equations is an important concept in mathematics. Likewise, linear systems described by ordinary differential equations partial differential equations and matrix equations will have unique solutions under the prescribed boundary conditions and the driving source terms. This is the way we solve a boundary value problem. But the uniqueness of a boundary value problem is not always guaranteed as we shall see unless additional conditions are stipulated. This issue is discussed in many math books and linear algebra books [85, 94]. The proof of uniqueness for Laplace and Poisson equations are given in [34, 59] which is slightly different from the electrodynamic case. To study uniqueness with Maxwell's equations is rather involved. But solving Maxwell's equations is analogous (or "homomorphic") to solving a system of linear algebra equations. Therefore, the uniqueness of Maxwell's equations is related to the uniqueness of linear algebra equations. You may not see it now, but eventually, you will get the epiphany that they are the same! Just imagine how bizzare it would be if there are more than one possible solution. One has to determine which is the real solution. To quote Star Trek, we need to know who the real McCoy is![1]

11.1 Different Solutions to Source-Free Maxwell's Equations

In this section, we will prove uniqueness theorem for electrodynamic problems under the prescribed boundary condition with unique sources in the system [35, 38, 55, 70, 92]. This is important, as when we solve Maxwell's equations, we are solving a set of partial differential equations as a boundary value problem with prescribed boundary conditions. We like to know the conditions under which such a problem has a unique solution.

Uniqueness for Maxwell's equations is only easily proved for linear Maxwell's equations, which are homomorphic to a system of linear algebraic equations. Therefore, to study the uniqueness theorem, we consider general linear anisotropic inhomogeneous media, where the tensors $\overline{\mu}$ and

[1]This phrase was made popular to the baby-boom generation, or the Trekkies by Star Trek. It actually refers to an African American inventor.

$\overline{\varepsilon}$ can be complex so that lossy media are included. In the frequency domain, let us assume two possible solutions with only one given set of sources $\mathbf{J}_i$ and $\mathbf{M}_i$.

First, let us assume that uniqueness is not guaranteed; there exist two solutions to Maxwell's equations in the presence of one set of common impressed sources $\mathbf{J}_i$ and $\mathbf{M}_i$.[2] Namely, these two solutions are $\mathbf{E}^a$, $\mathbf{H}^a$, and $\mathbf{E}^b$, $\mathbf{H}^b$. Both of them satisfy Maxwell's equations and the same boundary conditions, and also use a set of common impressed sources. What additional conditions do we need to impose to ensure that $\mathbf{E}^a = \mathbf{E}^b$ and $\mathbf{H}^a = \mathbf{H}^b$?

It follows that

$$\nabla \times \mathbf{E}^a = -j\omega\overline{\boldsymbol{\mu}} \cdot \mathbf{H}^a - \mathbf{M}_i \tag{11.1.1}$$

$$\nabla \times \mathbf{E}^b = -j\omega\overline{\boldsymbol{\mu}} \cdot \mathbf{H}^b - \mathbf{M}_i \tag{11.1.2}$$

$$\nabla \times \mathbf{H}^a = j\omega\overline{\boldsymbol{\varepsilon}} \cdot \mathbf{E}^a + \mathbf{J}_i \tag{11.1.3}$$

$$\nabla \times \mathbf{H}^b = j\omega\overline{\boldsymbol{\varepsilon}} \cdot \mathbf{E}^b + \mathbf{J}_i \tag{11.1.4}$$

By taking the difference of these two solutions, we have

$$\nabla \times (\mathbf{E}^a - \mathbf{E}^b) = -j\omega\overline{\boldsymbol{\mu}} \cdot (\mathbf{H}^a - \mathbf{H}^b) \tag{11.1.5}$$

$$\nabla \times (\mathbf{H}^a - \mathbf{H}^b) = j\omega\overline{\boldsymbol{\varepsilon}} \cdot (\mathbf{E}^a - \mathbf{E}^b) \tag{11.1.6}$$

Or alternatively, defining $\delta\mathbf{E} = \mathbf{E}^a - \mathbf{E}^b$ and $\delta\mathbf{H} = \mathbf{H}^a - \mathbf{H}^b$, we have

$$\nabla \times \delta\mathbf{E} = -j\omega\overline{\boldsymbol{\mu}} \cdot \delta\mathbf{H} \tag{11.1.7}$$

$$\nabla \times \delta\mathbf{H} = j\omega\overline{\boldsymbol{\varepsilon}} \cdot \delta\mathbf{E} \tag{11.1.8}$$

The difference solutions, $\delta\mathbf{E}$ and $\delta\mathbf{H}$, satisfy the original source-free Maxwell's equations. Source-free here implies that we are looking at the homogeneous solutions of the pertinent partial differential equations constituted by (11.1.7) and (11.1.8). They are also analogous to finding the null-space solution of a matrix equation in linear algebra.

To prove uniqueness, we would like to find a simplifying expression for $\nabla \cdot (\delta\mathbf{E} \times \delta\mathbf{H}^*)$. By using the product rule for divergence operator, and the vector identity (also called the scalar-tripple product), it can be shown that

$$\nabla \cdot (\delta\mathbf{E} \times \delta\mathbf{H}^*) = \delta\mathbf{H}^* \cdot \nabla \times \delta\mathbf{E} - \delta\mathbf{E} \cdot \nabla \times \delta\mathbf{H}^* \tag{11.1.9}$$

We need to simplify the right-hand side of the above with the goal of proving the uniqueness theorem. Then by taking the left dot product of $\delta\mathbf{H}^*$ with (11.1.7), and then the left dot product of $\delta\mathbf{E}$ with the complex conjugation of (11.1.8), we obtain

$$\begin{aligned} \delta\mathbf{H}^* \cdot \nabla \times \delta\mathbf{E} &= -j\omega\delta\mathbf{H}^* \cdot \overline{\boldsymbol{\mu}} \cdot \delta\mathbf{H} \\ \delta\mathbf{E} \cdot \nabla \times \delta\mathbf{H}^* &= -j\omega\delta\mathbf{E} \cdot \overline{\boldsymbol{\varepsilon}}^* \cdot \delta\mathbf{E}^* \end{aligned} \tag{11.1.10}$$

[2]It is not clear when the useful concept of impressed sources were first used in electromagnetics even though it was used in [95] in 1936. These are immutable sources that cannot be changed by the environment in which they are immersed. They are like current and voltage sources in circuit theory that are immutable irrespective of environmental loading, e.g., by changing the circuit elements in the circuit.

Now, taking the difference of the above, we get

$$\begin{aligned}\delta\mathbf{H}^* \cdot \nabla \times \delta\mathbf{E} - \delta\mathbf{E} \cdot \nabla \times \delta\mathbf{H}^* &= \nabla \cdot (\delta\mathbf{E} \times \delta\mathbf{H}^*) \\ &= -j\omega\delta\mathbf{H}^* \cdot \overline{\mu} \cdot \delta\mathbf{H} + j\omega\delta\mathbf{E} \cdot \overline{\varepsilon}^* \cdot \delta\mathbf{E}^* \end{aligned} \tag{11.1.11}$$

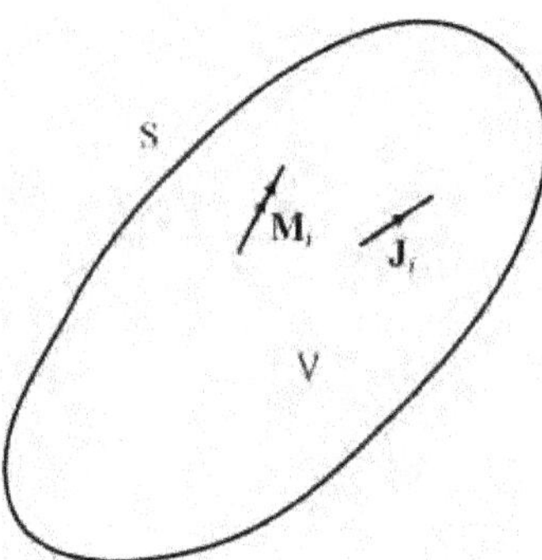

Figure 11.1: Geometry for proving the uniqueness theorem. We like to know the requisite boundary conditions on S plus the type of media inside V in order to guarantee the uniqueness of the solution in V.

Our goal is to find the conditions under which $\delta\mathbf{H}$ and $\delta\mathbf{E}$ are both zero, which will guarantee uniqueness of the solution. Next, we integrate the above equation (11.1.11) over a volume V bounded by a surface S as shown in Figure 11.1. After making use of Gauss's divergence theorem, we arrive at

$$\iint_V \nabla \cdot (\delta\mathbf{E} \times \delta\mathbf{H}^*) dV = \oint\!\!\!\oint_S (\delta\mathbf{E} \times \delta\mathbf{H}^*) \cdot d\mathbf{S} \tag{11.1.12}$$

or after equating with the right-hand side of (11.1.11), we have that

$$\oint\!\!\!\oint_S (\delta\mathbf{E} \times \delta\mathbf{H}^*) \cdot d\mathbf{S} = \iiint_V [-j\omega\delta\mathbf{H}^* \cdot \overline{\mu} \cdot \delta\mathbf{H} + j\omega\delta\mathbf{E} \cdot \overline{\varepsilon}^* \cdot \delta\mathbf{E}^*] dV \tag{11.1.13}$$

And next, we would like to know the kind of boundary conditions that would make the left-hand side equal to zero such that we can make a statement on $\delta\mathbf{H}$ and $\delta\mathbf{E}$ on the right-hand side. It is seen that the surface integral on the left-hand side will be zero if:[3]

1. If $\hat{n} \times \mathbf{E}$ is specified over S for the two possible solutions, so that $\hat{n} \times \mathbf{E}_a = \hat{n} \times \mathbf{E}_b$ on S, then $\hat{n} \times \delta\mathbf{E} = 0$, which is the PEC boundary condition for $\delta\mathbf{E}$, and then[4]

[3]In the following, please be reminded that PEC stands for "perfect electric conductor", while PMC stands for "perfect magnetic conductor". PMC is the dual of PEC. Also, a fourth case of impedance boundary condition is possible, which is beyond the scope of this course. Interested readers may consult Chew, Theory of Microwave and Optical Waveguides [92].

[4]We use the vector identity (also called the scalar-triple product) that $\mathbf{a} \cdot (\mathbf{b} \times c) = \mathbf{c} \cdot (\mathbf{a} \times \mathbf{b}) = \mathbf{b} \cdot (\mathbf{c} \times \mathbf{a})$. Also, from Section 1.3.3, $d\mathbf{S} = \hat{n} dS$.

$\oiint_S (\delta\mathbf{E} \times \delta\mathbf{H}^*) \cdot \hat{n} dS = \oiint_S (\hat{n} \times \delta\mathbf{E}) \cdot \delta\mathbf{H}^* dS = 0.$

2. If $\hat{n} \times \mathbf{H}$ is specified over S for the two possible solutions, so that $\hat{n} \times \mathbf{H}_a = \hat{n} \times \mathbf{H}_b$ on S, then $\hat{n} \times \delta\mathbf{H} = 0$, which is the PMC boundary condition for $\delta\mathbf{H}$, and then
$\oiint_S (\delta\mathbf{E} \times \delta\mathbf{H}^*) \cdot \hat{n} dS = -\oiint_S (\hat{n} \times \delta\mathbf{H}^*) \cdot \delta\mathbf{E} dS = 0.$

3. Let the surface S be divided into two mutually exclusive surfaces S_1 and S_2.[5] If $\hat{n} \times \mathbf{E}$ is specified over S_1, and $\hat{n} \times \mathbf{H}$ is specified over S_2. Then $\hat{n} \times \delta\mathbf{E} = 0$ (PEC boundary condition) on S_1, and $\hat{n} \times \delta\mathbf{H} = 0$ (PMC boundary condition) on S_2. Therefore, the left-hand side becomes
$\oiint_S (\delta\mathbf{E} \times \delta\mathbf{H}^*) \cdot \hat{n} dS = \iint_{S_1} + \iint_{S_2} = \iint_{S_1} (\hat{n} \times \delta\mathbf{E}) \cdot \delta\mathbf{H}^* dS$
$- \iint_{S_2} (\hat{n} \times \delta\mathbf{H}^*) \cdot \delta\mathbf{E} dS = 0.$

Thus, under the above three scenarios, the left-hand side of (11.1.13) is zero, and then the right-hand side of (11.1.13) becomes

$$\iiint_V [-j\omega\delta\mathbf{H}^* \cdot \overline{\mu} \cdot \delta\mathbf{H} + j\omega\delta\mathbf{E} \cdot \overline{\varepsilon}^* \cdot \delta\mathbf{E}^*] dV = 0 \tag{11.1.14}$$

For lossless media, $\overline{\mu}$ and $\overline{\varepsilon}$ are Hermitian tensors (or matrices[6]). It can then be seen, using the properties of Hermitian matrices or tensors, that $\delta\mathbf{H}^* \cdot \overline{\mu} \cdot \delta\mathbf{H}$ and $\delta\mathbf{E} \cdot \overline{\varepsilon}^* \cdot \delta\mathbf{E}^*$ are purely real. Taking the imaginary part of the above equation yields

$$\iiint_V [-\delta\mathbf{H}^* \cdot \overline{\mu} \cdot \delta\mathbf{H} + \delta\mathbf{E} \cdot \overline{\varepsilon}^* \cdot \delta\mathbf{E}^*] dV = 0 \tag{11.1.15}$$

The above two terms correspond to stored magnetic field energy and stored electric field energy in the difference solutions $\delta\mathbf{H}$ and $\delta\mathbf{E}$, respectively. The above being zero does not imply that $\delta\mathbf{H}$ and $\delta\mathbf{E}$ are zero since they can be negative of each other.

For resonant solutions, the stored electric energy and stored magnetic energy can balance (or cancel) each other, implying that $\delta\mathbf{H}$ and $\delta\mathbf{E}$ need not be zero when the above is zero. The above resonant solutions are those of the difference solutions satisfying PEC or PMC boundary condition or mixture thereof. Also, they are the solutions of the source-free Maxwell's equations (11.1.7).[7] Clearly, $\delta\mathbf{H}$ and $\delta\mathbf{E}$ need not be zero, even though (11.1.15) is zero. This happens when we encounter solutions that are the resonant modes of the volume V bounded by the surface S.

11.2 Conditions for Uniqueness

Uniqueness can only be guaranteed if the medium is lossy as shall be shown. It is also guaranteed if lossy impedance boundary conditions are imposed.[8] First we begin with the isotropic case which is simple.

[5] In math parlance, $S_1 \cup S_2 = S$.
[6] Tensors are a special kind of matrices that relate two physical vectors.
[7] They are also the homogeneous solution of a partial differential equation.
[8] See Chew, *Theory of Microwave and Optical Waveguides.* [96]

11.2.1 Isotropic Case

It is easier to see this for lossy isotropic media. Then (11.1.14) simplifies to

$$\iiint_V [-j\omega\mu|\delta\mathbf{H}|^2 + j\omega\varepsilon^*|\delta\mathbf{E}|^2]dV = 0 \tag{11.2.1}$$

For isotropic lossy media, $\mu = \mu' - j\mu''$ and $\varepsilon = \varepsilon' - j\varepsilon''$. Taking the real part of the above, we have from (11.2.1) that, "miraculously",

$$\iiint_V [-\omega\mu''|\delta\mathbf{H}|^2 - \omega\varepsilon''|\delta\mathbf{E}|^2]dV = 0 \tag{11.2.2}$$

Now the two terms in the integrand are of the same sign. Moreover, $|\delta\mathbf{H}|^2$ and $|\delta\mathbf{E}|^2$ are always positive definite. Thus the integrand in the above is always negative definite. The integral can be zero only if

$$\delta\mathbf{E} = 0, \quad \delta\mathbf{H} = 0 \tag{11.2.3}$$

everywhere in V, implying that $\mathbf{E}_a = \mathbf{E}_b$, and that $\mathbf{H}_a = \mathbf{H}_b$ everywhere in V. Consequently, it is seen that uniqueness is guaranteed only if the medium is lossy.

The physical reason is that when the medium is lossy, a homogeneous solution (also called a natural solution, a resonant solution, or a modal solution), which is a pure time-harmonic solution, cannot exist due to loss. The modes or the natural solutions, which are the source-free solutions of Maxwell's equations, are decaying sinusoids. But when we express equations (11.1.1) to (11.1.4) in the frequency domain, we are seeking time-harmonic solutions for which ω is real. Hence, (11.2.3) is true in order for (11.2.2) to be true.

Notice that the same conclusion can be drawn if we make μ'' and ε'' negative. This corresponds to active media, and uniqueness can be guaranteed for a time-harmonic solution if μ'' and ε'' are of the same sign. Again, here no natural time-harmonic solution can exist, since the resonant solution or the homogeneous solution is a growing sinusoid.

Therefore, uniqueness is guaranteed for active or passive media. However, if the medium is a mixture of active and passive media, uniqueness is not guaranteed again as the equation (11.2.2) can be satisfied with $\delta\mathbf{H}$ and $\delta\mathbf{E}$ being not zero if μ'' and ε'' are of the opposite signs.

11.2.2 General Anisotropic Case

The proof for general anisotropic media is more involved. For the lossless anisotropic media, we see that (11.1.14) is purely imaginary. However, when the medium is lossy, this same equation is not purely imaginary, and will have a real part. Hence, we need to find the real part of (11.1.14) for the general lossy case.

About taking the Real and Imaginary Parts of a Complicated Expression

To this end, we digress on taking the real and imaginary parts of a complicated expression. Here, we need to find the complex conjugate[9] of the integrand of (11.1.14), which is a scalar, and add it

[9]Also called Hermitian conjugate.

to itself to get its real part. To this end, we will find the conjugate of its integrand which is also a scalar number.

First, the complex conjugate of the first scalar term in the integrand of (11.1.14) is[10]

$$(-j\omega\delta\mathbf{H}^*\cdot\overline{\boldsymbol{\mu}}\cdot\delta\mathbf{H})^* = j\omega\delta\mathbf{H}\cdot\overline{\boldsymbol{\mu}}^*\cdot\delta\mathbf{H}^* = j\omega\delta\mathbf{H}^*\cdot\overline{\boldsymbol{\mu}}^\dagger\cdot\delta\mathbf{H} \tag{11.2.4}$$

The last equality follows because we are just taking the transpose of a scalar number. Similarly, the complex conjugate of the second scalar term in the same integrand is

$$(j\omega\delta\mathbf{E}\cdot\overline{\varepsilon}^*\cdot\delta\mathbf{E}^*)^* = -j\omega\delta\mathbf{E}^*\cdot\overline{\varepsilon}^\dagger\cdot\delta\mathbf{E} \tag{11.2.5}$$

But

$$j\omega\delta\mathbf{E}\cdot\overline{\varepsilon}^*\cdot\delta\mathbf{E}^* = j\omega\delta\mathbf{E}^*\cdot\overline{\varepsilon}^\dagger\cdot\delta\mathbf{E} \tag{11.2.6}$$

The above gives us the complex conjugate of the scalar quantity (11.1.14). Adding it to itself, we have

$$\iiint_V [-j\omega\delta\mathbf{H}^*\cdot(\overline{\boldsymbol{\mu}}-\overline{\boldsymbol{\mu}}^\dagger)\cdot\delta\mathbf{H} - j\omega\delta\mathbf{E}^*\cdot(\overline{\varepsilon}-\overline{\varepsilon}^\dagger)\cdot\delta\mathbf{E}]dV = 0 \tag{11.2.7}$$

For lossy media, $-j(\overline{\boldsymbol{\mu}}-\overline{\boldsymbol{\mu}}^\dagger)$ and $-j(\overline{\varepsilon}-\overline{\varepsilon}^\dagger)$ are Hermitian positive matrices. Hence the integrand is always positive definite, and the above equation cannot be satisfied unless $\delta\mathbf{H} = \delta\mathbf{E} = 0$ everywhere in V. Uniqueness is thus guaranteed in a lossy anisotropic medium.

A similar statement can be made for the anisotropic case if the medium is active. Then the integrand is positive definite, and the above equation cannot be satisfied unless $\delta\mathbf{H} = \delta\mathbf{E} = 0$ everywhere in V, thereby proving that uniqueness is satisfied.

11.3 Hindsight Using Linear Algebra

The proof of uniqueness for Maxwell's equations is very similar to the proof of uniqueness for a matrix equation [85]. The two problems are "homomorphic" to each other. As you will see, the proof using linear algebra is a lot simpler due to the simplicity of notations. To see this, consider a linear algebraic equation

$$\overline{\mathbf{A}}\cdot\mathbf{x} = \mathbf{b} \tag{11.3.1}$$

If a solution to a matrix equation exists without excitation, namely, when $\mathbf{b} = 0$, then the solution to (11.3.1) is the null-space solution [85], namely, $\mathbf{x} = \mathbf{x}_N$, or

$$\overline{\mathbf{A}}\cdot\mathbf{x}_N = 0 \tag{11.3.2}$$

For Maxwell's equations, $\mathbf{b}$ corresponds to the source terms. In a word, the solution in (11.3.2) is like the homogeneous solution of an ordinary differential equation or a partial differential equation [94].

[10] To arrive at these expressions, one makes use of the matrix algebra rule that if $\overline{\mathbf{D}} = \overline{\mathbf{A}}\cdot\overline{\mathbf{B}}\cdot\overline{\mathbf{C}}$, then $\overline{\mathbf{D}}^t = \overline{\mathbf{C}}^t\cdot\overline{\mathbf{B}}^t\cdot\overline{\mathbf{A}}^t$. This is true even for non-square matrices. But for our case here, $\overline{\mathbf{A}}$ is a 1×3 row vector, and $\overline{\mathbf{C}}$ is a 3×1 column vector, and $\overline{\mathbf{B}}$ is a 3×3 matrix. In vector algebra, the transpose of a vector is implied. Also, in our case here, $\overline{\mathbf{D}}$ is a scalar, and hence, its transpose is itself.

They are also analogous to the null-space solution of a matrix equation. In an enclosed region of volume V bounded by a surface S, homogeneous solutions are the resonant solutions (or the natural solutions) of this Maxwellian system. When these solutions exist, they give rise to nonuniqueness. (Note that these resonant solutions in the time domain exist for all time if the cavity is lossless.)

Also, notice that (11.1.7) and (11.1.8) for the difference solutions are Maxwell's equations without the source terms. When there are no source terms in a closed, lossless region V bounded by a surface S, **only** resonant solutions can exist for $\delta\mathbf{E} \neq 0$ and $\delta\mathbf{H} \neq 0$ with the relevant boundary conditions.

As previously mentioned, one way to ensure that these resonant solutions (or homogeneous solutions) are eliminated is to put in loss or gain. When loss or gain is present, then the resonant solutions are decaying sinusoids or growing sinusoids. Since we are looking for solutions in the frequency domain, or time-harmonic solutions, the solutions we are seeking are on the real ω axis on the complex ω plane. Thus the nonsinusoidal (non-time-harmonic) solutions are outside the solution space: They are not part of the time-harmonic solutions (which are on the real axis) that we are looking for.

We see that the sources of nonuniqueness are the homogeneous solutions or the resonant solutions of the system that persist for all time. These solutions are noncausal, and they are there in the system since the beginning of time and forever thereafter. One way to remove these resonant solutions is to set them to zero at the beginning by solving an initial value problem. However, this has to be done in the time domain. Therefore, one reason for nonuniqueness is because we are seeking the solutions in the frequency domain.

11.4 Connection to Poles of a Linear System

The output is the response to the input of a linear system. It can be represented by a transfer function $H(\omega)$ [57, 97]. If $H(\omega)$ has poles, and if the system is lossless, the poles are on the real axis. Therefore, when $\omega = \omega_{\text{pole}}$, the function $H(\omega)$ becomes undefined. In other words, one can add a constant term to the output, and the ratio between output to input is still infinity. This also is the reason for nonuniqueness of the output with respect to the input. Poles usually correspond to resonant solutions, and hence, the nonuniqueness of the solution is intimately related to the nonuniqueness of Maxwell's equations at the resonant frequencies of a structure. This is illustrated in the upper part of Figure 11.2.

If the input function is $f(t)$, with Fourier transform $F(\omega)$, then the output $y(t)$ is given by the following Fourier integral, namely,

$$y(t) = \frac{1}{2\pi} \int_{-\infty}^{\infty} d\omega e^{j\omega t} H(\omega) F(\omega) \tag{11.4.1}$$

where the Fourier inversion integral path is on the real axis on the complex ω plane. The Fourier inversion integral above is undefined or nonunique if poles exist on the real ω axis.

However, if loss is introduced, these poles will move away from the real axis as shown in the lower part of Figure 11.2. Then the transfer function is uniquely determined for all frequencies on the real axis. In this way, the Fourier inversion integral in (11.4.1) is well defined, and uniqueness of the solution is guaranteed.

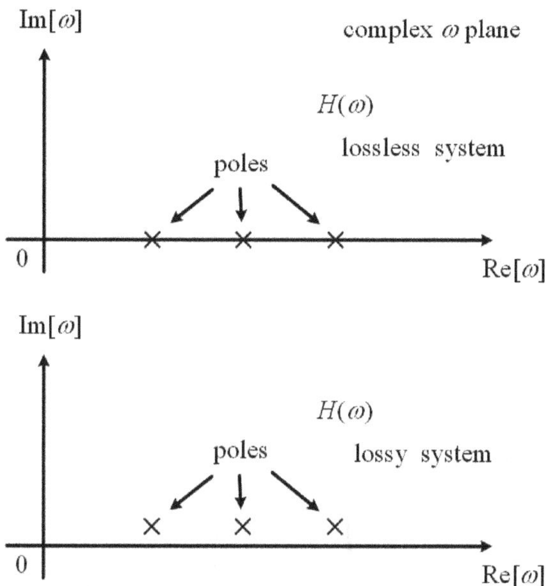

Figure 11.2: The nonuniqueness problem is intimately related to the locations of the poles of a transfer function being on the real axis, when one solves a linear system using Fourier transform techniques. because For a lossless system, the poles are located on the real axis, the Fourier inverse transform is not defined since the Fourier inversion contour is on the real axis. But when performing a Fourier inverse transform to obtain the solution in the time domain, then the Fourier inversion contour is undefined, and the solution cannot be uniquely determined.

When the poles are located on the real axis yielding possibly nonunique solutions, a remedy to this problem is to use the Laplace transform technique [57] when solving such problems. The Laplace transform technique allows the specification of initial values, which is similar to solving the problem as an initial value problem. As mentioned before, solving these problems as an initial value problem will remove nonuniqueness of the solution.

If you have trouble wrapping your head around this concept, it is good to connect back to the LC tank circuit example. The transfer function $H(\omega)$ is similar to the $Y(\omega) = j\omega C + \frac{1}{j\omega L}$. The transfer function has two poles at $\omega = \pm\sqrt{LC}$. If there is no loss, then the poles are located on the real ω axis, rendering the Fourier inversion contour undefined in (11.4.1). Hence, the solution is nonunique. However, if infinitesimal loss is introduced by adding a resistor, then the poles will migrate off the real axis making the Fourier integral in (11.4.1) well defined!

11.5 Radiation from Antenna Sources and Radiation Condition[11]

The above uniqueness theorem guarantees that if we have some antennas with prescribed current sources on them, the radiated field from these antennas are unique under certain conditions. To see how this can come about, we first study the radiation of sources into a region V bounded by a very large surface $S_{\text{large}} \to S_{\text{inf}}$ as shown in Figure 11.4 [38].

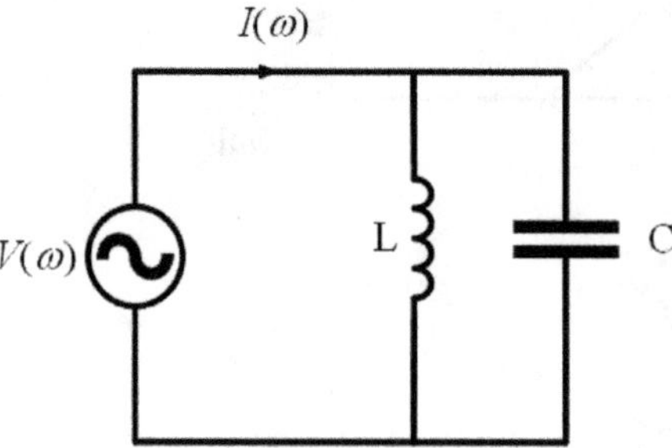

Figure 11.3: The transfer function of the LC tank circuit is $H(\omega) = Z(\omega)$ where the input is $I(\omega)$ and the output is $V(\omega) = H(\omega)I(\omega)$. One can show that the transfer function $H(\omega)$ has poles on the real ω axis, implying that the Fourier inverse transform in (11.4.1) where the integration is on the real ω axis does not exist. However, when a small resistor R is added in series with the inductor, the poles are shifted off from the real axis making the Fourier transform in (11.4.1) well defined.

When we solve a boundary value problem, $\hat{n} \times \mathbf{E}$ or $\hat{n} \times \mathbf{H}$ are specified on the surface at S_{large}. When the volume V bounded by S_{large} is large, the problem can have many resonant solutions. In fact, the region will be replete with resonant solutions as one makes S_{large} become very large. When we pick an operating frequency ω, the likelihood of ω coinciding with a resonant solution is high.

To gain more insight, we look at the resonant condition of a large rectangular cavity reproduced here as[12]

$$\beta^2 = \frac{\omega^2}{c^2} = \left(\frac{m\pi}{a}\right)^2 + \left(\frac{n\pi}{b}\right)^2 + \left(\frac{p\pi}{d}\right)^2 \tag{11.5.1}$$

The above is an equation of an Ewald sphere in a 3D mode space which is described by discrete points, or that the values of $\beta_x = \frac{m\pi}{a}$, $\beta_y = \frac{n\pi}{b}$, and $\beta_z = \frac{p\pi}{d}$ are discrete.

[11] May be skipped on first reading.

[12] This formula is usually covered in an undergraduate course in electromagnetics, and will be covered later in this course in section 25.2.1.

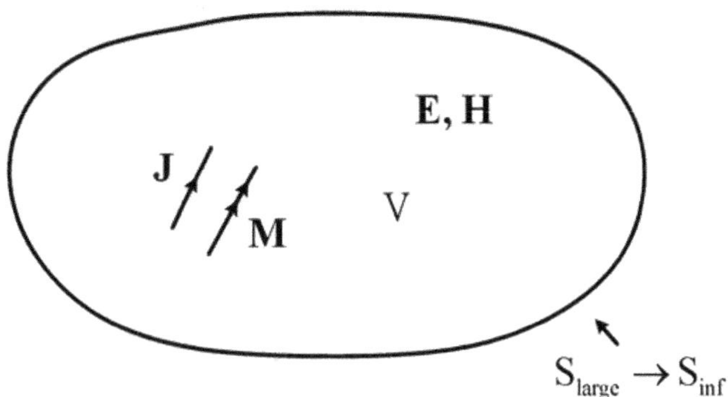

Figure 11.4: When $S_{\text{large}} \rightarrow S_{\text{inf}}$, the number of the resonant solution within the surface S becomes infinitely large. The chance of the operating frequency ω coinciding with the resonant frequency is 1. However, the solution can be made unique by assuming an infinitesimal loss. Therefore, the solution for antenna radiation in an infinite space can be made unique by imposing the Sommerfeld radiation condition. That is, we assume that the radiation wave travels to infinity, never to return. This is equivalent to assuming an infinitesimal loss when seeking the solution in V and later let $V \rightarrow \infty$.

Here, β_x, β_y, and β_z can be thought of the Fourier transform variables of x, y, and z. These points are known as reciprocal lattice points in solid state physics [98, 99].

We can continuously change the operating frequency ω above until the above equation is satisfied. When this happens, we encounter a resonant frequency of the cavity. At this operating frequency, the solution to Maxwell's equations inside the cavity is nonunique. As the dimensions of the cavity become large (or a, b, and d are large), then the number of ω's or resonant frequencies at which the above equation can be satisfied becomes very large. This is illustrated Figure 11.5 in 2D. Hence, the chance of the operating frequency ω coinciding with a resonant mode of the cavity is very high (with a probability of almost 1) giving rise to nonuniqueness. This above argument applies to cavities of other shapes as well.

The way to remove these resonant solutions is to introduce an infinitesimal amount of loss in region V. Then these resonant solutions will disappear from the real ω axis, where we seek a time-harmonic solution. Now we can take S_{large} to S_{inf}, which is infinitely large, and the solution will always remain unique even if the loss is infinitesimally small.

Notice that if $S_{\text{inf}} \rightarrow \infty$, the waves that leave the sources will never be reflected back because of the small amount of loss. The radiated field will just disappear into infinity. This is just what radiation loss is: power that propagates to infinity, never to return. In fact, one way of guaranteeing the uniqueness of the solution in region V when S_{inf} is infinitely large or when V is infinitely large, is to impose the radiation condition: the waves that radiate to infinity are outgoing waves only and never return. This is also called the Sommerfeld radiation condition [100]. Uniqueness of the field outside the sources is always guaranteed if we assume that the field radiates to infinity and never returns. This is equivalent to solving the cavity solutions with an infinitesimal loss, and then letting the size of the cavity become infinitely large.

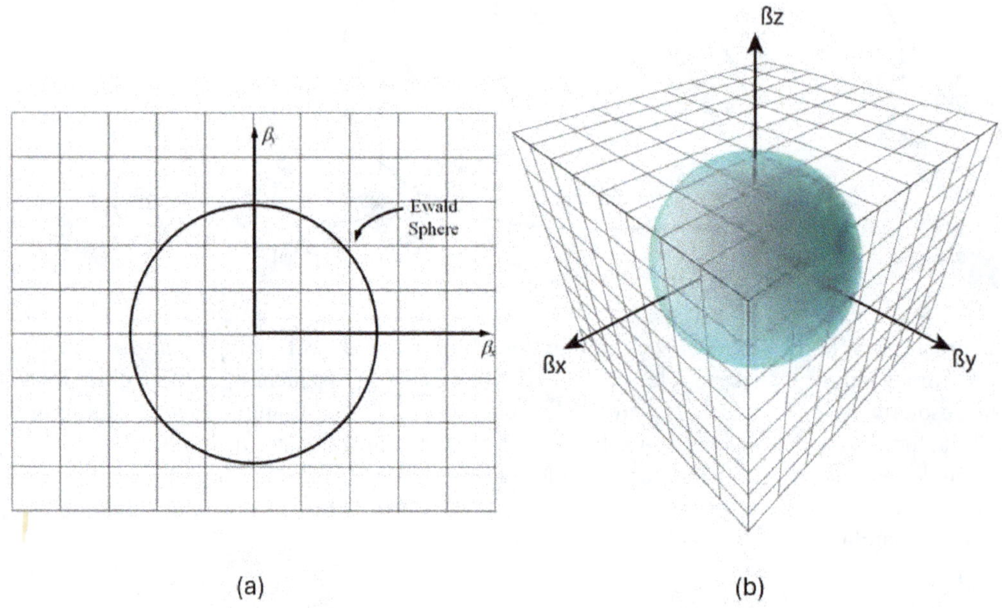

Figure 11.5: For a very large cavity, the grid spacing in the mode space (or Fourier space, also called the reciprocal lattice space [98, 99]) becomes very small. The radius of the sphere is given by β. Then the chance that the sphere surface encounters a resonant mode is very high. When this happens, the solution to the cavity problem is nonunique. (a) shows the 2D case while (b) shows the 3D case. The sphere defined by the equation $\beta^2 = \beta_x^2 + \beta_y^2 + \beta_z^2$ is known as the Ewald sphere. Here, k is equivalent to β.

Exercises for Lecture 11

Problem 11-1:

(i) Derive Equation (11.1.13) of the lecture notes.

(ii) Explain the conditions for uniqueness of the solutions to Maxwell's equations.

(iii) Explain why the nonuniqueness of Maxwell's equations is similar to the nonuniqueness of solving a matrix equation with a null space.

(iv) Explain why when one solves for the solutions of Maxwell's equations in a very large lossless cavity, the probability of encountering nonuniqueness is very high.

Lecture 12

Reciprocity Theorem

子貢問曰：“有一言而可以終身行之者乎？” 子曰：“其恕乎！己所不欲、勿施於人。”

Zi Gong [a disciple] asked: "Is there any one word that could guide a person throughout life?"

The Master replied: "How about 'reciprocity'! Never impose on others what you would not choose for yourself."

Analects XV.24, tr. David Hinton

Figure 12.1: (Left) A depiction of Confucius from a stone fresco from the Western Han dynasty (202 BC–9 AD). The emphasis of the importance of "reciprocity" by Confucius Analects as translated by D. Hinton [101]. (Right) A portrait of Jesus that is truer to its form. Jesus teaching from the New Testament says, "Do unto others as you would have them do unto you." Luke 6:31 and Matthew 7:12 [102]. You may recall what the Old Testament says. (Figure by: https://www.pinterest.com/pin/47217496082627759/)

Reciprocity theorem is one of the most important theorems in electromagnetics. With it we can develop physical intuition and sanity check to ascertain the correctness of a certain design or experiment. Often time, a lossless reciprocal system is also time reversible.

Reciprocity theorem is like a "tit-for-tat" relationship in humans: Goodwill is reciprocated with goodwill while ill will is countered with ill will. Both Confucius (551 BC–479 BC) and Jesus Christ

(4 BC–AD 30) espoused the concept, "Don't do unto others which you don't like others to do unto you." But in electromagnetics, this beautiful relationship can be expressed precisely and succinctly using mathematics. We shall see how this is done.

12.1 Mathematical Derivation

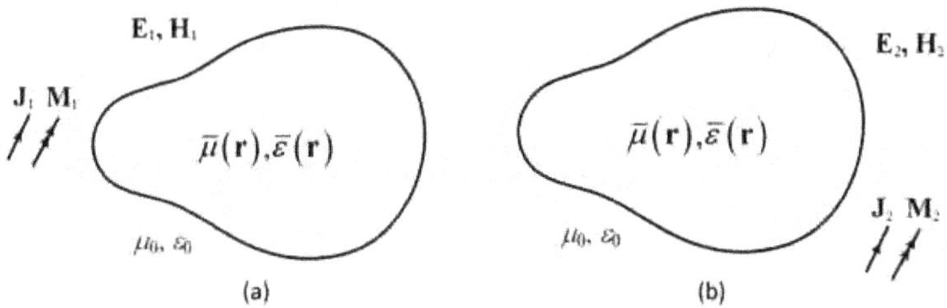

Figure 12.2: The geometry for proving reciprocity theorem. We perform two experiments on the same object or medium: (a) With sources $\mathbf{J}_1$ and $\mathbf{M}_1$ turned on, generating fields $\mathbf{E}_1$ and $\mathbf{H}_1$, and (b) With sources $\mathbf{J}_2$ and $\mathbf{M}_2$ turned on, generating fields $\mathbf{E}_2$ and $\mathbf{H}_2$. Magnetic currents, by convention, are denoted by double arrows.

Consider a general linear anisotropic inhomogeneous medium in the frequency domain where both $\overline{\mu}(\mathbf{r})$ and $\overline{\varepsilon}(\mathbf{r})$ are described by permeability tensor and permittivity tensor over a finite part of space as shown in Figure 12.2. In the phasor world, this representation of the medium is quite general, as it can include dispersive and conductive media. It can represent complex terrain, or complicated electronic circuit structures in circuit boards or microchips, as well as complicated antenna structures (see Figure 12.5).

We will do a Gedanken experiment[1] where a scatterer or an object is illuminated by fields from two different sets of sources which are turned on and off consecutively. This is illustrated in Figure 12.2: The geometry is not changed, but when only $\mathbf{J}_1$ and $\mathbf{M}_1$, as impressed sources, are turned on (case (a)), they generate fields $\mathbf{E}_1$ and $\mathbf{H}_1$ in this medium. On the other hand, when only $\mathbf{J}_2$ and $\mathbf{M}_2$ as impressed sources are turned on (case (b)), they generate $\mathbf{E}_2$ and $\mathbf{H}_2$ in this medium. Therefore, the pertinent equations in the frequency domain, for linear time-invariant systems, for these two cases are[2]

$$\nabla \times \mathbf{E}_1 = -j\omega\overline{\mu} \cdot \mathbf{H}_1 - \mathbf{M}_1 \tag{12.1.1}$$

$$\nabla \times \mathbf{H}_1 = j\omega\overline{\varepsilon} \cdot \mathbf{E}_1 + \mathbf{J}_1 \tag{12.1.2}$$

$$\nabla \times \mathbf{E}_2 = -j\omega\overline{\mu} \cdot \mathbf{H}_2 - \mathbf{M}_2 \tag{12.1.3}$$

$$\nabla \times \mathbf{H}_2 = j\omega\overline{\varepsilon} \cdot \mathbf{E}_2 + \mathbf{J}_2 \tag{12.1.4}$$

[1]Thought experiment in German.

[2]The current sources are impressed currents so that they are immutable, and not changed by the environment in which they are immersed [95, 55].

To prove reciprocity, we would like to find a simplifying expression for the divergence of the following quantity,

$$\nabla \cdot (\mathbf{E}_1 \times \mathbf{H}_2) = \mathbf{H}_2 \cdot \nabla \times \mathbf{E}_1 - \mathbf{E}_1 \cdot \nabla \cdot \mathbf{H}_2 \tag{12.1.5}$$

so that the divergence theorem can be invoked. We need to expand the right-hand side further so that reciprocity relationships can be derived. To this end, and from the above, we can show that (after left dot-multiply (12.1.1) with $\mathbf{H}_2$, and (12.1.4) with $\mathbf{E}_1$), we have

$$\mathbf{H}_2 \cdot \nabla \times \mathbf{E}_1 = -j\omega \mathbf{H}_2 \cdot \overline{\mu} \cdot \mathbf{H}_1 - \mathbf{H}_2 \cdot \mathbf{M}_1 \tag{12.1.6}$$

$$\mathbf{E}_1 \cdot \nabla \times \mathbf{H}_2 = j\omega \mathbf{E}_1 \cdot \overline{\varepsilon} \cdot \mathbf{E}_2 + \mathbf{E}_1 \cdot \mathbf{J}_2 \tag{12.1.7}$$

Then, using the above and subtracting them, and following the identity in (12.1.5), we get:

$$\begin{aligned}\nabla \cdot (\mathbf{E}_1 \times \mathbf{H}_2) &= \mathbf{H}_2 \cdot \nabla \times \mathbf{E}_1 - \mathbf{E}_1 \cdot \nabla \times \mathbf{H}_2 \\ &= -j\omega \mathbf{H}_2 \cdot \overline{\mu} \cdot \mathbf{H}_1 - j\omega \mathbf{E}_1 \cdot \overline{\varepsilon} \cdot \mathbf{E}_2 - \mathbf{H}_2 \cdot \mathbf{M}_1 - \mathbf{E}_1 \cdot \mathbf{J}_2\end{aligned} \tag{12.1.8}$$

By the same token,

$$\nabla \cdot (\mathbf{E}_2 \times \mathbf{H}_1) = -j\omega \mathbf{H}_1 \cdot \overline{\mu} \cdot \mathbf{H}_2 - j\omega \mathbf{E}_2 \cdot \overline{\varepsilon} \cdot \mathbf{E}_1 - \mathbf{H}_1 \cdot \mathbf{M}_2 - \mathbf{E}_2 \cdot \mathbf{J}_1 \tag{12.1.9}$$

If one assumes that

$$\overline{\mu} = \overline{\mu}^t, \qquad \overline{\varepsilon} = \overline{\varepsilon}^t \tag{12.1.10}$$

or when the tensors are symmetric, then it follows that $\mathbf{H}_1 \cdot \overline{\mu} \cdot \mathbf{H}_2 = \mathbf{H}_2 \cdot \overline{\mu} \cdot \mathbf{H}_1$ and $\mathbf{E}_1 \cdot \overline{\varepsilon} \cdot \mathbf{E}_2 = \mathbf{E}_2 \cdot \overline{\varepsilon} \cdot \mathbf{E}_1$.[3]

Upon subtracting (12.1.8) and (12.1.9), many terms not involving the currents cancel each other, and one gets a simplified equation

$$\nabla \cdot (\mathbf{E}_1 \times \mathbf{H}_2 - \mathbf{E}_2 \times \mathbf{H}_1) = -\mathbf{H}_2 \cdot \mathbf{M}_1 - \mathbf{E}_1 \cdot \mathbf{J}_2 + \mathbf{H}_1 \cdot \mathbf{M}_2 + \mathbf{E}_2 \cdot \mathbf{J}_1 \tag{12.1.11}$$

12.1.1 Lorentz Reciprocity Theorem

First, we imagine a surface S that is bounding a volume V. In this sense, S is arbitrary as long as it excludes or encloses the sources as shown in (a) and (b) in Figure 12.3. Now, integrating (12.1.11) over the volume V, and invoking Gauss's divergence theorem, we have the reciprocity relationship that

$$\begin{aligned}&\oiint_S d\mathbf{S} \cdot (\mathbf{E}_1 \times \mathbf{H}_2 - \mathbf{E}_2 \times \mathbf{H}_1) \\ &\qquad = -\iiint_V dV[\mathbf{H}_2 \cdot \mathbf{M}_1 + \mathbf{E}_1 \cdot \mathbf{J}_2 - \mathbf{H}_1 \cdot \mathbf{M}_2 - \mathbf{E}_2 \cdot \mathbf{J}_1]\end{aligned} \tag{12.1.12}$$

[3]It is to be noted that in matrix algebra, the dot product between two vectors is often written as $\mathbf{a}^t \cdot \mathbf{b}$, but in the physics literature, the transpose on $\mathbf{a}$ is implied. Therefore, the dot product between two vectors is just written as $\mathbf{a} \cdot \mathbf{b}$.

When the volume V contains no sources (see Figure 12.3), the reciprocity relationship reduces to

$$\oiint_S d\mathbf{S} \cdot (\mathbf{E}_1 \times \mathbf{H}_2 - \mathbf{E}_2 \times \mathbf{H}_1) = 0 \tag{12.1.13}$$

The above is also called Lorentz reciprocity theorem by some authors.[4] It can be used to prove reciprocal relations between modes in a source-free waveguide, since no sources are involved in this reciprocity theorem.

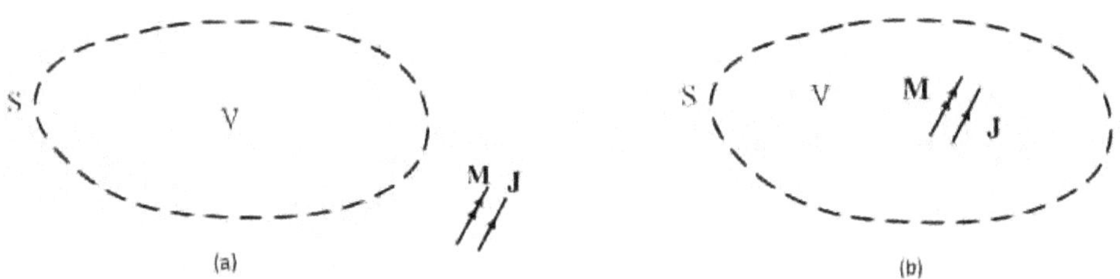

Figure 12.3: The geometry for proving reciprocity theorem when the surface S: (a) does not enclose the sources, and (b) encloses the sources. In the figure, the sources are supposed to be either $(\mathbf{M}_1, \mathbf{J}_1)$ producing fields $(\mathbf{E}_1, \mathbf{H}_1)$ or $(\mathbf{M}_2, \mathbf{J}_2)$ producing fields $(\mathbf{E}_2, \mathbf{H}_2)$.

12.1.2 Reaction Reciprocity Theorem

Next, when the surface S contains all the sources (see Figure 12.3), then the right-hand side of (12.1.12) will not be zero. On the other hand, when the surface $S \to \infty$, $\mathbf{E}_1$ and $\mathbf{H}_2$ become spherical waves which can be approximated by plane waves sharing the same $\boldsymbol{\beta}$ vector (also known as $\mathbf{k}$ vector or wave vector). Moreover, under the plane-wave approximation, in Maxwell's equations, we can replace ∇ with $-j\boldsymbol{\beta}$, and have $\omega\mu_0\mathbf{H}_2 = \boldsymbol{\beta} \times \mathbf{E}_2$, $\omega\mu_0\mathbf{H}_1 = \boldsymbol{\beta} \times \mathbf{E}_1$. Subsequently,

$$\mathbf{E}_1 \times \mathbf{H}_2 \sim \mathbf{E}_1 \times (\boldsymbol{\beta} \times \mathbf{E}_2) = \mathbf{E}_1(\boldsymbol{\beta} \cdot \mathbf{E}_2) - \boldsymbol{\beta}(\mathbf{E}_1 \cdot \mathbf{E}_2) \tag{12.1.14}$$

$$\mathbf{E}_2 \times \mathbf{H}_1 \sim \mathbf{E}_2 \times (\boldsymbol{\beta} \times \mathbf{E}_1) = \mathbf{E}_2(\boldsymbol{\beta} \cdot \mathbf{E}_1) - \boldsymbol{\beta}(\mathbf{E}_2 \cdot \mathbf{E}_1) \tag{12.1.15}$$

where the BAC-CAB formula has been used to simplify the above. But $\boldsymbol{\beta} \cdot \mathbf{E}_2 = \boldsymbol{\beta} \cdot \mathbf{E}_1 = 0$ in the far field since these far fields resemble plane waves. Furthermore, the $\boldsymbol{\beta}$ vectors are parallel to each other. Therefore, the two terms on the left-hand side of (12.1.12) cancel each other, and it vanishes when $S \to \infty$. (They cancel each other so that the remnant field vanishes faster than $1/r^2$. This is necessary[5] as the surface area S is growing larger and proportional to r^2.)

As a result, when $S \to \infty$, the left-hand side of (12.1.12) is zero, and it can be rewritten simply as

$$\int_V dV[\mathbf{E}_2 \cdot \mathbf{J}_1 - \mathbf{H}_2 \cdot \mathbf{M}_1] = \int_V dV[\mathbf{E}_1 \cdot \mathbf{J}_2 - \mathbf{H}_1 \cdot \mathbf{M}_2] \tag{12.1.16}$$

---

[4]Harrington, *Time-Harmonic Electric Field* [55].

[5]This is a mistake often committed by students of the course.

The inner product symbol is often used to rewrite the above as[6]

$$\langle \mathbf{E}_2, \mathbf{J}_1 \rangle - \langle \mathbf{H}_2, \mathbf{M}_1 \rangle = \langle \mathbf{E}_1, \mathbf{J}_2 \rangle - \langle \mathbf{H}_1, \mathbf{M}_2 \rangle \tag{12.1.17}$$

where the inner product[7] $\langle \mathbf{A}, \mathbf{B} \rangle = \int_V dV \mathbf{A}(\mathbf{r}) \cdot \mathbf{B}(\mathbf{r})$.

Thus, the above inner product is also called reaction or the reaction inner product, a concept introduced by Victor Rumsey [103]. The above is also called the Rumsey reaction theorem. Sometimes, the above is rewritten more succinctly or tersely as

$$\langle 2, 1 \rangle = \langle 1, 2 \rangle \tag{12.1.18}$$

where

$$\langle 2, 1 \rangle = \langle \mathbf{E}_2, \mathbf{J}_1 \rangle - \langle \mathbf{H}_2, \mathbf{M}_1 \rangle \tag{12.1.19}$$

The concept of inner product or reaction can be thought of as a kind of "measurement". The reciprocity theorem states as that the fields generated by sources 2 as "measured" by sources 1 is equal to fields generated by sources 1 as "measured" by sources 2. This measurement concept is clearer if we think of these sources as point sources or as Hertzian dipoles.

12.2 Conditions for Reciprocity

It is seen that the above proof hinges on (12.1.10) where $\overline{\mu} = \overline{\mu}^t$ and $\overline{\varepsilon} = \overline{\varepsilon}^t$. In other words, the anisotropic medium has to be described by symmetric tensors. They include conductive media, but not gyrotropic media that we studied previously; they are nonreciprocal. A ferrite biased by a magnetic field is often used in electronic circuits, and it corresponds to a gyrotropic, nonreciprocal medium.[8] Also, our starting equations (12.1.1) to (12.1.4) assume that the medium and the equations are linear time invariant so that Maxwell's equations can be written down in the frequency domain easily. Hence, the important conditions for reciprocity for a linear medium are

$$\overline{\mu} = \overline{\mu}^t, \qquad \overline{\varepsilon} = \overline{\varepsilon}^t \tag{12.2.1}$$

They include lossy anisotropic conductive media. Moreover, the medium is assumed stationary so that it is time-invariant.

---

[6]Previously, we used angular brackets to mean time average, but here, it means inner product.

[7]This inner product is quite different from those defined by mathematicians and physicists. We shall call this inner product the reaction inner product.

[8]Nonreciprocal media are important for making isolators in microwave. In an isolator, microwave signals can travel from Port 1 to Port 2, but not vice versa.

12.3 Application to a Two-Port Network and Circuit Theory

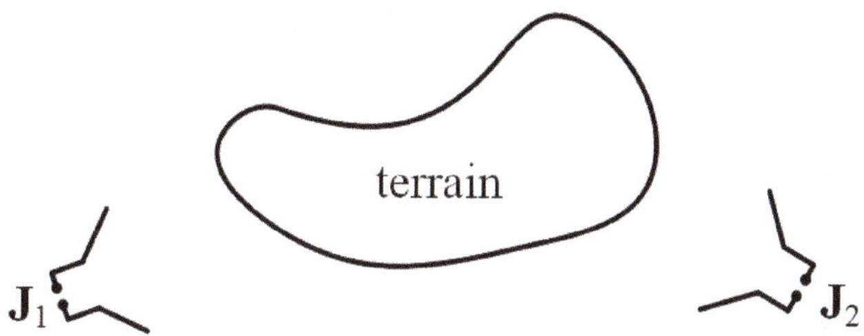

Figure 12.4: A geometry for proving the circuit relationship between two antennas using reciprocity theorem. Circuit relationship is possible when the ports of the antennas are small compared to wavelength. For these ports then, circuit theory prevails.

Figure 12.5: The terrain forming the media between the two antenna ports in Figure 12.4 can be as complicated as one can imagine, as long as the medium is reciprocal and stationary (time invariant). The one on the left shows a beautiful terrain with a whiff of civilization at the top of the hill, while the one on the right is a complicated, man-made printed circuit board with conductive elements. Circuit relationship is possible when the ports of the antennas are small compared to wavelength.

The reciprocity theorem can be used to distill and condense the interaction and relationship between two antennas over a complex terrain as long as the terrain comprises reciprocal media, namely, if $\overline{\mu} = \overline{\mu}^t$ and $\overline{\varepsilon} = \overline{\varepsilon}^t$ for these media.[9] In Figure 12.4, we assume that antenna 1 is driven by impressed current $\mathbf{J}_1$ while antenna 2 is driven by impressed current $\mathbf{J}_2$. It is assumed that the supports of these impressed currents are very small compared to wavelength so that circuit theory prevails at the antenna ports.[10] Further, it is assumed that the antennas are made from reciprocal

[9] It is to be noted that a gyrotropic medium considred in section 9.1 does not safisfy this reciprocity criteria, but it does satisfy the lossless criteria of section 10.3.2.

[10] It can be shown that when the frequency is low, or the wavelength is long, then one can replace electromagnetic theory with electro-quasistatic theory or magneto-quasistatic theory. Or, in short, with circuit theory.

media, such as conductive media. Since the system is linear time invariant, it can be written as the interaction between two ports as in circuit theory as shown in Figure 12.6. Since these two ports are small compared to wavelengths, in the neighborhood of the ports, then we can apply circuit concepts like potential theory by letting $\mathbf{E} = -\nabla\Phi$ (see (3.3.7)). Thus, we can define voltages and currents at these ports, and V-I relationships can be established in the manner of circuit theory.

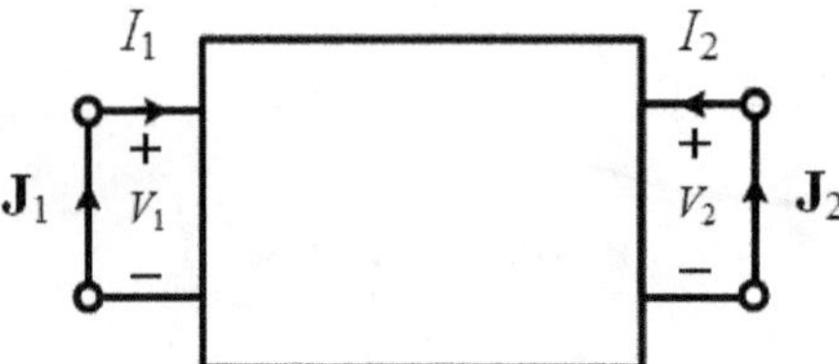

Figure 12.6: The interaction between two antennas, not necessarily in the far field of each other, can be reduced to a circuit theory description since the input and output ports of the antennas are small compared to the wavelength. But inside the box, any linear time-invariant medium can be there.

Focusing on a two-port network as shown in Figure 12.6, circuit theory implies that [104]

$$\begin{bmatrix} V_1 \\ V_2 \end{bmatrix} = \begin{bmatrix} Z_{11} & Z_{12} \\ Z_{21} & Z_{22} \end{bmatrix} \begin{bmatrix} I_1 \\ I_2 \end{bmatrix} \tag{12.3.1}$$

This form is permissible since we have a linear time-invariant system, and this is the most general way to establish a linear relationship between the voltages and the currents at the two ports. This is quite obvious in a network of circuit elements, but that it remains true for a general medium is harder to fathom. But it can be proved from electromagnetic theory. Furthermore, the matrix elements Z_{ij} can be obtained by performing a series of open-circuit and short-circuit measurements as in circuit theory.

Then assuming that the port 2 is turned on with $\mathbf{J}_2 \neq 0$, while port 1 is turned off with $\mathbf{J}_1 = 0$. In other words, port 1 is open circuit, and the source $\mathbf{J}_2$ is an impressed current source[11] that will produce an electric field $\mathbf{E}_2$ at port 1. Since the current at port 1 is turned off, or $\mathbf{J}_1 = 0$, the voltage measured at port 1 is the open-circuit voltage V_1^{oc}. Please note here that $\mathbf{J}_1$ and $\mathbf{J}_2$ are impressed currents and are only defined in their respective ports. Moreover, in the long-wavelength limit, the currents are constant in the wires that carry them. Consequently, the reaction between $\mathbf{E}_2$ and $\mathbf{J}_1$ is

$$\langle \mathbf{E}_2, \mathbf{J}_1 \rangle = \int_V dV(\mathbf{E}_2 \cdot \mathbf{J}_1) = I_1 \int_{\text{Port 1}} \mathbf{E}_2 \cdot d\mathbf{l} = -I_1 V_1^{oc} \tag{12.3.2}$$

In proving reciprocity, we perform two Gedanken experiments consecutively. Even though port 1 is assumed to be off with no current through it, the $\mathbf{J}_1$ above is the impressed current to be used when port 1 is turned on. You need to contemplate this a bit to wrap your head around this point!

[11]This is the same as the current source concept in circuit theory.

Since we assume the currents in the wire to be constant, then the current $\mathbf{J}_1$ is a constant current at the port when it is turned on. Or the current I_1 can be taken outside the integral. In slightly more detail, the current $\mathbf{J}_1 = \hat{l}I_1/A$ where A is the cross-sectional area of the wire, and $\hat{l}$ is a unit vector aligned with the axis of the wire. The volume integral $dV = Adl$, and hence the second equality follows in the derivation above, where $d\mathbf{l} = \hat{l}dl$. Since $\int_{\text{Port 1}} \mathbf{E}_2 \cdot d\mathbf{l} = -V_1^{oc}$, we arrive at the last equality above.

We can repeat the derivation with port 2 to arrive at the reaction

$$\langle \mathbf{E}_1, \mathbf{J}_2 \rangle = I_2 \int_{\text{Port 2}} \mathbf{E}_1 \cdot d\mathbf{l} = -I_2 V_2^{oc} \tag{12.3.3}$$

Reciprocity requires these two reactions to be equal, and therefore,

$$I_1 V_1^{oc} = I_2 V_2^{oc} \tag{12.3.4}$$

But from (12.3.1), we can set the pertinent currents to zero to find these open circuit voltages to be used in (12.3.4). Therefore, $V_1^{oc} = Z_{12}I_2$, $V_2^{oc} = Z_{21}I_1$. Since $I_1V_1^{oc} = I_2V_2^{oc}$ by the reaction concept or by reciprocity, then $Z_{12} = Z_{21}$. The above analysis can be easily generalized to an N-port network.

The simplicity of the above belies its importance. The above shows that the reciprocity concept in circuit theory is a special case of reciprocity theorem for electromagnetic theory. The terrain can also be replaced by complex circuits as in a circuit board, as long as the materials in the terrain or circuit board are reciprocal, linear, and time invariant. For instance, the complex terrain can also be replaced by complex antenna structures. Noted that even when the transmit and receive antennas are miles apart, as long as the transmit and receive ports of the linear time-invariant system can be characterized by a linear relation expounded by (12.3.1), and the ports small enough compared to wavelength so that circuit theory prevails at the ports, we can apply the above analysis! This relation that $Z_{12} = Z_{21}$ is true as long as the medium traversed by the fields is a reciprocal medium even though the ports may be far apart.

Before we conclude this section, it is to be mentioned that some researchers advocate the use of circuit theory to describe electromagnetic theory. Such is the case in the transmission line matrix (TLM) method [105], and the partial element equivalence circuit (PEEC) method [106]. Circuit theory is so simple that many people fall in love with it!

12.4 Voltage Sources in Electromagnetics

In the above discussions, we have used impressed current sources in reciprocity theorem to derive certain circuit concepts. (As mentioned before, impressed currents are immutable with respect to changes in thier environment, just as current sources are immutable in circuit theory.)

Before we end this section, it is prudent to mention how voltage sources are modeled in electromagnetic theory. They have to be immutable with respect to change in thier environment, and can be modeled by impressed magnetic currents.

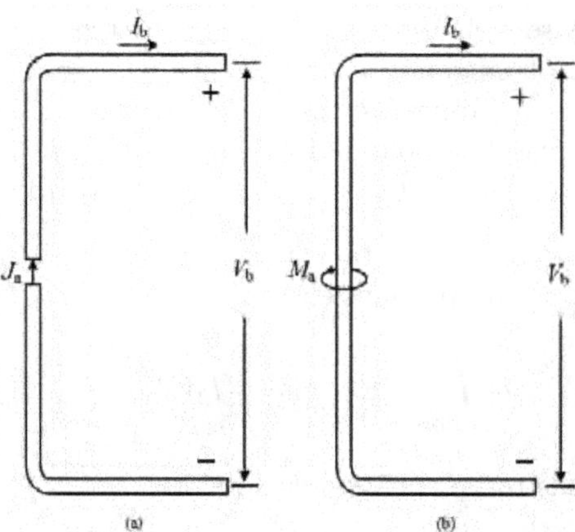

Figure 12.7: Two ways to model electromagnetic sources: (i) A current source modeled by an impressed current source $\mathbf{J}_a$ driving a very short antenna, and (ii) A voltage source modeled by an impressed magnetic frill source (loop source) $\mathbf{M}_a$ driving a very short antenna (Figure by: JIN AU KONG, mhnlakgilnojmhinhkckjpncpbhabphi/pages/pdf/web/viewer.html?file=httpsTheory [35]).

The use of the impressed currents so that circuit concepts can be applied is shown in Figure 12.7. The antenna in (a) is driven by a current source. But a magnetic current source (loop) can be used as a voltage source in circuit theory as shown by Figure 12.7b. By using duality concept, Faraday's law with magnetic current is similar to Ampère's law with electric current. An electric field has to curl around a magnetic current in Faraday's law, similar to the case of Ampère's law, where magnetic field curls around an electric current. This electric field will cause a voltage drop between the metal above and below the magnetic current loop making it behave like a voltage source.[12]

12.5 Hindsight

The proof of reciprocity theorem for Maxwell's equations is very deeply related to the symmetry of the operator involved. We can elucidate this from linear algebra. Given a linear system which can be modeled by a matrix equation driven by two right-hand sides, $\mathbf{b}_1$ and $\mathbf{b}_2$ with solutions $\mathbf{x}_1$ and $\mathbf{x}_2$, they can be written succinctly as

$$\overline{\mathbf{A}} \cdot \mathbf{x}_1 = \mathbf{b}_1 \tag{12.5.1}$$

$$\overline{\mathbf{A}} \cdot \mathbf{x}_2 = \mathbf{b}_2 \tag{12.5.2}$$

[12]More discussions can be found in Jordain and Balmain, Electromagnetic Waves and Radiation Systems [59].

We can left dot multiply the first equation with $\mathbf{x}_2$ and do the same with the second equation with $\mathbf{x}_1$ to arrive at

$$\mathbf{x}_2^t \cdot \overline{\mathbf{A}} \cdot \mathbf{x}_1 = \mathbf{x}_2^t \cdot \mathbf{b}_1 \tag{12.5.3}$$

$$\mathbf{x}_1^t \cdot \overline{\mathbf{A}} \cdot \mathbf{x}_2 = \mathbf{x}_1^t \cdot \mathbf{b}_2 \tag{12.5.4}$$

If $\overline{\mathbf{A}}$ is symmetric, the left-hand side of both equations are equal to each other.[13] Therefore, we can equate their right-hand sides to arrive at

$$\mathbf{x}_2^t \cdot \mathbf{b}_1 = \mathbf{x}_1^t \cdot \mathbf{b}_2 \tag{12.5.5}$$

The above is analogous to the statement of the reciprocity theorem which is

$$\langle \mathbf{E}_2, \mathbf{J}_1 \rangle = \langle \mathbf{E}_1, \mathbf{J}_2 \rangle \tag{12.5.6}$$

where the reaction inner product is $\langle \mathbf{E}_i, \mathbf{J}_j \rangle = \int_V dV \mathbf{E}_i(\mathbf{r}) \cdot \mathbf{J}_j(\mathbf{r})$ as mentioned before. The inner product in linear algebra is that of dot product in matrix theory, but the inner product for reciprocity theorem is that of infinite dimensional spaces.[14] So if the operators in Maxwell's equations are symmetrical, then reciprocity theorem applies.

It is prudent to mention that in linear algebra, for two vectors $\mathbf{a}$ and $\mathbf{b}$, there are two kinds of dot products or inner products. They are written as

$$\mathbf{a}^t \cdot \mathbf{b} = \sum_i a_i b_i$$

and

$$\mathbf{a}^\dagger \cdot \mathbf{b} = \sum_i a_i^* b_i$$

When $\mathbf{a} = \mathbf{b}$, the second inner product ensures that it is the positive definite. We shall call the second inner product energy inner product, while the first inner product, the reaction inner product used in reciprocity.

In the infinite dimensional continuum space (or Hilbert space), analogously, the above inner products between vector functions $\mathbf{a}(\mathbf{r})$ and $\mathbf{b}(\mathbf{r})$ are

$$\langle \mathbf{a}(\mathbf{r}), \mathbf{b}(\mathbf{r}) \rangle = \int dV \mathbf{a}(\mathbf{r}) \cdot \mathbf{b}(\mathbf{r})$$

and

$$\langle \mathbf{a}^*(\mathbf{r}), \mathbf{b}(\mathbf{r}) \rangle = \int dV \mathbf{a}^*(\mathbf{r}) \cdot \mathbf{b}(\mathbf{r})$$

In math literature, the complex conjugation is implied in the inner products, but not in electromagnetics literature. Notice that now, $\mathbf{r}$ replaces the role of i in the discrete case.

[13]This can be easily proven by taking the transpose of a scalar, and taking the transpose of the product of matrices.

[14]Such spaces are called Hilbert space.

Exercises for Lecture 12

Problem 12-1: Show that the left-hand side of (12.1.12) is in fact zero by showing that (12.1.14) and (12.1.15) are true in the far field.

Problem 12-2:

(i) Explain the difference in reciprocity expressed via the Lorentz reciprocity theorem and the Rumsey reaction theorem.

(ii) Show that in a reciprocity linear circuit containing N ports which can be modeled by an $N \times N$ impedance matrix $\overline{\mathbf{Z}}$, reciprocity means that $Z_{ij} = Z_{ji}$ where Z_{ij} is the ij-element of the matrix $\overline{\mathbf{Z}}$.

(iii) Show that the reciprocity theorem is related to a symmetric matrix system in linear algebra.

Problem 12-3: Explain why a magnetic loop around a metal rod can be used to model a voltage source.

Lecture 13

Equivalence Theorems, Huygens's Principle

Electromagnetic equivalence theorems are useful for simplifying solutions to many problems. The rule of physics and engineering is that if we encounter a problem that is hard to fathom, we break it down to a combination of smaller problems that are easier to solve. In this manner, highly complex problems can be solved. With the help of computers, highly complicated problems can be solved. Also, they offer physical insight into the behaviour of the electromagnetic fields of a Maxwellian system. They are closely related to Huygens's Principle. One application is their use in studying the radiation from an aperture antenna or from the output of a lasing cavity. These theorems are discussed in many textbooks [55, 59, 70, 35, 107]. Some authors also call them Love's equivalence principles [108] and Figure by has been given to Sergei Schelkunoff as well [95].

You may have heard of another equivalence theorem in special relativity. It was postulated by Einstein to explain why a light ray should bend around a star. The equivalence theorem in special relativity is very different from that in electromagnetics.

13.1 Equivalence Theorems or Equivalence Principles[1]

13.1.1 Inside-Out Case

In this lecture, we will consider three equivalence theorems: (1) The inside-out case, (2) The outside in case, (3) The general case. As mentioned above, we will derive these theorems using thought experiments or Gedanken experiments. As shall be shown later, they can also be derived mathematically using Green's theorem.

We will examine case (1) first. In this case, as shown in Figure 13.1(a), we let **J** and **M** be the

[1]This lecture has enough subtleties and depth that one should read it with care. It will not be until the end of the lecture book that such subtleties are encountered.

time-harmonic radiating sources inside a surface S radiating into a region $V = V_o \cup V_i$.[2] They produce $\mathbf{E}$ and $\mathbf{H}$ everywhere. These fields $\mathbf{E}$ and $\mathbf{H}$ are unique provided that the Sommerfeld radiation condition is satisfied.

We call these fields and sources Maxwellian since they satisfy Maxwell's equations. We postulate an equivalence problem indicated in Figure 13.1(b) by first constructing an imaginary surface S. In this equivalence problem, the fields outside S in V_o are the same in both (a) and (b). But in (b), the fields inside S in V_i are zero. Despite this, the fields and sources in (b) are Maxwellian, which will be explained next.

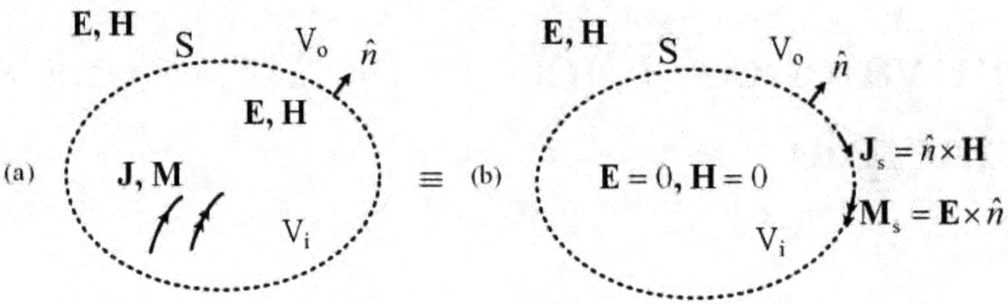

Figure 13.1: The inside-out problem where the two cases in (a) and (b) are equivalent. In (b), equivalence currents are impressed on the surface S so as to produce the same fields outside in V_o in both cases (a) and (b).

To explain further, in Figure 13.1(b), the tangential components of the fields are discontinuous at S. This is not possible for Maxwellian fields unless surface currents are impressed on the surface S. We have learned from electromagnetic boundary conditions that electromagnetic fields are discontinuous across a current sheet. Then we ask ourselves what surface currents are needed on surface S so that the boundary conditions (or jump conditions) for field discontinuities are satisfied on S. Clearly, surface currents are needed for these field discontinuities. Since both the $\mathbf{E}$ and the $\mathbf{H}$ fields are discontinuous, we need impressed $\mathbf{J}_s$ and $\mathbf{M}_s$ at the interface to account for the jump discontinuities. This will make the fields and sources in (b) Maxwellian-that is, they are also solutions to Maxwell's equations driven by different sources.

By virtue of the boundary conditions and the jump conditions in electromagnetics, these surface currents to be impressed on S are

$$\mathbf{J}_s = \hat{n} \times \mathbf{H}, \qquad \mathbf{M}_s = \mathbf{E} \times \hat{n} \tag{13.1.1}$$

We have learned from section 4.3.3 that an electric current sheet in Ampère's law produced a jump discontinuity in the magnetic field. By the same token, fictitious magnetic current is added to Faraday's law in section 5.3 for mathematical symmetry. Then by duality, a magnetic current

[2]This is the math notation for "union", the parlance for "sum".

sheet induces a jump discontinuity in the electric field. Because of the opposite polarity of the magnetic current $\mathbf{M}$ in Faraday's law compared to the electric current $\mathbf{J}$ in Ampère's law as is shown in section 5.3, this magnetic current sheet is proportional to $\mathbf{E} \times \hat{n}$ instead of $\hat{n} \times \mathbf{H}$.

Consequently, we can convince ourselves that $\hat{n} \times \mathbf{H}$ and $\mathbf{E} \times \hat{n}$ just outside S in both cases (a) and (b) are the same. Furthermore, we are persuaded that the above is a bona fide solution to Maxwell's equations. The case (a) in Figure 13.1 satisfies Maxwell's equations with current sources $\mathbf{J}$ and $\mathbf{M}$ inside V_i and for $\mathbf{E}$ and $\mathbf{H}$ everywhere.

In case (b), the $\mathbf{E}$ and $\mathbf{H}$ fields satisfy Maxwell's equations with the impressed surface current sources $\mathbf{J}_s$ and $\mathbf{M}_s$ on S, but with the original sources $\mathbf{J}$ and $\mathbf{M}$ removed in V_i. In case (b), the fields inside V_i are set to zero but $\mathbf{E}$ and $\mathbf{H}$ are the same as case (a) outside (or in V_0). Thus, there are discontinuous fields on the surface S, but the discontinuities are supported by the impressed currents $\mathbf{J}_s$ and $\mathbf{M}_s$. Hence, the fields and sources in (b) satisfy Maxwell's equations, or are Maxwellian.

Next, we have to convince ourselves that the fields outside S are the same in both cases. This follows from the uniqueness theorem: the fields in both cases satisfy the same boundary conditions and the radiation condition at infinity. It seems that there is some redundancy here, since both the boundary conditions for $\mathbf{E}$ and $\mathbf{H}$ are met here. But as long as these fields are consistent, that is okay. These fields are consistent since they are Maxwellian.

The above fact can be proved mathematically, as shall be shown later by a more mathematical manipulation. The fact that the fields are zero in V_i or inside S is known as the *extinction theorem*. This equivalence theorem and extinction theorem can also be proved mathematically.

13.1.2 Outside-In Case

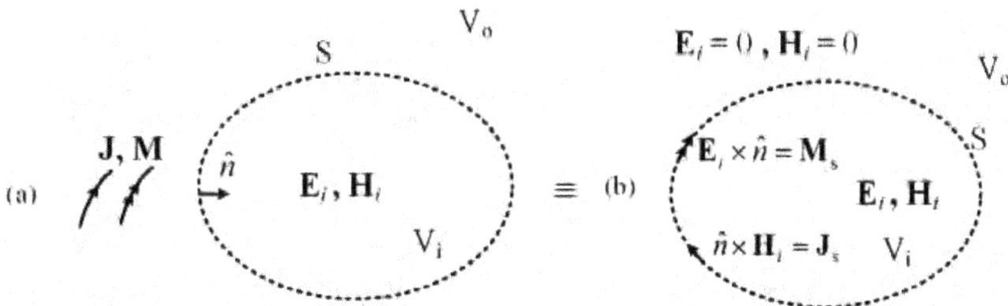

Figure 13.2: The outside-in problem where equivalence currents are impressed on the surface S to produce the same fields inside in both cases.

Similar to before, in Figure 13.2, we find an equivalence problem (b) where the fields inside S in V_i are the same as in (a), but the fields outside S in V_o in the equivalence problem is zero. The fields are discontinuous across the surface S, and hence, impressed surface currents $\mathbf{J}_s$ and $\mathbf{M}_s$ are needed to account for these discontinuities.

Then by the uniqueness theorem,[3] the fields $\mathbf{E}_i$, $\mathbf{H}_i$ inside V in both cases are the same. Again, by the *extinction theorem*, the fields produced by $\mathbf{E}_i \times \hat{n}$ and $\hat{n} \times \mathbf{H}_i$ are zero outside S.

It is to be noted that for both inside-out and outside-in cases, the field is extinct by the extinction theorem only in the volume or region that originally contains the sources. This will be clear when these equivalence problems are derived mathematically.

13.1.3 General Case

From these two cases, we can create a rich variety of equivalence problems. By linear superposition of the inside-out problem and the outside-in problem, then a third equivalence problem is shown in Figure 13.3:

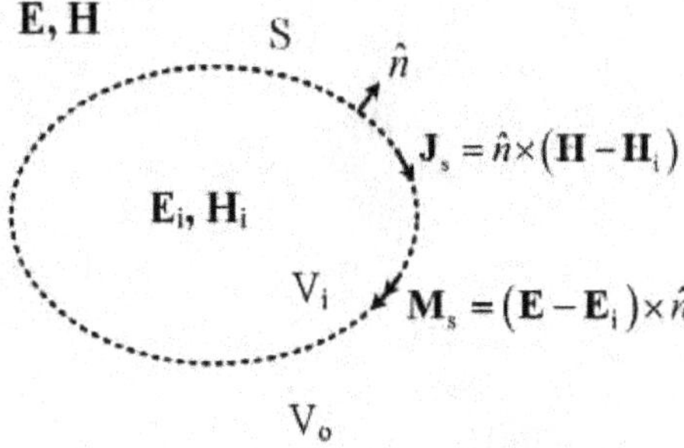

Figure 13.3: The general case where the fields are nonzero both inside and outside the surface S. Equivalence currents are needed on the surface S to support the jump discontinuities in the fields.

13.2 Electric Current on a PEC–Relation to Uniqueness Theorem

In this section, we show that new equivalence problems can be derived if the equivalence surface currents are radiating on the surface of a PEC. This can be obtained by using the equivalence problems in the previous section, to derive other corollaries of equivalence problems. We also need a result from reciprocity theorem, that impressed surface currents on the PEC cannot radiate.

We can start with the inside-out equivalence problem as shown in (b) of Figure 13.1. Since the fields inside S is zero for the inside-out problem, using a Gedanken experiment, one can insert a PEC object inside S without disturbing the fields $\mathbf{E}$ and $\mathbf{H}$ outside since the field is zero inside S. As the PEC object grows to snugly fit inside the surface S, then the electric current $\mathbf{J}_s = \hat{n} \times \mathbf{H}$, which is an impressed current source on top of a PEC, does not radiate by reciprocity theorem. Only one of the two impressed currents is radiating, namely, the magnetic current $\mathbf{M}_s = \mathbf{E} \times \hat{n}$ is

[3]We can add infinitesimal loss to ensure that uniqueness theorem is satisfied in this enclosed volume V_i.

radiating; and hence, $\mathbf{J}_s$ in Figure 13.4 can be removed. This is commensurate with the uniqueness theorem that only the knowledge of $\mathbf{E} \times \hat{n}$ plus the radiation condition, are needed to uniquely determine the fields outside S.

It is to be noted that $\mathbf{J}_s$, $\mathbf{M}_s$, $\mathbf{E}$ and $\mathbf{H}$ form a Maxwellian system before we insert a PEC object inside the surface S shown in (b) in Figure 13.1.

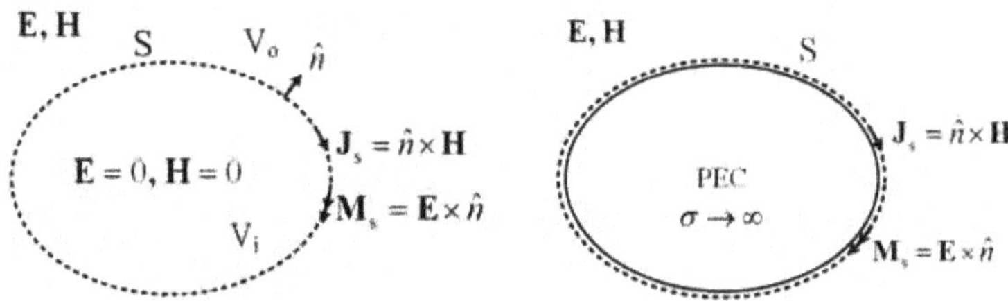

Figure 13.4: For the figure on the left, since the field is zero inside S, a PEC can be inserted inside S without disturbing the field outside as shown on the right. But the electric current does not radiate on a PEC surface. Thus only one of the two currents is needed.

13.3 Magnetic Current on a PMC–Relation to Uniqueness Theorem

Again, from reciprocity, an impressed magnetic current on a PMC cannot radiate. By the same token, we can perform the Gedanken experiment as before by inserting a PMC object inside S. It will not alter the field outside S, as the fields inside S is zero. As the PMC object grows to snugly fit the surface S, only the impressed electric current $\mathbf{J}_s = \hat{n} \times \mathbf{H}$ radiates, and the impressed magnetic current $\mathbf{M}_s = \mathbf{E} \times \hat{n}$ does not radiate and it can be removed. This is again commensurate with the uniqueness theorem that only the knowledge of the $\hat{n} \times \mathbf{H}$ is needed to uniquely determine the fields outside S.

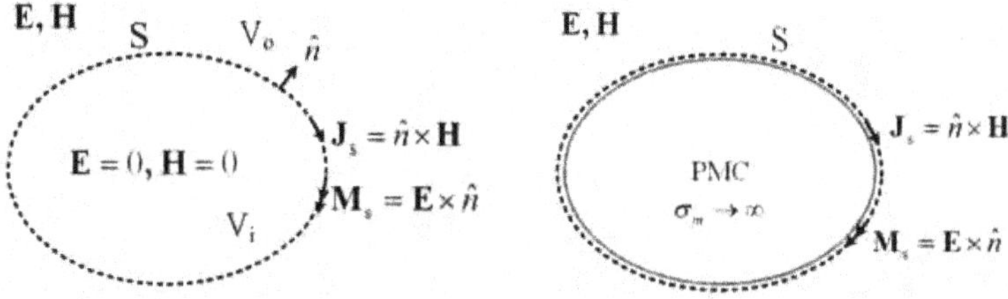

Figure 13.5: Similarly, a Gedanken experiment can be done by inserting a PMC inside S. By expanding the PMC surface to snugly fit inside S, only an electric current is needed to produce the field outside the surface S.

13.4 Huygens's Principle and Green's Theorem

Huygens's principle shows how a wave field on a surface determines the wave field outside the surface S. This concept was expressed by Christiaan Huygens heuristically in the 1600s [109]. But the mathematically precise expression of this idea was due to George Green[4] in the 1800s. This concept can be expressed mathematically for both scalar and vector waves. The derivation for the vector wave case is "homomorphic" to the scalar wave case. But the algebra in the scalar wave case is much simpler. Therefore, we shall study first the scalar wave case first using simpler algebra, followed by the electromagnetic vector wave case later where the vector algebra is more complex.

13.4.1 Scalar Waves Case

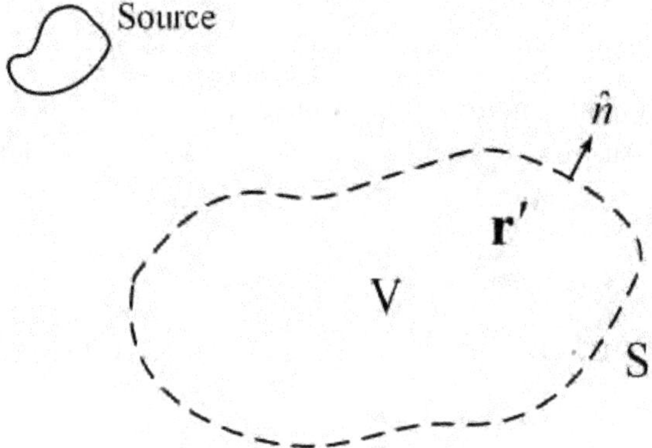

Figure 13.6: The geometry for deriving Huygens's principle for scalar wave equation. The dashed surface represents an "imagined" surface.

For a $\Psi(\mathbf{r})$ that satisfies the scalar wave equation inside V surrounded by an "imagined" surface S, then

$$(\nabla^2 + k^2)\,\Psi(\mathbf{r}) = 0, \qquad \mathbf{r} \in V \tag{13.4.1}$$

Notice that V does not contain the source that produces $\Psi(\mathbf{r})$ so that the right-hand side of (13.4.1) can be set to zero always where $\mathbf{r} \in V$. The corresponding scalar Green's function $g(\mathbf{r}, \mathbf{r}')$ for a homogeneous medium satisfies

$$(\nabla^2 + k^2)\,g(\mathbf{r}, \mathbf{r}') = -\delta(\mathbf{r} - \mathbf{r}'), \quad \forall\mathbf{r}, \quad \forall\mathbf{r}'. \tag{13.4.2}$$

[4]George Green (1793-1841) was self educated and the son of a miller in Nottingham, England [110]. If you visit Nottingham, you will see a windmill built in his honor.

which is the point source response in a homogeneous medium.[5] Next, we multiply (13.4.1) by $g(\mathbf{r}, \mathbf{r}')$ and (13.4.2) by $\Psi(\mathbf{r})$. Upon subtracting them, we have

$$\Psi(\mathbf{r})\delta(\mathbf{r}-\mathbf{r}') = g(\mathbf{r}, \mathbf{r}')\nabla^2\Psi(\mathbf{r}) - \Psi(\mathbf{r})\nabla^2 g(\mathbf{r}, \mathbf{r}') \tag{13.4.3}$$

Notice that the terms involving $gk^2\Psi$ cancel each other. As shall be shown, the right-hand side can be written as the divergence of a vector field, motivating the use of Gauss's divergence theorem. To this end, we integrate resultant equation over a volume V as shown in Figure 13.6.

There are two cases to consider: (1) when $\mathbf{r}'$ is in V, or (2) when $\mathbf{r}'$ is outside V. Thus, we have the dichotomous result on the left-hand side of the equation, namely,

$$\left.\begin{array}{ll}\text{if } \mathbf{r}' \in V, & \Psi(\mathbf{r}') \\ \text{if } \mathbf{r}' \notin V, & 0\end{array}\right\} = \int_V d\mathbf{r}\, [g(\mathbf{r}, \mathbf{r}')\nabla^2\Psi(\mathbf{r}) - \Psi(\mathbf{r})\nabla^2 g(\mathbf{r}, \mathbf{r}')], \tag{13.4.4}$$

The left-hand side evaluates to different values depending on where $\mathbf{r}'$ is due to the sifting property of the delta function $\delta(\mathbf{r}-\mathbf{r}')$. Since $g\nabla^2\Psi - \Psi\nabla^2 g = \nabla \cdot (g\nabla\Psi - \Psi\nabla g)$, the right-hand side of (13.4.4) can be rewritten using Gauss's divergence theorem, giving

$$\left.\begin{array}{ll}\text{if } \mathbf{r}' \in V, & \Psi(\mathbf{r}') \\ \text{if } \mathbf{r}' \notin V, & 0\end{array}\right\} = \oint_S dS\, \hat{n} \cdot [g(\mathbf{r}, \mathbf{r}')\nabla\Psi(\mathbf{r}) - \Psi(\mathbf{r})\nabla g(\mathbf{r}, \mathbf{r}')], \tag{13.4.5}$$

where S is the surface bounding V. The above is the Green's theorem, or the mathematical expression that once $\Psi(\mathbf{r})$ and $\hat{n} \cdot \nabla\Psi(\mathbf{r})$ are known on S, then $\Psi(\mathbf{r}')$ away from S may be found. This is similar to the expression of the equivalence principle where $\hat{n} \cdot \nabla\Psi(\mathbf{r})$ and $\Psi(\mathbf{r})$ are equivalence sources on the surface S. They can be used to find the fields inside and outside V, and the extinction theorem is also beautifully embodied in this equation, as shall be shown.

The first term on the right-hand side radiates via the Green's function $g(\mathbf{r}, \mathbf{r}')$ which radiates like a monopole source producing a spherically symmetric field. Since this is a monopole field, this source is also called a monolayer or single layer source. The second term radiates, on the other hand, via the normal derivative of the Green's function, namely $\hat{n} \cdot \nabla g(\mathbf{r}, \mathbf{r}')$. Since the derivative of a Green's function yields a dipole field (see Problem 4-2), with the dipole pointing normal to the surface S, the second term corresponds to sources that radiate like dipoles. These sources are also called double layer (or dipole layer) sources. (These terminologies are prevalent in the acoustics and mathematics communities.) As aforementioned, the above mathematical expression also embodies the *extinction theorem* that says if $\mathbf{r}'$ is outside V, or $\mathbf{r}' \notin V$, the left-hand side evaluates to zero. In this case, the monolayer source and the dipole layer source in (13.4.5) produce fields that cancel each other outside V.

---

[5]It can be shown that in a homogeneous medium, this scalar Green's function is given by $g(\mathbf{r}, \mathbf{r}') = 1/(4\pi|\mathbf{r} - \mathbf{r}'|)\exp(-jk|\mathbf{r} - \mathbf{r}'|)$. [1][p. 24-26]

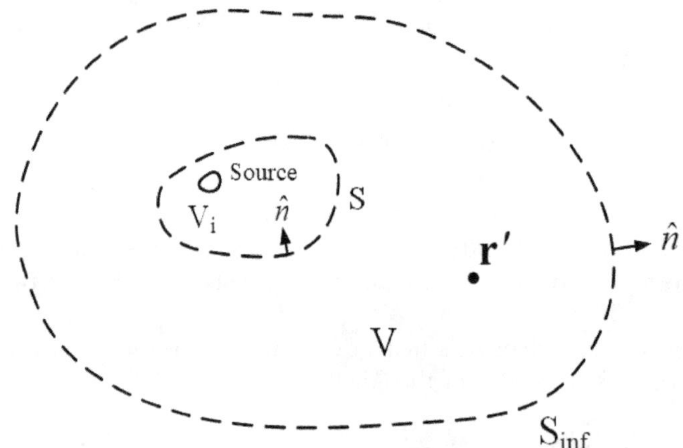

Figure 13.7: The geometry for deriving Huygens's principle for scalar wave. The radiation from the source can be replaced by equivalence sources on the surfaces S and S_{inf}, and the field inside V bounded by S and S_{inf} can be calculated using (13.4.5). Also, the field is zero inside S from (13.4.5). This is the extinction theorem. In other words, inside S, the field from the source is extinct by the field generated by the equivalent surface sources on S.

Let us assume that the volume V is bounded by two "imagined" surfaces S and S_{inf} (with $S_{\text{inf}} \to \infty$ eventually) as shown in Figure 13.7. Initially, the surface integral in (13.4.5) should include an integral over S_{inf}. But the integral over S_{inf} eventually vanishes as we shall explain next.

Remember that when $S_{\text{inf}} \to \infty$, all fields look like spherical waves which can be approximated by plane waves in the far field. Also, on the right-hand side of (13.4.5), the fields, $\Psi(\mathbf{r})$ on the surface S_{inf} in the integrand of (13.4.5) become plane waves, and $\nabla \to -\hat{r}jk$ on S_{inf}. Furthermore, for $g(\mathbf{r}-\mathbf{r}')$, $\mathbf{r}'$ remains finite, while $\mathbf{r}$ goes to infinity. Hence, $g(\mathbf{r}-\mathbf{r}') \sim O(1/r)$, when $r \to \infty$, which is a spherical wave morphing into a plane wave. Since $\Psi(\mathbf{r})$ is produced by finite sources in V_i, $\Psi(\mathbf{r}) \sim O(1/r)$, when $r \to \infty$.[6] It is also a spherical wave morphing into a plane wave, or $\nabla g \to -\hat{r}jkg$. Then, with some algebra, the integrand of the two terms on the right-hand side of (13.4.5) cancel each other on S_{inf}. With care, it can be shown that the leftover terms decay faster than $(1/r^2)$ integral over S_{inf} in (13.4.5) and eventually vanish.[7]

Therefore, the integral over S_{inf} becomes zero, and can be ignored, and (13.4.5) is valid for the case shown in Figure 13.7 as well but with the surface integral over surface S only. Physically, it implies that we are integrating over equivalence sources on S that radiate to infinity.

Here, the field outside S at $\mathbf{r}'$ is expressible in terms of the field on S. Therefore, this is similar to the inside-out equivalence principle we discussed previously in section 13.1.1, although this is for

[6]The symbol "O" means "of the order" in the math community. The computer science community uses a somewhat different notation.

[7]This decay rate faster than $(1/r^2)$ is necessary since the surface S_{inf} grows as r^2.

scalar wave case.

Notice that in deriving (13.4.5), $g(\mathbf{r}, \mathbf{r}')$ has only to satisfy (13.4.2) for both $\mathbf{r}$ and $\mathbf{r}'$ in V but no boundary condition has yet been imposed on $g(\mathbf{r}, \mathbf{r}')$. Therefore, if we further require that $g(\mathbf{r}, \mathbf{r}') = 0$ for $\mathbf{r} \in S$ when $\mathbf{r}'$ is in V, then (13.4.5) becomes

$$\Psi(\mathbf{r}') = -\oint_S dS\, \Psi(\mathbf{r})\, \hat{n} \cdot \nabla g(\mathbf{r}, \mathbf{r}'), \qquad \mathbf{r}' \in V. \tag{13.4.6}$$

On the other hand, if we require additionally that $g(\mathbf{r}, \mathbf{r}')$ satisfies (13.4.2) with the boundary condition $\hat{n} \cdot \nabla g(\mathbf{r}, \mathbf{r}') = 0$ for $\mathbf{r} \in S$, then (13.4.5) becomes

$$\Psi(\mathbf{r}') = \oint_S dS\, g(\mathbf{r}, \mathbf{r}')\, \hat{n} \cdot \nabla \Psi(\mathbf{r}), \qquad \mathbf{r}' \in V. \tag{13.4.7}$$

Equations (13.4.5), (13.4.6), and (13.4.7) are various forms of Huygens's principle, or the equivalence principle for scalar waves (acoustic waves) depending on the definition of $g(\mathbf{r}, \mathbf{r}')$. Equations (13.4.6) and (13.4.7) stipulate that only $\Psi(\mathbf{r})$ or $\hat{n} \cdot \nabla\Psi(\mathbf{r})$ need be known on the surface S in order to determine $\Psi(\mathbf{r}')$. The above are analogous to the PEC and PMC equivalence principles considered previously in sections 13.2 and 13.3.

13.4.2 Electromagnetic Waves Case

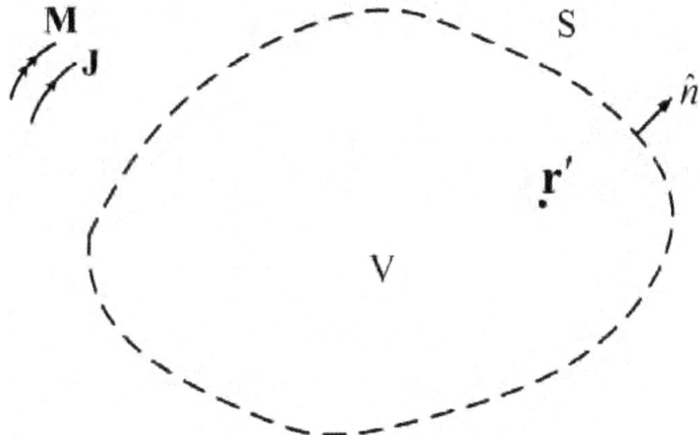

Figure 13.8: Derivation of the Huygens's principle for the electromagnetic case. One only needs to know the surface fields on surface S in order to determine the field at $\mathbf{r}'$ inside V.

The derivation of Huygens's principle and the Green's theorem for the electromagnetic case is more complicated than the scalar wave case. But fortunately, this problem is mathematically "homomorphic" to the scalar wave case. In dealing with the requisite vector algebra, we have to remember to cross the t's and dot the i's, to carry ourselves carefully through the somewhat laborious and complicated vector algebra! One can always refer back to the scalar wave case to keep our bearing straight.

In a source-free homogeneous region, an electromagnetic wave satisfies the vector wave equation

$$\nabla \times \nabla \times \mathbf{E}(\mathbf{r}) - k^2\,\mathbf{E}(\mathbf{r}) = 0, \qquad \mathbf{r} \in V. \tag{13.4.8}$$

The above also implies that the field $\mathbf{E}$ is Maxwellian. Again, we pick the volume V to contain no source so that the right-hand side of the above is zero when $\mathbf{r} \in V$. Comparing with the scalar wave case, the analogue of the scalar Green's function is the dyadic Green's function for the electromagnetic wave case [35, 2, 111, 112]. Moreover, the dyadic Green's function satisfies the equation[8]

$$\nabla \times \nabla \times \overline{\mathbf{G}}(\mathbf{r}, \mathbf{r}') - k^2\,\overline{\mathbf{G}}(\mathbf{r}, \mathbf{r}') = \overline{\mathbf{I}}\,\delta(\mathbf{r} - \mathbf{r}'). \tag{13.4.9}$$

In the above, the source point is given by $\mathbf{r}'$ while the field point is given by $\mathbf{r}$. It can be shown by direct back substitution that the dyadic Green's function in free space is [112]

$$\overline{\mathbf{G}}(\mathbf{r}, \mathbf{r}') = \left(\overline{\mathbf{I}} + \frac{\nabla\nabla}{k^2}\right) g(\mathbf{r}, \mathbf{r}') \tag{13.4.10}$$

The back substitution shows that as long as the scalar Green's function $g(\mathbf{r}, \mathbf{r}')$ satisfies (13.4.2), the above is the solution to (13.4.9). The above allows us to derive the vector Green's theorem [111, 2, 35].[9] We will use the insight we have gained in deriving the scalar Green's theorem to derive the Huygens's principle for the vector field.

Then, after post-multiplying (13.4.8) by $\overline{\mathbf{G}}(\mathbf{r}, \mathbf{r}') \cdot \mathbf{a}$, pre-multiplying[10] (13.4.9) by $\mathbf{E}(\mathbf{r})$, subtracting the resultant equations, the terms involving $k^2\mathbf{E}(\mathbf{r}) \cdot \overline{\mathbf{G}}(\mathbf{r}, \mathbf{r}') \cdot \mathbf{a}$ cancel out. We arrive at[11]

$$\mathbf{E}(\mathbf{r}) \cdot \mathbf{a}\delta(\mathbf{r} - \mathbf{r}') = \mathbf{E}(\mathbf{r}) \cdot \nabla \times \nabla \times \overline{\mathbf{G}}(\mathbf{r}, \mathbf{r}') \cdot \mathbf{a} - \nabla \times \nabla \times \mathbf{E}(\mathbf{r}) \cdot \overline{\mathbf{G}}(\mathbf{r}, \mathbf{r}') \cdot \mathbf{a} \tag{13.4.11}$$

We then integrate the difference over volume V, and using the sifting property of the delta function, considering two cases as we did for the scalar wave case, we have

$$\left.\begin{array}{ll} \text{if } \mathbf{r}' \in V, & \mathbf{E}(\mathbf{r}') \\ \text{if } \mathbf{r}' \notin V, & 0 \end{array}\right\} = \int_V dV\, \left[\mathbf{E}(\mathbf{r}) \cdot \nabla \times \nabla \times \overline{\mathbf{G}}(\mathbf{r}, \mathbf{r}') \right.$$
$$\left. - \nabla \times \nabla \times \mathbf{E}(\mathbf{r}) \cdot \overline{\mathbf{G}}(\mathbf{r}, \mathbf{r}')\right]. \tag{13.4.12}$$

[8] A dyad is an outer product between two vectors, and it behaves like a tensor, except that a tensor is more general than a dyad. A purist will call the above a tensor Green's function, as the above is not a dyad in its strictest definition.

[9] To keep our sanity, it is prudent to convert (13.4.9) by an arbitrary vector $\mathbf{a}$ to convert it from a dyad equation to a vector equation so that we can understand them better. The arbitrary vector $\mathbf{a}$ can be cancelled after you have gone through the somewhat insane manipulations.

[10] Since we are dealing with dyads which are tensors like matrices, order is very important here.

[11] For those less mathematical inclined, you may want to take a small diversion to derive the following.

Next, using the vector identity that[12]

$$-\nabla \cdot \left[\mathbf{E}(\mathbf{r}) \times \nabla \times \overline{\mathbf{G}}(\mathbf{r},\mathbf{r}') + \nabla \times \mathbf{E}(\mathbf{r}) \times \overline{\mathbf{G}}(\mathbf{r},\mathbf{r}')\right] \\ = \mathbf{E}(\mathbf{r}) \cdot \nabla \times \nabla \times \overline{\mathbf{G}}(\mathbf{r},\mathbf{r}') - \nabla \times \nabla \times \mathbf{E}(\mathbf{r}) \cdot \overline{\mathbf{G}}(\mathbf{r},\mathbf{r}'), \tag{13.4.13}$$

then the integrand of (13.4.12) can be written as a divergence. With the help of Gauss's divergence theorem, the right-hand side of (13.4.12) can be written as

$$\left.\begin{array}{ll}\text{if } \mathbf{r}' \in V, & \mathbf{E}(\mathbf{r}') \\ \text{if } \mathbf{r}' \notin V, & 0\end{array}\right\} = -\oint_S dS\, \hat{n} \cdot \left[\mathbf{E}(\mathbf{r}) \times \nabla \times \overline{\mathbf{G}}(\mathbf{r},\mathbf{r}') + \nabla \times \mathbf{E}(\mathbf{r}) \times \overline{\mathbf{G}}(\mathbf{r},\mathbf{r}')\right] \\ = -\oint_S dS\, \left[-\mathbf{E}(\mathbf{r}) \times \hat{n} \cdot \nabla \times \overline{\mathbf{G}}(\mathbf{r},\mathbf{r}') - j\omega\mu\, \hat{n} \times \mathbf{H}(\mathbf{r}) \cdot \overline{\mathbf{G}}(\mathbf{r},\mathbf{r}')\right]. \tag{13.4.14}$$

Notice that to invoke the Gauss's divergence theorem, the bounding surface S above is one that includes the surface at infinity as shown in the geometry for the scalar wave case in Figure 13.7. Again, the dichotomous value on the left-hand side of the above follows from the sifting property of the $\delta(\mathbf{r} - \mathbf{r}')$. It is simple and embodies the extinction theorem. It would not have been that simple if the right-hand side of (13.4.8) had not been made zero with a proper choice of V. Hence, the field is extinct for $\mathbf{r}' \notin V$ or in the volume that contains the sources that generate the fields.

The above is just the vector analogue of (13.4.5). We have used the cyclic relation of scalar-triple product to rewrite the last expression. Since $\hat{n} \times \mathbf{E}$ and $\hat{n} \times \mathbf{H}$ are associated with impressed surface magnetic current $\mathbf{M}_s$ and impressed surface electric current $\mathbf{J}_s$, respectively, the above can be thought of having these equivalence impressed surface currents radiating via the dyadic Green's function.

Again, notice that (13.4.14) is derived via the use of (13.4.9), but no boundary condition has yet been imposed on $\overline{\mathbf{G}}(\mathbf{r},\mathbf{r}')$ on S even though we have given a closed form solution for the free-space case previously. The above is similar to the outside-in equivalence theorem we have derived in section 13.1.2 using a Gedanken experiment . Now we have a mathematical derivation of the same theorem!

Next, if we require the additional boundary condition that $\hat{n} \times \overline{\mathbf{G}}(\mathbf{r},\mathbf{r}') = 0$ for $\mathbf{r} \in S$, this corresponds to a point source located at $\mathbf{r}'$ radiating via the dyadic Green's function, producing a field at $\mathbf{r}$, in the presence of a PEC surface. Then (13.4.14) becomes

$$\mathbf{E}(\mathbf{r}') = -\oint_S dS\, \hat{n} \times \mathbf{E}(\mathbf{r}) \cdot \nabla \times \overline{\mathbf{G}}(\mathbf{r},\mathbf{r}'), \qquad \mathbf{r}' \in V \tag{13.4.15}$$

for, using the scalar-triple product rule, it could be shown that $\hat{n} \times \mathbf{H} \cdot \overline{\mathbf{G}} = \mathbf{H} \cdot \hat{n} \times \overline{\mathbf{G}}$ implying that the second term in (13.4.14) is zero on a PEC surface due to the boundary condition we impose

[12]This identity can be established by using the identity $\nabla \cdot (\mathbf{A} \times \mathbf{B}) = \mathbf{B} \cdot \nabla \times \mathbf{A} - \mathbf{A} \cdot \nabla \times \mathbf{B}$. We will have to let (13.4.13) act on an aribitrary constant vector to convert the dyad into a vector before applying this identity. The equality of the volume integral in (13.4.12) to the surface integral in (13.4.14) is also known as the vector Green's theorem [111, 35]. Earlier form of this theorem was known as Franz formula [113].

on $\hat{n} \times \overline{\mathbf{G}}$. On the other hand, if we have a PMC surface, and we require that $\hat{n} \times \nabla \times \overline{\mathbf{G}}(\mathbf{r}, \mathbf{r}') = 0$ for $\mathbf{r} \in S$, then (13.4.14) becomes

$$\mathbf{E}(\mathbf{r}') = j\omega\mu \oint_S dS\, \hat{n} \times \mathbf{H}(\mathbf{r}) \cdot \overline{\mathbf{G}}(\mathbf{r}, \mathbf{r}'), \qquad \mathbf{r}' \in V \tag{13.4.16}$$

Equations (13.4.15) and (13.4.16) state that $\mathbf{E}(\mathbf{r}')$ is determined if either $\hat{n} \times \mathbf{E}(\mathbf{r})$ or $\hat{n} \times \mathbf{H}(\mathbf{r})$ is specified on S. This is in agreement with the uniqueness theorem. These are the mathematical expressions of the PEC and PMC equivalence problems we have considered previously in sections 13.2 and 13.3.

In (13.4.14), (13.4.15), and (13.4.16), the closed bounding surface S still includes the surface S_{inf}. Therefore, the dyadic Green's functions in (13.4.15) and (13.4.16) are for a closed cavity since boundary conditions are imposed on S for them. We need to prove next that for a homogeneous open region, the integration over S_{inf} vanishes.

The equations so far derived do not require the dyadic Green's function $\overline{\mathbf{G}}(\mathbf{r}, \mathbf{r}')$ to be translational invariant (or function of only $\mathbf{r} - \mathbf{r}'$), but it has to be a solution of (13.4.9). Thus, the equation for $\overline{\mathbf{G}}$ used in (13.4.15), using (13.4.12), can also be written as

$$\overline{\mathbf{G}}(\mathbf{r}, \mathbf{r}') = \frac{1}{k^2}[\nabla \times \nabla \times \overline{\mathbf{I}}\, g(\mathbf{r}, \mathbf{r}') - \overline{\mathbf{I}}\, \delta(\mathbf{r} - \mathbf{r}')], \tag{13.4.17}$$

In the above, we can use the BAC-CAB formula to simplify the double-curl operator $\nabla \times \nabla \times$

Taking the curl of the above, we have

$$\nabla \times \overline{\mathbf{G}}(\mathbf{r}, \mathbf{r}') = \nabla \times \overline{\mathbf{I}}\, g(\mathbf{r}, \mathbf{r}'). \tag{13.4.18}$$

In the back-substitution exercise we did earlier, the dyadic Green's function we have used so far is kosher if the scalar Green's function satisfies the following equation

$$(\nabla^2 + k^2) g(\mathbf{r}, \mathbf{r}') = -\delta(\mathbf{r} - \mathbf{r}') \tag{13.4.19}$$

without having to satisfy the radiation condition.[13] Hence, if we can solve the above equation for both bounded and unbounded regions, the dyadic Green's function defined in (13.4.12) is valid both for a homogeneous region that is bounded as well as unbounded. For an unbounded, homogeneous medium, the scalar Green's function, satisfying the radiation condition, has a closed form and is simply

$$g(\mathbf{r}, \mathbf{r}') = \frac{e^{-jk|\mathbf{r}-\mathbf{r}'|}}{4\pi|\mathbf{r} - \mathbf{r}'|}$$

as previously derived. Then, (13.4.14), for $\mathbf{r}' \in V$ and $\mathbf{r}' \neq \mathbf{r}$, after going through the detailed derivation, we can show that

$$\mathbf{E}(\mathbf{r}') = -\oint_S dS\ \hat{n} \times \mathbf{E}(\mathbf{r}) \cdot (\nabla \times g(\mathbf{r}, \mathbf{r}')\overline{\mathbf{I}}) + \frac{j}{\omega\epsilon} \oint_S dS\, \hat{n} \times \mathbf{H}(\mathbf{r}) \cdot \nabla \times (\nabla \times g(\mathbf{r}, \mathbf{r}')\overline{\mathbf{I}}). \tag{13.4.20}$$

[13]For instance, $g = g_h + g_s$, where g_h is the homogeneous solution and g_s satisfies the above equation with the singularity on the right-hand side.

Next, we can assume that $g(\mathbf{r}, \mathbf{r}')$ is translationally invariant and can be replaced by $g(\mathbf{r} - \mathbf{r}')$. Then using that $\nabla g(\mathbf{r} - \mathbf{r}') = -\nabla' g(\mathbf{r} - \mathbf{r}')$, the above can be rewritten as

$$\mathbf{E}(\mathbf{r}') = \nabla' \times \oint_S dS\, g(\mathbf{r} - \mathbf{r}')\, \hat{n} \times \mathbf{E}(\mathbf{r}) + \frac{j}{\omega\epsilon} \nabla' \times \nabla' \times \oint_S dS\, g(\mathbf{r} - \mathbf{r}')\, \hat{n} \times \mathbf{H}(\mathbf{r}). \tag{13.4.21}$$

The above can be applied to the geometry in Figure 13.7 where $\mathbf{r}'$ is enclosed in S and S_{inf}. However, the integral over S_{inf} vanishes by virtue of the radiation condition as for (13.4.5).[14] Then, (13.4.21) relates the field outside S at $\mathbf{r}'$ in terms of only the equivalence surface currents $\mathbf{M}_s = \mathbf{E} \times \hat{n}$ and $\mathbf{J}_s = \hat{n} \times \mathbf{H}$ on S. This is similar to the inside-out problem in the equivalence theorem (see section 13.1.1). It is also related to the fact that if the radiation condition is satisfied, then the field outside of the source region is uniquely satisfied. Therefore, this is also related to the uniqueness theorem.

13.5 Some Math Details

In the above, the equation (13.4.20) follows from (13.4.14). If we post-multiply it by an arbitrary vector $\mathbf{a}$, we can extract a term that resembles

$$\mathbf{A} \cdot (\nabla \times \bar{\mathbf{I}} g) \cdot \mathbf{a} \tag{13.5.1}$$

where $\mathbf{A} = \hat{n} \times \mathbf{E}$. Using the definition that $\bar{\mathbf{I}} g \cdot \mathbf{a} = g\mathbf{a}$, then we have

$$\mathbf{A} \cdot (\nabla \times \bar{\mathbf{I}} g) \cdot \mathbf{a} = -\mathbf{A} \cdot \nabla' \times (g\mathbf{a}) = -(\nabla' \times g\mathbf{a}) \cdot \mathbf{A} = -(\nabla' \times g\mathbf{A}) \cdot \mathbf{a} \tag{13.5.2}$$

The first term of the above equation (13.4.21) follows from the use of the above identity after cancelling $\mathbf{a}$ in the above equality.

For the second term, we obtain a term from (13.4.14) and (13.4.20) that resembles

$$\mathbf{B} \cdot (\nabla \times \nabla \times g\bar{\mathbf{I}}) \cdot \mathbf{a} \tag{13.5.3}$$

where $\mathbf{B} = \hat{n} \times \mathbf{H}$. It can be shown to be equivalent to

$$\mathbf{B} \cdot (\nabla \times \nabla \times g\bar{\mathbf{I}}) \cdot \mathbf{a} = \mathbf{B} \cdot (\nabla \times \nabla \times (g\mathbf{a}) = \nabla' \times \nabla' \times (g\mathbf{a}) \cdot \mathbf{B} = \nabla' \times \nabla' \times g\mathbf{B} \cdot \mathbf{a} \tag{13.5.4}$$

After canceling the vector $\mathbf{a}$ from the above equality, we obtain that $\mathbf{B} \cdot (\nabla \times \nabla \times g\bar{\mathbf{I}}) = \nabla' \times \nabla' \times g\mathbf{B}$. Using the above, we can get the second term in (13.4.21).

[14] It is to be noted that the integral over S_{inf} does not vanish because the field is vanishingly small, but the cancellation of the two terms in (13.4.21).

Exercises for Lecture 13

Problem 13-1:

(i) Using reciprocity theorem, show that an impressed current source on a PEC surface cannot radiate any field.

(ii) A dyad is defined in physics as a juxtaposition of two 3 vectors–for example, $\mathbf{ab}$ where $\mathbf{a}$ and $\mathbf{b}$ are three-component vectors called 3 vectors in physics. In matrix algebra, this is written as $\mathbf{a} \cdot \mathbf{b}^t$ and is known as an outer product (this outer product is denoted $\mathbf{a} \bigotimes \mathbf{b}$ in the math literature as well), whereas $\mathbf{a}^t \cdot \mathbf{b}$ is called an inner product. In physics, an inner product is just written $\mathbf{a} \cdot \mathbf{b}$. Assuming that $\mathbf{a}$ and $\mathbf{b}$ are independent of each other but not orthogonal, even though the dyad $\mathbf{ab}$ behaves like a 3×3 matrix, it only has one nonzero eigenvector with one nonzero eigenvalue. Find this eigenvector.

It can be shown that a dyad has at most one nonzero eigenvector. The number of nonzero eigenvalues of a matrix is the rank of the matrix. A dyad has a rank of at most 1. (This exercise teaches you what a dyad is.)

(iii) On the other hand, a dyadic is a superposition of dyads. The dyadic Green's function is "homomorphic" to the scalar Green's function, albeit with more complicated vector algebra. By direct back substitution, show that the solution to the following equation

$$\nabla \times \nabla \times \overline{\mathbf{G}}(\mathbf{r}, \mathbf{r}') - k^2 \, \overline{\mathbf{G}}(\mathbf{r}, \mathbf{r}') = \overline{\mathbf{I}} \, \delta(\mathbf{r} - \mathbf{r}'). \tag{E13.1}$$

is given by

$$\overline{\mathbf{G}}(\mathbf{r}, \mathbf{r}') = \left(\overline{\mathbf{I}} + \frac{\nabla\nabla}{k^2} \right) g(\mathbf{r}, \mathbf{r}') \tag{E13.2}$$

In the above, the dyadic $\overline{\mathbf{I}}$ is similar to an identity matrix because $\overline{\mathbf{I}} \cdot \mathbf{a} = \mathbf{a}$. Discuss how you would write $\overline{\mathbf{I}}$ in a three-component vector space. You can write $\overline{\mathbf{I}}$ in Cartesian, cylindrical, or spherical coordinates.

(iv) Given the vector wave equation for the electric field has a source term, namely that

$$\nabla \times \nabla \times \mathbf{E}(\mathbf{r}) - k^2 \, \mathbf{E}(\mathbf{r}) = -j\omega\mu \mathbf{J} \tag{E13.3}$$

By the principle of linear superposition, show that the solution to the above equation can be written as

$$\mathbf{E}(\mathbf{r}) = -j\omega\mu \iiint d\mathbf{r}' \overline{\mathbf{G}}(\mathbf{r}, \mathbf{r}') \cdot \mathbf{J}(\mathbf{r}') \tag{E13.4}$$

Problem 13-2: This problem is on Huygens's principle for electromagnetic fields in the lecture notes.

(i) Verify the identity (13.4.13) and then derive the expression (13.4.14).

(ii) Explain why (13.4.15) and (13.4.16) are the mathematical expressions of the equivalence problems shown in sections 13.2 and 13.3.

(iii) Going through the details, derive (13.4.20) for the free space case.

(iv) The surface integral n (13.4.21) is over the surface S and S_{inf} in Figure 13.7. Show that the surface integral over S_{inf} evaluates to zero by cancellation of two terms.

Part II

Transmission Lines, Waves in Layered Media, Waveguides, Cavity Resonators

Lecture 14

Circuit Theory Revisited

Circuit theory is one of the most successful and often used theories in electrical engineering. Its success is mainly due to its simplicity: it can capture the essence of the physics of highly complex circuits and structures, which is very important in the computer and microchip industry (or the integrated circuit [IC] design industry). In electrical engineering, we distill complicated concepts into the simplest form possible. Simplicity rules! Now, having understood electromagnetic theory in its full glory, it is prudent to revisit circuit theory and study its relationship to electromagnetic theory [59, 34, 70, 35]. Circuit theory can be regarded as the poor man's version of electromagnetics theory. It is simple, and yet, can be equally powerful in many ways.

The two most important laws in circuit theory are Kirchhoff's current law (KCL) and Kirchhoff's voltage law (KVL) [15, 57]. These two laws predate Maxwell's equations and were driven by the need for telegraphic technology then. But they are derivable from the current continuity equation, which follows from Ampère's law, and from Faraday's law.

14.1 Kirchhoff's Current Law (KCL)

KCL follows from the static Ampère's law that states that

$$\nabla \times \mathbf{H}(\mathbf{r}) = \mathbf{J}(\mathbf{r}) \tag{14.1.1}$$

By taking the divergence of the above, we arrive at

$$\nabla \cdot \mathbf{J}(\mathbf{r}) = 0 \tag{14.1.2}$$

from which KCL can be derived. The above is also the current continuity equation in the static limit. It implies that no net current flows into or out of a point $\mathbf{r}$ in space. First, we assume that all currents are flowing into a node as shown in Figure 14.1. By integrating the above current continuity equation over a volume containing the node, it is easy to show that

$$\sum_{i}^{N} I_i = 0 \tag{14.1.3}$$

which is the statement of KCL. This is shown for the schematic of Figure 14.1.

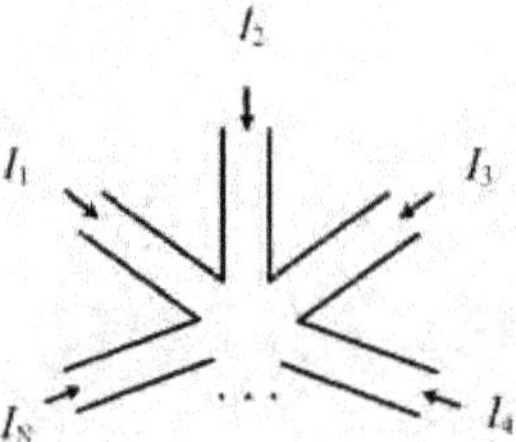

Figure 14.1: Schematic showing the derivation of KCL. All currents flowing into a node must add up to zero. This is reminiscent of the current continuity equation which is a statement of charge conservation. Viz., $\nabla \cdot \mathbf{J}(\mathbf{r}) = -\partial_t \varrho(\mathbf{r})$, which in the static limit, becomes $\nabla \cdot \mathbf{J}(\mathbf{r}) = 0$.

14.2 Kirchhoff's Voltage Law (KVL)

KCL is the consequence of Faraday's law in the static limit. For the truly static case when $\omega = 0$, it is

$$\nabla \times \mathbf{E} = 0 \tag{14.2.1}$$

The above implies that $\mathbf{E} = -\nabla\Phi$ or that scalar potential theory prevails. From this, we can deduce that

$$-\oint_C \mathbf{E} \cdot d\mathbf{l} = 0 \tag{14.2.2}$$

For statics, the statement that $\mathbf{E} = -\nabla\Phi$ also implies that we can define a voltage drop between two points, a and b to be

$$V_{ba} = -\int_a^b \mathbf{E} \cdot d\mathbf{l} = \int_a^b \nabla\Phi \cdot d\mathbf{l} = \Phi(\mathbf{r}_b) - \Phi(\mathbf{r}_a) = V_b - V_a \tag{14.2.3}$$

The equality $\int_a^b \nabla\Phi \cdot d\mathbf{l} = \Phi(\mathbf{r}_b) - \Phi(\mathbf{r}_a)$ can be understood by expressing this integral in one dimension along a straight line segment, or that

$$\int_a^b \frac{d}{dx}\Phi \cdot d\mathbf{x} = \Phi(\mathbf{r}_b) - \Phi(\mathbf{r}_a) \tag{14.2.4}$$

(If the contour a to b is a curved line, it can be thought of as a concatenation of many small straight line segments.) The above derivations of KCL and KVL from Maxwell's equations are surrealistically simple largely due to the work of Maxwell [114], and, later, the Maxwellians [43] who distilled Maxwell's work further so that it could be easily absorbed by us. As shall be shown later, to be exact, $\mathbf{E} = -\nabla\Phi - \partial/\partial t\mathbf{A}$, where the second term is due to the induction effect due to $\partial_t\mathbf{B}$ in Faraday's law. Therefore, when we ignore the induction effect, this is only valid in the low-frequency or long-wavelength limit,[1] or, in other words, that the dimension over which the above is applied is very small so that the retardation effect can be ignored.

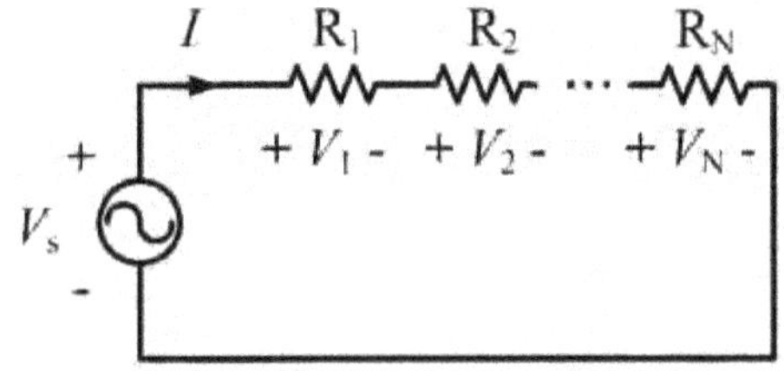

Figure 14.2: KVL where the sum of all voltages around a loop is zero, which is the consequence of static Faraday's law.

A good way to remember the above formula is that if $V_b > V_a$ then the potential at b is higher than that at a. Since $\mathbf{E} = -\nabla\Phi$, then the electric field points from point a to point b: Electric fields always point from the point of higher potential to point of lower potential. Faraday's law, when applied to the static case for a closed loop of resistors shown in Figure 14.2 gives KVL, or

$$\sum_{i}^{N} V_j = 0 \tag{14.2.5}$$

Notice that the voltage drop across a resistor is always positive, since the voltages to the left of the resistors in Figure 14.2 are always higher than the voltages to the right of the resistors. This implies that internal to the resistor, there is always an electric field that points from the left to the right. Therefore, the potential on the left side is always higher than that on the right side. A resistor impedes the flow of current, and hence, positive charges accumulate on the left side with negative charges on the right side. An electric field thus points from the left to the right as shown in Figure 14.3.

If one of the voltage drops is due to a voltage source, it can be modeled by a negative resistor as shown in Figure 14.4.[2] The voltage drop across a negative resistor is opposite to that of a positive resistor. As we have learned from the Poynting's theorem, negative resistors give out energy instead of dissipating energy. Remember that the complex power dissipated at one point in space is given by $\mathbf{E}\cdot\mathbf{J}^* = \sigma|\mathbf{E}|^2$ after letting $\mathbf{J} = \sigma\mathbf{E}$.

[1]These two concepts are synonymous in this course.

[2]This seemingly simple concept resulted in Esaki [54] being awarded the Nobel Prize for explaining, how a microwave transistor worked.

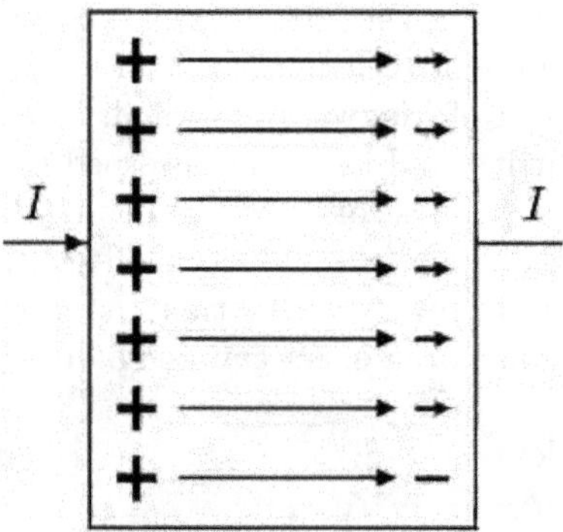

Figure 14.3: The schematic of the field inside a resistor. Due to charge accumulation, the potential on the left side is always higher than that on the right side. An electric field thus points from the left to the right inside the resistor.

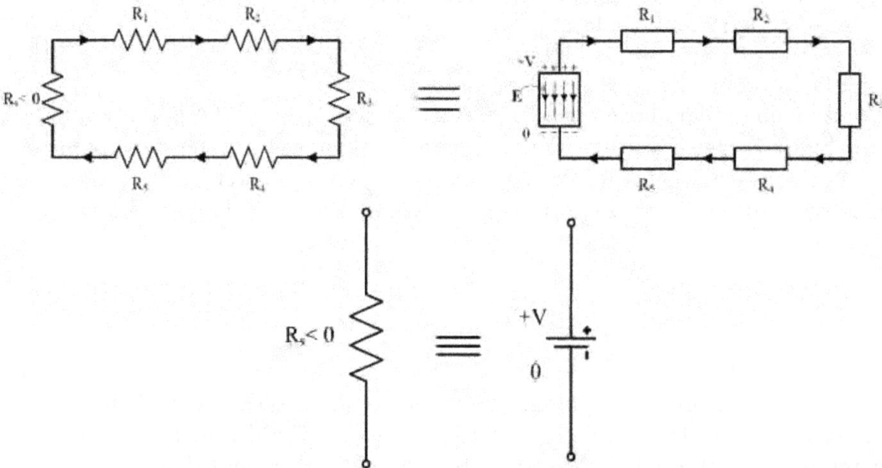

Figure 14.4: As shown above, a voltage source can also be modeled by a negative resistor.

14.2.1 Faraday's Law and the Flux Linkage Term

Faraday's law for the time-varying **B** flux case is

$$\nabla \times \mathbf{E} = -\frac{\partial \mathbf{B}}{\partial t} \tag{14.2.6}$$

In the static or long-wavelength limit, the right-hand side of the above can be ignored, but it can be amplified with the use of an inductor.

Writing the above in integral form, one gets

$$-\oint_C \mathbf{E} \cdot d\mathbf{l} = \frac{d}{dt} \int_s \mathbf{B} \cdot d\mathbf{S} \tag{14.2.7}$$

For simplicity, we apply the above to a loop shown in Figure 14.5, or a loop C that goes from a to b to c to d to a but the contour C does not go through the wires of the inductor. We can further assume that this loop is very small compared to wavelength so the potential theory that $\mathbf{E} = -\nabla\Phi$ applies.[3] Furthermore, we assume that the frequency is low such that this loop C has little or no magnetic flux through it so that the right-hand side of the above can be approximated by zero, or Faraday's law becomes

$$-\oint_C \mathbf{E} \cdot d\mathbf{l} = 0 \tag{14.2.8}$$

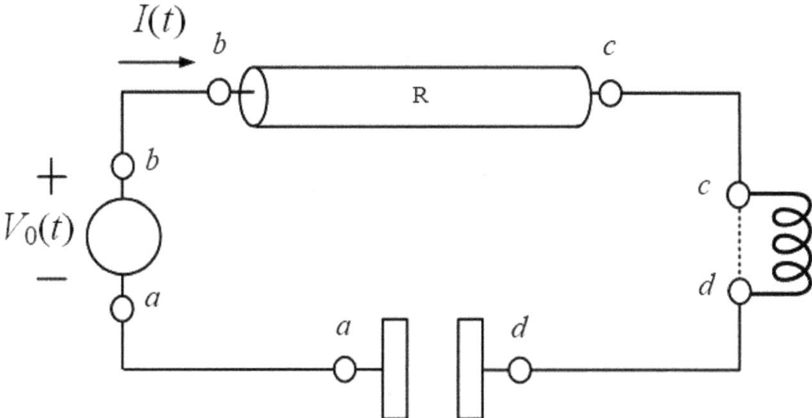

Figure 14.5: The KVL for a circuit loop consisting of resistor, inductor, and capacitor can also be derived from Faraday's law at low frequency. The KVL can be derived by integrating Faraday's law around the loop defined by *abcda*. This loop does not have any magnetic flux linkage. But the detail of V_{cd} and how it relates to the flux linkage in the inductor is discussed in the text. At low frequency, the flux linkage term in Faraday's law, and the displacement current term in the generalized Ampère's law are less important, unless their effects can be amplified by using inductors and capacitors. Thus the full glory of Maxwell's equation is not lost in circuit theory (redrawn from original Figure 4.2 in Ramo et al.) [34].

Since this loop does not go through the inductor, but goes directly from c to d, there is no flux linkage in this loop. Or $\nabla \times \mathbf{E} = 0$ and $\mathbf{E} = -\nabla\Phi$. In a word, scalar potential theory applies here

[3] You will later learn that the exact expression for the electric field is $\mathbf{E} = -\nabla\Phi - \partial_t \mathbf{A}$, where $\mathbf{A}$ is the magnetic vector potential. $\mathbf{A}$ is related to the magnetic field, but since the frequency is low, or there are no magnetic fields in the region considered, electric potential Φ suffices to describe the electric field $\mathbf{E}$.

to the electric field, and we have

$$-\int_a^b \mathbf{E}\cdot d\mathbf{l} - \int_b^c \mathbf{E}\cdot d\mathbf{l} - \int_c^d \mathbf{E}\cdot d\mathbf{l} - \int_d^a \mathbf{E}\cdot d\mathbf{l} = 0 \tag{14.2.9}$$

Inside the source or the battery, it is assumed that the electric field points opposite to the direction of integration **dl**, and hence the first term on the left-hand side of the above equation is positive and equal to $V_0(t)$, while the other terms are negative. Writing out the above more explicitly, after using (14.2.3),

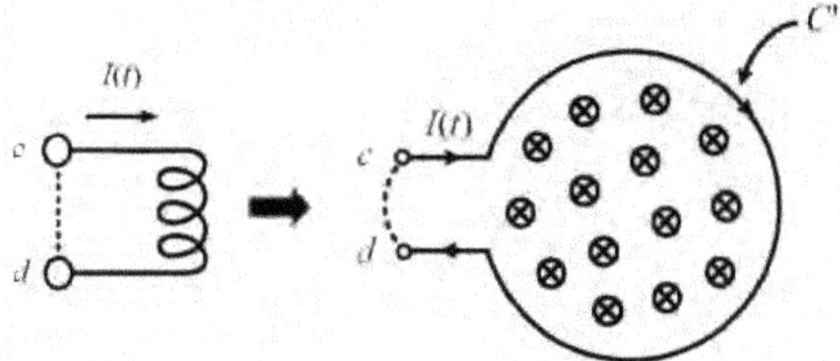

Figure 14.6: The voltage-current relation of an inductor can be obtained by unwrapping an inductor coil, and then calculating its flux linkage.

We have

$$V_0(t) + V_{cb} + V_{dc} + V_{ad} = 0 \tag{14.2.10}$$

Notice that in the above, in accordance with (14.2.3), $V_b > V_c$, $V_c > V_d$, and $V_d > V_a$. Therefore, V_{cb}, V_{dc}, and V_{ad} are all negative quantities but $V_0(t) > 0$. We will study the contributions to each of the terms, the inductor, the capacitor, and the resistor more explicitly next.

14.2.2 Inductor—A Flux Linkage Amplifier

An inductor is a lumped element that is used to amplify the flux linkage term in Faraday's law. To find the voltage current relation of an inductor, we apply Faraday's law to a closed loop C' formed by dc and the inductor coil shown in Figure 14.6 where, for simplicity, we have unwrapped the solenoid into a larger loop. Assume that the inductor wire is made of a perfect electric conductor (PEC), so that the electric field **E** in the wire is zero. Then the only contribution to the left-hand side of Faraday's law is the integration from point d to point c, the only place in the loop C' where **E** is not zero. We assume that outside the loop in the region between c and d, potential theory applies; namely, we can let $\mathbf{E} = -\nabla\Phi$. Now, we can connect V_{dc} in the previous equation to the flux linkage to the inductor. When the voltage source attempts to drive an electric current into the loop, Lenz's law (1834) comes into effect, essentially generating an opposing voltage.[4] The opposing

[4]Lenz's law can also be explained from Faraday's law (1831).

voltage gives rise to charge accumulation at d and c, and therefore, a low-frequency electric field at the gap at dc.

To this end, we form a new C' that goes from d to c, and then continues on to the wire that leads into the inductor. But this new loop will contain the flux $\mathbf{B}$ generated by the inductor current. Thus

$$\oint_{C'} \mathbf{E} \cdot \mathbf{dl} = \int_d^c \mathbf{E} \cdot \mathbf{dl} = -V_{dc} = -\frac{d}{dt} \int_{S'} \mathbf{B} \cdot \mathbf{dS} \tag{14.2.11}$$

As mentioned before, since the wire is a PEC, the integration around the loop C' is only nonzero from d to c. In the above, $\int_{S'} \mathbf{B} \cdot \mathbf{dS}$ is the flux linkage, which should be linearly proportional to I, the current. The inductance L is then defined as the flux linkage per unit current, or

$$L = \left[\int_{S'} \mathbf{B} \cdot \mathbf{dS} \right] / I \tag{14.2.12}$$

So the voltage in (14.2.11) is then

$$V_{dc} = \frac{d}{dt}(LI) = L\frac{dI}{dt} \tag{14.2.13}$$

since L is time independent.

Resistive Wire Case:

Had there been a finite resistance in the wire of the inductor, then the electric field is nonzero inside the wire. Taking this into account, we have

$$\oint_{C'} \mathbf{E} \cdot \mathbf{dl} = R_L I - V_{dc} = -\frac{d}{dt} \int_S \mathbf{B} \cdot \mathbf{dS} \tag{14.2.14}$$

Consequently,

$$V_{dc} = R_L I + L\frac{dI}{dt} \tag{14.2.15}$$

Thus, to account for the conductor loss of the coil, we can add a resistor in the equation. The above then becomes simpler in the frequency domain, namely

$$V_{dc} = R_L I + j\omega LI = (R_L + j\omega L)I = ZI \tag{14.2.16}$$

where $Z = R_L + j\omega L$.

14.2.3 Capacitance—A Displacement Current Amplifier

The capacitance is the proportionality constant between the charge Q stored in the capacitor, and the voltage V applied across the capacitor, or $Q = CV$. Then

$$C = \frac{Q}{V} \tag{14.2.17}$$

From the current continuity equation, one can easily show that in Figure 14.7,

$$I = \frac{dQ}{dt} \tag{14.2.18}$$

by charge conservation. Furthermore, using that $Q = CV_{da}$ where V_{da} is defined in Figure 14.5, we have

$$I = \frac{d}{dt}(CV_{da}) = C\frac{dV_{da}}{dt}$$

where C is time independent. Integrating the above equation, one gets

$$V_{da}(t) = \frac{1}{C}\int_{-\infty}^{t} I dt' \tag{14.2.19}$$

The above looks quite cumbersome in the time domain, but in the frequency domain, it becomes

$$I = j\omega CV_{da} = YV_{da} \tag{14.2.20}$$

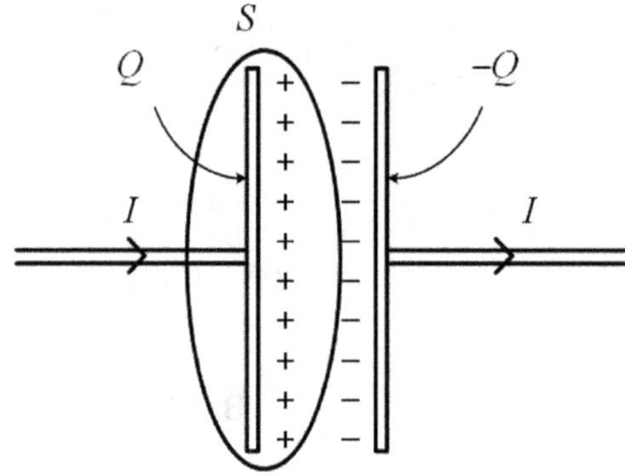

Figure 14.7: Schematic showing the calculation of the capacitance of a capacitor.

14.3 Resistor

The electric field is not zero inside the resistor as an electric field is needed to push electrons through it. As discussed in section 8.3, a resistor is a medium where collision of the electrons with the lattice dominates. As is well known, here Ohm's law prevails, and

$$\mathbf{J} = \sigma \mathbf{E} \tag{14.3.1}$$

where σ is the conductivity of the medium. From this, we deduce that $V_{cb} = V_c - V_b$ is a negative number given by (see also Figure 14.5)

$$V_{cb} = -\int_{b}^{c} \mathbf{E} \cdot \mathbf{dl} = -\int_{b}^{c} \frac{\mathbf{J}}{\sigma} \cdot \mathbf{dl} \tag{14.3.2}$$

where we assume a uniform current $\mathbf{J} = \hat{l}I/A$ in the resistor where $\hat{l}$ is a unit vector pointing in the direction of current flow in the resistor. We can assume that I is a constant along the length of the resistor, and thus, with $\mathbf{dl} = \hat{l}dl$, $\mathbf{J} \cdot \mathbf{dl} = Idl/A$, implying that

$$V_{cb} = -\int_b^c \frac{Idl}{\sigma A} = -I\int_b^c \frac{dl}{\sigma A} = -IR \tag{14.3.3}$$

where[5]

$$R = \int_b^c \frac{dl}{\sigma A} = \int_b^c \frac{\rho dl}{A} \tag{14.3.4}$$

Again, for simplicity, we assume long wavelength or low frequency in the above derivation. (The above formula is valid approximately even when the cross-sectional area A and resistivity ρ are functions of l.)

Inductors, capacitors, and resistors are known as lumped elements because they are usually much smaller than the wavelength, but interesting physics occurs inside them. They can be described by very simple voltage-current (VI) relationship. By so doing, circuit theory can be made very simple with far-ranging impact.

14.4 Generalized KCL and KVL, the Power of Phasors

Phasor technique allows us to generalize KCL and KVL to include the displacement current and induction voltage easily. To do this for generalized KCL, we rewrite Ampère's law in the frequency domain as

$$\nabla \times \mathbf{H} = j\omega\mathbf{D} + \mathbf{J} = \mathbf{J}_D + \mathbf{J} = \mathbf{J}_T \tag{14.4.1}$$

where the displacement current $\mathbf{J}_D = j\omega\mathbf{D}$. Taking the divergence of the above equation, we have

$$\nabla \cdot \mathbf{J}_T = 0 \tag{14.4.2}$$

where $\mathbf{J}_T = \mathbf{J}_D + \mathbf{J}$ includes the displacement current. KCL can be written easily for $\mathbf{J}_T$, which is very simple in the frequency domain. The above is the root to the generalized KCL. Note that here, the current $\mathbf{J}_T$ is complex.

Similarly, we rewrite Faraday's law in the frequency domain as

$$\nabla \times \mathbf{E} = -j\omega\mathbf{B} = -j\omega\nabla \times \mathbf{A} \tag{14.4.3}$$

where we have defined $\mathbf{B} = \nabla \times \mathbf{A}$ as before. From the above, we move all terms to the left-hand side, then we can define a new electric field $\underset{\sim}{\mathbf{E}} = \mathbf{E} + j\omega\mathbf{A}$ such that

$$\nabla \times \underset{\sim}{\mathbf{E}} = 0 \tag{14.4.4}$$

The above is the root to the generalized KVL. Now we can define $\underset{\sim}{\mathbf{E}} = \mathbf{E} + j\omega\mathbf{A} = -\nabla\Phi$. We can integrate the above around a closed loop to get

$$\oint_C \mathbf{E} \cdot d\mathbf{l} + j\omega\oint_C \mathbf{A} \cdot d\mathbf{l} = 0 \tag{14.4.5}$$

[5]The resisitivity $\rho = 1/\sigma$ where ρ has the unit of ohm-m, while σ has the unit of siemen/m.

Assume that the closed loop in the first term is made of the PEC wire C' and the air region is from d to c (see Figure 14.6). Thus the integration around the closed loop is nonzero only in the air region from d to c. Therefore,

$$V_{dc} = \int_d^c \mathbf{E} \cdot d\mathbf{l} = -j\omega \oint_C \mathbf{A} \cdot d\mathbf{l} = -j\omega \int_S (\nabla \times \mathbf{A}) \cdot d\mathbf{S} = -j\omega \int_S \mathbf{B} \cdot d\mathbf{S} = -j\omega LI \tag{14.4.6}$$

where we have used (14.2.12) to define an inductance L. The left-hand side is the voltage between points c and d, while the right-hand side is the flux linkage term responsible for generating this counter induction voltage. If we add a resistor in the loop, then the above can be converted to

$$V_{dc} = -(R + j\omega L)I \tag{14.4.7}$$

indicating the V_{dc} is a complex voltage.

14.5 Some Remarks

We have looked at the definition of inductor L and capacitor C in the above. But clever engineering is driven by heuristics: it is better, at times, to look at inductors and capacitors as energy storage devices, rather than flux linkage and charge storage devices.

Another important remark is that even though circuit theory, in its full glory is simpler than Maxwell's equations not all the physics is captured by it. The physics of the induction term in Faraday's law and the displacement current term in generalized Ampère's law are still retained and can be amplified by inductor and capacitor, respectively. In fact, wave physics are still retained in circuit theory: one can make slow-wave structure out of a series of inductors and capacitors. The lumped-element model of a transmission line is an example of a slow-wave structure design that will be studied in the next lecture. Since the wave is slow, it has a smaller wavelength, and as shall be seen, resonators can be made smaller: We see this in the LC tank circuit which is a much smaller resonator in wavelength with $L/\lambda \ll 1$ compared to a microwave cavity resonator that we will study later in the course. Therefore, circuit design is great for miniaturization. The shortcoming is that all inductors and capacitors are made from material media, and generally have higher losses than air or vacuum.

14.6 Energy Storage Method for Inductor and Capacitor

As aforementioned, oftentimes, it is more expedient to think of inductors and capacitors as energy storage devices. This enables us to identify stray (also called parasitic) inductances and capacitances more easily. This manner of thinking allows for an alternative way of calculating and understanding inductances and capacitances as well [34].

The energy stored in an inductor is due to its energy storage in the magnetic field, and it is alternatively written, according to circuit theory, as

$$W_m = \frac{1}{2}LI^2 \rightarrow L = \frac{2W_m}{I^2} \tag{14.6.1}$$

Therefore, it is simpler to think that a stray inductance exists whenever there is stray magnetic field to store magnetic energy. A piece of wire carries a current that produces a magnetic field enabling energy storage in the magnetic field. Hence, with this insight, we can understand why a piece of wire in fact behaves like a small inductor, which is non-negligible at high frequencies: Stray inductances occur whenever there are stray magnetic fields.

By the same token, a capacitor can be thought of as an electric energy storage device rather than a charge storage device. The energy stored in a capacitor, from circuit theory, is

$$W_e = \frac{1}{2}CV^2 \to C = \frac{2W_e}{V^2} \tag{14.6.2}$$

Therefore, whenever stray electric fields exist, one can think of stray capacitances as we have seen in the case of fringing field capacitances in a microstrip line.

14.7 Finding Closed-Form Formulas for Inductance and Capacitance

Closed-form formulas are important as they give us physical insight. Finding closed form solutions for inductors and capacitors is a difficult endeavor. As in solving Maxwell's equations or the waveguide problems, only certain simple geometries are amenable to closed-form solutions. Even a simple circular loop does not have a closed-form solution for its inductance L. If we assume a uniform current on a circular loop, in theory, the magnetic field can be calculated using the Biot-Savart law that we learned before, namely that

$$\mathbf{H}(\mathbf{r}) = \int \frac{I(\mathbf{r}')\mathbf{dl}' \times \hat{R}}{4\pi R^2} \tag{14.7.1}$$

But the above cannot be evaluated in closed-form except in terms of complicated elliptic integrals [115, 116]. Thus it is simpler to just measure the inductance.

However, if we have a solenoid as shown in Figure 14.8, an approximate formula for the inductance L can be found if the fringing field at the end of the solenoid can be ignored. The inductance can be found using the flux linkage method [34, 32]. Figure 14.9 shows the schematic used to find the approximate inductance of this inductor.

The capacitance of a parallel plate capacitor can be found by solving a boundary value problem (BVP) for electrostatics as shown in section 3.3.5. The electrostatic BVP for a capacitor involves Poisson's equation and Laplace's equation, which are scalar equations [50]. Finding the correct formula for the capacitor as shown in Figure 14.10 involving fringing field effect can be an exhausting exercise [117].

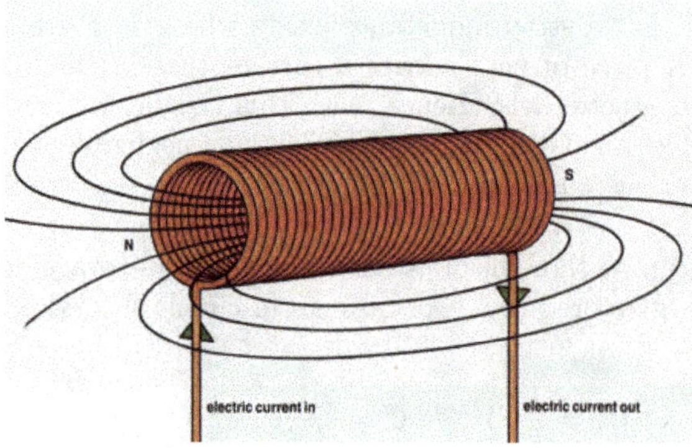

Figure 14.8: The flux-linkage method is used to estimate the inductance of a solenoid (Figure by SolenoidSupplier.Com).

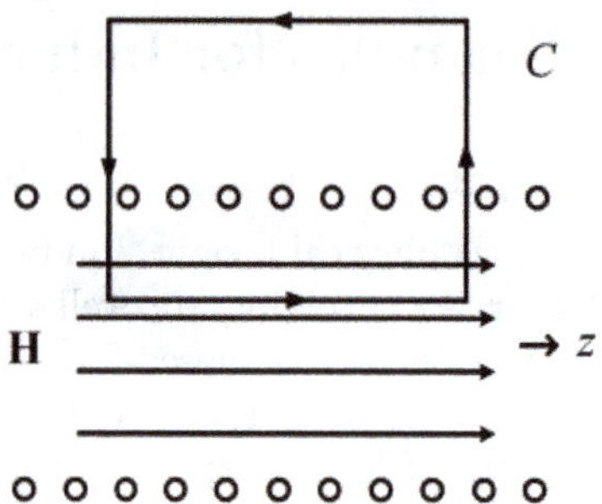

Figure 14.9: Finding the inductor flux linkage approximately by assuming the magnetic field is uniform inside a long solenoid. This approximation is more accurate as the solenoid becomes longer.

Alternatively, variational expressions can be used to find the lower and upper bounds of capacitors using, for example, Thomson's theorem [118] together with numerical methods.[6]

Assume a geometry of two conductors charged to $+V$ and $-V$ volts as shown in Figure 14.11. Surface charges will accumulate on the surfaces of the conductors. Using Poisson's equation, and a Green's function for Poisson's equation, one can express the potential in between the two conductors due to the surface charges density $\sigma(\mathbf{r})$. It can be expressed as (using (3.3.22) of lecture 3)

$$\Phi(\mathbf{r}) = \frac{1}{\varepsilon} \int_S dS' \frac{\sigma(\mathbf{r}')}{4\pi|\mathbf{r} - \mathbf{r}'|} \tag{14.7.2}$$

[6]There are many variational formulas for capacitance, some of which are discussed in [118][p. 53]. It seems that nature (or God) always tries to minimize something in seeking the solution. He is a minimalist.

where $S = S_1 + S_2$ is the union of two surfaces S_1 and S_2. Since Φ has values of $+V$ and $-V$ on the two conductors.

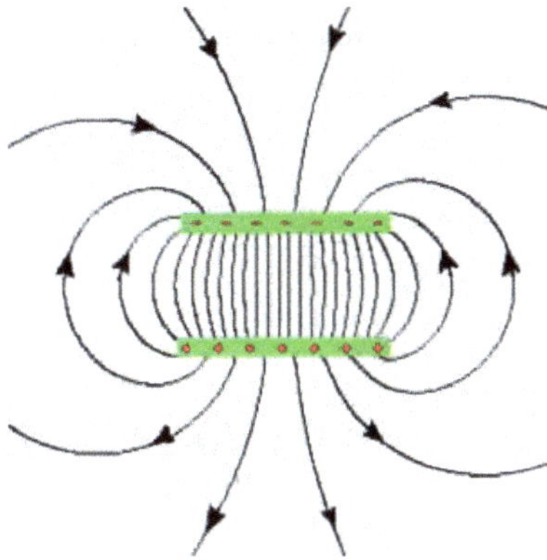

Figure 14.10: Nominally, the field in between two parallel plates in a capacitor is non-uniform. When the parallel plate is large, the ballpark value of the capacitor can be estimated by assuming a uniform field in between them. The correction to this simple formula incorporating fringing fields requires some tour-de-force analysis [117] (Figure by: https://electronics.stackexchange.com/questions/647189/what-are-changes-in-field-lines-after-two-plates-of-capacitors-are-connected).

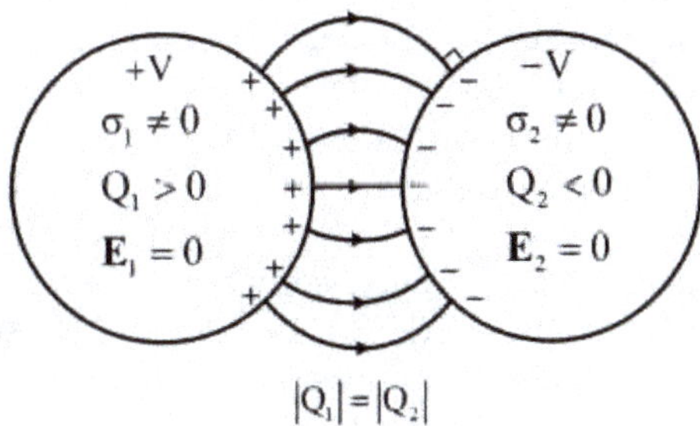

Figure 14.11: The capacitance between two charged conductors can be found by solving a BVP involving Laplace's equation as discussed in 3.3.5. If the conductors are odd shapes, then numerical methods are needed.

We require that

$$\Phi(\mathbf{r}) = \frac{1}{\varepsilon} \int_S dS' \frac{\sigma(\mathbf{r}')}{4\pi|\mathbf{r} - \mathbf{r}'|} = \begin{cases} +V, & \mathbf{r} \in S_1 \\ -V, & \mathbf{r} \in S_2 \end{cases} \tag{14.7.3}$$

In the above, $\sigma(\mathbf{r}')$, the surface charge density, is the unknown yet to be sought and it is embedded in an integral. But the right-hand side of the equation is known. Hence, this equation is also known as an integral equation where the unknown to be sought is embedded inside the integral. The integral equation can be solved by numerical methods that shall be discussed later.

Because we have found $\sigma(\mathbf{r})$, it can be now integrated to find Q, the total charge on one of the conductors. Since the voltage difference between the two conductors is known, the capacitance can be found as $C = Q/(2V)$. Here, $2V$ is assumed because it is the voltage difference between the two objects.

14.8 Importance of Circuit Theory in Integrated Circuit Design

The clock rate of computer circuits has peaked at about 3 GHz due to the resistive loss, or the I^2R loss. At this frequency, the wavelength is about 10 cm. Since transistors and circuit components are shrinking due to the compounding effect of Moore's law, most circuit components, which are measured in nanometers, are much smaller than the wavelength. Thus, most of the physics of electromagnetic signals in a microchip circuit can be captured by using circuit theory.

Figure 14.12 shows the schematic and the cross section of a computer chip at different levels: with the transistor at the bottommost level. The signals are taken out of a transistor by XY lines at the middle level that are linked to the ball-grid array at the topmost level of the chip. And then, the signal leaves the chip via a package. At the chip level, these structures are nanometer-sized and are much smaller than the wavelength,[7] they are usually modeled by lumped R, L, and C elements when retardation effect can be ignored. If retardation effect is needed, it can be modeled by a transmission line. The retardation effect is important at the package level where the dimensions of the components are sizeable compared to the wavelength.

A process of parameter extraction includes when computer software or field solvers (software that solves Maxwell's equations numerically–for example, HFSS [119, 120] and PEEC [106]) are used to extract these lumped-element parameters. Finally, a computer chip is modeled as a network involving a large number of transistors, diodes, and R, L, and C elements. Subsequently, a very useful and powerful commercial software called SPICE (Simulation Program with Integrated-Circuit Emphasis) [121] is a computer-aided software that helps to solve for the voltages and currents in this network.

Initially, SPICE software was written primarily to solve circuit problems. But the SPICE software now has many capabilities, including modeling of transmission lines (which shall be discussed next) for microwave engineering, which are important for modeling retardation effects. Figure 14.14 shows a graphical user interface (GUI) of an RF-SPICE that allows the modeling of transmission lines with a Smith chart interface.

[7]Circuit theory applies when the dimensions of a structure are much smaller than a wavelength.

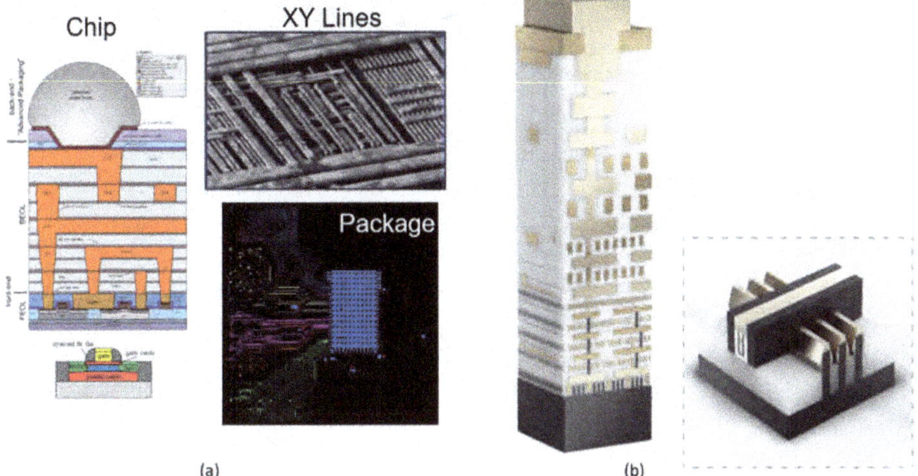

Figure 14.12: (a) Cross section of a chip (top left) and the XY lines in the chip (top right), and the interconnects in the package needed to take the signal out of the chip (bottom right). (b) A cross-sectional view of the modern design of a chip with Fin-FET transistors at the bottom, which switch faster and are smaller. Nowadays, transistors on a chip number in the billions. Clever engineering is needed to extract the humongous amount of data from these transistors (Figure by: Wikipedia, K. Radhakrishnan, Intel, https://www.svethardware.cz/intel-pro-vyrobu-tiger-lake-i-xe-vyuziva-10nmarchitekturu- superfin/52847).

A Smith chart, invented by P.H. Smith [122], is a graphical tool that we will discuss more fully in lecture 16. It is used in electrical engineering and telecommunications to analyze complex impedances in radio frequency (RF) and microwave circuits. It is indispensable in microwave engineering. It is particularly useful for matching impedances and visualizing how impedance changes along a transmission line. We will study this chart more rigorously in the next Lecture.

14.9 Decoupling Capacitors and Spiral Inductors

Decoupling capacitors are an important part of modern computer chip design. They can regulate voltage supply on the PDN (power delivery network) of the chip as they can remove high-frequency noise and voltage fluctuations from a circuit as shown in Figure 14.13. Figure 14.15 shows a 3D IC computer chip where decoupling capacitors are integrated into its design.

Inductors are also indispensable in IC design, as they can be used as a high-frequency choke. However, designing compact inductors on a chip is still a challenge. Spiral inductors are used because of their planar structure and ease of fabrication. However, miniaturizing inductors are a difficult frontier research topic [123].

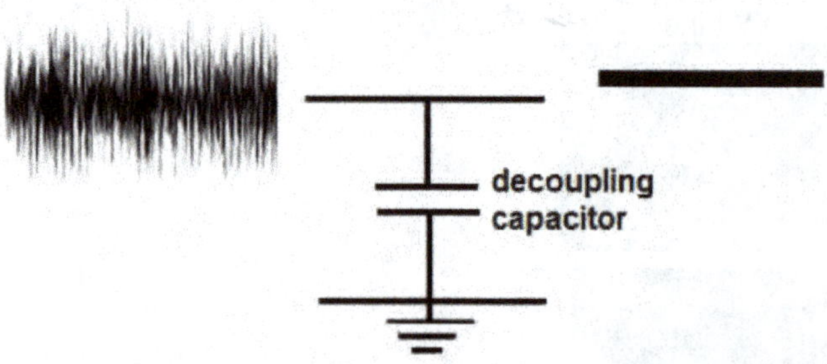

Figure 14.13: A decoupling capacitor is essentially a low-pass filter allowing low-frequency signals to pass through, while high-frequency signals are short-circuited (Figure by: https://www.ic-components.com/blog/a-comprehensive-guide-to-decoupling-and-bypass-capacitors-mechanisms,applications,and-differences.jsp).

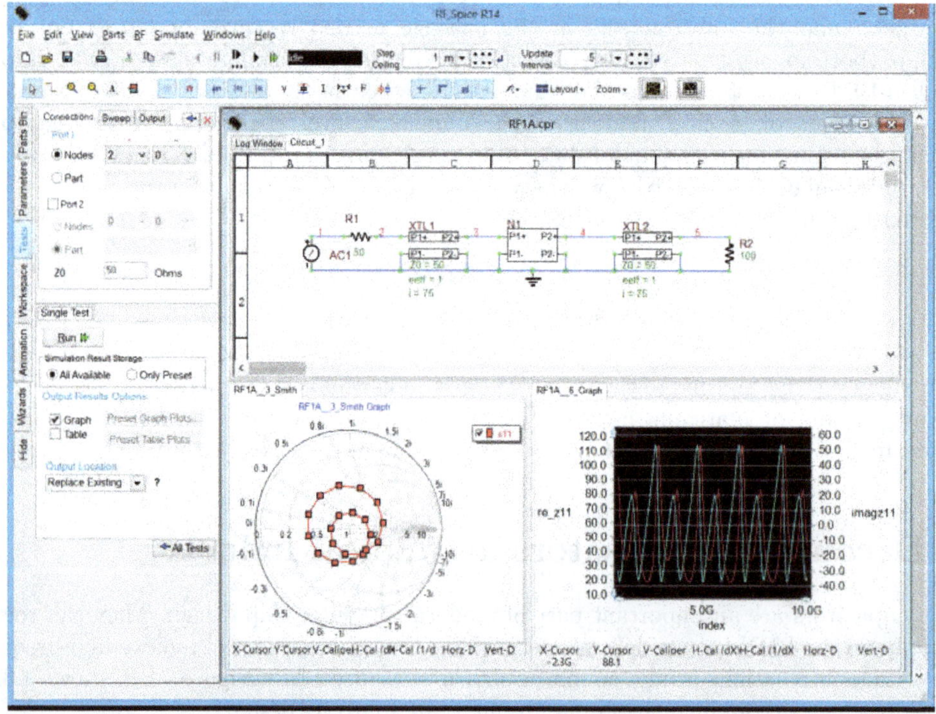

Figure 14.14: SPICE is also used to solve RF problems. A transmission line, to be discussed in the next lecture, is used in combination with circuit theory to account for retardation effects in a complex computer circuit (Figure by EMAG Technologies).

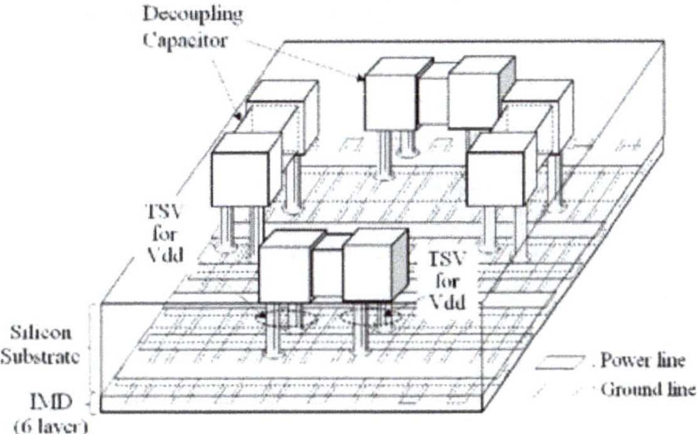

Figure 14.15: Modern computer chip design is 3D and is like an "urban jungle" like Manhattan or Hong Kong. There are different levels in the chip and they are connected by through-silicon vias (TSVs). IMD stands for intermetal dielectrics. One can see different XY lines serving as power and ground lines. Clever engineering has upheld the self-fulfilling prophesy of Moore's law. Because of the complexity of modern chip designs, and the common electromagnetic interference between the components, signal integrity (SI) and power integrity (PI) are important issues in their designs (Figure by: springer.com).

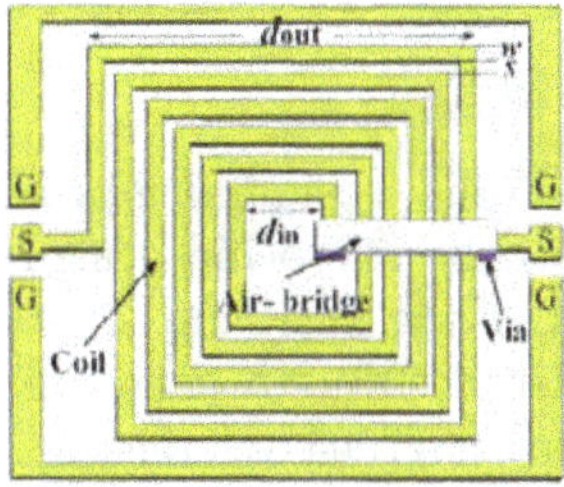

Figure 14.16: Spiral inductors are difficult to build on a chip, but by using laminal structure, it can be integrated into the IC fabrication process. The fabrication of circuit components has to be commensurate with the photolithography and chemical processes used (Figure by: researchgate.com, Quan Yuan).

14.10 Why the 3 GHz Barrier?

Moore's law projects that the packing density of transistors will double every 1.5 years. This trend has continued unabated until modern times. It has allowed each of us to carry the equivalent of the supercomputers of yesteryear in our pockets. Some of these cell phones have over 100 billion transistors in them. The most advanced have trillions of transistors in them. This unrelenting growth is truly the gift of God to us! It has given us a humongous amount of cloud storage and free email accounts. This growth is sometimes called the self-fulfilling prophecy of Moore's law. Because the semiconductor industry wants to see this growth, engineers become more ingenious in their chip designs. As you can see, chips are now three-dimensional via the use of TSVs (through-silicon vias).

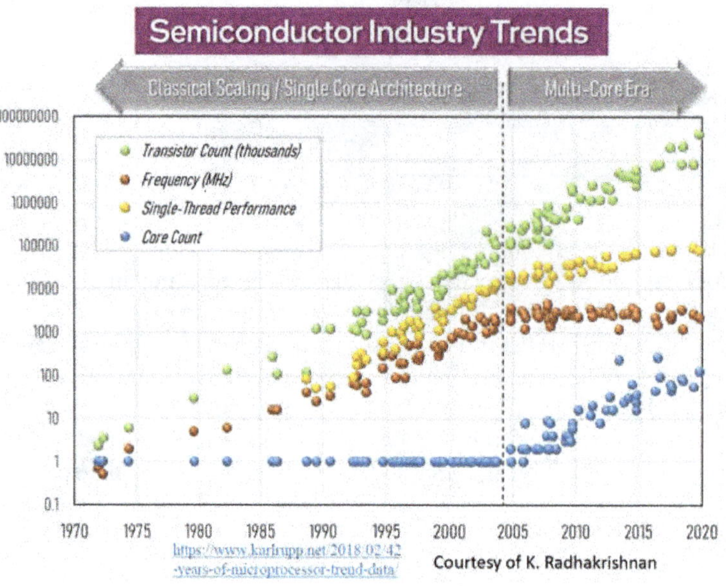

Figure 14.17: The semiconductor trends reveal the self-fulfilling prophesy of Moore's law. Ingenious engineering and 3D architecture of the microchip design allow the transistor and core counts to improve the performance of the microchip. Also, the clock rate (frequency) has not increased in the last two decades due to joule heating to be explained next.(Figure by: ieeexplore.ieee.org/document/9377004/references)

Despite the exciting growth of the microchip industry, the clock rate of computer chips has not increased very much: it has stagnated around 3 GHz. Thus, to improve the performance of microchips, core counts have increased to counter the stagnation in the clock rate (see Figure 14.17). There is a reason why the clock rate of CPU has not improved even though transistors now can switch at 800 GHz. The reason is joule heating, as explained simply by Figure 14.18.

The gate voltage of a MOSFET is used to switch it on and off. It can be modeled simply by a gate capacitor. The transistors of a microchip are connected to the outside world by interconnects. The PDN can be modeled by a voltage source connected to the gate by a wire as shown in Figure 14.18. The resistive loss on the wire can be modeled by a resistor R as shown. As the clock rate goes up, the frequency goes up, and the gate capacitance appears to be more like a short circuit, increasing the driving current I. Hence, the I^2R loss increases, giving rise to joule heating.[8] This

[8] I am indebted to Paul Y.S Cheung of HKU for sharing this insight with me.

joule heating, and the need to dissipate and manage this heat, is the reason why the clock rate has not risen in our computers in the last two decades.

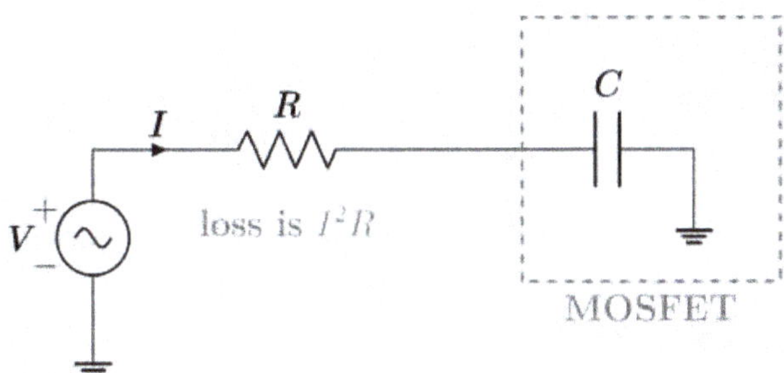

Figure 14.18: MOSFET driven by a power delivery network via the interconnects can be modeled simply as a gate capacitance that models the MOSFET transistor. The metal loss in the interconnects can be modeled as a resistor R in series with a voltage source V. As the frequency increases, more current flows through the capacitance, and hence, through the resistor, increasing the I^2R loss and giving rise to joule heating.

14.11 When Is Circuit Theory Valid?

Before we end this lecture, it will be good to ponder for a moment to ask, "When is circuit theory valid?" We will use dimensional analysis to help us answer this question. Faraday's law and Ampère's law are given as

$$\nabla \times \mathbf{E} = -\frac{\partial \mathbf{B}}{\partial t} \tag{14.11.1}$$

$$\nabla \times \mathbf{H} = \frac{\partial \mathbf{D}}{\partial t} + \mathbf{J} \tag{14.11.2}$$

Potential theory for which we assume that $\mathbf{E} = -\nabla\Phi$ follows from Faraday's law if we can drop the $\partial_t \mathbf{B}$ term in Faraday's law. Then KVL can be derived. Also, if we can drop the $\partial_t \mathbf{D}$ term in Ampère's law, then we have $\nabla \times \mathbf{H} = \mathbf{J}$, which implies that $\nabla \cdot \mathbf{J} = 0$ from which KCL can be derived. In a word, if we can drop the time derivative terms in the above, we can apply KVL and KCL.

For a geometry where the dimensions of the devices are of $O(L)$, the fields have to vary on the length scale of L in order to satisfy the boundary conditions around the devices. Thus we can argue that $\frac{\partial}{\partial x} \sim 1/L$ and similarly in all the other directions such as the y and z directions.

Thus we can argue that

$$\frac{\partial}{c\partial t} \ll \frac{\partial}{\partial x} \tag{14.11.3}$$

where c is the velocity of light.[9] For a sinusoidal signal, $\partial_t \sim O(\omega)$. The above is equivalent to

$$\frac{\omega}{c} \ll \frac{1}{L} \tag{14.11.4}$$

Since ω/c is the wavenumber which is $\frac{2\pi}{\lambda}$, the above is the same as

$$\frac{2\pi}{\lambda} \ll \frac{1}{L}, \text{ or } 2\pi L \ll \lambda \tag{14.11.5}$$

Therefore, if the dimensions L of the devices we are dealing with are much smaller than the wavelength λ, we can apply potential theory or circuit theory.

[9]We cannot compare apples with oranges, so we have to make the units on both sides of the inequality the same.

Exercises for Lecture 14

Problem 14-1: Circuit theory is often regarded as a subset of electromagnetic theory, but it is very simple, and hence popular. By amplifying the displacement current term using capacitors and flux linkage term using inductors, the full glory of electromagnetic theory is embedded in circuit theory.

(i) Derive Kirchhoff's current law and Kirchhoff's voltage law from electromagnetic theory.

(ii) Use the energy storage method: Find the inductance of the solenoid and show that it is the same as that obtained by the flux linkage method. Hint: Ramo et al. has a discussion of this.

(iii) Use the energy storage method: Find the capacitance of two parallel plate capacitors, and show that it is the same as solving Laplace's equation as a boundary value problem.

Problem 14-2:

(i) Using your undergraduate electrical engineering knowledge, show that the magnetic field energy storage in an inductor is given by

$$W_m = \frac{1}{2}LI^2$$

as in (14.6.1).

(ii) Similarly, show that the electric field energy storage in a capacitor is given by

$$W_e = \frac{1}{2}CV^2$$

as in (14.6.2).

Lecture 15

Transmission Lines

Transmission lines represent one of the most important electromagnetic technologies. The reason being that they can be described by simple theory, similar to circuit theory. As such, the theory is within the grasp of most practicing electrical engineers. Moreover, transmission line theory fills the gap in the physics of circuit theory: Circuit theory alone cannot describe wave phenomena, but when circuit theory is augmented with transmission line theory, wave phenomena with its corresponding wave physics such as retardation and time delay start to emerge.

Even though circuit theory has played an indispensable role in the development of the computer chip industry, eventually, circuit theory has to be embellished by transmission line theory so that high-speed circuits can be designed. Retardation effect, which causes time delay, clock skew, and phase shift, can be modeled simply using transmission lines. Nowadays, commercial circuit solver software programs such as SPICE[1] have the ability to include transmission lines as elements in their modeling.

15.1 Transmission Line Theory

Transmission lines were the first electromagnetic waveguides ever invented. They were driven by the needs of telegraphic technology. It is best to introduce transmission line theory from the viewpoint of circuit theory, which is elegant and one of the simplest theories of electrical engineering. This theory is also discussed in many textbooks and lecture notes.

Transmission lines are so important in modern-day electromagnetic engineering that most engineering electromagnetics textbooks would be incomplete without introducing the topics related to them [34, 59, 35, 37, 48, 70, 124].

[1]This is an acronym for a package "simulation program with integrated circuit emphasis" that came out of U. Cal., Berkeley [121].

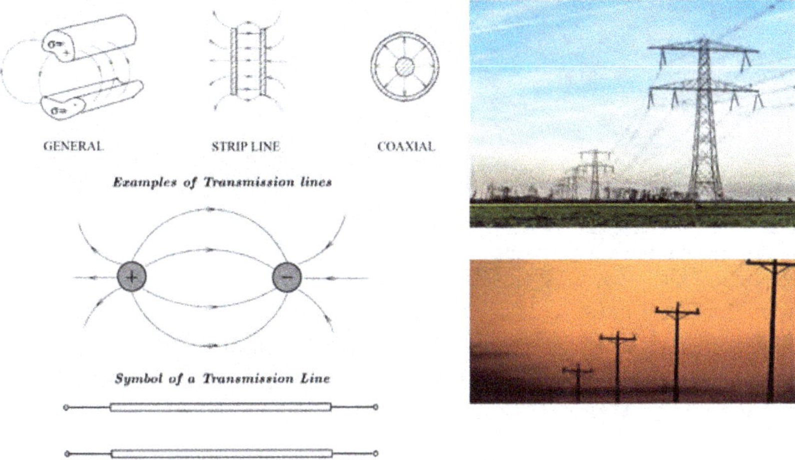

Figure 15.1: Various kinds of transmission lines. Schematically, all of them can be modeled by two parallel wires. On the right are pictures of a power transmission lines, and a telephone line. Due to technological advancements, telephone lines are a rare sight now (Figure by: Istockphoto.com).

Circuit theory is robust and is not sensitive to the detailed shapes of the components involved such as capacitors or inductors. Moreover, many transmission line problems cannot be analyzed simply when the full form of Maxwell's equations is used,[2] but approximate solutions can be obtained using circuit theory. We have seen previously that circuit theory is an approximation of electromagnetic field theory when the wavelength is very long (or the frequency is low): the longer the wavelength, the better the approximation [55]. Hence, in long-wavelength limit, transmission line theory can be approximated by circuit theory.

Examples of transmission lines are shown in Figure 15.1. The symbol for a transmission line is just two pieces of parallel wire, but in practice, these wires need not be parallel as shown in Figure 15.2.

Circuit theory also explains why waveguides such as transmission lines can be made sloppily when wavelength is long or the frequency low. For instance, in the long-wavelength limit, we can make twisted-pair waveguides with abandon, and yet they still work well (see Figure 15.2). Hence, it is simplest and best to first explain the propagation of electromagnetic signal on a transmission line using circuit analysis.

[2]Usually called full-wave analysis.

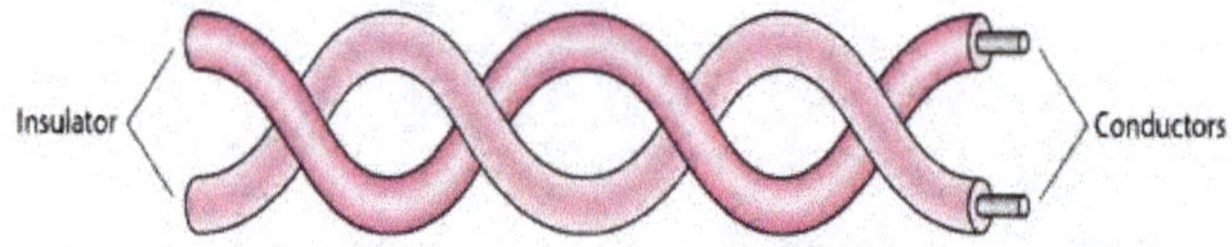

Figure 15.2: A twisted-pair transmission line where the two wires are not parallel to each other. The winding of the wires in series opposition cancels the linkage of the transmission line to low frequency external magnetic flux. (Figure by: https://stock.adobe.com/.

15.1.1 Time-Domain Analysis

We will start with performing the time-domain analysis of a simple, infinitely long transmission line. Remember that two pieces of metal can accumulate attractive positive and negative charges between them, giving rise to electric fields that start with positive charges and end with negative charges. The stored energy in the electric field gives rise to capacitive effect in the line which can be modeled by capacitances. Moreover, a piece of wire carrying a current generates a magnetic field, and hence, yields stored energy in the magnetic field. The stored magnetic field energy gives rise to inductive effect in the line which can be modeled by inductances. These stored energies are the sources of the capacitive and inductive effects.

But these capacitive and inductive effects are distributed over the spatial dimension of the transmission line. Therefore, it is helpful to think of the two pieces of metal as consisting of small segments of metal concatenated together. Each of these segments will have a small inductance, as well as a small capacitive coupling between them. Therefore, we can model two pieces of metal with a distributed lumped-element model as shown in Figure 15.3. For simplicity, we assume the other conductor to be a ground plane, so that it need not be approximated with lumped elements.[3]

In the transmission line, the voltage $V(z,t)$ and the current $I(z,t)$ are functions of both space z and time t, but we will model the space variation of the voltage and current with discrete step approximations. The voltage varies from node to node while the current varies from branch to branch of the lumped-element model (see Figure 15.3).

Telegrapher's Equations

First, we recall that the V-I relation of an inductor is

$$V_0 = L_0 \frac{dI_0}{dt} \tag{15.1.1}$$

where L_0 is the inductor, V_0 is the time-varying voltage drop across the inductor, and I_0 is the current through the inductor. Then using this relation between nodes 1 and 2 in Figure 15.3 (the

[3]Ground-plane physics can be modeled using image theorem that will be covered later in the course.

node numberings are in parenthesis), we have

$$V - (V + \Delta V) = L\Delta z \frac{\partial I}{\partial t} \tag{15.1.2}$$

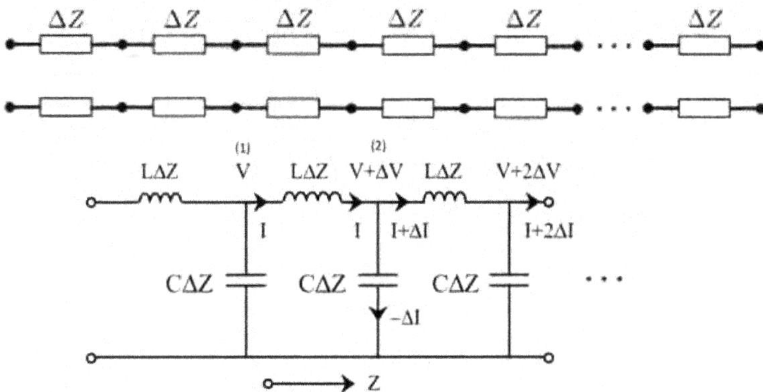

Figure 15.3: A long transmission line can be replaced by a concatenation of many short transmission lines. For each pair of short wires, there are capacitive couplings between them. Furthermore, when a current flows in the wires, a magnetic field is generated, making the wires behave like an inductor. Therefore, the transmission line can be replaced by a lumped-element approximation as shown. The lumped elements have inductances given by $L\Delta z$ and capacitances given by $C\Delta z$, distributed over the line. This is also known as the distributive model of the transmission line. There were many lumped-element models of a transmission line, but the final one that captures the physics correctly is due to Heaviside [43].

The left-hand side is the voltage drop across the series inductor, while the right-hand side follows from the aforementioned V-I relation of an inductor, but we have replaced $L_0 = L\Delta z$. Here, L is the inductance per unit length (line inductance) of the transmission line. And $L\Delta z$ is the incremental inductance due to the small segment of metal of length Δz. In the above, we assume that $V = V(z, t)$ and $I = I(z, t)$, so that time derivative is replaced by partial time derivative. Then the above (15.1.2) can be simplified to

$$\Delta V = -L\Delta z \frac{\partial I(z, t)}{\partial t} \tag{15.1.3}$$

where ΔV is the incremental voltage drop between the two nodes 1 and 2.

Next, we make use of the V-I relation for a single capacitor, which is

$$I_0 = C_0 \frac{dV_0}{dt} \tag{15.1.4}$$

where C_0 is the capacitor, I_0 is the current through the capacitor, and V_0 is a time-varying voltage drop across the capacitor. Thus, applying this relation at node 2 in Figure 15.3 gives the incremental shunt current to be

$$-\Delta I = C\Delta z \frac{\partial}{\partial t}(V + \Delta V) \approx C\Delta z \frac{\partial V}{\partial t} \tag{15.1.5}$$

where C is the capacitance per unit length, and $C\Delta z$ is the incremental capacitance between the small piece of metal and the ground plane. In the above, we have used Kirchhoff's current law to surmise that the current through the shunt capacitor is $-\Delta I$, where $\Delta I = I(z+\Delta z, t) - I(z, t)$. In the last approximation in (15.1.5), we dropped a term involving the product of Δz and ΔV, since it would be very small (or in the second order of magnitude in math parlance).

In the limit when $\Delta z \to 0$, one gets from (15.1.3) and (15.1.5) that

$$\frac{\partial V(z,t)}{\partial z} = -L\frac{\partial I(z,t)}{\partial t} \tag{15.1.6}$$

$$\frac{\partial I(z,t)}{\partial z} = -C\frac{\partial V(z,t)}{\partial t} \tag{15.1.7}$$

The above are the *telegrapher's equations.*[4] They are two coupled first-order equations, and can be converted into second-order equations easily by eliminating one of the two unknowns. Therefore,

$$\frac{\partial^2 V}{\partial z^2} - LC\frac{\partial^2 V}{\partial t^2} = 0 \tag{15.1.8}$$

$$\frac{\partial^2 I}{\partial z^2} - LC\frac{\partial^2 I}{\partial t^2} = 0 \tag{15.1.9}$$

The above are wave equations that we have previously studied (see section 3.2), where the velocity of the wave is given by

$$v = \frac{1}{\sqrt{LC}} \tag{15.1.10}$$

Furthermore, if we assume that

$$V(z,t) = V_0 f_+(z - vt), \qquad I(z,t) = I_0 f_+(z - vt) \tag{15.1.11}$$

corresponding to a right-traveling wave only, they can be verified to satisfy (15.1.6) and (15.1.7) as well as (15.1.8) and (15.1.9) by back-substitution.

Consequently, we can easily show, by back substitution to (15.1.6) and (15.1.7), that when the transmission line has only the right-traveling wave,

$$\frac{V(z,t)}{I(z,t)} = \frac{V_0}{I_0} = \sqrt{\frac{L}{C}} = Z_0 \tag{15.1.12}$$

[4]They can be thought of as the distillation of Faraday's law, $\nabla \times \mathbf{E} = -\partial_t \mathbf{B}$, and Ampère's law, $\nabla \times \mathbf{H} = \partial_t \mathbf{D}$, from Maxwell's equations without the source term. Their simplicity gives them an important role in engineering electromagnetics. One can think of them as the poor man's Maxwell's equations.

where Z_0 is a constant independent of z, which is the characteristic impedance of the transmission line. The above ratio is only true for a wave that travels in one direction, in this case, one that propagates in the $+z$ direction.

For a wave that travels in the negative z direction only, in a similar manner, we can let

$$V(z,t) = V_0 f_-(z+vt), \qquad I(z,t) = I_0 f_-(z+vt) \tag{15.1.13}$$

one can easily show by the same token that

$$\frac{V(z,t)}{I(z,t)} = \frac{V_0}{I_0} = -\sqrt{\frac{L}{C}} = -Z_0 \tag{15.1.14}$$

Again, Z_0 is a constant independent of z.

Time-domain analysis is very useful for transient analysis of transmission lines, especially when nonlinear elements are coupled to the transmission line.[5] Another major strength of the transmission line model is that it is a simple way to introduce time delay (also called retardation which is illustrated in Figure 3.1) in a simple circuit model.[6] As we saw when we studied the solution to the wave equation: solutions at different times are just the time-delayed version of the original solution.

Time Delay and Velocity of Light

Time delay is a wave propagation effect, and it is harder to incorporate into circuit theory or a pure circuit model consisting of R, L, and C only. In circuit theory, where the wavelength is assumed to be very long, Laplace's equation is usually solved, which is equivalent to Helmholtz's equation with infinite wave velocity, namely,

$$\lim_{c\to\infty} \nabla^2\Phi(\mathbf{r}) + \frac{\omega^2}{c^2}\Phi(\mathbf{r}) = 0 \quad \longrightarrow \quad \nabla^2\Phi(\mathbf{r}) = 0 \tag{15.1.15}$$

From the above, we see that Helmholtz's equation becomes Laplace's equation when the velocity of light c is infinite. Hence, events in Laplace's equation happen instantaneously. In other words, in circuit theory, where Laplace's equation is involved, we assume that the velocity of the wave is infinite, and there is no retardation effect. This is only true or a good approximation when the size of the structure is small compared to the wavelength.

15.1.2 Frequency-Domain Analysis—The Power of Phasor Technique Again!

As we have seen in previous lectures, the frequency-domain analysis greatly simplifies the analysis of many complicated phenomena. This was especially true in our analysis of conductive media, and

[5]Remember that we can only use frequency domain technique or Fourier transform technique for linear time-invariant systems from linear system theory [125].

[6]By a simple circuit model, we mean a model that has lumped elements such as R, L, and C as well as a transmission line element.

frequency-dispersive media as in the Drude-Lorentz-Sommerfeld model, etc. As such, frequency-domain analysis is very popular as it makes the transmission line equations very simple–one just replaces $\partial/\partial t \to j\omega$. Moreover, generalization to a lossy system is quite straightforward. Furthermore, for linear time-invariant systems, the time-domain signals can be obtained from the frequency-domain data by performing a Fourier inverse transform since phasors and Fourier transforms of a time variable are just related to each other by a constant (see section 6.2).

The telegrapher's equations (15.1.6) and (15.1.7) then in frequency-domain become

$$\frac{d}{dz}V(z,\omega) = -j\omega L I(z,\omega) \tag{15.1.16}$$

$$\frac{d}{dz}I(z,\omega) = -j\omega C V(z,\omega) \tag{15.1.17}$$

The above gives the notion that the change in the voltage $V(z,\omega)$ on a transmission line is proportional to the line impedance $j\omega L$ times the current $I(z,\omega)$. Similar notion can be said of the second equation above.

The corresponding 1D Helmholtz's equations can be derived from the above, and they are then (see section 7.2)

$$\frac{d^2V}{dz^2} + \omega^2 LCV = 0 \tag{15.1.18}$$

$$\frac{d^2I}{dz^2} + \omega^2 LCI = 0 \tag{15.1.19}$$

The above are second order ordinary differential equations (ODEs), and the general solutions to the above are (easily proved by back substitution)

$$V(z) = V_+ e^{-j\beta z} + V_- e^{j\beta z} \tag{15.1.20}$$

$$I(z) = I_+ e^{-j\beta z} + I_- e^{j\beta z} \tag{15.1.21}$$

where the wavenumber $\beta = \omega\sqrt{LC}$. This is similar to what we have seen previously for plane waves in the one-dimensional wave equation in free space, where

$$E_x(z) = E_{0+} e^{-jk_0 z} + E_{0-} e^{jk_0 z} \tag{15.1.22}$$

Here, $k_0 = \omega\sqrt{\mu_0\varepsilon_0}$. We see much similarity between (15.1.20), (15.1.21), and (15.1.22).

To see the solution in the time domain, we let the phasor $V_\pm = |V_\pm|e^{j\phi_\pm}$ in (15.1.20), and the time-harmonic voltage signal above can then be converted back to the time domain using the key formula in phasor technique as

$$V(z,t) = \Re e\{V(z,\omega)e^{j\omega t}\} \tag{15.1.23}$$

$$= |V_+|\cos(\omega t - \beta z + \phi_+) + |V_-|\cos(\omega t + \beta z + \phi_-) \tag{15.1.24}$$

As can be seen, the first term corresponds to a right-traveling wave, while the second term is a left-traveling wave.

Furthermore, if we assume only a one-way traveling wave, traveling to the right by letting $V_- = I_- = 0$, then it can be shown that, for a right-traveling wave, using (15.1.16) or (15.1.17)

$$\frac{V(z)}{I(z)} = \frac{V_+}{I_+} = \sqrt{\frac{L}{C}} = Z_0 \tag{15.1.25}$$

where Z_0 is the characteristic impedance. The above is the same as that in the time domain we have derived previously.[7] Since Z_0 is real, it implies that the phasors[8] $V(z)$ and $I(z)$ are in phase.

Similarly, applying the same process for a left-traveling wave only, by letting $V_+ = I_+ = 0$, then

$$\frac{V(z)}{I(z)} = \frac{V_-}{I_-} = -\sqrt{\frac{L}{C}} = -Z_0 \tag{15.1.26}$$

In other words, for the left-traveling waves, the voltage and current are 180° out of phase, while the right-traveling waves, they are in phase. This is similar to the time domain case we have before.

15.2 Lossy Transmission Line

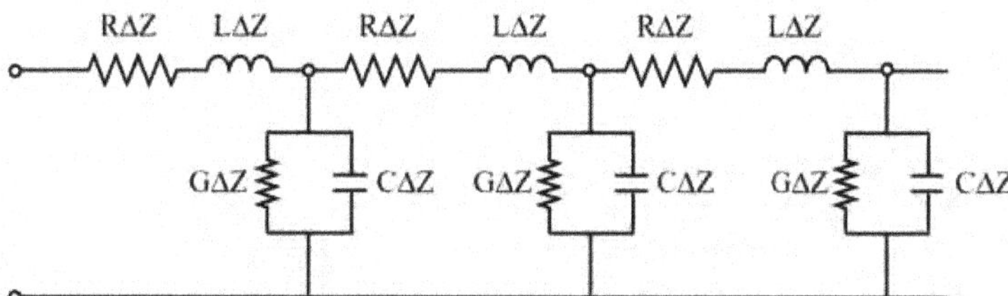

Figure 15.4: In a lossy transmission line, series resistance can be added to the series inductance, and the shunt conductance can be added to the shunt susceptance of the capacitor. This problem is "homomorphic" to the lossless case in the frequency domain (see Figure 15.3).

The phasor technique is empowered by, first, that the algebra for complex numbers is the same as that of real numbers. Second, resistors and conductances are replaced by impedances and admittances in the frequency domain. By exploiting "homomorphism", this makes the solution involving a network of impedances and admittances analogous to the network of resistances and conductances. The power of frequency-domain analysis is revealed in the study of lossy transmission lines. The previous analysis, which is valid for lossless transmission line, can be easily generalized to the lossy case in the frequency domain. In using frequency domain and phasor technique, impedances will become complex numbers, as shall be shown next.

[7] One can think that this ratio is independent of frequency, and hence, is valid both in the time domain and frequency domain.

[8] We will not denote phasors by under-tilde, as they are implied by the context.

To include loss, we use the lumped-element model as shown in Figure 15.4. One thing to note is that $j\omega L$ is actually the series line impedance of the lossless transmission lines, while $j\omega C$ is the shunt line admittance of the same line. First, we can rewrite the expressions for the telegrapher's equations in (15.1.16) and (15.1.17) in terms of series line impedance and shunt line admittance to arrive at

$$\frac{d}{dz}V = -ZI \tag{15.2.1}$$

$$\frac{d}{dz}I = -YV \tag{15.2.2}$$

where $Z = j\omega L$ and $Y = j\omega C$. The above can be easily generalized to the lossy case as shall be shown next.

The geometry in Figure 15.4 is topologically similar to, or "homomorphic" to the lossless case in Figure 15.3. Hence, when lossy elements are added in the geometry, we can surmise that the corresponding telegrapher's equations are similar to those above. But to include loss, we need only to generalize the series line impedance and shunt admittance from the lossless case to lossy case as follows:

$$Z = j\omega L \to Z = j\omega L + R \tag{15.2.3}$$

$$Y = j\omega C \to Y = j\omega C + G \tag{15.2.4}$$

where R is the series line resistance, and G is the shunt line conductance. Here, R can be used to model copper loss on a metallic transmission line and G, the dielectric leakage loss in the shunt capacitance. Thus, now Z and Y are the series impedance and shunt admittance (which are complex numbers rather than being pure imaginary numbers), respectively. We will further exploit the fact that the algebra of complex numbers is the same as the algebra of real numbers. We will refer to this as mathematical "homomorphism". Then, the corresponding Helmholtz's equations are

$$\frac{d^2V}{dz^2} - ZYV = 0 \tag{15.2.5}$$

$$\frac{d^2I}{dz^2} - ZYI = 0 \tag{15.2.6}$$

or

$$\frac{d^2V}{dz^2} - \gamma^2 V = 0 \tag{15.2.7}$$

$$\frac{d^2I}{dz^2} - \gamma^2 I = 0 \tag{15.2.8}$$

where $\gamma^2 = ZY$, or that one can also think of $\gamma^2 = -\beta^2$ by comparing with (15.1.18) and (15.1.19). Then the above is "homomorphic" to the lossless case except that now, β is a complex number, indicating that the field is decaying and oscillating as it propagates. As before, the above are second order 1D Helmholtz equations where the general solutions are

$$V(z) = V_+ e^{-\gamma z} + V_- e^{\gamma z} \tag{15.2.9}$$

$$I(z) = I_+ e^{-\gamma z} + I_- e^{\gamma z} \tag{15.2.10}$$

and

$$\gamma = \sqrt{ZY} = \sqrt{(j\omega L + R)(j\omega C + G)} = j\beta \tag{15.2.11}$$

Here, $\beta = \beta' - j\beta''$ is now a complex number. In other words,

$$e^{-\gamma z} = e^{-j\beta' z - \beta'' z}$$

is an oscillatory and decaying wave. Or focusing on the voltage case, we have

$$V(z) = V_+ e^{-\beta'' z - j\beta' z} + V_- e^{\beta'' z + j\beta' z} \tag{15.2.12}$$

Again, letting $V_\pm = |V_\pm| e^{j\phi_\pm}$, the above can be converted back to the time domain as

$$V(z,t) = \Re e\{V(z,\omega)e^{j\omega t}\} \tag{15.2.13}$$

$$= |V_+| e^{-\beta'' z} \cos(\omega t - \beta' z + \phi_+) + |V_-| e^{\beta'' z} \cos(\omega t + \beta' z + \phi_-) \tag{15.2.14}$$

The first term corresponds to a decaying wave moving to the right while the second term is also a decaying wave but moving to the left. When there is no loss, or $R = G = 0$, and from (15.2.11), we deduce that $j\beta = j\omega\sqrt{LC}$ is pure imaginary, or that $\beta = \beta'$ and that $\beta'' = 0$.

Notice that for the lossy case, the characteristic impedance, which is the ratio of the voltage to the current for a one-way wave, can similarly be derived using "homomorphism", viz.,

$$Z_0 = \frac{V_+}{I_+} = -\frac{V_-}{I_-} = \sqrt{\frac{L}{C}} = \sqrt{\frac{j\omega L}{j\omega C}} \to Z_0 = \sqrt{\frac{Z}{Y}} = \sqrt{\frac{j\omega L + R}{j\omega C + G}} \tag{15.2.15}$$

The above Z_0 is manifestly a complex number. Here, Z_0 is the ratio of the phasors of the one-way waves, and apparently, their current phasor and the voltage phasor will not be in phase for lossy transmission lines.

In the absence of loss, the above again becomes

$$Z_0 = \sqrt{\frac{L}{C}} \tag{15.2.16}$$

the characteristic impedance for the lossless case previously derived.

Exercises for Lecture 15

Problem 15-1:

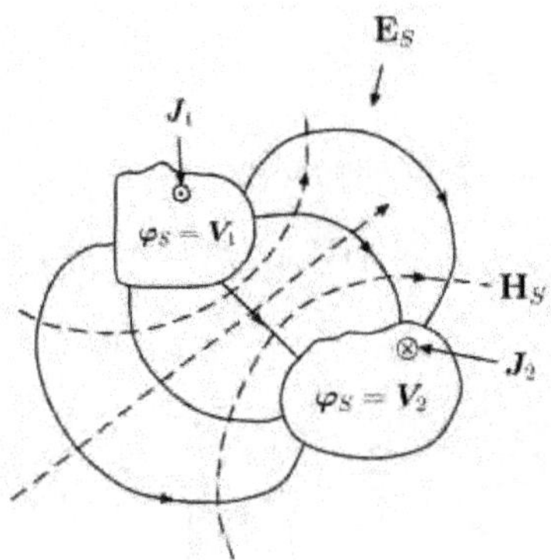

Figure 15.5: The TEM $\mathbf{E}_s$ and $\mathbf{H}_s$ fields of a transmission line.

The TEM field of the transmission line is as shown in Figure 15.5. The electromagnetic fields are purely transverse with no z component. The telegrapher's equations for transmission lines have been derived using circuit theory in the text. Hence, a transmission line can be twisted and bent with abandon, and yet it still works. But the telegrapher's equations can also be derived using field theory, which is more elegant to some people, as shall be shown below:

(i) The fields of a guided TEM (transverse electromagnetic) mode in a transmission line can be written as

$$\mathbf{E} = \mathbf{E}_s e^{-j\beta z}, \qquad \mathbf{H} = \mathbf{H}_s e^{-j\beta z} \tag{15.2.1}$$

where $\mathbf{E}_s$ and $\mathbf{H}_s$ are purely transverse fields with no z components. Show that in order for the above to satisfy Maxwell's equations, then

$$\nabla_s \times \mathbf{E}_s = 0, \qquad \nabla_s \times \mathbf{H}_s = 0 \tag{15.2.2}$$

The subscript s implies transverse to z, and it is also true for the del operator.

(ii) With your knowledge of potential theory, how would you solve the first equation to obtain $\mathbf{E}_s$? You can assume that one of the conductors has voltage V while the other conductor is at ground with zero voltage. For simplicity, you may repeat this exercise for the coax cable as has been done in the earlier part of this course.

(iii) Show from Maxwell's equations that once $\mathbf{E}_s$ is known, $\mathbf{H}_s$ can be obtained via the following equation:

$$\frac{\partial}{\partial z}\hat{z} \times \mathbf{E}_s = -j\omega\mu\mathbf{H}_s, \qquad \frac{\partial}{\partial z}\hat{z} \times \mathbf{H}_s = j\omega\mu\mathbf{E}_s \tag{15.2.3}$$

(iv) Can you derive the telegrapher's equation from the above equations? (Hint: Cross product one of the above with $\hat{z}$ to simplify it and then use line integral involving $\mathbf{E}_s$ and $\mathbf{H}_s$ to define the voltage and the current on the conductors (see also [92][p. 40-41]).

Problem 15-2: For the geometry shown in Figure 15.6, it actually does not support a TEM mode, a fact that you shall learn later in the course also. It actually does not have a closed form solution but we can use a circuit, or quasi-static approximation in finding the transmission line parameters. Namely, in the long-wavelength limit, we can still define a line capacitance and a line inductance. We can then use circuit theory concept to find the phase velocity of a low frequency wave on such a line. The line admittance and impedance of a transmission line can then be found by solving an electrostatic problem and a magnetostatic problem, respectively.

(i) The coaxial cable geometry below has inner radius a and outer radius b. It has a lossy dielectric medium between the inner and outer conductor. Find the admittance per unit length that you can substitute into the telegrapher's equations. Since the problem does not have a closed-form solution, an approximate solution can be obtained by assuming uniform radial electric field inside the coax. (Also, you may want to think about the lossless problem first where $\sigma_1 = 0$, solve the electrostatics problem, and then use math "homomorphism" between statics and dynamics and phasor technique to obtain the quasi-static dynamic solution.)

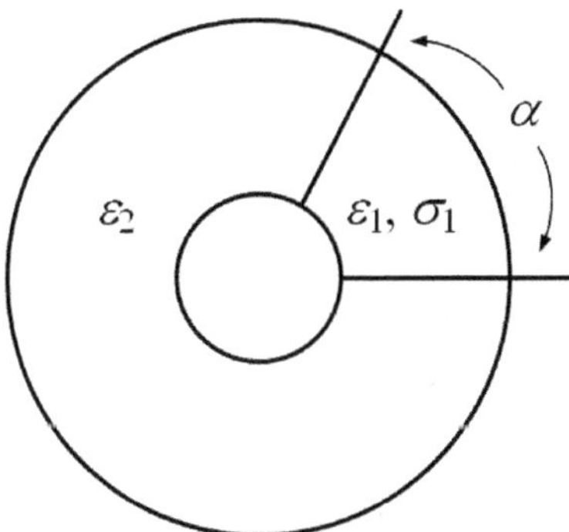

Figure 15.6: A coaxial cable with inhomogeneous, conductive medium inside.

(ii) Since in the static limit the magnetostatic problem is decoupled from the electrostatic problem, we can solve the magnetostatics problem first. Assume a current I that flows in the inner conductor (or $-I$ in the outer conductor), and that the conductors are lossless. Using the magnetic field you have found before, find the magnetic energy stored per unit length. Knowing that the energy storage for an inductor is $\frac{1}{2}LI^2$, find the inductance per unit length.

(iii) Use your results to find the characteristic impedance of this lossy transmission line, and also the propagation constant γ.

Lecture 16

More on Transmission Lines

As mentioned before, transmission line theory is indispensable in microwave engineering. The theory is the necessary augmentation of circuit theory for higher frequency analysis, and it is also indispensable to integrated circuit designers as computer clock rates become faster. Over the years, engineers have developed some very useful tools and measurement techniques to expand the design space of circuit designers. We will learn some of these tools in this lecture.[1]

As seen in the previous lecture, the telegrapher's equations are similar to the one-dimensional form of Maxwell's equations, and can be thought of as Maxwell's equations in their simplest form. Therefore, they entail a subset of the physics seen in the full Maxwell's equations. Transmission line is a poor man's way of incorporating wave physics into engineering designs without invoking the full bounty of Maxwell's equations.

16.1 Terminated Transmission Lines

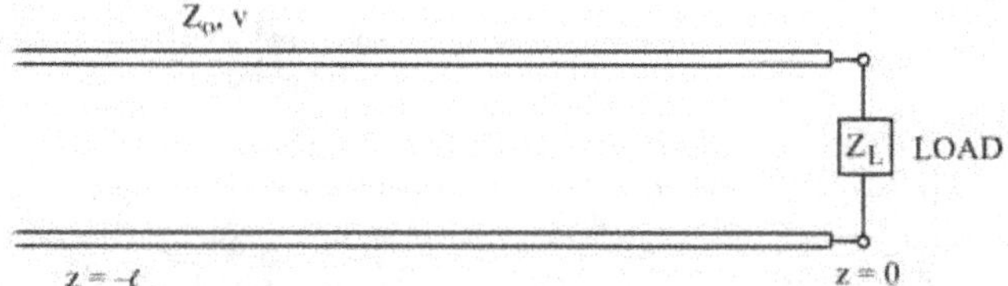

Figure 16.1: A schematic for a transmission line terminated with an impedance load Z_L at $z = 0$.

[1]Some of you may have studied this topic in your undergraduate electromagnetics course. However, this topic is important, and you will have to muster your energy to master this knowledge again.:)

For an infinitely long transmission line, the solution consists of the linear superposition of a wave traveling to the right plus a wave traveling to the left. If transmission line is terminated by a load as shown in Figure 16.1, a right-traveling wave will be reflected by the load to generate a left-traveling wave, and in general, the wave on the transmission line will be a linear superposition of the left- and right-traveling waves.

To simplify the analysis, we will assume that the line is lossless. The generalization to the lossy case is quite straightforward as shall be shown later. Thus, we assume that

$$V(z) = a_+ e^{-j\beta z} + a_- e^{j\beta z} = V_+(z) + V_-(z) \tag{16.1.1}$$

where $V_+(z)$ and $V_-(z)$ are right and left-traveling waves, respectively. Here $\beta = \omega/v = \omega\sqrt{LC}$. In the above, in general, $a_+ \neq a_-$. Besides, we assume that the system is linear; or we can define the amplitude of the left-going reflected wave a_- to be linearly related to the amplitude of the right-going or incident wave a_+. In other words, at $z = 0$, we can let

$$V_-(z = 0) = \Gamma_L V_+(z = 0) \tag{16.1.2}$$

where Γ_L is independent of $V_+(z = 0)$, making this relation a linear one. Thus, using the definition of $V_+(z)$ and $V_-(z)$ as implied in (16.1.1), we have

$$a_- = \Gamma_L a_+ \tag{16.1.3}$$

where Γ_L is termed the reflection coefficient. Hence, (16.1.1) becomes

$$V(z) = a_+ e^{-j\beta z} + \Gamma_L a_+ e^{j\beta z} = a_+ \left(e^{-j\beta z} + \Gamma_L e^{j\beta z}\right) \tag{16.1.4}$$

The corresponding current $I(z)$ on the transmission line is derived using the telegrapher's equations as previously defined. By recalling that

$$\frac{\mathrm{d}V}{\mathrm{d}z} = -j\omega L I$$

then for the general case,

$$I(z) = \frac{a_+}{Z_0}\left(e^{-j\beta z} - \Gamma_L e^{j\beta z}\right) \tag{16.1.5}$$

Notice the sign change in the second term of the above expression.

From (16.1.1) and (16.1.3), one sees that the left-going wave and the right-going wave are linearly proportional to each other. In other words, if we double the right-going wave, the left-going wave doubles. Thus, similar to Γ_L, a general or local reflection coefficient $\Gamma(z)$ (which is a function of z) at z can be defined by relating the left-traveling and right-traveling wave at location z such that

$$\Gamma(z) = \frac{V_-(z) = a_- e^{j\beta z}}{V_+(z) = a_+ e^{-j\beta z}} = \frac{a_- e^{j\beta z}}{a_+ e^{-j\beta z}} = \Gamma_L e^{2j\beta z} \tag{16.1.6}$$

where (16.1.3) has been used to relate a_- and a_+ to Γ_L. Of course, $\Gamma(z = 0) = \Gamma_L$. Furthermore, due to the V-I relation at an impedance load, we must have

$$\frac{V(z = 0)}{I(z = 0)} = Z_L \tag{16.1.7}$$

or that using (16.1.4) and (16.1.5) with $z = 0$, the left-hand side of the above can be rewritten, and we have

$$\frac{1+\Gamma_L}{1-\Gamma_L} Z_0 = Z_L, \quad \text{or} \quad \frac{1+\Gamma_L}{1-\Gamma_L} = \frac{Z_L}{Z_0} = Z_{nL} \tag{16.1.8}$$

where Z_{nL} is the normalized load. From the above, we can solve for Γ_L in terms of Z_{nL} to get

$$\Gamma_L = \frac{Z_{nL}-1}{Z_{nL}+1} = \frac{Z_L - Z_0}{Z_L + Z_0} \tag{16.1.9}$$

Thus, given the termination load Z_L and the characteristic impedance Z_0, the reflection coefficient Γ_L can be found, or vice versa. Or given Γ_L, the normalized load impedance, $Z_{nL} = Z_L/Z_0$, can be found.

It is seen that $\Gamma_L = 0$ if $Z_L = Z_0$. Thus a right-traveling wave will not be reflected and the left-traveling wave is absent. This is the case of a *matched load.* When there is no reflection, all energy of the right-traveling wave will be totally absorbed by the load.

In general, we can define a generalized impedance at $z \neq 0$ to be

$$\begin{aligned} Z(z) = \frac{V(z)}{I(z)} &= \frac{a_+(e^{-j\beta z} + \Gamma_L e^{j\beta z})}{\frac{1}{Z_0} a_+(e^{-j\beta z} - \Gamma_L e^{j\beta z})} \\ &= Z_0 \frac{1+\Gamma_L e^{2j\beta z}}{1-\Gamma_L e^{2j\beta z}} = Z_0 \frac{1+\Gamma(z)}{1-\Gamma(z)} \end{aligned} \tag{16.1.10}$$

where $\Gamma(z)$, the general reflection coefficient defined in (16.1.6) is used in the above. Since the voltage $V(z)$ and $I(z)$ are proportional to the electric field and the magnetic field at location z, this generalized impedance is the ratio of the electric field to the magnetic field at the location z. The above can also be normalized and written as

$$Z_n(z) = Z(z)/Z_0 = \frac{1+\Gamma(z)}{1-\Gamma(z)} \tag{16.1.11}$$

where $Z_n(z)$ is the normalized generalized impedance. Conversely, one can write the above as

$$\Gamma(z) = \frac{Z_n(z)-1}{Z_n(z)+1} = \frac{Z(z)-Z_0}{Z(z)+Z_0} \tag{16.1.12}$$

From (16.1.10) above, one gets

$$Z(z) = Z_0 \frac{1+\Gamma_L e^{2j\beta z}}{1-\Gamma_L e^{2j\beta z}} \tag{16.1.13}$$

One can show that by setting $z = -l$, using (16.1.9) for Γ_L, and after some algebra,

$$Z(-l) = Z_0 \frac{Z_L + jZ_0 \tan \beta l}{Z_0 + jZ_L \tan \beta l} \tag{16.1.14}$$

16.1.1 Short-Circuited Terminations

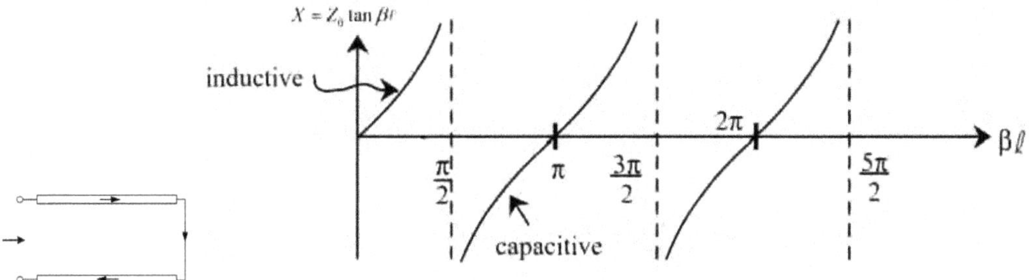

Figure 16.2: The input reactance (X) of a short-circuited transmission line as a function of its length l. The input impedance of the transmission line alternates between being inductive and capacitive as the length varies. A standing wave develops on this short-circuit terminated transmission line. This can make the input impedance look inductive or capacitive depending on if the input port is current dominated or voltage dominated, respectively. A large current flows on a short-circuited line to generate stored magnetic energy and, hence, the impedance becomes inductive depending on the value of l.

From (16.1.14) above, when we have a short such that $Z_L = 0$, then

$$Z(-l) = jZ_0 \tan(\beta l) = jX \tag{16.1.15}$$

On the short-circuited transmission line, a standing wave develops where the voltage and the current are out of phase with respect to each other. When the current is stronger than the voltage, the magnetic energy stored is larger than the electric energy stored. As in the case of an inductor, the input impedance is inductive. However, if one varies the length such that at the input port, the voltage is stronger than the current, the input impedance becomes capacitive. When $\beta l \ll 1$, then $\tan(\beta l) \approx \beta l$, and (16.1.15) becomes

$$Z(-l) \cong jZ_0\beta l \tag{16.1.16}$$

After using that $Z_0 = \sqrt{L/C}$ and that $\beta = \omega\sqrt{LC}$, (16.1.16) becomes

$$Z(-l) \cong j\omega Ll = j\omega L_{\text{eff}} \tag{16.1.17}$$

The above implies that a short length of transmission line connected to a short as a load looks like an inductor with $L_{\text{eff}} = Ll$, since much current will pass through this short producing a strong magnetic field with stored magnetic energy. Remember here that L is the line inductance, or inductance per unit length.

16.1.2 Open-Circuited Terminations

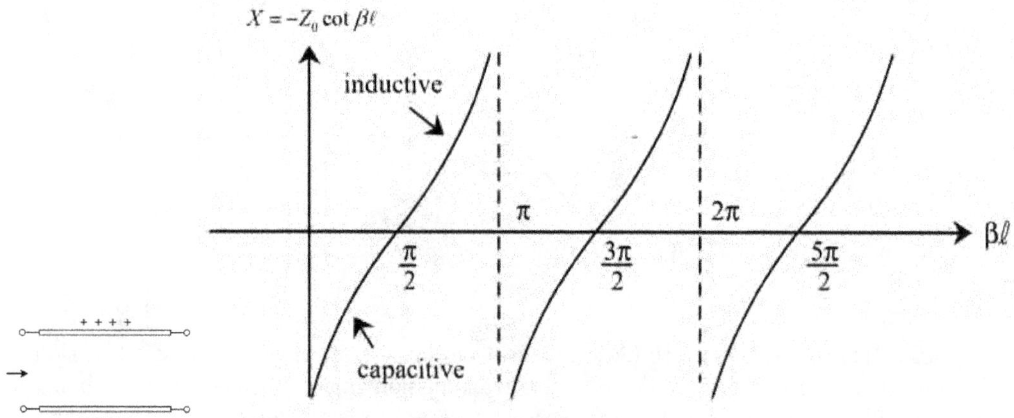

Figure 16.3: The input reactance (X) of an open-circuited transmission line as a function of its length l. Again, a standing wave develops on this open-circuit terminated transmission line. This can make the input impedance look capacitive or inductive depending on the length l.

When we have an open circuit such that $Z_L = \infty$, then from (16.1.14) above

$$Z(-l) = -jZ_0 \cot(\beta l) = jX \tag{16.1.18}$$

Then, when $\beta l \ll 1$, $\cot(\beta l) = 1/\tan(\beta l) \approx 1/\beta l$

$$Z(-l) \approx -j\frac{Z_0}{\beta l} \tag{16.1.19}$$

And then, again using $\beta = \omega\sqrt{LC}$, $Z_0 = \sqrt{L/C}$

$$Z(-l) \approx \frac{1}{j\omega Cl} = \frac{1}{j\omega C_{\text{eff}}} \tag{16.1.20}$$

Hence, an open-circuited terminated short length of transmission line appears like an effective capacitor with $C_{\text{eff}} = Cl$. Again, remember here that C is the line capacitance or capacitance per unit length of the transmission line.

As shown in Figure 16.3, the impedance at $z = -l$ is purely reactive when there is no loss, and goes through positive and negative values due to the standing wave set up on the transmission line. Therefore, by changing the length of l, one can make a short-circuited or an open-circuited line look like an inductor or a capacitor depending on its length l. This effect is shown in Figures 16.2 and 16.3. Moreover, the reactance X becomes infinite or zero with the proper choice of the length l. These are resonances or anti-resonances of the transmission line, very much like the resonance of an LC tank circuit. An LC circuit can look like an open or short circuit at different resonances and depending on whether they are connected in parallel or in series.

16.2 Smith Chart

In general, from (16.1.13) and (16.1.14), a length of transmission line can transform a load Z_L to a range of possible complex values $Z(-l)$. To understand this range of values better, we can use the Smith chart (thanks to the genius of Phillip H. Smith who invented it in 1939 before the advent of the computer) [122]. The Smith chart is essentially a graphical calculator for solving transmission line problems. It was used so much by microwave engineers during the early days that its use left a strong imprint and legacy on these engineers: it also has become an indispensable visual and mental aid for understanding and solving microwave engineering problems. This chart occupies an important place in the hearts and minds of many microwave engineers; many of them can manipulate and predict the outcome of a design visually or mentally in their minds.

Equation (16.1.12) indicates that there is a unique, one-to-one map between the normalized impedance $Z_n(z) = Z(z)/Z_0$ and reflection coefficient $\Gamma(z)$. In the normalized impedance form where $Z_n = Z/Z_0$, from (16.1.11) and (16.1.12)

$$\Gamma = \frac{Z_n - 1}{Z_n + 1}, \qquad Z_n = \frac{1+\Gamma}{1-\Gamma} \tag{16.2.1}$$

Equations in (16.2.1) are known as bilinear transforms in complex variables [94]: Conformal maps that map circles to circles. Such a map is shown in Figure 16.4, where lines on the right half of the complex Z_n plane are mapped to the circles on the complex Γ plane. Since straight lines on the complex Z_n plane are circles with infinite radii, they are mapped to circles on the complex Γ plane. The Smith chart shown on Figure 16.10 allows one to obtain the corresponding Γ given Z_n and vice versa as indicated in (16.2.1), but using a graphical calculator or the Smith chart. They can be done visually or mentally in many engineer's minds.

Notice that the imaginary axis on the complex Z_n plane maps to the circle of unit radius on the complex Γ plane. All points on the right-half plane are mapped to within the unit circle. The reason being that the right-half plane of the complex Z_n plane corresponds to passive impedances such that $R_n > 0$ that will absorb energy. Hence, by energy conservation, such an impedance load will have a reflection coefficient with an amplitude less than one, which are points within the unit circle as shown in Figure 16.4.

On the other hand, the left-half of the complex Z_n plane corresponds to impedances with negative resistances. These will be active elements that can generate energy, and hence, yielding $|\Gamma| > 1$; they correspond to points outside the unit circle on the complex Γ plane.

Another point to note is that points at infinity on the complex Z_n plane map to the point at $\Gamma = 1$ on the complex Γ plane, while the point zero on the complex Z_n plane maps to $\Gamma = -1$ on the complex Γ plane. These are the reflection coefficients of an open-circuit load and a short-circuit load, respectively. For a matched load, $Z_n = 1$ and $\Gamma = 0$ which maps to the zero point or the origin on the complex Γ plane implying no reflection. It looks like we are inundating you with a large amount of data here, and you may want to stop and contemplate these points a bit!

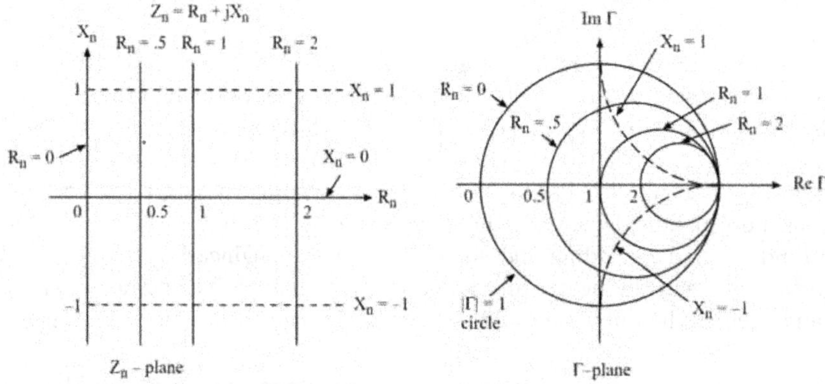

Figure 16.4: Bilinear map of the formula $\Gamma = \frac{Z_n - 1}{Z_n + 1}$, and $Z_n = \frac{1+\Gamma}{1-\Gamma}$. It maps circles to circles. The chart on the right, called the Smith chart, allows the values of Z_n to be determined quickly once given Γ, and vice versa.

The Smith chart also allows one to quickly evaluate the expression[2]

$$\Gamma(-l) = \Gamma_L e^{-2j\beta l} \tag{16.2.2}$$

and its corresponding Z_n, not by using (16.2.1) via a calculator, but by using a graphical calculator–the Smith chart. Since $\beta = 2\pi/\lambda$, it is more convenient to write $\beta l = 2\pi l/\lambda$ and measure the length of the transmission line in terms of wavelength. To this end, the above becomes

$$\Gamma(-l) = \Gamma_L e^{-4j\pi(l/\lambda)} \tag{16.2.3}$$

In the above, the phase of the general reflection coefficient is changing in the unit l/λ, or that our yardstick is wavelength here.[3] For increasing l, one moves away from the load to the generator (or source). As l increases, the phase is decreasing because of the negative sign. So given a point for Γ_L on the Smith chart, one has a negative phase or a decreasing phase by rotating the point clockwise. Also, due to the $\exp(-4j\pi l/\lambda)$ dependence of the phase, when $l = \lambda/4$, the reflection coefficient acquires a phase factor of $\exp(-j\pi)$, which rotates it a half circle around the chart, or to a point diametrically opposite on the Smith chart. And when $l = \lambda/2$, the reflection coefficient acquires a phase factor of $\exp(-j2\pi)$, which will rotate it full circle, or back to the original point.

Therefore, on the edge of the Smith chart, there are indications as to which direction one should rotate if one were to move toward the generator or toward the load. Again, there is a large amount of information here, but it is not rocket science. And it will be good if you can stop and contemplate a moment to wrap your head around these concepts.

---

[2]The factor of $2l$ in the exponent comes about because the wave has to travel a distance of $2l$ because the reflection coefficient is defined to be the ratio of the reflected wave to the incident wave at the location $z = -l$.

[3]Remember in electromagnetics, the yardstick is always the wavelength!

For two points diametrically opposite to each other on the Smith chart, Γ changes sign, and it can be shown easily from (16.2.1) that the normalized impedances are reciprocal of each other. Hence, the Smith chart can also be used to find the reciprocal of a complex number quickly. This is because if

$$\Gamma = \frac{Z_n - 1}{Z_n + 1} \tag{16.2.4}$$

and we replace $Z_n \to 1/Z_n$, we see that $\Gamma \to -\Gamma$, which is a point diametrically opposite on the Smith chart.

A full blown Smith chart is shown in Figure 16.10.

16.3 Voltage Standing Wave Ratio (VSWR)

From the previous section, one sees that the voltage and current are not constant on a transmission line. Therefore, one surmises that measuring the impedance of a device at microwave frequency is a tricky business. At low frequency, one can use an ohm meter with two wire probes to do such a measurement. But at microwave frequency, two pieces of wire become inductors, and two pieces of metal become capacitors. More sophisticated ways to measure the impedance are needed as described below.

Due to the interference between the forward-traveling wave and the backward-traveling wave, $V(z)$ is a function of position z on a terminated transmission line and it is given as

$$\begin{aligned} V(z) &= V_0 e^{-j\beta z} + V_0 e^{j\beta z} \Gamma_L \\ &= V_0 e^{-j\beta z} \left(1 + \Gamma_L e^{2j\beta z}\right) \\ &= V_0 e^{-j\beta z} \left(1 + \Gamma(z)\right) \end{aligned} \tag{16.3.1}$$

where we have used (16.1.6) for $\Gamma(z)$. Hence, $V(z)$ is not a constant but dependent on z, or

$$|V(z)| = |V_0||1 + \Gamma(z)| \tag{16.3.2}$$

For lack of a better name, this is called the standing wave, even though it is not truly a standing wave: it is the interference pattern formed by two traveling waves in opposite directions.

In Figure 16.5, the relationship between variations of $1 + \Gamma(z)$ as z varies is shown.

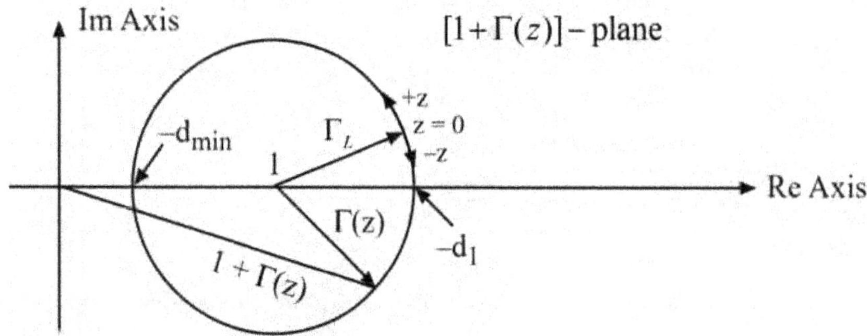

Figure 16.5: The voltage amplitude on a transmission line depends on $|V(z)|$, which is proportional to $|1+\Gamma(z)|$ per equation (16.3.2). This figure shows how $|1+\Gamma(z)|$ varies as z varies on a transmission line. Adding complex numbers on the complex plane is like adding vectors in 2D plane. $|1+\Gamma(z)|$ is the length of the "vector" that corresponds to the complex number $1+\Gamma(z)$.

Using the triangular inequality, one gets the lower and upper bounds, or

$$|V_0|(1-|\Gamma(z)|) \le |V(z)| \le |V_0|(1+|\Gamma(z)|) \tag{16.3.3}$$

But from (16.1.6) and that β is pure real for a lossless line, then $|\Gamma(z)| = |\Gamma_L|$; hence

$$V_{\text{min}} = |V_0|(1-|\Gamma_L|) \le |V(z)| \le |V_0|(1+|\Gamma_L|) = V_{\text{max}} \tag{16.3.4}$$

VSWR, or voltage standing wave ratio, is defined to be[4]

$$\text{VSWR} = \frac{V_{\text{max}}}{V_{\text{min}}} = \frac{1+|\Gamma_L|}{1-|\Gamma_L|} \tag{16.3.5}$$

Conversely, one can invert the above formula to get

$$|\Gamma_L| = \frac{\text{VSWR}-1}{\text{VSWR}+1} \tag{16.3.6}$$

Hence, the knowledge of voltage standing wave pattern (VSWP) or its VSWR, as shown in Figure 16.6, yields the knowledge of $|\Gamma_L|$, the amplitude of Γ_L. Notice that the relations between VSWR and $|\Gamma_L|$ are homomorphic to those between Z_n and Γ. Therefore, the Smith chart can also be used to evaluate the above equations.

[4]This word has no vowel, and you pronounce it like a word in Hebrew.

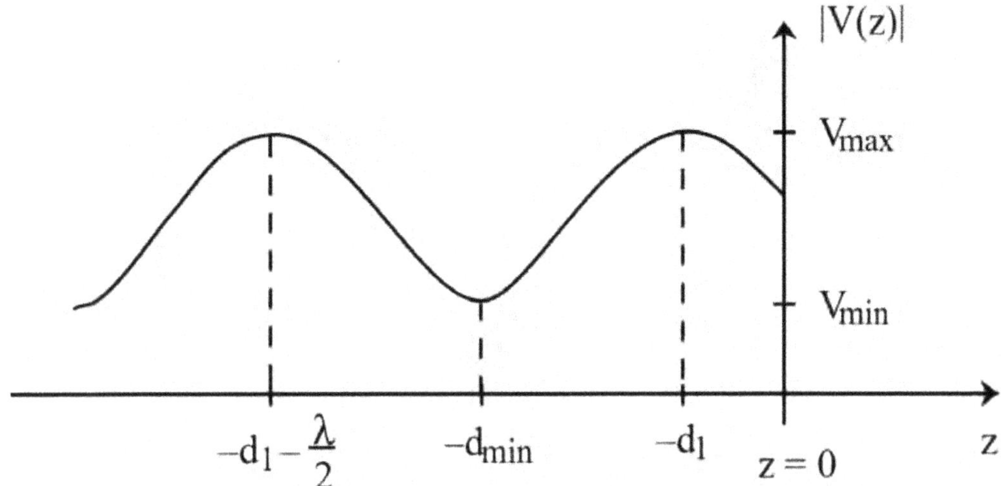

Figure 16.6: The voltage standing wave pattern (VSWP) as a function of z on a load-terminated transmission line.

The phase of Γ_L can also be determined from the measurement of the VSWP. The location of Γ_L in the complex Γ plane in Figure 16.5 is determined by the phase of Γ_L. Hence, the value of d_1 in Figure 16.5 is determined by the phase of Γ_L as well. The length of the transmission line waveguide needed to cancel the original phase of Γ_L to bring the voltage standing wave pattern to a maximum value at $z = -d_1$ is shown in Figure 16.6. Thus, d_1 is the value where the following equation is satisfied:

$$|\Gamma_L|e^{j\phi_L}e^{-4\pi j(d_1/\lambda)} = |\Gamma_L| \tag{16.3.7}$$

Therefore, by measuring the voltage standing wave pattern, one deduces both the amplitude and phase of Γ_L. From the complex value Γ_L, one can determine Z_L, the load impedance from the Smith chart.

In the old days, the voltage standing wave pattern was measured by a slotted-line instrument which consists of a coaxial waveguide with a slot opening as shown in Figure 16.7. A field probe can be inserted into the slotted-line to determine the strength of the electric field inside the coax waveguide. A typical experimental setup for a slotted line measurement is shown in Figure 16.8. A generator source, with low frequency modulation, feeds microwave energy into the coaxial waveguide. The isolator, allowing only the unidirectional propagation of microwave energy, protects the generator. The attenuator protects the slotted-line instrument. The wavemeter is an adjustable resonant cavity. When the wavemeter is tuned to the frequency of the microwave, it siphons off some energy from the source, giving rise to a dip in the signal of the SWR meter (a shorthand for voltage standing wave ratio meter). Hence, the wavemeter measures the frequency of the microwave.

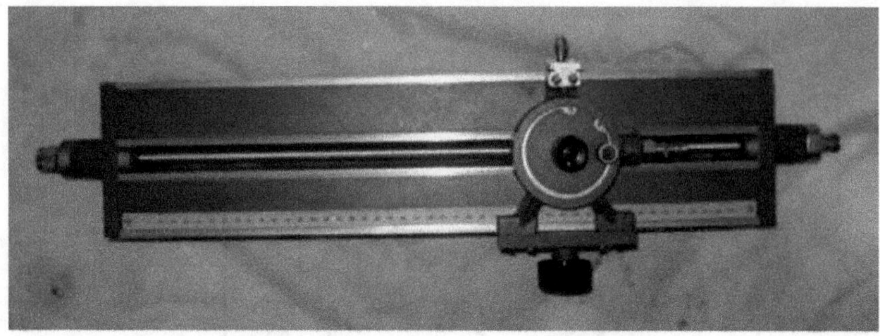

Figure 16.7: A slotted-line instrument which consists of a coaxial waveguide with a slot opening at the top to allow the measurement of the field strength, and hence, the voltage standing wave pattern in the waveguide (Figure by: Microwave101.com).

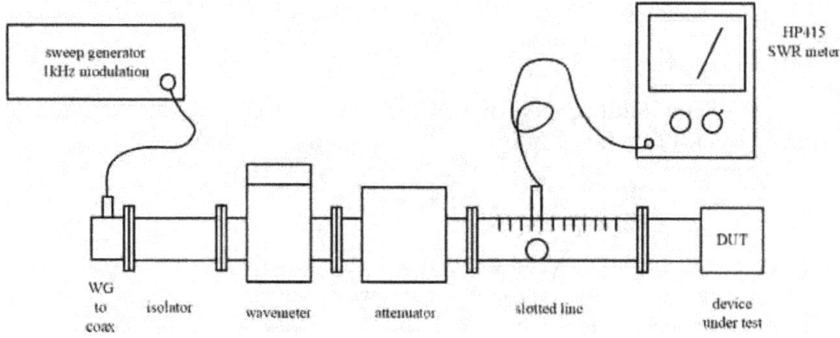

Figure 16.8: An experimental setup for a slotted-line measurement (Figure by: Pozar and Knapp, U. Mass [126]).

The slotted-line probe is usually connected to a square law detector with a rectifier that converts the microwave signal to a low-frequency signal. In this manner, the amplitude of the voltage in the slotted line can be measured with some low-frequency equipment, such as the SWR meter. Low-frequency equipment is a lot cheaper to make and maintain. That is also the reason why the source is modulated with a low-frequency signal. At low frequencies, circuit theory prevails, making engineering and design a lot simpler.

The above describes how the impedance of the device under test (DUT) can be measured at microwave frequencies. Nowadays, automated network analyzers (ANAs) make these measurements a lot simpler in a microwave laboratory. A picture of an ANA is shown in Figure 16.9. A schematic of how it works is shown in Figure 16.11. More resources on microwave measurements can be found on the web, such as in [127].

Notice that the above is based on the interference of the two traveling waves on a terminated transmission line. Such interference experiments are increasingly difficult in optical frequencies because of the much shorter wavelengths. As such, many experiments are easier to perform at microwave frequencies rather than at optical frequencies.

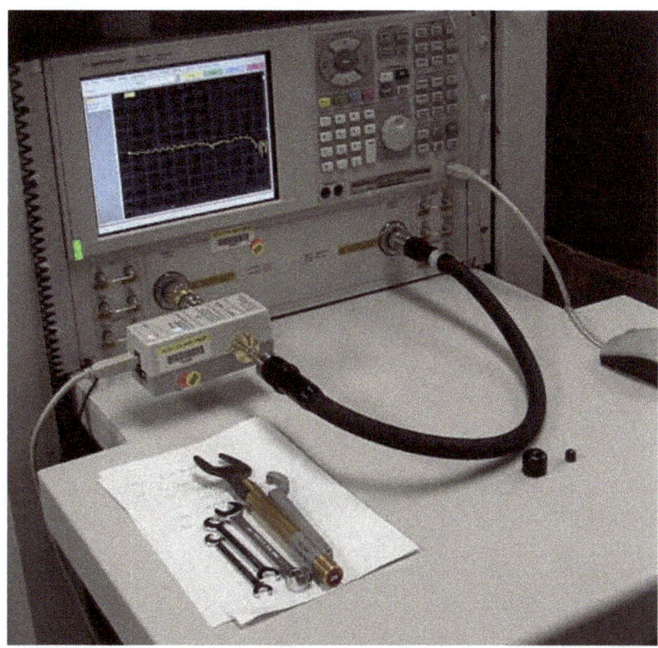

Figure 16.9: Microwave metrology technologies have progressed by leaps and bounds. I had to use a slotted line to measure the impedance of a device when I was a student. Now, a vector (measures both phase and amplitude) automated network analyzer is used and shown. It makes the measurements of microwave parameters a lot easier (Figure by: NIST).

Many electromagnetics technologies were first developed at microwave frequency, and later developed at optical frequency. Examples are phase imaging, optical coherence tomography, and beam steering with phase array sources. Another example is that quantum information and quantum computing can be done at optical frequency, but the recent trend is to use artificial atoms working at microwave frequencies. Engineering with longer wavelengths and larger components is easier; hence, microwave engineering is employed. For instance, the recent Sycamore quantum computer made by Google and the Chinese Zuchongzhi quantum computer use hoards of microwave engineering concepts [128, 129].[5] Another new frontier in the electromagnetic spectrum is in the

[5]Due to the humanly size of microwave components, microwave technologies have been used to test many new technologies such as MASER [130], metamaterials [131, 132], phase imaging [133], and various quantum technologies as cited above.

terahertz range. Due to the dearth of sources in the terahertz range, and the added difficulty in having to engineer with smaller components, this is an exciting and a largely untapped frontier in electromagnetic technology.

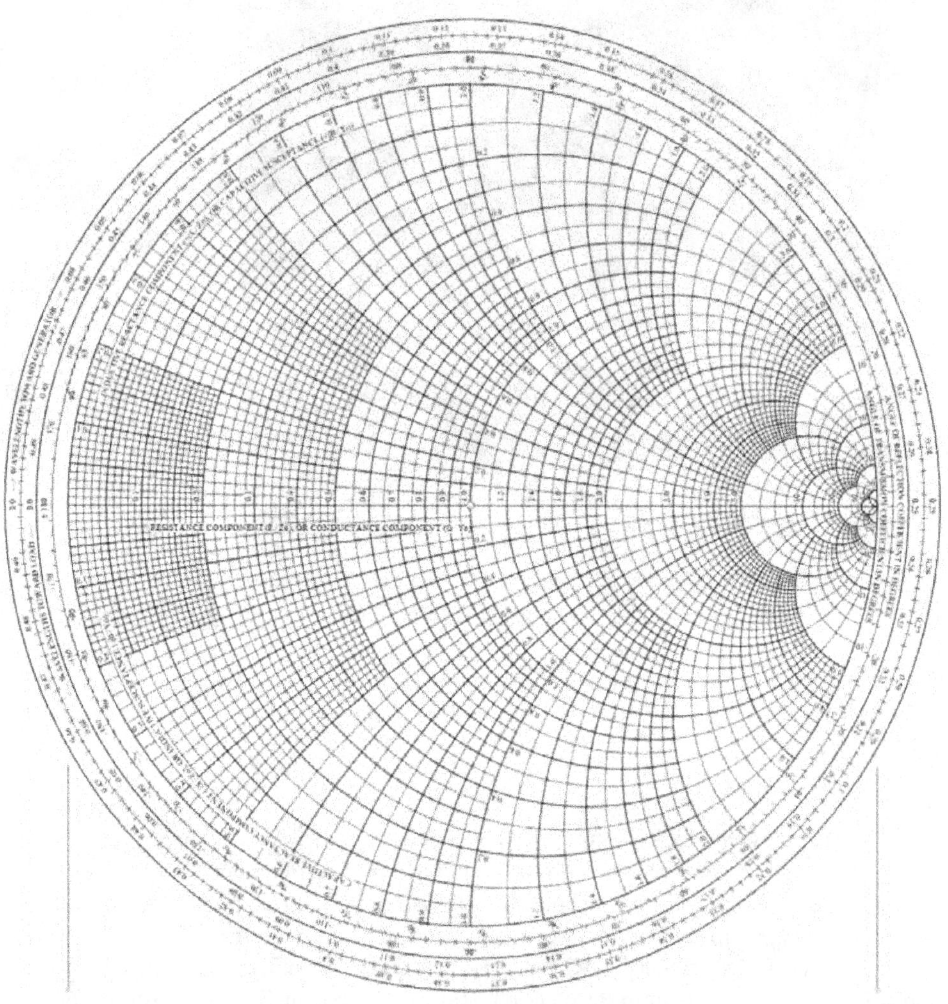

Figure 16.10: The Smith chart in its full glory. It was invented in 1939 before the age of digital computers, but it still allows microwave engineers to do mental estimations, gymnastics, and rough predictions with it. Because of its simplicity, it has an indelible influence on microwave engineering. (Figure by: https://en.wikipedia.org)

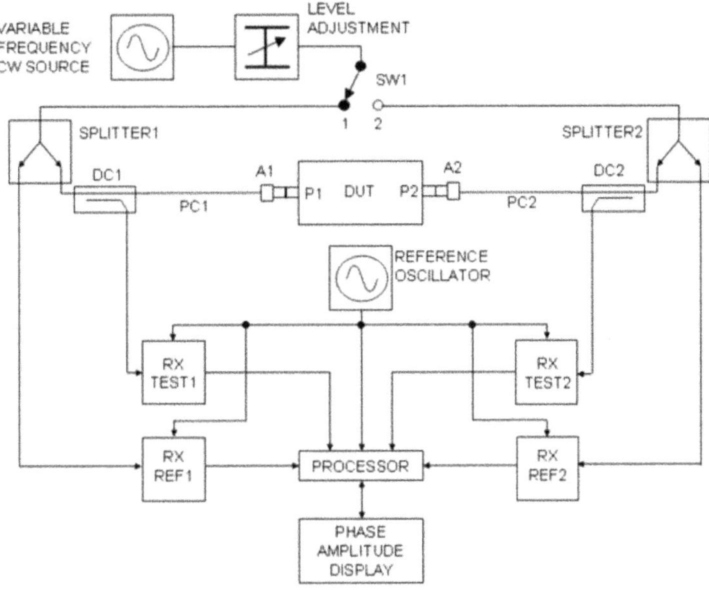

Figure 16.11: A schematic of the automated network analyzer. You can get the definitions of the acronyms from [134] (Figure by, en.wikipedia.org).

Exercises for Lecture 16

Problem 16-1: Look at lecture 8 of the notes from ECE350x:-
https://engineering.purdue.edu/wcchew/ece350/ee350-08.pdf

(i) Walk yourself through the example. For part (a), change the load to

$$Z_L = 20 + j30\ \Omega$$

and find the new answer.

(ii) For part (b), change

$$d_{\text{min}} = 3\lambda/16$$

and find the new answer.

Lecture 17

Multijunction Transmission Lines, Duality Principle

A simple extension of the transmission line theory of the previous lecture is to the case when transmission lines of different characteristic impedances and wave velocities are concatenated together. Myriads of devices can be designed using such admixtures of transmission lines, transistors, and diodes. Microwave engineering is a vibrant field because much of it can be described by transmission line theory, a poor man's Maxwell's equations. The wisdom of our predecessors is that the simpler the concepts are, the easier they can be engineered. The folklore is that when Maxwell completed his treatise in electricity and magnetism [114], few could read beyond the first fifty pages of his tome. It is through decades of regurgitation, distillation, and simplification that we can now teach this knowledge to undergraduate students. Also, many microwave components are thumb sized or hand sized, making the engineering of their wave physics components easier compared to radio waves and optics.

Therefore, due to the symmetry of Maxwell's equations between the electric field and the magnetic field, once a set of solutions has been found for Maxwell's equations, new solutions can be found by symmetry arguments. This is known as duality principle in electromagnetics. Recently, the use of symmetry of Maxwell's equations has given rise to the field of metamaterials. This field holds promise for new materials that can offer new physical phenomena [135, 136].

17.1 Multijunction Transmission Lines

The real world is usually more complex than the world of our textbooks. However, we need to distill problems in the real world into simpler problems so that we can explain them with our textbook examples. Figure 17.1 shows many real-world technologies, but they can be approximated with transmission line models, as shall be seen.

An area where multi-junction transmission lines play an important role is in the microwave integrated circuits (MICs) and the monolithic microwave integrated circuits (MMICs). An MMIC circuit is shown in Figure 17.2. Many of the components can be approximated by multijunction transmission lines. Thus, they are, a clear motivation for studying this topic here.

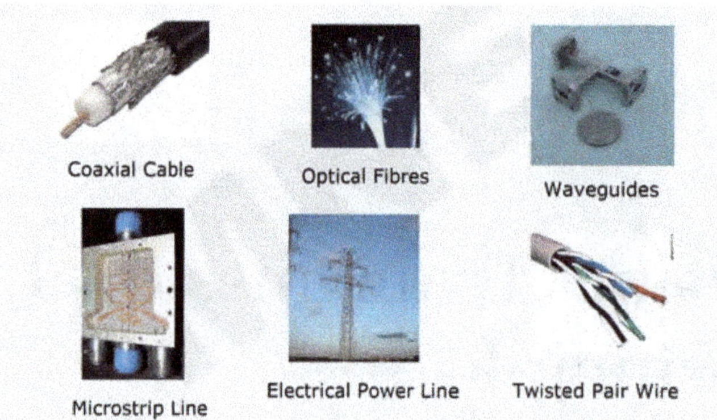

Figure 17.1: Different kinds of waveguides operating in different frequencies in power lines, RF circuits, microwave circuits, and optical fibers. Their salient physics or features can be captured or approximated by transmission lines (Figure by: Owen Casha).

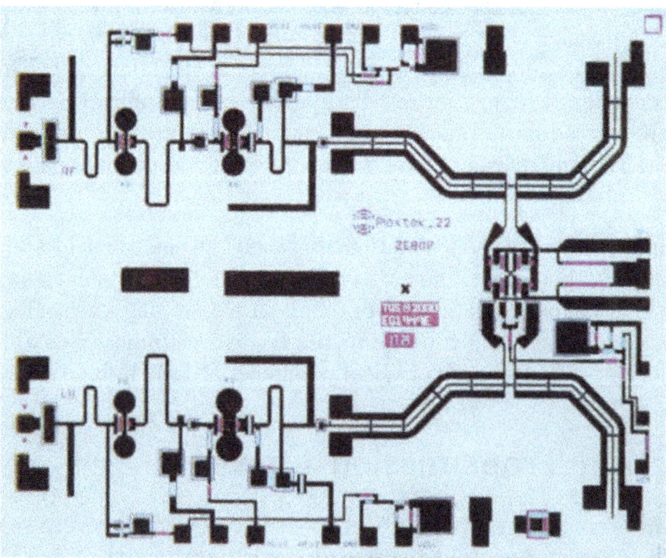

Figure 17.2: A generic GaAs MMIC. They are the motivation for studying multijunction transmission lines (Figure by: https://commons.wikimedia.org).

By concatenating sections of transmission lines of different characteristic impedances, a large variety of devices such as resonators, filters, radiators, and matching networks can be formed. We will start with a single-junction transmission line first. Good references for such a problem are the books by Collin [137] and Pozar [138], but much of the treatment here is not found in any textbook.

17.1.1 Single-Junction Transmission Lines

Consider two transmission lines connected at a single junction as shown in Figure 17.3. For simplicity, we assume that the transmission line to the right is infinitely long so that a right-traveling wave is not reflected; namely, there is no reflected wave. And we assume that the two transmission lines have different characteristic impedances, Z_{01} and Z_{02}. As such, due to the impedance mismatch at junction 1, a right-traveling wave in line 1 is reflected at junction 1.

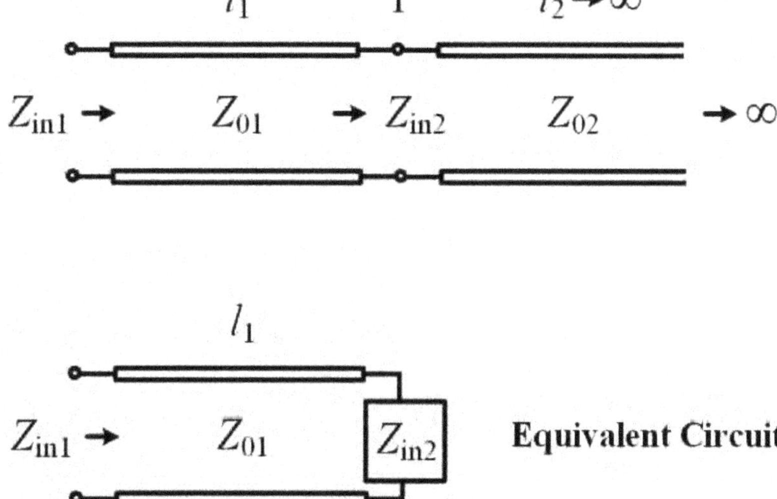

Figure 17.3: A single-junction transmission line can be modeled by an equivalent transmission line terminated in a load Z_{in2}.

The impedance of the transmission line at junction 1 looking to the right (see Figure 17.3), using the formula from previously derived,[1] is

$$Z_{in2} = Z_{02}\frac{1+\Gamma_{L,\infty}e^{-2j\beta_2 l_2}}{1-\Gamma_{L,\infty}e^{-2j\beta_2 l_2}} = Z_{02} \tag{17.1.1}$$

[1]We should always remember that the relations between the reflection coefficient Γ and the normalized impedance Z_n are $\Gamma = \frac{Z_n-1}{Z_n+1}$ and $Z_n = \frac{1+\Gamma}{1-\Gamma}$.

We have let $\Gamma_{L,\infty} = 0$ since no reflected wave exists on the last section. Thus, the above is just Z_{02}. As a result, transmission line 1 sees a load of $Z_L = Z_{in2} = Z_{02}$ hooked to its end. The equivalent circuit is shown in Figure 17.3 as well. Hence, we deduce that the reflection coefficient at junction 1 between line 1 and line 2, using the knowledge from the previous lecture, is Γ_{12}, and is given by

$$\Gamma_{12} = \frac{Z_L - Z_{01}}{Z_L + Z_{01}} = \frac{Z_{in2} - Z_{01}}{Z_{in2} + Z_{01}} = \frac{Z_{02} - Z_{01}}{Z_{02} + Z_{01}} \tag{17.1.2}$$

where we have assumed that $Z_L = Z_{in2} = Z_{02}$.

17.1.2 Two-Junction Transmission Lines—Generalized Reflection Coefficient[2]

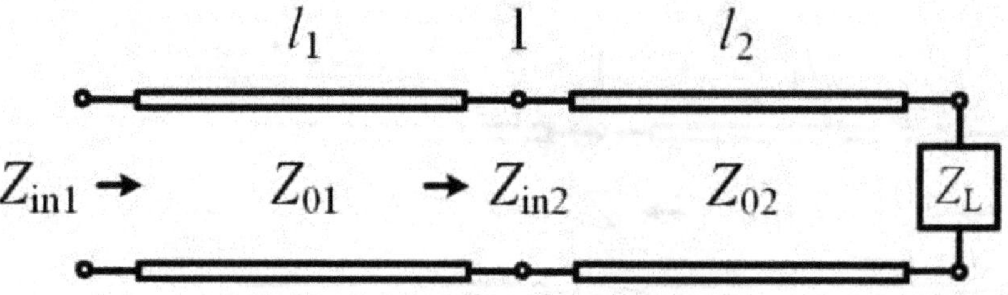

Figure 17.4: A two-junction transmission line can be modeled by a single-junction transmission line with a load. The last section (third section) is infinitely long and can be replaced with a load Z_L at the far (load) end of the second line. But it can be reduced to the equivalent circuit shown in the bottom of Figure 17.3.

Now, we look at the two-junction case. To this end, we first look at when line 2 is terminated by a load Z_L at its load end as shown in Figure 17.4. Then, using the formula derived in the previous lecture,

$$Z_{in2} = Z_{02}\frac{1 + \Gamma(-l_2)}{1 - \Gamma(-l_2)} = Z_{02}\frac{1 + \Gamma_{L2}e^{-2j\beta_2 l_2}}{1 - \Gamma_{L2}e^{-2j\beta_2 l_2}} \tag{17.1.3}$$

where we have used the fact that $\Gamma(-l_2) = \Gamma_{L2}e^{-2j\beta_2 l_2}$. It is to be noted that here, using knowledge from the previous lecture, the reflection coefficient at the load end of line 2 is

$$\Gamma_{L2} = \frac{Z_L - Z_{02}}{Z_L + Z_{02}} \tag{17.1.4}$$

[2]This term was used by me in [1]. It seems to be used by other authors later to mean other things, but I will continue to use this term in this course.

Now, line 1 sees a load of Z_{in2} hooked at its end. The equivalent circuit is the same as that shown in Figure 17.3. The generalized reflection coefficient at junction 1, which includes all the reflected waves from its right, is now[3]

$$\tilde{\Gamma}_{12} = \frac{Z_{in2} - Z_{01}}{Z_{in2} + Z_{01}} \tag{17.1.5}$$

To elaborate further, we substitute (17.1.3) into (17.1.5) to arrive at

$$\tilde{\Gamma}_{12} = \frac{Z_{02}(\frac{1+\Gamma}{1-\Gamma}) - Z_{01}}{Z_{02}(\frac{1+\Gamma}{1-\Gamma}) + Z_{01}} \tag{17.1.6}$$

where $\Gamma = \Gamma_{L2} e^{-2j\beta_2 l_2}$. The above can be rearranged to give

$$\tilde{\Gamma}_{12} = \frac{Z_{02}(1+\Gamma) - Z_{01}(1-\Gamma)}{Z_{02}(1+\Gamma) + Z_{01}(1-\Gamma)} \tag{17.1.7}$$

Finally, by further rearranging terms, after somewhat tedious algebra, it can be shown that the above becomes

$$\tilde{\Gamma}_{12} = \frac{\Gamma_{12} + \Gamma}{1 + \Gamma_{12}\Gamma} = \frac{\Gamma_{12} + \Gamma_{L2} e^{-2j\beta_2 l_2}}{1 + \Gamma_{12}\Gamma_{L2} e^{-2j\beta_2 l_2}} \tag{17.1.8}$$

where Γ_{12}, the local reflection coefficient at the junction between line 1 and line 2, is given by (17.1.2), and $\Gamma = \Gamma_{L2} e^{-2j\beta_2 l_2}$ is the local reflection coefficient[4] at $z = -l_2$ due to the load Z_L. In other words, referring to Figure 17.4, with knowledge of the local reflection coefficient Γ_{12} at the junction 1 in accordance to (17.1.2) and the reflection coefficient Γ_{L2} at the load end according to (17.1.4), one can use the above formula to find the generalized reflection coefficient $\tilde{\Gamma}_{12}$ at junction 1. The generalized reflection coefficient includes all reflections of the wave to the right of junction 1, while the local reflection coefficient Γ_{12} is due only to the impedance difference between line 1 and line 2 at junction 1. It is the locally reflected wave off the (1,2) junction.

17.1.3 Recursive Formula for Generalized Reflection Coefficient

Equation (17.1.8) is a *powerful formula* for multijunction transmission lines. Imagine now that we add another section of transmission line as shown in Figure 17.5. We can use the aforementioned method to first find $\tilde{\Gamma}_{23}$, the generalized reflection coefficient at junction 2. Using formula (17.1.8), it is given by

$$\tilde{\Gamma}_{23} = \frac{\Gamma_{23} + \Gamma_{L3} e^{-2j\beta_3 l_3}}{1 + \Gamma_{23}\Gamma_{L3} e^{-2j\beta_3 l_3}} \tag{17.1.9}$$

[3] I used to call this reflection coefficient "generalized reflection coefficient" [1], but this name seems to have been hijacked by some authors in the community for what we call "the general reflection coefficient" in this course.

[4] We will use the term "local reflection coefficient" at location z to mean the ratio between the amplitudes of the left-traveling wave and the right-traveling wave on a transmission line at z due to the discontinuity in the impedance at the (1,2) junction.

where Γ_{L3} is the load reflection coefficient due to the load Z_L hooked to the end of transmission line 3 as shown in Figure 17.5. Here, it is given as

$$\Gamma_{L3} = \frac{Z_L - Z_{03}}{Z_L + Z_{03}} \tag{17.1.10}$$

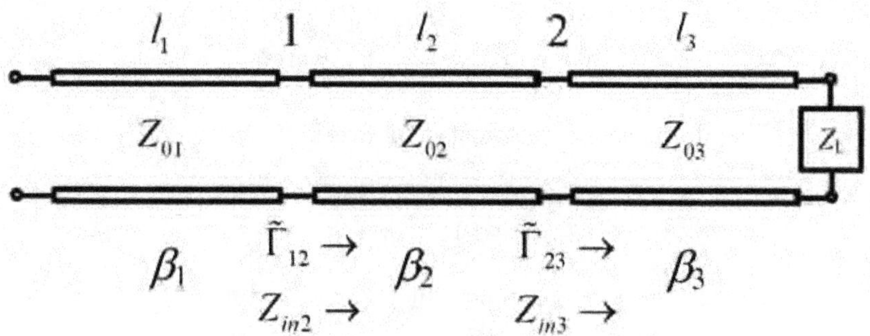

Figure 17.5: A two-junction transmission line with a load Z_L at the far end. The input impedance looking in from the far left can be found recursively using the formula (17.1.11) and (17.1.12).

Given the knowledge of $\tilde{\Gamma}_{23}$, we can use (17.1.8) again to find the new $\tilde{\Gamma}_{12}$ at junction 1. It is now

$$\tilde{\Gamma}_{12} = \frac{\Gamma_{12} + \tilde{\Gamma}_{23} e^{-2j\beta_2 l_2}}{1 + \Gamma_{12}\tilde{\Gamma}_{23} e^{-2j\beta_2 l_2}} \tag{17.1.11}$$

The equivalent circuit is again that shown in Figure 17.3. Therefore, we can use (17.1.11) recursively to find the generalized reflection coefficient for a multijunction transmission line. Once the reflection coefficient is known, the impedance at that location can also be found.[5] For instance, at junction 1, the impedance is now given by

$$Z_{in2} = Z_{01} \frac{1 + \tilde{\Gamma}_{12}}{1 - \tilde{\Gamma}_{12}} \tag{17.1.12}$$

instead of (17.1.3). In the above, Z_{01} is used because the generalized reflection coefficient $\tilde{\Gamma}_{12}$ is the total reflection coefficient for an incident wave from transmission line 1 that is sent toward the junction 1. Previously, Z_{02} was used in (17.1.3) because the reflection coefficients in that equation were for an incident wave sent from transmission line 2.

[5]We have found from the previous lecture that there is a one-to-one map between the reflection coefficient and the normalized impedance at location z.

If the incident wave were to have come from line 2, then one can write Z_{in2} as

$$Z_{in2} = Z_{02} \frac{1 + \tilde{\Gamma}_{23} e^{-2j\beta_2 l_2}}{1 - \tilde{\Gamma}_{23} e^{-2j\beta_2 l_2}} \tag{17.1.13}$$

With some algebraic manipulation, it can be shown that (17.1.12) and (17.1.13) are identical. Therefore, it is important to envision in our minds an incident wave and a reflected wave, and the line on which these waves are traveling. But (17.1.12) is closer to an experimental scenario where one measures the reflection coefficient by sending a wave from line 1 with no knowledge of what is to the right of junction 1.

Transmission lines can be made easily in microwave integrated circuits (MICs) by etching or milling. A picture of a microstrip line waveguide or transmission line is shown in Figure 17.6.

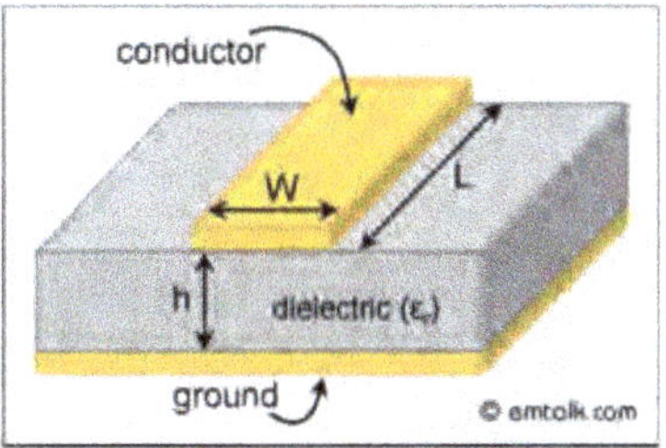

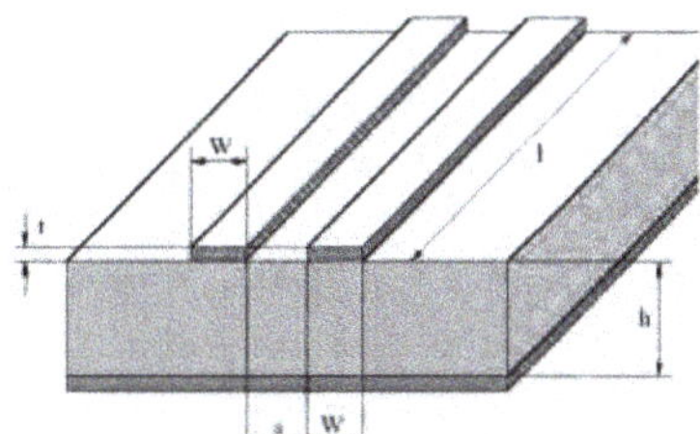

Figure 17.6: (Left) Schematic of a microstrip line with the signal line above and a ground plane below. (Right) A strip line with each strip carrying currents of opposite polarity. A ground plane is not needed in the second case. Unlike the coaxial transmission line, there is no closed-form solution for these geometries. Thanks to efficient computational electromagnetics (CEM) algorithms, numerical solutions to these problems exist (Figure by: emtalk.com, qucs.sourceforge.net).

17.1.4 Stray Capacitance and Inductance

The junction between two transmission lines is not as simple as we have assumed. In the real world, or in MIC, the waveguide junction has discontinuities in line width, or shape. This can give rise to excess charge accumulation or constricted current flow. Excess charge gives rise to an excess electric field that corresponds to excess electric stored energy. This can be modeled by a stray or parasitic capacitance C_s as shown in Figure 17.8.

Alternatively, there could be a constricted current flow that gives rise to stronger magnetic field.[6] Excess magnetic field compared to normal gives rise to excess magnetic stored energy. This can be modeled by stray or parasitic inductances L_{s1} and L_{s2}. Hence, a junction can be approximated by a circuit model as shown in Figure 17.8 to account for these effects. The Smith chart or the method we have outlined above can still be used to solve for the input impedances of a transmission circuit when these parasitic circuit elements are added.

[6]The magnetic field around a thinner wire is stronger than that around a thicker wire.

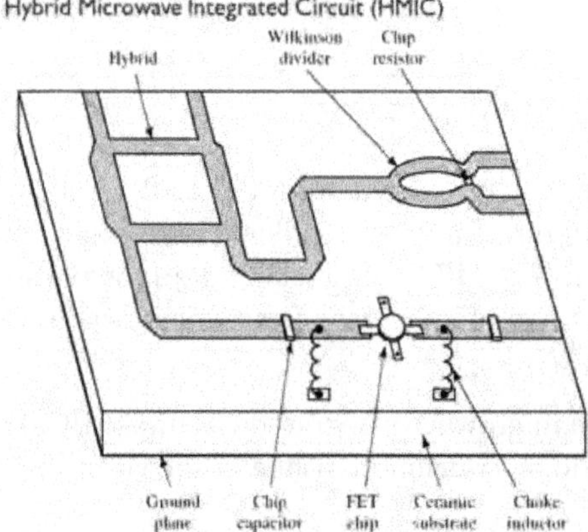

Figure 17.7: A general microwave integrated circuit with different kinds of elements (Figure by: Slideplayer.com).

Notice that when the frequency is zero or low, these stray capacitances and inductances are negligible. We retrieve the simple junction model. But since their impedance and admittance are $j\omega L_s$ and $j\omega C_s$, respectively, they are non-negligible and are instrumental in modeling high frequency circuits.

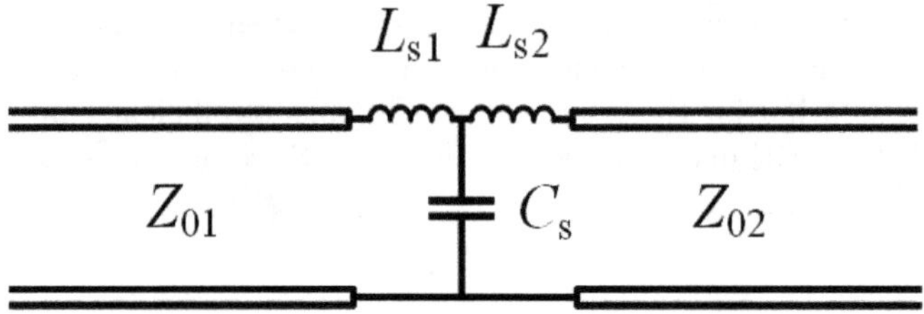

Figure 17.8: A junction between two microstrip lines can be modeled with a stray junction capacitance and stray inductances. The capacitance is used to account for excess charges at the junction, while the inductances model the excess current at the junction. They are important as the frequency increases.

17.1.5 Multiport Network

A section of transmission line can be thought of as a two-port network as in circuit theory, but with the difference that the inputs are the incident waves at each of these ports, and the outputs are the reflected waves at the ports. This concept can be generalized to a multi-port network with N ports easily. The inputs are the incident voltage waves (on a transmission line), and the outputs are the reflected voltage waves at each of the ports. Since the system is linear, in general, the inputs and outputs are linearly related by a scattering matrix $\overline{\mathbf{S}}$, or

$$\mathbf{V}^{-} = \overline{\mathbf{S}} \cdot \mathbf{V}^{+} \tag{17.1.14}$$

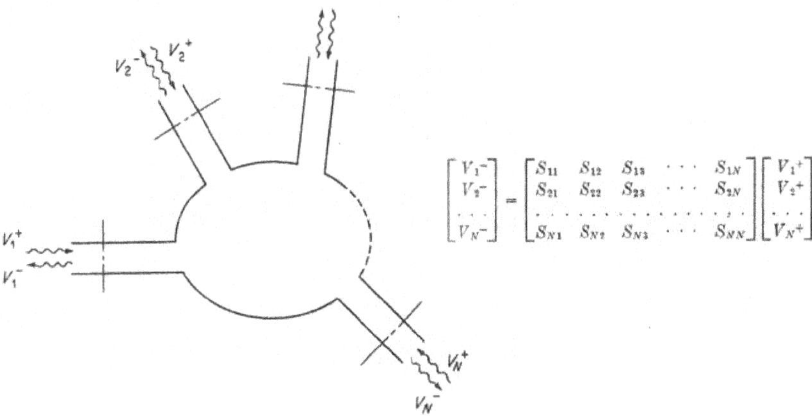

Figure 17.9: A multiport network can be characterized by its scattering matrix and the scattering parameters. Once known, microwave circuits can be concatenated together to form larger complex circuits. If the scattering parameters are known over a broad bandwidth, the behavior of the circuit in the time domain can also be obtained by Fourier transform (Figure by: Collin [104]).

The scattering matrix is a very useful and important microwave engineering concept. Here, $\mathbf{V}^+$ is a column vector containing all the incident voltage wave amplitudes $[V_1^+, V_2^+, \ldots, V_N^+]$, and similarly, $\mathbf{V}^-$, contains all the reflected voltage wave amplitudes (see Figure 17.9). Since the system is linear, they are related by the scattering matrix $\overline{\mathbf{S}}$. It encapulates or characterizes the properties of a complex microwave circuit with complex numbers called the scattering parameters. The scattering parameters can be measured or calculated. Once the scattering matrix of a microwave circuit is known, it can be concatenated with other microwave circuits similarly characterized.

17.2 Duality Principle

Duality principle exploits the inherent symmetry of Maxwell's equations. Once a set of $\mathbf{E}$ and $\mathbf{H}$ fields have been found to solve Maxwell's equations for a certain geometry, another set for a similar geometry can be found by invoking this principle. Maxwell's equations in the frequency domain, including the fictitious magnetic sources, are

$$\nabla \times \mathbf{E}(\mathbf{r},\omega) = -j\omega\mathbf{B}(\mathbf{r},\omega) - \mathbf{M}(\mathbf{r},\omega) \tag{17.2.1}$$
$$\nabla \times \mathbf{H}(\mathbf{r},\omega) = j\omega\mathbf{D}(\mathbf{r},\omega) + \mathbf{J}(\mathbf{r},\omega) \tag{17.2.2}$$
$$\nabla \cdot \mathbf{B}(\mathbf{r},\omega) = \varrho_m(\mathbf{r},\omega) \tag{17.2.3}$$
$$\nabla \cdot \mathbf{D}(\mathbf{r},\omega) = \varrho(\mathbf{r},\omega) \tag{17.2.4}$$

One way to make Maxwell's equations invariant is to perform the following substitutions (or transformations).

$$\mathbf{E} \rightarrow \mathbf{H}, \quad \mathbf{H} \rightarrow -\mathbf{E}, \quad \mathbf{D} \rightarrow \mathbf{B}, \quad \mathbf{B} \rightarrow -\mathbf{D} \tag{17.2.5}$$

$$\mathbf{M} \rightarrow -\mathbf{J}, \quad \mathbf{J} \rightarrow \mathbf{M}, \quad \varrho_m \rightarrow -\varrho, \quad \varrho \rightarrow \varrho_m \tag{17.2.6}$$

The above swaps retain the right-hand rule for plane waves. When material media is included, such that $\mathbf{D} = \overline{\varepsilon} \cdot \mathbf{E}$, $\mathbf{B} = \overline{\mu} \cdot \mathbf{H}$ for anisotropic media, Maxwell's equations become

$$\nabla \times \mathbf{E} = -j\omega\overline{\mu} \cdot \mathbf{H} - \mathbf{M} \tag{17.2.7}$$

$$\nabla \times \mathbf{H} = j\omega\overline{\varepsilon} \cdot \mathbf{E} + \mathbf{J} \tag{17.2.8}$$

$$\nabla \cdot \overline{\mu} \cdot \mathbf{H} = \varrho_m \tag{17.2.9}$$

$$\nabla \cdot \overline{\varepsilon} \cdot \mathbf{E} = \varrho \tag{17.2.10}$$

In addition to the above swaps, one needs further to swap the material parameters, namely,

$$\overline{\mu} \rightarrow \overline{\varepsilon}, \quad \overline{\varepsilon} \rightarrow \overline{\mu} \tag{17.2.11}$$

17.2.1 Unusual Swaps[7]

There are other swaps where seemingly the right-hand rule is not preserved (see Figure 17.10), for example,

$$\mathbf{E} \rightarrow \mathbf{H}, \mathbf{H} \rightarrow \mathbf{E}, \mathbf{M} \rightarrow -\mathbf{J}, \mathbf{J} \rightarrow -\mathbf{M}, \tag{17.2.12}$$
$$\varrho_m \rightarrow -\varrho, \varrho \rightarrow -\varrho_m, \overline{\mu} \rightarrow -\overline{\varepsilon}, \overline{\varepsilon} \rightarrow -\overline{\mu} \tag{17.2.13}$$

The above swaps will leave Maxwell's equations invariant, but when applied to a plane wave, the right-hand rule seems violated.

The deeper reason is that solutions to Maxwell's equations are not unique, since there is a time-forward as well as a time-reverse solution. In the frequency domain, this shows up in the choice of

[7]This section can be skipped on first reading.

the sign of the $\mathbf{k}$ vector[8] where in a plane wave $k = \pm\omega\sqrt{\mu\varepsilon}$. When one does a swap of $\mu \to -\varepsilon$ and $\varepsilon \to -\mu$, k is still indeterminate, and one can always choose a root where the right-hand rule is retained.

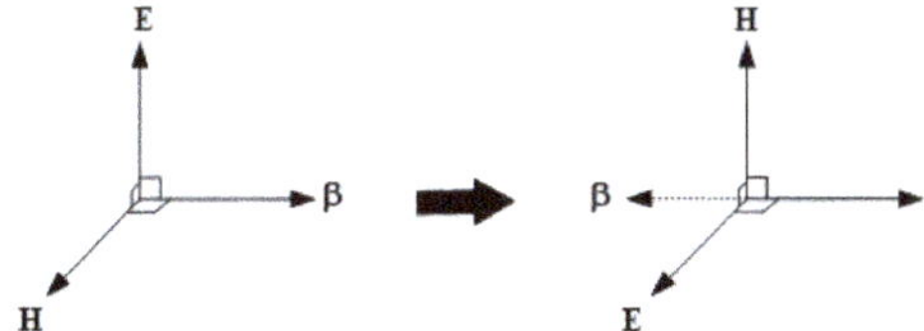

Figure 17.10: The unusual swap expounded in (17.2.12), though leaving Maxwell's equations unchanged, seems to disobey the right-hand rule for $\mathbf{E}$, $\mathbf{H}$, and $\boldsymbol{\beta}$. But one can change the sign of β (as it is not unique) to obey the right-hand rule again.

17.2.2 Left-Handed Materials and Double-Negative Materials

The above unusual swap reminds us of the double-negative (DNG) materials or left-handed materials (LHM) that have inspired some recent works in metamaterials in electromagnetics [135, 136]. Assuming a simple source-free homogeneous-medium case where we have let $\mu \to -\mu$ and $\varepsilon \to -\varepsilon$ to arrive at

$$\nabla \times \mathbf{E} = j\omega\mu\mathbf{H} \tag{17.2.14}$$

$$\nabla \times \mathbf{H} = -j\omega\varepsilon\mathbf{E} \tag{17.2.15}$$

If we further assume a plane-wave solution in the above and let the space dependence of the solution be $\exp(-j\boldsymbol{\beta}\cdot\mathbf{r})$, such a plane-wave solution obeys the left-hand rule rather than the right-hand rule. Hence, such a material, first proposed by Veselago [135] and later promulgated by Pendri [136], had been a hot topic of research. Since $\mathbf{E}(\mathbf{r}, t) = \Re e\left\{\mathbf{E}(\mathbf{r}, \omega)e^{-j\omega t}\right\}$, the above equations can also be obtained by letting $t \to -t$, or by letting $j \to -j$. The above can be thought of as a left-handed solution traveling forward in time, or a right-handed solution traveling backward in time.

17.3 Fictitious Magnetic Currents

Even though magnetic charges or monopoles do not exist, magnetic dipoles do. For instance, a magnet can be regarded as a magnetic dipole. Also, it is believed that electrons have spins, and these spins make electrons behave like tiny magnetic dipoles in the presence of a magnetic field.

Also, if we form electric current into a loop, it produces a magnetic field that looks like the electric field of an electric dipole.

[8] In this course, we use $\mathbf{k}$ and $\boldsymbol{\beta}$ interchangeably.

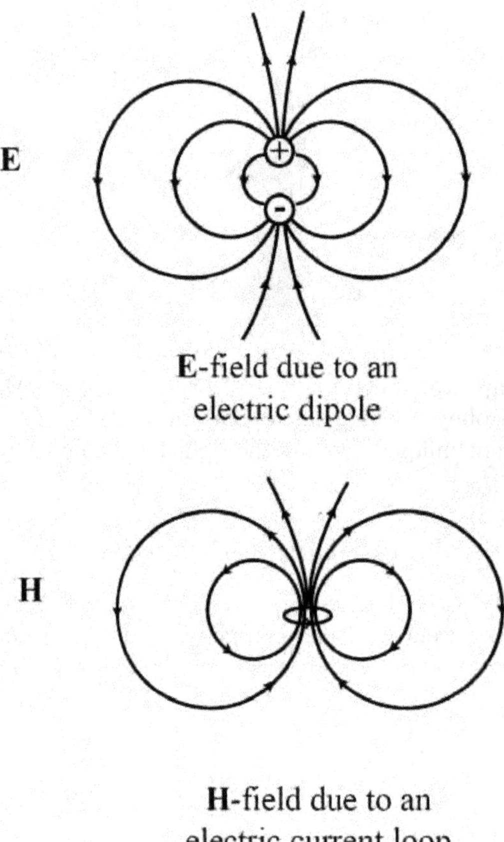

Figure 17.11: Sketches of the electric field **E** due to an electric dipole and the magnetic field **H** due to a electric current loop. The **E** and **H** fields have the same pattern and can be described by the same formula. Hence, the magnetic field of a current loop resembles that of a magnetic dipole. As such, a current loop is a good mimicry of a magnetic dipole.

This resembles a magnetic dipole field. Hence, a magnetic dipole can be "mimicked" using a small electric current loop (see Figure 17.11). The magnetic field external to the current loop is essentially that of a magnetic dipole. Because of these similarities, it is common to introduce fictitious magnetic charges and magnetic currents into Maxwell's equations. One can think that these magnetic charges always occur in pairs. Thus, they do not contradict the absence of magnetic monopole.

The electric current loops can be connected in series to make a toroidal antenna as shown in Figure 17.12. The toroidal antenna is used to drive a current in an electric dipole. Notice that the toroidal antenna resembles the primary winding of a transformer circuit. In essence, the toroidal loops, which mimic a magnetic current loop, produces an electric field that will drive current through the cylinder forming a long electric dipole. This is dual to the fact that an electric current loop produces a magnetic field. This is the working principle behind the measurement-while-drilling tool in the oil industry [139]. The entire drill stem inside a well bore for well logging can be used as an antenna. It also serves as a Goubau line [140].

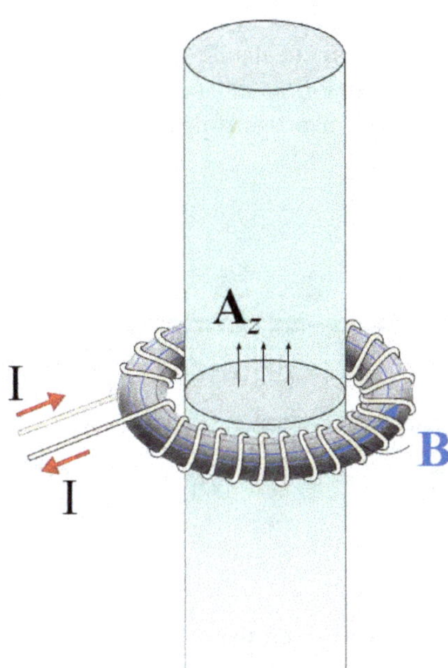

Figure 17.12: A toroidal antenna used to drive an electric current through a conducting cylinder of a dipole. It works similarly to a transformer: one can think of them as the primary and secondary turns (windings) of a transformer. Alternatively, one can think of the sequence of small electric current loops as forming a larger magnetic current loop. This larger magnetic current loop produces an electric field that drives an electric current through the conducting cylinder (Figure by Q. S. Liu [141]).

Exercises for Lecture 17

Problem 17-1: The multisection (or junction) transmission line is as shown in the figure below.

(i) Using the generalized reflection coefficient derived in class, find $\tilde{\Gamma}_{23}$ and Z_{in3}.

(ii) Then find $\tilde{\Gamma}_{12}$ and Z_{in2}.

(iii) What is the value of Z_{02} you can choose to have zero reflection at Junction 1? (Note: This problem can also be solved using the graphical calculator, the Smith chart, but the close-form formulas allow one to calculate the reflection coefficients and the impedances exactly. Part (iii) of this problem is that of a quarter-wave transformer matching, which can be found in many textbooks.)

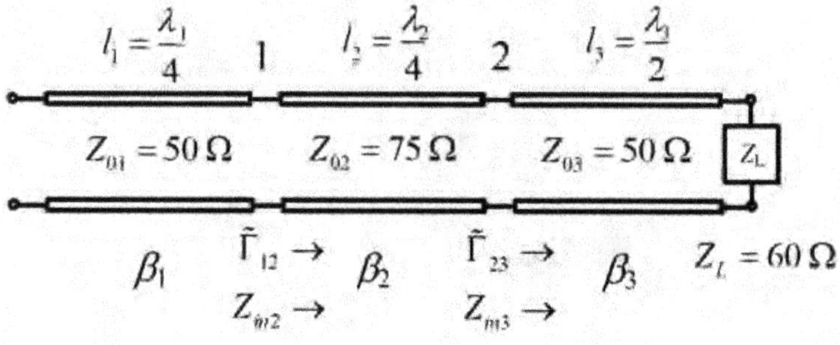

Figure 17.13: A transmission-line matching using quarter-wave transformer.

Lecture 18

Reflection, Transmission, Interesting Physical Phenomena

We have seen the derivation of a reflection coefficient in a transmission line that relates the amplitude of the reflected wave to that of the incident wave. By doing so, we have used a simplified form of Maxwell's equations, the telegrapher's equations, which are equations in one dimension. Here, we will solve Maxwell's equations in their full glory, but in order to do so, we will look at a very simple problem of plane-wave reflection and transmission at a single plane interface.

This study will result in the Fresnel reflection and transmission coefficients, and embedded in them are interesting physical phenomena as shall be shown. We shall elucidate the physics of these interesting phenomena as well. This is a rare example of a simple closed-form solution to Maxwell's equations that offers scientists physical insight to the vast number of electromagnetic solutions.[1] Even though this is only the tip of the iceberg, it offers physical insight into the interaction of a wave field with a simple medium or geometry.

(Much of the contents of this lecture can be found in Kong, and also the ECE 350X lecture notes. They can be found in many textbooks, even though the notations can be slightly different [34, 35, 37, 48, 55, 59, 70, 87, 92, 124].)

18.1 Reflection and Transmission–Single-Interface Case

We will derive the plane-wave reflection and transmission coefficients in closed form for the single interface case between two dielectric media. These reflection and transmission coefficients are the Fresnel reflection and transmission coefficients because they were first derived by Austin-Jean

[1]One notable point is that glass spectacles were made in China as early as the Eastern Han dynasty (AD 25–220) [142] and a law of refraction was known in the Islamic world around AD 900 [143].

Fresnel (1788-1827).[2]

The single-plane interface, plane-wave reflection, and transmission problem, with complicated mathematics, are "homomorphic" to the transmission line problem. Here, the complexity comes because we have to keep track of the 3D polarizations of the electromagnetic fields in this case, as well as finding a solution in 3D space. We shall learn later that mathematical "homomorphism" can be used to exploit the simplicity of transmission line theory in seeking the solutions to the multiple dielectric interface problems.

18.1.1 TE Polarization (Perpendicular or E Polarization)[3]

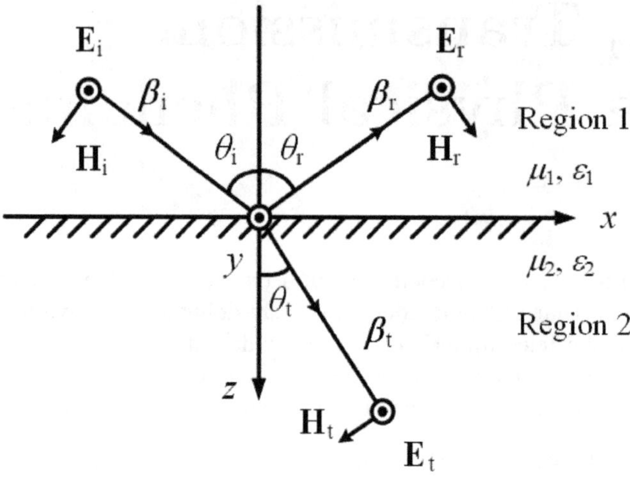

Figure 18.1: A schematic showing the reflection of the TE polarization wave impinging on a dielectric interface.

To set up the above problem, motivated by physical insight, the wave in region 1 can be written as the superposition or sum of the incident and reflected plane waves. Here, $\mathbf{E}_i(\mathbf{r})$ and $\mathbf{E}_r(\mathbf{r})$ are the incident and reflected plane waves, respectively, with different $\boldsymbol{\beta}$ vectors. The total field is $\mathbf{E}_i(\mathbf{r})+\mathbf{E}_r(\mathbf{r})$ which are the phasor representations of the fields. We assume a plane wave polarized in the y direction where the wave vectors are $\boldsymbol{\beta}_i = \hat{x}\beta_{ix} + \hat{z}\beta_{iz}$, $\boldsymbol{\beta}_r = \hat{x}\beta_{rx} - \hat{z}\beta_{rz}$, respectively for the incident, reflected waves. In region 2, we assume only a transmitted wave with wave vector given by $\boldsymbol{\beta}_t = \hat{x}\beta_{tx} + \hat{z}\beta_{tz}$, respectively for the transmitted waves. Then, from section 7.3, we have,

[2]Note that he lived before the completion of Maxwell's equations in 1865. But when Fresnel derived these coefficients in 1823, they were based on the elastic theory of light; hence, the formulas are not exactly the same as what we are going to derive (see Born and Wolf, *Principles of Optics*, p. 40 [63]).

[3]These polarizations are also variously know as TE_z, or the s and p polarizations, a descendent from the notations for acoustic waves where s and p stand for shear and pressure waves, respectively.

in full mathematical form, an incident plane wave given by

$$\mathbf{E}_i = \hat{y}E_0 e^{-j\boldsymbol{\beta}_i\cdot\mathbf{r}} = \hat{y}E_0 e^{-j\beta_{ix}x - j\beta_{iz}z} \tag{18.1.1}$$

which represents a uniform incident plane wave. For the reflected plane wave, we have

$$\mathbf{E}_r = \hat{y}R^{TE}E_0 e^{-j\boldsymbol{\beta}_r\cdot\mathbf{r}} = \hat{y}R^{TE}E_0 e^{-j\beta_{rx}x + j\beta_{rz}z} \tag{18.1.2}$$

which is a uniform reflected wave with R^{TE} being the Fresnel reflection coefficient. By the same token, the plane-wave solution in region 2 only has a transmitted plane wave; hence

$$\mathbf{E}_t = \hat{y}T^{TE}E_0 e^{-j\boldsymbol{\beta}_t\cdot\mathbf{r}} = \hat{y}T^{TE}E_0 e^{-j\beta_{tx}x - j\beta_{tz}z} \tag{18.1.3}$$

with T^{TE} being the Fresnel transmission coefficient. In the above, we assume that the incident wave is known and hence, E_0 is known. From (18.1.2) and (18.1.3), R^{TE} and T^{TE} are two unknowns yet to be sought.

To find them, we need two equations, and they can be found by imposing boundary conditions at the interface.[4] These boundary conditions are tangential **E** field continuous and tangential **H** field continuous, which become $\hat{n} \times \mathbf{E}$ continuous and $\hat{n} \times \mathbf{H}$ continuous conditions at the interface.

Imposing $\hat{n} \times \mathbf{E}$ continuous at $z = 0$, we get

$$E_0 e^{-j\beta_{ix}x} + R^{TE}E_0 e^{-j\beta_{rx}x} = T^{TE}E_0 e^{-j\beta_{tx}x}, \quad \forall x \tag{18.1.4}$$

where $\forall$ means "for all". In order for the above to be valid for all x, it is necessary that $\beta_{ix} = \beta_{rx} = \beta_{tx}$, which is also known as the phase-matching condition.[5] From the above, by letting $\beta_{ix} = \beta_{rx} = \beta_1 \sin\theta_i = \beta_1 \sin\theta_r$, we obtain that $\theta_r = \theta_i$ or that the law of reflection stating that the angle of reflection is equal to the angle of incidence.

By letting $\beta_{ix} = \beta_1 \sin\theta_i = \beta_{tx} = \beta_2 \sin\theta_t$, we obtain Snell's law of refraction that $\beta_1 \sin\theta_i = \beta_2 \sin\theta_t$. (This law of refraction was also known in the Islamic world in the 900 AD to Ibn Sahl [143].)

The exponential terms or the phase terms on both sides of (18.1.4) are the same. Now, canceling common terms on both sides of the equation (18.1.4), the above simplifies to

$$1 + R^{TE} = T^{TE} \tag{18.1.5}$$

Next, to impose $\hat{n} \times \mathbf{H}$ continuous at the interface, one needs to find the **H** field using $\nabla \times \mathbf{E} = -j\omega\mu\mathbf{H}$. Since these are plane waves, we can replace ∇ with $-j\boldsymbol{\beta}$. Then we have $\mathbf{H} = -j\boldsymbol{\beta} \times$

[4]Here, we will treat this problem as a boundary value problem where the unknowns are sought from equations obtained from boundary conditions.

[5]The phase-matching condition can also be proved by taking the Fourier transform of the equation with respect to x. Among the physics community, this is also known as momentum matching, as the wavenumber of a wave is related to the momentum of the particle. Remember that the momentum of a particle is proportional to $\hbar k$.

$\mathbf{E}/(-j\omega\mu) = \boldsymbol{\beta} \times \mathbf{E}/(\omega\mu)$. By so doing,[6]

$$\mathbf{H}_i = \frac{\boldsymbol{\beta}_i \times \mathbf{E}_i}{\omega\mu_1} = \frac{\boldsymbol{\beta}_i \times \hat{y}}{\omega\mu_1} E_0 e^{-j\boldsymbol{\beta}_i \cdot \mathbf{r}} = \frac{\hat{z}\beta_{ix} - \hat{x}\beta_{iz}}{\omega\mu_1} E_0 e^{-j\boldsymbol{\beta}_i \cdot \mathbf{r}} \tag{18.1.6}$$

$$\mathbf{H}_r = \frac{\boldsymbol{\beta}_r \times \mathbf{E}_r}{\omega\mu_1} = \frac{\boldsymbol{\beta}_r \times \hat{y}}{\omega\mu_1} R^{TE} E_0 e^{-j\boldsymbol{\beta}_r \cdot \mathbf{r}} = \frac{\hat{z}\beta_{rx} + \hat{x}\beta_{rz}}{\omega\mu_1} R^{TE} E_0 e^{-j\boldsymbol{\beta}_r \cdot \mathbf{r}} \tag{18.1.7}$$

$$\mathbf{H}_t = \frac{\boldsymbol{\beta}_t \times \mathbf{E}_t}{\omega\mu_2} = \frac{\boldsymbol{\beta}_t \times \hat{y}}{\omega\mu_2} T^{TE} E_0 e^{-j\boldsymbol{\beta}_t \cdot \mathbf{r}} = \frac{\hat{z}\beta_{tx} - \hat{x}\beta_{tz}}{\omega\mu_2} T^{TE} E_0 e^{-j\boldsymbol{\beta}_t \cdot \mathbf{r}} \tag{18.1.8}$$

Imposing $\hat{n} \times \mathbf{H}$ continuous or H_x continuous at $z = 0$, we have

$$-\frac{\beta_{iz}}{\omega\mu_1} E_0 e^{-j\beta_{ix}x} + \frac{\beta_{rz}}{\omega\mu_1} R^{TE} E_0 e^{-j\beta_{rx}x} = -\frac{\beta_{tz}}{\omega\mu_2} T^{TE} E_0 e^{-j\beta_{tx}x} \tag{18.1.9}$$

As mentioned before, the phase-matching condition requires that $\beta_{ix} = \beta_{rx} = \beta_{tx}$. The dispersion relation for plane waves requires that in their respective media,

$$\beta_{ix}^2 + \beta_{iz}^2 = \beta_{rx}^2 + \beta_{rz}^2 = \omega^2\mu_1\varepsilon_1 = \beta_1^2, \qquad \text{Region 1} \tag{18.1.10}$$

$$\beta_{tx}^2 + \beta_{tz}^2 = \omega^2\mu_2\varepsilon_2 = \beta_2^2, \qquad \text{Region 2} \tag{18.1.11}$$

Since

$$\beta_{ix} = \beta_{rx} = \beta_{tx} = \beta_x \tag{18.1.12}$$

because of phase matching. The above implies that, if β_{rx} and β_{ix} are in the same region, by subtracting (18.1.10) with β_x^2, then it becomes

$$\beta_{iz}^2 = \beta_{rz}^2 = \beta_1^2 - \beta_x^2 = \beta_{1z}^2 \tag{18.1.13}$$

The above implies that $\beta_{iz} = \beta_{rz} = \beta_{1z}$.

Moreover, since $\beta_1^2 \neq \beta_2^2$, from (18.1.10) and (18.1.11), subtracting (18.1.11) with β_x^2, it follows that $\beta_{tz}^2 = \beta_2^2 - \beta_x^2 = \beta_{2z}^2 \neq \beta_{1z}^2$ usually.

[6]Compared to transmission line theory, we note here that field theory is a lot more complicated and will drive you daffy. That is the reason for the triumph of transmission line theory as well. Thank goodness that we have transmission line theory and circuit theory, which can capture the major physics of electromagnetic theory!

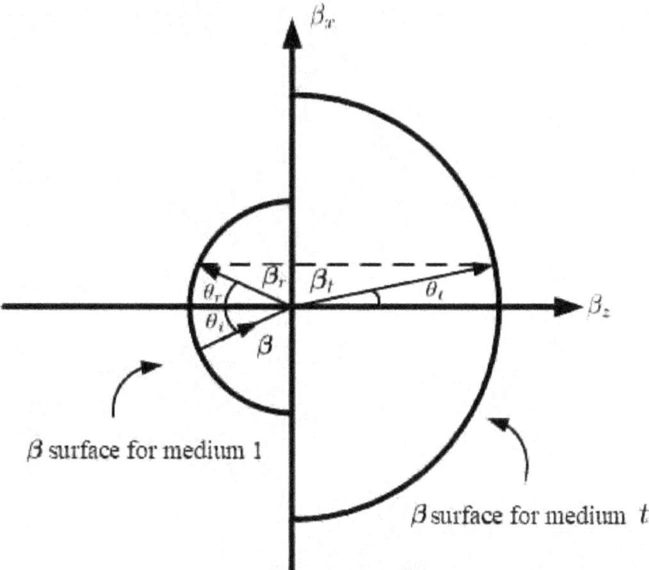

Figure 18.2: Wavenumber (β or k) surfaces in two regions showing the phase-matching condition. In this case, the wavenumber in medium t is larger (with larger $\boldsymbol{\beta}$ vector) than the wavenumber in medium 1. Then the wave vectors for the incident wave, reflected wave, and transmitted wave have to be aligned in such a way that their components parallel to the interface are equal because of the phase-matching condition. Also, one can see that Snell's law is satisfied as required by the phase-matching condition.

The above can be easily understood from the phase-matching diagram shown in Figure 18.2. In this figure, depending on the region that the $\boldsymbol{\beta}$ is in, the length of the $\boldsymbol{\beta}$ vector, which is $\sqrt{\beta_x^2 + \beta_z^2} = \beta = \omega\sqrt{\mu\varepsilon}$ changes from medium to medium. The circles in Figure 18.2 decide on the length of the $\boldsymbol{\beta}$ vectors. As shown in Figure 18.2, because of the dispersion relation that $\beta_{rx}^2 + \beta_{rz}^2 = \beta_{ix}^2 + \beta_{iz}^2 = \beta_1^2$, $\beta_{tx}^2 + \beta_{tz}^2 = \beta_2^2$, they are equations of two circles in 2D whose radii are β_1 and β_2, respectively. (The tips of the $\boldsymbol{\beta}$ vectors for regions 1 and 2 have to be on a spherical surface [also called β or k surface] in the β_x, β_y, and β_z) space in the general 3D case, but in this figure, we only show a cross section of the sphere assuming that $\beta_y = 0$.

Phase matching implies that the x-component of these $\boldsymbol{\beta}$ vectors are equal to each other as shown. One sees immediately that $\theta_i = \theta_r$ in Figure 18.2, and also as θ_i increases, θ_t increases. Then (18.1.9) simplifies to

$$\frac{\beta_{1z}}{\mu_1}\left(1 - R^{TE}\right) = \frac{\beta_{2z}}{\mu_2}T^{TE} \tag{18.1.14}$$

where $\beta_{1z} = \sqrt{\beta_1^2 - \beta_x^2}$, and $\beta_{2z} = \sqrt{\beta_2^2 - \beta_x^2}$.

Solving (18.1.5) and (18.1.14) for R^{TE} and T^{TE} yields the Fresnel coefficients to be[7]

$$R^{TE} = \left(\frac{\beta_{1z}}{\mu_1} - \frac{\beta_{2z}}{\mu_2}\right) \Big/ \left(\frac{\beta_{1z}}{\mu_1} + \frac{\beta_{2z}}{\mu_2}\right) \tag{18.1.15}$$

$$T^{TE} = 2\left(\frac{\beta_{1z}}{\mu_1}\right) \Big/ \left(\frac{\beta_{1z}}{\mu_1} + \frac{\beta_{2z}}{\mu_2}\right) \tag{18.1.16}$$

18.1.2 TM Polarization (Parallel or H Polarization)[8]

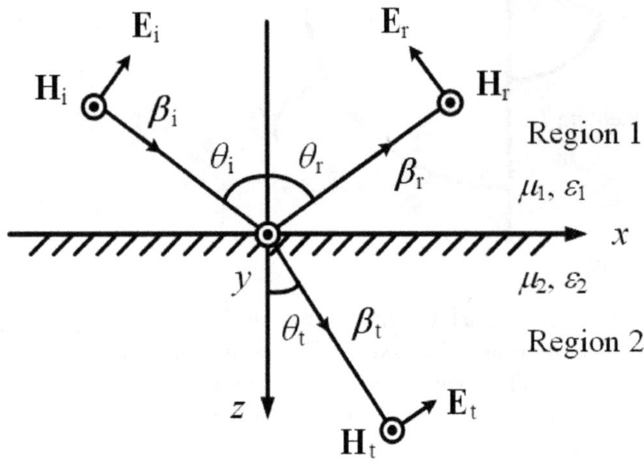

Figure 18.3: A similar schematic showing the reflection of the TM polarization wave impinging on a dielectric interface. The solution to this problem can be easily obtained from the solution for TE polarization by invoking the duality principle.

The solution to the TM polarization case can be obtained by invoking the duality principle where we do the substitution $\mathbf{E} \to \mathbf{H}$, $\mathbf{H} \to -\mathbf{E}$, and $\mu \rightleftharpoons \varepsilon$ as shown in Figure 18.3. The reflection coefficient for the TM magnetic field is then

$$R^{TM} = \left(\frac{\beta_{1z}}{\varepsilon_1} - \frac{\beta_{2z}}{\varepsilon_2}\right) \Big/ \left(\frac{\beta_{1z}}{\varepsilon_1} + \frac{\beta_{2z}}{\varepsilon_2}\right) \tag{18.1.17}$$

$$T^{TM} = 2\left(\frac{\beta_{1z}}{\varepsilon_1}\right) \Big/ \left(\frac{\beta_{1z}}{\varepsilon_1} + \frac{\beta_{2z}}{\varepsilon_2}\right) \tag{18.1.18}$$

Please remember now that R^{TM} and T^{TM} are reflection and transmission coefficients for the magnetic fields, whereas R^{TE} and T^{TE} are those for the electric fields. Some textbooks may define

[7]For mnemonics, we can also call these coefficients R_{12}^{TE} and T_{12}^{TE} with the understanding that the incident wave is in region 1 impinging in the region 1 and 2 interfaces.

[8]Also known as TM_z polarization.

these reflection coefficients based on electric field only, and they will look different, and duality principle cannot be applied.

18.1.3 Lens Optics and Ray Tracing

The Fresnel coefficients are derived for an infinitely flat surface. But when the wavelength is very short, a curved surface resembles a flat surface to the plane wave,[9] and the Fresnel coefficients can be used to estimate the reflected and transmitted waves. This is the fundamental principle behind lens optics.[10]

When the electromagnetic wave is described by a ray, the field is known also as ray optics. In this case, Maxwell's equations are not solved in their full glory, but approximately. The approximation is a very good one when the frequency is high and the wavelength is short. We will learn more in high-frequency methods later in the course. When the geometry is simple, solving the ray optics problem is similar to solving a geometry problem since one needs to know how the rays intersect with the geometry (see Figure 18.4). But when the geometry is highly complex, ray-tracing methods are used for tracking the light rays as they propagate through a complex environment. Computer codes have been written to do ray tracing. Ray tracing has been used to enhance the fidelity of a graphical picture in the movie industry. This has in turn fueled the growth of the graphics processing units (GPUs) in the computer industry [144].

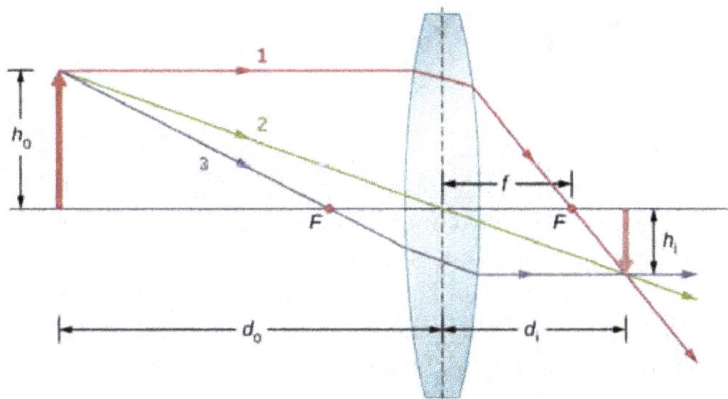

Figure 18.4: Ray tracing, based on Snell's law of refraction, can be used to solve many optics problems when the wavelength is small compared the to the size of the geometry (Figure by Steven Mellema, The Cosmic Universe).

[9]This is very much akin to the notion that the Earth is flat to people who do not venture outside the vicinity of their neighborhood.

[10]Lenses were made in China since the Han dynasty in ancient times [142].

18.2 Interesting Physical Phenomena

Three interesting physical phenomena emerge from the solutions of the single-interface problem. They are *total internal reflection*, *Brewster angle effect*, and *surface plasmonic resonance*. We will examine them next.

18.2.1 Total Internal Reflection

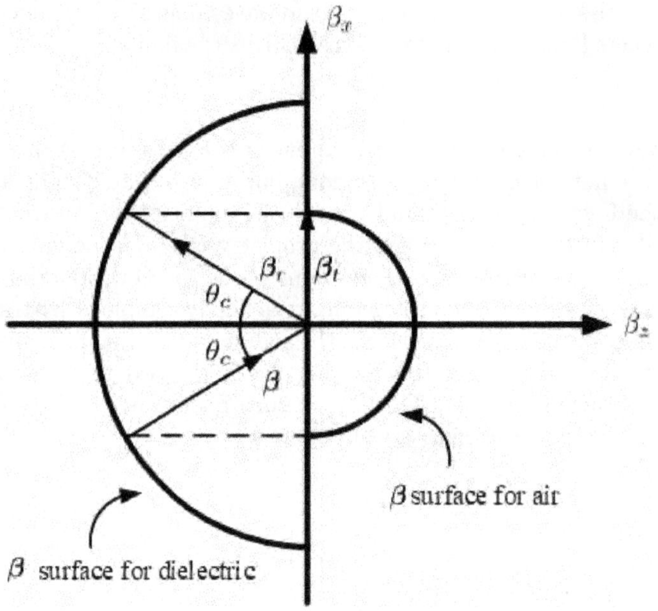

Figure 18.5: Phase-matching diagram for the total internal reflection case. Wavenumber surfaces in two regions show the phase-matching condition. In this case, the wavenumber in medium t is smaller than the wavenumber in medium 1. The figure shows an incident wave vector coming in at the critical angle. Then the transmitted wave vector is parallel to the interface as shown with $\beta_z = 0$. When the incident angle is larger than the critical angle, β_z becomes an imaginary number as the wave vector in region t is complex and cannot be drawn.

How total internal reflection can come about is also explainable by phase matching via the phase-matching diagram. This phase-matching condition can be illustrated using β-surfaces as shown in Figure 18.5. It turns out that because of the phase-matching condition, for interfaces where the left-hand side is optically more dense than the right-hand side, and for certain value of the incident angle, β_{2z} becomes pure imaginary.

For an optically less dense medium for which $\beta_2 < \beta_1$, according to the Snell's law of refraction, the transmitted β will refract away from the normal, as seen in the figure. Therefore, eventually the vector $\boldsymbol{\beta}_t$ becomes parallel to the x-axis when $\beta_{ix} = \beta_{rx} = \beta_2 = \omega\sqrt{\mu_2\varepsilon_2}$ and $\theta_t = \pi/2$. The incident angle at which this happens is termed the critical angle θ_c (see Figure 18.5).

At the critical angle, since $\beta_{ix} = \beta_1 \sin\theta_i = \beta_{rx} = \beta_1 \sin\theta_r = \beta_2$, or

$$\sin\theta_r = \sin\theta_i = \sin\theta_c = \frac{\beta_2}{\beta_1} = \frac{\sqrt{\mu_2\varepsilon_2}}{\sqrt{\mu_1\varepsilon_1}} = \frac{n_2}{n_1} \tag{18.2.1}$$

where n_1 is the refractive index defined as $c_0/v_i = \sqrt{\mu_i\varepsilon_i}/\sqrt{\mu_0\varepsilon_0}$ where v_i is the phase velocity of the wave in region i. Hence,

$$\theta_c = \sin^{-1}(n_2/n_1) \tag{18.2.2}$$

When $\theta_i > \theta_c$. $\beta_x > \beta_2$ and $\beta_{2z} = \sqrt{{\beta_2}^2 - {\beta_x}^2}$ becomes pure imaginary. When β_{2z} becomes pure imaginary, the wave cannot propagate in region 2, or $\beta_{2z} = -j\alpha_{2z}$, and the wave becomes evanescent. The physical reason for the decaying nature of the evanescent wave is quite different from that of a decaying wave in a lossy medium. The former is due to phase matching, and the need for the field to satisfy the boundary condition, while the latter is due to the loss of energy to the lossy medium. In a lossless medium, one can also show that the evanescent wave does not carry real power, but only reactive power using the complex Poynting's theorem.

The reflection coefficient (18.1.15) becomes of the form

$$R^{TE} = (A - jB)/(A + jB) \tag{18.2.3}$$

Since the numerator is the complex conjugate of the denominator. It is clear that $|R^{TE}| = 1$ always, and that $R^{TE} = e^{j\theta_{TE}}$. Therefore, a total internally reflected wave suffers a phase shift. A phase shift in the frequency domain corresponds to a time delay in the time domain. Such a time delay is achieved by the wave traveling laterally in region 2 before being refracted back to region 1. Such a lateral shift is called the Goos-Hänchen shift as shown in Figure 18.6 [63].[11] A wave that travels laterally along the surface of two media is also known as a lateral wave [145, 146].

(Please be reminded that total inter nal reflection comes about entirely due to the phase-matching condition when region 2 is a faster medium than region 1. Hence, it will occur with all manner of waves, such as elastic waves, sound waves, seismic waves, quantum waves, etc., and even waves in a cylindrical fiber.)

[11]You may be perplexed by our use of finite beam width of the plane wave for our physical argument. But you will learn later that a finite beam width can be approximated by a bundle of plane waves with similar wave numbers.

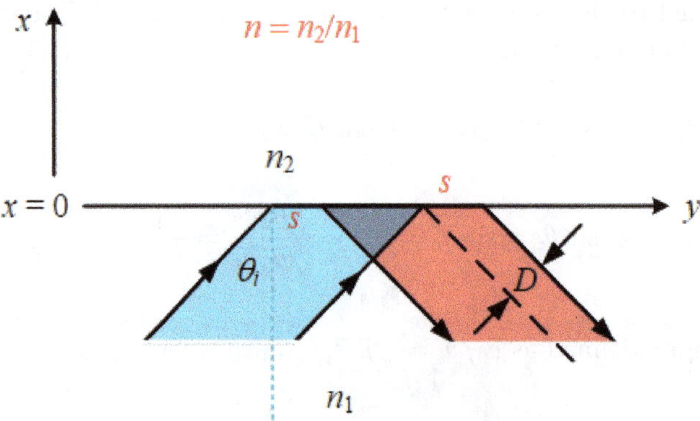

Figure 18.6: Goos-Hänchen Shift. A phase delay is equivalent to a time delay. In the time domain, one imagines a pulse traveling along a ray. When the ray hits the interface, it travels laterally on the interface for a distance s before being refracted back. The extra distance s traveled gives rise to the time delay (Figure by Paul R. Berman (2012), Scholarpedia [147]).

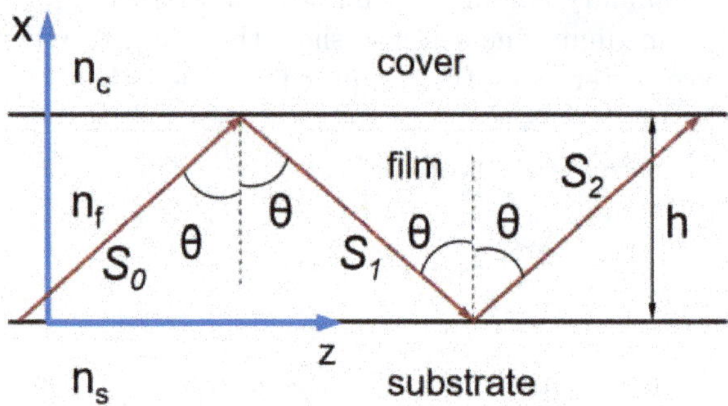

Figure 18.7: The total internal reflections (TIR) at the two interfaces of a thin-film waveguide can be used to guide an optical wave (Figure by E.N. Glytsis, NTUA, Greece [148]).

The guidance of a wave in a dielectric slab waveguide is due to total internal reflection at the dielectric-to-air interface. The wave bounces between the two interfaces of the slab and creates evanescent waves outside, as shown in Figure 18.7. The guidance of waves in an optical fiber works by a similar mechanism of total internal reflection, as shown in Figure 18.8. Optical light waves can provide carrier frequencies that are extremely high compared to radio or microwaves, giving rise to the tremendous bandwidth increase. Due to the tremendous impact the optical fiber has had on modern-day communications, Charles Kao, the father of the optical fiber, was awarded the Nobel Prize in 2009. His work was first published in [149].

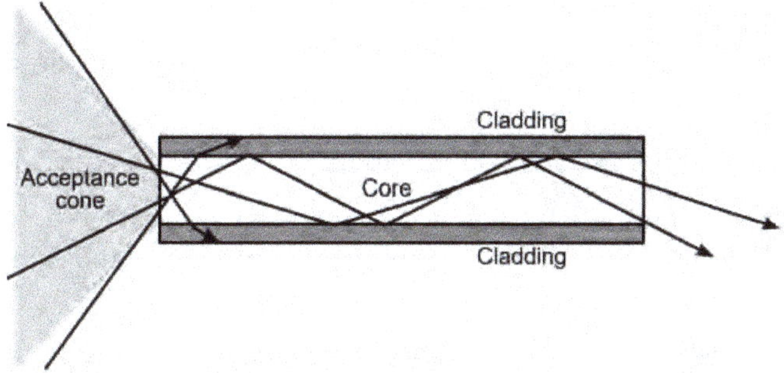

Figure 18.8: An optical fiber consists of a core and a cladding region. Total internal reflections (TIRs), due to phase-matching condition, can occur at the core-cladding interface. TIR can guide an optical wave in the fiber, which has a tremendous impact on the communication industry due to the low loss of the optical fiber (Figure by: https://www.researchgate.net [150]).

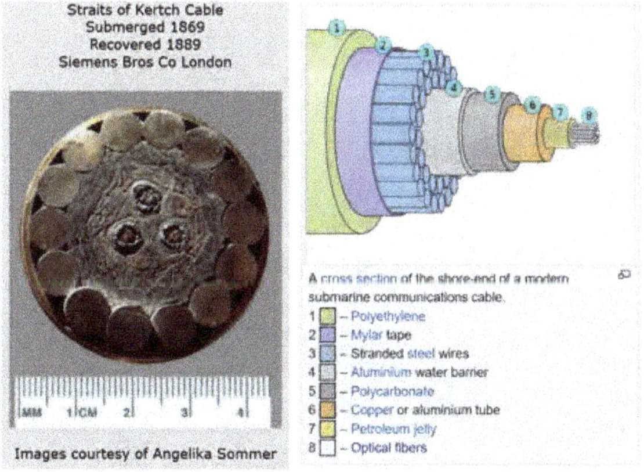

Figure 18.9: An old 1869 submarine cable made of coaxial cables (left) and modern submarine cable made of optical fibers (right) (Figure by Atlantic-Cable [151], https://en.wikipedia.org [152]). Comparing them to the optical fiber now, we have come a long way!

Waveguides have affected international communications for over a hundred years now. Since telegraphy was in place before the full advent of Maxwell's equations, submarine cables for global communications were laid as early as the 1850s. Figure 18.9 shows a submarine cable from 1869 using coaxial cables, compared to the one used in the modern world using optical fibers!

Exercises for Lecture 18

Problem 18-1:

(i) Derive the Fresnel reflection coefficients for TE and TM plane waves for the single interface problem.

(ii) Explain how Snell's law follows from the phase-matching condition.

(iii) Find the condition for total internal reflection and explain why the wave is evanescent on one side of the interface. Under what condition can total internal reflection occur, and in which medium is the wave evanescent?

(iv) For the lossless medium case, what happens to the amplitude of the reflection coefficient when total internal reflection occurs?

(v) Using the complex Poynting's theorem, show that the evanescent wave can only carry reactive power in a lossless medium.

Lecture 19

Brewster Angle, Surface Plasmon Polariton, Homomorphism with Transmission Lines

As simple as it looks, embedded in the TM Fresnel reflection coefficient are a few more interesting physical phenomena. We have looked at the physics of total internal reflection, which has inspired many interesting technologies such as waveguides, the most important of which is the optical fiber. In this lecture, we will look at other physical phenomena. These are the phenomena of Brewster's angle [153, 154] and of surface plasmon resonance, or polariton [155, 156].

Even though transmission line theory and the theory of plane-wave reflection and transmission look quite different, they are very similar in their underlying mathematical structures. For lack of a better name, we call this mathematical "homomorphism" (math analogy).[1] Later, to simplify the mathematics of waves in layered media, we will draw upon this mathematical "homomorphism" between multisection transmission line theory and plane-wave theory in layered media.

19.1 Brewster's Angle

First, we will continue with understanding some interesting phenomena associated with the single-interface problem starting with the Brewster's angle. Brewster's angle was discovered in 1815 [153, 154]. Furthermore, most materials at optical frequencies have $\varepsilon_2 \neq \varepsilon_1$, but $\mu_2 \approx \mu_1$. In other words, it is hard to obtain magnetic materials at optical frequencies or in optics. Therefore, even though the TE and TM polarizations are dual to each other, the TM polarization for light behaves differently from TE polarization. As such, we shall focus on the reflection and transmission of the

[1]The use of this term could be to the chagrin of a mathematician, but it has also been used in a subject called homomorphic encryption or computing [157].

TM polarization of light: The previously derived TM reflection coefficients are given here, namely,

$$R^{TM} = \left(\frac{\beta_{1z}}{\varepsilon_1} - \frac{\beta_{2z}}{\varepsilon_2}\right) \Big/ \left(\frac{\beta_{1z}}{\varepsilon_1} + \frac{\beta_{2z}}{\varepsilon_2}\right) \tag{19.1.1}$$

The transmission coefficient is easily gotten by the formula $T^{TM} = 1 + R^{TM}$.

Observe that for R^{TM}, it is possible that $R^{TM} = 0$ if

$$\varepsilon_2\beta_{1z} = \varepsilon_1\beta_{2z} \tag{19.1.2}$$

Squaring the above, making the note that $\beta_{iz} = \sqrt{\beta_i^2 - \beta_x^2}$, one gets[2]

$$\varepsilon_2{}^2(\beta_1{}^2 - \beta_x{}^2) = \varepsilon_1{}^2(\beta_2{}^2 - \beta_x{}^2) \tag{19.1.3}$$

where β_x is the same for all regions because of phase matching. Solving the above, assuming $\mu_1 = \mu_2 = \mu$, gives

$$\beta_x = \omega\sqrt{\mu}\sqrt{\frac{\varepsilon_1\varepsilon_2}{\varepsilon_1 + \varepsilon_2}} = \underbrace{\beta_1 \sin\theta_1}_{\beta_{1x}} = \underbrace{\beta_2 \sin\theta_2}_{\beta_{2x}} \tag{19.1.4}$$

The latter two equalities in the above come from phase matching. Therefore, at the Brewster's angle, by letting $\beta_i = \omega\sqrt{\mu\varepsilon_i}$ in the above, we get

$$\sin\theta_1 = \sqrt{\frac{\varepsilon_2}{\varepsilon_1 + \varepsilon_2}}, \quad \sin\theta_2 = \sqrt{\frac{\varepsilon_1}{\varepsilon_1 + \varepsilon_2}} \tag{19.1.5}$$

or squaring the above and adding them, we get

$$\sin^2\theta_1 + \sin^2\theta_2 = 1, \tag{19.1.6}$$

Then, assuming that θ_1 and θ_2 are less than $\pi/2$, and using the identity that $\cos^2\theta_1 + \sin^2\theta_1 = 1$, after subtracting this identity from the above, we infer that $\cos^2\theta_1 = \sin^2\theta_2$. It follows that

$$\sin\theta_2 = \cos\theta_1 \tag{19.1.7}$$

In other words, because $\sin\theta_2 = \cos(\frac{\pi}{2} - \theta_2)$, the above implies that

$$\theta_1 + \theta_2 = \pi/2 \tag{19.1.8}$$

as shown in Figure 19.1 where $\theta_b = \theta_1$. The above formula can be used to explain why at Brewster's angle, no light is reflected back to region 1. Figure 19.1 shows that the induced polarization dipoles in region 2 always have their axes aligned in the direction of a reflected wave. A dipole does not radiate along its axis, a fact that we will learn later in the course. But at this point, for your lack of knowledge, we can verify heuristically by field sketch and looking at the Poynting's vector. Therefore, these induced dipoles in region 2 do not radiate in the direction of the reflected wave.

[2]Notice that if $\varepsilon_2 = \varepsilon_1$, which is possible for the TE polarization where these will be for permeability for nonmagnetic materials, then the solution is $\beta_x = \infty$, which is an uninteresting solution.

(Notice that when the contrast is very weak meaning that $\varepsilon_1 \cong \varepsilon_2$, then $\theta_1 \cong \theta_2$. And from (19.1.8), we infer that $\theta_1 \cong \theta_2 \cong \pi/4$.)

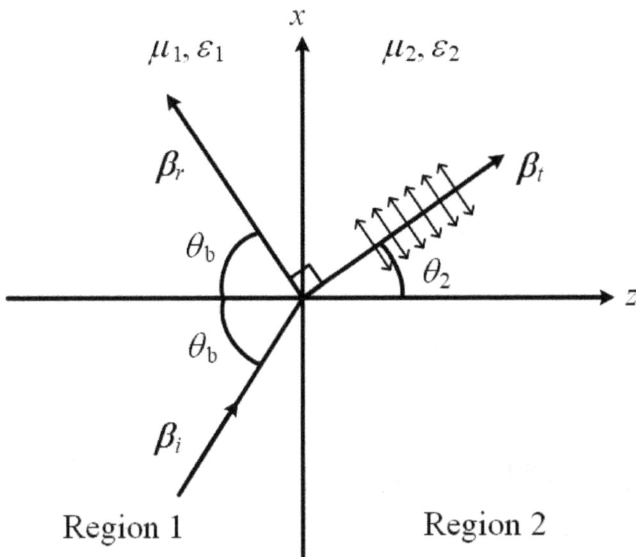

Figure 19.1: A figure showing a plane wave being reflected and transmitted at the Brewster's angle. In region t (or region 2), the transmitted field induces small dipoles to produce polarization density $\mathbf{P}$ with polarization current $j\omega\mathbf{P}$. The polarization current or dipoles are all pointing in the β_r direction, and hence, there is no radiation in that direction. In this figure, $\theta_b = \theta_1$, and hence, from (19.1.8), $\theta_b + \theta_2 = \pi/2$.

Because of the Brewster's angle effect for TM polarization when $\varepsilon_2 \neq \varepsilon_1$, $|R^{TM}|$ has to go through a null when $\theta_i = \theta_b$. Therefore in general, $|R^{TM}| \leq |R^{TE}|$ is as shown in the plots in Figure 19.2. Thus when a randomly (or arbitrarily) polarized light is incident on a surface, the polarization where the electric field is parallel to the surface (TE polarization) is reflected more than the polarization where the magnetic field is parallel to the surface (TM polarization). This phenomenon is used to design sunglasses to reduce road surface glare for drivers. For light reflected off a road surface, they are predominantly horizontally polarized with respect to the surface of the road. When sunglasses are made with vertical polarizers,[3] they will filter out or mitigate the reflected rays from the road surface to reduce road glare. This phenomenon can also be used to improve the quality of photography by using a polarizer filter as shown in Figure 19.3.

[3]Defined as one that will allow vertical polarization to pass through.

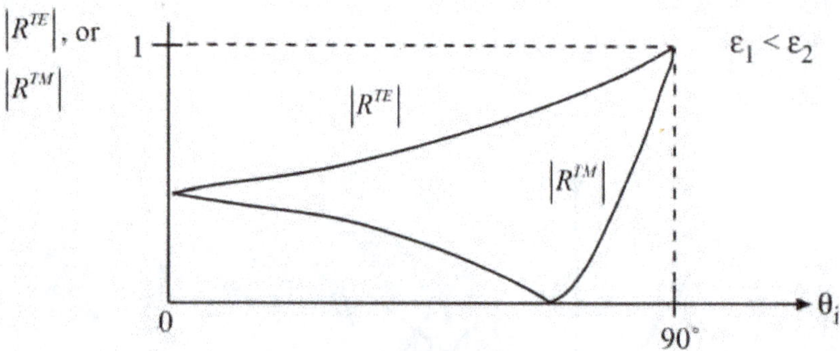

Figure 19.2: Because $|R^{TM}|$ has to go through a null when $\theta_i = \theta_b$; therefore, $|R^{TM}| \leq |R^{TE}|$ for all θ_i as shown in the above plots.

Figure 19.3: Because TM and TE lights will be reflected differently, polarizer filter can produce remarkable effects on the quality of the photograph by reducing glare (Figure by: https://physics.stackexchange.com/questions/264846/why-the-reflection-of-randomly-polarized-light-is-not-random) [154].

19.1.1 Surface Plasmon Polariton (SPP)

Surface plasmon polariton (SPP) occurs for the same mathematical reason as the Brewster's angle effect, but the physical mechanism is quite different. Many papers and textbooks will introduce this phenomenon from a different angle. But here, we will see it from the Fresnel reflection coefficient for the TM waves. When the denominator of the reflection coefficient R^{TM} is zero, it can become infinite. (This heralds the presence of some interesting physical phenomena, and in this case, a resonance behavior.) This is possible if $\varepsilon_2 < 0$, which is possible if medium 2 is a plasma medium. In this case, the criterion for the denominator to be zero is

$$-\varepsilon_2\beta_{1z} = \varepsilon_1\beta_{2z} \tag{19.1.9}$$

When R^{TM} becomes infinite, it implies that a reflected wave exists with no incident wave.[4] Hence, there is a plasmonic resonance or guided mode existing at the interface without the presence of an incident wave. It is a self-sustaining wave propagating in the x direction; and hence, it is a guided mode propagating in the x direction.

Solving (19.1.9) after squaring it, as in the Brewster's angle case, yields

$$\beta_x = \omega\sqrt{\mu}\sqrt{\frac{\varepsilon_1\varepsilon_2}{\varepsilon_1+\varepsilon_2}} \tag{19.1.10}$$

This is exactly the same equation for the Brewster's angle except now that ε_2 is negative.[5] Even though $\varepsilon_2 < 0$, but $\varepsilon_1 + \varepsilon_2 < 0$ is still possible, so that the expression under the square root sign in (19.1.10) is positive. Thus, β_x can be pure real. The corresponding β_{1z} and β_{2z} in (19.1.9) can be pure imaginary as explained below, and (19.1.9) can still be satisfied.

We shall show that the above corresponds to a guided wave propagating in the x direction. When this happens,

$$\beta_{1z} = \sqrt{{\beta_1}^2 - {\beta_x}^2} = \omega\sqrt{\mu}\left[\varepsilon_1\left(1 - \frac{\varepsilon_2}{\varepsilon_1+\varepsilon_2}\right)\right]^{1/2} = \omega\sqrt{\mu}\left[\frac{\varepsilon_1^2}{\varepsilon_1+\varepsilon_2}\right]^{1/2} \tag{19.1.11}$$

Since $\varepsilon_1 + \varepsilon_2 < 0$, it is seen that β_{1z} becomes pure imaginary. Moreover, $\beta_{2z} = \sqrt{{\beta_2}^2 - {\beta_x}^2}$ and ${\beta_2}^2 < 0$, making β_{2z} become an even larger imaginary number. This corresponds to a trapped wave (or a bound state) at the interface. The wave decays exponentially in both directions away from the interface and they are both evanescent waves.[6] The physical characteristic of this mode is shown in Figure 19.4, and is the only case in electromagnetics where a single interface can guide a surface wave, while such phenomenon abounds for elastic waves [158, 159].

When one operates close to the resonance of the mode so that the denominator in (19.1.10) is almost zero, then β_x can be very large. The wavelength in the x direction becomes very short in this case. And since $\beta_{iz} = \sqrt{\beta_i^2 - \beta_x^2}$, then β_{1z} and β_{2z} become even larger imaginary numbers. Thus the mode becomes tightly confined or bound to the interface, making the confinement of the mode very tight. This evanescent wave is much more rapidly decaying than that encountered in total internal reflection, at an interface between medium 1 and medium 2. There, $\beta_{2z} = \sqrt{\beta_2^2 - \beta_x^2}$ where β_x is no larger than β_1. On the other hand, because β_x from (19.1.10) can be very large when $\varepsilon_1 + \varepsilon_2 \approx 0$ close to plasmonic resonance; hence, (19.1.11) can be large imaginary causing the SPP mode is tightly confined to the interface. It portends its use in tightly packed optical components where cross-talk between components can be an issue. As such, this has stirred up some excitement in the optics community.

---

[4]In other words, a solution exists without the excitation term. This is often encountered in a resonance system like an LC tank circuit. Current flows in the tank circuit despite the absence of an exciting or driving voltage. In an ordinary differential equation or partial differential equation without a driving term (source term), such solutions are known as homogeneous solutions (to clarify the potpourri of math terms, homogeneous solutions here refer to solutions with zero source term). In a matrix equation $\overline{\mathbf{A}} \cdot \mathbf{x} = \mathbf{b}$ without a right-hand side or $\mathbf{b} = 0$, it is known as a null-space solution. These are all equivalent ways of saying a resonance solution.

[5]We see that it is often a dangerous proposition to square and square-root a function. We have to be guided by physical insight to see if we are finding a sane or insane solution!

[6]We have learned about evanescent waves in Lecture18. These waves do not carry real power.

An ordinary waveguide like a transmission line requires the exchange between stored electric energy and magnetic energy for the guided wave. But for this SPP waveguide, the magnetic field is conspicuously absent: it is the kinetic energy of the electrons in the plasma that provides the other stored energy component; and hence, the waveguide can be made small. The resonance is due to the exchange of the stored electric field energy and the kinetic energy of the electrons. Since the kinetic energy of the electrons replaces the stored energy of the magnetic field in an ordinary waveguide, this effect is also known as the kinetic inductance.

https://en.wikipedia.org/wiki/Surface_plasmon

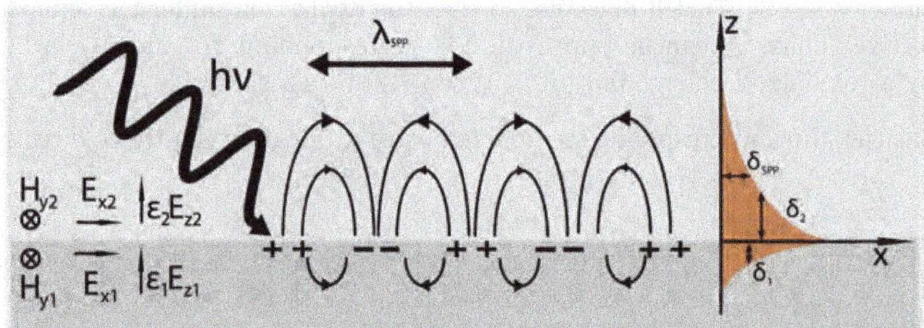

Figure 19.4: Figure showing a surface plasmonic mode propagating at an air-plasma interface. As in all resonant systems, a resonant mode entails the exchange of energies. In the case of surface plasmonic resonance, the energy is exchanged between the kinetic energy of the electrons (also known as kinetic inductance) and the energy stored in the electric field (Figure by: https://en.wikipedia.org [160]).

19.2 "Homomorphism" of Uniform Plane Waves and Transmission Lines Equations

Transmission line theory is very simple due to its one-dimensional nature. And the problem of reflection and transmission of plane waves at a planar interface is actually "homormophic" to that of the transmission line problem. Therefore, the problem of plane waves through layered media can be mapped into the multisection transmission line problem due to this mathematical "homomorphism" between the two problems. Hence, we can kill two birds with one stone: apply all the trappings of transmission line theory that we have learned to solve for the solutions of waves through layered medium problems.[7] Transmission line theory is simple and well loved by engineers.

For uniform plane waves, since they are proportional to $\exp(-j\boldsymbol{\beta}\cdot\mathbf{r})$, we know that with $\nabla \rightarrow -j\boldsymbol{\beta}$,

[7]This treatment is not found elsewhere and is unique to these lecture notes.

Maxwell's equations for a general isotropic homogeneous medium become

$$\boldsymbol{\beta} \times \mathbf{E} = \omega\mu\mathbf{H} \tag{19.2.1}$$

$$\boldsymbol{\beta} \times \mathbf{H} = -\omega\varepsilon\mathbf{E} \tag{19.2.2}$$

We will specialize these equations for different polarizations.

19.2.1 TE or TE$_z$ Waves

For this, one assumes a TE wave traveling in the z direction with an electric field polarized in the y direction, or $\mathbf{E} = \hat{y}E_y$. The corresponding magnetic field can be derived, and is seen in Figure 18.1, $\mathbf{H} = \hat{x}H_x + \hat{z}H_z$. Then we have from (19.2.1) that

$$\beta_z E_y = -\omega\mu H_x \tag{19.2.3}$$

$$\beta_x E_y = \omega\mu H_z \tag{19.2.4}$$

From (19.2.2), we deduce that

$$\beta_z H_x - \beta_x H_z = -\omega\varepsilon E_y \tag{19.2.5}$$

The above equations involve three variables, E_y, H_x, and H_z. Our goal is to make these equations homomorphic to the telegrapher's equations. But there are only two variables in the telegrapher's equations which are V and I. To this end, we will eliminate one of the variables from the above three equations. Using (19.2.4), we can express H_z in terms of E_y. Then, we can show from (19.2.5) that

$$\begin{aligned}\beta_z H_x &= -\omega\varepsilon E_y + \beta_x H_z = -\omega\varepsilon E_y + \frac{\beta_x^2}{\omega\mu} E_y \\ &= -\omega\varepsilon(1 - \beta_x^2/\beta^2)E_y = -\omega\varepsilon\cos^2\theta E_y\end{aligned} \tag{19.2.6}$$

where $\beta_x = \beta\sin\theta$ has been used.

Since we still have a plane wave with $\exp(-j\beta_z z)$ dependence, (19.2.3) and (19.2.6) can be written to look like the telegrapher's equations by letting $-j\beta_z \to d/dz$. Thus, we have from (19.2.3) and (19.2.6) that

$$\frac{d}{dz}E_y = j\omega\mu H_x \tag{19.2.7}$$

$$\frac{d}{dz}H_x = j\omega\varepsilon\cos^2\theta E_y \tag{19.2.8}$$

If we let $E_y \to V$, $H_x \to -I$, $\mu \to L$, $\varepsilon\cos^2\theta \to C$, the above is exactly analogous to the telegrapher's equations. The equivalent characteristic impedance of these equations above is then

$$Z_0 = \sqrt{\frac{L}{C}} = \sqrt{\frac{\mu}{\varepsilon}\frac{1}{\cos\theta}} = \sqrt{\frac{\mu}{\varepsilon}\frac{\beta}{\beta_z}} = \frac{\omega\mu}{\beta_z} \tag{19.2.9}$$

The above $\omega\mu/\beta_z$ is also known as the wave impedance for a propagating plane wave with propagation direction or the $\boldsymbol{\beta}$ vector inclined with an angle θ respect to the z-axis. It is analogous to the characteristic impedance Z_0 of a transmission line. When $\theta = 0$, as can be shown, the wave impedance $\omega\mu/\beta_z$ becomes the intrinsic impedance of space (see section 7.3).

A two-region, single-interface reflection problem can then be mathematically mapped to a single-junction connecting two-transmission-lines problem discussed in section 17.1.1. The equivalent characteristic impedances of these two regions are then

$$Z_{01} = \frac{\omega\mu_1}{\beta_{1z}}, \qquad Z_{02} = \frac{\omega\mu_2}{\beta_{2z}} \tag{19.2.10}$$

We can use the above to find Γ_{12} as given by

$$\Gamma_{12} = \frac{Z_{02} - Z_{01}}{Z_{02} + Z_{01}} = \frac{(\mu_2/\beta_{2z}) - (\mu_1/\beta_{1z})}{(\mu_2/\beta_{2z}) + (\mu_1/\beta_{1z})} \tag{19.2.11}$$

The above is the same as the Fresnel reflection coefficient we have previously derived for TE waves or R^{TE} after some simple rearrangement, but they are derived differently here.

Assuming that we have a single-junction transmission line, one can define a transmission coefficient given by

$$T_{12} = 1 + \Gamma_{12} = \frac{2Z_{02}}{Z_{02} + Z_{01}} = \frac{2(\mu_2/\beta_{2z})}{(\mu_2/\beta_{2z}) + (\mu_1/\beta_{1z})} \tag{19.2.12}$$

The above is similar to the continuity of the voltage across the junction, which is the same as the continuity of the tangential electric field across the interface. It is also the same as the Fresnel transmission coefficient T^{TE}.

19.2.2 TM or TM$_z$ Waves

For TM polarization, by invoking duality principle, the corresponding equations are, from (19.2.7) and (19.2.8),

$$\frac{d}{dz} H_y = -j\omega\varepsilon E_x \tag{19.2.13}$$

$$\frac{d}{dz} E_x = -j\omega\mu \cos^2\theta H_y \tag{19.2.14}$$

Just for consistency of units, since electric field is in V m^{-1} and magnetic field is in A m$^{-1,}$ we may chose the following map to convert the above into the telegrapher's equations, namely;

$$E_y \to V, \quad H_y \to I, \quad \mu\cos^2\theta \to L, \quad \varepsilon \to C \tag{19.2.15}$$

Then, the equivalent characteristic impedance is now

$$Z_0 = \sqrt{\frac{L}{C}} = \sqrt{\frac{\mu}{\varepsilon}} \cos\theta = \sqrt{\frac{\mu}{\varepsilon}} \frac{\beta_z}{\beta} = \frac{\beta_z}{\omega\varepsilon} \tag{19.2.16}$$

The above is also termed the wave impedance of a TM propagating wave making an inclined angle θ with respect to the z-axis. Notice that this wave impedance again becomes the intrinsic impedance of space when $\theta = 0$.

Now, using the reflection coefficient for a single-junction transmission line, and the appropriate characteristic impedances for the two lines as given in (19.2.16), we arrive at

$$\Gamma_{12} = \frac{(\beta_{2z}/\varepsilon_2) - (\beta_{1z}/\varepsilon_1)}{(\beta_{2z}/\varepsilon_2) + (\beta_{1z}/\varepsilon_1)} \tag{19.2.17}$$

Notice that (19.2.17) has a sign difference from the definition of R^{TM} derived earlier in the last lecture. The reason is that R^{TM} defined previously is for the reflection coefficient of magnetic field while Γ_{12} above is for the reflection coefficient of the voltage or the electric field. This difference is also seen in the definition for transmission coefficients.[8] A voltage transmission coefficient can be defined to be

$$T_{12} = 1 + \Gamma_{12} = \frac{2(\beta_{2z}/\varepsilon_2)}{(\beta_{2z}/\varepsilon_2) + (\beta_{1z}/\varepsilon_1)} \tag{19.2.18}$$

But this will be the transmission coefficient for the voltage, which is not the same as T^{TM} which is the transmission coefficient for the magnetic field or the current. Different textbooks may define different transmission coefficients for this polarization.

[8]This is often a source of confusion for these reflection and transmission coefficients.

Exercises for Lecture 19

Problem 19-1:

(i) Explain the criterion for the Brewster's angle, and explain why it is prevalent for TM waves.

(ii) Explain heuristically why an electric dipole does not radiate in its axial directions, or along the axis of the dipole. Hint: Sketch the electric and magnetic fields in the vicinity of an electric dipole and look at the Poynting's vector.

(iii) Explain why the condition for surface plasmonic polariton (SPP) is the same as the condition for the Brewster's angle. How does one distinguish an SPP case from the Brewster's angle case?

(iv) Explain why the SPP mode is tightly confined to the interface, more so than the evanescent wave due to total internal reflection.

(v) Explain why the reflection and transmission from a single interface problem can be made "homomorphic" to a single-junction transmission line problem.

Lecture 20

Waves in Layered Media

Waves in layered media are an important topic in electromagnetics. Many media can be approximated by planarly layered media. For instance, the propagation of radio wave on the earth's surface was of interest and first tackled by Sommerfeld in 1909 [161]. Other researchers have worked on the half-space problem to layered-medium problem case [146, 162, 145, 163].

The earth can be approximated by planarly layered media to capture the important physics behind the wave propagation. As such, many geophysics problems can be understood by studying waves in layered media. Many microwave components are made by planarly layered structures or laminated structures such as microstrip and coplanar waveguides. Layered media are also important in optics: they can be used to make optical filters such as Fabry-Pérot filters. As technologies and fabrication techniques become better, there is an increasing need to understand the interaction of waves with layered structures.

20.1 Waves in Layered Media

20.1.1 Generalized Reflection Coefficient for Layered Media

Because of the "homomorphism" between the transmission line problem and the plane-wave reflection by interfaces, we will exploit the simplicity of the transmission line theory to arrive at formulas for plane-wave reflection by layered media. We can capitalize on using the multisection transmission line formulas for the generalized reflection coefficient, which is

$$\tilde{\Gamma}_{12} = \frac{\Gamma_{12} + \tilde{\Gamma}_{23} e^{-2j\beta_2 l_2}}{1 + \Gamma_{12}\tilde{\Gamma}_{23} e^{-2j\beta_2 l_2}} \tag{20.1.1}$$

In the above, Γ_{12} is the local reflection at the 1,2 junction, whereas $\tilde{\Gamma}_{ij}$ is the generalized reflection coefficient at the i, j interface; here, $\tilde{\Gamma}_{12}$ includes multiple reflections from behind the 1,2 junction. It can be used to study electromagnetic waves in layered media, which are shown in Figures 20.1 and 20.2.

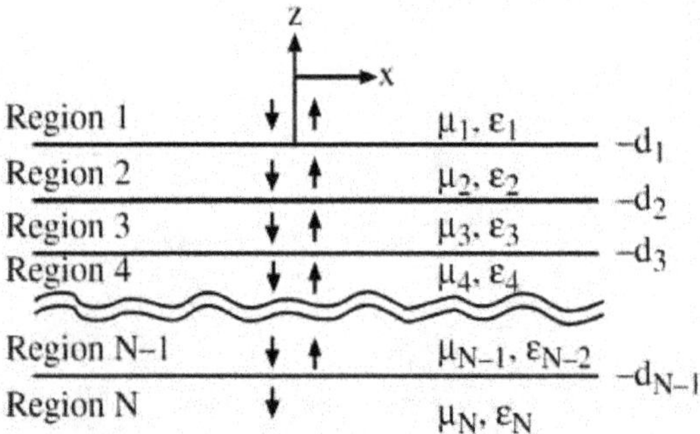

Figure 20.1: Schematic of waves in layered media. A wave entering the medium from above can be multiply reflected before emerging from the top again or transmitted to the bottommost medium.

Using the result from the multijunction transmission line, by analogy where characteristic impedances are equivalent to wave impedances only when the region is infinitely large, we can write down the generalized reflection coefficient for a layered medium with an incident wave at the 1,2 interface, including multiple reflections from behind the interface. In addition to the wave-impedance to characteristic-impedance replacements,[1] we do the following replacements: $\Gamma_{12} \rightarrow R_{12}$, $\tilde{\Gamma}_{23} \rightarrow \tilde{R}_{23}$, $\tilde{\Gamma}_{12} \rightarrow \tilde{R}_{12}$, and $\beta_2 \rightarrow \beta_{2z}$. Then we have

$$\tilde{R}_{12} = \frac{R_{12} + \tilde{R}_{23} e^{-2j\beta_{2z} l_2}}{1 + R_{12}\tilde{R}_{23} e^{-2j\beta_{2z} l_2}} \tag{20.1.2}$$

where R_{12} is the local Fresnel reflection coefficient and $\tilde{R}_{ij}$ is the generalized reflection coefficient at the i, j interface that includes reflections from behind the i, j interface. Here, l_2 is now the thickness of region 2. In the above, we assume that the wave is incident from medium (region) 1, which is semi-infinite, and the generalized reflection coefficient $\tilde{R}_{12}$ above is defined at the media 1 and 2 interface. It is assumed that there are multiple reflections coming from the right of the 2,3 interface, so that the 2,3 reflection coefficient is the generalized reflection coefficient $\tilde{R}_{23}$.

Figure 20.2 shows the case of a normally incident wave entering a layered medium. For this case, the wave impedance, defined in the previous lecture, becomes the intrinsic impedance of the homogeneous medium filling the whole space.

[1]Remember from the previous lecture that the wave-impedance formulas are different for TM and TE polarizations.

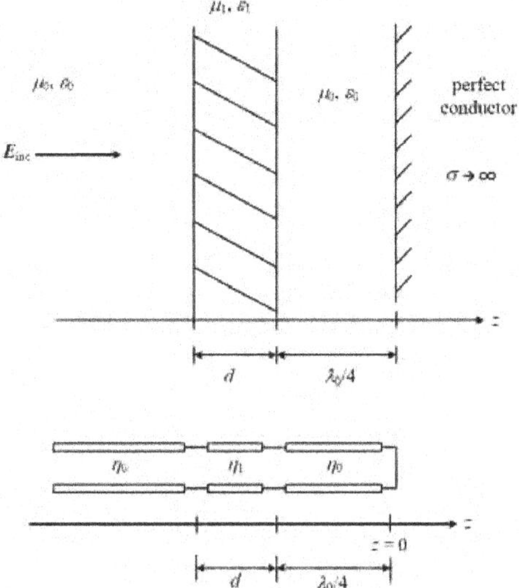

Figure 20.2: The equivalence of a layered medium problem to a transmission line problem. This equivalence is possible even for oblique incidence where a wave impedance is defined as in the previous lecture. The wave impedance is analogous to the characteristic impedance of a transmission line. For normal incidence, the wave impedance defined in the previous lecture becomes an intrinsic impedance (Figure by J.A. Kong, *Electromagnetic Wave Theory*).

20.1.2 Ray Series Interpretation of Generalized Reflection Coefficient

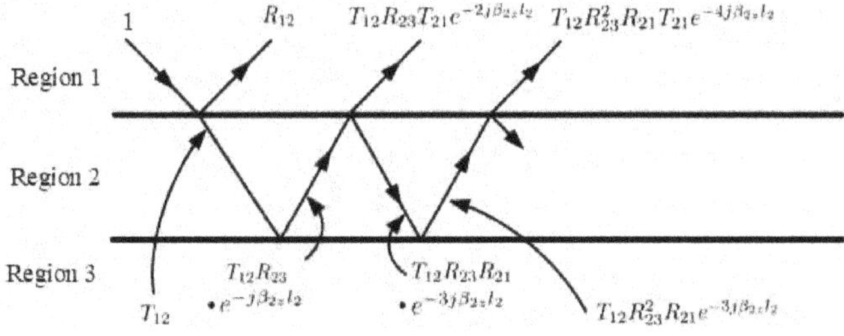

Figure 20.3: The expression of the generalized reflection coefficient decomposed into a ray series. Here, $l_2 = d_2 - d_1$ is the thickness of the slab (Figure by W.C. Chew [164]).

For simplicity, we will assume that $\tilde{R}_{23} = R_{23}$ in this section. By manipulation, one can convert the generalized reflection coefficient $\tilde{R}_{12}$ into a form that has a ray physics interpretation. By adding and subtracting the term

$$R_{12}^2 R_{23} e^{-2j\beta_{2z} l_2}$$

on the numerator of (20.1.2), and rearranging terms later, (20.1.2) can be shown to become

$$\tilde{R}_{12} = R_{12} + \frac{R_{23} e^{-2j\beta_{2z} l_2}(1 - R_{12}^2)}{1 + R_{12} R_{23} e^{-2j\beta_{2z} l_2}} \tag{20.1.3}$$

by using the fact that $R_{12} = -R_{21}$, which can be easily seen from the Fresnel reflection coefficient formula. Now we define a transmission coefficient such that $T_{ij} = 1 + R_{ij}$. Making $1 - R_{12}^2 = (1 - R_{12})(1 + R_{12}) = (1 - R_{12})(1 - R_{21}) = T_{12}T_{21}$, the above can be rewritten as

$$\tilde{R}_{12} = R_{12} + \frac{T_{12} T_{21} R_{23} e^{-2j\beta_{2z} l_2}}{1 + R_{12} R_{23} e^{-2j\beta_{2z} l_2}} \tag{20.1.4}$$

Next, using the fact that $(1 - x)^{-1} = 1 + x + x^2 + \ldots +$, the denominator of the second term above can be expanded as a power series, and the above can further be rewritten as

$$\tilde{R}_{12} = R_{12} + T_{12} R_{23} T_{21} e^{-2j\beta_{2z} l_2} + T_{12} R_{23}^2 R_{21} T_{21} e^{-4j\beta_{2z} l_2} + \cdots . \tag{20.1.5}$$

In the above, the series helps us to elucidate the physics of the generalized reflection coefficient $\tilde{R}_{12}$ that includes multiple reflections from below the first interface. The first term in the above is just the result of a single reflection off the first interface. The n-th term above is the consequence of the n-th reflection from the three-layer medium (see Figure 20.3). Hence, the expansion of (20.1.2) into (20.1.5) renders a lucid physical interpretation for the generalized reflection coefficient. Consequently, the series in (20.1.5) can be thought of as a *ray* series or a *geometrical optics* series. It is the consequence of multiple reflections and transmissions in region 2 of the three-layer medium, obtained by expanding the denominator of the second term in (20.1.4). Hence, the denominator of the second term in (20.1.4) is physically interpreted as a consequence of multiple reflections within region 2.

20.2 Phase Velocity and Group Velocity

We have seen that a single interface can guide a wave, as in the case of the SPP (surface plasmon polariton) (see section 19.1.1). Later, you will see that multiple interfaces can also be used to guide waves as in the dielectric slab waveguide. These guided waves will have a phase velocity that we have seen before. But in order to generate a localized wave that has a beginning and an end (like a pulse), it has to be multifrequency or polychromatic. If a localized pulse is sent along these waveguides, the it will not travel at the phase velocity, but instead, it travels at the group velocity, as we shall see.

A monochromatic signal will travel with a phase velocity. On the other hand, a polychromatic signal will propagate with a the phase velocity also if the medium is non-dispersive.. The case is more complicated if the medium is frequency dispersive, as in the Drude-Lorentz-Sommerfeld (DLS) model, meaning that different frequency components will travel with different phase velocity. As we shall see, the localized pulse can propagate with a velocity that is different from the phase velocity, which is called the group velocity.

We are ready to investigate the difference between the phase velocity and the group velocity. In this course, we will use k and β interchangeably to represent wavenumber.[2]

20.2.1 Phase Velocity

The phase velocity is the velocity of the phase of a wave. It is only well defined for a monochromatic signal (also called a time-harmonic, or sinusoidal signal) at one given frequency. Given a sinusoidal wave signal–for example, the voltage signal on a transmission line–using phasor technique, its simplest representation in the time domain can be easily found and takes the form

$$\begin{aligned} V(z,t) &= V_0\cos(\omega t - kz + \alpha) \\ &= V_0\cos\left[k\left(\frac{\omega}{k}t - z\right) + \alpha\right] \end{aligned} \tag{20.2.1}$$

The above signal has no beginning nor end, and it is of infinite extend. We have seen before that a function $f(vt - z)$ moves with a velocity of v to the right. Thus the sinusoidal signal moves *en block* to the right. Thus, this sinusoidal signal moves with a well-defined velocity ω/c, which is the phase velocity or

$$v_{ph} = \frac{\omega}{k} \tag{20.2.2}$$

where, for example, $k = \omega\sqrt{\mu\varepsilon}$ inside a simple coax. Hence, its phase velocity is

$$v_{ph} = 1/\sqrt{\mu\varepsilon} \tag{20.2.3}$$

It is the velocity of the phase of the wave which is well-defined for a monochromatic signal.

But a dielectric medium can be frequency dispersive, in which case $\varepsilon(\omega)$ is not a constant but a function of ω, as has been shown with the DLS model. A function of time such as $f(vt - z)$, for a fixed z, can be Fourier decomposed into a linear superposition of many time-harmonic signals. Therefore, signals with different ω's will travel with different phase velocities, making the definition of the velocity of the signal complicated.

More bizarre still is when the coax is filled with a plasma medium where

$$\varepsilon = \varepsilon_0\left(1 - \frac{{\omega_p}^2}{\omega^2}\right) \tag{20.2.4}$$

Then, $\varepsilon < \varepsilon_0$ in the above means that the phase velocity given by (20.2.3) can be larger than the velocity of light in vacuum (assuming $\mu = \mu_0$). Also, $\varepsilon = 0$ when $\omega = \omega_p$, implying that $k = 0$; then, in accordance with (20.2.2) and (20.2.3), $v_{ph} = \infty$.

These ludicrous observations can be avoided only if we can show that information in a dispersive medium can only be sent by using a wave packet.[3] The same goes for energy: it can only be sent by

[2] As mentioned before, the microwave community prefers β, while the optics community prefers k.

[3] In information theory, according to Shannon, the basic unit of information is a bit, which can only be sent by a digital signal, wave packet, or broadband signal. A wave packet cannot be monochromatic or has a very narrow band.

wave packets but not by a CW signal; only in this manner can a finite amount of energy be sent. Therefore, it is prudent for us to study the velocity of a wave packet, which is not a monochromatic signal but a polychromatic signal. As shall be seen, these wave packets in a dispersive medium can only travel at the group velocity, which is physical and always less than the velocity of light.

20.2.2 Group Velocity

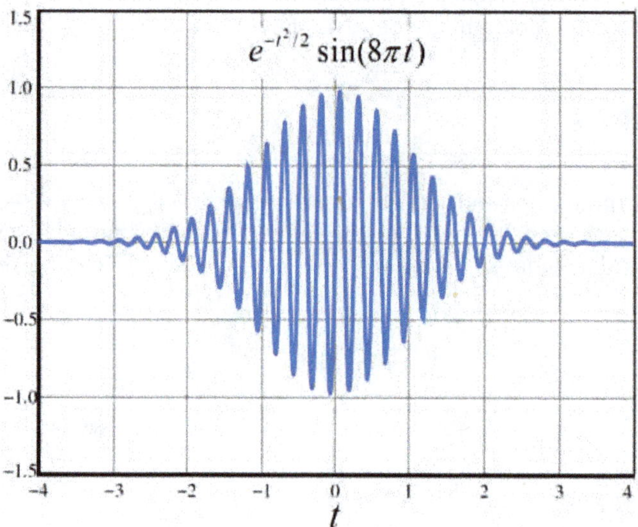

Figure 20.4: A Gaussian wave packet as a function of t for a fixed z. It can be thought of as a linear superposition of many monochromatic waves with slightly different frequencies. If one Fourier transforms the above signal, it will become a narrowband signal centered about certain ω_0 (Figure by Wikipedia).

Now, consider a narrowband wave packet as shown in Figure 20.4. It cannot be monochromatic, but can be written as a linear superposition of many frequencies. One way to express this is to write this wave packet as an integral in terms of Fourier transform, or a summation over many frequencies. Namely, at a fixed z,[4]

$$V(z,t) = \int_{-\infty}^{\infty} d\omega \underset{\sim}{V}(z,\omega) e^{j\omega t} \tag{20.2.5}$$

To make $V(z,t)$ be related to a traveling wave, we assume that $\underset{\sim}{V}(z,\omega)$ is the solution to the one-dimensional Helmholtz equation. We have encountered this equation in solving one-dimensional

[4]The Fourier transform technique is akin to the phasor technique but different. For simplicity, we will use $\underset{\sim}{V}(z,\omega)$ to represent the Fourier transform of $V(z,t)$.

Maxwell's equations in (7.2.6) and in the transmission line theory as shown in (15.2.5) and (15.2.6). There, μ, ε, L, and C can be functions of frequency, and yet the equations are still valid when phasor technique is used. In general,

$$\frac{d^2}{dz^2}\underset{\sim}{V}(z,\omega) + k^2(\omega)\underset{\sim}{V}(z,\omega) = 0 \tag{20.2.6}$$

for both Maxwell's equations and the TEM mode of a transmission line, $k^2 = \omega^2\mu\varepsilon$. The previous derivations have been for dispersionless media. But one can easily extend the derivation in section 7.2 to a dispersive medium where $V(z,\omega) = E_x(z,\omega)$. Alternatively, one can generalize the derivation in section 15.2 to the case of dispersive transmission lines, where, for instance, the coaxial transmission line is filled with a dispersive material–then $k^2 = \omega^2\mu_0\varepsilon(\omega)$. Thus, upon solving the above ordinary differential equation (20.2.6), for a fixed ω, one obtains

$$V(z,\omega) = V_0(\omega)e^{-jkz}$$

where $k = \omega\sqrt{\mu_0\varepsilon(\omega)}$, and

$$V(z,t) = \int_{-\infty}^{\infty} d\omega V_0(\omega)e^{j(\omega t - kz)} \tag{20.2.7}$$

Here, since we have to connect a broadband time-domain signal to its Fourier transform, we will use Fourier transform technique rather than phasor technique as phasor technique is more for a monochromatic signal. At this point, it is prudent to notice that if the medium is dispersionless, then $k = \omega\sqrt{\mu_0\varepsilon} = \omega/v_{ph}$ where $v_{ph} = 1/\sqrt{\mu_0\varepsilon}$ and ε is independent of frequency.[5] Then all Fourier components travel with the same phase velocity. Thus, $V(z,t) = f(v_{ph}t - z)$, which is an arbitrary function traveling with velocity v_{ph} without distortion.

We shall next look at the dispersive medium case. For this medium, different Fourier components will travel with different phase velocities. If these phase velocities are all different, then the pulse traveling through it will be distorted and become quite complicated. To simplify matters, we will assume a narrowband pulse and find the velocity of this narrowband pulse. As shall be shown, if the bandwidth is narrow enough, the envelope of the pulse will travel with a different velocity called the group velocity, which is different from the phase velocity. This means that the group velocity can only be defined for narrowband pulse (also called a quasi-monochromatic pulse), which is an asymptotic concept.

In the above, $V(z,t)$ is a real value. As such, the negative frequency components of the above integral have to be complex conjugates of the positive frequency components. Then, we can also rewrite the above as

$$V(z,t) = \int_{-\infty}^{0} d\omega V_0(\omega)e^{j(\omega t - kz)} + \int_{0}^{\infty} d\omega V_0(\omega)e^{j(\omega t - kz)} \tag{20.2.8}$$

[5]There is no truly dispersionless medium except for vacuum.

For real-valued signal, then $V_0(-\omega) = V_0^*(\omega)$ and that $k(-\omega) = k^*(\omega)$, we can write the above as a sum over only the $+\omega$ part of the integral and take twice the real part of the integral.

$$V(z,t) = 2\Re e \int_0^\infty d\omega V_0(\omega) e^{j(\omega t - kz)} \tag{20.2.9}$$

In the general case, k may be a complicated function of ω as shown in Figure 20.5.

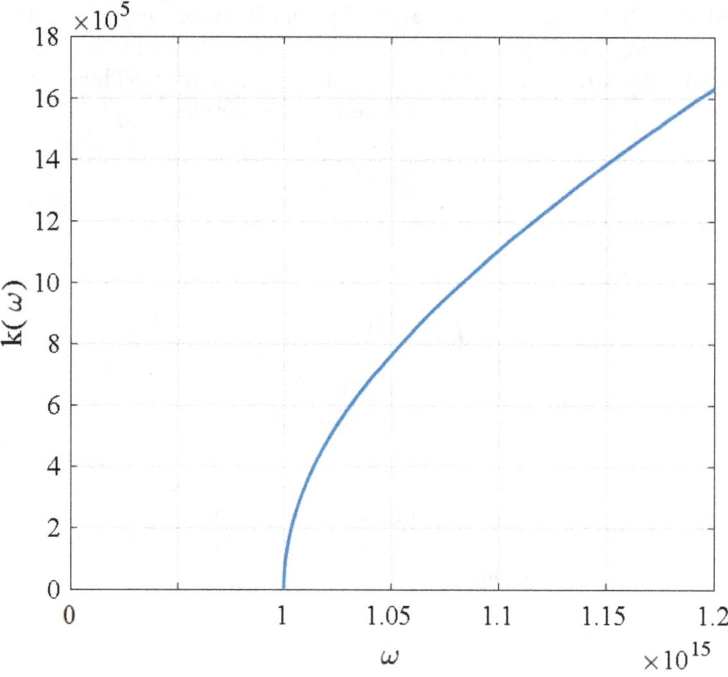

Figure 20.5: A typical frequency dependent $k(\omega)$ for a plasma medium, albeit the frequency dependence can be more complicated than shown here.

Since this is a wave packet, we assume that $V_0(\omega)$ is narrowband (quasi-monochromatic or polychromatic) and centered about a frequency ω_0, the carrier frequency as shown in Figure 20.6. Therefore, when the integral in (20.2.7) is performed, we need only to sum over a narrow range of frequencies in the vicinity of ω_0. Henceforth, we can approximate $k(\omega)$ in the integrand in the vicinity of $\omega = \omega_0$ by Taylor series expansion, and let

$$k(\omega) \cong k(\omega_0) + (\omega - \omega_0)\frac{dk(\omega_0)}{d\omega} + \frac{1}{2}(\omega - \omega_0)^2 \frac{d^2k(\omega_0)}{d\omega^2} + \cdots \tag{20.2.10}$$

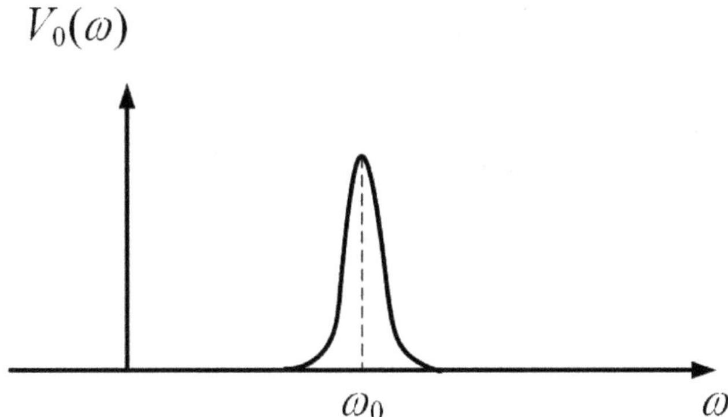

Figure 20.6: The frequency spectrum of $V_0(\omega)$ is the Fourier transform of $V(z=0,t)$ (see (20.2.7)). As can be seen, it is a narrowband signal. Only with a narrowband signal can we define what a group velocity is. It is the velocity of the envelope of a pulse function.

Since we need to integrate over $\omega \approx \omega_0$, we can substitute (20.2.10) into (20.2.9) and rewrite it as

$$V(z,t) \cong 2\Re e\left[e^{j[\omega_0 t - k(\omega_0)z]} \underbrace{\int_0^\infty d\omega V_0(\omega) e^{j(\omega-\omega_0)t} e^{-j(\omega-\omega_0)\frac{dk}{d\omega}z}}_{F(t-\frac{dk}{d\omega}z)}\right] \tag{20.2.11}$$

where more specifically, in the above,

$$F\left(t - \frac{dk}{d\omega}z\right) = F\left(t - z/v_g\right) = \int_0^\infty d\omega V_0(\omega) e^{j(\omega-\omega_0)t} e^{-j(\omega-\omega_0)\frac{dk}{d\omega}z} \tag{20.2.12}$$

where $v_g = \frac{d\omega}{dk}$ is the velocity of the function $F\left(t - z/v_g\right)$. Moreover, as shall be shown, $F\left(t - z/v_g\right)$ is a slowly varying function moving with v_g given by

$$v_g = \frac{d\omega}{dk} \tag{20.2.13}$$

The above is known as the group velocity. Notice that for a dispersionless medium, the group velocity is the same as the phase velocity.

We shall next show that $F\left(t - z/v_g\right)$ is a slowly varying function. Since $V_0(\omega) \neq 0$ over a small range of frequency near ω_0 (see Figure 20.6), by a change of variable by letting $\Omega = \omega - \omega_0$, it becomes

$$F\left(t - z/v_g\right) \cong \int_{-\Delta}^{+\Delta} d\Omega V_0(\Omega + \omega_0) e^{j\Omega(t - z/v_g)} \tag{20.2.14}$$

When Ω ranges from $-\Delta$ to $+\Delta$ in the above integral, the value of ω ranges from $\omega_0 - \Delta$ to $\omega_0 + \Delta$. We assume that outside this range of ω, as shown in Figure 20.6. $V_0(\omega)$ is sufficiently small so that its value can be ignored.

The above itself is a Fourier transform integral: its physical meaning is that this function $F\left(t - z/v_g\right)$ involves only the low frequencies of the Fourier spectrum where $e^{j\Omega(t-z/v_g)}$ is evaluated over small Ω values. Hence, $F\left(t - z/v_g\right)$ is a slowly varying envelope function of its argument or $(t - z/v_g)$, or wave packet that moves with the velocity v_g as shown Figure 20.4.

In (20.2.11), the envelope function $F(t - \frac{z}{v_g})$ is multiplied by the rapidly varying function

$$e^{j[\omega_0 t - k(\omega_0)z]} \tag{20.2.15}$$

before one takes the real part of the entire function. Hence, this rapidly varying part represents the rapidly varying carrier frequency shown in Figure 20.4. More importantly, this carrier, the rapidly varying part of the signal, as shown in (20.2.15), moves with the velocity

$$v_{ph} = \frac{\omega_0}{k(\omega_0)} \tag{20.2.16}$$

which is the phase velocity. Thus we have elucidated the wave physics that a narrowband pulse has an envelope that travels with the group velocity v_g, while the "innards" of the pulse, which include the wave inside the envelope, travel with the phase velocity v_{ph}.

20.3 Wave Guidance in a Layered Medium

Now that we understand phase and group velocity, we are at ease with studying the propagation of a guided wave in a layered medium. We have seen that in the case of a surface plasmonic resonance, the value of β_x, given by (19.1.10), and reproduced here is

$$\beta_x = \omega\sqrt{\mu}\sqrt{\frac{\varepsilon_1\varepsilon_2}{\varepsilon_1 + \varepsilon_2}} \tag{20.3.1}$$

If β_x is linearly proportional to ω, or that μ, ε_1 and ε_2 are frequency independent corresponding to the dispersionless case. Then a pulse, as previously discussed, will not be distorted, as all Fourier components travel with the same phase velocity. However, if medium 2 is a plasma medium, ε_2 can be a strong function of frequency. Then different frequencies travel with a different phase velocity. The cluster of frequencies, as a group will travel at the group velocity, which is very different from the phase velocity.

The surface plasmon polariton (SPP) wave is guided by an interface because the Fresnel reflection coefficient becomes infinite. This physically means that a reflected wave exists even if an incident wave is absent or vanishingly small, which is a resonance behaviour. This condition can be used to find a guided mode in a layered medium–namely, to find the condition under which the generalized reflection coefficient (20.1.2) becomes infinite.[6]

[6]As mentioned in section 19.1.1, this is equivalent to finding a solution to a problem with no driving term (forcing function) or finding the homogeneous solution to an ordinary differential equation or partial differential equation. It is also equivalent to finding the null-space solution of a matrix equation.

20.3.1 Transverse Resonance Condition

Therefore, to have a guided mode exist in a layered medium due to multiple bounces, the generalized reflection coefficient becomes infinite when the denominator of (20.1.2) is zero, or

$$1 + R_{12}\tilde{R}_{23}e^{-2j\beta_{2z}l_2} = 0 \tag{20.3.2}$$

where l_2 is the thickness of the dielectric slab. Since $R_{12} = -R_{21}$, the above can be written as

$$1 = R_{21}\tilde{R}_{23}e^{-2j\beta_{2z}l_2} \tag{20.3.3}$$

The above has the physical meaning that the wave, after going through two reflections at the two interfaces, the 21 and 23 interfaces, which are R_{21} and $\tilde{R}_{23}$, respectively, plus a phase delay given by $e^{-2j\beta_{2z}l_2}$, becomes itself again. When this happens, the wave interfere constructively. This is also known as the transverse resonance condition. When specialized to the case of a dielectric slab with two interfaces and three regions, the above becomes

$$1 = R_{21}R_{23}e^{-2j\beta_{2z}l_2} \tag{20.3.4}$$

The above can be generalized to finding the guided mode in a general layered medium. It can also be specialized to finding the guided mode of a dielectric slab.

Exercises for Lecture 20

Problem 20-1: This problem is about physical interpretation of the generalized reflection coefficient for a layered medium.

(i) By using $T_{ij} = 1 + R_{ij}$ and $R_{ij} = -R_{ji}$, show that the generalized reflection coefficient can be written as

$$\tilde{R}_{12} = \frac{R_{12} + \tilde{R}_{23}e^{-2j\beta_{2z}l_2}}{1 + R_{12}\tilde{R}_{23}e^{-2j\beta_{2z}l_2}} = R_{12} + \frac{T_{12}T_{21}R_{23}e^{-2j\beta_{2z}l_2}}{1 + R_{12}R_{23}e^{-2j\beta_{2z}l_2}}$$

where l_2 is the thickness of region 2.

(ii) Using the geometric series expansion that $1/(1-x) = 1 + x + x^2 + x^3 + \cdots$, show that the generalized reflection coefficient can be rewritten as

$$\tilde{R}_{12} = R_{12} + T_{12}R_{23}T_{21}e^{-2j\beta_{2z}l_2} + T_{12}R_{23}^2R_{21}T_{21}e^{-4j\beta_{2z}l_2} + \cdots.$$

(iii) Give the physical meanings of each of the terms above, and the meaning of the phase delay term.

(iv) Explain what phase and group velocities are.

(v) Derive (20.2.11) of the lecture; explain why $F\left(t - \frac{dk}{d\omega}z\right)$ is a slowly varying function, and explain what a group velocity is.

Lecture 21

Dielectric Slab Waveguides

As mentioned before, the dielectric slab waveguide shares many salient features with the optical fiber waveguide, one of the most important waveguides of this century.[1] The analysis of the optical fiber requires the Maxwellian solution in cylindrical coordinates, which is beyond our scope, but it can be found in [2, 92, 96]. Before we embark on studying dielectric slab waveguides, we will revisit transverse resonance again. The transverse resonance condition allows one to derive the guidance conditions for a dielectric slab waveguide easily without having to match the boundary conditions at the interfaces again; the boundary conditions are already embedded in the derivation of the Fresnel reflection coefficients. Much of the material in this lecture can also be found in [92, 35, 48]

21.1 Generalized Transverse Resonance Condition

The generalized transverse resonance condition is a powerful condition that can be used to derive the guidance conditions of modes in a layered medium. To derive this condition, we first have to realize that a guided mode in a waveguide is due to the coherent or constructive interference of the waves. This implies that if a plane wave starts at position 1 (see Figure 21.1)[2] and is multiply reflected as shown, it will regain its original phase in the x direction at position 5. Since this mode progresses in the z direction, all these waves (also known as partial waves) are in phase in the z direction by the phase-matching condition. Otherwise, the boundary conditions cannot be satisfied. (That is, waves at 1 and 5 will gain the same phase in the z direction.) But for them to add coherently or interfere coherently in the x direction, the transverse phase in the x direction at 5 must be the same as 1.

Assuming that the wave starts with amplitude 1 at position 1 and that the distance between 1 and

[1]The optical fiber also has very low loss in the order of 0.1 dB/km, which is much lower than that achievable with metal waveguide.

[2]The waveguide convention is to assume the direction of the propagation of a mode to be in the z direction. Since we are analyzing a guided mode in a layered medium, the z-axis is as shown in this figure, which is parallel to the interfaces. This is different from the previous lecture where z is an axis that is normal to the layers.

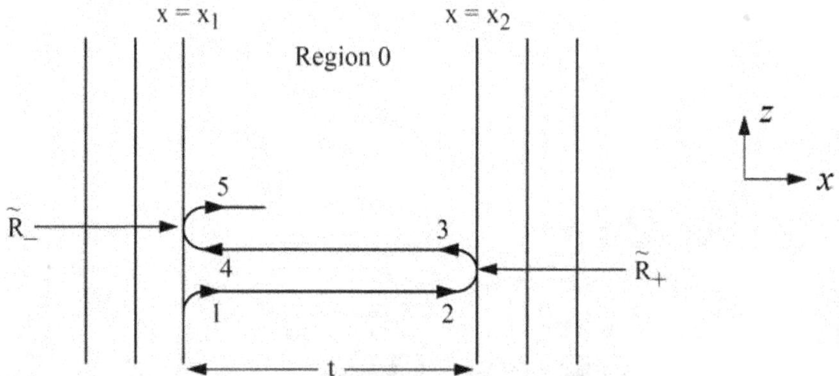

Figure 21.1: For constructive interference to occur, the phase of the wave at position 5 should be equal to the transverse phase at position 1. The above is an illustration, as the ray does not bend as shown.

2 is t, the wave will gain a transverse phase of $e^{-j\beta_{0x}t}$ when it reaches position 2. Upon reflection at $x = x_2$, at position 3, the wave becomes $\tilde{R}_+ e^{-j\beta_{0x}t}$ where $\tilde{R}_+$ is the generalized reflection coefficient at the right interface of region 0. Finally, at position 5, it becomes $\tilde{R}_- \tilde{R}_+ e^{-2j\beta_{0x}t}$ where $\tilde{R}_-$ is the generalized reflection coefficient at the left interface of region 0. For constructive interference to occur or for the mode to exist, we require that the transverse phase of the wave at position 5 is the same as that at position 1, or

$$\tilde{R}_- \tilde{R}_+ e^{-2j\beta_{0x}t} = 1 \tag{21.1.1}$$

The above is the generalized transverse resonance condition for the guidance condition for a plane-wave mode traveling in a layered medium.

Alternatively, we can look at the generalized reflection coefficient of the previous lecture. Internal modes exist in the layered medium when the generalized reflection coefficient goes to infinity. For a three-layered medium, looking at (20.3.3), the denominator of the equation will be zero when

$$R_{21} R_{23} e^{-2j\beta_{2x}t} = 1 \tag{21.1.2}$$

Here we assume the wave-number in the middle region is β_{2x} for a wave propagating in the x direction.

21.1.1 Parallel Plate Waveguide

Note that for (21.1.1), when we have two parallel metallic plates, where the metallic plates are assumed to be PEC, then $R_{\pm}^{TM} = 1$, and $R_{\pm}^{TE} = -1$,[3] the guidance condition becomes

$$1 = e^{-2j\beta_{0x}t} \quad \Rightarrow \beta_{0x} = \frac{m\pi}{t}, \qquad m = 0, 1, 2, \ldots, \tag{21.1.3}$$

[3]This can be seen from the Fresnel reflection coefficient by setting one of the media to be PEC.

These are just the guidance conditions for parallel plate waveguides. In this waveguide, the modes are guided by total reflections at the air-metallic interface due to the impenetrability of a PEC surface. Details of this waveguide are given in the ECE 350X lecture notes as well as in Kong [35, 48].

21.2 Dielectric Slab Waveguide

The most important dielectric waveguide of the modern world is the optical fiber , whose invention was credited to Charles Kao [149]. He was awarded the Nobel Prize in 2009 [165]. However, the closed-form analysis of the optical fiber requires the use of cylindrical coordinates and special functions such as Bessel functions. In order to capture the essence of the optical fiber or different dielectric waveguides without the use of special functions such as Bessel functions, one can study the slab dielectric waveguide, which shares many salient wave-physics features. We start with analyzing the TE modes in this waveguide. (This waveguide is also used as a thin-film optical waveguide (see Figure 21.2).

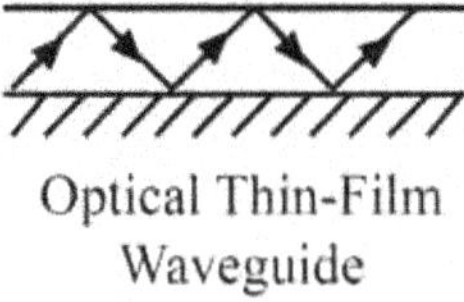

Figure 21.2: An optical thin-film waveguide is made by coating a thin dielectric film or sheet on a metallic surface. The wave is guided by total internal reflection at the top interface, and by metallic reflection at the bottom interface where the wave is totally reflected as well.

21.2.1 TE Case

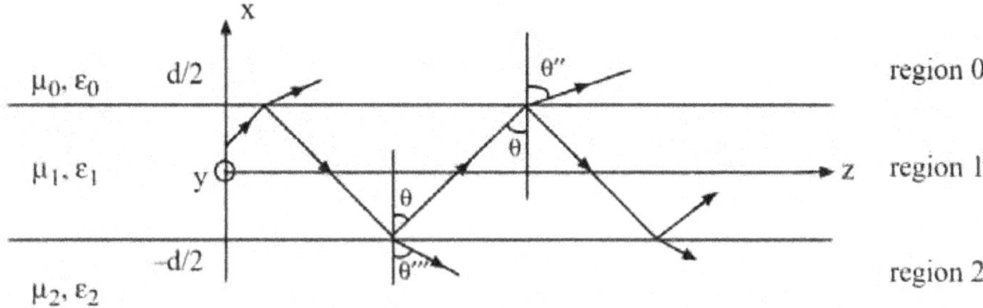

Figure 21.3: Schematic for the analysis of a guided mode in the dielectric waveguide. Total internal reflections occur at the top and bottom interfaces. If the waves add coherently, the wave is guided along the dielectric slab.

We shall look at the application of the transverse resonance condition to a TE wave guided in a dielectric slab waveguide. Again, in accordance with convention, we assume the direction of propagation of the guided mode to be in the z direction. Specializing the above equation for the dielectric slab waveguide shown in Figure 21.3, we have the guidance condition, similar to the transverse resonance condition as

$$1 = R_{10}R_{12}e^{-2j\beta_{1x}d} \tag{21.2.1}$$

where d is the thickness of the dielectric slab. Guidance of a mode is due to total internal reflection, and, hence, we expect region 1 to be optically more dense (in terms of optical refractive indices) than regions 0 and 2.[4]

To simplify the analysis further, we assume region 2 to be the same as region 0 so that $R_{12} = R_{10}$. The new guidance condition is then

$$1 = R_{10}^2 e^{-2j\beta_{1x}d} \tag{21.2.2}$$

In the above,

$$R_{10} = \frac{\mu_0\beta_{1x} - \mu_1\beta_{0x}}{\mu_0\beta_{1x} + \mu_1\beta_{0x}} \tag{21.2.3}$$

By phase matching, β_z is the same in all the three regions of Figure 21.3. In the above, $\beta_{ix} = \sqrt{\beta_i^2 - \beta_z^2}$, and by expressing all of the β_{ix} in terms of the variable β_z, the above (21.2.2) becomes an implicit equation to be solved for β_z. Also, we assume that $\varepsilon_1 > \varepsilon_0$ so that total internal reflection occurs at both interfaces as the wave bounces around so that $\beta_{0x} = \beta_{2x} = -j\alpha_{0x}$. In other words, for TE polarization, the local, single-interface or Fresnel reflection coefficient is

$$R_{10} = \frac{\mu_0\beta_{1x} - \mu_1\beta_{0x}}{\mu_0\beta_{1x} + \mu_1\beta_{0x}} = \frac{\mu_0\beta_{1x} + j\mu_1\alpha_{0x}}{\mu_0\beta_{1x} - j\mu_1\alpha_{0x}} = e^{j\theta_{TE}} \tag{21.2.4}$$

where θ_{TE} is the Goos-Hänchen shift for total internal reflection. It is given by[5]

$$\theta_{TE} = 2\tan^{-1}\left(\frac{\mu_1\alpha_{0x}}{\mu_0\beta_{1x}}\right) \tag{21.2.5}$$

The guidance condition for constructive interference according to (21.2.1) and (21.2.2), after using (21.2.4) and (21.2.5), is such that

$$2\theta_{TE} = 2\beta_{1x}d + 2l\pi \tag{21.2.6}$$

From the above, in view of (21.2.5), we divide the above by four, and taking its tangent, we get

$$\tan\left(\frac{\theta_{TE}}{2}\right) = \tan\left(\frac{l\pi}{2} + \frac{\beta_{1x}d}{2}\right) \tag{21.2.7}$$

[4]Optically more dense means higher optical refractive index, or higher dielectric constant. As a reminder, the refractive index of a medium with permittivity ε and permeability μ is $n = \sqrt{\mu\varepsilon/(\mu_0\varepsilon_0)}$ where ε_0 and μ_0 are the permittivity and permeability, respectively, of vacuum.

[5]A phase shift in the frequency domain is equivalent to a time delay in the time domain. This effect can be observed in time-domain measurements.

or using (21.2.5) for the left-hand side,

$$\frac{\mu_1 \alpha_{0x}}{\mu_0 \beta_{1x}} = \tan\left(\frac{l\pi}{2} + \frac{\beta_{1x} d}{2}\right) \tag{21.2.8}$$

The above gives rise to

$$\mu_1 \alpha_{0x} = \mu_0 \beta_{1x} \tan\left(\frac{\beta_{1x} d}{2}\right), \qquad l \text{ even} \tag{21.2.9}$$

$$-\mu_1 \alpha_{0x} = \mu_0 \beta_{1x} \cot\left(\frac{\beta_{1x} d}{2}\right), \qquad l \text{ odd} \tag{21.2.10}$$

It can be shown that when l is even, the mode profile is even , whereas when l is odd, the mode profile is odd . The above can also be rewritten as

$$\frac{\mu_0}{\mu_1} \frac{\beta_{1x} d}{2} \tan\left(\frac{\beta_{1x} d}{2}\right) = \frac{\alpha_{0x} d}{2}, \qquad \text{even modes} \tag{21.2.11}$$

$$-\frac{\mu_0}{\mu_1} \frac{\beta_{1x} d}{2} \cot\left(\frac{\beta_{1x} d}{2}\right) = \frac{\alpha_{0x} d}{2}, \qquad \text{odd modes} \tag{21.2.12}$$

Again, the above equations can be expressed in the β_z variable, but they do not have closed-form solutions, except for graphical solutions (or numerical solutions). We shall discuss their graphical solutions next.[6]

To solve the above graphically, it is best to plot the equations in terms of one common variable. It turns out that β_{1x} is the simplest common variable to use for graphical solutions since the left-hand side of (21.2.11) and (21.2.12) are simple functions of β_{1x}. To this end, using the fact that

$$-\alpha_{0x}^2 = \beta_0^2 - \beta_z^2$$

and that

$$\beta_{1x}^2 = \beta_1^2 - \beta_z^2$$

and eliminating β_z from these two equations by subtraction, one can show that α_{0x} on the right-hand side becomes

$$\alpha_{0x} = [\omega^2(\mu_1\epsilon_1 - \mu_0\epsilon_0) - \beta_{1x}^2]^{\frac{1}{2}} \tag{21.2.13}$$

By regarding β_{1x} as the abscissa and α_{0x} as the ordinate, we notice that the above is the equation

[6]This technique was put together by the community of scholars in the optical waveguide field.

of a circle. Thus the right-hand side of (21.2.11) and (21.2.12) can be simplified to become

$$\frac{\mu_0}{\mu_1}\frac{\beta_{1x}d}{2}\tan\left(\frac{\beta_{1x}d}{2}\right) = \frac{\alpha_{0x}d}{2}$$
$$= \sqrt{\omega^2(\mu_1\epsilon_1 - \mu_0\epsilon_0)\frac{d^2}{4} - \left(\frac{\beta_{1x}d}{2}\right)^2}, \quad \text{even modes} \tag{21.2.14}$$

$$-\frac{\mu_0}{\mu_1}\frac{\beta_{1x}d}{2}\cot\left(\frac{\beta_{1x}d}{2}\right) = \frac{\alpha_{0x}d}{2}$$
$$= \sqrt{\omega^2(\mu_1\epsilon_1 - \mu_0\epsilon_0)\frac{d^2}{4} - \left(\frac{\beta_{1x}d}{2}\right)^2}, \quad \text{odd modes} \tag{21.2.15}$$

We can solve the above graphically by plotting

$$y_1 = \frac{\mu_0}{\mu_1}\frac{\beta_{1x}d}{2}\tan\left(\frac{\beta_{1x}d}{2}\right), \quad \text{even modes} \tag{21.2.16}$$

$$y_2 = -\frac{\mu_0}{\mu_1}\frac{\beta_{1x}d}{2}\cot\left(\beta_{1x}\frac{d}{2}\right), \quad \text{odd modes} \tag{21.2.17}$$

$$y_3 = \left[\omega^2(\mu_1\epsilon_1 - \mu_0\epsilon_0)\frac{d^2}{4} - \left(\frac{\beta_{1x}d}{2}\right)^2\right]^{\frac{1}{2}} = \frac{\alpha_{0x}d}{2} \tag{21.2.18}$$

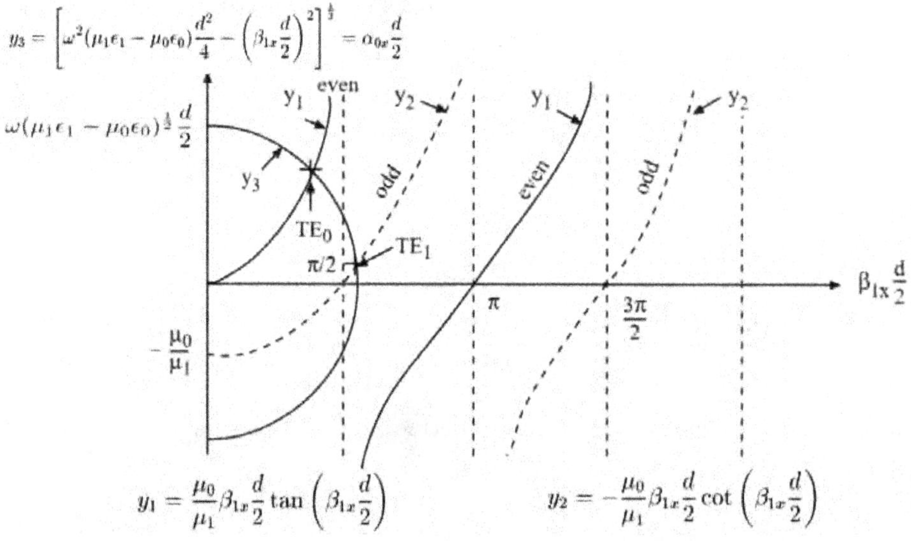

Figure 21.4: A way to solve (21.2.14) and (21.2.15) is via a graphical method. In this method, both the right-hand side and the left-hand side of the equations are plotted on the same plot. The solutions are at the intersection points of these plots.

In the above, y_3 is the equation of a circle in terms of $\beta_{1x}\frac{d}{2}$. The radius of the circle is given by

$$\omega(\mu_1\epsilon_1 - \mu_0\epsilon_0)^{\frac{1}{2}}\frac{d}{2}. \tag{21.2.19}$$

The solutions to (21.2.14) and (21.2.15) are given by the intersections of y_3 with y_1 and y_2. To increase the number of solutions, we can increase the radius of the circle defined by y_3. We note from (21.2.1) that the radius of the circle can be increased in three ways: (i) by increasing the frequency ω, (ii) by increasing the contrast $\frac{\mu_1\epsilon_1}{\mu_0\epsilon_0}$, and (iii) by increasing the thickness d of the slab.[7] By increasing these three parameters, the number of trapped modes (or guided modes) inside the slab waveguide increases. The mode profiles of the first two modes are shown in Figure 21.5.

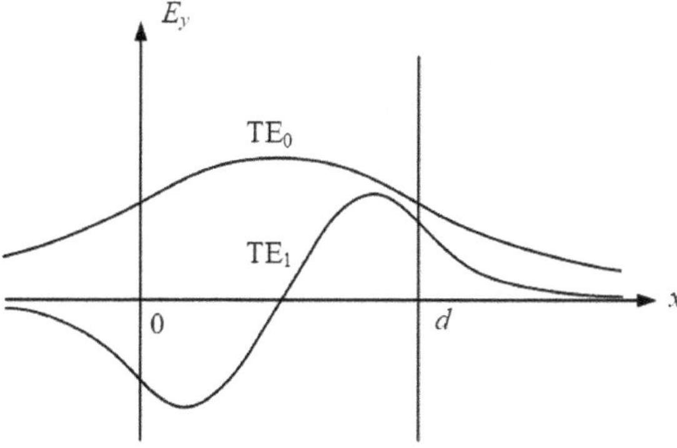

Figure 21.5: Mode profiles of the TE_0 and TE_1 modes of a dielectric slab waveguide (Figure by J.A. Kong [35]).

When $\beta_{0x} = -j\alpha_{0x}$, the reflection coefficient for total internal reflection is

$$R_{10}^{TE} = \frac{\mu_0\beta_{1x} + j\mu_1\alpha_{0x}}{\mu_0\beta_{1x} - j\mu_1\alpha_{0x}} = \exp\left[+2j\tan^{-1}\left(\frac{\mu_1\alpha_{0x}}{\mu_0\beta_{1x}}\right)\right] \tag{21.2.20}$$

and $\left|R_{10}^{TE}\right| = 1$. Hence, the wave is guided by total internal reflections at the two interfaces.

Cutoff happens when the total internal reflection ceases to occur–that is, when the frequency decreases such that $\alpha_{0x} = 0$. This is also the point where total internal reflection stops to occur. When this happens, the mode is not trapped or confined anymore or the wave is not evanescent outside the slab. From Figure 21.4, we see that $\alpha_{0x} = 0$ when

$$\omega(\mu_1\epsilon_1 - \mu_0\epsilon_0)^{\frac{1}{2}}\frac{d}{2} = \frac{m\pi}{2}, \qquad m = 0, 1, 2, 3, \ldots \tag{21.2.21}$$

[7] These are important salient features of a dielectric waveguide. These features are also shared by the optical fiber.

or

$$\omega_{mc} = \frac{m\pi}{d(\mu_1\epsilon_1 - \mu_0\epsilon_0)^{\frac{1}{2}}}, \qquad m = 0, 1, 2, 3, \ldots \tag{21.2.22}$$

where ω_{mc} is the cutoff frequency of the m-th mode. The mode that corresponds to the m-th cutoff frequency above is labeled the TE_m mode. Thus the TE_0 mode is the mode that has no cutoff or propagates at all frequencies. This is shown in Figure 21.6, where the TE mode profiles are similar since they are dual to each other. The boundary conditions at the dielectric interface is that the tangential E and H fields have to be continuous. The TE_0 or TM_0 mode can satisfy this boundary condition at all frequencies. At the cutoff frequency, the field outside the slab is not evanescent but has to become flat, implying that $\alpha_{0x} = 0$. Therefore, the mode is not confined to the vicinity of the waveguide, and there is no guidance.

Next, we will elucidate more physics of the dielectric slab guided mode. At cutoff, $\alpha_{0x} = 0$, and from the dispersion relation that $\alpha_{0x}^2 = \beta_z^2 - \beta_0^2$, we conclude that

$$\beta_z = \beta_0 = \omega\sqrt{\mu_0\epsilon_0},$$

for all the modes. Hence, the phase velocity, ω/β_z, and the group velocity, $d\omega/d\beta_z$, are those of the outer region. This is because when $\alpha_{0x} = 0$, the wave is not evanescent outside, and the energy of the mode is predominantly carried by the exterior field.

When ω becomes increasingly larger, the radius of the circle in the plot of y_3 also becomes increasingly larger. As seen from Figure 21.4, the solution for $\beta_{1x} \to \frac{l\pi}{d}$ for all the modes. In other words, the semicircle in Figure 21.4 will intersect all the vertical dotted lines as ω increases.

From the dispersion relation for region 1, since $\omega^2\mu_1\epsilon_1 \gg \beta_{1x}^2 \approx (l\pi/d)^2$, we have

$$\beta_z = \sqrt{\omega^2\mu_1\epsilon_1 - \beta_{1x}^2} \approx \sqrt{\omega^2\mu_1\epsilon_1 - (l\pi/d)^2} \approx \omega\sqrt{\mu_1\epsilon_1}, \qquad \omega \to \infty \tag{21.2.23}$$

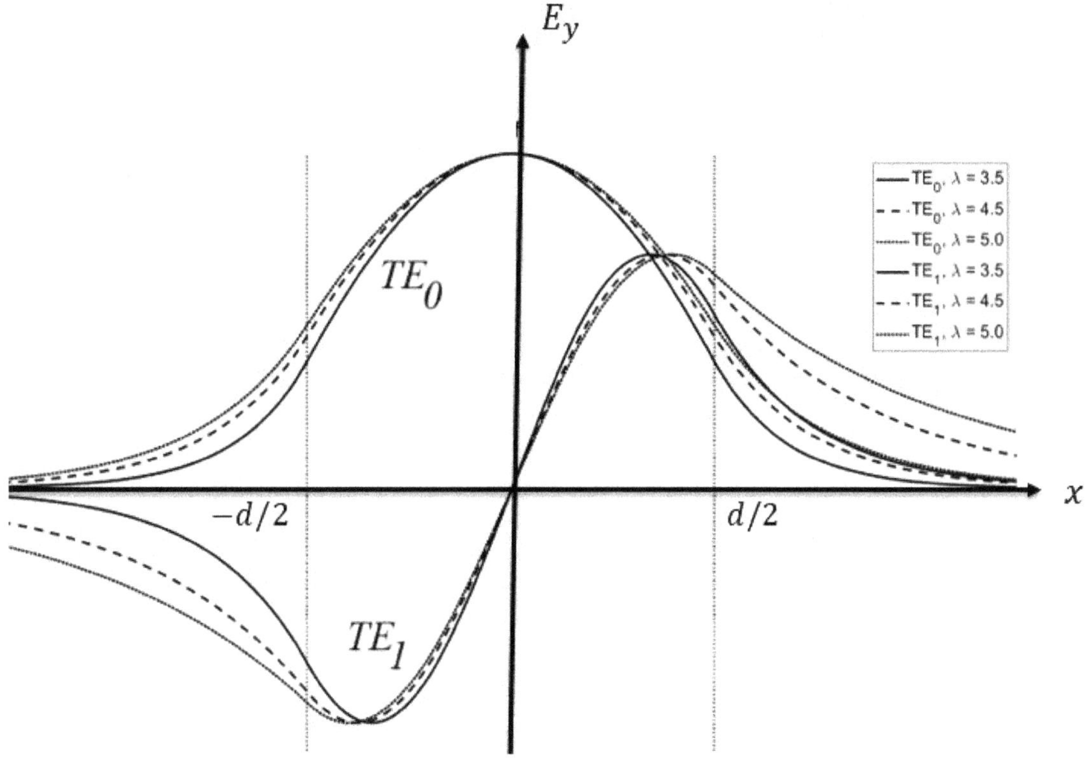

Figure 21.6: Mode profiles of the TM modes of a dielectric slab. The TE modes are dual to the TM modes and have similar mode profiles.

Hence, the group and phase velocities approach that of the dielectric slab. This is because when $\omega \to \infty$, $\alpha_{0x} \to \infty$, implying the rapid exponential decay of the fields outside the waveguide. Therefore, the fields are trapped or confined in the slab and propagating within it. Because of this, the dispersion diagram of the different modes appears as shown in Figure 21.7.

In Figure 21.7,[8] k_{c1}, k_{c2}, and k_{c3} are the cutoff wavenumber or frequency of the first three modes. In the same figure, if the wave is in free space, then $k_z = \omega\sqrt{\mu_0\varepsilon_0}$. This is also called the light line. However, in the k_z-k_1 plot, we see that

$$k_z = \left(\frac{\mu_0\varepsilon_0}{\mu_1\varepsilon_1}\right)^{1/2} k_1 = k_0 \tag{21.2.24}$$

Thus on the k_z versus k_1 plot, the slope of the light line is $\left(\frac{\mu_0\varepsilon_0}{\mu_1\varepsilon_1}\right)^{1/2}$ as shown in Figure 21.7.

[8]Please note again that in this course we will use β and k interchangeably for wavenumbers.

When the mode is at cutoff, the energy of the mode is in the outer region (or free space) and the dispersion curve hugs the light line (so that they have the same slope), so that both the guided mode and the free space wave have the same group velocity. As the frequency increases above the cutoff, the mode is better trapped or confined to the waveguide region. In this case, the group velocity of the mode approaches that of the waveguide medium which is $k_z = k_1$.

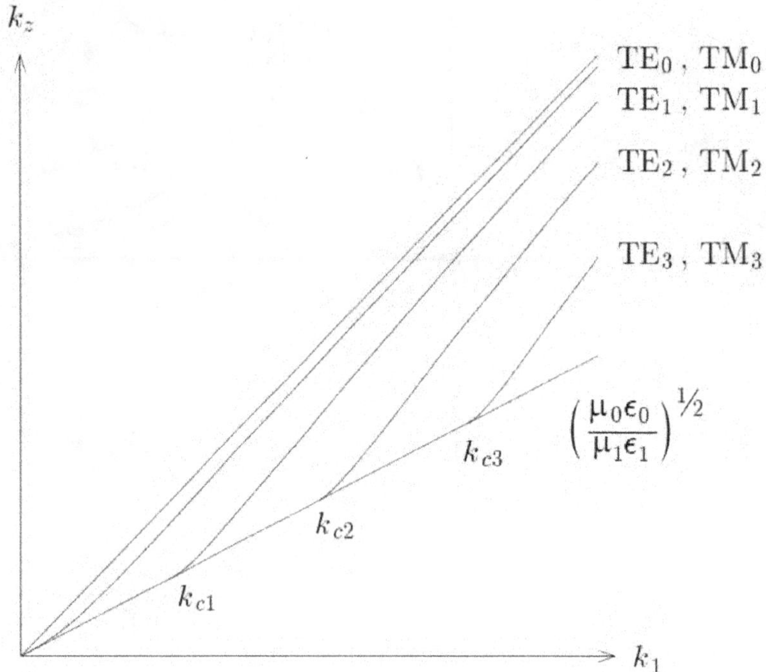

Figure 21.7: Here, we have k_z versus k_1 plots for dielectric slab waveguide. Near its cutoff, the energy of the mode is in the outer region, and hence its group velocity given by $\frac{d\omega}{dk_z}$ is close to that of the outer region. At high frequencies, the mode is tightly bound to the slab, and its group velocity approaches that of the dielectric slab (Figure by J.A. Kong [35]).

21.2.2 TM Case

For the TM case, a similar guidance condition analogous to (21.2.1) can be derived using the duality principle, and with the understanding that the reflection coefficients in (21.2.1) are now TM reflection coefficients. Similar derivations show that the above guidance conditions, for $\epsilon_2 = \epsilon_0$,

$\mu_2 = \mu_0$, and reduce to

$$\frac{\epsilon_0}{\epsilon_1}\beta_{1x}\frac{d}{2}\tan\beta_{1x}\frac{d}{2} = \sqrt{\omega^2(\mu_1\epsilon_1 - \mu_0\epsilon_0)\frac{d^2}{4} - \left(\beta_{1x}\frac{d}{2}\right)^2}, \qquad \text{even modes} \tag{21.2.25}$$

$$-\frac{\epsilon_0}{\epsilon_1}\beta_{1x}\frac{d}{2}\cot\beta_{1x}\frac{d}{2} = \sqrt{\omega^2(\mu_1\epsilon_1 - \mu_0\epsilon_0)\frac{d^2}{4} - \left(\beta_{1x}\frac{d}{2}\right)^2}, \qquad \text{odd modes} \tag{21.2.26}$$

21.2.3 A Note on CutOff of Dielectric Waveguides

The concept of cutoff in dielectric waveguides is quite different from that of hollow waveguides, which we shall learn later in the course. A mode is guided in a dielectric waveguide if the wave is trapped (or confined) inside the dielectric slab. The trapping is due to the total internal reflections at the top and the bottom interfaces of the waveguide. When total internal reflection ceases to occur at any of the two interfaces, the wave is not guided or trapped inside the dielectric slab anymore. This happens when $\alpha_{ix} = 0$ where i can indicate the topmost or the bottommost region. In other words, the wave ceases to be evanescent in either the topmost or the bottommost region.

21.2.4 Alternative Derivation of the Guidance Condition

The previous derivation does not allow us to determine the mode profiles as in Figure 21.5. It would not let us know that if l is even, the mode profile is even and so on. To get this information, we need an alternative derivation as is shown here. We focus on the TE case, because the TM case is easily arrived at by using the duality principle . For the TE case, E_y is a solution to the wave equation in each region. In region 0, we assume a solution of the form

$$E_{0y} = E_0 e^{-j\beta_{0x}x - j\beta_z z} \tag{21.2.27}$$

that corresponds to an upgoing traveling wave from the interface, and

$$\beta_{0x}^2 + \beta_z^2 = \omega^2\mu_0\epsilon_0 = \beta_0^2 \tag{21.2.28}$$

In region 1, we assume a two-way traveling wave, and the solution is of the form

$$E_{1y} = [A_1 e^{-j\beta_{1x}x} + B_1 e^{j\beta_{1x}x}]e^{-j\beta_z z} \tag{21.2.29}$$

where the first term is an upgoing wave while the second term is a downgoing wave, and

$$\beta_{1x}^2 + \beta_z^2 = \omega^2\mu_1\epsilon_1 = \beta_1^2 \tag{21.2.30}$$

In region 2, the solution is a downgoing traveling wave of the form

$$E_{2y} = E_2 e^{j\beta_{2x}x - j\beta_z z} \tag{21.2.31}$$

with

$$\beta_{2x}^2 + \beta_z^2 = \omega^2\mu_2\epsilon_2 = \beta_2^2 \tag{21.2.32}$$

We assume that all the solutions in the three regions have the same z variation of $e^{-j\beta_z z}$ or the same β_z by the phase-matching condition.

In region 1, we have an upgoing wave as well as a downgoing wave. The two waves have to be related by the reflection coefficient R^{TE} for the electric field at the boundaries. Therefore, at $x = \frac{d}{2}$, a downgoing wave is a consequence of the reflection of the upgoing wave. Therefore, we have

$$B_1 e^{j\beta_{1x}\frac{d}{2}} = R_{10}^{TE} A_1 e^{-j\beta_{1x}\frac{d}{2}} \tag{21.2.33}$$

where R_{10}^{TE} is the TE reflection coefficient at the region 1 and region 0 interface.

At $x = -\frac{d}{2}$, an upgoing wave is a consequence of the reflection of the downgoing wave. Or we have

$$A_1 e^{j\beta_{1x}\frac{d}{2}} = R_{12}^{TE} B_1 e^{-j\beta_{1x}\frac{d}{2}} \tag{21.2.34}$$

where R_{12}^{TE} is the reflection coefficient at the region 1 and region 2 interface. Multiplying equations (21.2.33) and (21.2.33) together, we have

$$A_1 B_1 e^{j\beta_{1x}d} = R_{12}^{TE} R_{10}^{T} E A_1 B_1 e^{-j\beta_{1x}d} \tag{21.2.35}$$

A_1 and B_1 are nonzero only if

$$1 = R_{12}^{TE} R_{10}^{TE} e^{-2j\beta_{1x}d} \tag{21.2.36}$$

The above is the exact guidance condition of a dielectric slab waveguide previously derived from the transverse resonance condition in (21.2.37). If medium 3 is equal to medium 1, then $R_{12}^{TE} = R_{10}^{TE}$, and the guidance condition becomes

$$1 = (R_{10}^{TE})^2 e^{-2j\beta_{1x}d} \tag{21.2.37}$$

Taking the square root of (21.2.37), we have

$$R_{10}^{TE} e^{-j\beta_{1x}d} = \pm 1 \tag{21.2.38}$$

When we choose the plus sign, $B_1 = A_1$ from (21.2.33), and from (21.2.29)

$$E_{1y} = 2A_1 \cos(\beta_{1x}x) e^{-j\beta_z z} \quad \Rightarrow \text{even in } x \tag{21.2.39}$$

When we choose the minus sign in (11) we have $B_1 = -A_1$, and

$$E_{1y} = -2jA_1 \sin(\beta_{1x}x) e^{-j\beta_z z} \quad \Rightarrow \text{odd in } x \tag{21.2.40}$$

It can be shown that the two solutions given by (21.2.38) map to the even and odd solutions in (21.2.11) and (21.2.12), and hence their names.

21.3 Photonic Integrated Circuits (PIC)

This is a rapidly emerging area due to rapid growth of lithography process and demand in bandwidth of data tranmission. Even though a slab waveguide is a one-dimensional waveguide, it can be made into a 3D waveguide quite easily by enabling inhomogeneity in the y and z directions in Figure 21.3. This is done approximately using the effective index method [92, 166, 167, 168, 169, 170, 171]. The method allows even 3D problems to be solved approximately [172]. But if the approximate method fails, more accurate numerical methods such as the rigorous coupled wave analysis (RCWA) [173] or the finite-difference time-domain method (FDTD) [174] can be used. Well documented copies of RCWA code is found in [175]. The spirit of RCWA is similar to that of the mode-matching method described in [1].

Exercises for Lecture 21

Problem 21-1:

(i) Explain what a general transverse resonance condition is.

(ii) By going through the lecture notes on alternative derivation in dielectric waveguides, explain why when n in (21.2.9) and (21.2.10) is even or odd, it corresponds to mode profiles that are even or odd, respectively.

(iii) Explain why the TE_0 mode of a dielectric slab waveguide has no cutoff frequency.

(iv) Show that the two solutions given by (21.2.38) map to the even and odd solutions in (21.2.11) and (21.2.12).

Lecture 22

Hollow Waveguides

Hollow waveguides are useful for high-power microwaves. Air has a higher breakdown voltage compared to most materials, and hence could be a good medium for propagating high power electromagnetic wave. In addition, air has low loss. Also, hollow metallic waveguides are sufficiently shielded from the rest of the world so that interference from other sources is minimized. Furthermore, for radio astronomy, they can provide a low-noise system immune to interference. Air generally has less loss than materials, and loss is often the source of thermal noise. Therefore, a low-loss waveguide is also a low-noise waveguide.[1]

22.1 General Information on Hollow Waveguides

It is to be noted that in microwave engineering, a metallic surface can be modeled with a conductive medium. This can be gotten from the discussion of the physics of the Drude-Lorentz-Sommerfeld model in Sect. 8.3.5. When collision is infrequent, then the medium becomes a conductive medium. When the skin depth of the conductive medium is small, because the conductivity is high, it can be approximated by a PEC.

Many waveguides can be approximated with PEC walls. Then these problems can be solved in closed-form. An example is the coaxial waveguide previously discussed with closed form solution. In addition, there are many other waveguide problems that have closed-form solutions. Closed form solutions are desirable and beautiful because they offer physical insight to the problems.

Closed-form solutions to Laplace's and Helmholtz's equations are obtained by the separation-of-variables method. The separation-of-variables method works only for separable coordinate systems. (There are 11 separable coordinates for Helmholtz's equation, but 13 for Laplace's equation.) Some examples of separable coordinate systems are Cartesian, cylindrical, and spherical coordinates.

[1]The fluctuation dissipation theorem [176, 177] says that when a system loses energy to the environment, it also receives the same amount of energy from the environment for energy conservation. In a word, a lossy system loses energy to its environment, but it also receives energy back from the environment in terms of thermal noise. Thus, the lossier a system is, the more thermal noise is needed for energy balance.

These three coordinates are about all we need to know for solving many engineering problems. For problems requiring coordinates other than these three coordinates, complex special functions need to be defined for their solutions, which are hard to compute. Therefore, more complicated cases are now handled with numerical methods using computers.

When a waveguide has a center conductor or two conductors like a coaxial cable , it can support a TEM wave (or TEM mode) where both the electric field and the magnetic field are orthogonal to the direction of propagation. The uniform plane wave is an example of a TEM wave (see section 7.3, (7.3.11)). However, when the waveguide is hollow or is filled completely with a homogeneous medium and lacks a center conductor, it cannot support a TEM mode, as we shall prove next. Much of the material of this lecture can be found in [35, 92, 137].

22.1.1 Absence of a TEM Mode in a Hollow Waveguide

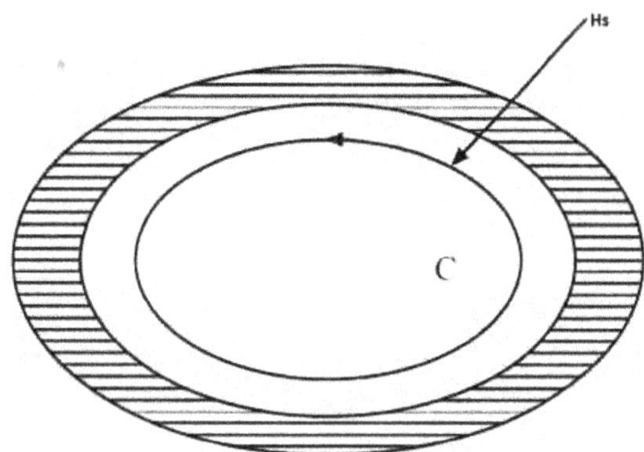

Figure 22.1: The absence of a TEM mode in a hollow waveguide enclosed by a PEC wall. The magnetic field lines form a closed loop inside the waveguide due to the absence of magnetic charges.

We would like to prove by contradiction (reductio ad absurdum) that a hollow waveguide as shown in Figure 22.1 (that is, without a center conductor) cannot support a TEM mode. The proof is as follows: If we assume that a TEM mode does exist, then the magnetic field has to end on itself due to the absence of magnetic charges on the waveguide wall. In this case, it is clear that $\oint_C \mathbf{H}_s \cdot d\mathbf{l} \neq 0$ for any closed contour following the magnetic field lines. But Ampère's law states that the above is equal to

$$\oint_C \mathbf{H}_s \cdot d\mathbf{l} = j\omega\varepsilon \int_S \mathbf{E} \cdot d\mathbf{S} + \int_S \mathbf{J} \cdot d\mathbf{S} \tag{22.1.1}$$

The left-hand side of the above equation is clearly nonzero by the above argument. But for a hollow waveguide, $\mathbf{J} = 0$ and the above becomes

$$\oint_C \mathbf{H}_s \cdot d\mathbf{l} = j\omega\varepsilon \int_S \mathbf{E} \cdot d\mathbf{S} = j\omega\varepsilon \int_S \mathbf{E} \cdot \hat{n} dS \tag{22.1.2}$$

where $\hat{n} = \hat{z}$. Hence, this equation cannot be satisfied unless there is a $E_z \neq 0$ component on the right-hand side. This implies that a TEM mode where both E_z and H_z are zero in a hollow waveguide is impossible without a center conductor.

By the above argument, in a hollow waveguide filled with a homogeneous medium, only TE_z (TE to z) or TM_z (TM to z) modes can exist. For a TE_z wave (or TE wave), $E_z = 0$, and $H_z \neq 0$ while for a TM_z wave (or TM wave), $H_z = 0$, $E_z \neq 0$. These classes of problems can be decomposed into two scalar problems like the layered medium case [2] by using the pilot potential method. However, when the hollow waveguide is filled with a center conductor, the TEM mode can exist in addition to TE and TM modes. (It is to be noted that in such a hollow waveguide, the TEM mode is the degenerate case of either the TE or the TM modes in the waveguide. Hence, the subsequent analyses for TE and TM modes are also valid for the TEM mode if it does exist.)

We begin by studying some simple waveguides, where closed-form solutions exist to hollow waveguides with PEC walls, such as the rectangular waveguides. These closed-form solutions offer physical insight into the propagation of waves in such a waveguide. Another waveguide with slightly more complicated closed-form solutions is the circular hollow waveguide. The solutions need to be sought in terms of Bessel functions. Another waveguide with very complicated closed-form solutions is the elliptical waveguide. However, the solutions are too complicated to be considered; these days, the preferred method of solving these complicated problems is via numerical methods.

22.1.2 TE Case ($E_z = 0$, $H_z \neq 0$, TE_z Case)

In this case, the field inside the waveguide is TE to z (or TE_z). To ensure such a TE field, we can write the $\mathbf{E}$ field as

$$\mathbf{E}(\mathbf{r}) = \nabla \times \hat{z}\Psi_h(\mathbf{r}) \tag{22.1.3}$$

By construction, (22.1.3) will guarantee that $E_z = 0$. Here, $\Psi_h(\mathbf{r})$ is a scalar potential and $\hat{z}$ is called the pilot vector.[3] (The subscript h is used because, as shall be shown, this scalar potential can be related to the z component of the $\mathbf{H}$ field.)

The waveguide is assumed source-free and filled with a lossless, homogeneous material. Eq. (22.1.3) also satisfies the source-free condition since, clearly, $\nabla \cdot \mathbf{E} = 0$. And hence, from Maxwell's equations we see that

$$\nabla \times \mathbf{E} = -j\omega\mu\mathbf{H} \tag{22.1.4}$$

$$\nabla \times \mathbf{H} = j\omega\varepsilon\mathbf{E} \tag{22.1.5}$$

Next, it can be shown that

$$\nabla \times \nabla \times \mathbf{E} - \omega^2\mu\varepsilon\mathbf{E} = 0 \tag{22.1.6}$$

Furthermore, using the appropriate vector identity, such as the BAC-CAB formula, and using the $\nabla \cdot \mathbf{E} = 0$ condition, it can be shown that the electric field $\mathbf{E}(\mathbf{r})$ satisfies the following Helmholtz

[2]Looking at the TE and TM waves in the half-space reflection problem (see Figures 18.1 and 18.3), we see that for TE waves, $H_z \neq 0$, and for TM waves, $E_z \neq 0$.

[3]It "pilots" the field so that it is transverse to z.

wave equation (or partial differential equation) that

$$(\nabla^2 + \beta^2)\mathbf{E}(\mathbf{r}) = 0 \tag{22.1.7}$$

where $\beta^2 = \omega^2\mu\varepsilon$. Substituting (22.1.3) into (22.1.7), we get

$$(\nabla^2 + \beta^2)\nabla \times \hat{z}\Psi_h(\mathbf{r}) = 0 \tag{22.1.8}$$

In the above, we can show that $\nabla^2\nabla \times \hat{z}\Psi = \nabla \times \hat{z}(\nabla^2\Psi)$, or that these operators commute.[4] Then it follows that

$$\nabla \times \hat{z}(\nabla^2 + \beta^2)\Psi_h(\mathbf{r}) = 0 \tag{22.1.9}$$

Thus, if $\Psi_h(\mathbf{r})$ satisfies the following Helmholtz wave equation or partial differential equation

$$(\nabla^2 + \beta^2)\Psi_h(\mathbf{r}) = 0 \tag{22.1.10}$$

then (22.1.9) is satisfied, and so is (22.1.7).[5] Hence, the **E** field constructed with (22.1.3) satisfies Maxwell's equations if $\Psi_h(\mathbf{r})$ satisfies (22.1.10).

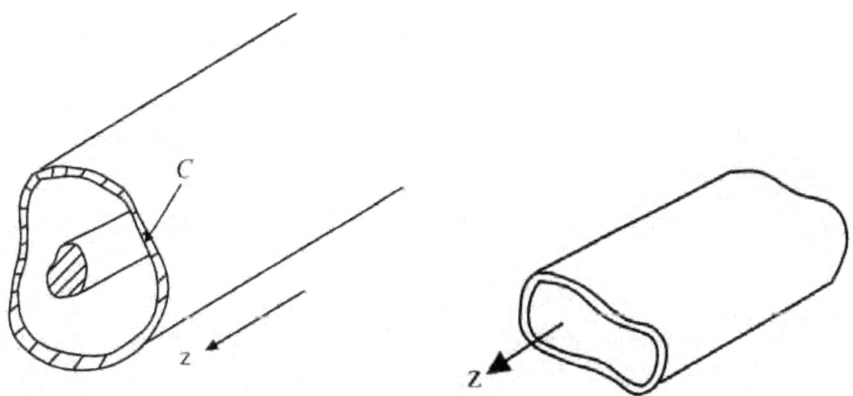

Figure 22.2: A hollow metallic waveguide with a center conductor (left), and without a center conductor (right).

Next, we look at the boundary condition for $\Psi_h(\mathbf{r})$ which is derivable from the boundary condition for **E**. The boundary condition for **E** is that $\hat{n} \times \mathbf{E} = 0$ on C, the PEC wall of the waveguide. But from (22.1.3), using the BAC-CAB formula, then

$$\hat{n} \times \mathbf{E} = \hat{n} \times (\nabla \times \hat{z}\Psi_h) = -\hat{n} \cdot \nabla\Psi_h = 0 \tag{22.1.11}$$

[4]This is a mathematical parlance, and a commutator is defined to be $[A, B] = AB - BA$ for two operators A and B. If these two operators commute, then $[A, B] = 0$.

[5](22.1.10) is a sufficient but unnecessary condition.

(In applying the BAC-CAB formula, one has to be mindful that ∇ operates on a function to its right, and the function Ψ_h should be placed to the right of the ∇ operator always.)

In the above $\hat{n} \cdot \nabla = \hat{n} \cdot \nabla_s$ where $\nabla_s = \hat{x}\frac{\partial}{\partial x} + \hat{y}\frac{\partial}{\partial y}$ (a 2D gradient operator) since $\hat{n}$ has no z component. The boundary condition (22.1.11) then becomes

$$\hat{n} \cdot \nabla_s \Psi_h = \partial_n \Psi_h = 0, \text{ on } C \tag{22.1.12}$$

where C is the waveguide wall where ∂_n is a shorthand notation for $\hat{n} \cdot \nabla_s$ operator which is a scalar operator for a normal derivative. The above is also known as the homogeneous Neumann boundary condition.

Furthermore, in a waveguide, just as in a transmission line case, we are looking for traveling wave solutions of the form $\exp(\mp j\beta_z z)$ for (22.1.10), or

$$\Psi_h(\mathbf{r}) = \Psi_{hs}(\mathbf{r}_s)e^{\mp j\beta_z z} \tag{22.1.13}$$

where $\mathbf{r}_s = \hat{x}x + \hat{y}y$,[6] or in short, $\Psi_{hs}(\mathbf{r}_s) = \Psi_{hs}(x, y)$ is a 2D function. Thus from the above, $\partial_n \Psi_h = 0$ implies that $\partial_n \Psi_{hs} = 0$ since ∂_n involves ∂_x and ∂_y, and only $\Psi_{hs}(x, y)$ is a function of x and y. With this assumption, $\frac{\partial^2}{\partial z^2} \to -{\beta_z}^2$, and (22.1.10) becomes even simpler, namely,

$$({\nabla_s}^2 + \beta^2 - {\beta_z}^2)\Psi_{hs}(\mathbf{r}_s) = ({\nabla_s}^2 + \beta_s^2)\Psi_{hs}(\mathbf{r}_s) = 0\,, \quad \partial_n \Psi_{hs}(\mathbf{r}_s) = 0, \text{ on } C \tag{22.1.14}$$

where $\nabla_s^2 = \partial^2/\partial x^2 + \partial^2/\partial y^2$ and $\beta_s^2 = \beta^2 - \beta_z^2$. The above is a boundary value problem (BVP) for a 2D waveguide problem. The above 2D wave equation (or partial differential equation) is also called the reduced wave equation.

Please notice that the above equation, rewritten as

$$-{\nabla_s}^2 \Psi_{hs}(\mathbf{r}_s) = +\beta_s^2 \Psi_{hs}(\mathbf{r}_s) \tag{22.1.15}$$

is "homomorphic" to the matrix eigenvalue problem $\overline{\mathbf{A}} \cdot \mathbf{x} = \lambda \mathbf{x}$ where ∇_s^2 is playing the role of the matrix operator $\overline{\mathbf{A}}$, while $\Psi_{hs}(\mathbf{r}_s)$ is analogous to the eigenvector $\mathbf{x}$ and β_s^2 is similar to the eigenvalue λ. The difference from the matrix operator is that the differential operator ∇_s^2 does not have a unique inverse unless boundary conditions are specified.

22.1.3 TM Case ($E_z \neq 0$, $H_z = 0$, TM$_z$ Case)

Repeating similar treatment for TM waves, the TM magnetic field is then

$$\mathbf{H} = \nabla \times \hat{z}\Psi_e(\mathbf{r}) \tag{22.1.16}$$

where

$$(\nabla^2 + \beta^2)\Psi_e(\mathbf{r}) = 0 \tag{22.1.17}$$

[6] In waveguide theory here, we will often use "s" to denote a quantity that is transverse to the z-axis, the direction of propagation of the traveling wave.

The subscript e is used for the pilot potential because Ψ_e can be related to the z component of the $\mathbf{E}$ field. We need to derive the boundary condition for $\Psi_e(\mathbf{r})$ from the fundamental boundary condition that $\hat{n} \times \mathbf{E} = 0$ on the PEC waveguide wall. To this end, we find the corresponding $\mathbf{E}$ field by taking the curl of the magnetic field in (22.1.16), and thus the $\mathbf{E}$ field is proportional to

$$\mathbf{E} \sim \nabla \times \nabla \times \hat{z}\Psi_e(\mathbf{r}) = \nabla\nabla \cdot (\hat{z}\Psi_e) - \nabla^2 \hat{z}\Psi_e = \nabla \frac{\partial}{\partial z}\Psi_e + \hat{z}\beta^2\Psi_e \tag{22.1.18}$$

where we have used the BAC-CAB formula to simplify the above. The tangential component of the above is $\hat{n} \times \mathbf{E}$ which is proportional to

$$\hat{n} \times \nabla \frac{\partial}{\partial z}\Psi_e + \hat{n} \times \hat{z}\beta^2\Psi_e$$

In the above, $\hat{n} \times \nabla$ is a tangential derivative, and it is clear that both the above terms will be zero if $\Psi_e = 0$ on the waveguide wall. Therefore, if

$$\Psi_e(\mathbf{r}) = 0 \text{ on } C, \tag{22.1.19}$$

where C is the waveguide wall, then

$$\hat{n} \times \mathbf{E}(\mathbf{r}) = 0 \text{ on } C \tag{22.1.20}$$

Equation (22.1.19) is also called the homogeneous Dirichlet boundary condition.

Next, we assume that

$$\Psi_e(\mathbf{r}) = \Psi_{es}(\mathbf{r}_s)e^{\mp j\beta_z z} \tag{22.1.21}$$

This will allow us to replace $\partial^2/\partial z^2 = -\beta_z^2$ in (22.1.17). With some manipulation, the boundary value problem (BVP) related to equation (22.1.17) reduces to a simpler partial differential equation for a 2D problem, namely, a reduced wave equation,

$$({\nabla_s}^2 + \beta_s^2)\Psi_{es}(\mathbf{r}_s) = 0 \tag{22.1.22}$$

with the homogeneous Dirichlet boundary condition that

$$\Psi_{es}(\mathbf{r}_s) = 0, \; \mathbf{r}_s \text{ on } C \tag{22.1.23}$$

The above summarizes the basic theory for hollow waveguides of arbitrary shape. To illustrate the above theory, we will solve some simple waveguide problems.

22.2 Rectangular Waveguides

Rectangular waveguides are among the simplest waveguides to analyze because closed-form solutions exist in Cartesian coordinates, the simplest of coordinate systems. One can imagine traveling waves in the x and y directions bouncing off the walls of the waveguide causing standing waves to exist inside the waveguide. We have already seen this wave physics in a transmission line: when a transmission line is terminated with a short, traveling waves in both z directions are observed, giving rise to a standing wave. But in a rectangular waveguide, we will see standing waves in the x and y directions, and a traveling wave in the z direction.

As shall be shown, it turns out that not all electromagnetic waves can be guided by a hollow PEC waveguide. Only when the wavelength is short enough, or the frequency is high enough that an electromagnetic wave can be guided by a hollow waveguide. When a waveguide mode cannot propagate in a waveguide, that mode is regarded as cutoff. The concept of cutoff for hollow waveguide is quite different from that of a dielectric waveguide we have studied previously.

22.2.1 TE Modes ($H_z \neq 0$, H Modes or TE$_z$ Modes)

For this mode, the scalar potential $\Psi_{hs}(\mathbf{r}_s)$ satisfies

$$(\nabla_s{}^2 + \beta_s{}^2)\Psi_{hs}(\mathbf{r}_s) = 0, \quad \frac{\partial}{\partial n}\Psi_{hs}(\mathbf{r}_s) = 0 \quad \text{on } C \tag{22.2.1}$$

where $\beta_s{}^2 = \beta^2 - \beta_z{}^2$. A viable solution using separation of variables[7] for $\Psi_{hs}(x, y)$ is then

$$\Psi_{hs}(x, y) = A\cos(\beta_x x)\cos(\beta_y y) \tag{22.2.2}$$

where $\beta_x{}^2 + \beta_y^2 = \beta_s^2$. One can see that the above is the representation of standing waves in the x and y directions. It is quite clear that $\Psi_{hs}(x, y)$ satisfies the boundary value problem (BVP) and boundary conditions defined by equation (22.2.1). Furthermore, cosine functions, rather than sine functions, are chosen with the hindsight that the above satisfies the homogenous Neumann boundary condition at $x = 0$, and $y = 0$ surfaces.

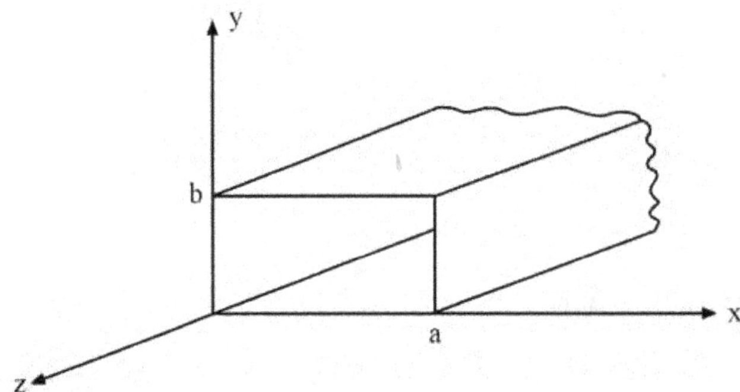

Figure 22.3: The schematic of a rectangular waveguide. By convention, the length of the longer side is usually named a.

To further satisfy the boundary condition at $x = a$ and $y = b$ surfaces, it is necessary that the boundary condition for eq. (22.1.12) is satisfied or that

$$\partial_x \Psi_{hs}(x, y)\big|_{x=a} \sim \sin(\beta_x a)\cos(\beta_y y) = 0, \tag{22.2.3}$$

[7]For those who are not familiar with this topic, please consult p. 385 of Kong [35].

$$\partial_y \Psi_{hs}(x, y)\big|_{y=b} \sim \cos(\beta_x x)\sin(\beta_y b) = 0, \tag{22.2.4}$$

The above puts constraints on the values of β_x and β_y, implying that $\beta_x a = m\pi$, $\beta_y b = n\pi$ where m and n are integers. Hence, (22.2.2) becomes

$$\Psi_{hs}(x, y) = A\cos\left(\frac{m\pi}{a}x\right)\cos\left(\frac{n\pi}{b}y\right) \tag{22.2.5}$$

after letting $\beta_x = \frac{m\pi}{a}$, $\beta_y = \frac{n\pi}{b}$. They can only take on these values in order for the boundary conditions to be satisfied. Consequently,

$$\beta_x^2 + \beta_y^2 = \left(\frac{m\pi}{a}\right)^2 + \left(\frac{n\pi}{b}\right)^2 = \beta_s^2 = \beta^2 - {\beta_z}^2 \tag{22.2.6}$$

Clearly now, (22.2.5) satisfies the requisite homogeneous Neumann boundary condition at the four waveguide walls.

At this point, it is prudent to stop and ponder on what we have done. Equation (22.2.1) is homomorphic to a matrix eigenvalue problem

$$\overline{\mathbf{A}} \cdot \mathbf{x}_i = \lambda_i \mathbf{x}_i \tag{22.2.7}$$

where $\mathbf{x}_i$ is the eigenvector and λ_i is the eigenvalue. Therefore, β_s^2 is actually an eigenvalue, and $\Psi_{hs}(\mathbf{r}_s)$ is an eigenfunction (or an eigenmode), which is analogous to an eigenvector. Here, the eigenvalue β_s^2 is indexed by m, n, and so is the eigenfunction in (22.2.5): it is also indexed by m and n. The corresponding eigenmode is also called the TE_{mn} mode.

The above condition on β_s^2 expressed by (22.2.6) is also known as the guidance condition for the modes in the waveguide. Furthermore, from (22.2.6),

$$\beta_z = \sqrt{\beta^2 - \beta_s^2} = \sqrt{\beta^2 - \left(\frac{m\pi}{a}\right)^2 - \left(\frac{n\pi}{b}\right)^2} \tag{22.2.8}$$

And from (22.2.8), when the frequency is low enough, then

$$\beta^2 = \omega^2\mu\varepsilon < \beta_s^2 = \left(\frac{m\pi}{a}\right)^2 + \left(\frac{n\pi}{b}\right)^2 \tag{22.2.9}$$

and β_z in (22.2.8) becomes pure imaginary and the mode cannot propagate or becomes evanescent in the z direction.[8] For fixed m and n, the frequency at which the above happens is called the cutoff frequency of the TE_{mn} mode of the waveguide. It is given by

$$\omega_{mn,c} = \frac{1}{\sqrt{\mu\varepsilon}}\sqrt{\left(\frac{m\pi}{a}\right)^2 + \left(\frac{n\pi}{b}\right)^2} \tag{22.2.10}$$

When $\omega < \omega_{mn,c}$, or the wavelength is longer than a certain critical value,[9] the TE_{mn} mode is evanescent and cannot propagate inside the waveguide. The corresponding cutoff wavelength is

[8] We saw this happening in a plasma medium earlier and also in total internal reflection.

[9] We use the formula $\beta = \omega/v = 2\pi/\lambda$ to find the wavelength, assuming a plane wave propagates in the same homogeneous medium that fills the waveguide.

then

$$\lambda_{mn,c} = \frac{2}{[\left(\frac{m}{a}\right)^2 + \left(\frac{n}{b}\right)^2]^{1/2}} \tag{22.2.11}$$

So when $\lambda > \lambda_{mn,c}$, the mode cannot propagate inside the waveguide or it cannot "enter" the waveguide.

Lowest Guided TE Mode in a Rectangular Waveguide

When $m = n = 0$, then $\Psi_h(\mathbf{r}) = \Psi_{hs}(x, y) \exp(\mp j\beta_z z)$ from (22.2.5) is a function independent of x and y. Then $\mathbf{E}(\mathbf{r}) = \nabla \times \hat{z}\Psi_h(\mathbf{r}) = \nabla_s \times \hat{z}\Psi_h(\mathbf{r}) = 0$. It turns out the only way for $H_z \neq 0$ is for $\mathbf{H}(\mathbf{r}) = \hat{z}H_0$ which is a static field in the waveguide. This is not a very interesting mode, and thus TE_{00} propagating mode is assumed not to exist and not useful. So the TE_{mn} modes cannot have both $m = n = 0$.

Thus, the TE_{10} mode, when $a > b$, is the mode with the lowest cutoff frequency or longest cutoff wavelength. Only when the frequency is above this cutoff frequency and the wavelength is shorter than this cutoff wavelength, can then be the TE_{10} mode propagate.

For the TE_{10} mode, for the mode to propagate, from (22.2.11), it is needed that

$$\lambda < \lambda_{10,c} = 2a \tag{22.2.12}$$

The above has the nice physical meaning that the wavelength has to be smaller than $2a$ in order for the mode to fit into the waveguide. As a mnemonic, we can think that photons have "sizes", corresponding to its wavelength. Only when its wavelength is small enough can the photons go into (or be guided by) the waveguide. The TE_{10} mode, when $a > b$, is also the mode with the lowest cutoff frequency or longest cutoff wavelength.

It is seen with the above analysis, that when the wavelength is short enough or the frequency is high enough, many modes can be guided. Each of these modes has a different group and phase velocity. But for most applications, only a single guided mode is desirable. Hence, it is important to know the cutoff frequencies of the fundamental mode (the mode with the lowest cutoff frequency) and what the next higher mode is. This allows one to pick a frequency window within which only a single mode can propagate in the waveguide.

It is to be noted that when a mode is cutoff, the field is evanescent, and there is no real power flow down the waveguide: only reactive power is carried by such a mode as the evanescent wave in total internal reflection (see exercise problem in lecture 18).

Exercises for Lecture 22

Problem 22-1: Find the group and phase velocity of a guided mode in a hollow waveguide.

(i) Show that their product is always a constant.

(ii) Explain why the group velocities are zero and the phase velocities are infinite right at cutoff of a mode.

(iii) What do the group and phase velocities become when the frequency is very high? Explain why. (Hint: Visualize the orientation of the $\boldsymbol{\beta}$ vector in a hollow waveguide as the frequency becomes very large.)

Lecture 23

More on Hollow Waveguides

We have seen that the hollow PEC waveguide is one of the simplest of waveguides other than the transmission line. Closed-form solutions exist for such waveguides as seen in the rectangular waveguide case. The solution is elegantly simple and beautiful, requiring only trigonometric functions. So we will continue with the study of the rectangular waveguide, and then address another waveguide, the circular waveguide where closed-form solutions also exist. However, the solution has to be expressed in terms of "Bessel functions", called special functions. As the name implies, these functions are seldom used outside the context of studying wave phenomena. Bessel functions in cylindrical coordinates are the close cousin of the sinusoidal functions in Cartesian coordinates. Whether Bessel functions are more complex or esoteric compared to sinusoidal functions is in the eyes of the beholder. Once we become familiar with them, they are simple! They are also the function that describes the concentric ripple wave that you see in your tea cup every morning (see Figure 23.1)!

Figure 23.1: The ripple wave (also called capillary wave) in your tea cup is describable by a Bessel function (Figure by dreamstime.com).

23.1 Rectangular Waveguides, Cont.

We have seen the mathematics for the TE modes of a rectangular waveguide in the previous lecture. We shall now study the TM modes and both the TE and TM modes of a circular waveguide.

23.1.1 TM Modes ($E_z \neq 0$, E Modes or TM$_z$ Modes)

These modes are not the exact dual of the TE modes because of the boundary conditions. The dual of a PEC (perfect electric conductor) wall is a PMC (perfect magnetic conductor) wall. However, the previous exercise for TE modes can be repeated for the TM modes with caution on the boundary conditions. The scalar wave function (or eigenfunction/eigenmode) for the TM modes, satisfying the homogeneous Dirichlet (instead of Neumann)[1] boundary condition with ($\Psi_{es}(\mathbf{r}_s) = 0$) on the entire waveguide wall is

$$\Psi_{es}(x, y) = A \sin\left(\frac{m\pi}{a}x\right) \sin\left(\frac{n\pi}{b}y\right) \tag{23.1.1}$$

In the above, remember that the meaning of the subscript e implies that this scalar potential Ψ is related to the z component of the $\mathbf{E}$ field, while the subscript s implies that this refers to a quantity that is transverse to the z-axis. Here, $\beta_x = \frac{m\pi}{a}$ and $\beta_y = \frac{n\pi}{b}$. Sine functions are chosen for the standing waves, and the chosen values of β_x and β_y ensure that the Dirichlet boundary condition is satisfied on the $x = a$ and $y = b$ walls. Neither of the m and n can be zero, lest $\Psi_{es}(x, y) = 0$, or the entire field is zero. Hence, both $m > 0$, and $n > 0$ are needed. Thus, the lowest TM mode is the TM$_{11}$ mode. Thinking of this as an eigenvalue problem, as mentioned in the previous lecture, then the eigenvalue is

$$\beta_s^2 = \beta_x^2 + \beta_y^2 = \left(\frac{m\pi}{a}\right)^2 + \left(\frac{n\pi}{b}\right)^2 \tag{23.1.2}$$

which is the same as the TE case. Therefore, the corresponding cutoff frequencies and cutoff wavelengths for the TM$_{mn}$ modes are the same as the TE$_{mn}$ modes. Also, these TE and TM modes are degenerate since they share the same eigevalues. In other words, the lowest mode, which is the TM$_{11}$ mode and the TE$_{11}$ have the same cutoff frequency. Figure 23.2 shows the dispersion curves for different modes of a rectangular waveguide. Notice that the group velocities of all the modes are zero at cutoff, and then the group velocities approach that of the waveguide medium inside as frequency increases. These observations can be explained physically when we study the bouncing wave picture next.

23.1.2 Bouncing Wave Picture

We have seen that the transverse variation of a mode in a rectangular waveguide can be expanded in terms of sine and cosine functions, which represent standing waves and are a superposition of two traveling waves, or they are

$$\begin{aligned}&\left[e^{-j\beta_x x} \pm e^{j\beta_x x}\right]\left[e^{-j\beta_y y} \pm e^{j\beta_y y}\right]\\ &\qquad = e^{-j\beta_x x - j\beta_y y} + e^{j\beta_x x + j\beta_y y} \pm e^{j\beta_x x - j\beta_y y} \pm e^{-j\beta_x x + j\beta_y y}\end{aligned} \tag{23.1.3}$$

[1]Again, "homogeneous" here for the math community means "zero".

Each term on the right-hand side corresponds to a plane wave traveling in different directions. When the above is expanded and together with the $\exp(-j\beta_z z)$, the mode is propagating in the z direction in addition to being the standing waves in the transverse direction. Or we see four waves bouncing around in the x and y directions and propagating in the z direction. The picture of this bouncing wave is depicted in Figure 23.3

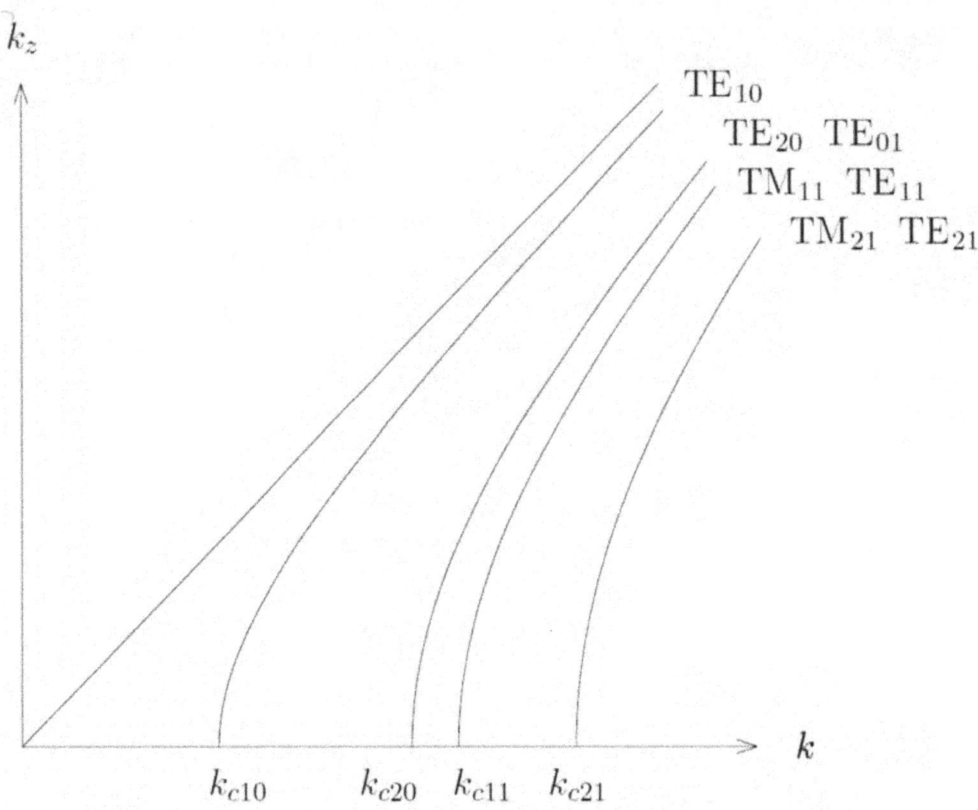

Figure 23.2: Dispersion curves for a rectangular waveguide (Figure by J.A. Kong [35]). Notice that the lowest TM mode is the TM_{11} mode, and that the TM_{mn} modes and the TE_{mn} modes have the same cutoff frequencies if they exist. Here, k is equivalent to β in this course. At cutoff, $\beta_z = k_z = 0$, or the guided mode does not propagate in the z direction, and as shown above, the group velocity is zero. But when $\omega \to \infty$, the mode propagates in a direction almost parallel to the axis of the waveguide (this is called the "paraxial wave" in the parlance of wave physics), and hence, the group velocity approaches that of the waveguide medium.

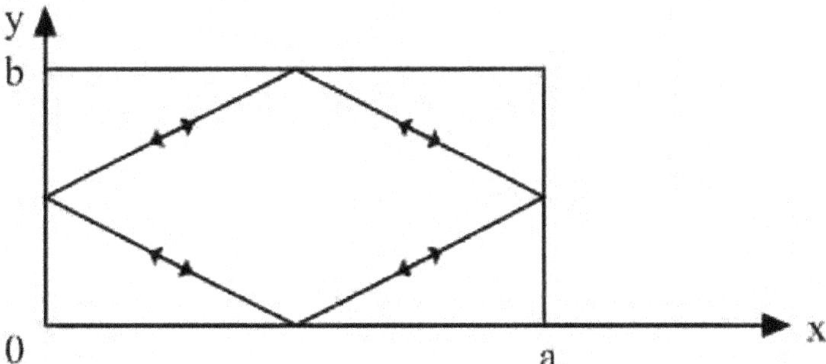

Figure 23.3: The waves in a rectangular waveguide can be thought of as bouncing waves off the four walls as they propagate in the z direction out of the paper.

23.1.3 Field Plots

Given the knowledge of the vector pilot potential of a waveguide, one can derive all the field components. For example, for the TE modes, if we know $\Psi_h(\mathbf{r})$, then

$$\mathbf{E} = \nabla \times \hat{z}\Psi_h(\mathbf{r}), \qquad \mathbf{H} = -\nabla \times \mathbf{E}/(j\omega\mu) \tag{23.1.4}$$

Then all the electromagnetic fields of a waveguide mode can be found and computed for these modes, and similarly for TM modes.

Plots of the fields of different rectangular waveguide modes are shown in Figure 23.4. Notice that for higher m's and n's, with $\beta_x = m\pi/a$ and $\beta_y = n\pi/b$, the corresponding β_x and β_y are larger with higher spatial frequencies.

Thus, the transverse spatial wavelengths are getting shorter. Also, since $\beta_z = \sqrt{\beta^2 - \beta_x^2 - \beta_y^2} = \sqrt{\beta^2 - (m\pi/a)^2 - (n\pi/b)^2}$, higher frequencies where $\beta^2 = \omega^2\mu\varepsilon$ are larger than $(m\pi/a)^2 + (n\pi/b)^2$ to make β_z real in order to propagate the higher order modes or the high m and n modes in a rectangular waveguide.

Notice also how the electric field and magnetic field curl around each other. Since $\nabla \times \mathbf{H} = j\omega\varepsilon\mathbf{E}$ and $\nabla \times \mathbf{E} = -j\omega\mu\mathbf{H}$, they do not curl around each other "immediately" in the time domain, but with a $\pi/2$ phase delay because of the $j\omega$ factor corresponding to a time delay in the time domain. Therefore, in a snapshot of the $\mathbf{E}$ and $\mathbf{H}$ fields shown in Figure 23.5: they do not curl around each other at one location, but at a displaced location due to the $\pi/2$ phase difference due to $j\omega$ giving rise to a time delay.

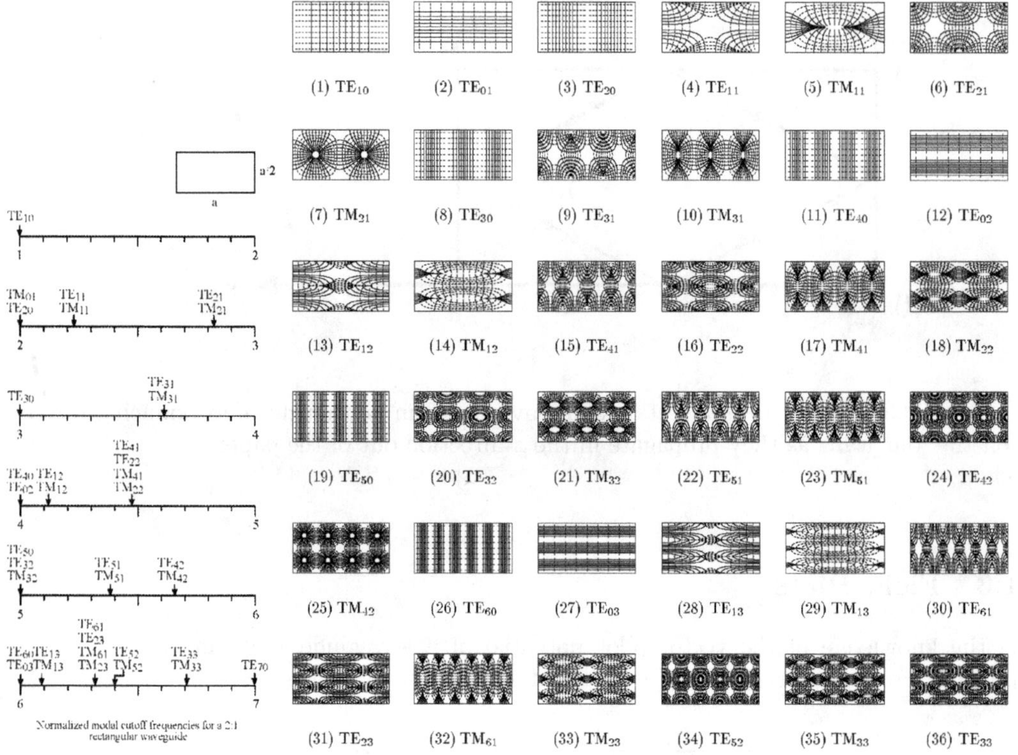

Figure 23.4: Transverse field plots of different modes in a rectangular waveguide (Figure by Andy Greenwood. Original plots published in Lee, Lee, and Chuang, IEEE T-MTT, 33.3 (1985): pp. 271-274. [178]).

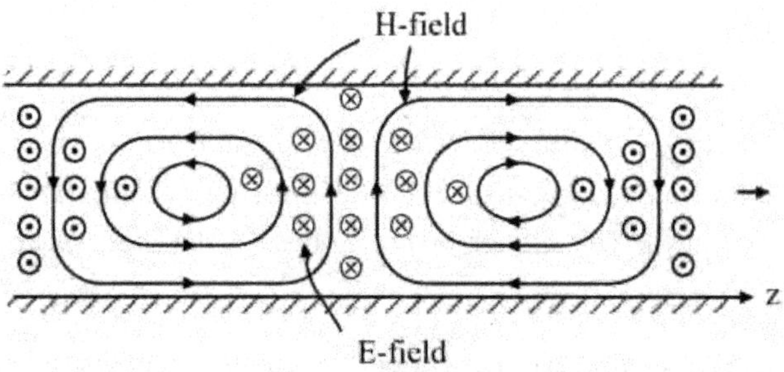

Figure 23.5: A snapshot of the field plot of a TE_{10} mode propagating in the z direction of a rectangular waveguide. Notice that the **E** and **H** fields do not exactly curl around each other because of the time delay.

23.2 Circular Waveguides

Another waveguide whose closed-form solutions can be easily obtained is the circular hollow waveguide as shown in Figure 23.6. Now they involve the use of Bessel functions, which are special functions. They are different from the trigonometric functions which are used pervasively.

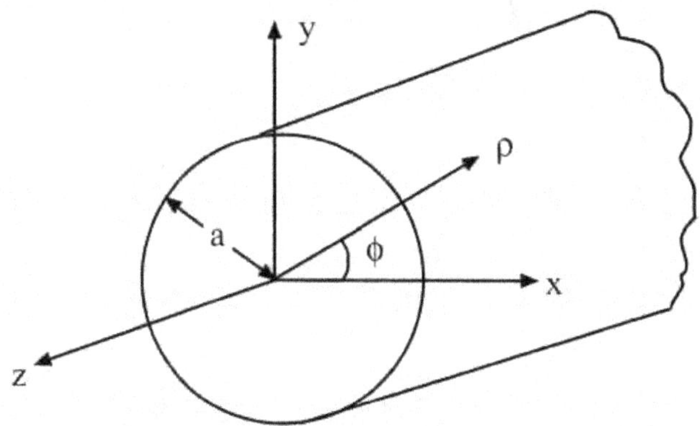

Figure 23.6: Schematic of a circular waveguide in cylindrical coordinates. It is one of the separable coordinate systems.

23.2.1 TE Case

For a circular waveguide, it is best first to express the Laplacian operator, ${\nabla_s}^2 = \nabla_s \cdot \nabla_s$, in cylindrical coordinates. The second term ∇_s is a gradient operator while the first term $\nabla_s \cdot$ is a divergence operator: they have different physical meanings, and look quite different in cylindrical coordinates. Formulas for grad and div operators in cylindrical coordinates are given in many text books [35, 179]. Doing a table lookup,

$$\nabla_s \Psi = \hat{\rho}\frac{\partial}{\partial \rho}\Psi + \hat{\phi}\frac{1}{\rho}\frac{\partial}{\partial \phi}$$

$$\nabla_s \cdot \mathbf{A} = \frac{1}{\rho}\frac{\partial}{\partial \rho}\rho A_\rho + \frac{1}{\rho}\frac{\partial}{\partial \phi}A_\phi$$

Then using the definitionn above for the evaluation of ${\nabla_s}^2 = \nabla_s \cdot \nabla_s$, we have

$$\left({\nabla_s}^2 + {\beta_s}^2\right)\Psi_{hs} = \left(\frac{1}{\rho}\frac{\partial}{\partial \rho}\rho\frac{\partial}{\partial \rho} + \frac{1}{\rho^2}\frac{\partial^2}{\partial \phi^2} + {\beta_s}^2\right)\Psi_{hs}(\rho, \phi) = 0 \tag{23.2.1}$$

The above is the partial differential equation for field in a circular waveguide. It is an eigenvalue problem where β_s^2 is the eigenvalue, and $\Psi_{hs}(\mathbf{r}_s)$ is the eigenfunction (equivalence of an eigenvector). Here, $\mathbf{r}_s = \hat{\rho}\rho + \hat{\phi}\phi$ is the transverse to z position vector in cylindrical coordinates in 2D. Please be reminded that the above is "homomorphic" to a matrix eigenvalue problem as discussed at the end of section 22.1.2 in the previous lecture.

Using separation of variables, we let

$$\Psi_{hs}(\rho, \phi) = B_n(\beta_s \rho) e^{\pm jn\phi} \tag{23.2.2}$$

Then $\frac{\partial^2}{\partial \phi^2} \to -n^2$, and (23.2.1) simplifies to an ordinary differential equation which is

$$\left(\frac{1}{\rho}\frac{d}{d\rho}\rho\frac{d}{d\rho} - \frac{n^2}{\rho^2} + {\beta_s}^2\right) B_n(\beta_s\rho) = 0 \tag{23.2.3}$$

Here, dividing the above equation by β_s^2, we can let $\beta_s\rho$ in (23.2.2) and (23.2.3) be x. Then the above can be rewritten as

$$\left(\frac{1}{x}\frac{d}{dx}x\frac{d}{dx} - \frac{n^2}{x^2} + 1\right) B_n(x) = 0 \tag{23.2.4}$$

The above is known as Bessel's equation, whose solutions are special functions denoted as $B_n(x)$.[2]

These special functions, also called cylinder functions, are $J_n(x)$, $N_n(x)$, $H_n^{(1)}(x)$, and $H_n^{(2)}(x)$ which are called the Bessel, Neumann, and Hankel functions of the first kind and the Hankel function of the second kind, respectively, where n is their order and x is their arguments.[3] Since this is a second order ordinary differential equation, it has only two independent solutions. Therefore, only two of the four commonly encountered solutions of Bessel's equation are independent: They can be expressed in terms of each other. Their relationships are shown below:[4]

Bessel, $$J_n(x) = \frac{1}{2}[{H_n}^{(1)}(x) + {H_n}^{(2)}(x)] \tag{23.2.5}$$

Neumann, $$N_n(x) = \frac{1}{2j}[{H_n}^{(1)}(x) - {H_n}^{(2)}(x)] \tag{23.2.6}$$

Hankel–First Kind, $${H_n}^{(1)}(x) = J_n(x) + jN_n(x) \tag{23.2.7}$$

Hankel–Second Kind, $${H_n}^{(2)}(x) = J_n(x) - jN_n(x) \tag{23.2.8}$$

When x is large, using the asymptotic formulas for these special functions [115], it can be shown that

$${H_n}^{(1)}(x) \sim \sqrt{\frac{2}{\pi x}} e^{jx - j(n+\frac{1}{2})\frac{\pi}{2}}, \quad x \to \infty \tag{23.2.9}$$

$${H_n}^{(2)}(x) \sim \sqrt{\frac{2}{\pi x}} e^{-jx + j(n+\frac{1}{2})\frac{\pi}{2}}, \quad x \to \infty \tag{23.2.10}$$

---

[2]This equation was studied by Friedrich Wilhelm Bessel, 1784–1846. It can also be written as $\left(\frac{d^2}{dx^2} + \frac{1}{x}\frac{d}{dx} - \frac{n^2}{x^2} + 1\right) B_n(x) = 0$.

[3]Some textbooks use $Y_n(x)$ for Neumann functions.

[4]Their relations with each other are similar to those between $\exp(\pm jx)$, $\sin(x)$, and $\cos(x)$. The marked difference is that the trigonometric functions are all regular, but some of the cylinder special functions are singular at $x = 0$.

They correspond to traveling wave solutions when $x = \beta_s\rho \to \infty$.

Since $J_n(x)$ and $N_n(x)$ are linear superpositions of these traveling wave solutions, they correspond to standing wave solutions. Moreover, $N_n(x)$, ${H_n}^{(1)}(x)$, and ${H_n}^{(2)}(x)$ are singular at $x = 0$. Or in a word, they tend to ∞ when $x \to 0$. Since the field has to be regular when $\rho \to 0$ at the center of the waveguide shown in Figure 23.6, the only viable solution for the hollow waveguide, to be chosen from (23.2.5) to (23.2.9), is that $B_n(\beta_s\rho) = AJ_n(\beta_s\rho)$, which is regular at $\rho = 0$. Thus, for a circular hollow waveguide, the eigenfunction or mode has to be of the form

$$\Psi_{hs}(\rho, \phi) = AJ_n(\beta_s\rho)e^{\pm jn\phi} \tag{23.2.11}$$

To ensure that the eigenfunction and the eigenvalue are unique, a boundary condition for the partial differential equation is needed. The homogeneous Neumann boundary condition,[5] or that $\partial_n\Psi_{hs} = 0$, on the PEC waveguide wall then translates to

$$\frac{d}{d\rho}J_n(\beta_s\rho) = 0, \quad \rho = a \tag{23.2.12}$$

Defining ${J_n}'(x) = \frac{d}{dx}J_n(x)$,[6] the above is the same as

$${J_n}'(\beta_s a) = 0 \tag{23.2.13}$$

The above are the zeros of the derivative of Bessel functions and they are tabulated in many textbooks and handbooks.[7] The m-th zero of ${J_n}'(x)$ is denoted here to be ξ_{nm}, a dimensionless number. Plots of Bessel functions and their derivatives are shown in Figure 23.8, and some zeros of Bessel functions and their derivatives are also shown in Tables 23.2.1 and 23.2.2. With this knowledge, the guidance condition for a waveguide mode is then

$$\beta_s = \xi_{nm}/a \tag{23.2.14}$$

for the TE$_{nm}$ mode. From the above, β_s^2 can be obtained which is the eigenvalue of (23.2.1) and (23.2.3). It is a constant independent of frequency.

Using the fact that $\beta_z = \sqrt{\beta^2 - \beta_s^2}$, then β_z will become pure imaginary if $\beta = \omega\sqrt{\mu\varepsilon}$ is small. Here, β can be made small by lowering the frequency so that $\beta^2 < \beta_s^2$. From this, the corresponding cutoff frequency of the TE$_{nm}$ mode is

$$\omega_{nm,c} = \frac{1}{\sqrt{\mu\varepsilon}}\frac{\xi_{nm}}{a} \tag{23.2.15}$$

When $\omega < \omega_{nm,c}$, the corresponding mode cannot propagate in the waveguide as β_z becomes pure imaginary. The corresponding cutoff wavelength, using $\omega\sqrt{\mu\varepsilon} = \beta = 2\pi/\lambda$, is then

$$\lambda_{nm,c} = \frac{2\pi}{\xi_{nm}}a \tag{23.2.16}$$

By the same token, when $\lambda > \lambda_{nm,c}$, the corresponding mode cannot be guided by the waveguide. It is not exactly precise to say this, but this gives us the heuristic notion that if the wavelength or "size" of the wave or photon is too big, it cannot "fit" inside the waveguide.

[5]Note that "homogeneous" here means "zero" in math.

[6]Note that this is a standard math notation, which has a different meaning in some engineering texts.

[7]Notably, Abramowitz and Stegun, *Handbook of Mathematical Functions* [115]. An online version is available [180].

23.2.2 TM Case

The corresponding partial differential equation and boundary value problem for this case is[8]

$$\left(\frac{1}{\rho}\frac{\partial}{\partial\rho}\rho\frac{\partial}{\partial\rho}+\frac{1}{\rho^2}\frac{\partial^2}{\partial\phi^2}+{\beta_s}^2\right)\Psi_{es}(\rho,\phi)=0 \tag{23.2.17}$$

with the homogeneous Dirichlet boundary condition, $\Psi_{es}(a,\phi)=0$, on the waveguide wall.[9] The eigenfunction solution is

$$\Psi_{es}(\rho,\phi)=AJ_n(\beta_s\rho)e^{\pm jn\phi} \tag{23.2.18}$$

with the boundary condition that $J_n(\beta_s a)=0$. The m-th zeros of $J_n(x)$ are labeled as α_{nm} here [115, 35], as well as in Tables 23.2.1 and 23.2.2; and hence, the guidance condition for the TM_{nm} mode is that

$$\beta_s=\frac{\alpha_{nm}}{a} \tag{23.2.19}$$

where the eigenvalue for (23.2.17) is β_s^2 which is a constant independent of frequency, analogous to the TE case. With $\beta_z=\sqrt{\beta^2-\beta_s^2}$, the corresponding cutoff frequency is

$$\omega_{nm,c}=\frac{1}{\sqrt{\mu\varepsilon}}\frac{\alpha_{nm}}{a} \tag{23.2.20}$$

or when $\omega<\omega_{nm,c}$, the mode cannot be guided. The cutoff wavelength is then

$$\lambda_{nm,c}=\frac{2\pi}{\alpha_{nm}}a \tag{23.2.21}$$

with the notion that when $\lambda>\lambda_{nm,c}$, the mode cannot be guided.

Tables 23.2.1 and 23.2.2 show the zeros ξ_{nm} and α_{nm}. They are important for figuring out the cutoff frequencies of the TE and TM modes of a circular hollow waveguide.

It turns out that the lowest mode in a circular waveguide is the TE_{11} mode. It is actually a close cousin of the TE_{10} mode of a rectangular waveguide. This can be gathered by comparing their field plots: these modes morph into each other as we deform the shape of a rectangular waveguide into a circular waveguide.

[8]Again, this is analogous to the matrix eigenvalue problem where differential operators are matrix operators, β_s^2 is the eigenvalue, and the eigenfunction Ψ_{es} is the eigenvector.

[9]Again, note that this problem is analogous to the matrix eigenvalue problem. The matrix has a unique inverse (or is full rank) only if a boundary condition is stipulated for the differential operator.

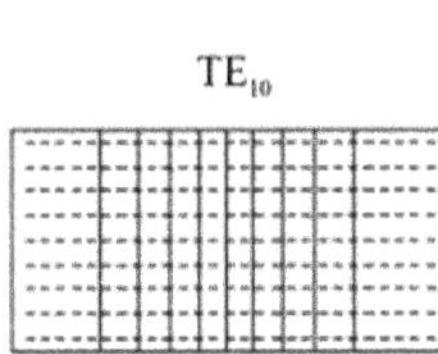

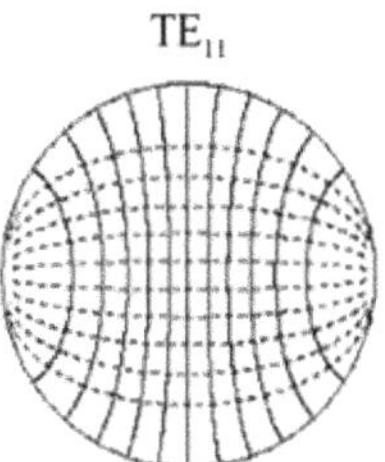

Figure 23.7: Side-by-side comparison of the field plots of the TE_{10} mode of a rectangular waveguide versus that of the TE_{11} mode of a circular waveguide. If one is imaginative enough, one can see that the field plot of TE_{10} mode morphs into that of TE_{11} mode as we change the waveguide shape. Electric fields are those that have to end on the waveguide walls with $\hat{n} \times \mathbf{E} = 0$. The E field is normal to the waveguide all.

Table 23.2.1. Roots of $J_n'(x) = 0$.

n	ξ_{n1}	ξ_{n2}	ξ_{n3}	ξ_{n4}
0	3.832	7.016	10.174	13.324
1	1.841	5.331	8.536	11.706
2	3.054	6.706	9.970	13.170
3	4.201	8.015	11.346	14.586
4	5.318	9.282	12.682	15.964
5	6.416	10.520	13.987	17.313

Table 23.2.2. Roots of $J_n(x) = 0$.

n	α_{n1}	α_{n2}	α_{n3}	α_{n4}
0	2.405	5.520	8.654	11.792
1	3.832	7.016	10.174	13.324
2	5.135	8.417	11.620	14.796
3	6.380	9.761	13.015	16.223
4	7.588	11.065	14.373	17.616
5	8.771	12.339	15.700	18.980

Figure 23.8 shows the plots of the Bessel function $J_n(x)$ and its derivative $J_n'(x)$. Tables 23.2.1 and 23.2.2 show the roots of $J_n'(x)$ and $J_n(x)$, which are important for determining the cutoff frequencies of the TE and TM modes of circular waveguides. They are useful for determining the guidance conditions of the TE_{nm} mode and TM_{nm} mode of a circular waveguide [92].

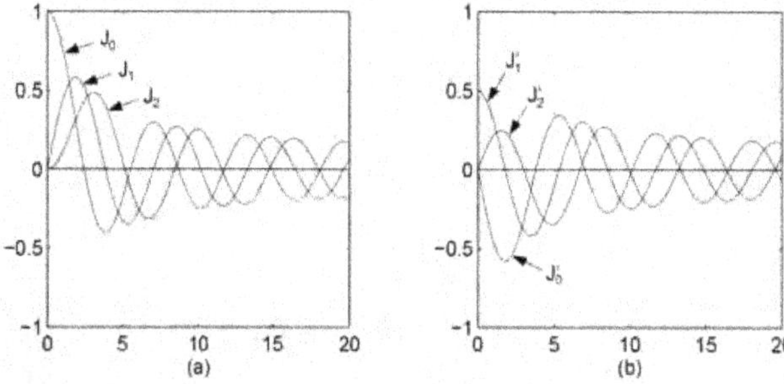

Figure 23.8: Plots of the Bessel functions, $J_n(x)$, of different orders, and their derivatives $J_n'(x)$. The zeros of these functions are used to find the eigenvalue β_s^2 of the problem, and hence, the guidance condition. The left figure is for TM modes, while the right figure is for TE modes. Here, $J_n'(x) = dJ_n(x)/dx$ [92].

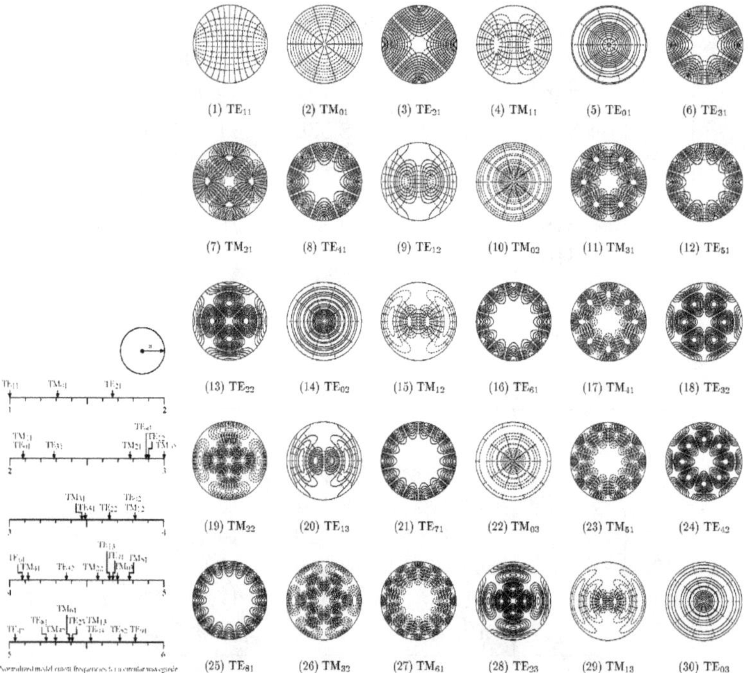

Figure 23.9: Transverse field plots of different modes in a circular waveguide (Figure by Andy Greenwood. [178]). The axially symmetric TE_{01} mode has the lowest loss and hence, finds a number of real-world applications, such as in radio astronomy.

Exercises for Lecture 23

Problem 23-1:

(i) For a 2 cm by 1 cm rectangular waveguide, and a 1 cm radius circular waveguide, find the first three propagating modes starting with the lowest cutoff frequencies.

(ii) For the TM modes, when we impose homogeneous Dirichlet boundary condition on the waveguide wall. Show that all components of the tangential electric field will be zero.

(iii) For the TM mode, starting with that

$$\mathbf{H}(\mathbf{r}) = \nabla \times \hat{z}\Psi_e(\mathbf{r})$$

Find all the components of electromagnetic fields in a rectangular waveguide.

(iv) Starting with a TM mode in a waveguide where $H_z = 0$, and $E_z \neq 0$ initially, show that if this mode becomes a TEM mode so that $H_z = 0$, $E_z = 0$ then $\beta_z = \beta$. (This can happen in a coaxial waveguide, for instance.) What happens to β_s for the TEM mode, and what happens to the Helmholtz equations that $\Psi_e(\mathbf{r})$ and $\Psi_h(\mathbf{r})$ originally satisfy? Explain why the fields of the TEM mode of a waveguide, if they exist, are electrostatic and magnetostatic in nature.

Lecture 24

More on Waveguides and Transmission Lines

Waveguides are a fundamental component of microwave circuits and systems. The study of closed-form solutions offers us physical insight. One can use such insight to design more complex engineering systems. In this lecture, we will use heuristics to understand some complex systems whose designs follow from the physical insight of simpler systems.

In addition, we will show that the waveguide problem is homomorphic to the transmission line problem. Here again, many transmission line techniques can be used to solve some complex waveguide problems approximately encountered in microwave and optical engineering by adding junction capacitances and inductances from Figure 17.8 in lecture 17.

24.1 Circular Waveguides, Cont.

The scalar potential (or pilot potential) for the modes in the circular waveguide is expressible as

$$\Psi_{\alpha s}(\rho, \phi) = AJ_n(\beta_s \rho)e^{\pm jn\phi} \tag{24.1.1}$$

where $\alpha = h$ for TE waves and $\alpha = e$ for TM waves.[1] The Bessel function or wave is expressible in terms of Hankel functions as in (23.2.5). Since Hankel functions are inward and outward-traveling waves, Bessel functions represent standing waves. Therefore, the Bessel waves can be thought of as bouncing traveling waves, as found in the rectangular waveguide case. In the azimuthal direction, one can express $e^{\pm jn\phi}$ as traveling waves in the ϕ direction, or they can be expressed as $\cos(n\phi)$ and $\sin(n\phi)$, which are standing waves in the ϕ direction.

[1]As mentioned before, the pilot potentials are related to the H_z and E_z components of the fields, as reflected in the subscripts of the potential. Also, the potential is regular at the center of the waveguide; and hence, the Bessel function is chosen as the solution.

24.1.1 An Application of Circular Waveguides

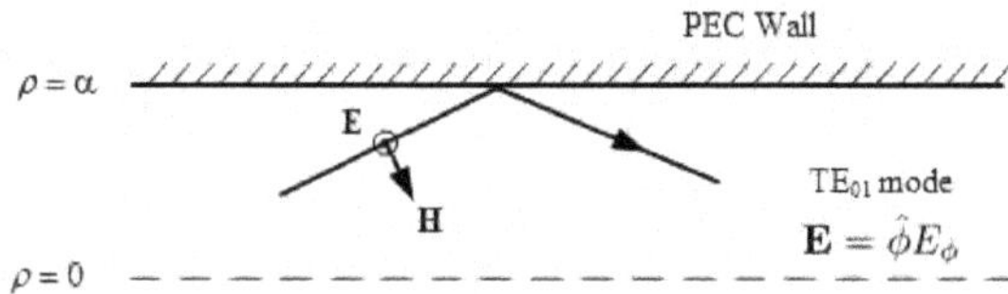

Figure 24.1: Bouncing wave picture of the Bessel wave inside a circular waveguide for the TE_{01} mode. One can also explain the low-loss physics [181] using the TE mode of a parallel-plate waveguide (see related text).

When a real-world waveguide is made, the wall of the metal waveguide is not made of a perfect electrical conductor, but with some metal of finite conductivity. Hence, tangential $\mathbf{E}$ field is not zero on the waveguide wall, implying that $\hat{n}\cdot(\mathbf{E}\times\mathbf{H}^*) \neq 0$. Since $\mathbf{E}\times\mathbf{H}^*$ is the complex Poynting's vector, it being not zero implies that energy can dissipate (or power can flow) into the waveguide wall.

It turns out that due to symmetry, the TE_{01} mode of a circular waveguide has the lowest loss of all the waveguide modes, including rectangular waveguide modes of a rectangular waveguide. Hence, this waveguide mode is of great interest to astronomers who are interested in building low-loss, low-noise systems.[2].

The TE_{01} mode has electric field given by $\mathbf{E} = \hat{\phi}E_\phi$. Furthermore, looking at the magnetic field, the current is mainly circumferential (azimuthal) flowing in the ϕ direction. By looking at a bouncing wave picture of the guided waveguide mode, this mode, similar to the parallel waveguide mode, has a small component of tangential magnetic field on a waveguide wall: It becomes increasingly smaller as the frequency increases (see Figure 24.1). The reason is that the wave vector for the waveguide becomes increasingly parallel to the axis of the waveguide with a large β_z component compared to the β_s component.[3] In a word, the wave becomes paraxial in the high-frequency limit.

The tangential magnetic field needs to be supported by a surface current on the waveguide wall. This implies that the surface current on the waveguide wall becomes smaller as the frequency increases. Consequently, the wall loss (or copper loss or eddy current loss) of the waveguide becomes smaller for higher frequencies. In fact, for high frequencies, the TE_{01} mode has the smallest copper loss of all waveguide modes (including waveguides of other shapes): It becomes the mode of choice (see Figure 24.2). Waveguides supporting the TE_{01} modes are used to connect the antennas of the very large array (VLA) for detecting extraterrestrial signals in radio astronomy [185] as shown in Figure 24.3. These signals are in general quite weak. The low wall loss gives rise to a good SNR (signal-to-noise ratio). [4]

[2]Low-loss systems are also low-noise due to energy conservation and the fluctuation dissipation theorem [176, 177, 182]. This is similar to Johnson-Nyquist noise [183, 184]

[3]Recall that for a fixed mode, β_s^2 is the eigenvalue of the system, and, hence, β_s is independent of frequency.

[4]This follows from the fluctuation dissipation theorem, which is an energy-conserving theorem that says that whatever electromagneticenergy is absorbed by the environment is returned to the environment as thermal radiation.

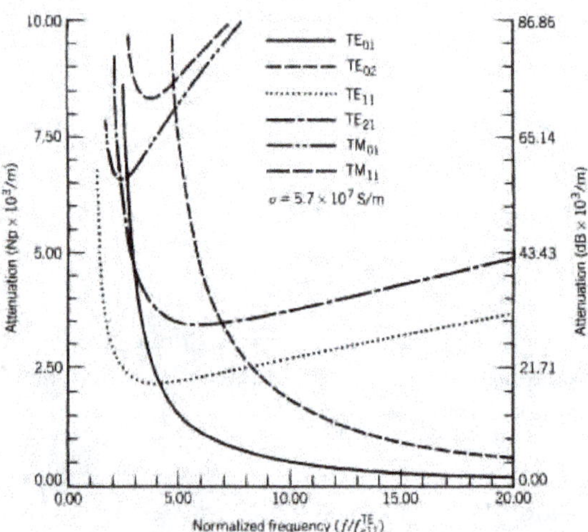

Figure 24.2: Plots of losses of different modes in a circular waveguide with a radius of 1.5 cm. It is seen that at high frequencies, the TE_{01} mode has the lowest loss (Figure by: C. A. Balanis and E. Holzman, "Circular waveguides," Encyclopedia of RF and Microwave Engineering, 2005.).

Figure 24.3: Picture of the VLA in New Mexico, USA (Figure by: Wikipedia). The low loss of the circular waveguide gives good SNR according to the fluctuation dissipation theorem [177], which makes the system good for detecting weak radio astronomy signals from outer space.

Figure 24.4 shows two ways of engineering a circular waveguide so that the TE_{01} mode is enhanced: (i) by using a mode filter that discourages the guidance of other modes except the TE_{01} mode, and (ii) by designing a corrugated waveguide wall to discourage the flow of axial current but encourage the flow of circumferential (azimuthal) current. Thus it discourages the propagation of the non-TE_{01} mode. (The TE_{01} mode only has azimuthal current in the ϕ direction.) More details of circular waveguides can be found in [181]. Typical loss of a circular waveguide can be as low as 2 dB/km.[5]

As shall be shown, an open circular waveguide can be made into an aperture antenna quite easily because the fields of the aperture are axially symmetric. To this end, the axially symmetric TE_{01} mode is enhanced by design. Because of this, the radiation pattern of such an antenna is axially symmetric, which can be used to produce axially symmetric circularly polarized (CP) waves. Such antenna is called a horn antenna. Ways to enhance the TE_{01} mode of the horn antenna are also desirable [186] as shown in Figure 24.5.

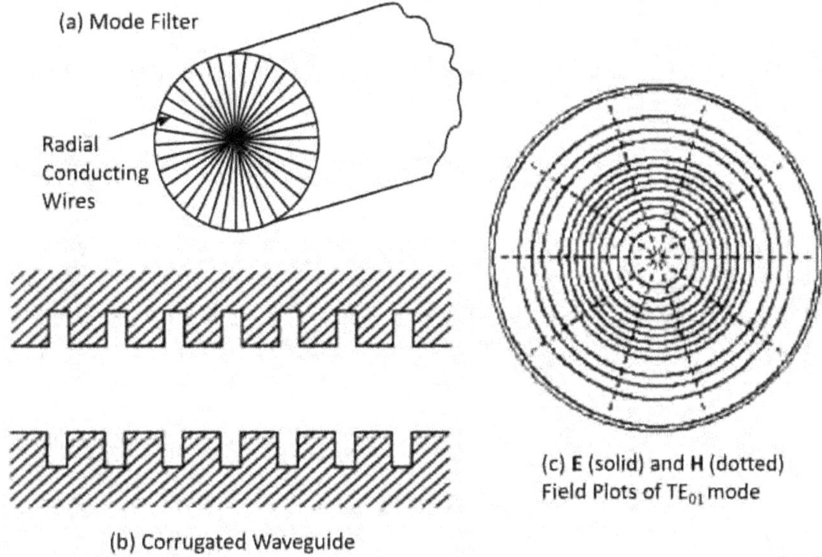

Figure 24.4: The TE_{01} has $\mathbf{E} = \hat{\phi}E_\phi$, and the tangential $\mathbf{H}$ field only has a z component on the waveguide wall. Hence, the wall current is purely circumferential (azimuthal) for this mode. There are ways to enhance the TE_{01} mode in a circular waveguide: (a) use a mode filter that only allows the TE_{01} mode to go through or (b) use a corrugated waveguide to discourage axial current flow from the other modes but encourage the circumferential (azimuthal) current flow from this TE_{01} mode. The field plot of the mode is shown in (c). Such a waveguide is used in radio astronomy to design the communication links between antennas in a very large array (VLA [185]), or it is used in a circular horn antenna [186].

[5]For optical fiber, this figure of merit (FOM) can be as low as 0.1 dB/km, making the optical fiber a darling for long-distance communication.

Figure 24.5: Picture of a circular horn antenna where corrugated walls are used to enhance the TE_{01} mode, which is axially symmetric. It discourages the other nonaxially symmetric modes. Therefore, the antenna has an axially symmetric field producing a radiation pattern (to be addressed later) that is axially symmetric (Figure by: Wikipedia).

24.2 Remarks on Quasi-TEM Modes, Hybrid Modes, Surface Plasmonic Modes

We have analyzed some simple structures where closed-form solutions are available. These simple elegant solutions offer physical insight as to how waves are guided and how they are cutoff from guidance. As has been shown, for some simple waveguides, the modes can be divided into TEM, TE, and TM modes. However, most waveguides are not simple. We will remark on various complexities that arise in real-world applications.

24.2.1 Quasi-TEM Modes

Many waveguides cannot support a pure TEM mode even when two conductors are present. For example, two pieces of metal make a transmission line, and in the case of a circular coax, a TEM mode can propagate in the waveguide. But most two-metal transmission lines do not support a pure TEM mode. Instead, they support a quasi-TEM mode. In the optical fiber case, when the index contrast of the fiber is very small, the mode is also quasi-TEM as it has to degenerate to the

TEM case when the contrast is absent.

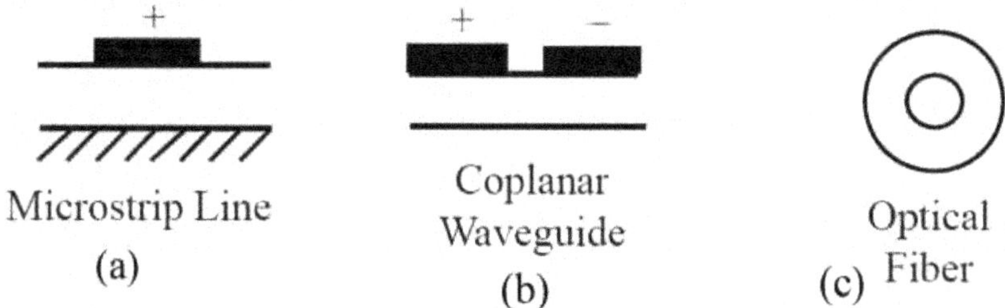

Figure 24.6: Some examples of practical coaxial-like waveguides are a microstrip line and a coplanar waveguide (left). For the microstrip line (a), the signal line (denoted with a + sign) mimics the center conductor of a coax, while the ground plane (hashed lines) represents the outer conductor of a coax. The coplanar waveguide (b) needs no ground plane and operates like a twin-ax, where the ± line indicates the signal line and ground line of the twin-ax. The optical fiber is indicated in (c). It operates by total-internal-reflection at the interface between the center (core) of the waveguide and the cladding (outside the core). The environments of these waveguides are inhomogeneous media, and hence, a pure TEM mode cannot propagate on these waveguides.

Absence of TEM Modes in Inhomogeneously-Filled Waveguides

In the following, we will give physical arguments as to why a pure TEM mode cannot exist in a microstrip line, a coplanar waveguide, or an optical fiber. When a wave is TEM, it is necessary that the wave propagates with the phase velocity of the medium in which it propagates with $\exp(-j\beta_i z)$ dependence where β_i is the wavenumber of the medium. But when a uniform waveguide has inhomogeneity in between, as shown in Figure 24.6, this is not possible anymore. We can prove this assertion by reductio ad absurdum. Very simply put, if the waves are TEM in all the regions, they will have the respective phase velocity of the regions, each with a different $\exp(-j\beta_i z)$ dependence. Thus, phase matching is impossible at the interfaces between these regions.

We shall study this point in greater detail. Assume only TM wave in a piecewise homogeneous region, then using the vector pilot potential approach, the $\mathbf{E}$ field is

$$\mathbf{E} = \frac{1}{j\omega\varepsilon_i} \nabla \times \nabla \times (\hat{z}\Psi_e) \tag{24.2.1}$$

where ε_i is the permittivity of the region. By doing some algebra and assuming that the field is a waveguide mode such that Ψ_e has $e^{-j\beta_z z}$ dependence, then using the BAC-CAB formula, one can show that the above simplifies to

$$\mathbf{E} = \frac{1}{j\omega\varepsilon_i} \left[\nabla\nabla \cdot \hat{z}\Psi_e - \nabla^2 \hat{z}\Psi_e\right] \tag{24.2.2}$$

Or

$$E_z = \frac{1}{j\omega\varepsilon_i}\left[\frac{\partial^2}{\partial z^2}\Psi_e - \nabla^2\Psi_e\right] \tag{24.2.3}$$

Therefore, E_z is given by

$$E_z = \frac{1}{j\omega\varepsilon_i}(\beta_i^2 - \beta_z^2)\Psi_e \tag{24.2.4}$$

The above derivation is certainly valid in a piecewise homogeneous region. But each of the piecewise homogeneous media can be made arbitrarily small, and hence, it is also valid for inhomogeneous media. If this mode becomes TEM, then $E_z = 0$, and this is possible only if $\beta_z = \beta_i$. In other words, the phase velocity of the waveguide mode is the same as a plane TEM wave in the same medium.

Now, if we assume that a TEM wave exists in both inhomogeneous regions of the microstrip line or all three dielectric regions of the optical fiber in Figure 24.6 then the phase velocities in the z direction, determined by ω/β_z of each region, will be ω/β_i of the respective region. As a consequence, phase matching is not possible and the boundary condition cannot be satisfied at the dielectric interfaces.

Nevertheless, the lumped element circuit model of the transmission line is still a very good model for such a waveguide. If the line capacitance and line inductance of such lines can be estimated, β_z can still be estimated. As has been shown before, circuit theory is valid when the frequency is low or the wavelength is large compared to the size of the structures.

24.2.2 Hybrid Modes–Inhomogeneously Filled Waveguides

It turns out that when such inhomogeneity is present, as in the case of the optical fiber, both TE and TM waves are needed in the waveguide. This will allow the boundary conditions to be satisfied at the interface of different regions [96]. These modes are called hybrid modes. Sometimes, these hybrid modes are called EH or HE modes, as in an optical fiber. Nevertheless, the guidance is still via a bouncing wave picture, where the bouncing waves are reflected off the boundaries of the waveguides. In the case of an optical fiber or a dielectric waveguide, the reflection is due to total internal reflection. But in the case of a PEC metalic waveguides, the reflection is due to the PEC metal walls, which are a perfect reflector.

24.2.3 Guidance of Modes

Propagation of a plane wave in free space is by the exchange of electric stored energy and magnetic stored energy . The same physics also happen in a waveguide. For example, in the transmission line, the guidance is by the exchange of electric and magnetic stored energy via the coupling between the line capacitance and the line inductance of the line. In this case, the waveguide size, like the cross-section of a coaxial cable, can be made much smaller than the wavelength and the wave will still be guided.

In the case of hollow waveguides, the **E** and **H** fields are coupled through their space and time variations representing a bouncing wave inside the waveguide. Namely,

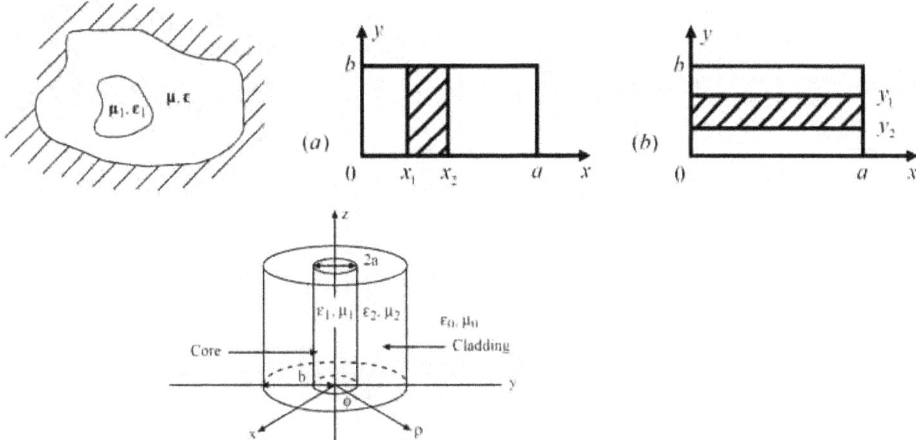

Figure 24.7: Some examples of inhomogeneously filled waveguides where hybrid modes exist: (top-left) A general inhomogeneously filled waveguide, (top-right) two examples of slab-loaded rectangular waveguides, and (bottom) an optical fiber with core and cladding.

$$\nabla \times \mathbf{E} = -j\omega\mu\mathbf{H}, \qquad \nabla \times \mathbf{H} = j\omega\varepsilon\mathbf{E} \tag{24.2.5}$$

Hence, in the frequency domain, the fields are coupled through their spatial derivative, and hence, the exchange of the energies stored is via the space that stores these energies, like that of a plane wave. These waveguides work only when these plane waves can "enter" the waveguide. Therefore, the size of these waveguides has to be about half a wavelength.

The surface plasmonic waveguide is an exception in that the exchange is between the electric field energy stored with the *kinetic energy* stored in the moving electrons in the plasma instead of magnetic energy stored. This form of energy stored is sometimes referred to as coming from *kinetic inductance*. Therefore, the dimension of the waveguide can be very small compared to wavelength, and yet the surface plasmonic mode can be guided.

24.3 "Homomorphism" of Hollow Waveguides and Transmission Lines

Previously, we have demonstrated mathematical "homomorphism" between plane waves in layered medium and transmission lines. Such "homomorphism" can be further extended to hollow waveguides and transmission lines.[6] But unlike the plane wave in the layered medium case, we cannot

[6]For a waveguide with a center conductor like the coaxial cable, the fundamental mode with no cutoff is the TEM mode, and it actually is the tranmission line mode. The higher-order modes in a coax can be classified into TE and TM modes, just like a hollow waveguide.

replace the ∇ operator with $-j\boldsymbol{\beta}$ in a hollow waveguide. Hence, the mathematics is slightly more elaborate. We can show this first for TE modes in a hollow waveguide, and the case for TM modes can be established by invoking duality principle.[7]

24.3.1 TE Case

For this case, $E_z = 0$, and from Maxwell's equations

$$\nabla \times \mathbf{H} = j\omega\varepsilon\mathbf{E} \tag{24.3.1}$$

First we let $\nabla = \nabla_s + \nabla_z$, $\mathbf{H} = \mathbf{H}_s + \mathbf{H}_z$ where $\nabla_z = \hat{z}\frac{\partial}{\partial z}$, and $\mathbf{H}_z = \hat{z}H_z$, and the subscript s implies the transverse of the z components. Then,

$$(\nabla_s + \nabla_z) \times (\mathbf{H}_s + \mathbf{H}_z) = \nabla_s \times \mathbf{H}_s + \nabla_z \times \mathbf{H}_s + \nabla_s \times \mathbf{H}_z = j\omega\varepsilon\mathbf{E} \tag{24.3.2}$$

where it is understood that $\nabla_z \times \mathbf{H}_z = 0$. Notice that the first term on the right-hand side of the above is pointing in the z direction. By letting $\mathbf{E} = \mathbf{E}_s + \mathbf{E}_z$ and equating transverse components in (24.3.1), we have[8]

$$\nabla_z \times \mathbf{H}_s + \nabla_s \times \mathbf{H}_z = j\omega\varepsilon\mathbf{E}_s \tag{24.3.3}$$

To further simplify the above equation, we shall relate $\mathbf{H}_z$ from the above with the other field components. To this end, we look at Faraday's law from which we have

$$\nabla \times \mathbf{E} = -j\omega\mu\mathbf{H} \tag{24.3.4}$$

Again, by letting $\mathbf{E} = \mathbf{E}_s + \mathbf{E}_z$, we can let (24.3.4) be written as

$$\nabla_s \times \mathbf{E}_s + \nabla_z \times \mathbf{E}_s + \nabla_s \times \mathbf{E}_z = -j\omega\mu(\mathbf{H}_s + \mathbf{H}_z) \tag{24.3.5}$$

Equating z components of the above, we have

$$\nabla_s \times \mathbf{E}_s = -j\omega\mu\mathbf{H}_z \tag{24.3.6}$$

The above allows us to express $\mathbf{H}_z$ in terms of $\mathbf{E}_s$. Thus (24.3.3) can be rewritten as

$$\nabla_z \times \mathbf{H}_s + \nabla_s \times \frac{1}{-j\omega\mu}\nabla_s \times \mathbf{E}_s = +j\omega\varepsilon\mathbf{E}_s \tag{24.3.7}$$

The above can be further simplified by noting that

$$\nabla_s \times \nabla_s \times \mathbf{E}_s = \nabla_s(\nabla_s \cdot \mathbf{E}_s) - \nabla_s \cdot \nabla_s\mathbf{E}_s = -\nabla_s^2\mathbf{E}_s \tag{24.3.8}$$

But since $\nabla \cdot \mathbf{E} = 0$, and $E_z = 0$ for TE modes, it also implies that $\nabla_s \cdot \mathbf{E}_s = 0$, which justifies the last equality in the above. From Maxwell's equations, we have previously shown that for a homogeneous source-free medium,

$$(\nabla^2 + \beta^2)\mathbf{E} = 0, \qquad \text{or} \qquad (\nabla^2 + \beta^2)\mathbf{E}_s = 0 \tag{24.3.9}$$

[7] I have not seen exposition of such mathematical homomorphism elsewhere except in very simple cases [35].

[8] And from the above, it is obvious that $\nabla_s \times \mathbf{H}_s = j\omega\varepsilon\mathbf{E}_z$, but this equation will not be used in the subsequent derivation.

since $E_z = 0$ for TE mode. Furthermore, assuming $e^{\mp j\beta_z z}$ for the z dependence of the waveguide modes, (24.3.9) then becomes

$$(\nabla_s{}^2 + \beta^2 - \beta_z{}^2)\mathbf{E}_s = 0 \tag{24.3.10}$$

or that $\mathbf{E}_s$ satisfies the reduced wave equation. Thus,

$$(\nabla_s{}^2 + \beta_s{}^2)\mathbf{E}_s = 0 \tag{24.3.11}$$

where $\beta_s^2 = \beta^2 - \beta_z^2$ is the transverse wavenumber. Consequently, from (24.3.8), we arrive at the simplification, or that

$$\nabla_s \times \nabla_s \times \mathbf{E}_s = -\nabla_s^2 \mathbf{E}_s = \beta_s^2 \mathbf{E}_s \tag{24.3.12}$$

As such, using this in (24.3.7), it becomes

$$\begin{aligned} \nabla_z \times \mathbf{H}_s &= j\omega\varepsilon \mathbf{E}_s + \frac{1}{j\omega\mu}\beta_s{}^2 \mathbf{E}_s \\ &= j\omega\varepsilon \left(1 - \frac{\beta_s{}^2}{\beta^2}\right)\mathbf{E}_s = j\omega\varepsilon \frac{\beta_z{}^2}{\beta^2}\mathbf{E}_s \end{aligned} \tag{24.3.13}$$

Then, letting $\beta_z = \beta\cos\theta$, the above can further be rewritten as

$$\nabla_z \times \mathbf{H}_s = j\omega\varepsilon \cos^2\theta \mathbf{E}_s \tag{24.3.14}$$

Now, the above resembles one of the two telegrapher's equations that we seek.

Again, looking at (24.3.4), assuming $E_z = 0$, equating transverse components, we have

$$\nabla_z \times \mathbf{E}_s = -j\omega\mu \mathbf{H}_s \tag{24.3.15}$$

More explicitly, we can rewrite (24.3.14) and (24.3.15) in the above as

$$\frac{\partial}{\partial z}\hat{z} \times \mathbf{H}_s = j\omega\varepsilon \cos^2\theta \mathbf{E}_s \tag{24.3.16}$$

$$\frac{\partial}{\partial z}\hat{z} \times \mathbf{E}_s = -j\omega\mu \mathbf{H}_s \tag{24.3.17}$$

Visibly, the above resembles the telegrapher's equations . We can multiply (24.3.17) by $\hat{z}\times$ to get

$$\frac{\partial}{\partial z}\mathbf{E}_s = j\omega\mu \hat{z} \times \mathbf{H}_s \tag{24.3.18}$$

Now (24.3.16) and (24.3.18) are a set of coupled equations that look even more like the telegrapher's equations. We can have $\mathbf{E}_s \to V$, $\hat{z} \times \mathbf{H}_s \to -I$, $\mu \to L$, $\varepsilon\cos^2\theta \to C$, and the above resembles the telegrapher's equations, or that the waveguide problem is homomorphic to the transmission line problem. The characteristic impedance of this equivalent line is then

$$Z_0 = \sqrt{\frac{L}{C}} = \sqrt{\frac{\mu}{\varepsilon\cos^2\theta}} = \sqrt{\frac{\mu}{\varepsilon}}\frac{1}{\cos\theta} = \frac{\omega\mu}{\beta_z} \tag{24.3.19}$$

which is the wave impedance of a hollow waveguide. Consequently, the TE modes of a waveguide can be mapped into a transmission problem. This can be done, for instance, for the TE_{mn} mode of a rectangular waveguide. Then, in the above

$$\beta_z = \sqrt{\beta^2 - \left(\frac{m\pi}{a}\right)^2 - \left(\frac{n\pi}{b}\right)^2} \tag{24.3.20}$$

Here, each TE_{mn} mode will be represented by a different equivalent characteristic impedance Z_0, since β_z is different for different TE_{mn} modes.

When this is used in practice, and when multimodes are present, the number of equivalent transmission lines can become unwieldy. Engineers like to use simpler systems. Simplicity rules! A simpler alternative is usually sought, such as the mode-matching method [96].

24.3.2 TM Case

This case can be derived using duality principle. Invoking duality, and after some algebra, then the equivalences of (24.3.16) and (24.3.18) become

$$\frac{\partial}{\partial z}\mathbf{E}_s = j\omega\mu\cos^2\theta\hat{z}\times\mathbf{H}_s \tag{24.3.21}$$

$$\frac{\partial}{\partial z}\hat{z}\times\mathbf{H}_s = j\omega\varepsilon\mathbf{E}_s \tag{24.3.22}$$

To keep the dimensions commensurate, we let $\mathbf{E}_s \to V$, $\hat{z}\times\mathbf{H}_s \to -I$, $\mu\cos^2\theta \to L$, $\varepsilon \to C$. Again, the above resembles the telegrapher's equations. We can thus let

$$Z_0 = \sqrt{\frac{L}{C}} = \sqrt{\frac{\mu\cos^2\theta}{\varepsilon}} = \sqrt{\frac{\mu}{\varepsilon}}\cos\theta = \frac{\beta_z}{\omega\varepsilon} \tag{24.3.23}$$

Please note that (24.3.19) and (24.3.23) are very similar to those in the plane-wave case, which are the wave impedance for the TE and TM modes, respectively.

The above implies that if we have a waveguide of an arbitrary cross section filled with layered media, the problem can be mapped to a multisection transmission line problem. Then it can be solved with transmission line methods. When V and I are continuous at a transmission line junction, $\mathbf{E}_s$ and $\mathbf{H}_s$ will also be continuous. Therefore, the transmission line solution would also imply continuous $\mathbf{E}_s$ and $\mathbf{H}_s$ field solutions. No mode conversion happens at such junctions, as shall be explained later.

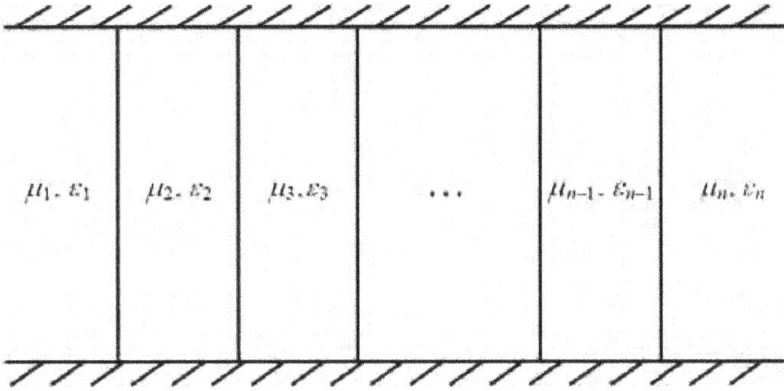

Figure 24.8: A waveguide filled with layered medium is mathematically homomorphic to a multisection transmission line problem. In this problem, there is no mode conversion and a single mode can propagate through the waveguide. And yet, the boundary conditions at the interfaces can be satisfied, like plane waves in layered media. Hence, the transmission-line method can be used to solve this problem.

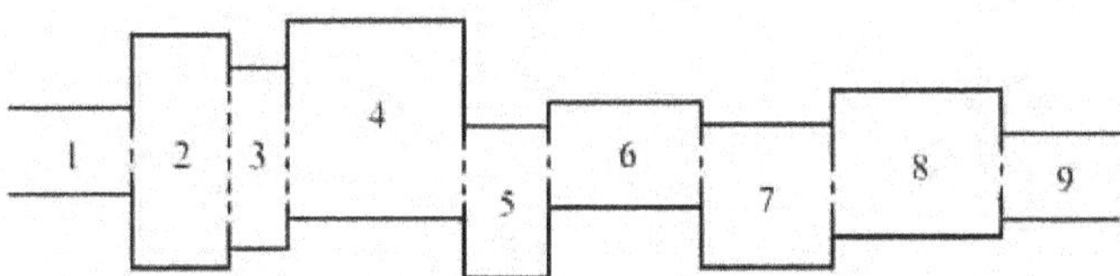

Figure 24.9: A multisection waveguide is not exactly homormorphic to a multisection transmission line problem when the cross section of the waveguides are not equal to each other. The boundary conditions at the junction cannot be met by assuming a single mode. Therefore, multimodes are needed to match the boundary conditions [96]. Circuit elements are needed at the junctions to approximately capture the physics at the waveguide junctions as shown in the next figure.

24.3.3 Mode Conversion

In the waveguide shown in Figure 24.8, there is no mode conversion at the junction interface. Assuming a rectangular waveguide as an example, what this means is that if we send a TE_{10} mode into the waveguide, this same mode will propagate throughout the length of the waveguide. The reason is that only this mode alone is sufficient to satisfy the boundary condition at the junction interface. The mode profile does not change throughout the length of the waveguide.

To elaborate further, from our prior knowledge, the transverse fields of the waveguide–for example, for the TM mode–can be derived to be

$$\mathbf{H}_s = \nabla \times \hat{z}\Psi_{es}(\mathbf{r}_s)e^{\mp j\beta_z z} \tag{24.3.24}$$

$$\mathbf{E}_s = \frac{\mp\beta_z}{\omega\varepsilon}\nabla_s\Psi_{es}(\mathbf{r}_s)e^{\mp j\beta_z z} \tag{24.3.25}$$

In the above, β_s^2 and $\Psi_{es}(\mathbf{r}_s)$ are eigenvalue and eigenfunction, respectively, that depend only on the geometrical cross-sectional shape of the waveguide, but not the materials filling the waveguide. These eigenfunctions are the same throughout different sections of the waveguide and only $\beta_z = \sqrt{\beta_i^2 - \beta_s^2}$ changes from section to section. Therefore, with the incident mode at the junction, the boundary conditions can be easily satisfied at the junctions with the inclusion of reflected and transmitted waves of the same mode.

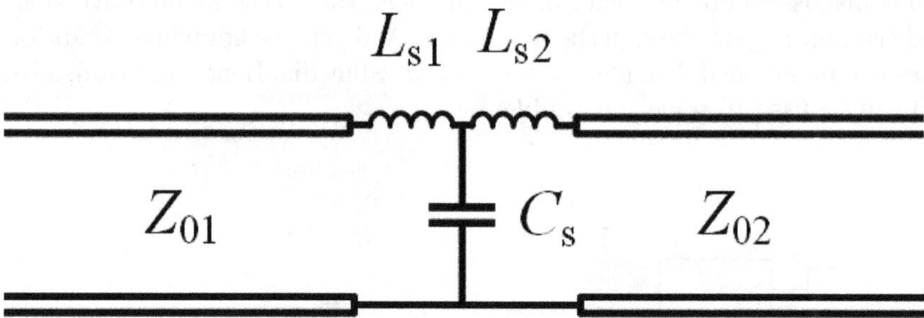

Figure 24.10: When the single-mode approximation holds, one assumes that away from the waveguide junction, there is only one single mode. Close to the waveguide junction, however, the evanescent, higher-order modes give rise to localize stored electric and magnetic energies at the waveguide junction. Circuit elements are used to account for these stored energies at the waveguide junctions. They also account for the fact that the currents and voltages are not continuous across the junctions anymore, as the fields of the dominant modes in each section (as shown in Figure 24.9) are not continuous anymore. The moral of the story is that engineers love to replace complicated theories with simple ones in order to solve complex problems.

However, for a multijunction waveguide as show in Figure 24.9, the tangential **E** and **H** continuous condition cannot be satisfied by a single mode in each waveguide alone: V and I continuity at a transmission line junction will not guarantee the continuity of the mode-matching tangential **E** and tangential **H** fields at the waveguide junction.

Multimodes have to be assumed on both sides of the junction at each section in order to match boundary conditions at the junction [92]. Moreover, the mode-matching method for multiple modes has to be used at each junction. Typically, a single mode incident at a junction will give rise to multiple modes reflected and multiple modes transmitted. The multiple modes give rise to the phenomenon of *mode conversion* at a junction. Instead, the mode-matching method is used to solve this problem [96].

However, the operating frequency can be chosen so that only one mode is propagating at each section of the waveguide, and the other modes are cutoff or evanescent . Then one can assume a single-mode approximation of the waveguide modes in each section of the waveguide. In this case, the higher-order modes are evanescent away from the waveguide junction, giving rise to localized energy storage at a junction in the **E** and **H** fields. These energies can be either inductive or capacitive. The junction effects may be modeled by a simple circuit model as shown in Figure 24.10. These junction elements also account for the physics that the currents and voltages are not continuous anymore across the junction. Moreover, these junction lumped circuit elements account for the stored electric and magnetic energies at the junction.

Exercises for Lecture 24

Problem 24-1:

(i) Explain when you would have quasi-TEM modes and hybrid modes.

(ii) Show that the rectangular or circular waveguide problem can be made homomorphic to a transmission line problem.

(iii) Explain how you would map the waveguide parameters to transmission line parameters for the TE and TM modes of a hollow waveguide.

(iv) Derive (24.3.24) and (24.3.25). Explain why mode conversion occurs at a waveguide junction.

Lecture 25

Cavity Resonators

Cavity resonators are important components of microwave and optical systems. They work by constructive and destructive interference of bouncing waves in an enclosed region. Just as LC resonators in circuits, they can be used as filters or as devices to enhance certain physical interactions. These can happen in radiation antennas or electromagnetic sources such as magnetrons or lasers. They can also be used to make high sensitivity sensors. We will study a number of them. For many of them, we will discuss them only heuristically in this lecture. Their full solutions are usually obtained using numerical methods.

25.1 Transmission Line Model of a Resonator

The simplest cavity resonator is formed by using a transmission line. The source end can be terminated by Z_S and the load end can be terminated by Z_L. When Z_S and Z_L are non dissipative, such as when they are reactive loads (capacitive or inductive), the reflection coefficient from (16.1.9), for $Z_L = jX$ where X is the reactance of the load, is

$$\Gamma_L = \frac{jX - Z_0}{jX + Z_0} \tag{25.1.1}$$

Like the Fresnel reflection coefficient for the total internal reflection case as shown in (18.2.3), it has a magnitude of 1, indicating that the energy is totally reflected. No energy is dissipitated as a wave is totally reflected off them. Therefore, if the wave can bounce and interfere constructively between the two ends, a coherent solution or a resonant solution can exist due to constructive inference.

The resonant solution exists even when the source is turned off. In mathematical parlance, this is a homogeneous solution to a partial differential equation or ordinary differential equation, since the right-hand side of the pertinent equation is zero. The right-hand side of these equations usually corresponds to a source term or a driving term. In physics parlance, this is a natural solution since it exists naturally without the need for a driving or exciting source. In linear algebra, when the matrix equation $\overline{\mathbf{A}} \cdot \mathbf{x} = 0$ has a non trivial solution, it is the null-space solution.

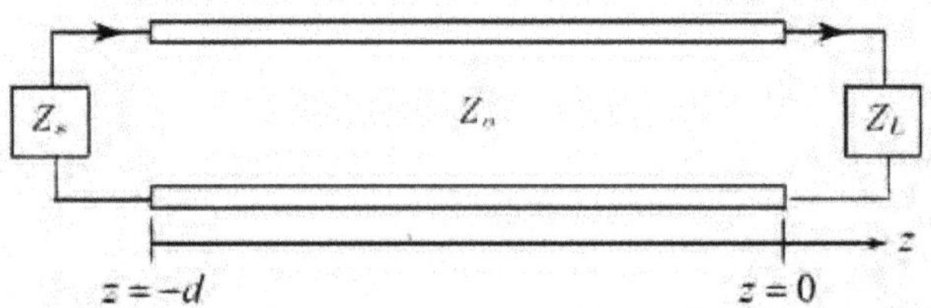

Figure 25.1: A simple ideal resonator is made by terminating a transmission line with two reactive loads at its two ends, the source end with Z_S and the load end with Z_L.

The transverse resonance condition for 1D problem can be used to derive the resonance frequency, namely that

$$1 = \Gamma_S \Gamma_L e^{-2j\beta_z d} \tag{25.1.2}$$

where Γ_S and Γ_L are the reflection coefficients at the source and the load ends, respectively; β_z is the wavenumber of the wave traveling in the z direction; and d is the length of the transmission line.

For a TEM mode in the transmission line, as in a coax filled with homogeneous medium, $\beta_z = \beta$, where usually, β is the wavenumber for the homogeneous medium. Otherwise, for a quasi-TEM mode, $\beta_z = \beta_e$ where β_e is some effective wavenumber for a z-propagating wave in an inhomogeneous (heterogeneous) medium. In general,

$$\beta_e = \omega / v_e \tag{25.1.3}$$

where v_e is the effective phase velocity of the wave in the heterogeneous structure like a microstrip line.

When the source and load impedances are replaced by short or open circuits, then the reflection coefficients are simply -1 for a short and $+1$ for an open circuit. The (25.1.2) above then becomes

$$\pm 1 = e^{-2j\beta_e d} \tag{25.1.4}$$

The $\pm$ signs correspond to different combinations of open and short circuits at the two ends of the transmission lines. When a "+" sign is chosen, it indicates that either both ends are short circuit, or are open circuit. Then the resonance condition is such that

$$\beta_e d = p\pi, \quad p = 0, 1, 2, \ldots, \quad \text{or integer} \tag{25.1.5}$$

For a TEM or a quasi-TEM mode in a transmission line, $p = 0$ is corresponds to the case when the voltage is uniformly zero on the transmission line, or the voltage is a constant on the line. The latter is possible when the line consists of two pieces of metal with uniform charge on each piece of metal: this is the electrostatic mode similar to the field of a charged capacitor. This two cases are not interesting as they correspond to modes of zero frequency.

The first interesting mode is when $p = 1$, corresponding to a half wavelength on the transmission line. The voltage distribution is shown in Figure 25.2.

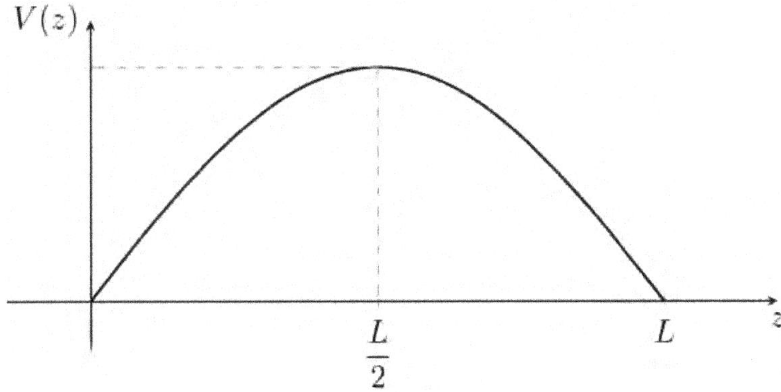

Figure 25.2: The voltage distribution on a half-wave transmission line resonator at its resonant frequency. Since it needs a half wavelength to resonate, it is not the smallest resonator, and hence, it is not employed in cell phone antennas.

When the line is open at one end and shorted at the other end in (25.1.2), the resonance condition corresponds to the "−" sign in (25.1.4), which gives rise to

$$e^{-2j\beta_e d} = e^{-jp\pi} = -1, \quad p \quad \text{odd integer} \tag{25.1.6}$$

The above implies that

$$\beta_e d = p\pi/2, \quad p \quad \text{odd integer} \tag{25.1.7}$$

As shown in Figure 25.3 , the lowest mode is when $p = 1$ corresponding to a quarter wavelength on the transmission line, which is smaller than a half-wavelength transmission line terminated with short or open at both ends. Designing a small resonator using a quarter-wave resonator is a prerogative in modern day electronic design. For example, miniaturization in cell phones calls for smaller components that can be packed into smaller spaces.

A quarter wavelength resonator has a voltage distribution shown in Figure 25.3. One that is made with a with a coax is shown in Figure 25.4. It is easier to make a short approximately as indicated at the left end: a good metallic conductor like copper suffices because it is close to a perfect electric conductor (PEC) at microwave frequency.[1] But it is hard to make a true open circuit as shown

[1]A PEC has a skin depth of zero. A good conductor like copper will have skin depth much less than the wavelength at microwave frequencies, and, hence, it behaves essentially like a PEC.

at the right end. A true open circuit means that the current has to be zero. But when a coax is terminated with an open, the electric current does not end abruptly. The fringing field at the right end gives rise to stray capacitance through which displacement current can flow in accordance with the generalized Ampère's law. Hence, we have to model the right end termination with a small stray or fringing field capacitance as (see Figure 25.4). This makes the transmission line slightly longer.

To design a true open circuit, one needs to terminate the right end of the transmission line with a perfect magnetic conductor (PMC) in theory. By going through a Gedanken experiment, since $H_\phi = 0$ at the termination, one can show that the current at the right termination has to be zero.

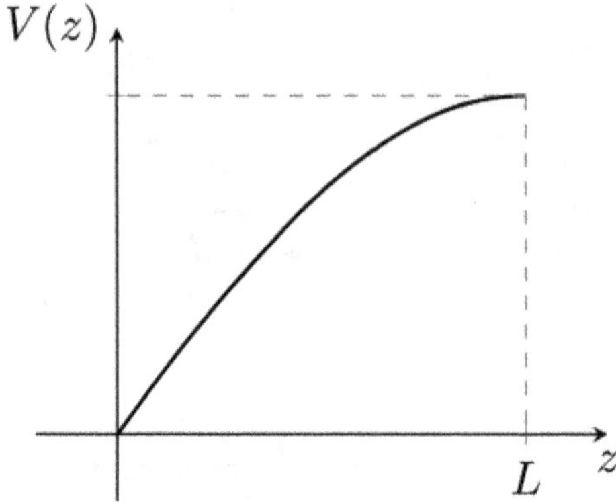

Figure 25.3: The voltage distribution on a quarter-wave transmission line resonator at its lowest resonant mode. Since it needs only a quarter wavelength to resonate, it can be made small compared to wavelength, and hence, it is a very popular resonator used in cell phone antennas.

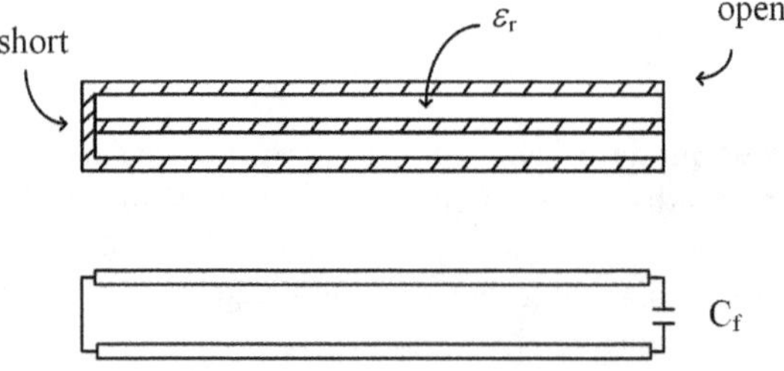

Figure 25.4: A short and open circuited transmission line can be a resonator, but the open end has to be modeled with a fringing field capacitance C_f since there is no exact or true open circuit. The resonance condition will have to be derived from (25.1.2), which will usually give a transcendental equation. Graphical method can be used to solve the transcendental equation.

25.2 Cylindrical Waveguide Resonators

Since a cylindrical waveguide[2] is "homomorphic" to a transmission line, we can model a mode in this waveguide as a transmission line. Then the termination of the waveguide with either a short or an open circuit at its end makes it into a resonator.

Again, there is no true open circuit in an open-ended waveguide, as there will be fringing fields at its open ends. If the aperture is large enough, the open end of the waveguide radiates and may be used as an antenna as shown in Figure 25.5.

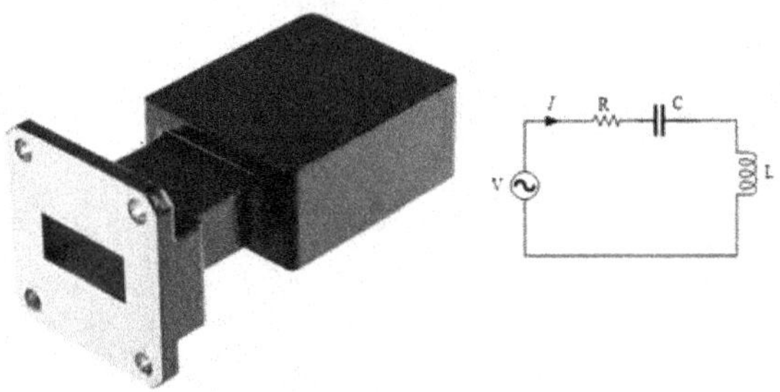

Figure 25.5: A rectangular waveguide terminating with a short at one end and an open circuit at the other end. The open end can also act as an antenna as it radiates. When the cavity is injected with electromagnetic fields coinciding with its resonance frequency, the fields inside the cavity become large, and so do the fields at the aperture, making it a better radiator. This is a cavity-backed antenna: it uses resonance tunneling to enhance its radiation capability. (Resonance tunneling phenomenon describes the ease with which photons tunnel through a barrier when the photons are close to the resonance frequency of the barrier structure). The RLC circuit can approximate the physics of the structure (Figure by RFcurrent.com).

25.2.1 Rectangular Cavity Resonator

As previously shown, single-section waveguide resonators can be modeled with a transmission line using "homomorphism" with the appropriately chosen β_z. Then, $\beta_z = \sqrt{\beta^2 - \beta_s^2}$ where β_s can be found by first solving a 2D waveguide problem corresponding to the reduced-wave equation.

For a rectangular waveguide, for example, from the previous lecture,

$$\beta_z = \sqrt{\beta^2 - \left(\frac{m\pi}{a}\right)^2 - \left(\frac{n\pi}{b}\right)^2} \tag{25.2.1}$$

for both TE_{mn} and TM_{mn} modes.[3] If the waveguide terminates with two shorts (which are easy

[2]Both rectangular and circular waveguides are cylindrical waveguides.

[3]Note that for a certain mn mode, with a choice of frequency, $\beta_z = 0$, which does not happen in a transmission line.

to make) at its ends, then the resonance condition is that

$$\beta_z = p\pi/d, \quad p \text{ integer} \tag{25.2.2}$$

Together, using (25.2.1), we have the condition that

$$\beta^2 = \frac{\omega^2}{c^2} = \left(\frac{m\pi}{a}\right)^2 + \left(\frac{n\pi}{b}\right)^2 + \left(\frac{p\pi}{d}\right)^2 \tag{25.2.3}$$

The above can only be satisfied by certain select frequencies: the resonant frequencies of the rectangular cavity. The corresponding mode is called the TE_{mnp} mode or the TM_{mnp} mode depending on if these modes are TE to z or TM to z. One can think of these modes as a consequence of the TE_{mn} or TM_{mn} modes in the rectangular waveguide bouncing back and forth in the z direction.

The entire electromagnetic fields of the cavity can be found from the pilot scalar potentials previously defined, namely that

$$\mathbf{E} = \nabla \times \hat{z}\Psi_h, \quad \mathbf{H} = \nabla \times \mathbf{E}/(-j\omega) \tag{25.2.4}$$

$$\mathbf{H} = \nabla \times \hat{z}\Psi_e, \quad \mathbf{E} = \nabla \times \mathbf{H}/(j\omega\varepsilon) \tag{25.2.5}$$

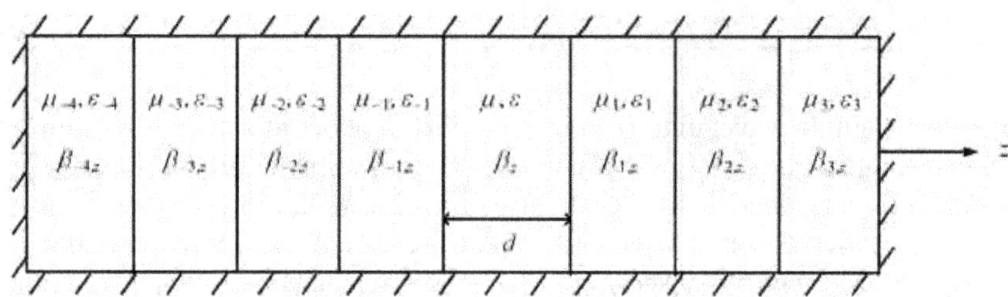

Figure 25.6: A waveguide filled with layered dielectrics can also become a resonator. The transverse resonance condition can be used to find the resonant modes. This can be obtained by exploiting the mathematical "homomorphism" between the waveguide problem and the transmission line problem.

25.2.2 Layered Medium Cavity

Since the layered medium problem in a waveguide is the same as the layered medium problem in open space, we can use the generalized transverse resonance condition to find the resonant frequencies and hence the modes of a waveguide cavity loaded with layered medium as shown in Figure 25.6. This condition is repeated below as:

$$\tilde{R}_-\tilde{R}_+e^{-2j\beta_z d} = 1 \tag{25.2.6}$$

where d is the length of the waveguide section and the above condition is applied. Here, $\tilde{R}_-$ and $\tilde{R}_+$ are the generalized reflection coefficients to the left and right of the center waveguide section. The above is similar to the resonant condition using the transmission line model in equation (25.1.2), except that now, we have replaced the transmission line reflection coefficient with TE or TM generalized reflection coefficients.

25.2.3 Lowest Mode of a Rectangular Cavity

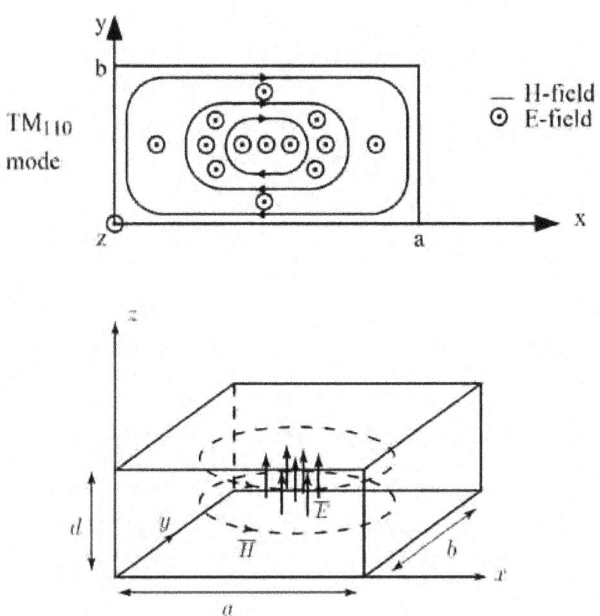

Figure 25.7: The top and 3D views of the E and H fields of the TM_{110} mode of a rectangular resonant cavity. Since this is a snapshot view of two sinusoidal fields that are 90 degrees out of phase, the relative signs of E and H fields are immaterial.

To find the lowest mode of a rectangular cavity is quite tricky. We assume that $a > b > c$, and find a mode where the combination of m, n, p in (25.2.3) will make β the smallest. We can first try to make one of them zero so that β will be greatly reduced. It is not possible to make both of them zero, as this will reduce it to a 1D problem. We begin by assuming a guided TM to z mode bouncing between two ends of a rectangular cavity. The lowest TM mode is the TM_{11} mode. At the cutoff of this mode, the $\beta_z = 0$ or $p = 0$, implying no variation of the field in the z direction. Let us call this mode the TM_{110} mode. The fields of this mode are visualized in Figure 25.7. It is a mode that bounces in the x and y directions, but does not propagate in the z direction. The electric field is pointing in the z direction in the cavity. Now, we do a Gedankenexperiment by putting two metallic shorts above and below the cavity. The boundary conditions at these two end caps are still satisfied since the tangential **E** field is zero. Therefore, this mode can exist and satisfy the boundary conditions on the six walls of the cavity with one of β_i, $i = x, y, z$ being zero. This is the lowest resonant mode that makes the right-hand side of (25.2.3) as small as possible by setting $p = 0$.

The top and side views of the fields of this mode are shown in Figure 25.7. The corresponding resonant frequency of this mode, from (25.2.3), satisfies the equation

$$\frac{\omega_{110}^2}{c^2} = \left(\frac{\pi}{a}\right)^2 + \left(\frac{\pi}{b}\right)^2 \tag{25.2.7}$$

Looking at the TE to z modes more carefully, it is required that $p \neq 0$, otherwise; the field is zero in the cavity. For example, it is possible to have the TE_{101} mode with a nonzero **E** field. The resonant frequency of this mode, from (25.2.3), is

$$\frac{\omega_{101}^2}{c^2} = \left(\frac{\pi}{a}\right)^2 + \left(\frac{\pi}{d}\right)^2 \tag{25.2.8}$$

Clearly, this mode has a higher resonant frequency compared to the TM_{110} mode if $d < b$.

25.2.4 Circular, Cylindrical, and Spherical Cavity Cases

The above analysis can be applied to circular and other cylindrical waveguides with β_s determined differently. For instance, for a circular waveguide, β_s is determined differently using Bessel functions, and for a general arbitrarily shaped waveguide as shown in Figure 25.9, β_s may have to be determined numerically.

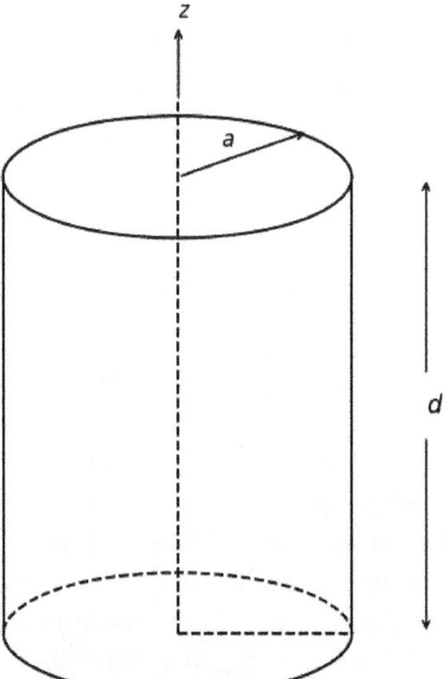

Figure 25.8: A circular resonant cavity made by terminating a circular waveguide at both ends.

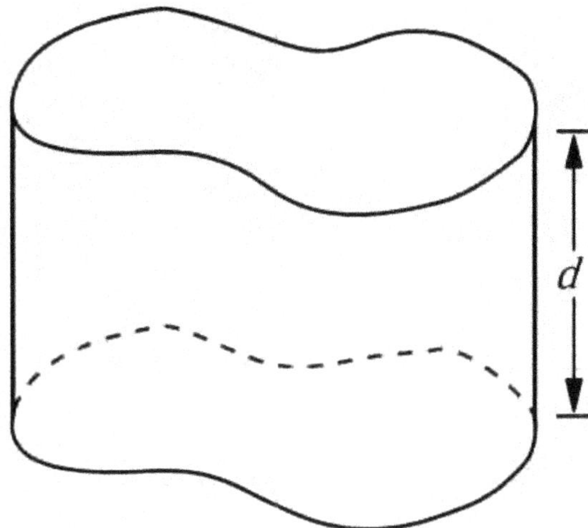

Figure 25.9: An arbitrary cylinder resonant cavity made by terminating an arbitrary waveguide at both ends.

For a spherical cavity, one would have to analyze the problem in spherical coordinates. The equations will have to be solved by the separation of variables using spherical harmonics. Details are given on p. 468 of Kong [35]. These days, when the cavity is of arbitrary shape, numerical methods can be used to find its resonant frequencies [187].

25.2.5 A General 3D Cavity

For a source-free cavity, the electromagnetic fields inside the cavity satisfy

$$(\nabla^2 + \omega\mu\varepsilon)\mathbf{E}(\mathbf{r}) = 0 \tag{25.2.9}$$

A generalization of this to 3D cavity is given in [187]. The details of this method is beyond the scope of this course, but interested readers can follow this paper more deeply. A beautiful theorem that all electromagneticists should know is Helmholtz theorem. It says that an arbitrary field inside a cavity can be decomposed into a divergence-free part plus a curl-free part. Detail proofs can be found in [50, 104, 188].

25.3 Some Applications of Resonators

Resonators in microwaves and optics can be used for designing filters, energy trapping devices, and antennas. As filters, they are used like LC resonators in circuit theory. A concatenation of them can be used to narrow or broaden the bandwidth of a filter. As an energy trapping device, a resonator can build up a strong field inside the cavity if it is excited with energy close to its resonance frequency, similar to an LC tank circuit. They can be used in klystrons and magnetrons as microwave sources, as a laser cavity for optical sources, or as a wavemeter to measure the frequency of the electromagnetic field at microwave frequencies. An antenna is a radiator that we will discuss more fully later. The use of a resonator can help in resonance tunneling to enhance the radiation efficiency of an antenna.

25.3.1 Filters

Circuit theory plays an important role in the design of microwave filters, as circuit theory is simple. An LC tank circuit can be used as a simple filter in electronic circuits. A concatenation of a number of LC tank circuits can be used to design a broadband filter. By the same token, microstrip line resonators, and a concatenation of them, are often used to make filters [189].

Transmission lines are often used to model microstrip lines in a complex microwave integrated circuits (MIC) or monolithic MIC (MMIC). In these circuits, due to the etching process, it is a lot easier to make an open circuit rather than a short circuit. But a true open circuit is hard to make, as an open-ended microstrip line has fringing field at its end as shown in Figure 25.11 [190, 191]. The fringing field gives rise to fringing field capacitance as shown in Figure 25.4. Then the appropriate Γ_S and Γ_L can be used to model the effect of fringing field capacitance. Figure 25.10 shows a concatenation of several microstrip resonators to make a microstrip filter. This is like using a concatenation of LC tank circuits to design filters in circuit theory [192]. Another rapidly emerging MIC technology is the substrate integrated waveguide (SIW) circuit [193] due to its ease of fabrication and increased confinement of the waves on the substrate. As a result, myriads of MIC can be fabricated from it. A sample paper is [194] with a picture of the SIW in Figure 25.12.

Optical filters can be made with an optical etalon such as a Fabry-Pérot resonator, or concatenation of them. This is shown in Figure 25.13.

Figure 25.10: A microstrip filter designed using concatenated resonators. The connectors to the coax cable are the SMA (sub-miniature type A) connectors (Figure by aginas.fe.up.pt).

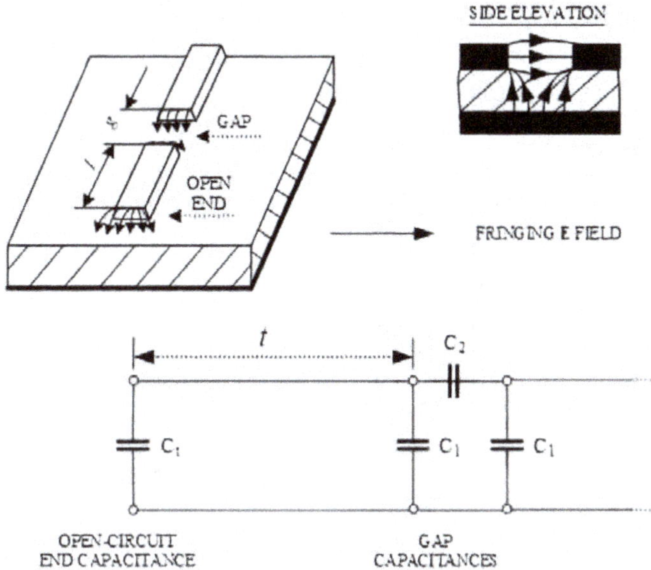

Figure 25.11: End effects and junction effects in a microwave integrated circuit are important [190, 191] (Figure by Microwave Journal). The fringing fields at the end of a microstrip line or between two microstrip lines can be modeled with parasitic capacitances. The reduction of a complex microwave circuit to a simple lumped-element approximation is a process known as parameter extraction.

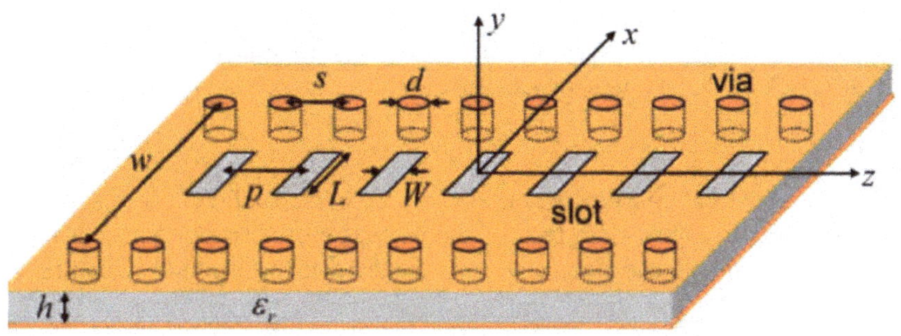

Figure 25.12: A substrate integrated waveguide from [194]. The shorting pins help to confine the fields, and they are also easily fabricated as part of the microwave integrated circuit.

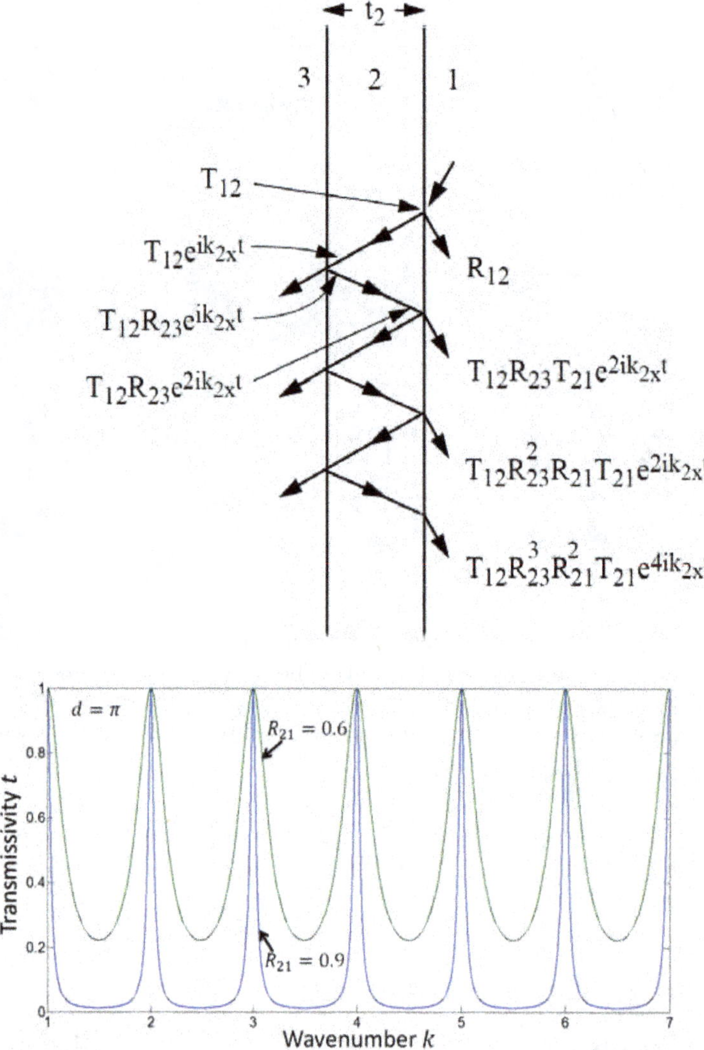

Figure 25.13: Design of a Fabry-Pérot resonator [61, 195, 92, 196]. As the magnitude of the reflection coefficient becomes close to one, the wave is better trapped inside the slab. A resonant mode exists inside the slab, providing a means for resonance tunneling.

25.3.2 Electromagnetic Sources (Heuristically)

Microwave sources are often made by transferring kinetic energy from an electron beam to microwave energy. Klystrons, magnetrons, and traveling wave tubes are examples of such devices. However, the cavity resonator in a klystron enhances the interaction of the electrons with the microwave field allowing for more effective energy transfer, causing the field to grow in amplitude as shown in Figure 25.14.

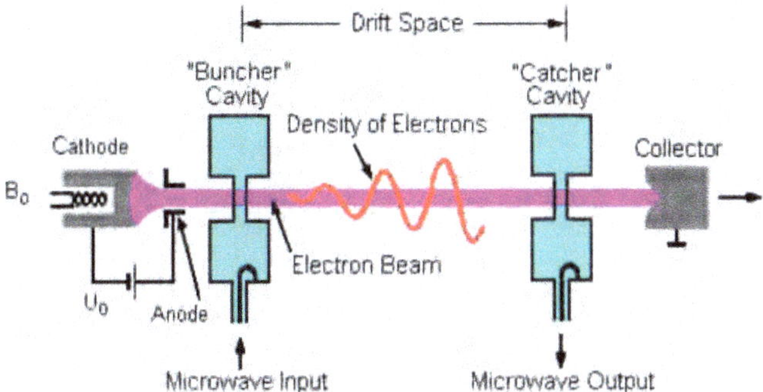

Figure 25.14: A klystron works by converting the kinetic energy of an electron beam into the energy of a traveling microwave next to the beam. As the microwave rides on the electron beam, it absorbs energy from the kinetic energy of the electrons, making its amplitude grow as it propagates. The amplified microwave can be collected by the "catcher" cavity (Figure by: https://en.wikipedia.org/wiki/Kly [197]).

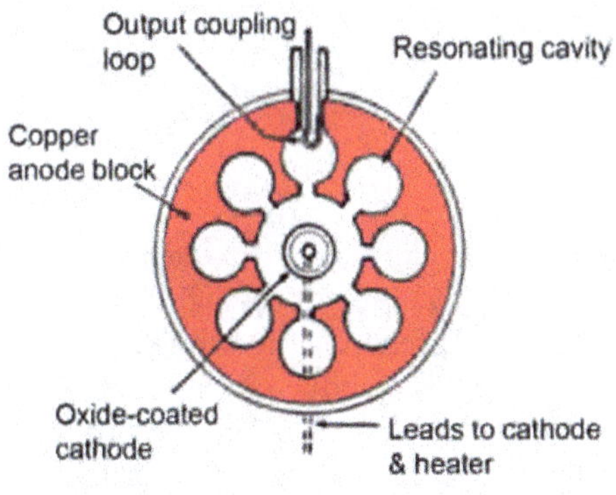

Figure 25.15: A magnetron works by having a high-Q microwave cavity resonator. When the cavity is injected with energetic electrons from the cathode to the anode, the kinetic energy of the electron feeds into the energy of the microwave. The cavity resonance amplifies this field-electron interaction, causing energy transfer from the kinetic energy of the electrons to the electromagnetic field energy (Figure by https://en.wikipedia.org/wiki/Cavity magnetron [198]).

Magnetron cavity works also by transferring the kinetic energy of the electron into microwave energy. By injecting hot energetic electrons into the magnetron cavity, the electromagnetic cavity resonance is magnified by the absorption of kinetic energy from the hot electrons, giving rise to amplified microwave energy as shown in Figure 25.15.

Figure 25.16 shows a laser cavity resonator used to enhance light wave interaction with material media. By using stimulated emission of electronic transition (the physics is beyond the purview of this course), light energy can be amplified.

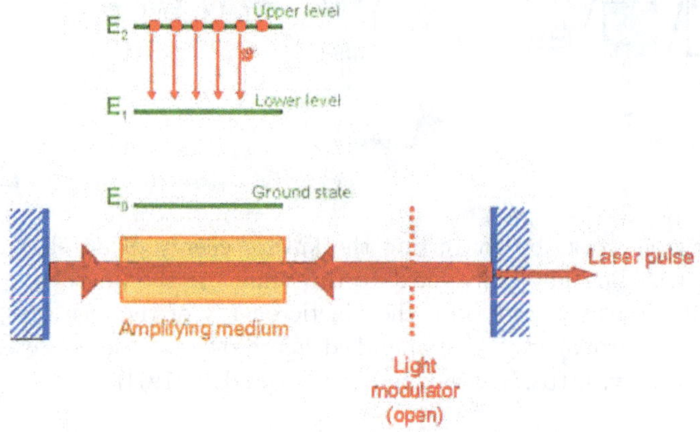

Figure 25.16: A simple view of the physical principle behind the workings of the laser. The cavity again enhances the interaction of the photons with the amplifying medium (Figure by: www.optique-ingenieur.org).

Energy trapping of a waveguide or a resonator can be used to enhance the efficiency of a semiconductor laser as shown in Figure 25.17. The trapping of the light energy by the heterojunctions as well as the index profile allows the light to interact more strongly with the lasing medium or the active medium of the laser. This enables a semiconductor laser to work at room temperature. In 2000, Z. I. Alferov and H. Kroemer, together with J.S. Kilby, were awarded the Nobel Prize for information and communication technology: Alferov and Kroemer for the invention of a room-temperature semiconductor laser, and Kilby for the invention of the electronic integrated circuit (IC) or the chip.

25.3.3 Frequency Sensor

A cavity resonator can be used as a frequency sensor. It acts as an energy trap, because it will siphon off energy from a microwave waveguide when the microwave frequency hits the resonance frequency of the cavity resonator. This can be used to determine the frequency of the passing wave. Wavemeters are shown in Figures 25.19 and 25.18. As seen in the picture, there is an entry microwave port for injecting microwaves into the cavity, and another exit port for the taking the microwaves out of the cavity sensor. The resonant frequency of the cavity resonator can be continuously tuned by changing the location of the plunger. The passing microwave, when it hits the resonance frequency of the cavity, will create a large field inside it due to resonance coupling. The larger field will dissipate more energy on the cavity metallic wall and will cause less energy to leave the cavity. This dip in energy transmission at the resonant freqeuncy of the cavity reveals the frequency of the microwave.

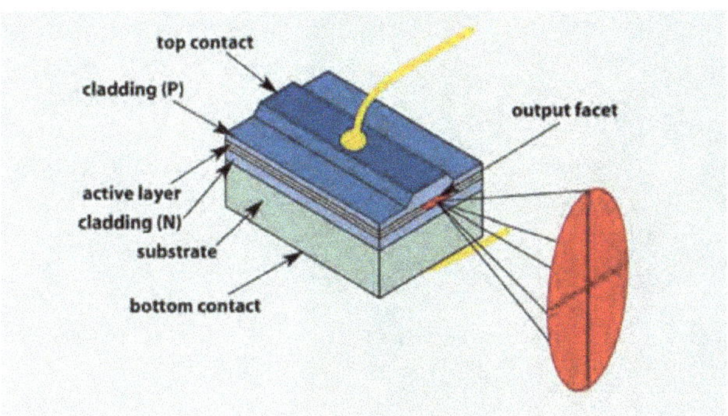

Figure 25.17: A semiconductor laser (also known as a laser diode) at work. Room temperature lasing is possible due to the tight confinement of both light photons and electron-hole pair carriers. Their close proximity causes the diode to lase at room temperature (Figure by: Photonics.com).

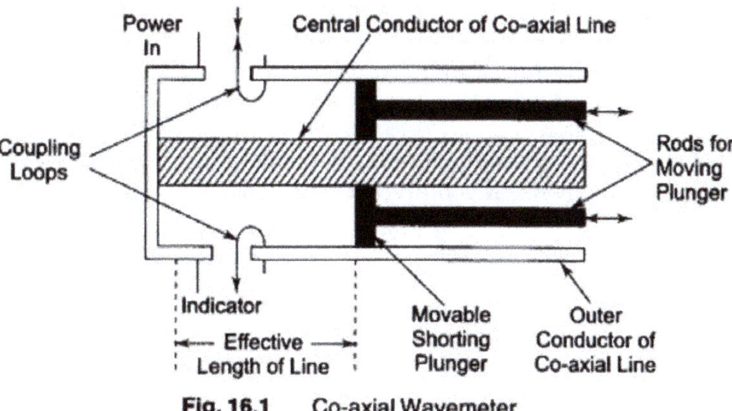

Figure 25.18: The innardsx of a wavemeter. The location of the plunger short can be continuously moved by rotating the cap of the cavity shown in the previous figure (Figure by: eeeguide.com).

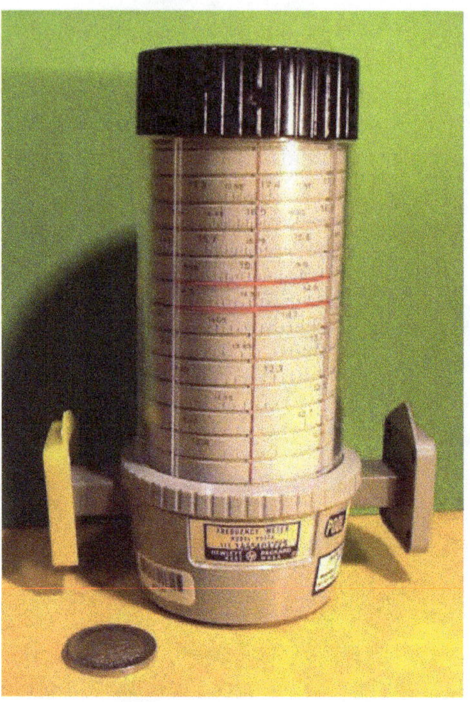

Figure 25.19: An absorption wave meter can be used to measure the frequency of a microwave. If the microwave energy enters the cavity at its resonant frequency, strong field buildup inside the cavity causes increased loss and absorption of the microwave energy by the cavity. A dip in energy level of the transmitted signal indicates the coincidence of the resonant frequency of the microwave with the frequency of the passing microwave (Figure By: https://en.wikipedia.org/wiki/Absorption wavemeter).

Due to the complexity of waveguides and resonators, most modern waveguide and resonator problems are solved with numerical methods. A good recent review paper is given in [199].

Exercises for Lecture 25

Problem 25-1:

(i) Explain what happens at the junction of two hollow waveguides if they do not share the same cross section.

(ii) Explain why the lowest resonant mode of a rectangular cavity is the TM_{110} mode when $a > b > d$.

(iii) Name some applications of a cavity resonator.

(iv) Using an LC tank circuit model for cavity-backed antenna, explain why the fields are strong inside the cavity when one operates near the resonant frequency of the cavity.

Lecture 26

Quality Factor of Cavities, Mode Orthogonality

Cavity resonators are important for making narrowband filters. The bandwidth of a filter is related to the Q or the quality factor of the cavity. As shown previously, a concatenation of cavity resonators can be used to engineer different filter designs. Resonators can also be used to design various sensing systems, as well as measurement systems. We will study the concept of Q in this lecture.

Also, before we leave the lectures on waveguides and resonators, it will be prudent to discuss mode orthogonality. Since this concept is very similar to eigenvector orthogonality found in matrix or linear algebra, we will relate mode orthogonality in waveguides and cavities to eigenvector orthogonality.

26.1 The Quality Factor of a Cavity—General Concept

The quality factor of a cavity or its Q measures how ideal or lossless a cavity resonator is. An ideal lossless cavity resonator will sustain free oscillations forever, while most resonators sustain free oscillations for a finite time due to loss or damping of the oscillation. This is because of losses coming from radiation, dissipation in the dielectric material filling the cavity, or resistive loss of the metallic part of the cavity. When loss is incurred from radiation of the structure, the term "radiation damping" is often used.

26.1.1 Analogue with an LC Tank Circuit

Much physical insight can be gotten by studying a simple resonator. One of the simplest resonators imaginable is the LC tank circuit. By using it as an analogue, we can better understand the resonance of a cavity. In theory, when there is no loss in an LC tank circuit, it can oscillate forever. Moreover, if we turn off the source as in Figure 26.1(b), a free oscillation solution persists in the circuit.[1]

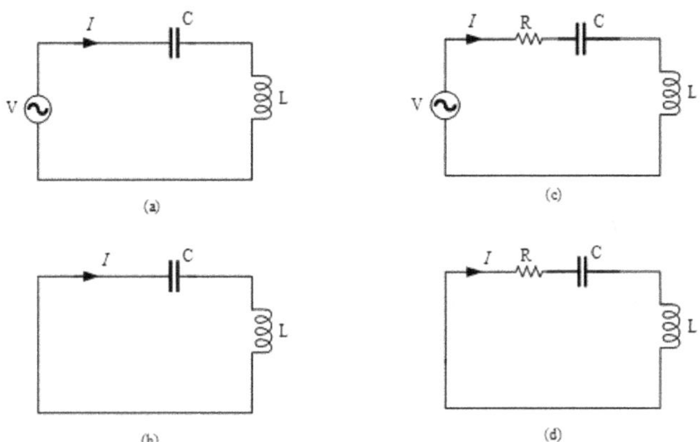

Figure 26.1: For the circuit on the left, it will resonate forever even if the voltage source is turned off. But for the circuit on the right, if the source is turned off, the current in the circuit will decay with time due to dissipation in the resistor and damping due to loss.

One can write the voltage-current relation in the lossless circuit as

$$I(\omega) = \frac{V(\omega)}{j\omega L + 1/(j\omega C)} = V(\omega)Y(\omega) \tag{26.1.1}$$

where

$$Y(\omega) = \frac{1}{j\omega L + 1/(j\omega C)} \tag{26.1.2}$$

The above $Y(\omega)$ in (26.1.2) can be thought of as the transfer function of the linear time-invariant system where the input is $V(\omega)$ and the output is $I(\omega)$. When the voltage is zero or turned off, a nonzero current exists or persists at the resonance frequency of the oscillator. The resonant frequency is when the denominator in the above equation is zero, so that I is finite despite $V = 0$.[2] This resonant frequency, obtained by setting the denominator of Y to zero, is given by $\omega_R = 1/\sqrt{LC}$.

When a small resistor is added in the circuit to give rise to loss, the voltage-current relation becomes

$$I(\omega) = \frac{V(\omega)}{j\omega L + R + 1/(j\omega C)} = V(\omega)Y(\omega) \tag{26.1.3}$$

where

$$Y(\omega) = \frac{1}{j\omega L + R + 1/(j\omega C)} \tag{26.1.4}$$

[1]This is analogous to the homogeneous solution of an ordinary differential equation or the natural solution of a physical system.

[2]We take advantage of the fact that zero divided by zero is undefined.

Now, the denominator of the above functions can never go to zero for real ω. But there exists complex ω that will make Y become infinite. These are the complex resonant frequencies of the circuit. Thus, the homogeneous solution (also called the natural solution, or free oscillation) can only exist at the complex resonant frequencies for this lossy circuit. With complex resonances, the voltage and the current are decaying sinusoids.

To further convince yourself, $Y(\omega)$ is the impulse response of the system which is the response of the system when it is excited with an impulse function. One can easily find the response by taking the inverse Fourier transform of (26.1.4). The inverse Fourier transform of such a function, for instance, is found in many textbooks on signals and systems [57] [Table 7.2]. It can be shown then that the response of the system or $i(t)$, which is the inverse Fourier transform of $I(\omega)$ above, is of the form

$$i(t) \sim I_0 e^{-\alpha t} \cos(\omega_0 t + \phi) u(t) \tag{26.1.5}$$

where $u(t)$ is the unit step function.

By the same token, the impulse response of an electromagnetic cavity can be represented by poles, each of which represents the resonant mode of the cavity. Each of these modes corresponds to the free oscillation of the cavity. (To simplify matters, we choose the operating frequency in the vicinity of the resonant frequency, and then the impulse response can be approximated by a single pole system.) Because of losses, the free oscillation has a complex frequency given by $\omega_R = \omega_0 + j\alpha$. Upon Fourier inverse transforming, the cavity has an electromagnetic field with time dependence as follows:[3]

$$\mathbf{E} \propto e^{-\alpha t} \cos(\omega t + \phi_1), \quad \mathbf{H} \propto e^{-\alpha t} \cos(\omega t + \phi_2) \tag{26.1.6}$$

That is, they are decaying sinusoids. The total stored energy of the system is proportional to $\frac{1}{4}\varepsilon |\mathbf{E}|^2 + \frac{1}{4}\mu |\mathbf{H}|^2$. But by assuming the loss to be small, the stored energy can be time-averaged similar to sinusoidal fields. Thus, the time-average stored energy has the form

$$\langle W_T \rangle = \langle W_E \rangle + \langle W_H \rangle \cong W_0 e^{-2\alpha t} \tag{26.1.7}$$

If there is no loss, $\langle W_T \rangle$ will remain constant. However, with loss, the average stored energy will decrease to $1/e$ of its original value at $t = \tau = \frac{1}{2\alpha}$. The Q of a cavity is defined as the number of free oscillations in radians that the field undergoes before the energy stored decreases to $1/e$ of its original value (see Figure 26.2). In a time interval $\tau = \frac{1}{2\alpha}$, the number of free oscillations in radians is $\omega\tau$ or $\frac{\omega}{2\alpha}$; hence, the Q is defined to be [35]

$$Q \cong \frac{\omega}{2\alpha} \tag{26.1.8}$$

Here, Q is an approximate and asympototic concept; it makes sense only if the system has low loss and narrow band.

[3]Remember from your signals and systems course that poles have to occur in conjugate pairs due to the real-value nature of the signal.

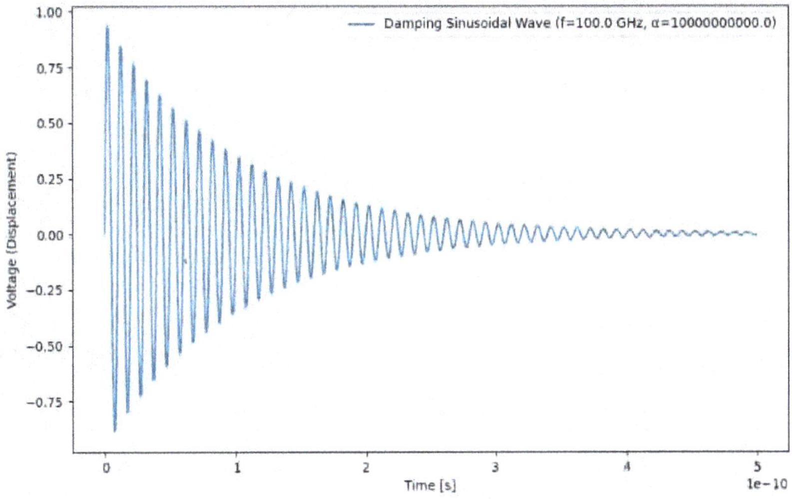

Figure 26.2: A typical time domain response of a high Q system.

The cavity field diminishes with time because of losses in the cavity. These losses can come about because of imperfect conductor on the cavity wall (also called metal loss or copper loss), materials can have complex permittivity or a loss tangent, or due to radiation if the cavity has a hole. Furthermore, by energy conservation, the decrease in stored energy per unit time must be equal to the total power dissipated in the losses of a cavity. In other words,

$$\langle P_D \rangle = -\frac{d\langle W_T \rangle}{dt} \tag{26.1.9}$$

By further assuming that $\langle W_T \rangle$ has to be of the form in (26.1.7), then

$$-\frac{d\langle W_T \rangle}{dt} \cong 2\alpha W_0 e^{-2\alpha t} = 2\alpha \langle W_t \rangle \tag{26.1.10}$$

From the above, we can estimate the decay constant

$$\alpha \cong \frac{\langle P_D \rangle}{2\langle W_T \rangle} \tag{26.1.11}$$

Hence, we can rewrite equation (26.1.8) for the Q as

$$Q \cong \frac{\omega \langle W_T \rangle}{\langle P_D \rangle} \tag{26.1.12}$$

By further letting $\omega = 2\pi/T$, we lend further physical interpretation to express Q as

$$Q \cong 2\pi \frac{\langle W_T \rangle}{\langle P_D \rangle T} = 2\pi \frac{\text{Total Energy Stored}}{\text{Energy Dissipated/Cycle}} \tag{26.1.13}$$

In the above, Q is a concept defined at the resonant frequency of the resonator, and hence, we evaluate $\omega = \omega_R$.

In a cavity, the energy can dissipate in either the dielectric loss or the wall loss of the cavity due to the finiteness of the conductivity. It has to be re-emphasized that the Q is a low-loss, asymptotic concept, and hence, the above formulas are only approximately true.

26.1.2 Relation to Bandwidth and Pole Location

As seen from above, the resonance of a system is related to the pole of the transfer function. For instance, in our previous example of the RLC tank circuit, the admittance $Y(\omega)$ can be thought of as a transfer function in linear system theory: The input is the voltage $V(\omega)$, while the output is the current $I(\omega)$. If we encounter the resonance of the system at a particular frequency, the transfer function becomes infinite. This infinite value can be modeled by a pole of the transfer function in the complex ω plane.

For a system with a simple pole, the transfer function $Y(\omega)$ in (26.1.4) of the system can be written as

$$Y(\omega) = \frac{A}{j\omega - j\omega_R} + \frac{A}{-j\omega - j\omega_R^*} \tag{26.1.14}$$

where A is real. The above can be obtained by performing the partial fraction of (26.1.4).

We can easily convince ourselves that, when ω is real,

$$Y(-\omega) = Y^*(\omega) \tag{26.1.15}$$

so that $y(t)$ is real. Here, $y(t)$ is the Fourier transform of $Y(\omega)$.

The second term is added to give this transfer function conjugate symmetry [57], a necessity for real-valued signals. The transfer function is the Fourier transform of an impulse response, which is real. And hence, the Fourier transform of $Y(\omega)$ has to be real value.

In the above, where we have assumed that $\omega_R = \omega_0 + j\alpha$, the resonant frequency is complex. In general, to find the current $i(t)$, we take the inverse Fourier transform of $I(\omega)$, or that

$$i(t) = \frac{1}{2\pi}\int_{-\infty}^{\infty} Y(\omega)V(\omega)e^{j\omega t}d\omega \tag{26.1.16}$$

The above does guarantee the real-valuedness of $i(t)$ because of the way we have defined $Y(\omega)$. For the other real-valuedness requirements, we have $V(-\omega) = V^*(\omega)$. The path of integration (also called Fourier inversion contour/path) above is along the real ω axis.[4] Because of the loss in the system, the poles are never encountered on the real ω axis, and hence, the above Fourier inverse transform is well defined and unique.[5]

[4]If one derives Fourier transform from a Fourier series, the frequencies are all real to begin with, and, hence, the final Fourier inversion path is on the real axis [57]

[5]Does this sound familiar? We have to introduce loss in the electromagnetic system in order to prove uniqueness of the solution!

In principle, when $\omega = \omega_R$, the transfer function $Y(\omega)$ becomes infinite, but this does not happen in practice because ω_R is complex, and ω, the variable of integration (or the operating frequency) is real. In other words, when the pole is displaced slightly off the real axis to account for loss, it gives rise to a well-defined Fourier inversion path, yielding uniqueness to the solution.

For frequencies close to ω_R, we can use a single-pole approximation to evaluate the integrand. It is quite clear that $|Y(\omega)|$ would peak at $\omega = \omega_0$. At $\omega = \omega_0 \pm \alpha$, the magnitude of $|Y(\omega)|$ would be $1/\sqrt{2}$ smaller, or that the power, which is proportional to $|Y(\omega)|^2$, would be half as small. Therefore, the half-power points compared to the peak are at $\omega = \omega_0 \pm \alpha$. We can surmise this behavior by studying the magnitude of $|1/(\omega - \omega_R)|$ with respect to Figure 26.3.[6]

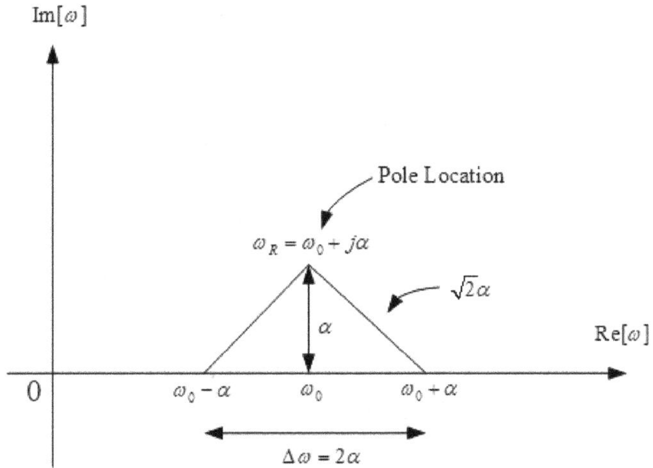

Figure 26.3: From this figure, it is seen that $\omega - \omega_R$ is smallest when $\omega = \omega_0$. Hence, $1/(\omega-\omega_R)$ is largest when $\omega = \omega_0$ on the real axis. Furthermore, its magnitude diminishes to $1/\sqrt{2}$ of its peak value when $\omega = \omega_0 \pm \alpha$. These are the half-power points; hence, the full-width half-maximum (FWHM) bandwidth is when $\Delta\omega = \alpha$. This is also the half-power bandwidth.

Thus, the full-width half-maximum (FWHM) bandwidth is defined to be $\Delta\omega = 2\alpha$. And the Q can be written as in terms of the half-power bandwidth $\Delta\omega$ of the system, namely,

$$Q = \omega/(2\alpha) \cong \omega_0/\Delta\omega \tag{26.1.17}$$

Again, we bear in mind that Q is defined at the resonance of the problem and it is a low-loss concept, and hence, $\omega \cong \omega_0$. The above implies that the narrower the bandwidth, the higher the Q of the system. Typical plots of transfer function versus frequency for a system with different Qs are shown in Figure 26.4.

[6]In many undergraduate electrical engineering textbooks, Bode plots are introduced to illustrate this point.

The physics of the Q of an electromagnetic cavity is similar to that of an RLC tank circuit. It is much harder to derive the input-output relation for a general electromagnetic system, other than with the use of impedance matrix $\overline{\mathbf{Z}}$ system. In this case, the input/output of the linear system is given by

$$\mathbf{I} = \overline{\mathbf{Y}} \cdot \mathbf{V} \tag{26.1.18}$$

where $\overline{\mathbf{Y}} = (\overline{\mathbf{Z}})^{-1}$, the inverse of the impedance matrix.

It is beyond the scope of this course, but it can be shown that the resonant modes of the system correspond to the zeroes of the determinant of the impedance matrix $\overline{\mathbf{Z}}$ or $f(\omega) = \det(\overline{\mathbf{Z}})$. Here, $f(\omega)$ will exhibit poles as in the RLC tank circuit. In fact, there are some scholars who use circuit theory to model Maxwell's equations called the TLM (transmission-line matrix) method [105].

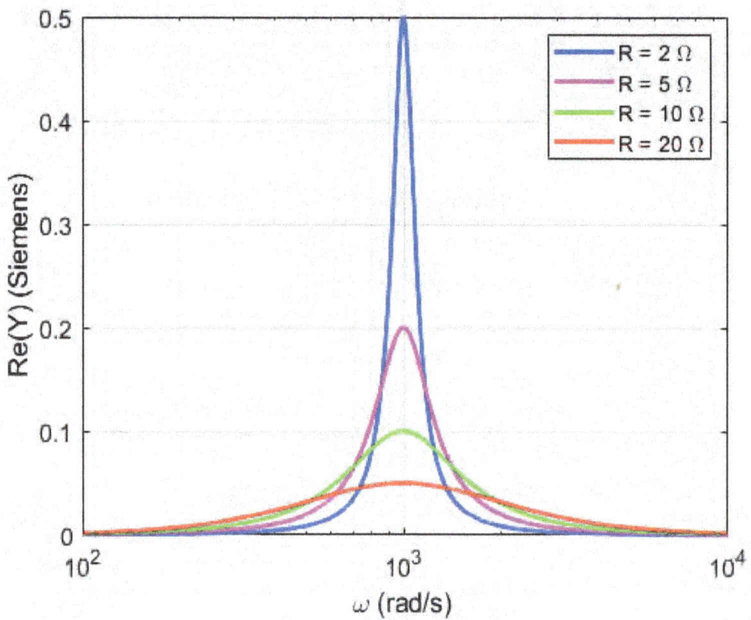

Figure 26.4: A typical system response versus frequency for different Qs using (26.1.4). The Q is altered by changing the resistor R in the circuit. Only the real part of Y is plotted.

26.1.3 Wall Loss and Q for a Metallic Cavity—A Perturbation Concept

To estimate the Q of a cavity, we will need to calculate the loss inside the cavity as well as the energy stored according to (26.1.12). We can use perturbation concept to estimate the Q. First, we assume a lossless cavity so that the cavity wall is made from perfect electric conductor (PEC). In this case, $\hat{n} \times \mathbf{E} = 0$ and no power can be absorbed by the waveguide wall. Then we assume a small loss by assuming that the cavity wall is now made of imperfect conductors and hence, $\hat{n} \times \mathbf{E} \neq 0$ but small. We can assume that the magnetic field $\mathbf{H}$ remains largely unchanged before and after we have introduced loss: for a PEC surface, a nonzero $\hat{n} \times \mathbf{H}$ is needed to support a surface current. But $\hat{n} \times \mathbf{E}$ is zero for a PEC and becomes nonzero for imperfect conductors.

If the cavity is filled with air, then the loss comes mainly from the metallic loss or copper loss from the cavity wall. In this case, the time-average power dissipated on the wall is given by [35]

$$\langle P_D \rangle = \frac{1}{2}\Re e \oint_S (\mathbf{E} \times \mathbf{H}^*) \cdot \hat{n}\, dS = \frac{1}{2}\Re e \oint_S (\hat{n} \times \mathbf{E}) \cdot \mathbf{H}^*\, dS \tag{26.1.19}$$

where we have used the scalar-triple product identity to rewrite the integrand, and S is the surface of the cavity wall.[7] Here, $(\hat{n} \times \mathbf{E})$ is the tangential component of the electric field ,which would have been zero if the cavity wall were made of an ideal PEC. Also, $\hat{n}$ is taken to be the outward-pointing normal at the surface S. The β (or k) vector in the transmitted medium is very large due to the high conductivity of the wall. Due to the phase-matching condition, the transmitted wave vector β is almost normal to the interface.[8] Therefore, we can approximate the transmitted wave as a wave propagating normal to the interface. In the metal, it decays predominantly in the direction of propagation, which is normal to the surface as well (see section 8.1).

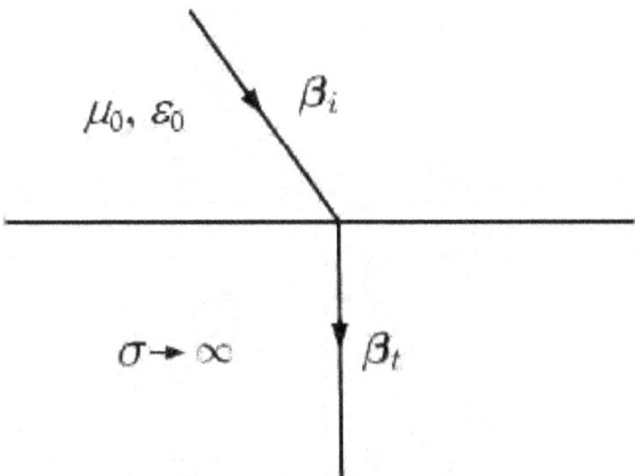

Figure 26.5: Due to high σ in the metal and large $|\boldsymbol{\beta}_t|$, phase-matching condition requires the transmitted wave vector to be almost normal to the surface. Then the field can be assumed to be a normally incident plane-wave field in the transmitted or the metal region. The tangential electric field is almost zero, and the tangential magnetic field remains almost unchanged using this perturbative concept.

For such a wave, we can approximate $\hat{n} \times \mathbf{E} = \mathbf{H}_t Z_m$ where Z_m is the intrinsic impedance for the metallic conductor, as shown in section 8.1, which is $Z_m = \sqrt{\frac{\mu}{\varepsilon_m}} \approx \sqrt{\frac{\mu}{-j\frac{\sigma}{\omega}}} = \sqrt{\frac{\omega\mu}{2\sigma}}(1+j)$,[9] where

[7] We have used the cyclic identity that $\mathbf{a} \cdot (\mathbf{b} \times \mathbf{c}) = \mathbf{b} \cdot (\mathbf{c} \times \mathbf{a}) = \mathbf{c} \cdot (\mathbf{a} \times \mathbf{b})$ in the above (see Some Useful Mathematical Formulas, in Appendix A of [1]).

[8] The horizontal components of the wave vectors $\boldsymbol{\beta}_i$ and $\boldsymbol{\beta}_t$ have to be real for phase matching. Hence, the $\boldsymbol{\beta}_t$ is only complex in the normal direction.

[9] When an electromagnetic wave enters a conductive region with a large β, it can be shown that the wave is

we have assumed that $\varepsilon_m \approx -\frac{j\sigma}{\omega}$ and $\mathbf{H}_t$ is the tangential magnetic field. From these equations, we can see that the tangential $\mathbf{E}$ field is small, but the tangential $\mathbf{H}$ field is not small. It is to be noted that the above approximation is only valid on a flat surface, which is mostly true on a large part of the cavity wall.

This relation between $\mathbf{E}$ and $\mathbf{H}$ will ensure that power is flowing into the metallic surface. Therefore,

$$\langle P_D \rangle = \frac{1}{2}\Re e \oint_S \sqrt{\frac{\omega\mu}{2\sigma}}(1+j)\left|\mathbf{H}_t\right|^2 dS = \frac{1}{2}\sqrt{\frac{\omega\mu}{2\sigma}}\oint_S \left|\mathbf{H}_t\right|^2 dS \tag{26.1.20}$$

By further assuming that the stored electric and magnetic energies of a cavity are equal to each other at resonance,[10] the stored energy can be obtained by

$$\langle W_T \rangle = \frac{1}{2}\mu\int_V |\mathbf{H}|^2 dV = \frac{1}{2}\varepsilon\int_V |\mathbf{E}|^2 dV \tag{26.1.21}$$

Written explicitly, the $Q \cong \frac{\omega\langle W_T\rangle}{\langle P_D\rangle}$ becomes

$$Q = \sqrt{2\omega\mu\sigma}\frac{\int_V \left|\mathbf{H}\right|^2 dV}{\oint_S \left|\mathbf{H}_t\right|^2 dS} = \frac{2}{\delta}\frac{\int_V \left|\mathbf{H}\right|^2 dV}{\oint_S \left|\mathbf{H}_t\right|^2 dS} \tag{26.1.22}$$

In the above, δ is the skin depth of the metallic wall. Hence, the more energy stored there with respect to the power dissipated, the higher the Q of a resonating system. Also, the lower the metal loss, or the smaller the skin depth, the higher the Q would be.

Notice that in (26.1.22), the numerator is a volume integral and hence is proportional to volume, while the denominator is a surface integral and is proportional to surface. Thus, the Q, a dimensionless quantity, is roughly proportional to

$$Q \sim \frac{V}{S\delta}A \tag{26.1.23}$$

where V is the volume of the cavity, while S is its surface area and A is a dimensionless constant yet to be determined. From the above, it is noted that a large cavity compared to its skin depth has a larger Q than a small cavity. Thus, in microwave engineering, a large cavity (compared to wavelength) is needed to obtain a high Q resonance. The above analysis is good for microwave engineering where in the metal, the skin-depth of the electromagnetics wave is much smaller than the wavelength.

---

refracted to propagate normally to the surface as shown in Figure 26.5, and, hence, this formula can be applied. Moreover, the Fresnel reflection and transmission coefficients still apply here.

[10]They are definitely equal to each other for the lossless case. For the lossy case, they are approximately equal.

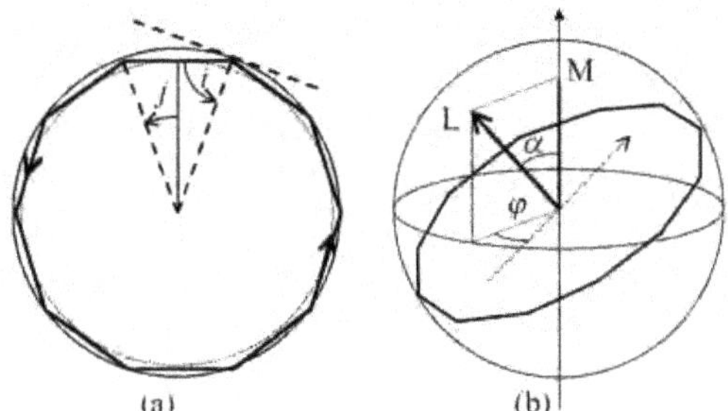

Figure 26.6: A dielectric resonator can be made by using total internal reflection of the bouncing wave within the resonator [200]. Such a mode is called a whispering gallery mode. It has high Q because the glass, from which the cavity is made, has little loss. Furthermore, the cavity can be made very large compared to wavelength, increasing the stored energy.

26.1.4 Example: The Q of TM$_{110}$ Mode

For the TM$_{110}$ mode, as can be seen from the previous lecture, the only electric field is $\mathbf{E} = \hat{z}E_z$, with $\partial/\partial z = 0$. Then with $m = 1$, $n = 1$, and $p = 0$, we have

$$E_z = E_0 \sin\left(\frac{\pi x}{a}\right) \sin\left(\frac{\pi y}{b}\right) \tag{26.1.24}$$

The magnetic field can be derived from the electric field using Maxwell's equation or Faraday's law, giving

$$H_x = \frac{j\omega\epsilon}{\omega^2\mu\epsilon}\frac{\partial}{\partial y}E_z = \frac{j\left(\frac{\pi}{b}\right)}{\omega\mu}E_0 \sin\left(\frac{\pi x}{a}\right)\cos\left(\frac{\pi y}{b}\right) \tag{26.1.25}$$

$$H_y = \frac{-j\omega\epsilon}{\omega^2\mu\epsilon}\frac{\partial}{\partial x}E_z = -\frac{j\left(\frac{\pi}{a}\right)}{\omega\mu}E_0 \cos\left(\frac{\pi x}{a}\right)\sin\left(\frac{\pi y}{b}\right) \tag{26.1.26}$$

Therefore[11]

$$\begin{aligned}
\int_V |\mathbf{H}|^2\, dV &= \int_{-d}^{0}\int_0^b\int_0^a dx\,dy\,dz \left[|H_x|^2 + |H_y|^2\right] \\
&= \frac{|E_0|^2}{\omega^2\mu^2}\int_{-d}^{0}\int_0^b\int_0^a dx\,dy\,dz \\
&\quad \left[\left(\frac{\pi}{b}\right)^2 \sin^2\left(\frac{\pi x}{a}\right)\cos^2\left(\frac{\pi y}{b}\right) + \left(\frac{\pi}{a}\right)^2 \cos^2\left(\frac{\pi x}{a}\right)\sin^2\left(\frac{\pi y}{b}\right)\right] \\
&= \frac{|E_0|^2}{\omega^2\mu^2}\frac{\pi^2}{4}\left[\frac{a}{b} + \frac{b}{a}\right] d
\end{aligned} \tag{26.1.27}$$

[11]Since the electric field is simpler than the magnetic field, it is easier to find the energy stored using the electric field. Like an LC tank circuit, the magnetic field energy stored and the electric field energy stored are equal to each other.

A cavity has six faces, finding the tangential exponent at each face and integrating it, gives the following:

$$\begin{aligned}\oint_S |\mathbf{H}_t|\, dS &= 2\int_0^b \int_0^a dx\, dy \left[|H_x|^2 + |H_y|^2\right] \\ &\quad + 2\int_{-d}^0 \int_0^a dx\, dz\, |H_x(y=0)|^2 + 2\int_{-d}^0 \int_0^b dy\, dz\, |H_y(x=0)|^2 \\ &= \frac{2|E_0|^2}{\omega^2\mu^2}\frac{\pi^2 ab}{4}\left[\frac{1}{a^2}+\frac{1}{b^2}\right] + \frac{2\left(\frac{\pi}{b}\right)^2}{\omega^2\mu^2}|E_0|^2\frac{ad}{2} + \frac{2\left(\frac{\pi}{a}\right)^2}{\omega^2\mu^2}|E_0|^2\frac{bd}{2} \\ &= \frac{\pi^2|E_0|^2}{\omega^2\mu^2}\left[\frac{b}{2a}+\frac{a}{2b}+\frac{ad}{b^2}+\frac{bd}{a^2}\right]\end{aligned} \tag{26.1.28}$$

Hence the Q from (26.1.22) is finally given as

$$Q = \frac{1}{2\delta}\frac{\left(\frac{ad}{b}+\frac{bd}{a}\right)}{\left(\frac{b}{2a}+\frac{a}{2b}+\frac{ad}{b^2}+\frac{bd}{a^2}\right)} \tag{26.1.29}$$

The result shows that the larger the cavity, the higher the Q. This is because the Q, as mentioned before, is the ratio of the energy stored in a volume to the energy dissipated over the surface of the cavity.

It is easy to make large cavities in optics as the wavelength is small. But it is imperative to note that at optical frequencies, the loss mechanism is quite different. The electrons in a piece of metal behave more like a sea of electrons with little restoring force. Hence, it is more like a plasma medium where the inertial force dominates over the collision force (see section 8.3.4, (8.3.14)) with very little restoring force. We can set $\kappa = 0$ and the equation of motion can be approximated by

$$m_e\frac{d^2x}{dt^2} + m_e\Gamma\frac{dx}{dt} = -eE \tag{26.1.30}$$

Also, a dielectric resonator cavity can be made out of glass where the primary loss will be from the material. Quality factor $\approx 10^8 \sim 10^9$ is possible [200]. A dielectric resonator using total internal reflection to trap the wave as it bounces around is called a whispering gallery mode resonator. The glass can be made with very low loss.[12] This together with the large size of the cavity compared to wavelength gives the resonator very high Q. The large cavity size increases the stored energy as well, which is good for increasing its Q as is evident from (26.1.23).

26.2 Mode Orthogonality and Matrix Eigenvalue Problem

It turns out that the modes of a waveguide or a resonator are orthogonal to each other. This is intimately related to the orthogonality of eigenvectors of a matrix operator.[13] Thus, it is best to

[12] Silica is a gift of God. With it, we have the optical fiber with little loss and the semiconductor devices that have revolutionized our modern world.

[13] This mathematical "homomorphism" is not discussed in any other electromagnetic textbooks. This "homomorphism" is not what is defined in the math literature.

understand this by the homomorphism between the electromagnetic mode problem and the matrix eigenvalue problem. Because of this similarity, electromagnetic modes are also called eigenmodes. Thus it is prudent that we revisit the matrix eigenvalue problem (EVP) here.

26.2.1 Hermiticity of an Operator

We can use inner product to define the Hermiticity and the symmetry of an operator. This is great because it can be easily extended to infinite dimensional Hilbert spaces.

Transposes of a Matrix Operator

The transpose of a matrix operator is defined with matrix operator notation, to be such that[14]

$$\mathbf{f}^t \cdot \overline{\mathbf{G}} \cdot \mathbf{g} = \mathbf{g}^t \cdot \overline{\mathbf{G}}^t \cdot \mathbf{f} \tag{26.2.1}$$

If $\overline{\mathbf{G}}^t = \overline{\mathbf{G}}$, then $\overline{\mathbf{G}}$ is symmetric. In other words, its transpose is itself.

The conjugate (Hermitian) transpose of a matrix operator is defined, such that

$$\mathbf{f}^\dagger \cdot \overline{\mathbf{G}} \cdot \mathbf{g} = \mathbf{g}^\dagger \cdot \overline{\mathbf{G}}^\dagger \cdot \mathbf{f} \tag{26.2.2}$$

If $\overline{\mathbf{G}}^\dagger = \overline{\mathbf{G}}$, then $\overline{\mathbf{G}}$ is Hermitian. In other words, its conjugate transpose is itself. These operators are also called self-adjoint operators.

The above definition is convenient because it can be extended to infinite dimensional spaces, which are usually called Hilbert spaces. Then the inner product between two vectors becomes the functional inner product.

$$\mathbf{f}^t \cdot \mathbf{g} \to \langle f, g \rangle \tag{26.2.3}$$

where the inner product on the right-hand side is defined as

$$\langle f, g \rangle = \int f(\mathbf{r}) g(\mathbf{r}) d\mathbf{r} \tag{26.2.4}$$

The above inner product exists if the integral converges.

26.2.2 Matrix Eigenvalue Problem (EVP)

It is known in matrix theory that if a matrix is Hermitian, then its eigenvalues are all real. Furthermore, their eigenvectors with distinct eigenvalues are orthogonal to each other [85]. Assume that an eigenvalue and an eigenvector exists for the Hermitian matrix $\overline{\mathbf{A}}$. Then

$$\overline{\mathbf{A}} \cdot \mathbf{v}_i = \lambda_i \mathbf{v}_i \tag{26.2.5}$$

[14]The transpose of a scalar is a scalar, and then we use the rule for the transpose of product of matrix operators [201].

To prove the real value of λ_i, we dot multiply the above from the left by $\mathbf{v}_i^\dagger$ where $\dagger$ indicates conjugate transpose. Then the above becomes

$$\mathbf{v}_i^\dagger \cdot \overline{\mathbf{A}} \cdot \mathbf{v}_i = \lambda_i \mathbf{v}_i^\dagger \cdot \mathbf{v}_i \tag{26.2.6}$$

Since $\overline{\mathbf{A}}$ is Hermitian, or $\overline{\mathbf{A}}^\dagger = \overline{\mathbf{A}}$, the quantity $\mathbf{v}_i^\dagger \cdot \overline{\mathbf{A}} \cdot \mathbf{v}_i$ is purely real. Moreover, the quantity $\mathbf{v}_i^\dagger \cdot \mathbf{v}_i$ is positive real.[15] So in order for the above to be satisfied, λ_i has to be real.

To prove orthogonality of eigenvectors, now assume that $\overline{\mathbf{A}}$ has two eigenvectors with distinct eigenvalues such that

$$\overline{\mathbf{A}} \cdot \mathbf{v}_i = \lambda_i \mathbf{v}_i \tag{26.2.7}$$

$$\overline{\mathbf{A}} \cdot \mathbf{v}_j = \lambda_j \mathbf{v}_j \tag{26.2.8}$$

When one left dot multiplies the first equation with $\mathbf{v}_j^\dagger$ and does the same to the second equation with $\mathbf{v}_i^\dagger$, one gets

$$\mathbf{v}_j^\dagger \cdot \overline{\mathbf{A}} \cdot \mathbf{v}_i = \lambda_i \mathbf{v}_j^\dagger \cdot \mathbf{v}_i \tag{26.2.9}$$

$$\mathbf{v}_i^\dagger \cdot \overline{\mathbf{A}} \cdot \mathbf{v}_j = \lambda_j \mathbf{v}_i^\dagger \cdot \mathbf{v}_j \tag{26.2.10}$$

Taking the conjugate transpose of (26.2.9) in the above, and since $\overline{\mathbf{A}}$ is Hermitian, their left-hand sides (26.2.9) and (26.2.10) are the same. Subtracting the two equations, we arrive at

$$0 = (\lambda_i - \lambda_j) \mathbf{v}_j^\dagger \cdot \mathbf{v}_i \tag{26.2.11}$$

For distinct eigenvalues, $\lambda_i \neq \lambda_j$, the only way for the above to be satisfied is if

$$\mathbf{v}_j^\dagger \cdot \mathbf{v}_i = C_i \delta_{ij} \tag{26.2.12}$$

Hence, eigenvectors of a Hermitian matrix with distinct eigenvalues are orthogonal to each other. The eigenvalues are also real.

26.2.3 Power Orthogonality

The subject of power orthogonality is intimately related to mode orthogonality [104]. Because of mode orthogonality, one can show that each individual mode in a waveguide carries power separately from each other, which is power orthogonality. This power orthogonality can be generalized to plane-wave modes in free space.

Given an incident plane wave impinging on a scatterer, only the plane wave with plane-wave **k** vector that is parallel to the **k** vector of the incident plane wave can carry energy away from the incident plane wave due to power orthogonality. Hence, the total power scattered by a scatterer with plane wave impinging on it is proportional to the power scattered in the forward direction. This is the statement of the optical theorem [50].

[15] Convince yourself that this is the case if you are skeptical.

Exercises for Lecture 26

Problem 26-1: For a rectangular metallic cavity, show that the cavity modes can also be derived using a scalar potential approach.

(i) Show that for:

$$\text{TM modes,}\quad \mathbf{H}(\mathbf{r}) = \nabla \times \hat{z}\Psi_e(\mathbf{r})$$

$$\text{TE modes,}\quad \mathbf{E}(\mathbf{r}) = \nabla \times \hat{z}\Psi_h(\mathbf{r})$$

(ii) Find the scalar wave equations that $\Psi_e(\mathbf{r})$ and $\Psi_h(\mathbf{r})$ satisfy.

(iii) Find the boundary conditions on the six walls of the cavity for $\Psi_e(\mathbf{r})$ and $\Psi_h(\mathbf{r})$.

(iv) Show that the modes of a rectangular cavity are orthogonal to each other (nothing that this is not restricted to rectangular cavities only). Show that the scalar wave modes found in (ii) and (iii) are orthogonal to each other. Namely, that

$$\int_V dV \Psi_{mnl}(\mathbf{r})\Psi_{m'n'l'}(\mathbf{r}) = C\delta_{mm'}\delta_{nn'}\delta_{ll'}$$

where Ψ above can be either Ψ_e or Ψ_h. Show that because of this, the fields are also orthogonal, namely that

$$\int_V dV \mathbf{H}_{mnl}(\mathbf{r}) \cdot \mathbf{H}_{m'n'l'}(\mathbf{r}) = C'\delta_{mm'}\delta_{nn'}\delta_{ll'}$$

The above volume integrals are over the volume of the cavity.

Part III

Radiation, High-Frequency Approximation, Computational Electromagnetics, Quantum Electromagnetics

Lecture 27

Scalar and Vector Potentials

27.1 Scalar and Vector Potentials for Time-Harmonic Fields

Now that we have studied the guidance of waves by waveguides, and the trapping of electromagnetic waves by cavity resonators, it will be interesting to consider how electromagnetic waves radiate from sources. This is best done via the scalar and vector potential formulation.

Previously, we have studied the use of scalar potential Φ for electrostatic problems. Then we learned the use of vector potential $\mathbf{A}$ for magnetostatic problems. Now, we will study the combined use of both scalar and vector potentials concurrently for solving time-harmonic (electrodynamic) problems.

This is important for bridging the gap between the static regime where the frequency is zero or low, and the electrodynamic regime where the frequency is not low. For the electrodynamic regime, it is important to understand the radiation of electromagnetic fields, which has a plethora of advanced applications. Electrodynamic regime is important for studying antennas, communications, sensing, wireless power transfer applications, and much more. High-frequency electromagnetics is important in THz technologies, optics, nanolithography, and quantum technologies. Hence, it is imperative that we understand how time-varying electromagnetic fields radiate from sources.

It is also crucial to understand when static or circuit (quasi-static) regimes are important. The circuit theory has been used to solve many highly complex problems. These solutions have fueled the microchip, the integrated circuit design (ICD) industry, millimeter wave integrated circuits, and the antenna design industry. There is also the emerging area of antenna on a chip (AoC) and antenna in a package (AiP) [202, 203]. Therefore, it is beneficial to understand when electromagnetic problems can be approximated with simple circuit problems and solved using simple laws such as Kirchhoff's current law (KCL), and Kirchhoff's voltage law (KVL).

27.2 Scalar and Vector Potentials for Statics—A Review

Previously, we have studied scalar and vector potentials for electrostatics and magnetostatics where the frequency ω is identically zero. The four Maxwell's equations for a homogeneous medium are then

$$\nabla \times \mathbf{E} = 0 \tag{27.2.1}$$

$$\nabla \times \mathbf{H} = \mathbf{J} \tag{27.2.2}$$

$$\nabla \cdot \varepsilon \mathbf{E} = \varrho \tag{27.2.3}$$

$$\nabla \cdot \mu \mathbf{H} = 0 \tag{27.2.4}$$

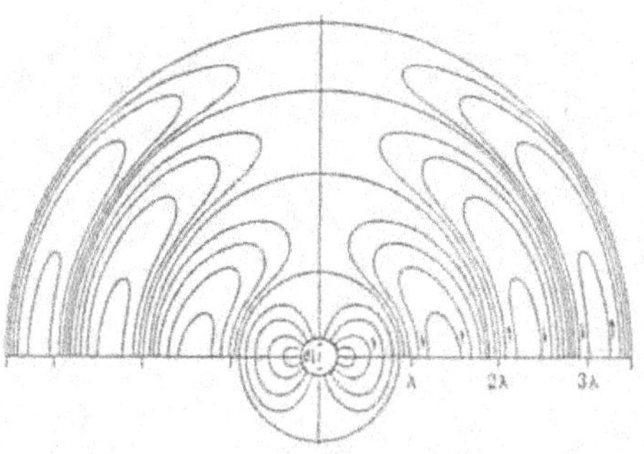

Figure 27.1: Plot of a snapshot of the time-harmonic electric field around a dipole source which is time-varying, where only the field in the upper half-space is shown. Close to the source, the field resembles that of a static electric dipole, but far away from the source, the electromagnetic field is detached from the source: due to the finite velocity of light, the electromagnetic field cannot keep up with the changing dipole field near the source. In other words, the source starts to shed energy to the far region and radiate.

Looking at the first equation above, and using the knowledge that $\nabla \times (\nabla \Phi) = 0$, we can construct a solution to (27.2.1) easily. Thus, in order to satisfy the first of Maxwell's equations or Faraday's law above, we let

$$\mathbf{E} = -\nabla \Phi \tag{27.2.5}$$

The above $\mathbf{E}$ implies that (27.2.1) is satisfied. It also implies that the static field is conservative, or that $\oint \mathbf{E} \cdot d\mathbf{l} = 0$, implying that no net work is done if one moves a charge in a closed loop. Using the (27.2.5) in (27.2.3), we get

$$\nabla \cdot \varepsilon \nabla \Phi = -\varrho \tag{27.2.6}$$

Then for a homogeneous medium where ε is a constant, $\nabla \cdot \varepsilon \nabla \Phi = \varepsilon \nabla \cdot \nabla \Phi = \varepsilon \nabla^2 \Phi$. Thus, we have

$$\nabla^2 \Phi = -\frac{\varrho}{\varepsilon} \tag{27.2.7}$$

which is the Poisson's equation for electrostatics for a homogeneous medium.

Now looking at (27.2.4) where $\nabla \cdot \mu \mathbf{H} = 0$, we let

$$\mu \mathbf{H} = \nabla \times \mathbf{A} \tag{27.2.8}$$

Since $\nabla \cdot (\nabla \times \mathbf{A}) = 0$, the last of Maxwell's equations (27.2.4) is automatically satisfied. Using the above in the second of Maxwell's equations from the previous page, we get

$$\nabla \times \nabla \times \mathbf{A} = \mu \mathbf{J} \tag{27.2.9}$$

Now, using the fact that[1] $\nabla \times \nabla \times \mathbf{A} = \nabla(\nabla \cdot \mathbf{A}) - \nabla^2 \mathbf{A}$ and Coulomb gauge indicates that $\nabla \cdot \mathbf{A} = 0$, we arrive at

$$\nabla^2 \mathbf{A} = -\mu \mathbf{J} \tag{27.2.10}$$

which is the vector Poisson's equation for a homogeneous medium. We will repeat the above derivation when $\omega \neq 0$.

27.2.1 Scalar and Vector Potentials for Electrodynamics

Since dynamic or time-varying problems are of the utmost importance in electromagnetics, we will study them next. To this end, assuming linearity, we will start with frequency domain Maxwell's equations with sources $\mathbf{J}$ and ϱ included, and later, see how these sources $\mathbf{J}$ and ϱ can radiate electromagnetic fields. Maxwell's equations in the frequency domain, or in time-harmonic electromagnetics, are[2]

$$\nabla \times \mathbf{E} = -j\omega\mu \mathbf{H} \tag{27.2.11}$$

$$\nabla \times \mathbf{H} = j\omega\varepsilon \mathbf{E} + \mathbf{J} \tag{27.2.12}$$

$$\nabla \cdot \varepsilon \mathbf{E} = \varrho \tag{27.2.13}$$

$$\nabla \cdot \mu \mathbf{H} = 0 \tag{27.2.14}$$

We can view the above as a linear system of equations where $\mathbf{J}$ and ϱ are inputs and $\mathbf{E}$ and $\mathbf{H}$ are outputs.

In order to satisfy the last Maxwell's equation, as before, we let

$$\mu \mathbf{H} = \nabla \times \mathbf{A} \tag{27.2.15}$$

Now, using (27.2.15) in (27.2.11), we have

$$\nabla \times (\mathbf{E} + j\omega \mathbf{A}) = 0 \tag{27.2.16}$$

Since $\nabla \times (\nabla \Phi) = 0$, the above implies that $\mathbf{E} + j\omega \mathbf{A} = -\nabla \Phi$, or that

$$\mathbf{E} = -j\omega \mathbf{A} - \nabla \Phi \tag{27.2.17}$$

[1] The following simplification can also be derived from the BAC-CAB formula.

[2] The time-domain field is a linear superposition of many frequency domain fields. We can get the frequency domain of Maxwell's equations by using either phasor technique or Fourier transform technique.

The above indicates that the electrostatic theory of letting $\mathbf{E} = -\nabla\Phi$, which we learned previously in section 3.3.1, is not exactly correct when $\omega \neq 0$. The $-j\omega\mathbf{A}$ term above, in accordance to Faraday's law, is the contribution to the electric field from the time-varying magnetic field, and hence, is termed the induction term.[3]

Furthermore, the above shows that once $\mathbf{A}$ and Φ are known, one can determine the fields $\mathbf{H}$ and $\mathbf{E}$ assuming that $\mathbf{J}$ and ϱ are given. To this end, we will derive equations for $\mathbf{A}$ and Φ in terms of the sources $\mathbf{J}$ and ϱ. Substituting (27.2.15) and (27.2.17) into (27.2.12) gives

$$\nabla \times (\nabla \times \mathbf{A}) = j\omega\mu\varepsilon(-j\omega\mathbf{A} - \nabla\Phi) + \mu\mathbf{J} \tag{27.2.18}$$

after using that $\nabla \times (\nabla \times \mathbf{A}) = \nabla\nabla \cdot \mathbf{A} - \nabla \cdot \nabla\mathbf{A}$, and upon rearrangement of terms, we have

$$\nabla^2\mathbf{A} + \omega^2\mu\varepsilon\mathbf{A} = -\mu\mathbf{J} + j\omega\mu\varepsilon\nabla\Phi + \nabla\nabla \cdot \mathbf{A} \tag{27.2.19}$$

Moreover, using (27.2.17) in (27.2.13), we have

$$\nabla \cdot (j\omega\mathbf{A} + \nabla\Phi) = -\frac{\varrho}{\varepsilon} \tag{27.2.20}$$

In the above, (27.2.19) and (27.2.20) represent two equations for the two unknowns $\mathbf{A}$ and Φ, expressed in terms of the known quantities, the sources $\mathbf{J}$ and ϱ. But these equations are coupled to each other. They look complicated and are rather unwieldy to solve at this point.

Fortunately, the above can be simplified! As in the magnetostatic case, the vector potential $\mathbf{A}$ in (27.2.15) is not unique. To show this, one can always construct a new $\mathbf{A}' = \mathbf{A} + \nabla\Psi$ that produces the same magnetic field $\mu\mathbf{H}$ via (27.2.15), since $\nabla \times (\nabla\Psi) = 0$. It is quite clear that $\mu\mathbf{H} = \nabla \times \mathbf{A} = \nabla \times \mathbf{A}'$. Moreover, one can further show that Φ is also nonunique [50]. Namely, with

$$\mathbf{A}' = \mathbf{A} + \nabla\Psi \tag{27.2.21}$$

$$\Phi' = \Phi - j\omega\Psi \tag{27.2.22}$$

it can be shown that the new $\mathbf{A}'$ and Φ' produce the same $\mathbf{E}$ and $\mathbf{H}$ field. The above is known as gauge transformation [50], clearly showing the nonuniqueness of $\mathbf{A}$ and Φ.

To make them unique, in addition to specifying what $\nabla \times \mathbf{A}$ should be in (27.2.15), we need to specify its divergence or $\nabla \cdot \mathbf{A}$ as in the electrostatic case.[4] A clever way to specify the divergence of $\mathbf{A}$ is to choose it to simplify the complicated equations above in (27.2.19). We choose a gauge so that the last two terms on the right-hand side in equation (27.2.19) cancel each other. In other words, we let

$$\nabla \cdot \mathbf{A} = -j\omega\mu\varepsilon\Phi \tag{27.2.23}$$

[3] Notice that in electrical engineering, most concepts related to magnetic fields are inductive!

[4] When given a vector $\mathbf{A}$ and an arbitrary vector $\mathbf{k}$, in addition to specifying what $\mathbf{k} \times \mathbf{A}$ is, it is also necessary to specify what $\mathbf{k} \cdot \mathbf{A}$ is to uniquely specify $\mathbf{A}$.

The above is judiciously chosen so that the pertinent equations (27.2.19) and (27.2.20) will be simplified and decoupled. With the use of (27.2.23) in (27.2.19) and (27.2.20), they now become

$$\nabla^2 \mathbf{A} + \omega^2 \mu\varepsilon \mathbf{A} = -\mu \mathbf{J} \tag{27.2.24}$$

$$\nabla^2 \Phi + \omega^2 \mu\varepsilon \Phi = -\frac{\varrho}{\varepsilon} \tag{27.2.25}$$

Equation (27.2.23) is known as the Lorenz gauge[5] and the above equations are Helmholtz equations with source terms. These are the equations that we can solve for the scalar potential Φ and the vector potential $\mathbf{A}$ given the sources $\mathbf{J}$ and ϱ. Not only are these equations simplified, but they also can be solved independently of each other since they are decoupled from each other.

Equations (27.2.24) and (27.2.25) can be solved using the Green's function method we learned previously. Equation (27.2.24) in Cartesian coordinates (only in Cartesian coordinates) actually constitutes three scalar equations for the three x, y, and z components, namely that

$$\nabla^2 A_i + \omega^2 \mu\varepsilon A_i = -\mu J_i \tag{27.2.26}$$

where i above can be x, y, or z. Therefore, (27.2.24) and (27.2.25) together constitute four scalar equations similar to each other. Hence, we need only to solve their point-source response, or the Green's function of these equations, by solving

$$\nabla^2 g(\mathbf{r}, \mathbf{r}') + \beta^2 g(\mathbf{r}, \mathbf{r}') = -\delta(\mathbf{r} - \mathbf{r}') \tag{27.2.27}$$

where $\beta^2 = \omega^2 \mu\varepsilon$. (This Green's function is sometimes referred to as the fundamental solution or the canonical solution of the problem.)

Previously, we have shown that when $\beta = 0$,

$$g(\mathbf{r}, \mathbf{r}') = g(|\mathbf{r} - \mathbf{r}'|) = \frac{1}{4\pi|\mathbf{r} - \mathbf{r}'|}$$

When $\beta \neq 0$, the correct solution is

$$g(\mathbf{r}, \mathbf{r}') = g(|\mathbf{r} - \mathbf{r}'|) = \frac{e^{-j\beta|\mathbf{r}-\mathbf{r}'|}}{4\pi|\mathbf{r} - \mathbf{r}'|} \tag{27.2.28}$$

which can be verified by back substitution or derived [35] [1, p. 26].

By using the principle of linear superposition, or convolution, the solutions to (27.2.24) and (27.2.25) are then

$$\mathbf{A}(\mathbf{r}) = \mu \iiint_V d\mathbf{r}' \mathbf{J}(\mathbf{r}') g(|\mathbf{r} - \mathbf{r}'|) = \mu \iiint_V d\mathbf{r}' \mathbf{J}(\mathbf{r}') \frac{e^{-j\beta|\mathbf{r}-\mathbf{r}'|}}{4\pi|\mathbf{r} - \mathbf{r}'|} \tag{27.2.29}$$

$$\Phi(\mathbf{r}) = \frac{1}{\varepsilon} \iiint_V d\mathbf{r}' \varrho(\mathbf{r}') g(|\mathbf{r} - \mathbf{r}'|) = \frac{1}{\varepsilon} \iiint_V d\mathbf{r}' \varrho(\mathbf{r}') \frac{e^{-j\beta|\mathbf{r}-\mathbf{r}'|}}{4\pi|\mathbf{r} - \mathbf{r}'|} \tag{27.2.30}$$

In the above, $d\mathbf{r}'$ is the shorthand notation for $dx'\,dy'\,dz'$: they are volume integrals. The above are three-dimensional convolutional integrals in space. Since $\mathbf{A}(\mathbf{r})$ is linearly proportional to $\mathbf{J}(\mathbf{r})$, and that $\Phi(\mathbf{r})$ is linearly proportional to $\varrho(\mathbf{r})$, it is a hallmark of a linear system.

[5] Please note this "Lorenz" is not the same as "Lorentz".

27.2.2 Degree of Freedom in Maxwell's Equations

From (27.2.24) and (27.2.25), if $\mathbf{J}$ and ϱ are independent, these two equations are independent. Howevever, $\mathbf{J}$ and ϱ are related by the current continuity equation, $\nabla \cdot \mathbf{J} = -j\omega\varrho$. Therefore, using the Lorenz gauge (27.2.23), (27.2.25) can be derived from (27.2.24). Hence, solving (27.2.24) suffices to solve Maxwell's equations. Therefore, only three degrees of freedom are needed to solve Maxwell's equations fully for a homogeneous linear medium. Once $\mathbf{A}$ is found, Φ can be found via the Lorenz gauge. But at statics with $\omega = 0$, Φ cannot be derived from $\mathbf{A}$ via the Lorenz gauge. In this case, it appears that four degrees of freedom are needed to fully solve Maxwell's equations. But the following argument will show that three degrees of freedom are still needed.

For the static case, since $\nabla \cdot \mathbf{J} = 0$ via (27.2.2), there are only two degrees of freedom in (27.2.10). This together with (27.2.7), only three degrees of freedom are needed to solve Maxwell's equations as in the dynamic case. Note that when the scalar and vector potential approach is used to solve Maxwell's equations, all four of Maxwell's equations, (27.2.11) to (27.2.14), are involved. When a numerical scheme is developed to solve Maxwell's equations using this approach, the solutions are stable down to very low frequencies [204].

27.2.3 More on Scalar and Vector Potentials

Note that Maxwell's equations are mathematical symmetrical if we add a magnetic current $\mathbf{M}$ to Maxwell's equations and magnetic charge ϱ_m to Gauss's law.[6] Thus, the equations become

$$\nabla \times \mathbf{E} = -j\omega\mu\mathbf{H} - \mathbf{M} \tag{27.2.31}$$

$$\nabla \times \mathbf{H} = j\omega\varepsilon\mathbf{E} + \mathbf{J} \tag{27.2.32}$$

$$\nabla \cdot \mu\mathbf{H} = \varrho_m \tag{27.2.33}$$

$$\nabla \cdot \varepsilon\mathbf{E} = \varrho \tag{27.2.34}$$

The above can be solved in two stages, using the principle of linear superposition because the above is a linear time invariant system. Thus, the sources of the system can be turned on and off consecutively to obtain different solutions to the system. First, we can set $\mathbf{M} = 0$, $\varrho_m = 0$, and $\mathbf{J} \neq 0$, $\varrho \neq 0$, and solve for the fields as we have done before. Second, we can set $\mathbf{J} = 0$, $\varrho = 0$ but let $\mathbf{M} \neq 0$, $\varrho_m \neq 0$ and solve for the fields next. Then the total general solution, by linearity, is just the linear superposition of these two solutions.

For the second case, we set $\mathbf{J} = 0$, $\varrho = 0$ and $\mathbf{M} \neq 0$, $\varrho_m \neq 0$. Then, we can define an electric vector potential $\mathbf{F}$ such that [42, 52]

$$\mathbf{D} = -\nabla \times \mathbf{F} \tag{27.2.35}$$

and a magnetic scalar potential Φ_m such that

$$\mathbf{H} = -\nabla\Phi_m - j\omega\mathbf{F} \tag{27.2.36}$$

[6]In fact, Maxwell himself exploited this symmetry [42].

By invoking duality principle (see section 17.2), one gathers that [52]

$$\mathbf{F}(\mathbf{r}) = \varepsilon \iiint d\mathbf{r}'\mathbf{M}(\mathbf{r}')g(|\mathbf{r}-\mathbf{r}'|) = \varepsilon \iiint d\mathbf{r}'\mathbf{M}(\mathbf{r}')\frac{e^{-j\beta|\mathbf{r}-\mathbf{r}'|}}{4\pi|\mathbf{r}-\mathbf{r}'|} \tag{27.2.37}$$

$$\Phi_m(\mathbf{r}) = \frac{1}{\mu} \iiint d\mathbf{r}'\varrho_m(\mathbf{r}')g(|\mathbf{r}-\mathbf{r}'|) = \frac{1}{\mu} \iiint d\mathbf{r}'\varrho_m(\mathbf{r}')\frac{e^{-j\beta|\mathbf{r}-\mathbf{r}'|}}{4\pi|\mathbf{r}-\mathbf{r}'|} \tag{27.2.38}$$

Again, the above is the hallmark of a linear system, since the fields $\mathbf{F}(\mathbf{r})$ and $\Phi_m(\mathbf{r})$ are linearly proportional to the sources $\mathbf{M}(\mathbf{r})$ and $\varrho_m(\mathbf{r})$.

As mentioned before, even though magnetic sources do not exist, they can be engineered. For instance, an electric current loop antenna resembles a magnetic dipole. Another example is the toroidal antenna shown in Figure 17.12, which resembles a magnetic current loop antenna. In many engineering designs, one can use fictitious magnetic sources to enrich the diversity of electromagnetic technologies.

27.3 When Is Static Electromagnetic Theory Valid?

Static electromagnetic theory is often simpler than dynamic electromagnetic theory. It will be prudent to know when we can apply static theory instead of the more complicated electrodynamic theory.

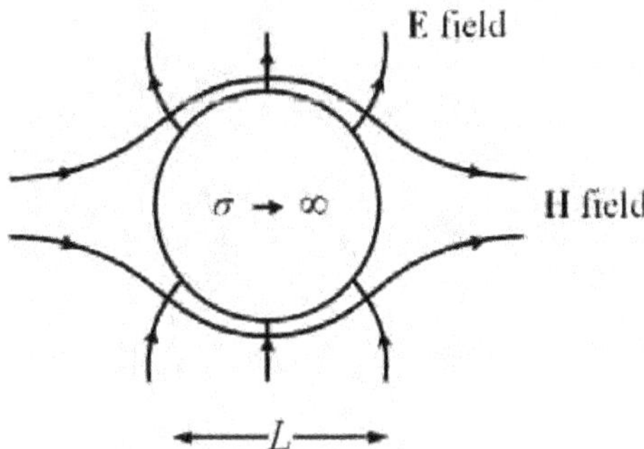

Figure 27.2: The electric and magnetic fields are great contortionists around a perfectly conducting particle. They deform themselves to satisfy the boundary conditions, $\hat{n}\times\mathbf{E} = 0$, and $\hat{n}\cdot\mathbf{H} = 0$ on the PEC surface, even when the particle is very small. In other words, the fields vary on the length-scale of L. Approximately, $\nabla \sim 1/L$, which is large when L is small.

27.3.1 Cutting through The Chaste

To see when static electromagnetics can be used to approximate electrodynamics, we consider the Helmholtz equations previously derived and ask when they can be replaced by Poisson's/Laplace's equations. They are reproduced here as

$$\nabla^2 \mathbf{A} + \omega^2 \mu\varepsilon \mathbf{A} = -\mu \mathbf{J} \tag{27.3.1}$$

$$\nabla^2 \Phi + \omega^2 \mu\varepsilon \Phi = -\frac{\varrho}{\varepsilon} \tag{27.3.2}$$

By looking at Figure 27.2, in order to satisfy the requisite boundary conditions on the wall of an object, the electromagnetic fields have to contort themselves around the object (as contortionists do in a circus). In order for this to happen, by dimensional analysis,[7] $\partial/\partial x$, $\partial/\partial y$, and $\partial/\partial z$ are of the order of $1/L$, or ∇^2 is of the order of $1/L^2$. Now we compare the Laplacian operator term, which is of order $1/L^2$, and the $\beta^2 = \omega^2\mu\varepsilon$ term. If $\beta^2 L^2 \ll 1$, then the Helmholtz equation is dominated by the Laplacian operator, and it can be replaced by Laplace's equations. In this case, static electromagnetics applies.

In the above, $\beta = 2\pi/\lambda$ where λ is the wavelength. Thus, if $L/\lambda \ll 1$, static theory applies. Therefore, in electromagnetics, the yardstick is the wavelength. When the object size is much smaller than the wavelength, we are in the static or quasi-static regime, whereas if the object size is about the wavelength or larger, we are in the electrodynamic regime or the wave-physics regime [205].

27.3.2 An Example

In a word, one can solve a problem using static analysis even in optics where the frequency is high, or the wavelength is short. This is possible when the size of the object L is much smaller than the optical wavelength, which is about 400 nm for blue light. This is true even in optics, where ω is humongous or the wavelength is very short. Nowadays, plasmonic nanoparticles of about 10 nm can be fabricated. If the particle is small enough compared to the wavelength of the light, electrostatic analysis can be used to study their interaction with light. Hence, static electromagnetic theory can be used to analyze the wave-particle interaction. This was done in one of the homework assignments. This kind of scattering is also referred to as Rayleigh scattering [206].

Figure 27.3 shows an incident light whose wavelength is much longer than the size of the particle. The incident field induces an electric dipole moment on the particle, whose scattered field can be written as

$$\mathbf{E}_s = (\hat{r} 2\cos\theta + \hat{\theta}\sin\theta)\left(\frac{a}{r}\right)^3 E_s \tag{27.3.3}$$

while the incident field $\mathbf{E}_0$ and the interior field $\mathbf{E}_i$ to the particle can be expressed as

$$\mathbf{E}_0 = \hat{z}E_0 = (\hat{r}\cos\theta - \hat{\theta}\sin\theta)E_0 \tag{27.3.4}$$

$$\mathbf{E}_i = \hat{z}E_i = (\hat{r}\cos\theta - \hat{\theta}\sin\theta)E_i \tag{27.3.5}$$

[7] Which is often used in fluid analysis [39].

By matching boundary conditions, as was done in the homework, it can be shown that

$$E_s = \frac{\varepsilon_s - \varepsilon}{\varepsilon_s + 2\varepsilon} E_0 \tag{27.3.6}$$

$$E_i = \frac{3\varepsilon}{\varepsilon_s + 2\varepsilon} E_0 \tag{27.3.7}$$

For a plasmonic nanoparticle, for some materials, the particle medium behaves like a plasma (see lecture 8), and ε_s in the above can be approximately negative, making the denominators of the above expression very close to zero. This is the hallmark of a resonance phenomenon as we have seen in the surface plasmonic polariton, the transverse resonance condition, LC tank circuit, and, now, the plasmon nanoparticle. Therefore, the amplitude of the internal and scattered fields can be very large when this happens, and the nanoparticles will glitter in the presence of light. Even the ancient Romans realized this! (See section 8.3.7.)

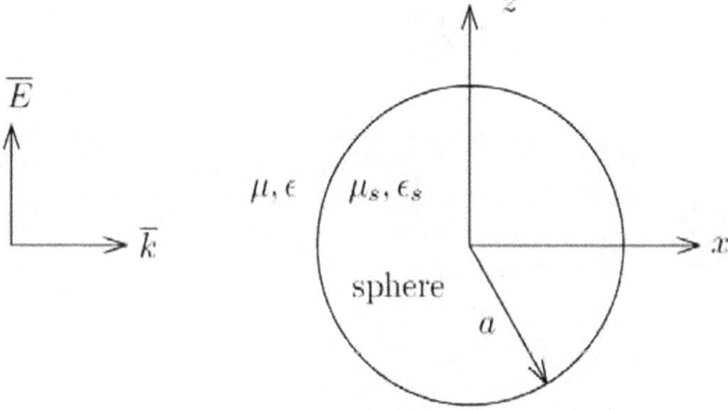

Figure 27.3: A plane electromagnetic wave incident on a particle. When the particle size is small compared to the wavelength, electrostatic analysis can be used to solve this problem (Figure by J. A. Kong [35]).

Figure 27.4 shows a nanoparticle induced in plasmonic oscillation by a light wave. Figure 27.5 shows that different color fluids can be obtained by immersing nanoparticles in fluids with different background permittivity (ε in (27.3.6) and (27.3.7)) causing the plasmonic particles to resonate at different frequencies. This is because the resonance frequency of the plasmonic nanoparticle is obtained by solving $\varepsilon_s(\omega) + 2\varepsilon = 0$, which depends on the background medium, ε.

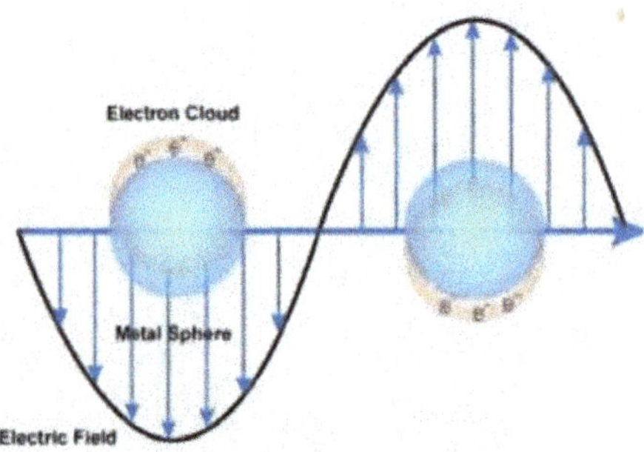

Figure 27.4: A nanoparticle undergoes electromagnetic oscillation when an electromagnetic wave impinges on it. The oscillation is inordinately large when the incident wave's frequency coincides with the resonance frequency of the plasmonic particle (Figure by sigmaaldric.com).

Figure 27.5: Different color fluids containing nanoparticles can be obtained by changing the permittivity ε of the background fluids in (27.3.6) and (27.3.7) (Figure by nanocomposix.com).

Exercises for Lecture 27

Problem 27-1: The equations for vector and scalar potentials have been found using the Lorenz gauge in the lecture notes.

(i) Now, find these equations using Coulomb's gauge. **Hint: This is described in J.D. Jackson's book and many other physics texts.**

(ii) Verify by back substitution that (27.2.28) is in fact a solution of (27.2.27).

(iii) Describe how you would use the electric vector potential and the magnetic scalar potential to find the electromagnetic field if you only have magnetic current and magnetic charge as sources.

(iv) Explain when the static electromagnetic theory can be used to approximate a time-varying electromagnetic field.

Lecture 28

Radiation by a Hertzian Dipole

Radiation of an electromagnetic field is of ultimate importance for wireless communication systems. The first demonstration of the wave like nature of an electromagnetic field was by Heinrich Hertz in 1888 [20], some 23 years after Maxwell's equations were fully established. Guglielmo Marconi, after much dogged perserverence with a series of experiments, successfully transmitted a wireless radio signal from Cornwall, England, to Newfoundland, Canada, in 1901 [207]. The experiment was serendipitous since he did not know that the ionosphere was on his side: The ionosphere helped to bounce the radio wave back to earth from outer space. Marconi's success ushered in the age of wireless communication, which is omnipresent in our present daily lives. Hence, radiation by arbitrary sources is an important topic for antennas and wireless communications. We will start with studying the Hertzian dipole, which is the simplest of radiation sources we can think of.

28.1 History

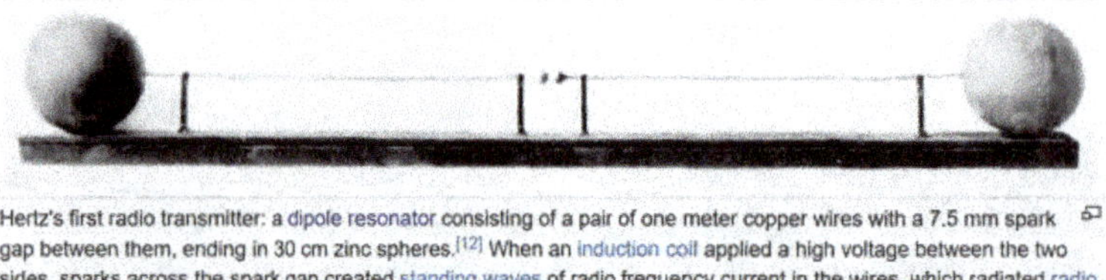

Figure 28.1: Hertz's original experiment on a small dipole (Figure by: H. Hertz, "On the finite velocity of propagation of electromagnetic actions," Electric Waves, vol. 110, 1888).

The original historic Hertzian dipole experiment is shown in Figure 28.1. It was done in 1887 by Heinrich Hertz [20]. The schematic for the original experiment is also shown in Figure 28.2.

A metallic sphere has a capacitance which can be found in closed form with respect to infinity or a ground plane. He could also estimate the inductance of the leads that were attached to the dipole. Therefore, the resonance frequency of his antenna can be calculated. As we shall see, the radiation strength of the dipole is proportional to $p = ql$, the dipole moment. Here, q is the charge at the ends of the dipole and l is its length.

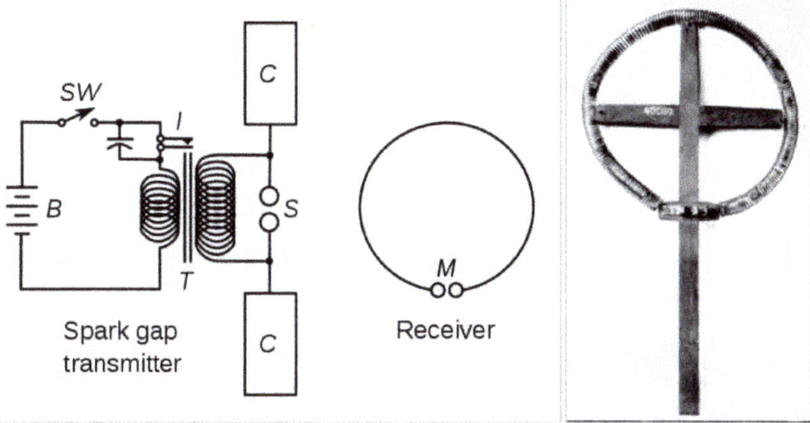

Hertz's 1887 apparatus for generating and detecting radio waves: a spark transmitter *(left)* consisting of a dipole antenna with a spark gap *(S)* powered by high voltage pulses from a Ruhmkorff coil *(T)*, and a receiver *(right)* consisting of a loop antenna and spark gap.

One of Hertz's radio wave receivers: a loop antenna with an adjustable micrometer spark gap *(bottom)*.[12]

Figure 28.2: More on Hertz's original experiment on a small dipole. The antenna was powered by a transformer. The radiated electromagnetic field was detected by a loop receiver antenna that generated a spark at its gap M (Figure by H. Hertz, "On the finite velocity of propagation of electromagnetic actions," Electric Waves, vol. 110, 1888).

28.2 Approximation by a Point Source

Figure 28.3 is the schematic of a small Hertzian dipole resembling the original dipole that Hertz made. Assuming that the spheres at the ends store charges of value q, and l is the effective length of the dipole, then the dipole moment $p = ql$. The charge q is varying time-harmonically because it is driven by the generator. Since

$$\frac{dq}{dt} = I,$$

we have the current moment

$$Il = \frac{dq}{dt}l = j\omega ql = j\omega p \tag{28.2.1}$$

for this Hertzian dipole.

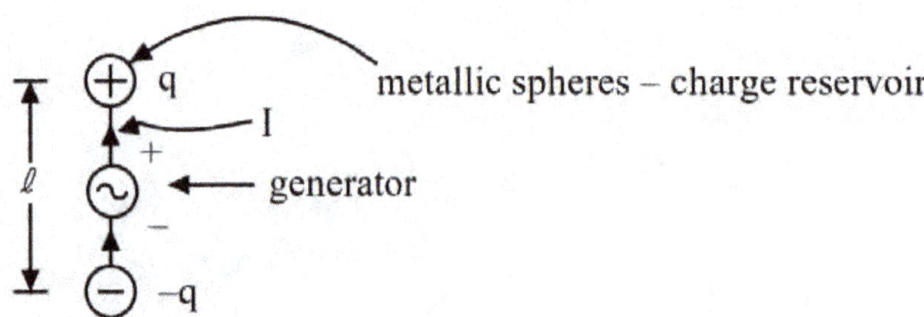

Figure 28.3: Schematic of a small Hertzian dipole, which is a close approximation of that first proposed by Hertz.

A Hertzian dipole is a dipole which is much smaller than the wavelength under consideration so that we can approximate it by a point current distribution, or a current density. Mathematically, it is given by [35, 48]

$$\mathbf{J}(\mathbf{r}) = \hat{z}Il\delta(x)\delta(y)\delta(z) = \hat{z}Il\delta(\mathbf{r}) \tag{28.2.2}$$

The dipole is as shown in Figure 28.3 schematically. As long as we are not too close to the dipole so that it does not look like a point source anymore, the above is a good mathematical model and approximation for describing a Hertzian dipole.

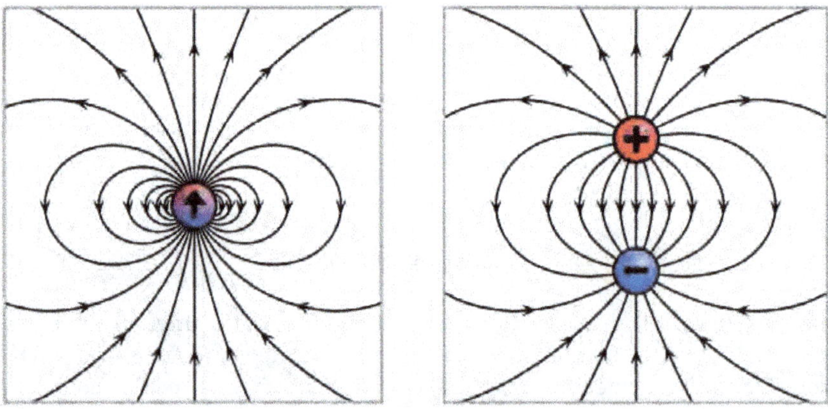

Figure 28.4: The field of a point dipole field where the separation of the two charges is infinitesimally small versus that of a dipole field with a finite separation between the charges. When one is far away from the dipole sources, their fields are similar to each other (Figure By https://en.m.wikipedia.org).

We learned previously that the vector potential is related to the current as follows:

$$\mathbf{A}(\mathbf{r}) = \mu \iiint_V d\mathbf{r}'\mathbf{J}(\mathbf{r}')\frac{e^{-j\beta|\mathbf{r}-\mathbf{r}'|}}{4\pi|\mathbf{r}-\mathbf{r}'|} \tag{28.2.3}$$

Since the current is a 3D delta function in space, using the sifting property of a delta function, and the equation $\delta(\mathbf{r}) = \delta(x)\delta(y)\delta(y)$, the above integral can be evaluated in closed form. The corresponding vector potential is then given by

$$\mathbf{A}(\mathbf{r}) = \hat{z}A_z = \hat{z}\frac{\mu Il}{4\pi r}e^{-j\beta r} \tag{28.2.4}$$

Since the vector potential $\mathbf{A}(\mathbf{r})$ is cylindrically symmetric (also called axisymmetric or axially symmetric), the corresponding magnetic field is obtained by using that $\mathbf{B} = \nabla \times \mathbf{A}$. Thus, using the curl operator in cylindrical coordinates ,

$$\mathbf{H} = \frac{1}{\mu}\nabla \times \mathbf{A} = \frac{1}{\mu}\left(\hat{\rho}\frac{1}{\rho}\frac{\partial}{\partial\phi}A_z - \hat{\phi}\frac{\partial}{\partial\rho}A_z\right) \tag{28.2.5}$$

Due to axisymmetry, $\frac{\partial}{\partial\phi} = 0$. Here, $r = \sqrt{\rho^2 + z^2}$. In the above, we can use the chain rule that

$$\frac{\partial}{\partial\rho} = \frac{\partial r}{\partial\rho}\frac{\partial}{\partial r} = \frac{\rho}{\sqrt{\rho^2 + z^2}}\frac{\partial}{\partial r} = \frac{\rho}{r}\frac{\partial}{\partial r}$$

to find the $\hat{\phi}$ component of $\mathbf{H}$, since A_z is a function of r only. As a result,

$$\mathbf{H} = -\hat{\phi}\frac{\rho}{r}\frac{Il}{4\pi}\left(-\frac{1}{r^2} - j\beta\frac{1}{r}\right)e^{-j\beta r} \tag{28.2.6}$$

In spherical coordinates, $\frac{\rho}{r} = \sin\theta$ and (28.2.6) becomes [35]

$$\mathbf{H} = \hat{\phi}H_\phi = \hat{\phi}\frac{Il}{4\pi r^2}(1 + j\beta r)e^{-j\beta r}\sin\theta \tag{28.2.7}$$

Notice that H_ϕ is a function of r and θ only in spherical coordinates. The electric field can be derived using Maxwell's equations in spherical coordinates, namely that

$$\begin{aligned}\mathbf{E} &= \frac{1}{j\omega\epsilon}\nabla \times \mathbf{H} = \frac{1}{j\omega\epsilon}\left(\hat{r}\frac{1}{r\sin\theta}\frac{\partial}{\partial\theta}\sin\theta H_\phi - \hat{\theta}\frac{1}{r}\frac{\partial}{\partial r}rH_\phi\right)\\ &= \frac{Ile^{-j\beta r}}{j\omega\epsilon 4\pi r^3}\left[\hat{r}2\cos\theta(1 + j\beta r) + \hat{\theta}\sin\theta(1 + j\beta r - \beta^2 r^2)\right]\end{aligned} \tag{28.2.8}$$

The above expression is rather complicated and it is hard to elucidate the physics from the math (in other words, we cannot see forest for the trees). However, we can see that as $r \to \infty$, there are terms that decay as $1/r^3$, $1/r^2$, and $1/r$. The complex Poynting's vector is proportional to $\mathbf{E}\times\mathbf{H}^*$, and the term that will convect power to infinity has to decay as $1/r^2$ for energy conservation, since the energy is radiating into a spherical surface whose area grows as r^2.

Terms that decay faster than $1/r^2$ cannot convect (carry) energy to infinity, but store reactive power that can exchange with the power from the source. As in the case of a capacitor or an inductor, this reactive power takes energy from the source and then returns it to the source again.

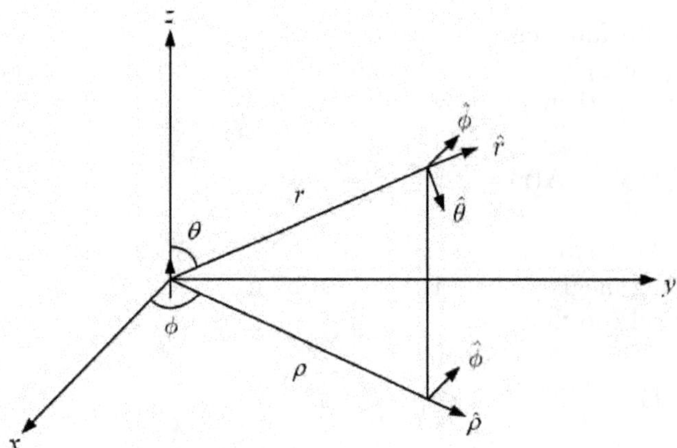

Figure 28.5: Spherical coordinates are used to calculate the fields of a Hertzian dipole.

28.2.1 Case I. Near Field, $\beta r \ll 1$

As we have seen, the yardstick in electromagnetics is the wavelength. Since $\beta = \frac{\omega}{c} = \frac{2\pi}{\lambda}$, $\beta r \ll 1$ implies that one is close to the dipole source in terms of wavelength. We are in the near field of the antenna. In the near field of an antenna, we have mainly reactive power: it is characterized by that the energy flows from the source to the medium in one cycle, and the converse in another cycle. Since $\beta r \ll 1$,[1] retardation effect within this short distance from the point dipole can be ignored. Also, we let $\beta r \to 0$, and keeping the largest terms (or leading order terms in math parlance), then from (28.2.8), with $Il = j\omega p$, we have

$$\mathbf{E} \cong \frac{p}{4\pi\epsilon r^3}(\hat{r}2\cos\theta + \hat{\theta}\sin\theta), \qquad \beta r \ll 1 \tag{28.2.9}$$

For the **H** field, from (28.2.7), with $\beta r \ll 1$, then

$$\mathbf{H} \cong \hat{\phi}\frac{j\omega p}{4\pi r^2}\sin\theta \tag{28.2.10}$$

or

$$\eta_0\mathbf{H} \cong \hat{\phi}\frac{j\beta r p}{4\pi\varepsilon r^3}\sin\theta \tag{28.2.11}$$

[1]Please be reminded that β and k are synonymous in this course: β for microwave engineers and k for optical engineers.

Now comparing (28.2.11) and (28.2.9), we see that (28.2.11) is $O(\beta r)$ smaller than (28.2.9). Since $\mathbf{E}$ and $\eta_0\mathbf{H}$ have the same unit, we can now compare them. Thus, it is seen that

$$\eta_0\mathbf{H} \ll \mathbf{E}, \qquad \text{when } \beta r \ll 1 \tag{28.2.12}$$

where $p = ql$ is the dipole moment.[2] The above implies that in the near field, the electric field $\mathbf{E}$ dominates over the magnetic field $\mathbf{H}$. Moreover, the $\mathbf{E}$ and the $\mathbf{H}$ fields are out of phase, meaning that the power that corresponds to a complex Poynting's vector $\mathbf{E}\times\mathbf{H}^*$ is almost purely imaginary or reactive (see section 10.3.1). This reactive power pumps stored energy into the near field, and siphons energy from the near field as well (similar to a capacitor).

In the above, βr could be made very small by making $\frac{r}{\lambda}$ small or by making $\omega \to 0$. The above becomes like the static field of a dipole. Another viewpoint is that in the near field, the field varies rapidly, and space derivatives are much larger than the time derivative.[3]

For instance,

$$\frac{\partial}{\partial x} \gg \frac{\partial}{c\partial t}$$

Alternatively, we can say that the above is equivalent to

$$\frac{\partial}{\partial x} \gg \frac{\omega}{c}, \qquad \frac{\partial^2}{\partial x^2} \gg \frac{\omega^2}{c^2}, \qquad \frac{\partial^2}{\partial x^2} \gg \frac{1}{c^2}\frac{\partial^2}{\partial t^2}$$

In other words,

$$\nabla^2 - \frac{1}{c^2}\frac{\partial^2}{\partial t^2} \approx \nabla^2$$

which is just the Laplacian operator. In other words, static theory prevails over dynamic theory when $\beta r \ll 1$. The above approximations are consistent with that the retardation effect is negligible over this length scale.

28.2.2 Case II. Far Field (Radiation Field), $\beta r \gg 1$

This is an interesting zone where we can see electromagnetics power being convected (carried) to infinity from the source. This is also known as the far zone.[4] In this case, electromagnetic retardation effect is important. In other words, phase delay cannot be ignored. Thus, keeping the leading order terms in (28.2.8), we then have

$$\mathbf{E} \cong \hat{\theta} j\omega\mu \frac{Il}{4\pi r} e^{-j\beta r} \sin\theta \tag{28.2.13}$$

[2]Here, $\eta_0 = \sqrt{\mu/\varepsilon}$. We multiply $\mathbf{H}$ by η_0 so that the quantities we are comparing have the same unit.

[3]This is in agreement with our observation that electromagnetic fields are great contortionists: they will deform themselves to match the boundary first before satisfying Maxwell's equations. Since the source point is very small, the fields will deform themselves so as to satisfy the boundary conditions near the source region. If this region is small compared to wavelength, the fields will vary rapidly and spatially over a small length scale compared to the wavelength.

[4]This is also called the Fraunhofer zone in German .

and similarly from (28.2.7)

$$\mathbf{H} \cong \hat{\phi} j\beta \frac{Il}{4\pi r} e^{-j\beta r} \sin\theta \tag{28.2.14}$$

Note that $\frac{E_\theta}{H_\phi} = \frac{\omega\mu}{\beta} = \sqrt{\frac{\mu}{\epsilon}} = \eta_0$ which is similar to the intrinsic impedance relationship of a plane wave. Here, $\mathbf{E}$ and $\mathbf{H}$ are orthogonal to each other and they are both orthogonal to the direction of propagation, as in the case of a plane wave. Or in a word, the $\mathbf{E}$ and the $\mathbf{H}$ fields of a spherical wave resembles a plane wave in the far-field approximation.

The *radiation field pattern* of a Hertzian dipole is the plot of $|\mathbf{E}|$ as a function of θ at a constant $\mathbf{r}$ when $\beta r \gg 1$ or in the far field of the antenna. Hence, it is proportional to $\sin\theta$, and it can be proved that it is a circle as shown in Figure 28.6. Also, notice that in the above, the $\mathbf{E}$ field and the $\mathbf{H}$ field are in phase in the far field, meaning that the complex Poynting's vector $\mathbf{S} = \mathbf{E} \times \mathbf{H}^*$ is purely real and that the $\mathbf{S}$ decays as $1/r^2$. This implies energy conservation as the wave propagates, and the power can travel to infinity.

28.3 Radiation Power of a Hertzian Dipole

The time-averaged power flow in the far field, after using (28.2.13) and (28.2.14), is given by

$$\langle \mathbf{S} \rangle = \frac{1}{2}\Re e[\mathbf{E} \times \mathbf{H}^*] = \hat{r}\frac{1}{2}\eta_0 |H_\phi|^2 = \hat{r}\frac{\eta_0}{2}\left(\frac{\beta Il}{4\pi r}\right)^2 \sin^2\theta = \hat{r}\langle S_r \rangle \tag{28.3.1}$$

$$\langle S_r \rangle = \frac{\eta_0}{2}\left(\frac{\beta Il}{4\pi r}\right)^2 \sin^2\theta \tag{28.3.2}$$

The *radiation power pattern* is the plot of $r^2\langle S_r \rangle$ as $r \to \infty$ or in the far field as shown in Figure 28.7.

The total power radiated by a Hertzian dipole is thus given by

$$P_T = \int d\Omega \langle S_r \rangle = \int_0^{2\pi} d\phi \int_0^{\pi} d\theta r^2 \sin\theta \langle S_r \rangle = 2\pi \int_0^{\pi} d\theta \frac{\eta_0}{2}\left(\frac{\beta Il}{4\pi}\right)^2 \sin^3\theta \tag{28.3.3}$$

Since

$$\int_0^{\pi} d\theta \sin^3\theta = -\int_1^{-1} (d\cos\theta)[1 - \cos^2\theta] = \int_{-1}^{1} dx(1 - x^2) = \frac{4}{3} \tag{28.3.4}$$

then the total power becomes

$$P_T = \frac{4}{3}\pi\eta_0 \left(\frac{\beta Il}{4\pi}\right)^2 = \frac{\eta_0(\beta Il)^2}{12\pi} \tag{28.3.5}$$

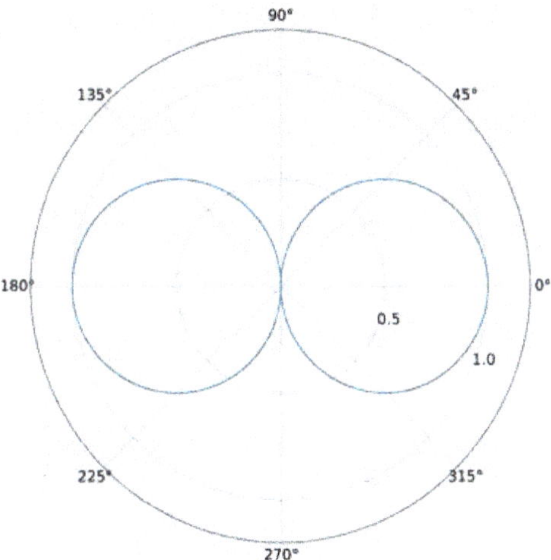

Figure 28.6: Radiation field pattern of a Hertzian dipole. It can be shown that the pattern is a circle.

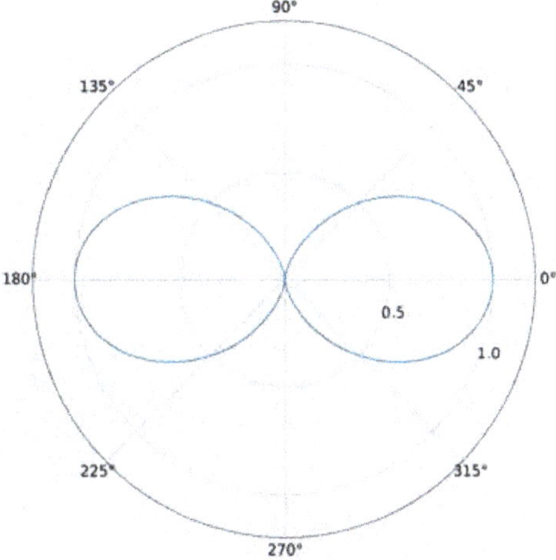

Figure 28.7: Radiation power pattern of a Hertzian dipole, which is also the same as the directive gain pattern.

28.3.1 Radiation Resistance—Circuit Equivalence of a Hertzian Dipole

Engineers love to replace complex systems with simpler systems. Simplicity rules again! For example, the voltage or current sources driving an antenna are usually made from electronic circuits.

Hence, it will be expedient to replace an antenna with a circuit equivalent so that it can interface with the driving circuit components. A raw Hertzian dipole, when driven by a voltage source, essentially looks like a capacitor due to the preponderance of electric field energy stored in the dipole field. But at the same time, the dipole radiates, giving rise to radiation loss. Thus, a simple circuit equivalence of a Hertzian dipole is a capacitor in series with a resistor. The resistor accounts for radiation loss of the dipole.

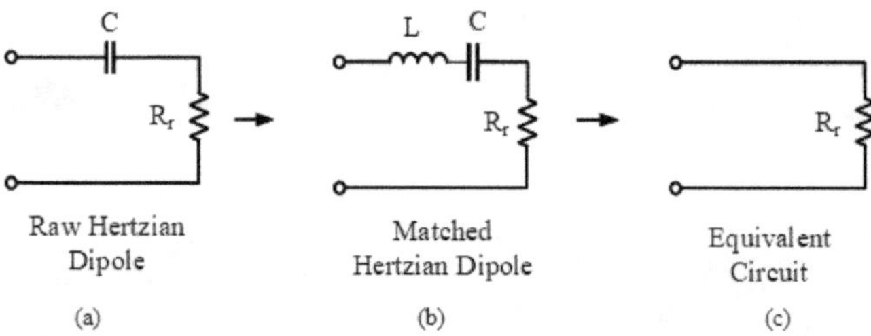

Figure 28.8: A short Hertzian dipole, shown in Figure 28.1, resembles an open circuit with a capacitor connected to it because the two spheres shown in Figure 28.1 will have charges stored in them, giving rise to electric energy stored. The radiation loss of the antenna can be modeled by a radiation resistance R_r. Hence, we have (a) an equivalent circuit of a raw Hertzian dipole without matching; (b) an equivalent circuit of a matched Hertzian dipole (using maximum power transfer theorem [57, sections 2.5 and 4.3], [125]); and (c) an equivalent circuit of a matched dipole at the resonance frequency of the LC tank circuit.

Hence, the way to drive the Hertzian dipole effectively is to use a matching network according to the maximum power transfer theorem.[5] Or an inductor has to be added in series with the intrinsic capacitance of the Hertzian dipole to cancel it at the resonance frequency of the tank circuit. Eventually, after matching, the Hertzian dipole can be modeled as just a resistor. Then the power absorbed by the Hertzian dipole from the driving source is $P_T = \frac{1}{2}I^2 R_r$. Here, the *radiation resistance* R_r is the effective resistance that will dissipate the same power as the total radiation power P_T when a current I flows through the resistor. It is defined by [35]

$$R_r = \frac{2P_T}{I^2} = \eta_0 \frac{(\beta l)^2}{6\pi} \approx 20(\beta l)^2, \qquad \text{where } \eta_0 = 377 \approx 120\pi \ \Omega \tag{28.3.6}$$

In the above, we have used (28.3.5) for the total power radiated by the Hertzian dipole. For example, for a Hertzian dipole with $l = 0.1\lambda$, $R_r \approx 8\Omega$.

The above assumes that the current is uniformly distributed over the length of the Hertzian dipole. This is approximately true if there are two charge reservoirs at its two ends, as in Hertz's original dipole antenna. For a small dipole with no charge reservoir at the two ends, the currents have

[5]The maximum power transfer theorem has to be used with caution, because with the recent advances in power electronics and transformers, the limit of the maximum power transfer theorem can be exceeded.

to vanish at the tips of the dipole as shown in Figure 28.9. The effective length of an equivalent Hertzian dipole for the dipole with triangular distribution is *half* of its actual length due to the manner in which the currents are distributed.[6] Such a formula can be used to estimate the radiation resistance of a small/short dipole.

For a filamental wire current with a nonuniform current, it can be approximated with a current given by

$$\mathbf{J}(\mathbf{r}) = \hat{z}\delta(x)\delta(y)I(z) \tag{28.3.7}$$

where $I(z)$ is the current distribution on the filamental wire. When such a current is used in the integral in (28.2.3), the integral does not have a closed form solution, unless we make further approximations like ignoring the phase term $\exp(-j\beta|\mathbf{r}-\mathbf{r}'|)$, which accounts for retardation effect. Then in (28.2.3), if $\mathbf{r}'$ is assumed to be small always so that the current on the antenna is equiphase. In a word, the retardation effect between different currents on the antennas is negligible. Then it can be approximated as

$$\mathbf{A}(\mathbf{r}) \cong \mu \iiint_V d\mathbf{r}'\mathbf{J}(\mathbf{r}')\frac{e^{-j\beta|\mathbf{r}|}}{4\pi|\mathbf{r}|} = \mu\frac{e^{-j\beta r}}{4\pi r}\iiint_V d\mathbf{r}'\mathbf{J}(\mathbf{r}') \tag{28.3.8}$$

where the phase term has been taken outside the integral. Then the radiation field is proportional to the integral summation of the radiation current alone.

For example, a half-wave dipole does not have a triangular current distribution but a sinusoidal one as shown in Figure 28.11. Nevertheless, we approximate the current distribution of a half-wave dipole with a triangular distribution and apply the above formula (28.3.6) with equiphase assumption. Then, picking $a = \frac{\lambda}{2}$, and letting $l_{\text{eff}} = \frac{\lambda}{4}$ in (28.3.6), we have

$$R_r \approx 50\Omega \tag{28.3.9}$$

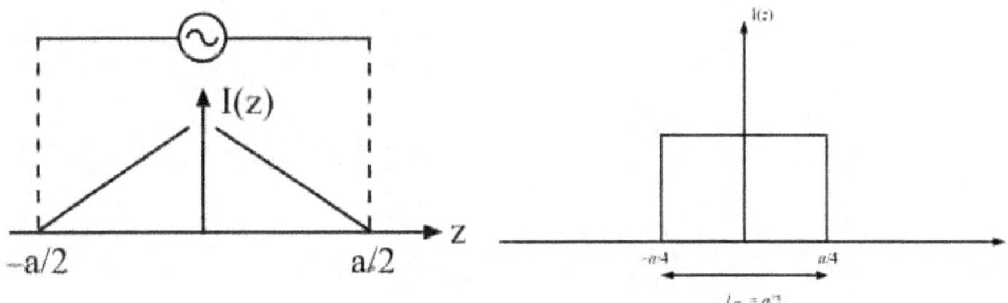

Figure 28.9: The current pattern on a short dipole can be approximated by a triangle since the current has to vanish at the end points of the short dipole. Furthermore, this dipole can be approximated by an effective Hertzian dipole half its length with uniform current. A more accurate current distribution on a wire dipole antenna can be found by using numerical methods, such as the method of moments [208, 209, 210].

[6]As shall be shown, when the dipole is short, the details of the current distribution are inessential in determining the radiation field. It is the area under the current distribution that is important.

The true current distribution on a half-wave dipole resembles that shown in Figure 28.11. The current is zero at the end points, but the current has a more sinusoidal-like distribution as in a transmission line. Hence, a half-wave dipole is not much smaller than a wavelength and does not qualify to be a Hertzian dipole. Furthermore, the current distribution on the half-wave dipole is not triangular in shape as above. Moreover, when we calculate (28.2.4), we assume that there is no phase delay (equiphase) between different parts of the current.[7] This is not true when the dipole antenna is not short compared to wavelength. This retardation effect, which comes from the $\exp(-j\beta|\mathbf{r}-\mathbf{r}'|)$ in (28.2.3), has to be accounted for. The calculation of such radiation integrals will be discussed in the next lecture.

A more precise calculation using the semianalytic method shows that $R_r = 73\ \Omega$ for a [59][p. 332].[8] This also implies that a half-wave dipole with sinusoidal current distribution is a better radiator than a dipole with just a triangular current distribution. More precise values can be obtained using numerical methods.

In fact, one can think of a half-wave dipole as a flared, open transmission line. In the beginning, this flared open transmission line came in the form of biconical antennas which are shown in Figure 28.10 [211]. If we recall that the characteristic impedance of a transmission line is $\sqrt{L/C}$, then as the spacing of the two metal pieces becomes bigger, the equivalent characteristic impedance becomes bigger. Therefore, the impedance can gradually transform from a small impedance like 50 Ω to that of free space, which is 377 Ω. This impedance matching helps mitigate reflection from the ends of the flared transmission line and enhances radiation. Because of the matching nature of bicone antennas, they are better radiators with higher radiaton loss and lower Q. Thus they have a broader bandwidth, and are important in UWB (ultra-wide band) antennas [212].

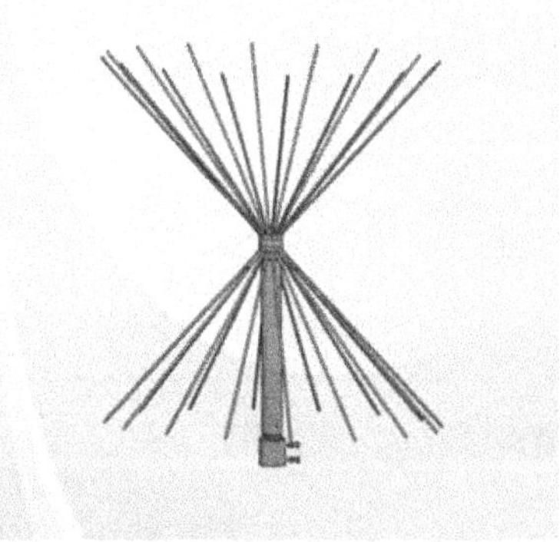

Figure 28.10: A bicone antenna can be thought of as a flared transmission line with gradually changing characteristic impedance. This enhances impedance matching and the radiation of the antenna (Figure by: antennasproduct.com).

[7] A more precise calculation of the current distribution requires a numerical method like the method of moments.

[8] This explains why the characteristic impedance of some transmission lines is 75 Ω.

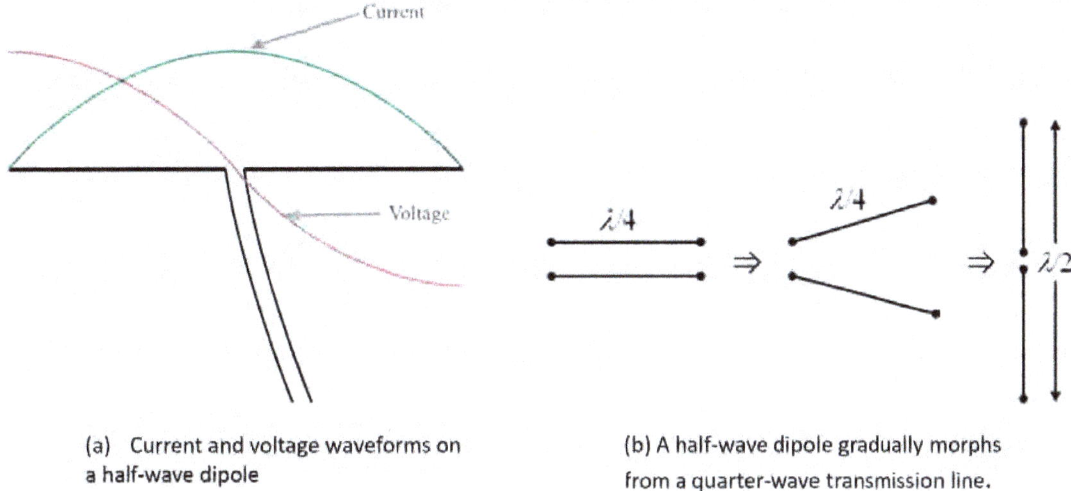

Figure 28.11: Approximate current distribution on a half-wave dipole (Figure by: electronics-notes.co). The currents are zero at the two end tips due to the current continuity equation, or KCL. A more precise calculation of its input impedance is given in [59].

Exercises for Lecture 28

Problem 28-1:

(i) Derive (28.2.7) and (28.2.8).

(ii) Verify (28.3.5).

(iii) Derive (28.3.6) and (28.3.9).

(iv) Explain why the current distribution on a short dipole can be approximated by a triangle.

(v) Explain why the retardation effect is ignored in such calculations as (28.3.8).

Lecture 29

Radiation Fields, Directive Gain, Effective Aperture

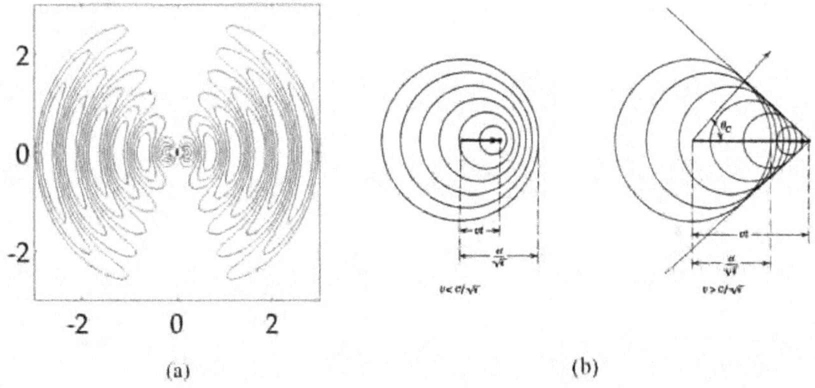

Figure 29.1: (a) Electric field around a time-oscillating dipole (Figure by physics stack exchange). (b) Equipotential lines around a moving charge that gives rise to Cherenkov radiation (Figure by J.D. Jackson [50]). We will not study Cherenkov (Cerenkov) radiation in this course, but it is written up in [35] and [50]. Its discovery and its explanation led to a Nobel Prize.

The reason why charges radiate is because they move or accelerate. In the case of a dipole antenna, the charges move back and forth between the two poles of the antenna. Near to the dipole source, quasi-static physics prevails, and the field resembles that of a static dipole. If the dipole is flipping its signs constantly due to the change in the direction of the current flow, the field would also have to flip its signs constantly. But electromagnetic waves travel with a finite velocity. The field from the source ultimately cannot keep up with the sign change of the source field: it has to be "torn" away from the source field and radiate. Another interesting radiation is the Cherenkov (also spelled Cerenkov) radiation. It is due to a charge moving faster than the velocity of light. As an electron cannot move faster than the speed of light in a vacuum, this can only happen in the material media or plasma, where the velocity of the electron can be faster than the group velocity of wave in the medium. Ultimately, the electric field from the particle is torn off from the charge and radiates. These two kinds of radiation are shown in the Figure 29.1.

For a system in time-harmonic oscillation, we have shown how to connect the vector and scalar potentials to the sources $\mathbf{J}$ and ϱ of an electromagnetic system in our previous lectures. This is a very important connection: it implies that once we know the sources, we know how to find the fields $\mathbf{A}$ and Φ, and then $\mathbf{E}$ and $\mathbf{H}$. But the relation between the fields and the sources are in general rather complex. In this lecture, we will simplify this relation by making a radiation field or far-field approximation. To this end, we assume that the point where the field is observed is very far from the source location in terms of wavelength. This approximation is very useful for understanding the physics of the radiation field from a source such as an antenna. It is also important for understanding the far field of an optical system. As shall be shown, this radiation field carries the energy generated by the sources to infinity.

29.1 Radiation Fields or Far-Field Approximation

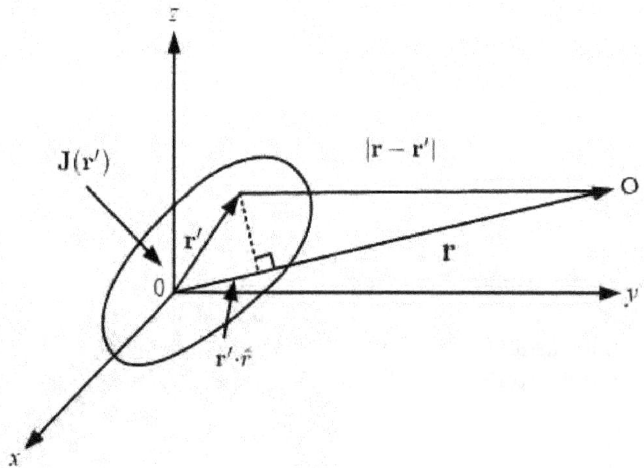

Figure 29.2: The relation of the observation point located at $\mathbf{r}$ to the source location at $\mathbf{r}'$. The distance of the observation point $\mathbf{r}$ to the source location $\mathbf{r}'$ is $|\mathbf{r} - \mathbf{r}'|$.

In the previous lecture, we derived the relation of the vector and scalar potentials to the sources $\mathbf{J}$ and ϱ as shown in (27.2.29) and (27.2.30).[1] They are given by

$$\mathbf{A}(\mathbf{r}) = \mu \iiint_V d\mathbf{r}' \mathbf{J}(\mathbf{r}') \frac{e^{-j\beta|\mathbf{r}-\mathbf{r}'|}}{4\pi|\mathbf{r}-\mathbf{r}'|} \tag{29.1.1}$$

$$\Phi(\mathbf{r}) = \frac{1}{\varepsilon} \iiint_V d\mathbf{r}' \varrho(\mathbf{r}') \frac{e^{-j\beta|\mathbf{r}-\mathbf{r}'|}}{4\pi|\mathbf{r}-\mathbf{r}'|} \tag{29.1.2}$$

where $\beta = \omega\sqrt{\mu\varepsilon} = \omega/c$ is the wavenumber. The integrals in (29.1.1) and (29.1.2) are normally untenable, but when the observation point is far from the source, approximation to the integrals can be made, giving them a nice physical interpretation.

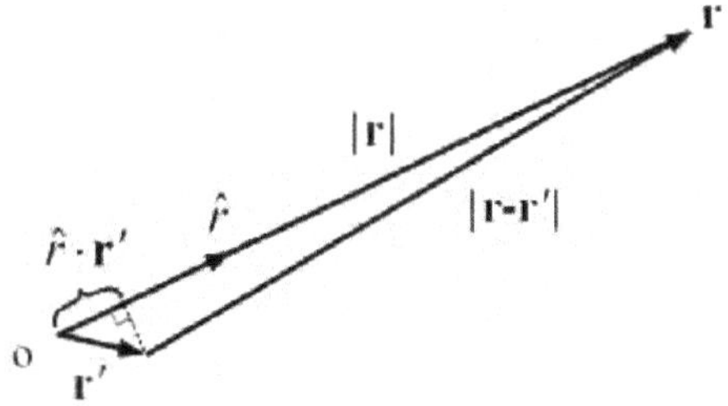

Figure 29.3: The relation between $|\mathbf{r}|$ and $|\mathbf{r}-\mathbf{r}'|$ using the parallax method, or that $|\mathbf{r}-\mathbf{r}'| \approx |\mathbf{r}| - \mathbf{r}' \cdot \hat{r}$. It is assumed that $\mathbf{r}$ is almost parallel to $\mathbf{r}-\mathbf{r}'$.

29.1.1 Far-Field Approximation

When $|\mathbf{r}| \gg |\mathbf{r}'|$, then $|\mathbf{r}-\mathbf{r}'| \approx r - \mathbf{r}' \cdot \hat{r}$, where $r = |\mathbf{r}|$. This approximation can be shown algebraically[2] or by geometrical argument as shown in Figure 29.3. Thus (29.1.1) above becomes

$$\mathbf{A}(\mathbf{r}) \approx \frac{\mu}{4\pi} \iiint_V d\mathbf{r}' \frac{\mathbf{J}(\mathbf{r}')}{r - \mathbf{r}' \cdot \hat{r}} e^{-j\beta r + j\beta\mathbf{r}'\cdot\hat{r}} \approx \frac{\mu e^{-j\beta r}}{4\pi r} \iiint_V d\mathbf{r}' \mathbf{J}(\mathbf{r}') e^{j\beta\mathbf{r}'\cdot\hat{r}} \tag{29.1.3}$$

In the above, $\mathbf{r}' \cdot \hat{r}$ is small compared to r. Hence, we have made use of that $1/(1-\Delta) \approx 1$ when Δ is small, so that $1/(r - \mathbf{r}' \cdot \hat{r})$ can be approximated by $1/r$. Also, we assume that the frequency is sufficiently high such that $\beta\mathbf{r}' \cdot \hat{r}$ is not necessarily small. Thus, $e^{j\beta\mathbf{r}'\cdot\hat{r}} \neq 1$. Hence, we keep the exponential term in (29.1.3) but simplify the denominator to arrive at the last expression above.

If we let $\boldsymbol{\beta} = \beta\hat{r}$, which is the $\boldsymbol{\beta}$ vector (or $\mathbf{k}$ vector in optics), and let $\mathbf{r}' = \hat{x}x' + \hat{y}y' + \hat{z}z'$, then we can convert the factor in the integrand into one that looks like a plane wave or 3D Fourier

[1] This topic is found in many standard textbooks in electromagnetics [59, 52, 35]. It is also found in lecture notes [48, 213].

[2] To show this algebraically, we let $|\mathbf{r}-\mathbf{r}'|^2 = (\mathbf{r}-\mathbf{r}')\cdot(\mathbf{r}-\mathbf{r}') = |\mathbf{r}|^2 + |\mathbf{r}'|^2 - 2\mathbf{r}\cdot\mathbf{r}' \cong |\mathbf{r}|^2 - 2\mathbf{r}\cdot\mathbf{r}'$. By taking the square root of this approximation and using the algebraic approximation that $(1-x)^{0.5} \cong 1 - 0.5x$, we can also arrive at the given approximation.

transform. Namely,

$$e^{j\beta\mathbf{r}'\cdot\hat{r}} = e^{j\boldsymbol{\beta}\cdot\mathbf{r}'} = e^{j\beta_x x' + j\beta_y y' + j\beta_z z'} \tag{29.1.4}$$

The above is the expression for a plane wave propagating in the $\boldsymbol{\beta} = \beta\hat{r}$ direction or $\hat{r}$ direction. Therefore, (29.1.3) resembles a 3D Fourier transform integral,[3] namely, the above integral becomes

$$\mathbf{A}(\mathbf{r}) \approx \frac{\mu e^{-j\beta r}}{4\pi r} \iiint_V d\mathbf{r}'\mathbf{J}(\mathbf{r}')e^{j\boldsymbol{\beta}\cdot\mathbf{r}'} \tag{29.1.5}$$

and (29.1.5) can be rewritten as

$$\mathbf{A}(\mathbf{r}) \cong \frac{\mu e^{-j\beta r}}{4\pi r}\mathbf{F}(\boldsymbol{\beta}) \tag{29.1.6}$$

where

$$\mathbf{F}(\boldsymbol{\beta}) = \iiint_V d\mathbf{r}'\mathbf{J}(\mathbf{r}')e^{j\boldsymbol{\beta}\cdot\mathbf{r}'} \tag{29.1.7}$$

The above is a special kind of 3D Fourier transform of $\mathbf{J}(\mathbf{r}')$ with the Fourier transform variable $\boldsymbol{\beta} = \hat{r}\beta$. In a word, the Fourier data is restricted to be on a sphere surface with radius β, which is not usual, or that $|\boldsymbol{\beta}|^2 = {\beta_x}^2 + {\beta_y}^2 + {\beta_z}^2 = \beta^2$ which is a constant for a fixed frequency. In other words, the length of the vector $\boldsymbol{\beta}$ is fixed to be β, whereas in a usual 3D Fourier transform, β_x, β_y, and β_z are independent variables only. Or the value of ${\beta_x}^2 + {\beta_y}^2 + {\beta_z}^2$ ranges from zero to infinity, but in (29.1.7), it is different.

The above is the 3D "Fourier transform" of the current source $\mathbf{J}(\mathbf{r}')$, with Fourier variables β_x, β_y, and β_z restricted to lying on a sphere of radius β and $\boldsymbol{\beta} = \beta\hat{r}$. This spherical surface in the Fourier space is also called the *Ewald sphere.*

29.1.2 Locally Plane Wave Approximation

We can write $\hat{r}$ or $\boldsymbol{\beta}$ in terms of direction cosines in spherical coordinates or that

$$\hat{r} = \hat{x}\cos\phi\sin\theta + \hat{y}\sin\phi\sin\theta + \hat{z}\cos\theta \tag{29.1.8}$$

Hence,

$$\mathbf{F}(\boldsymbol{\beta}) = \mathbf{F}(\beta\hat{r}) = \mathbf{F}(\beta, \theta, \phi) \tag{29.1.9}$$

It is not truly a 3D function in space, since β, the length of the vector $\boldsymbol{\beta}$, is fixed; thus it is a function of θ and ϕ space variables. Or it is a 3D Fourier transform with data restricted on a spherical surface.

In (29.1.6), when $r \gg \mathbf{r}'\cdot\hat{r}$, and when the frequency is high or β is large, $e^{-j\beta r}$ is now a rapidly varying function of r while $\mathbf{F}(\boldsymbol{\beta})$ is only a slowly varying function of $\hat{r}$ or of θ and of ϕ, the

[3] Except that the vector $\boldsymbol{\beta}$ is of fixed length.

observation angles. In other words, the prefactor in (29.1.6), $\exp(-j\beta r)/r$, resembles a spherical wave which is modulated by a slowly varying function of θ and ϕ, namely, $\mathbf{F}(\boldsymbol{\beta})$.

Therefore, in the far field, the wave radiated by a finite source resembles a spherical wave. Moreover, a spherical wave resembles a plane wave when one is sufficiently far from the source such that $\beta r \gg 1$, or $2\pi r/\lambda \gg 1$. This happens when r is many wavelengths away from the source. This is obviated if we write $e^{-j\beta r} = e^{-j\boldsymbol{\beta}\cdot\mathbf{r}}$ where $\boldsymbol{\beta} = \hat{r}\beta$ and $\mathbf{r} = \hat{r}r$ so that a spherical wave resembles a plane wave locally. This phenomenon is shown in Figure 29.4 and Figure 29.5.

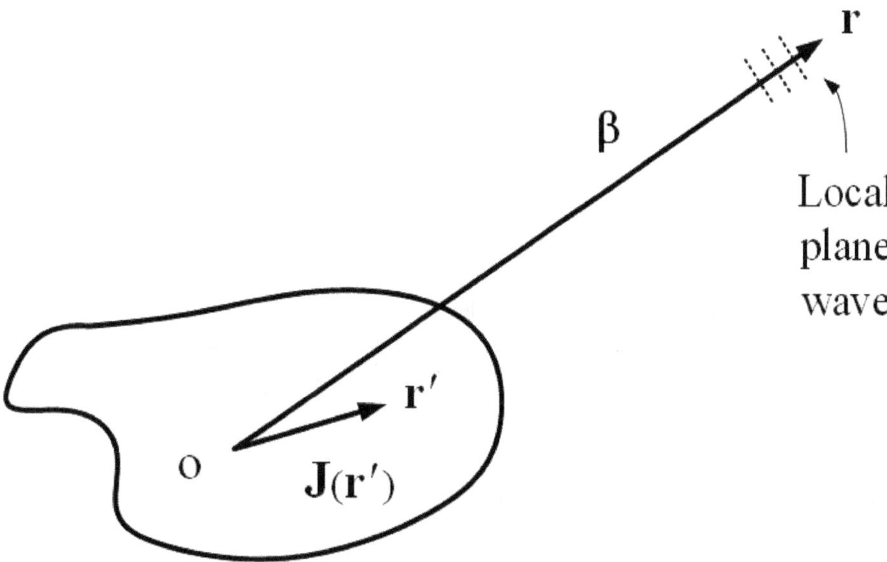

Figure 29.4: A finite source radiates a field that resembles a spherical wave far from the source. In the vicinity of the observation point $\mathbf{r}$, when β is large, the field is strongly dependent on r via $\exp(-j\beta r)$ but weakly dependent on $\boldsymbol{\beta}$ ($\boldsymbol{\beta}$ hardly changes direction in the vicinity of the observation point $\mathbf{r}$). Hence, the field becomes locally like a plane wave in the far field.

Then it is clear that with the local plane-wave approximation, we can let $\nabla \to -j\boldsymbol{\beta} = -j\beta\hat{r}$, and with this approximation, we have

$$\mathbf{H} = \frac{1}{\mu}\nabla \times \mathbf{A} \approx -j\frac{\beta}{\mu}\hat{r} \times (\hat{\theta}A_\theta + \hat{\phi}A_\phi) = j\frac{\beta}{\mu}(\hat{\theta}A_\phi - \hat{\phi}A_\theta) \tag{29.1.10}$$

Similarly [48, 213],

$$\mathbf{E} = \frac{1}{j\omega\varepsilon}\nabla \times \mathbf{H} \cong -j\frac{\beta}{\omega\epsilon}\hat{r} \times \mathbf{H} \cong -j\omega(\hat{\theta}A_\theta + \hat{\phi}A_\phi) \tag{29.1.11}$$

Notice that $\boldsymbol{\beta} = \beta\hat{r}$, the direction of propagation of the local plane wave, is orthogonal to $\mathbf{E}$ and $\mathbf{H}$ in the far field. This is a hallmark of a plane wave since the wave is locally a plane wave.

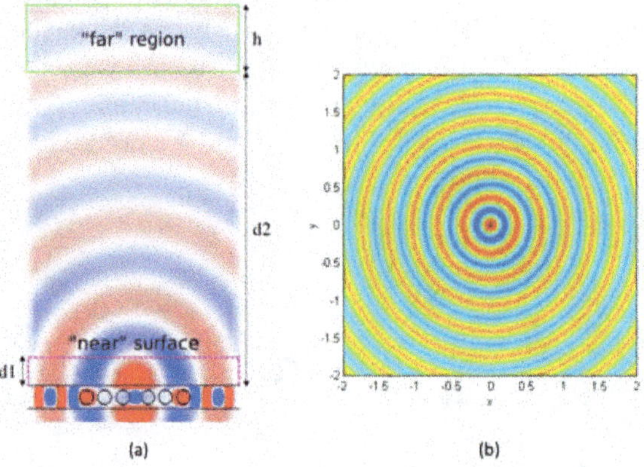

Figure 29.5: We can use numerical simulations to show the physical attributes of a spherical wave. (a) A leaky hole in a waveguide leaks a spherical wave (Figure by MEEP, MIT). (b) A point source radiates a spherical wave (Figure by ME513, Purdue Engineering). Most of these simulations are done with the FDTD (finite-difference time-domain) method that we will learn later in the course. When the wavelength is short, or the frequency high, a spherical wave looks locally like a plane wave. This is similar to the notion of humans, who are small, thinking that the earth is flat around us. To this day, some people still believe that the earth is flat:)

Moreover, there is more than one way to derive the electric field $\mathbf{E}$. For instance, using (29.1.10) for the magnetic field, the electric field can also be written as

$$\mathbf{E} = \frac{1}{j\omega\mu\varepsilon}\nabla \times (\nabla \times \mathbf{A}) \tag{29.1.12}$$

Using the BAC-CAB formula for the double-curl operator, plus that $\nabla \simeq -j\boldsymbol{\beta}$, for a plane wave, the above can be rewritten as

$$\mathbf{E} = \frac{1}{j\omega\mu\varepsilon}\left(\nabla\nabla\cdot\mathbf{A} - \nabla^2\mathbf{A}\right) \cong \frac{1}{j\omega\mu\varepsilon}\left(-\boldsymbol{\beta}\boldsymbol{\beta} + \beta^2\bar{\mathbf{I}}\right)\cdot\mathbf{A} \tag{29.1.13}$$

where we have used that $\nabla^2\mathbf{A} = -\beta^2\mathbf{A} = -\beta^2\bar{\mathbf{I}}\cdot\mathbf{A}$, where $\bar{\mathbf{I}}$ is the identity dyad.[4]

Alternatively, we bring $\beta^2 = \omega^2\mu\epsilon$ out of the parentheses, and rewrite the above as

$$\mathbf{E} \cong -j\omega\left(-\hat{\beta}\hat{\beta} + \bar{\mathbf{I}}\right)\cdot\mathbf{A} = -j\omega\left(-\hat{r}\hat{r} + \bar{\mathbf{I}}\right)\cdot\mathbf{A} \tag{29.1.14}$$

[4]Note that $\nabla\cdot\mathbf{A} \neq 0$ here.

where we have assume that $\hat{\beta} = \hat{r}$. Since we can express the identity dyad $\bar{\mathbf{I}} = \hat{r}\hat{r} + \hat{\theta}\hat{\theta} + \hat{\phi}\hat{\phi}$ in spherical coordinates,[5] then the above becomes

$$\mathbf{E} \cong -j\omega\left(\hat{\theta}\hat{\theta} + \hat{\phi}\hat{\phi}\right) \cdot \mathbf{A} = -j\omega(\hat{\theta}A_\theta + \hat{\phi}A_\phi) \tag{29.1.15}$$

which is the same as previously derived. It also shows that the electric field is transverse to the $\boldsymbol{\beta}$ vector.[6]

Furthermore, it can be shown that in the far field, using the locally plane-wave approximation,

$$|\mathbf{E}|/|\mathbf{H}| \approx \eta \tag{29.1.16}$$

where η is the intrinsic impedance of free space, which is a property of a plane wave. Moreover, one can show that the time-averaged Poynting's vector, or the power density flow, in the far field is

$$\langle \mathbf{S} \rangle = \frac{1}{2}\Re e\left(\mathbf{E} \times \mathbf{H}^*\right) \approx \frac{1}{2\eta}|\mathbf{E}|^2\hat{r} = \langle S_r \rangle \hat{r} \tag{29.1.17}$$

which again resembles the property of a plane wave.[7] In a word, the radiated field is a spherical wave, and the Poynting's vector is radial. Therefore,

$$\langle \mathbf{S} \rangle = \hat{r}\langle S_r(\theta, \phi) \rangle, \quad \text{where} \quad \langle S_r(\theta, \phi) \rangle = \frac{1}{2\eta}|\mathbf{E}|^2 \tag{29.1.18}$$

and $\langle S_r \rangle$ is the time-averaged radial power density. The plot of $|\mathbf{E}(\theta, \phi)|$ is termed the far-field pattern or the radiation pattern of an antenna or the source, while the plot of $|\mathbf{E}(\theta, \phi)|^2$ is its far-field power pattern.

29.1.3 Directive Gain Pattern

A directive gain pattern that characterizes the radiation pattern of a general source or an antenna can be defined. Once the far-field radiation power pattern or the radial power density $\langle S_r \rangle$ is known, the total power radiated by the antenna in the far field is found by integrating over all angles, namely,

$$P_T = r^2 \int d\Omega \langle S_r(\Omega) \rangle = r^2 \int_0^\pi \int_0^{2\pi} \sin\theta d\theta d\phi \langle S_r(\theta, \phi) \rangle \tag{29.1.19}$$

In the above, $d\Omega = \sin\theta d\theta d\phi$ In free space, the above evaluates to a constant independent of r due to energy conservation.

Now assume that this same antenna is radiating isotropically in all directions; then the average power density of this fictitious isotropic radiator as $r \to \infty$ is

$$S_{\text{av}} = \frac{P_T}{4\pi r^2} \tag{29.1.20}$$

[5]Easily verified by a sanity check.

[6]We can also arrive at the above by letting $\mathbf{E} = -j\omega\mathbf{A} - \nabla\Phi$ and using the appropriate formula for the scalar potential. There is more than one road that leads to Rome!

[7]To avoid confusion, we will use $\mathbf{S}$ to denote the instantaneous Poynting's vector and $\underset{\sim}{\mathbf{S}}$ to denote the complex Poynting's vector (see 10.3.1).

A dimensionless directive gain pattern can be defined such that [35, 213]

$$G(\theta,\phi) = \frac{\langle S_r(\theta,\phi)\rangle}{S_{\text{av}}} = \frac{4\pi r^2 \langle S_r(\theta,\phi)\rangle}{P_T} \tag{29.1.21}$$

Now we can use the Hertzian dipole to illustrate this concept. Using P_T from (28.3.5), and $\langle S_r\rangle$ from (28.3.1), we arrive at

$$G(\theta,\phi) = \frac{\frac{\eta_0}{2}\left(\frac{\beta Il}{4\pi r}\right)^2 \sin^2\theta}{\frac{1}{4\pi r^2}\frac{4}{3}\eta_0\pi\left(\frac{\beta Il}{4\pi}\right)^2} = \frac{3}{2}\sin^2\theta \tag{29.1.22}$$

This directive gain pattern is a measure of the radiation power pattern of the antenna or source compared to when it radiates isotropically. The above function is independent of r in the far field since $S_r \sim 1/r^2$ in the far field, but it is a function of (θ,ϕ).

The directivity of an antenna $D = \max(G(\theta,\phi))$ is the maximum value of the directive gain. This is 1.5 for the Hertzian dipole. It is to be noted that by its mere definition in (29.1.21),

$$\int d\Omega G(\theta,\phi) = 4\pi \tag{29.1.23}$$

where $\int d\Omega = \int_0^{2\pi}\int_0^{\pi} \sin\theta d\theta d\phi$. It is seen that since the directive gain pattern is normalized, when the radiation power is directed to the main lobe of the antenna, the corresponding side lobes and back lobes will be diminished. This follows from energy conservation.

29.2 Effective Aperture and Directive Gain

An antenna also has an effective area or aperture A_e, such that if a plane wave carrying power density denoted by $\langle S_{\text{inc}}\rangle$ impinges on the antenna, then the power received by the antenna, P_{received}, is given by

$$P_{\text{received}} = \langle S_{\text{inc}}\rangle A_e \tag{29.2.1}$$

Here, the transmit antenna and the receive antenna are in the far field of each other. Hence, we can approximate the field from the transmit antenna to be a plane wave when it reaches the receive antenna. The receive antenna, if it is a good antenna, will siphon off some energy from the plane wave to power the local electronics within the antenna. Therefore, the antenna absorbs some energy to generate a voltage at the receiver load to deliver the power received by the antenna. This receiving system can be linearly modeled and the received power is linearly proportional to the incident power density $\langle S_{\text{inc}}\rangle$ with the proportionality constant which is the effective aperture of the antenna.[8]

[8]This concept can be extended to an effective cross section or aperture of an atom that has the dimension of area.

A wonderful relationship exists between the directive gain pattern $G(\theta, \phi)$ and the effective aperture, namely that

$$A_e(\theta, \phi) = \frac{\lambda^2}{4\pi} G(\theta, \phi) \tag{29.2.2}$$

Therefore, the effective aperture of an antenna is also direction dependent. Different plane waves incident on the antenna at different angles will see different effective apertures. The above implies that the radiation property of an antenna is related to its receiving property. (This is a beautiful consequence of reciprocity theorem!) The constant of proportionality, $\lambda^2/(4\pi)$, is a universal constant that is valid for all antennas satisfying the reciprocity theorem. The derivation of this constant for a Hertzian dipole is given in Kong [35], or using blackbody radiation law [213, 214].)

The directivity and the effective aperture can be enhanced by designing antennas with different gain patterns. When the radiative power of the antenna can be directed to be in a certain direction, then the directive gain and the effective aperture (for that given direction) of the antenna is improved. This is shown in Figure 29.6. Such focusing of the radiation fields of the antenna can be achieved using reflector antennas or array antennas. Array antennas, as shall be shown, work by constructive and destructive wave field of the antenna.

Being able to do point-to-point communications at high data rate is an important modern application of an antenna array. Figure 29.7 shows the gain pattern of a sophisticated antenna array design for 5G applications. Now, 5G antennas operating in the high band usually range in frequencies from 24 GHz to 50 GHz, putting them in the millimeter wave range, given the antennas much higher data throughput or bandwidth.

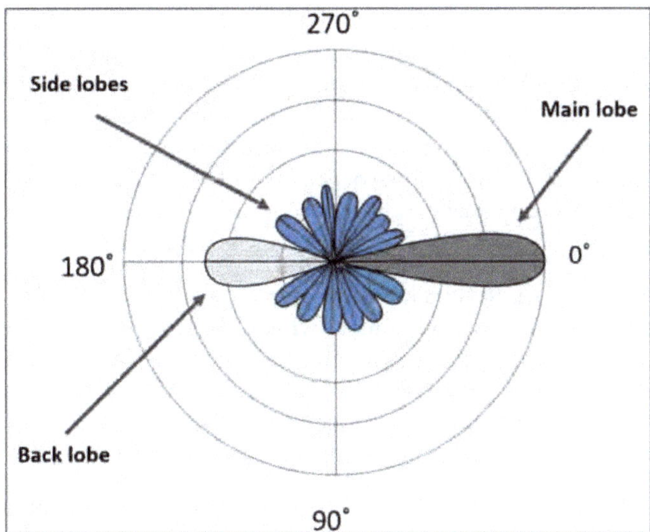

Figure 29.6: The directive gain pattern of an array antenna. The directivity is increased by constructive interference among the elements of the array antenna (Figure by Wikimedia).

Figure 29.7: The directive gain pattern of a sophisticated array antenna for 5G applications. Presently, 5G antennas work from 24 GHz to 50 GHz, which are in the millimeter range (Figure by: Ozeninc.com).

29.2.1 The Electromagnetic Spectrum

Now that we have learned more about the physics of electromagnetic waves, it is meaningful to discuss the electromagnetic spectrum. The electromagnetic field has been used from very low frequencies to very high frequencies. At very low frequencies–, ultra-low frequencies (ULFs) of less than 3 Hz, extremely low frequencies (ELFs) of 3Hz to 3,000 Hz, and very low frequencies (VLFs) of 3 KHz to 30 KHz– have been used to probe the Earth's surface and in submarine communications because of their deeper penetration depths. This is because a saline solution, which is primarily a conductive medium, has larger skin depth the lower the frequency. Remember that the skin depth is given by the formula $\delta = \sqrt{2/(\omega\mu\sigma)}$!

The AM radio stations, operating in the several hundred KHz have wavelengths of several hundreds meters. FM radio stations are in the 100 MHz range, while TV stations operate in the several hundreds MHz. Microwaves have wavelengths in the order of cm, and infrared light ranges from 1000 μm to 1 μm. A perfect electric conductor (PEC) has zero skin depth, but when the skin depth is much smaller than the wavelength and the size of the object, then one can approximate a conductive medium as a PEC–for example, copper at microwave frequencies.

The visible spectrum ranges from 700 nm to 400 nm. There is no PEC at these optical frequencies. As we learned from the Drude-Lorentz-Sommerfeld model, when the frequency is high, the inertial term in (8.3.14) in the lecture 8, which is proportional to $m_e d^2x/dt^2$, is important. When this term dominates, the medium resembles a plasma medium [215]. Hence, a metallic medium is often modeled as a plasma medium.

Ultraviolet (UV) light ranges from 400 nm to 100 nm, while X-rays are generally below 100 nm to

1 nm. Gamma rays are generally below 1 nm. UV light of 193 nm and EUV light of 13.5 nm are now used for nanolithography. X-rays are important for imaging, while gamma rays are used for some medical applications. The lights above UV are ionizing, and thus generally harmful to the human body. At high frequencies, it is expedient to think of electromagnetic waves as consisting of streams of particles (photons).

The new frontier of the electromagnetic spectrum, at the present moment, is in the terahertz range, from 0.1 THz to 10 THz. The dearth of THz sources has made the technology in this area in its infancy. THz signals can be generated from frequency doubling from lower semiconductor devices, nonlinear mixing of optical signals, or the use of quantum cascade lasers.

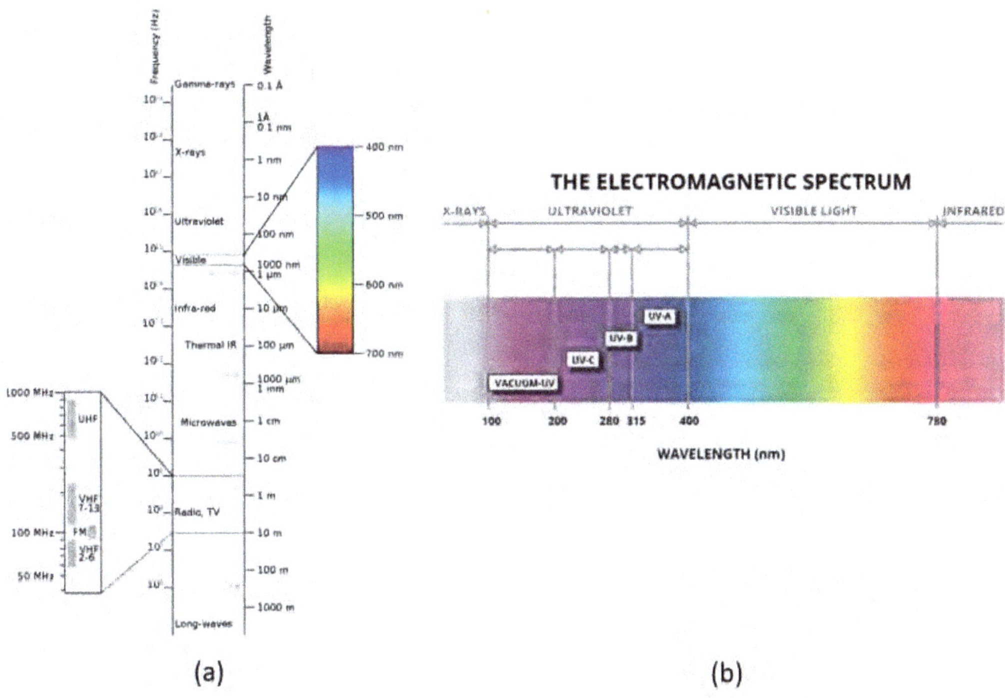

Figure 29.8: (a) The electromagnetic spectrum from static to gamma ray. (b) The electromagnetic spectrum with emphasis around the visible/optical spectrum. EUV with wavelength of 13.5 nm is used for nanolithography now, while ELF with a frequency of around 3 Hz is used for submarine communications (Figure by Wikimedia).

Exercises for Lecture 29

Problem 29-1:

(i) Show that in the far field, the ratio of $|\mathbf{E}|/|\mathbf{H}| = \eta$. Give the physical reason for this.

(ii) In this lecture, it was shown that

$$\mathbf{A}(\mathbf{r}) \cong \frac{\mu e^{-j\beta r}}{4\pi r}\mathbf{F}(\boldsymbol{\beta})$$

Verify the above expression and give the expression for $\mathbf{F}(\boldsymbol{\beta})$.

(iii) Explain why when we are in the far field such that $\beta r \gg 1$, in the neighborhood of an observation around $\mathbf{r}$, the variation of the field is dominated by $e^{-j\beta r}$, and hence, why we can make a local plane wave approximation. Is it necessary that the radius of curvature of the wavefront be much larger than the wavelength for this approximation? Explain why and why not?

(iv) For an antenna, give the physical meaning for the directive gain pattern $G(\theta, \phi)$ defined in (29.1.21) and the effective aperture A_e defined in (29.2.1).

Lecture 30

Array Antennas, Fresnel Zone, Rayleigh Distance

Our world is beset with the dizzying impact of wireless communication. It has greatly impacted our lives.[1] Wireless communication is impossible without using antennas. Hence it is important to design these communication systems with the proper antennas so that they operate with utmost efficiency and sensitivity. We have seen that a simple Hertzian dipole has low directivity in section 29.1.3. The radiation pattern looks like a donut, and the directivity of the Hertzian dipole antenna is only 1.5. Hence, for point-to-point communications, much power is wasted. However, the directivity of antennas can be improved if a group or array of dipoles can work cooperatively together by using constructive and destructive interferences. They can be made to constructively interfere in the desired direction, and destructively interfere in other directions to enhance the directivity of the array of antennas. Since the far-field approximation of the radiation field can be made, and the relationship between the far field and the source is a Fourier transform relationship,[2] clever engineering can be done by borrowing knowledge from the signal processing area. After understanding the far-field physics, one can also understand many optical phenomena, such as how a laser pointer works. Many textbooks have been written about array antennas, some of which are in the reference list [216, 217].

We like to emphasize that the world of electromagnetics morphs from circuit physics to wave physics when the wavelength becomes small and the structure size is of the order of the wavelength. If the frequency increases such that the wavelength becomes much smaller than the feature sizes, then ray physics prevails, and electromagnetic waves behave like particles. We will study the wave physics nature of electromagnetic fields in this lecture. Later, we will address the ray nature of electromagnetic fields.

[1] I cannot imagine a day that I do not receive messages from my relatives and friends in Malaysia and Singapore. It has made the world feel smaller and more transparent.

[2] It actually is a Fourier transform relationship on an Ewald sphere.

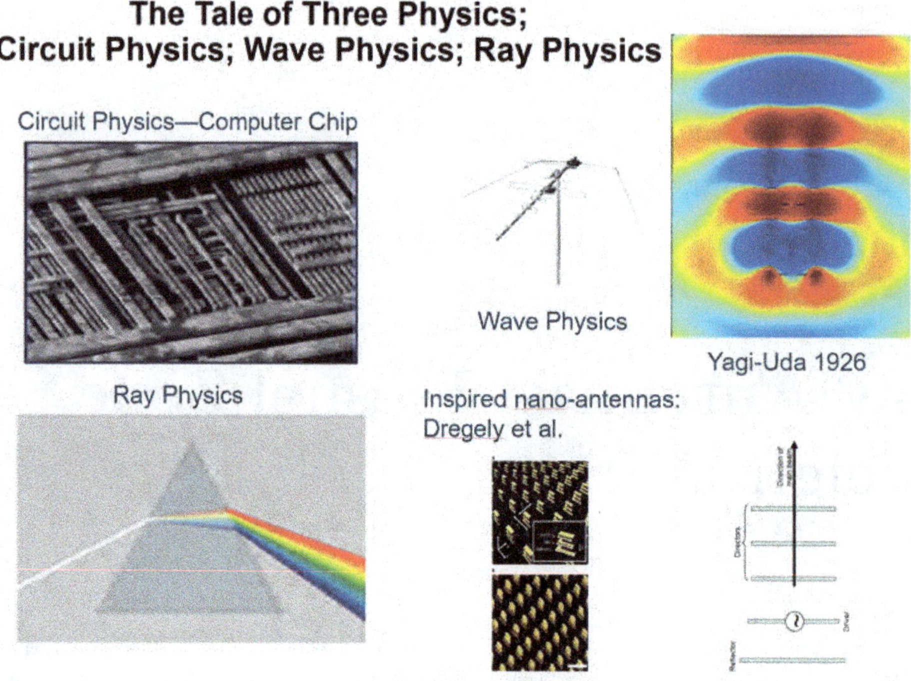

Figure 30.1: The world of electromagnetics morphs from circuit physics to wave physics, and then ray physics. The plethora of physics that emerges in electromagnetics is fascinating. Different mathematical tools are devised to solve electromagnetics problems in these regimes. At low frequencies, KCL, KVL, and potential theory can be use to describe the workings of most devices. At mid-frequencies, full Maxwell's equations have to be solved to capture the wave physics correctly. At high frequencies, electromagnetic waves behave like particles and rays. Ray-tracing algorithms are often used to capture their wave physics. What is missing here is the quantum nature of electromagnetic fields that will be discussed at the end of this course (Figure by wikipedia).

30.1 Array Pattern–Unit Pattern and Array Factor

The fact that the far field of an antenna is proportional to the Fourier transform of the current on the antenna can be exploited for further insight in many areas of wave physics. This fact is important, and hence, (29.1.3) is reproduced below:

$$\mathbf{A}(\mathbf{r}) \approx \frac{\mu e^{-j\beta r}}{4\pi r} \iiint_V d\mathbf{r}' \mathbf{J}(\mathbf{r}') e^{j\boldsymbol{\beta}\cdot\mathbf{r}'} \cong \frac{\mu e^{-j\beta r}}{4\pi r} \mathbf{F}(\boldsymbol{\beta}) \tag{30.1.1}$$

In the above,

$$\mathbf{F}(\boldsymbol{\beta}) = \iiint_V d\mathbf{r}'\mathbf{J}(\mathbf{r}')e^{j\boldsymbol{\beta}\cdot\mathbf{r}'} \tag{30.1.2}$$

The above is the Fourier transform of the radiation current restricted to the Ewald sphere.

In an antenna array, we make cookie-cutter replicas of a primary antenna and design them to radiate in unison. By appropriately adjusting their mutual phase relations and amplitudes, we can engineer and produce a rich diversity of radiation patterns by constructive and destructive interference. For simplicity, we begin with two antenna elements, which are replicas of each other, but they have unequal amplitudes and phases. Mathematically, this can be expressed as

$$\mathbf{J}(\mathbf{r}') = \mathbf{J}_u(\mathbf{r}') + B\mathbf{J}_u(\mathbf{r}' - \mathbf{r}_T) \tag{30.1.3}$$

where B is a complex amplitude and, hence, the two currents are not equi-phase. Moreover, the second current is being translated to a new location $\mathbf{r}_T$.

Using that the Fourier transform of $\mathbf{J}_u(\mathbf{r}')$ is

$$\mathbf{F}_u(\beta) \tag{30.1.4}$$

then the Fourier transform of $\mathbf{J}_u(\mathbf{r}' - \mathbf{r}_T)$ is

$$\mathbf{F}_u(\boldsymbol{\beta})e^{-j\boldsymbol{\beta}\cdot\mathbf{r}_T} \tag{30.1.5}$$

Consequently, the Fourier transform (30.1.3) is

$$\mathbf{F}(\boldsymbol{\beta}) = \mathbf{F}_u(\boldsymbol{\beta})(1 + Be^{-j\boldsymbol{\beta}\cdot\mathbf{r}_T}) \tag{30.1.6}$$

And then, $\mathbf{A}(\mathbf{r})$ in (30.1.1), which is a Fourier transform integral, becomes

$$\mathbf{A}(\mathbf{r}) \simeq \underbrace{\frac{\mu e^{-j\beta r}}{4\pi r}\mathbf{F}_u(\boldsymbol{\beta})}_{\mathbf{A}_u(\mathbf{r})}\underbrace{(1 + Be^{-j\boldsymbol{\beta}\cdot\mathbf{r}_T})}_{A.F.} \tag{30.1.7}$$

where

$$\mathbf{A}_u(\mathbf{r}) = \frac{\mu e^{-j\beta r}}{4\pi r}\mathbf{F}_u(\boldsymbol{\beta}) \tag{30.1.8}$$

In the above, $\mathbf{A}_u(\mathbf{r})$ is the field due to a unit element of the array, and $A.F.$ is the array factor given as

$$A.F. = 1 + Be^{-j\boldsymbol{\beta}\cdot\mathbf{r}_T} \tag{30.1.9}$$

Using that in the far electric field,

$$\mathbf{E} \cong -j\omega(\hat{\theta}A_\theta + \hat{\phi}A_\phi) = -j\omega\left(\hat{\theta}\hat{\theta} + \hat{\phi}\hat{\phi}\right)\cdot\mathbf{A} = -j\omega\left(\hat{\theta}\hat{\theta} + \hat{\phi}\hat{\phi}\right)\cdot\mathbf{A}_u A.F. \tag{30.1.10}$$

the far electric field due to a unit element of the array is thus

$$\mathbf{E}_u \cong -j\omega\left(\hat{\theta}\hat{\theta} + \hat{\phi}\hat{\phi}\right)\cdot\mathbf{A}_u \tag{30.1.11}$$

Then, using (30.1.10) and (30.1.11), the far-field pattern can be written as

$$|\mathbf{E}| = |\mathbf{E}_u| A.F. \tag{30.1.12}$$

Therefore, the far-field pattern is the product of the unit pattern with the array factor $A.F.$ The array factor above is for a two-element array. But $A.F.$ can be generalized to an N-element array to give

$$A.F. = 1 + B_1 e^{-j\boldsymbol{\beta}\cdot\mathbf{r}_{T1}} + B_2 e^{-j\boldsymbol{\beta}\cdot\mathbf{r}_{T2}} + \cdots = \sum_{i=0}^{N} B_i e^{-j\boldsymbol{\beta}\cdot\mathbf{r}_{Ti}} \tag{30.1.13}$$

where we take $B_0 = 1$.

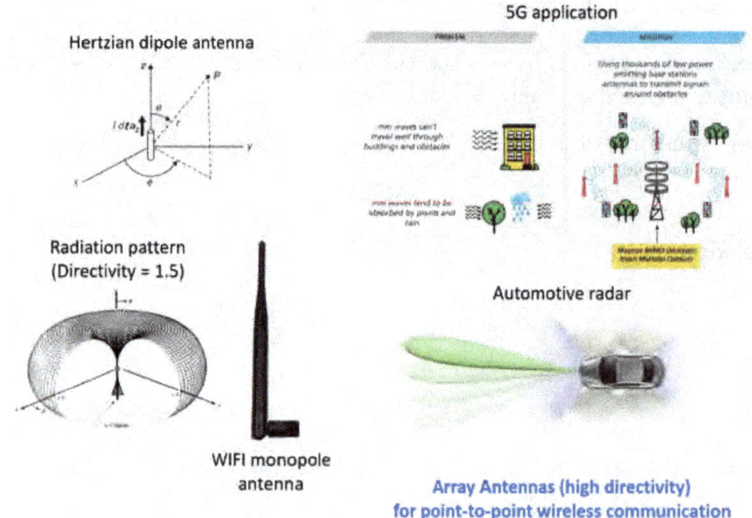

Figure 30.2: The left figures show the low directivity of a Hertzian dipole. The right figures show applications for which array antennas are needed to improve directivity and better communication links of a 5G communication system, which can operate as high as 54 GHz. The bottom right figure shows the collision avoidance system (CAS) of a car. The CAS of cars typically operate around 77 GHz so that the electromagnetic signals can penetrate through fog and rain (assembled by D.Y. Na).

30.2 Linear Array of Dipole Antennas

An antenna array can be designed so that the constructive and destructive interference in the far field can be used to steer the direction of radiation pattern of the antenna, or the far-field radiation pattern of an antenna array. This is because the far field of a source is related to the source by a Fourier transform relationship as shown in the previous lecture (see subsection 29.1.1). The relative phases of the array elements can be changed slowly over time with respect to the operating frequency so that the beam of an array antenna can be steered in real time. This has important applications in, for example, air traffic control. Note that if the current sources are impressed current sources, they can be regarded as the input to Maxwell's equations. Then the fields are the output of the system, and we are dealing with a linear time-invariant system for which linear system theory can be used. For instance, we can use Fourier transform to analyze the problem in the frequency domain. The time domain response then can be obtained by inverse Fourier transform. This is provided that the current sources are impressed and time-harmonic. "Impressed sources" means they are unchanging as they radiate.

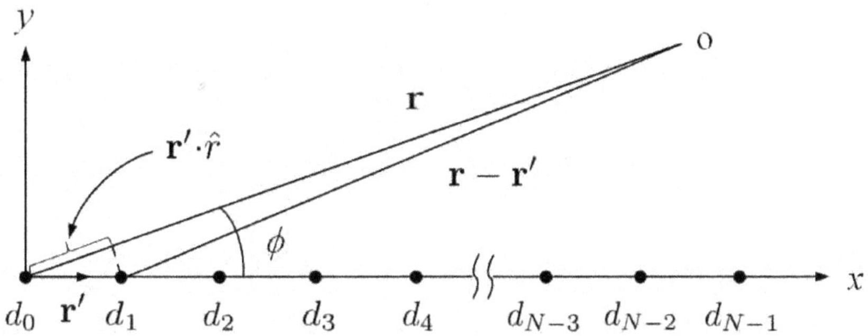

Figure 30.3: Schematic of a dipole array where the dipoles are aligned on the x-axis and we observe the field on the xy plane. To simplify the math, the far-field approximation can be used to find its far field or radiation field.

To gain physical insight into how constructive and destructive interference works for an antenna array, we assume a simple linear dipole array as shown in Figure 30.3. First, without loss of generality and for simplicity to elucidate the physics, we assume that this is a linear array of point Hertzian dipoles aligned on the x-axis. The current can then be described mathematically as follows:

$$\begin{aligned}\mathbf{J}(\mathbf{r}') = \hat{z}Il[&A_0\delta(x') + A_1\delta(x'-d_1) + A_2\delta(x'-d_2) + \cdots \\ &+ A_{N-1}\delta(x'-d_{N-1})]\delta(y')\delta(z')\end{aligned} \tag{30.2.1}$$

Again for simplicity, all the dipoles are pointing in the z-axis. The far field can be found using the approximate formula derived in the previous lecture, namely, (29.1.3), reproduced below:

$$\mathbf{A}(\mathbf{r}) \approx \frac{\mu e^{-j\beta r}}{4\pi r} \iiint_V d\mathbf{r}'\mathbf{J}(\mathbf{r}')e^{j\boldsymbol{\beta}\cdot\mathbf{r}'} \tag{30.2.2}$$

To reiterate, the above implies that the far field is related to the Fourier transform of the current source $\mathbf{J}(\mathbf{r}')$. [3]

30.2.1 Far-Field Approximation of a Linear Array

The vector potential on the xy plane in the far field, using the sifting property of delta function, yields the following equation for $\mathbf{A}(\mathbf{r})$ using (30.2.2),

$$\mathbf{A}(\mathbf{r}) \cong \hat{z}\frac{\mu Il}{4\pi r}e^{-j\beta r}\iiint d\mathbf{r}'[A_0\delta(x') + A_1\delta(x'-d_1) + \cdots]\delta(y')\delta(z')e^{j\beta\mathbf{r}'\cdot\hat{r}} \tag{30.2.3}$$

In the above, for simplicity, we will assume that the observation point is only on the xy plane, or that $\mathbf{r} = \boldsymbol{\rho} = \hat{x}x + \hat{y}y$ where $\boldsymbol{\rho}$ is the position vector of the observation point in the xy plane. Thus, $\hat{r} = \hat{x}\cos\phi + \hat{y}\sin\phi$. Also, since the sources are aligned on the x-axis, then $\mathbf{r}' = \hat{x}x'$ and $\mathbf{r}'\cdot\hat{r} = x'\cos\phi$. Consequently, $e^{j\beta\mathbf{r}'\cdot\hat{r}} = e^{j\beta x'\cos\phi}$. By so doing, the far field of a linear array is

$$\mathbf{A}(\mathbf{r}) \cong \hat{z}\frac{\mu Il}{4\pi r}e^{-j\beta r}[A_0 + A_1e^{j\beta d_1\cos\phi} + A_2e^{j\beta d_2\cos\phi} + \cdots + A_{N-1}e^{j\beta d_{N-1}\cos\phi}] \tag{30.2.4}$$

Next, for further simplification, we let $d_n = nd$, implying an equally spaced array with distance d between adjacent elements. We next let $A_n = e^{jn\Psi}$, assuming a progressively increasing phase shift between different elements. Such an antenna array is called a *linear phase array*. Thus, in (30.2.4), we can identify the array factor $A.F.$ in the above as

$$\begin{aligned} A.F. \cong [1 &+ e^{j(\beta d\cos\phi+\Psi)} + e^{j2(\beta d\cos\phi+\Psi)} + \cdots \\ &+ e^{j(N-1)(\beta d\cos\phi+\Psi)}] \end{aligned} \tag{30.2.5}$$

With the simplifying assumptions, the above series can be summed in closed form because it is a series of the form $1 + x + x^2 + x^3 + \cdots + x^{N-1}$. Again, the aggregate pattern of the array is the product of the unit pattern of a lone antenna element multiplied by the array factor.

30.2.2 Radiation Pattern of an Array

The array factor $A.F.$ above (30.2.5) can be summed in closed form using the formula

$$\sum_{n=0}^{N-1} x^n = \frac{1-x^N}{1-x} \tag{30.2.6}$$

Then in the far field,

$$\mathbf{A}(\mathbf{r}) \cong \hat{z}\frac{\mu Il}{4\pi r}e^{-j\beta r}\frac{1-e^{jN(\beta d\cos\phi+\Psi)}}{1-e^{j(\beta d\cos\phi+\Psi)}} \tag{30.2.7}$$

[3]Again, this is an unusual kind of Fourier transform where the length of the $\boldsymbol{\beta}$ vector is fixed. Nevertheless, one can gain a lot of insight by using one's knowledge of Fourier transform to understand a number of physical phenomena. For instance, it tells us that our eyes actually see the Fourier tranform of an object, and it is lens optics that does the inverse Fourier transform for us. Furthermore, we know that the Fourier transform of an impulse train is an impulse train in our signal-processing course. We can extend this to 3D space to ascertain that the Fourier transform of sources located at lattice points become points located at the reciprocal lattice points in the Fourier space [99]. Physicists also call the Fourier space the momentum space because of the association of $\hbar\mathbf{k}$ with momentum.

Ordinarily, as shown previously in (29.1.11), $\mathbf{E} \approx -j\omega(\hat{\theta}A_\theta + \hat{\phi}A_\phi)$. But since $\mathbf{A}$ is $\hat{z}$ directed, we have $A_\phi = 0$. Furthermore, on the xy plane, $E_\theta \approx -j\omega A_\theta = j\omega A_z$. As a consequence,

$$|E_\theta| \cong |E_0| \left| \frac{1 - e^{jN(\beta d \cos\phi + \Psi)}}{1 - e^{j(\beta d \cos\phi + \Psi)}} \right|,$$

$$= |E_0| \left| \frac{\sin\left(\frac{N}{2}(\beta d \cos\phi + \Psi)\right)}{\sin\left(\frac{1}{2}(\beta d \cos\phi + \Psi)\right)} \right|, \quad \mathbf{r} \to \infty \tag{30.2.8}$$

The above can be used to plot the far-field pattern of an antenna array.

Equation (30.2.8) has an array factor that is of the form

$$\frac{|\sin(Nx)|}{|\sin x|} \tag{30.2.9}$$

This function, which appears in digital signal processing frequently, is also called the digital sinc function [218]. Again, the reason why this is so is because the far field is proportional to the Fourier transform of the current. The current in this case is a finite array of Hertzian dipole, which is the product of a box function and an infinite array of Hertzian dipole. The Fourier transform of such a current, as is well known, is the digital sinc.[4]

Plots of $|\sin(3x)|$ and $|\sin x|$ are shown as an example and the resulting $\frac{|\sin(3x)|}{|\sin x|}$ is also shown in Figure 30.4. The function peaks (also called the principal maximum) when both the numerator and the denominator of the digital sinc vanish. Their values can be found using the Taylor series, expanding both the numerator and the denominator at the point where they both vanish. This happens when $x = n\pi$ for integer n. (Such an apparent singularity is called a removable singularity.)

In equation (30.2.9), $x = \frac{1}{2}(\beta d \cos\phi + \Psi)$. We notice that the *maximum* of the array radiation pattern in (30.2.8) would occur if $x = n\pi$, or if

$$\beta d \cos\phi + \Psi = 2n\pi, \qquad n = 0, \pm1, \pm2, \pm3, \cdots \tag{30.2.10}$$

Solving the above gives the location of the maxima of the radiation pattern. The *zeros* or *nulls* of the radiation pattern will occur at $Nx = n\pi$, or

$$\beta d \cos\phi + \Psi = \frac{2n\pi}{N}, \qquad n = \pm1, \pm2, \pm3, \cdots, \quad n \neq mN \tag{30.2.11}$$

[4]All good electrical engineers should know that the Fourier transform of an infinite impulse train in the time domain is an impulse train in the frequency domain. Hence, the Fourier transform of an infinitely long array of point sources is an infinitely long array of delta functions in the spectral Fourier space. The Fourier transform of a box function is a sinc function. Hence, the Fourier transform of a finite size array of dipoles is actually the convolution of the sinc function with an infinitely long array of delta function, yielding the digital sinc.

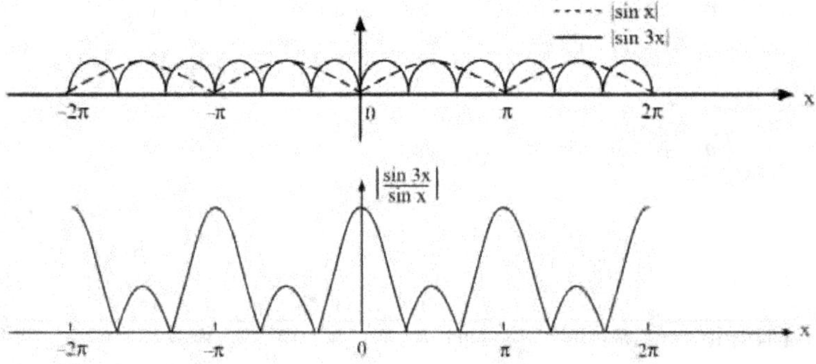

Figure 30.4: Plots of $|\sin x|$, $|\sin 3x|$ of the digital sinc, $\frac{|\sin 3x|}{|\sin x|}$. It peaks at points where both the numerator and denominator vanish.

For example,

Case I. $\Psi = 0, \beta d = \pi$, principal maxima are at $\phi = \pm\frac{\pi}{2}$. If $N = 5$, nulls are at $\phi = \pm\cos^{-1}\left(\frac{2n}{5}\right)$, or $\phi = \pm 66.4°, \pm 36.9°, \pm 113.6°, \pm 143.1°$. The radiation pattern is seen to form *lobes* of the antenna radiation pattern. The largest lobe is called the main lobe, while the smaller lobes are called side lobes. Since $\Psi = 0$, the radiated fields in the y direction are in phase and the peak of the radiation lobe is in the y direction or in the broadside direction (see Figure 30.5 for the definition of broadside and endfire). Hence, this is called a *broadside array.* The radiation pattern of such an array is shown in Figure 30.5.

Case II. $\Psi = \pi, \beta d = \pi$, principal maxima are at $\phi = 0, \pi$. If $N = 4$, nulls are at $\phi = \pm\cos^{-1}\left(\frac{n}{2} - 1\right)$, or $\phi = \pm 120°, \pm 90°, \pm 60°$. Since the sources are out of phase by 180°, and $N = 4$ is even, the radiation fields cancel each other in the broadside, but add in the x direction or the endfire direction. This is called the *endfire array.* Figure 30.6 shows the radiation pattern of such an array.

From the above examples, it is seen that the interference effects between the different antenna elements of a linear array focus the power in a given direction. We can use an antenna array to increase the directivity of antennas. Moreover, it is shown that the radiation patterns can be changed by adjusting the spacings of the elements as well as the relative phase shift between them. The idea of antenna array design is to make the main lobe of the pattern much higher than the side lobes so that the radiated power of the antenna is directed along the main lobe rather than the side lobes. So side-lobe-level suppression is an important goal of designing a highly directive antenna. Also, by changing the phase of the antenna elements in real time, the beam of the antenna can be steered in real time with no moving parts. The antenna array we have studied is of the simplest kind. For instance, the antenna element current amplitudes can be nonuniform, and we will have a nonuniform array.

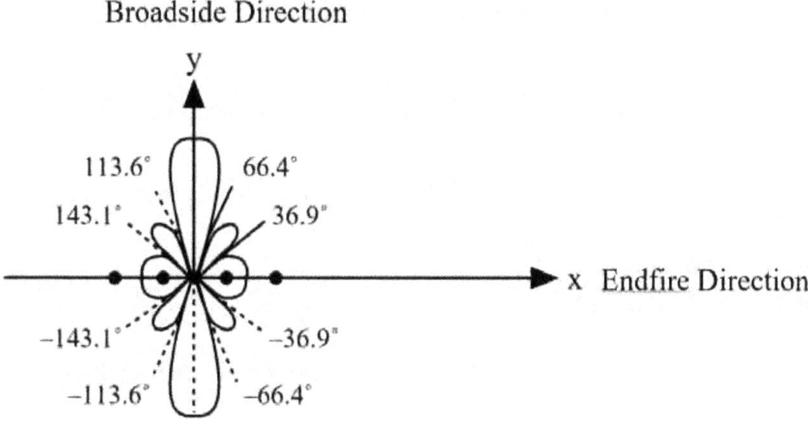

Figure 30.5: The radiation pattern of a five-element broadside array. These terms have descended from war ships. Assuming that the ship is traveling in the x direction, then the y direction is the broadside of the ship. The broadside and endfire directions of the array are also labeled.

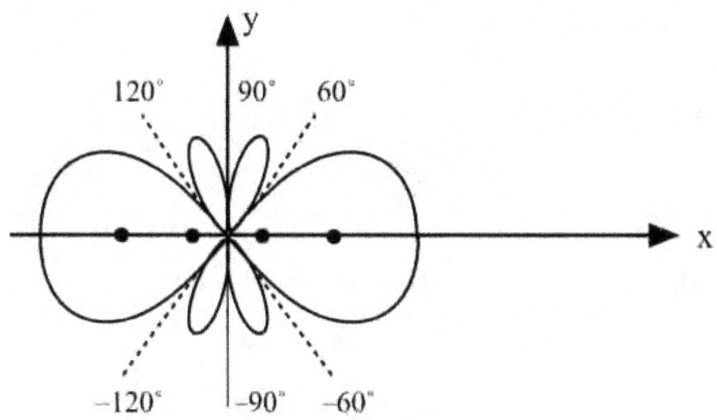

Figure 30.6: By changing the phase of the linear array, the radiation pattern of the antenna array can be changed to become an endfire array as shown.

It is to be noted that the series in (30.2.4) can be thought of as a polynomial if the array is equally spaced, and the phases for A_n properly chosen. A typical term in (30.2.4) is of the form

$$A_n e^{j\beta d_n \cos\phi} = A_n e^{j\beta n d \cos\phi} = A_n \left(e^{j\beta d \cos\phi}\right)^n \tag{30.2.12}$$

or $A_n x^n$.

Sometimes, this polynomial can be summed in closed form. These are the cases in the binomial arrays and Chebyshev arrays, and their radiation patterns can be easily found, and are amenable to engineering designs.

There have been much written about antenna arrays in recent years where an incomplete list is as follows: [219, 220, 221]

30.3 Validity of the Far-Field Approximation

In making the far-field approximation in (30.2.3), it will be interesting to ponder when the far-field approximation is valid? That is, when we can approximate

$$e^{-j\beta|\mathbf{r}-\mathbf{r}'|} \approx e^{-j\beta r + j\beta \mathbf{r}'\cdot\hat{r}} \tag{30.3.1}$$

to arrive at (30.2.2) and (30.2.3). This is especially important because when we integrate over $\mathbf{r}'$, it can range over large values especially for a large array. In this case, $\mathbf{r}'$ can be as large as $(N-1)d$. The above approximation is important also because it tells when the field generated by an array antenna becomes a spherical wave. All finite sources generate fields that eventually become spherical waves. How far does one have to go before these fields become spherical waves?

To answer this question, we need to study the approximation in (30.3.1) more carefully. First, we have

$$|\mathbf{r}-\mathbf{r}'|^2 = (\mathbf{r}-\mathbf{r}')\cdot(\mathbf{r}-\mathbf{r}') = r^2 - 2\mathbf{r}\cdot\mathbf{r}' + r'^2 \tag{30.3.2}$$

We can take the square root of the above to get

$$|\mathbf{r}-\mathbf{r}'| = r\left(1 - \frac{2\mathbf{r}\cdot\mathbf{r}'}{r^2} + \frac{r'^2}{r^2}\right)^{1/2} \tag{30.3.3}$$

The above is exact so far with no approximation. Next, we use the Taylor series expansion to get, for small x, that

$$(1+x)^n \approx 1 + nx + \frac{n(n-1)}{2!}x^2 + \cdots \tag{30.3.4}$$

or that

$$(1+x)^{1/2} \approx 1 + \frac{1}{2}x - \frac{1}{8}x^2 + \cdots \tag{30.3.5}$$

We can apply this approximation by letting

$$x \cong -\frac{2\mathbf{r}\cdot\mathbf{r}'}{r^2} + \frac{r'^2}{r^2} \tag{30.3.6}$$

in (30.3.3). To this end, we arrive at[5]

$$|\mathbf{r}-\mathbf{r}'| \approx r\left[1-\frac{\mathbf{r}\cdot\mathbf{r}'}{r^2}+\frac{1}{2}\frac{r'^2}{r^2}-\frac{1}{2}\left(\frac{\mathbf{r}\cdot\mathbf{r}'}{r^2}\right)^2+\cdots\right] \tag{30.3.7}$$

In the above, we have not kept every term of the x^2 terms by assuming that $r'^2 \ll \mathbf{r}'\cdot\mathbf{r}$, and terms much smaller than the last term in (30.3.7) can be neglected.

We can multiply out the right-hand side of the above to further arrive at

$$\begin{aligned}|\mathbf{r}-\mathbf{r}'| &\approx r-\frac{\mathbf{r}\cdot\mathbf{r}'}{r}+\frac{1}{2}\frac{r'^2}{r}-\frac{1}{2}\frac{(\mathbf{r}\cdot\mathbf{r}')^2}{r^3}+\cdots\\ &= r-\hat{r}\cdot\mathbf{r}'+\frac{1}{2}\frac{r'^2}{r}-\frac{1}{2r}(\hat{r}\cdot\mathbf{r}')^2+\cdots\end{aligned} \tag{30.3.8}$$

The last two terms in the last line of (30.3.8) are of the same order.[6] Moreover, their sum is bounded by $r'^2/(2r)$ since $\hat{r}\cdot\mathbf{r}'$ is always less than r'. Hence, using the above in (30.3.1), the far field approximation is valid if

$$\beta\frac{r'^2}{2r} \ll 1 \tag{30.3.9}$$

In the above, β is involved because the approximation has to be valid in the exponent of (30.3.1), namely, $\exp(-j\beta|\mathbf{r}-\mathbf{r}'|)$ where β multiplies $|\mathbf{r}-\mathbf{r}'|$ or its approximation. If (30.3.9) is valid, then

$$e^{j\beta\frac{r'^2}{2r}} \approx 1 \tag{30.3.10}$$

and thus, the first two terms on the right-hand side of (30.3.8) suffice to approximate $|\mathbf{r}-\mathbf{r}'|$ on the left-hand side, which are the two terms we have kept in the far-field approximation.

30.3.1 Rayleigh Distance

If we have an infinite time-harmonic current sheet in the xy plane, it can be shown that by matching boundary conditions, it will launch plane waves on both sides of the current sheet propagating in the z direction [35, p. 652].

[5]The art of making such approximation is called perturbation expansion [49].

[6]The math parlance for saying that these two terms are approximately of the same magnitude as each other.

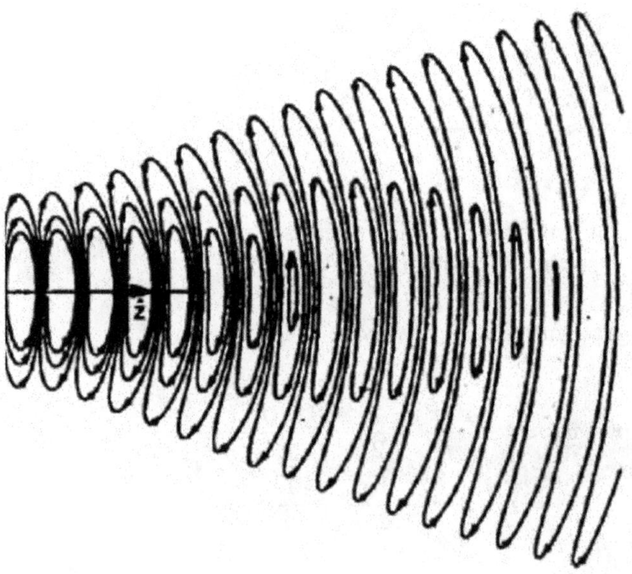

Figure 30.7: In the above figure, which is the right half of a Gaussian beam, it displays the physics of the near field, the Fresnel zone, and the far zone as one moves from the left to the right in the picture. In the far zone, the field behaves like a spherical wave (Figure by H. Haus, Electromagnetic Noise and Quantum Optical Measurements, [91]).

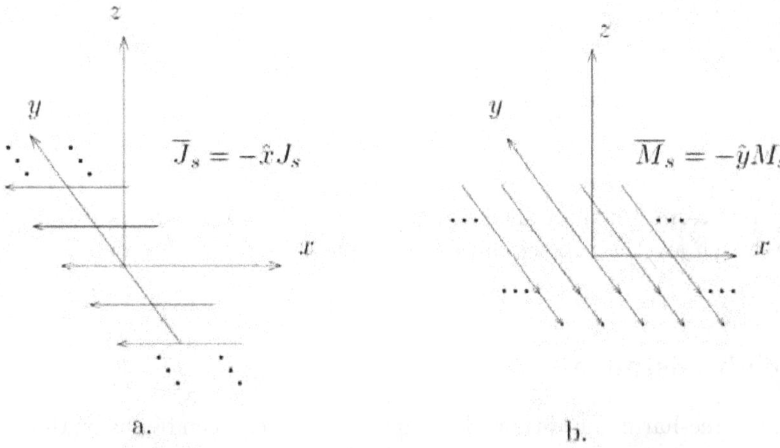

Figure 30.8: An infinite electric or magnetic current sheet, by the equivalence principle, can launch a perfect plane wave. The plane waves are propagating above and below the current sheet. [35, p. 653]

Thus, if we have an aperture antenna like the opening of a waveguide that is much larger than the wavelength, it will launch a wave that is almost like a plane wave from it. We can use Huygens's original hypothesis that the current wavefront can be predicted by putting point sources at the previous wavefront. This is also vindicated by our derivation of Huygens's principle in the previous lecture. Thus, when a wave field leaves a large aperture antenna, it can be approximately described by a plane wave, and later, by a Gaussian beam [91] (see Figure 30.7).

As seen in the above figure, near to the antenna aperture, or in the near zone, it is approximately a plane wave with wavefronts parallel to the aperture surface. Far from the antenna aperture, or in the far zone, the field behaves like a spherical wave, with its typical wavefront. In between, we are in the Fresnel zone. A Gaussian beam describes the morphing of a finite size plane wavefront to a spherical wavefront.

Consequently, after using that $\beta = 2\pi/\lambda$, for the far-field approximation to be valid, we need (30.3.9) to be valid, or that

$$r \gg \frac{\pi}{\lambda} r'^2 \tag{30.3.11}$$

If the aperture of the antenna is of radius W, then $r' < r_{\max} \cong W$ and the far-field approximation is valid if

$$r \gg \frac{\pi}{\lambda} W^2 = r_R \tag{30.3.12}$$

If r is larger than this distance, then the far-field approximation is valid, and an antenna beam behaves like a spherical wave: it starts to diverge. This distance r_R is also known as the Rayleigh distance. In other words, after this distance, the wave from a finite size source resembles a spherical wave which is diverging in all directions (see Figure 30.7). Also, notice that the shorter the wavelength λ, the larger is this distance before the far-field approximation can be made.

This also explains why a laser pointer works. A laser pointer light can be thought of as radiation from a finite size source located at the aperture of the laser pointer, as shall be shown using the equivalence theorem later. The laser pointer beam remains collimated for quite a distance before it becomes a divergent beam or a beam with a spherical wavefront.

In some textbooks [35], it is common to define acceptable phase error to be $\pi/8$. The Rayleigh distance is the distance beyond which the phase error is below this value. When the phase error of $\pi/8$ is put on the right-hand side of (30.3.9), one gets

$$\beta \frac{r'^2}{2r} \approx \frac{\pi}{8} \tag{30.3.13}$$

Using the approximation, the Rayleigh distance is defined to be

$$r_R = \frac{2D^2}{\lambda} \tag{30.3.14}$$

where $D = 2W$ is the diameter of the antenna aperture. Learning the demarcation distance between the near zone, Fresnel zone, and far zone is a concept that is important in both optics and microwaves.

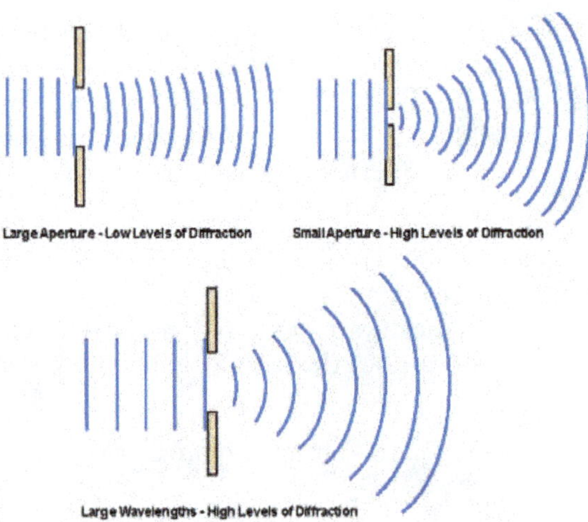

Figure 30.9: Diffraction of a plane wave by apertures of different sizes. The laser pointer works like diffraction of short wavelength plane waves by a large aperture. The beam remains like a plane wave after diffraction, and hence it is collimated for a Rayleigh distance before beam divergence sets in (Figure by imagen-estilo.com).

30.3.2 Near Zone, Fresnel Zone, and Far Zone

Therefore, when a source radiates, the radiation field is divided into the near zone, the Fresnel zone, and the far zone (also known as the radiation zone, or the Fraunhofer zone in optics). The Rayleigh distance is the demarcation boundary between the Fresnel zone and the far zone. The larger the aperture of an antenna array is, the further one has to be in order to reach the far zone of an antenna from (30.3.14). This distance becomes larger too when the wavelength is short. In the far zone, the far field behaves like a spherical wave and its radiation pattern is proportional to the Fourier transform of the current.

In some sources, like the Hertzian dipole, in the near zone, much reactive energy is stored in the electric field or the magnetic field near to the source. This near zone receives reactive power from the source, which corresponds to instantaneous power that flows from the source but is returned to the source after one time-harmonic cycle. Thus, reactive power corresponds to energy that sloshes back and forth between the source and the near field. Hence, a Hertzian dipole has input impedance that looks like that of a capacitor, because much of the near field of this dipole is the electric field.

The field in the far zone carries power that radiates to infinity. But the near field stores energy and carries reactive power that does not propagate to infinity. As a result, the field in the near zone decays rapidly. Furthermore, it sustains reactive power that needs to be exchanged with the source. On the other hand, the field in the far zone decays as $1/r$ for energy conservation. Moreover, the far field is no longer bound to the source, as it convects energy to infinity.

Exercises for Lecture 30

Problem 30-1:

(i) Go through the lecture notes and derive the expression (30.2.8).

(ii) Repeat Case II of the same lecture notes, but with $N = 5$, and plot the far field pattern of this new array.

(iii) Find the leading order approximation, up to the quadratic term, of the expression $|\mathbf{r} - \mathbf{r}'|$ when $r' \ll r$. In other words, rederive equation (30.3.8) and reconfirm the definition of the Rayleigh distance in (30.3.12). Also, derive the Rayleigh distance defined in (30.3.14).

(iv) Explain why a laser pointer beam remains collimated after the light beam has left the aperture of the laser pointer.

Lecture 31

Different Types of Antennas—Heuristics

We have studied different closed-form solutions and approximate solutions to Maxwell's equations. Examples of closed-form solutions are found in transmission lines, waveguides, resonators, and dipoles. Examples of approximate solutions are found in circuit theory and far-field approximations. These solutions offer us insights into the physical behaviour of electromagnetic fields, and also the physical mechanisms of how things work. These physical insights often inspire us to develop new designs.

Fortunately for us, Maxwell's equations are valid from subatomic lengthscales to galactic lengthscales. In vacuum, they have been validated to extremely high accuracy (see section 1.1). Furthermore, since the 1960s, many numerical solutions provided by commercial software have made it possible to solve Maxwell's equations of complex structures. This field of solving Maxwell's equations numerically is known as computational electromagnetics (CEM), which shall be discussed later in this course. Abundant commercial software is now available to solve Maxwell's equations to high fidelity. Therefore, design engineers these days do not require advanced knowledge of math and physics, and the solutions of Maxwell's equations can be obtained by learning how to use this commercial software. This is a boon to many design engineers: by running this software with cut-and-try engineering, wonderful systems can be designed. The art of electromagnetic design using simulation before the actual hardware is made is known as virtual prototyping. They have replaced expensive cut-and-try experiments.

It used to be said that if we lock 100 monkeys in a room and let them punch at 100 keyboards, they will never type out Macbeth nor Hamlet. But with 100 engineers trained with good physical insight, when locked up in a room with commercial software, with enough time and patience, they can come up with wonderful designs of different electromagnetic systems. Antenna design now is mainly driven by heuristics and cut-and-try engineering for virtual prototyping. Therefore, we will discuss the functions of different antennas heuristically in this lecture.

31.1 Resonance Tunneling in Antennas

We realized the power of resonance enhancement when we were young by playing on a swing in the park. By pumping the swing at its resonance frequency, we can cause it to swing at a large amplitude without a Herculean effort. A simple antenna like a short dipole behaves like a Hertzian dipole with an effective length. A short dipole has an input impedance resembling that of a capacitor. Hence, it is difficult to drive current into the antenna unless other elements are added. Hertz was clever by using two metallic spheres attached to the stem of the dipole to increase the current flow. A large current flow on the stem of the antenna makes the stem resemble an inductor. Thus, the end cap capacitances and the stem inductance together act like a resonator, enhancing the current flow on the antenna.

Some antennas are deliberately built to resonate with their structure to make **J** larger to enhance their radiation. A half-wave dipole is such an antenna, as shown in Figure 31.1 [211], [59, Fig. 11-30]. These antennas use resonance tunneling to increase the currents on them, and thus to enhance their radiation efficiencies. A half-wave dipole can also be thought of as a transmission line flared open in order to make it radiate. The transmission line can be gradually morphed from a quarter-wavelength transmission line, as shown in Figure 31.1. A transmission line is a poor radiator because the electromagnetic energy is trapped between two pieces of metal. But a flared transmission line can radiate its field to free space more easily. (The dipole antenna, though a simple device, has been extensively studied by King [222].[1])

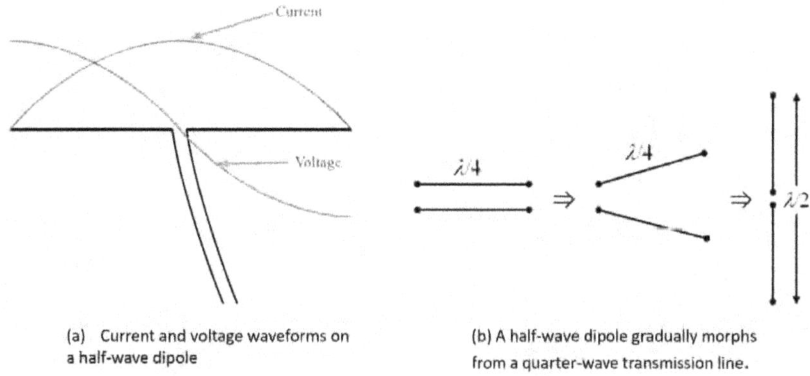

Figure 31.1: A half-wave dipole can be thought of as a resonator with radiation loss. It can be thought of as a quarter-wavelength transmission line that is gradually opened up or flared (Figure by: electronics-notes.com).

One can also think of a long piece of a wire as a waveguide. It is called a *Goubau line*, as shown in Figure 31.2, which can be thought of as the limiting case of a coaxial cable where the outer conductor is gradually moved to be infinitely far away [140]. The wave is weakly guided since it

[1]He is reputed to have graduated more than 100 PhD students studying the dipole antenna.

now can shed energy to infinity. The behavior of a wire as a Goubau line waveguide can be used to explain heuristically why a half-wave dipole resonates when it is about half a wavelength.

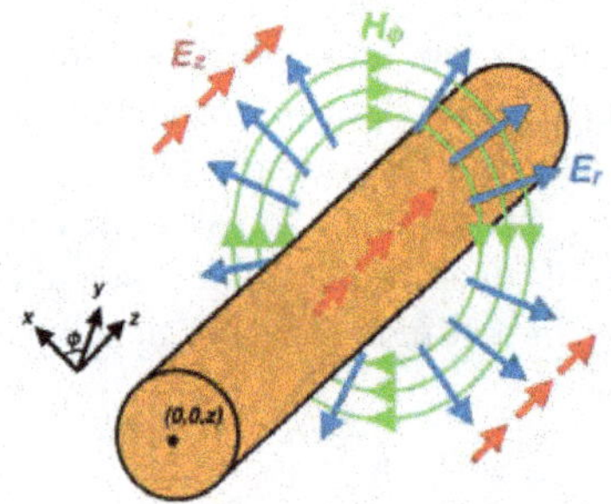

Figure 31.2: The electromagnetic field around a Goubau line (Figure by B. Vaughn and D. Peroulis, "An updated applied formulation for the Goubau transmission line," Journal of Applied Physics, vol. 126, no. 19, p. 194902, 2019). The field resembles that of a coaxial line with the outer conductor (or ground) gradually moved to infinity. It can be thought of as a quasi-transmission line. When applied to a half-wave dipole, the mode is quasi-guided on the wire and is torn away from the wire antenna as it reaches its ends. Then it is launched into space at the end of the line.

Figure 31.3: A Yagi-Uda antenna was invented by heuristics with great physical insight in 1926 even before we had computers. The principal element of the antenna is the folded dipole with four times the radiation resistance of a half-wave dipole. When an array of wire dipole antennas, each of which is less than half a wavelength, the array acts as a waveguide, or a director. When the wire antenna is slightly more than half a wavelength, the wire dipole ceases to be a waveguide, and it acts as a reflector [223]. Therefore, the antenna radiates predominantly in one direction (Figure by: https://www.wikiwand.com).

A folded dipole is often used to alter the input impedance of a dipole antenna [224]. Even though it can have a resonant frequency lower than that of a normal dipole, the lowest resonant mode does not radiate well due to destructive cancelation. The mode that radiates well has the same resonant length as an unfolded dipole. It has a radiation resistance about four times that of a half-wave dipole of similar length, which is about 300 ohms. This is equal to the characteristic impedance of a twin-lead transmission line [225]. Figure 31.3 shows a Yagi-Uda antenna driven by a folded dipole. This antenna was very popular and adorned the roof of every household before high-frequency cable modems brought broadband signals to our homes, which happened around the mid-1990s.

A Yagi-Uda antenna is also another interesting invention. It was invented in 1926 by Yagi and Uda in Japan by plainly using physical intuition [223]. Physical intuition was a tool of the engineers of yesteryears, while modern engineers prefer to use sophisticated computer-aided design (CAD) software. Nevertheless, physical intuition is still important. The principal driver element of the antenna is the folded dipole. Surprisingly, the array of dipole elements, whose lengths are slightly less than a half wavelength, in front of the driver element are acting like a waveguide in space, while the sole element at the back, slightly larger than a half wavelength, acts like a reflector. Therefore, the field radiated by the driver element will be directed toward the front of the antenna. Thus, this antenna has higher directivity than just a stand-alone dipole. Due to its simplicity, this antenna has been made into nanoantennas which operate at optical frequencies [226].

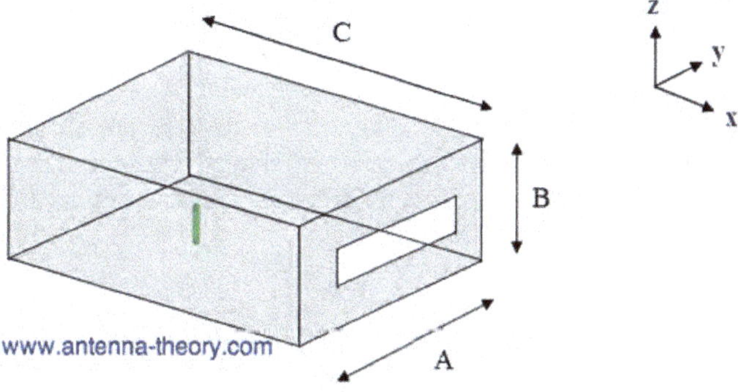

Figure 31.4: A cavity-backed slot antenna radiates well using the physics of resonant tunneling. When the small dipole radiates close to the resonant frequency of the cavity, the field strength is strongly enhanced inside the cavity, and hence around the slot. This makes the slot into a good radiator (Figure by antenna-theory.com).

A slot antenna is a simple antenna to make [227]. To improve the radiation efficiency of a slot antenna, it is made to radiate via a cavity. A cavity-backed slot antenna that uses such a concept is shown in Figure 31.4. A small dipole with poor radiation efficiency is placed inside the cavity. When the operating frequency is close to the resonant frequency of the cavity, the field strength inside the cavity becomes very strong, and much of the energy can leak from the cavity via the slot on its side. This makes the antenna radiate more efficiently into free space compared to just the small dipole alone, due to the physics of resonant tunneling.

Another antenna that resembles a cavity-backed slot antenna is the microstrip patch antenna (or just called a patch antenna). This is shown in in Figure 31.5. This antenna also radiates efficiently by resonant tunneling. Roughly, when L (see the left side of Figure 31.5) is half a wavelength, the patch antenna resonates. It is similar to the resonant frequency of a transmission line with open circuits at both ends. The current sloshes back and forth across the length of the patch antenna along the L direction. The fringing electric fields at the two ends can be thought of as equivalent magnetic currents that radiate in phase in the direction normal to the patch.

The second design (see the right side of Figure 31.5) has an inset feed. It allows the antenna to resonate at a lower frequency because the current has to go through a tortuous path, and has a longer path to slosh through when it is at resonance. Then it resonates at a lower frequency, meaning that the antenna can be made smaller.

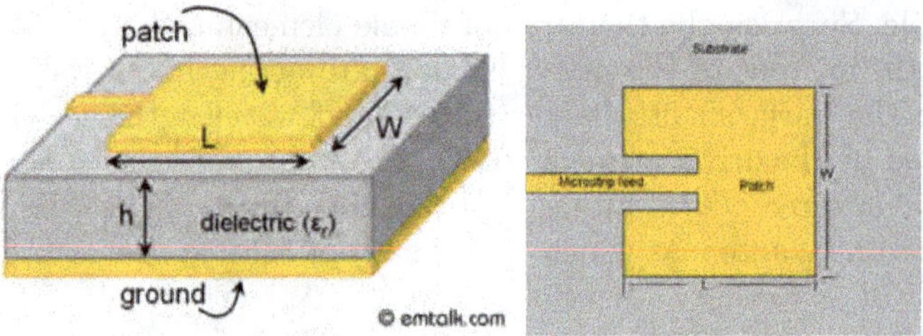

Figure 31.5: A microstrip patch antenna also radiates well when it resonates. The patch antenna resembles a cavity resonator with a magnetic wall [228]. Again, it uses the physics of resonant tunneling to enhance its radiation efficiency. The notch increases the length of the current flow and reduces the resonant frequency of the antenna (Figure by emtalk.com).

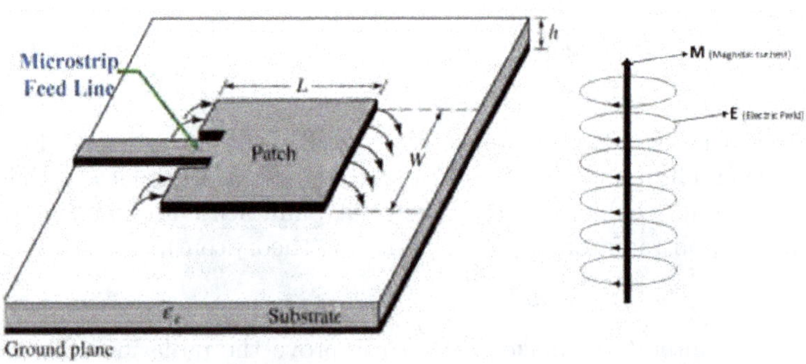

Figure 31.6: A microstrip patch antenna showing fringing electric fields at the two apertures. The fringing electric fields can be thought of as equivalent magnetic currents. The magnetic currents have the same E fields, and they can be used to model the radiation of the patch antenna (Figure by everthingRF.com).

31.2 Horn Antennas

The impedance of free space is 377 ohms while that of most transmission lines is about 50 ohms. This impedance mismatch can be mitigated by using a flared horn (see Figure 31.7) [229]. The gradual transition region allows the wave to travel from a region of low impedance to a region of high impedance with reduced reflection.

One can think that the characteristic impedance is $Z_0 = \sqrt{L/C}$ of a transmission line if it is made of two pieces of metal. As the horn flares, C becomes smaller, increasing its characteristic impedance to get close to that of free space, which is about 377 Ohms. This allows for better impedance matching from the source to free space. It is similar to the quarter-wave transformer for matching the characteristic impedance Z_0 of a line to a load with impedance Z_L. The requirement is that the quarter-wave transformer has an impedance given by $Z_T = \sqrt{Z_0 Z_L}$, which is the geometrical mean of the two impedances.

A corrugated horn in a circular waveguide, as we have discussed previously in section 24.1.1, discourages current flows in the nonaxial symmetric modes. The reason is that the axially symmetric modes have only circumferential currents while the nonaxial symmetric modes have axial currents. The corrugation impedes the flow of axial currents, and hence, discourages the propagation of the nonaxial symmetric modes. On the contrary, it encourages the propagation of the axially symmetric TE_{01} mode in the circular waveguide or the circular horn antenna. Because this mode is axially symmetric, this antenna can radiate fields that are axially symmetric, giving rise to an axially symmetric radiation pattern [230, 231, 232].

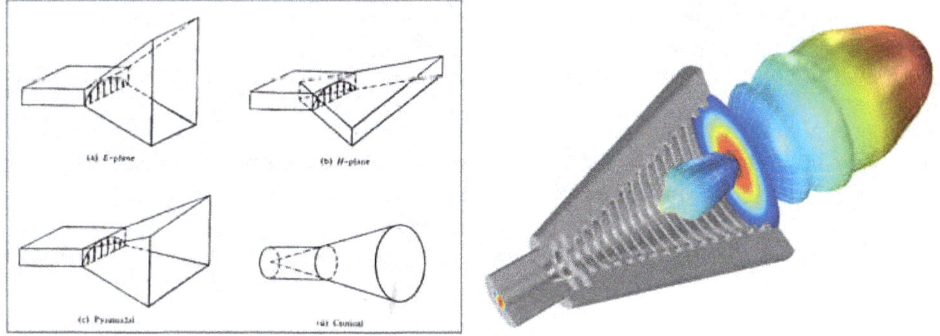

Figure 31.7: A horn antenna works under the same principle as the biconical antenna. Its flared horn changes the waveguide impedance to match the impedance of a waveguide to the impedance of free space. The right figure is that of a corrugated circular horn antenna. The corrugation enhances the propagation of the TE_{01} mode in the circular waveguide, and thus it enhances the cylindrical symmetry of the mode and the radiation field (Figure by: tutorialpoints.com and comsol.com).

A Vivaldi antenna (invented by P. Gibson in 1978 [233]) is shown in Figure 31.8.[2] It is also called a notched antenna, and the two pieces of metal act as a coplanar waveguide (see [96][p. 4]). It works by the same principle to gradually match the impedance of the source to that of free space. But such a gradually flared coplanar waveguide (flared horn) has the element of a frequency-independent antenna. The low-frequency component of the signal will radiate from the wide end of the flared notch, while the high frequency component will radiate from the narrow end of the notch. Thus, this antenna can radiate effectively over a broad range of frequencies, giving the antenna a broad bandwidth performance. It is good for transmitting a pulsed signal, which has a broad frequency spectrum.

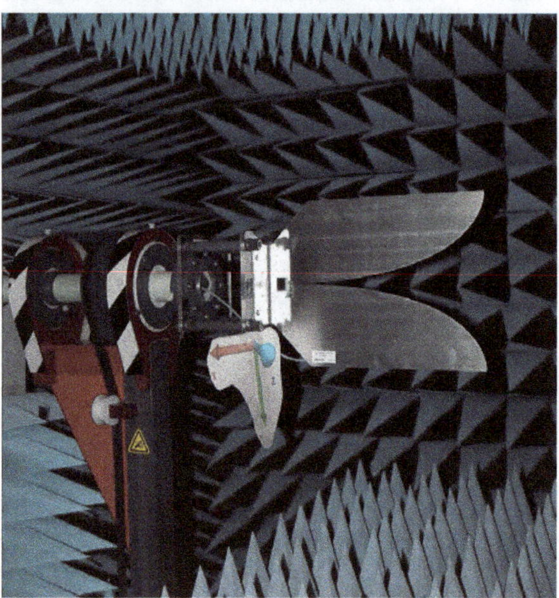

Figure 31.8: A Vivaldi antenna, also called a notched antenna, works like a horn antenna, but uses very little metal and is lightweight. Hence, it is cheap to build, and its flared notch makes it broadband. It is broadband because the high frequency signals can radiate from the narrow part of the notch, and the lower-frequency signals can radiate from the wider part of the notch (Figure by: https://en.wikipedia.org).

31.3 Quasi-Optical Antennas

High-frequency or short-wavelength electromagnetic fields behave like light rays in optics. Therefore, many high-frequency antennas are designed based on the principle of ray optics. A reflector antenna is such an antenna as shown in Figure 31.9. The reflector antenna in this case is a Cassegrain design [234][3] where a subreflector is present. This allows the antenna to be fed from

[2]He must have loved the musician Vivaldi so much.:)

[3]The name came from an optical telescope of similar design [235]

behind the parabolic dish, where the electronics can be stored and isolated as well. Reflector antennas [236] are prevalent in radio astronomy and space exploration due to the high directivity and sensitivity needed for low signals. Moreover, due to their large size compared to wavelength, they have a large effective aperture or area.

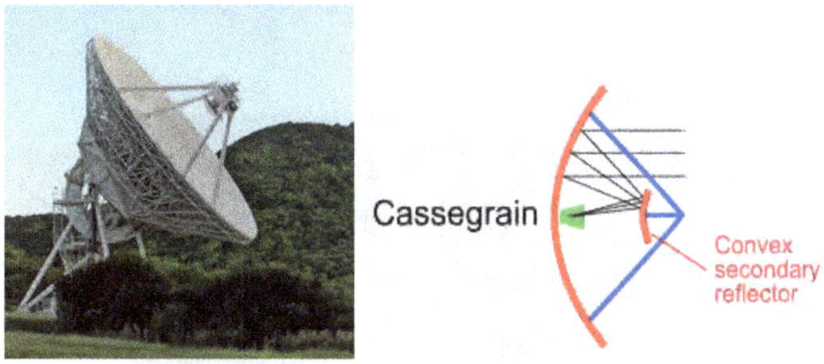

Figure 31.9: The left picture of an NRAO (National Radio Astronomy Observatory) radio telescope antenna of Cassegrain design in Virginia, USA (Figure by: Britannica.com). The right is the detail of the Cassegrain design (Figure by: rev.com). Its design is inspired by ray optics. In fact, the Cassegrain design was first used in an optical telescope.

Another recent invention is the reflectarray antenna [237, 238] which is very popular. One of them is shown in Figure 31.10. Due to recent advances in simulation technology, complicated structures can be simulated on a computer, including one with a complicated surface design. Patch elements can be etched onto a flat surface as shown, giving it an effective impedance that is spatially varying, making it reflect like a curved surface. Such a surface is known as a metasurface [239, 240]. Its flat structure can greatly economize on the space usage compared to a reflector antenna.

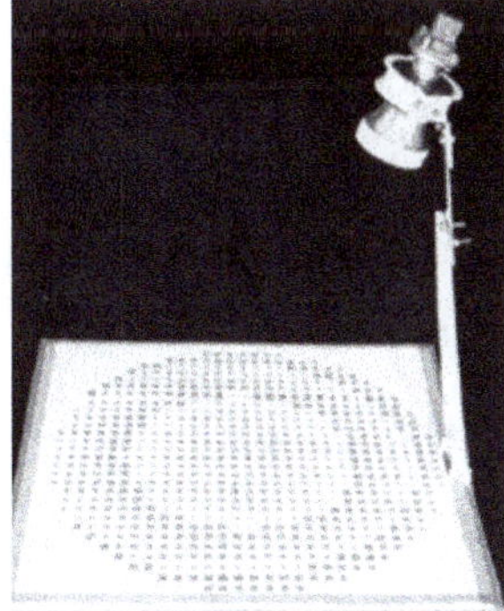

Figure 31.10: A reflectarray where the reflector is a flat surface. Patches are unequally spaced to give the array the focusing effect. No closed form solution exists for such a reflectarray, but due to advancement in computational electromagnetics (CEM), it can be simulated or virtual-prototyped in a computer (Figure by: antenna-theory.com).

Another quasi-optical antenna is the lens antenna as shown in Figure 31.11 [241]. The design of this antenna follows lens optics and is only valid when the wavelength is very short compared to the curvature of the surfaces. In this case, reflection and transmission of a curved surface are similar to that of a flat surface. This is called the tangent-plane approximation of a curved surface, and is valid at high frequencies.

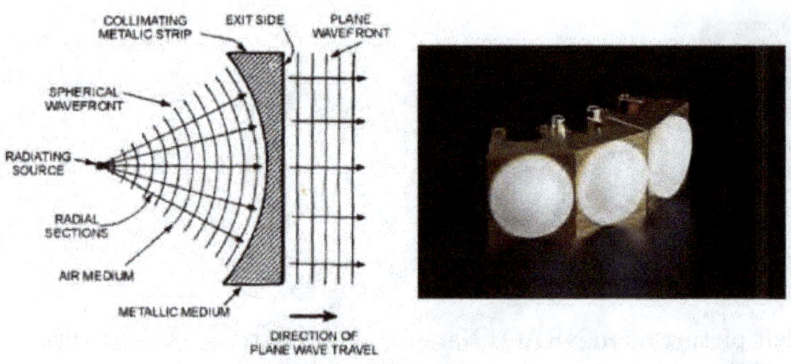

Figure 31.11: The left figure shows a lens antenna where the lens is made of artificial dielectrics made from metallic strips (Figure by: electriciantutoring.tpub.com). The right figure shows some dielectric lenses at the aperture of an open waveguide to focus the microwave exiting from the waveguide opening (Figure by: micro-radar.de). When the wavelength is much smaller than the size of the structure, a wave behaves like a ray, and quasi-optical concepts can be used to design the antenna.

31.4 Small Antennas

Small antennas are in vogue these days due to the advent of the cell phone, and the importance of economizing on the antenna size due to miniaturization requirements. Also, the antennas should have enough bandwidth to accommodate the signals from different cell phone companies, which use different carrier frequencies. An interesting small antenna is the PIFA (planar inverted F antenna because it is shaped like an F) shown in Figure 31.12 [242]. Because it is shorted at one end and open circuit at the other end, it acts like a quarter wavelength resonator, making it substantially smaller. But the resonator has a low Q because of the "slots" or "openings" around it from which energy can leak, giving rise to radiation loss. The low Q gives this antenna a broader bandwidth. Because it is shorted at one end, its driving point impedance can be changed by altering the feed location.

An interesting small antenna is the U-slot antenna shown in Figure 31.13 [243, 244]. Because the current is forced to follow a longer tortuous path by the U-slot, it can resonate with a longer wavelength (lower frequency) and hence, can be made smaller compared to wavelength. In order to give the antenna a larger bandwidth, its Q is made smaller by etching it on a thick dielectric substrate (shown as the dielectric material region in the figure). But feeding it with a longer

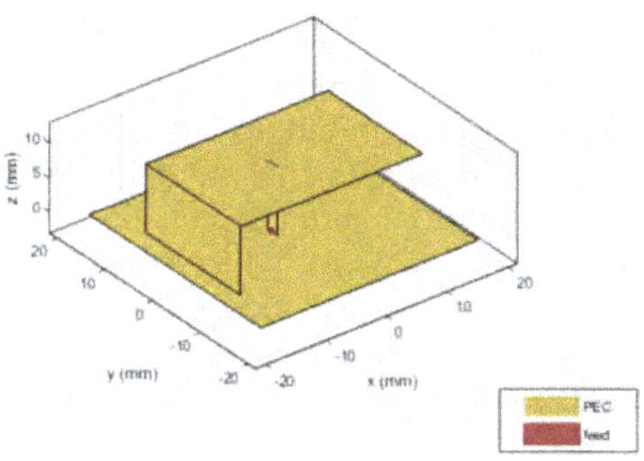

Figure 31.12: A PIFA (planar inverted F antenna) is compact, broadband, and easy to fabricate. It behaves like a quarter wavelength transmission line resonator. It is good for cell phone antennas due to its small size (Figure by: https://www.mathworks.com/help/antenna/ug/ai-for-rapid-analysis-and-design-of-microstrip-patch.html).

probe will make the bandwidth of the antenna smaller, due to the larger inductance of the probe.[4] An ingenious invention is to use an L probe [245]. The L probe has an inductive part as well as a capacitive part. Their reactances cancel each other, allowing the electromagnetic energy to "tunnel" through the antenna, making it a better radiator.

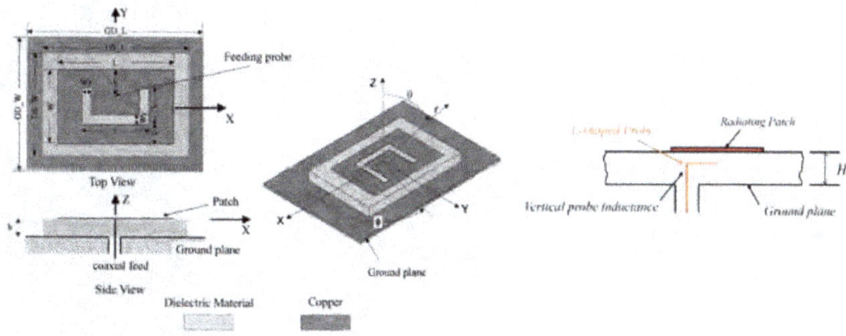

Figure 31.13: The left figure shows a U slot patch antenna design. The right figure shows a patch antenna fed by an L probe with significant increase in bandwidth. The capacitive coupling of the L probe together with its inductance could form an LC tank circuit to help with the impedance matching of the antenna (Figure by K.M. Luk) [245].

Another area where small antennas are needed is in RFID (radio frequency identification) tags [246].

[4]Remember that larger inductance implies more stored magnetic field energy, and, hence, the higher Q of the system.

Since tags are placed outside the packages of products–for example, in a warehouse–an RFID tag has a transmit-receive antenna that can communicate with the external world.[5] The communication is done through an RFID reader. The RFID reader can talk to a small computer chip embedded in the tag where data about the package can be stored. Thus, an RFID reader can quickly and remotely communicate with the RFID tag to retrieve information about the package. Such a small antenna design for an RFID tag is shown in Figure 31.14. It uses image theory (which we shall learn later) so that the antenna can be made half as small. Then slots are cut into the radiating patch so that the current follows a longer path. This lowers the resonant frequency of the antenna, allowing it to be made smaller. The take-home message here is that to make an antenna a few times smaller than a wavelength to resonate, the current on the antenna has to flow through a tortuous path. In this manner, the antenna can be made a few times smaller than the wavelength.

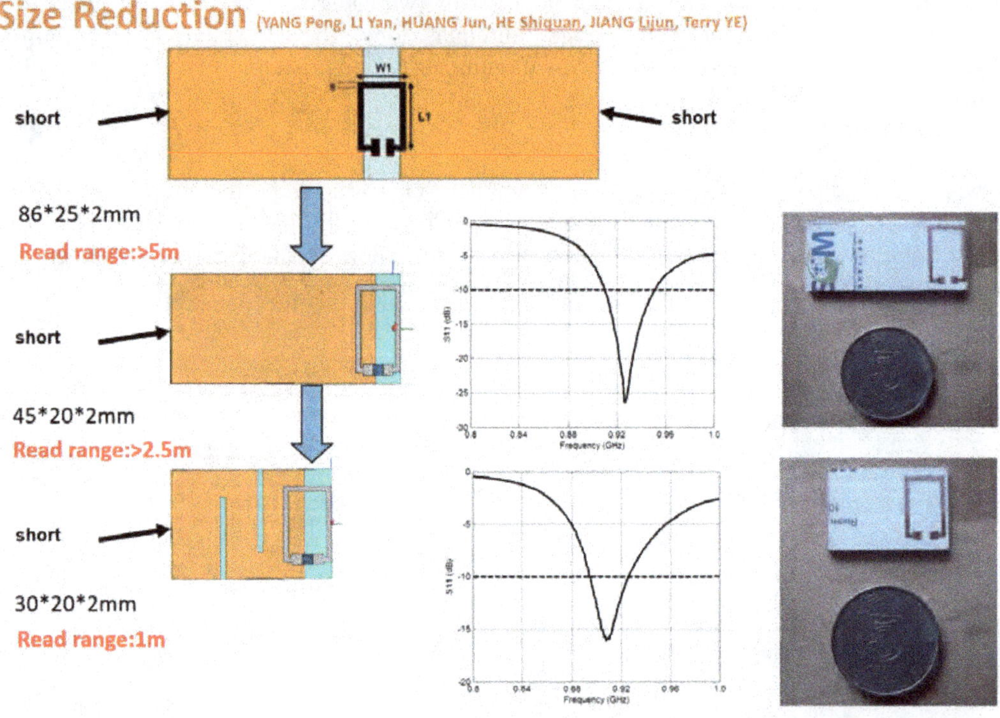

Figure 31.14: Some RFID antennas designed at the University of Hong Kong in collaboration with LSCM (Logistics and Supply Chain Multitech, R&D Center, Hong Kong). The image theory together with the use of tortuous current flow patterns are used to miniaturize the antennas (Figure by P. Yang, Y. Li, J. Huang, L.J. Jiang, S.Q. He, T. Ye, and W.C. Chew).

[5] A lower-frequency version of it is used in credit cards and ID cards.

An RFID reader can be designed to read the information from a batch of vials (or test tubes) containing different chemicals or medicine. Reading the info from a large batch of vials is important in the pharmaceutical industry. Hence, a large loop antenna is needed, but at a sufficiently high frequency (for large bandwidth). However, a loop antenna, if we look at a piece of wire as a Goubau transmission line [140], will have resonant frequencies. When a loop antenna resonates, the current is nonuniform on it just as in a transmission line. This happens at higher frequencies. (Fundamentally, this comes from the retardation effect of electromagnetic fields.) It will result in a nonuniform magnetic field inside the loop, defeating the design of the RFID reader.

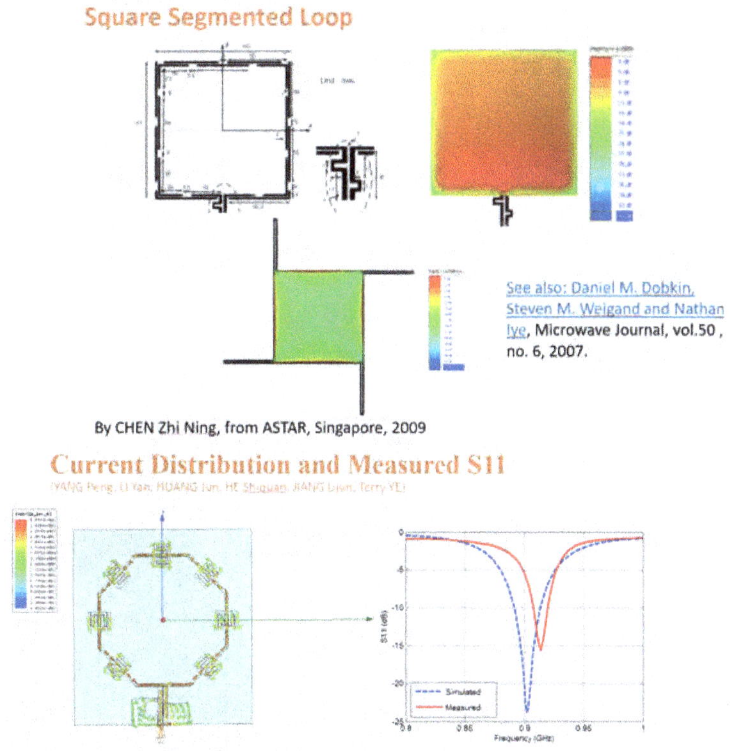

Figure 31.15: The top figure shows an RFID reader designed by [247] using a capacitively loaded inductor loop. The bottom figure shows simulation and measurement done at the University of Hong Kong and LSCM (Figure by Z.N. Chen [247] and P. Yang, Y. Li, J. Huang, L.J. Jiang, S.Q. He, T. Ye, and W.C. Chew). The series impedance is made to vanish to reduce the voltage drop across it and, hence, reduces the phase shift.

One way to view the physics of nonuniform current is that a line of wire can be viewed as a quasi-transmission line. It can be modeled with the lumped element model as shown in Figure 15.3. The series inductance in the lumped element model can be series-loaded with capacitance to reduce its series impedance, to increase the phase velocity of a wave on such a line. This will help the current around a loop to be equi-phase and hence more uniform [248]. In this way, the voltage and phase are equalized between points on the loop and become uniform across the loop. Therefore, one way to enable a uniform current in a large loop is to capacitively load the loop. This will ensure a constant phase, or a more uniform current around the loop, and hence, a more efficient reader. Such a design is shown in Figure 31.15.

Exercises for Lecture 31

Problem 31-1:

(i) Explain why a folded dipole has a radiation resistance four times that of an unfolded dipole **(Hint: The answer can be found on wikipedia. But please don't regurgitate it; understand the logic before you provide an answer.)**

(ii) Explain how a cavity-backed slot antenna works.

(iii) Explain why the corrugated horn antenna produces an axially symmetric radiation pattern.

(iv) Explain how a lens antenna works.

(v) Explain why a PIFA is smaller than a half-wave dipole.

(vi) Explain how you would make the current uniform on a large loop antenna. Use the lumped element model and show that the phase velocity can go to infinity. You can use (15.2.11) to find the phase velocity of a capacitively loaded transmission line, and explain how you would choose the capacitive loading so that the phase velocity is infinite. You can assume the lossless case so that R and G in (15.2.11) are zero.

Lecture 32

Shielding, Image Theory

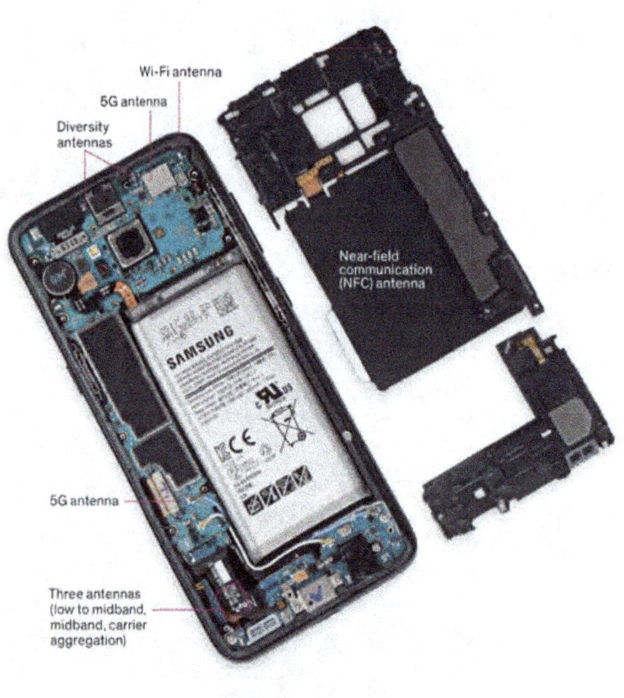

Figure 32.1: Because cell phone components are very compact, EMC/EMI (electromagnetic compatibility/electromagnetic interferece) knowledge is instrumental in the design of cell phones. Clever use of shielding is necessary to prevent interferences between different components. Compatibility means that when each component works well in isolation, the components continue to work well when brought together (Figure By: https://spectrum.ieee.org/building-smartphone-antennas-that-play-nice-together).

The physical mechanism of electromagnetic shielding and electromagnetic image theory (also called image theory) go hand in hand. They work by the moving of charges around so as to cancel the impinging fields. By understanding simple cases of shielding and image theory, we can gain enough insight to solve some real-world problems. For instance, the art of shielding is very important in the fields of electromagnetic compatibility (EMC) and electromagnetic interference (EMI). In the modern age where we have more electronic components working side by side in a very compact environment–for example inside a cell phone (see Figure 32.1)–EMC and EMI become an increasingly challenging issue. Due to the complexity of these problems, they have to be solved using heuristics with a high dosage of physical insight and experience.

32.1 Shielding

We can understand shielding by understanding how electric charges move around in a conductive medium. They move around to shield out the electric field, or cancel the impinging field inside the conductor. There are two cases to consider: the static case and the dynamic case. The physical arguments needed to understand these two cases are very different. Moreover, since there are no magnetic charges around, the shielding of a magnetic field is very different from the shielding of an electric field, as shall be seen below.

32.1.1 A Note on Electrostatic Shielding

We begin with the simple case of electrostatic shielding. For electrostatic problems, a conductive medium suffices to produce surface charges that shield out the electric field from the conductive medium. If the electric field is not zero, then since $\mathbf{J} = \sigma\mathbf{E}$, the electric current inside the conductor will keep flowing. The current will produce charges on the surface of the conductor to cancel the impinging field until inside the conductive medium, $\mathbf{E} = 0$. In this case, electric current ceases to flow in the conductor. This is the quiescent limit.

In other words, when the field reaches the quiescent state, the charges have to redistribute themselves so as to shield out the electric field, and that the total internal electric field, $\mathbf{E} = 0$ at equilibrium. And from Faraday's law that tangential $\mathbf{E}$ field is continuous, then $\hat{n} \times \mathbf{E} = 0$ on the conductor surface since $\hat{n} \times \mathbf{E} = 0$ inside the conductor. Hence, the electric field has to be normal to the conductor surfaces. Figure 32.2 shows the static electric field, in the quiescent state, between two conductors (even though they are not PECs). Moreover, since $\mathbf{E} = 0$ inside the conductor, $\nabla\Phi = 0$ implying that the potential is a constant inside a conductor at equilibrium.

32.1.2 Relaxation Time

The time it takes for the charges to move around until they reach their quiescent distribution such that $\mathbf{E} = 0$ is called the relaxation time. It is very similar to the RC time constant of an RC circuit consisting of a resistor in series with a capacitor (see Figure 32.3). It can be proven that this relaxation time τ is related to ε/σ, but the proof is beyond the scope of this course at this point. It is worthwhile to note that this constant has the same unit as the RC time constant of an RC circuit where a charged capacitor relaxes as $\exp(-t/\tau)$ where the relaxation time $\tau = RC$. Note that when $\sigma \to \infty$, the relaxation time is zero. In other words, in a perfect conductor or a superconductor, the charges reorient themselves instantaneously if the external field is time-varying so that $\mathbf{E}(t) = 0$ always inside the conductor.[1]

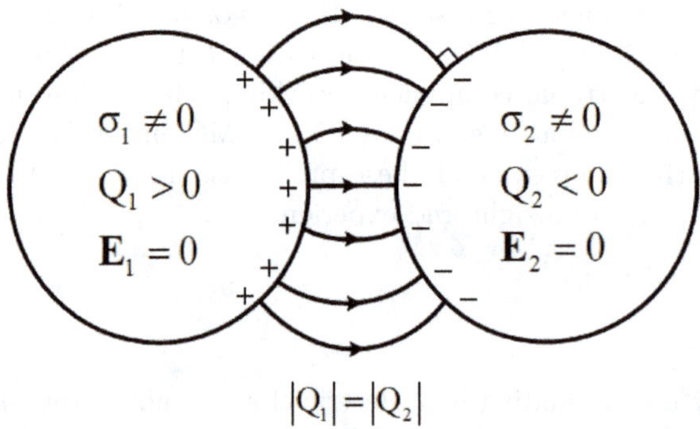

Figure 32.2: The objects can just be conductors, and in the quiescent state (static state), the tangential electric field will be zero on their surfaces. Also, $\mathbf{E} = 0$ inside the conductor, or $\nabla\Phi = 0$, or Φ is a constant inside.

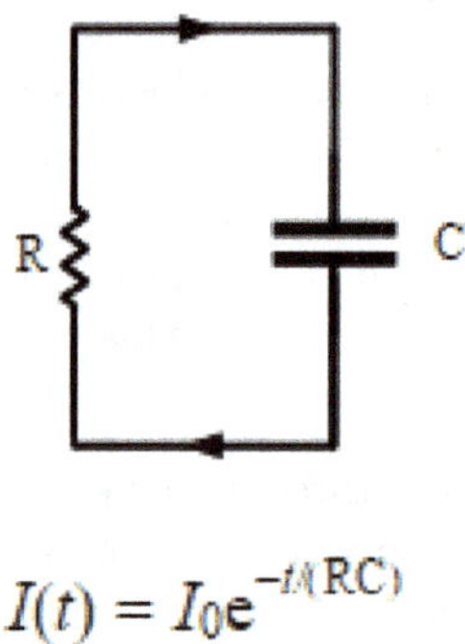

Figure 32.3: The relaxation (or disappearance of accumulated charges) in a conductive object is similar to the relaxation of charges from a charged capacitance in an RC circuit as shown.

Electrostatic shielding or low-frequency shielding is important at low frequencies. The Faraday cage or Faraday shield is an important application of such a shielding (see Figure 32.4). By grounding the Faraday cage, the potential inside the cage is set to zero[2] [249]. Earnshaw's theorem says that for Laplaces's problems, the maximum or minimum of a region V bounded by the surface S has to

[1]Note that a PEC is a fictitious medium and the closest real-world equivalent to it is the superconductor.

[2]Since potential is a relative concept, we can set any constant potential to zero. But in electrical engineering, it is customary to assume the ground potential to be zero.

be on the surface S. Thus, for a volume V bounded by a surface S with equipotential, the potential inside the volume V has to be a constant. Thus, the electric field $\mathbf{E} = -\nabla\Phi$ has to be zero.

Figure 32.4: Faraday cage demonstration on volunteers in the Palais de la Découverte in Paris (Figure by: https://en.wikipedia.org). When the cage is grounded, the potential at the surface of the cage is zero. By the solution to Laplace's equation, the potential inside the cage is a constant. Hence, the electric field inside the cage, proportional to $\nabla\Phi$, is zero. Charges will surge from the ground to the cage surface to ensure zero potential inside the cage. Therefore, a grounded Faraday cage effectively shields the external fields from entering the cage.

However, if the conductor charges are induced by an external electric field that is time varying, then the charges have to constantly redistribute/reorient themselves to try to shield out the incident time-varying electric field. Currents have to be constantly flowing around the conductor. Then the electric field cannot be zero inside the conductors as shown in Figure 32.5. In other words, an object with finite conductivity cannot shield out completely a time-varying electric field. It can be shown that the depth of penetration of the field into the conductive object is about a skin depth $\delta = \sqrt{2/(\omega\mu\sigma)}$. The lower the frequency ω or the conductivity σ, the larger the penetration depth.

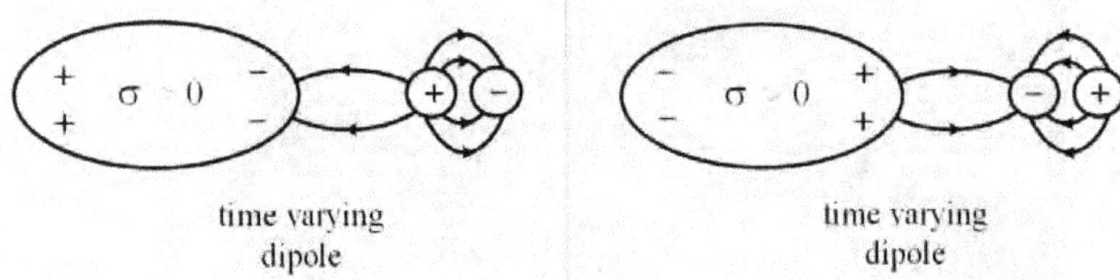

Figure 32.5: If the source that induces the charges on the conductor is time varying, the current in the conductor is always nonzero because the charges have to move around to respond to the external time-varying charges. The two figures above show the orientation of the charges for two snapshots in time. In other words, a time-varying field can penetrate the conductor to approximately within a skin-depth $\delta = \sqrt{2/(\omega\mu\sigma)}$ from a plane-surface/plane-wave approximation.

Thus, for the PEC, the charges can reorient themselves instantaneously on the surface when the inducing (incident or impinging) electric fields from outside are time varying. In other words, the relaxation time $\tau = \varepsilon/\sigma$ is zero. As a consequence, the time-varying electric field $\mathbf{E}$ is always zero inside PEC, with $\hat{n} \times \mathbf{E} = 0$ on the surface of the PEC, even for time-varying fields.

It is to be noted that when a point charge density is placed in a lossy conductive medium, the closed-form solution for the potential exists in the time-domain as well [1, p. 232-235]. In fact, $\tau = 0$ when $\sigma \to \infty$, implying that a point charge density disappears instantly when placed in a PEC medium. But this is an idealization. As we have seen from the Drude-Lorentz-Sommerfeld model, nothing is truly instantaneous in the real world!

32.2 Image Theory

The image theory here in electromagnetics is quite different from that in optics. As mentioned before, when the frequency of the fields is high, the waves associated with the fields can be described by rays. Therefore, ray optics can be used to solve many high-frequency problems. We can use ray optics to understand how an image is generated in a mirror. But the image theory in electromagnetics is quite different from that in ray optics.

Image theory can be used to derived closed-form solutions to boundary value problems when the geometry is simple with a lot of symmetry, especially for electrostatics and magnetostatics. These closed-form solutions in turn offer physical insight into more complicated problems. This theory or theorem is also discussed in many textbooks [2, 47, 59, 70, 87, 107, 250].

32.2.1 Electric Charges and Electric Dipoles

Image theory for a flat conductor surface or a half-space is quite easy to derive. To see that, we can start with electrostatic theory of putting a positive charge above a flat plane. As mentioned before, for electrostatics, the plane or half-space does not have to be a perfect conductor, but only a conductor (or a metal). From the discussion in the previous section 32.1.1, the tangential static electric field on the surface of the conductor has to be zero.

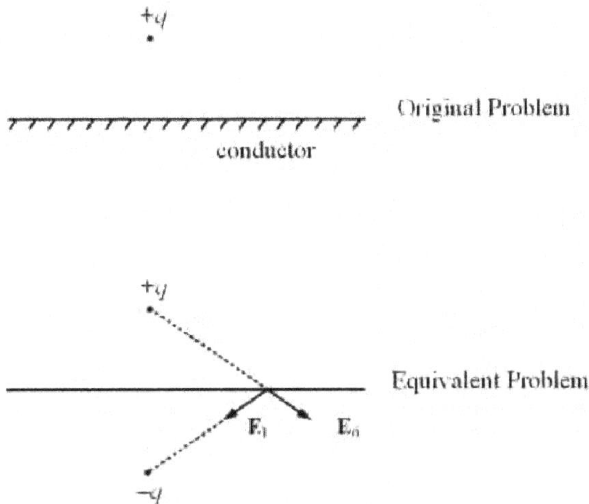

Figure 32.6: The use of image theory to solve the BVP (boundary value problem) of a point charge on top of a conductor. The boundary condition is that $\hat{n} \times \mathbf{E} = 0$ on the conductor surface. By placing a negative charge judiciously with respect to the original charge, by the principle of linear superposition, both of them produce a total field with no tangential component at the interface.

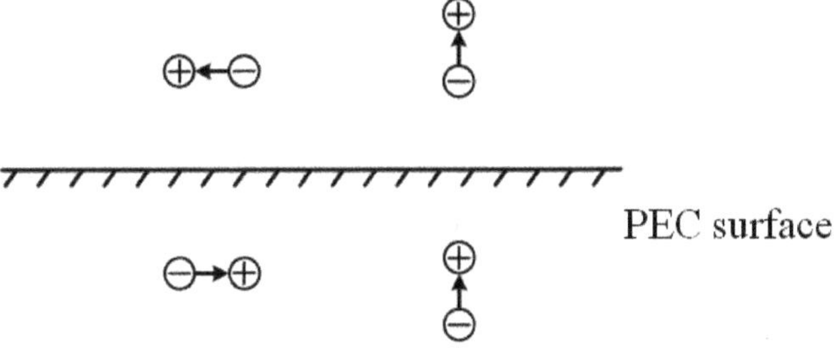

Figure 32.7: By image theory, on a conductor surface, a horizontal static dipole reflects to one of opposite polarity, while a static vertical dipole reflects to one of the same polarity. If the dipoles are time-varying, then a PEC will have a same reflection rule.

By the principle of linear superposition, the tangential static electric field can be canceled by putting an image charge of opposite sign at the symmetric mirror image location of the original charge. This is shown in Figure 32.6. Now we can mentally add the total field due to these two charges. When the total static electric field due to the original charge and image charge is sketched, it will look like that in Figure 32.8. It is seen that the static electric field satisfies the boundary condition that $\hat{n} \times \mathbf{E} = 0$ at the conductor interface due to symmetry.

An electric dipole is made from a positive charge placed in close proximity to a negative charge. Using the fact that an electric charge reflects to an electric charge of opposite polarity above a conductor, one can easily see that a static horizontal electric dipole reflects to a static horizontal electric dipole of opposite polarity. By the same token, a static vertical electric dipole reflects to static vertical electric dipole of the same polarity as shown in Figure 32.7.

If this electric dipole is a Hertzian dipole whose field is time-varying, then one needs a PEC surface to shield out the electric field. Also, the image charges will follow the original dipole charges instantaneously except for the retardation effect. Thus, the image theory for static electric dipoles over a half-space still holds true if the dipoles now become Hertzian dipoles, but over a PEC surface.

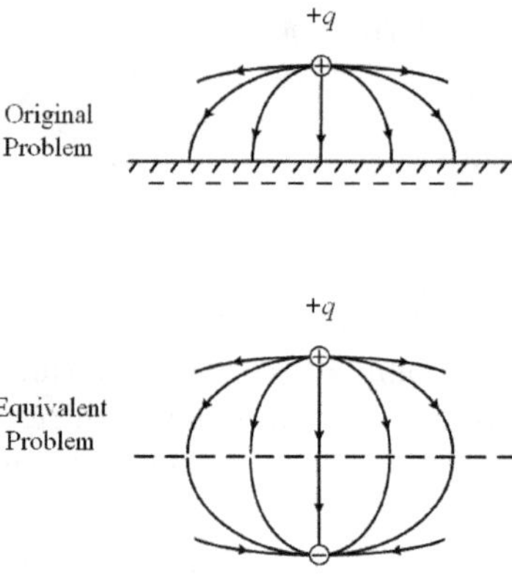

Figure 32.8: By image theory, the total electric field of the original problem and the equivalent problem when we add the total electric field due to the original charge and the image charge. It is to be noted that if we take the top half of the bottom figure, it is equivalent to the field of the top figure. The field for the bottom figure is easily obtained by superposing the fields of two opposite charges symmetrically.

32.2.2 Magnetic Charges and Magnetic Dipoles

A static magnetic field can penetrate a conductive medium. This is apparent from our experience when we play with a bar magnet over a copper sheet: the magnetic field from the magnet can still be experienced by iron filings put on the other side of the copper sheet.

However, this is not the case for a time-varying magnetic field. Inside a conductive medium, a time-varying magnetic field will produce a time-varying electric field, which in turn produces the conduction current via $\mathbf{J} = \sigma\mathbf{E}$. This is termed an eddy current, which by Lenz's law produces an opposing magnetic field that repels the magnetic field from entering the conductive medium.[3]

Now, consider a static magnetic field penetrating into a perfect electric conductor. A minute amount of time variation will produce an electric field, which in turn produces an infinitely large eddy current. So the stable state for a static magnetic field inside a PEC is for it to be expelled from the perfect electric conductor. This in fact is what we observe when a magnetic field is brought near a superconductor. Therefore, for the static magnetic field, where $\mathbf{B} = 0$ inside the PEC, then $\hat{n} \cdot \mathbf{B} = 0$ on the PEC surface (see Figure 32.9).

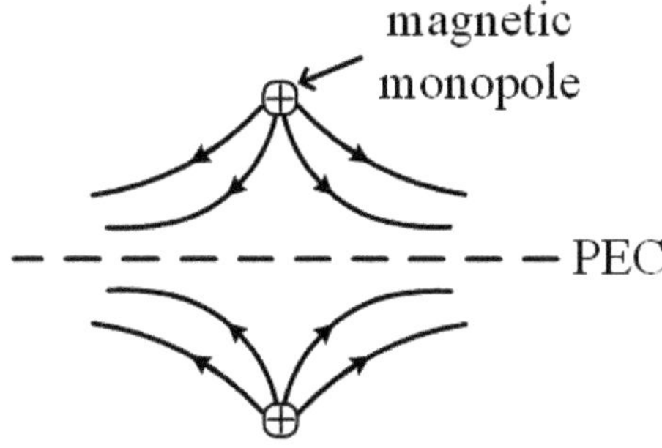

Figure 32.9: On a PEC surface, the requisite boundary condition is $\hat{n} \cdot \mathbf{B} = 0$. Hence, a static magnetic monopole on top of a PEC surface will have a magnetic field distributed as shown. By image theory, such a distribution of the **B** field can be obtained by adding a magnetic monopole of the same polarity at its image point.

Now, assuming that a magnetic monopole exists, it will reflect to itself on a PEC surface so that $\hat{n} \cdot \mathbf{B} = 0$ as shown in Figure 32.9. Therefore, a magnetic charge reflects to a charge of similar polarity on the PEC surface.

By extrapolating this to magnetic dipoles, they will reflect themselves to the magnetic dipoles as shown in Figure 32.10. A horizontal magnetic dipole reflects to a horizontal magnetic dipole of the same polarity, and a vertical magnetic dipole reflects to a vertical magnetic dipole of opposite polarity. Then, a vertical dipolar bar magnet near a superconducting half-space reflects to a vertical bar magnet of opposite polarity: it can be levitated by a superconductor half-space when this magnet is placed close to it. This is also known as the Meissner effect [251], which is shown in Figure 32.11.

A time-varying magnetic dipole can be made from an electric current loop. Over a PEC, a time-

[3]The repulsive force occurs by virtue of energy conservation. Since "work done" is needed to set the eddy current in motion in the conductor, or to impart kinetic energy to the electrons forming the eddy current, a repulsive force is felt in Lenz's law so that work is done in pushing the magnetic field into the conductive medium.

varying magnetic dipole will reflect the same way as a static magnetic dipole as shown in Figure 32.10.

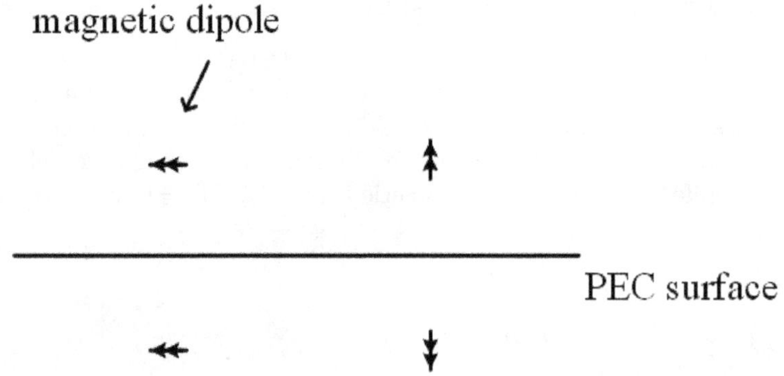

Figure 32.10: Using the rule of how a magnetic monopole reflects itself on a PEC surface, the reflection rules for magnetic dipoles can be ascertained. (Magnetic dipoles are often denoted by double arrows).

Figure 32.11: On a PEC (superconducting) surface, a static vertical magnetic dipole above it (formed by a small permanent bar magnet here) reflects to an image of a bar magnet of opposite polarity. Hence, the magnetic dipoles repel each other, displaying the Meissner effect. The magnet, because of the repulsive force from its image, levitates above the superconductor (Figure by Wikipedia).

32.2.3 Perfect Magnetic Conductor (PMC) Surfaces

Magnetic conductors do not appear naturally in this world since there are no free-moving magnetic charges around. Magnetic monopoles are yet to be discovered. On a PMC surface, by duality, $\hat{n} \times \mathbf{H} = 0$. At low frequency, it can be mimicked by a high-μ material. One can see that for magnetostatics, at the interface of a high-μ material and air, the magnetic flux is approximately normal to the surface, resembling the $\mathbf{H}$ field near a PMC surface (see Figure 32.12).

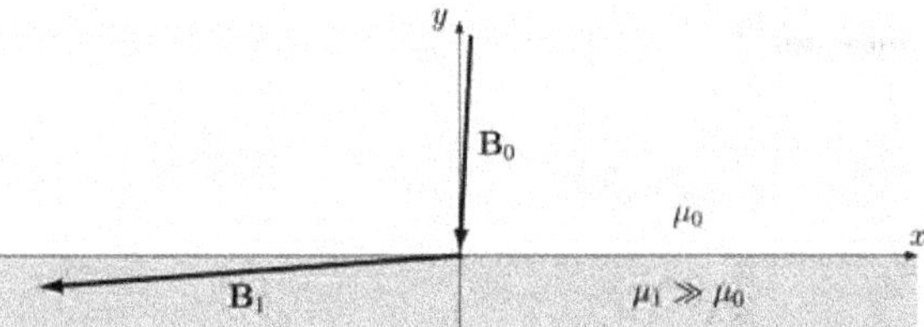

Figure 32.12: At the interface between free space and a high-μ material, one can see that magnetic field in the free-space region is almost normal to the surface in order for normal **B** and tangential **H** to be continuous. Hence, looking from the free-space side, the high-μ material resembles a PMC.

High-μ materials are hard to find at higher frequencies. Since $\hat{n} \times \mathbf{H} = 0$ on such a surface, no electric current can flow on such a surface. Hence, a PMC can be mimicked by a surface where no surface electric current can flow. This has been achieved in microwave engineering with a mushroom surface as shown in Figure 32.13 [252]. The mushroom structure, consisting of a wire and an endcap, can be thought of as forming an LC tank circuit. Close to the resonance frequency of this tank circuit, the surface of mushroom structures essentially become open circuits with little or no current flowing on the surface, or $\mathbf{J}_s \cong 0$. In other words, $\hat{n} \times \mathbf{H} \cong 0$. This resembles a PMC surface, because when no surface electric current can flow on such a surface, the tangential magnetic field is small, the hallmark of a good magnetic conductor, by using the duality principle.

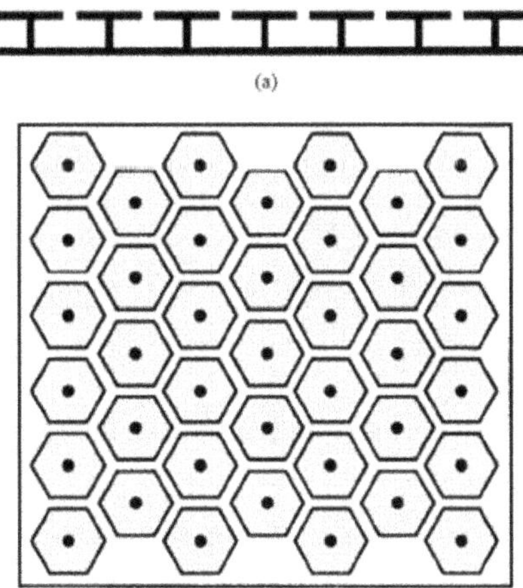

Figure 32.13: A mushroom structure operates like an LC tank circuit. At the right resonant frequency, the surface resembles an open-circuit surface where no current can flow. Hence, tangential magnetic field is zero, resembling a perfect magnetic conductor (Figure by Sievenpiper [252]).

Mathematically, a surface that is dual to the PEC surface is the perfect magnetic conductor (PMC) surface. The magnetic dipole is also dual to the electric dipole. Thus, over a PMC surface, these electric and magnetic dipoles will reflect differently as shown in Figure 32.14. One can go through Gedanken experiments and verify that the reflection rules are as shown in the figure.

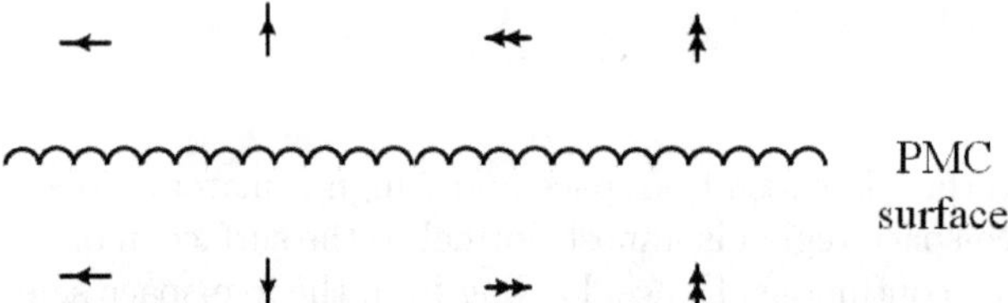

Figure 32.14: Reflection rules for electric and magnetic dipoles over a PMC surface. Magnetic dipoles are denoted by double arrows.

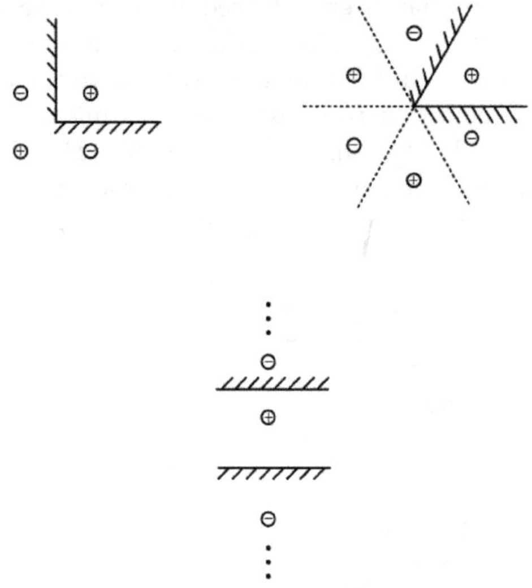

Figure 32.15: Image theory for multiple images [34].

32.2.4 Multiple Images

For the geometry shown in Figure 32.15, using symmetry argument, one can start with electrostatic theory and convince oneself that $\hat{n} \times \mathbf{E} = 0$ on the metal surface with the placement of charges as shown. For conducting media, the charges will relax to the quiescent distribution after the relaxation time. For PEC surfaces, one can extend these cases to time-varying dipoles because the charges in the PEC medium can reorient instantaneously (that is, with zero relaxation time) to shield out or expel the $\mathbf{E}$ and $\mathbf{H}$ fields. Again, one can repeat the above exercise for magnetic charges, magnetic dipoles, and PMC surfaces.

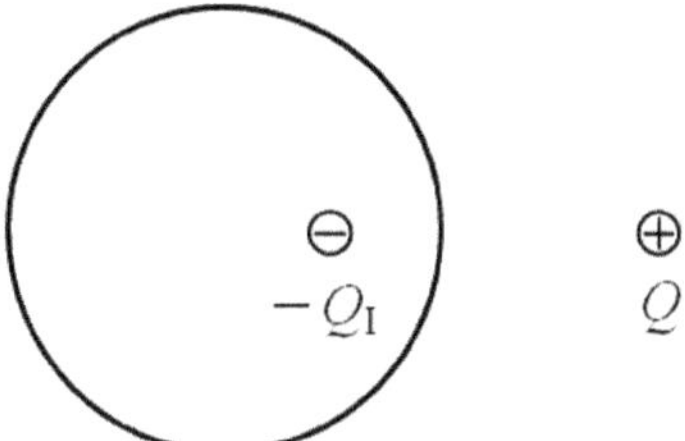

Figure 32.16: Image theory for a point charge near a cylinder or a sphere can be found in closed form. Details are given in [34].

32.2.5 Some Special Cases—Spheres, Cylinders, and Dielectric Interfaces

One curious case is for a static charge placed near a conductive sphere (or cylinder) as shown in Figure 32.16.[4] A charge of $+Q$ reflects to a charge of $-Q_I$ inside the sphere where $Q_I \neq Q$. For electrostatics, the sphere (or cylinder) need only be a conductor. However, this cannot be generalized to electrodynamics or a time-varying problem because of the retardation effect: A time-varying dipole or charge will be felt at different points asymmetrically on the surface of the sphere from the original and image charges. Exact cancelation of the tangential electric field on the surface of the sphere or cylinder cannot occur for a time-varying field.

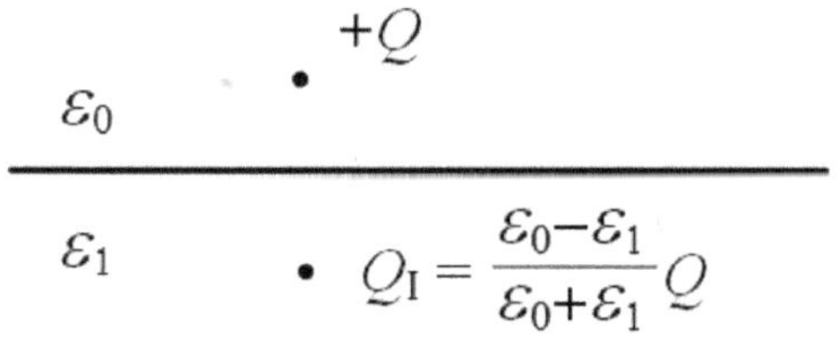

ε_0 $+Q$

ε_1

$$Q_\mathrm{I} = \frac{\varepsilon_0 - \varepsilon_1}{\varepsilon_0 + \varepsilon_1} Q$$

Figure 32.17: A static charge over a dielectric interface can be found in closed form using Fourier transform technique, to be discussed later. The solution is beyond the scope of this lecture at this point.

When a static charge is placed over a dielectric interface, image theory can be used to find the closed-form solution. This solution can be derived using Fourier transform technique, which we shall learn later [38]. It can also be extended to multiple interfaces. But image theory cannot be used in a simple manner for the electrodynamic case due to the different speeds of light in different media, giving rise to different retardation effects.

[4]This is worked out in detail in p. 48 and p. 49, Ramo et al. [34].

Exercises for Lecture 32

Problem 32-1:

(i) Explain why in the case of a conductor with finite conductivity, the electric field is completely shielded from the inside of the conductor when the electric field is static.

(ii) Explain why a conductor with finite conductivity cannot shield out a magnetic field in the static limit.

(iii) Explain why a bar magnet can levitate on top of a superconductor.

(iv) Using image theory, explain why the boundary conditions are satisfied in the examples shown in Figure 32.15.

(v) For Figure 32.17, show by using the principle of linear superposition that by image theory the normal flux and tangential field are continuous when they are produced by the original source plus its image source.

Lecture 33

High-Frequency Solutions, Gaussian Beams

When the frequency is very high, the wavelength of electromagnetic wave becomes very short. In this limit, many solutions to Maxwell's equations can be found approximately. These solutions offer a very different physical picture of electromagnetic waves, and they are often used in optics where the wavelength is short. So it was no surprise that for a while, optical fields were thought to satisfy a very different set of equations from those of electricity and magnetism. Therefore, it came as a surprise when it was later revealed that in fact, optical fields satisfy the same Maxwell's equations as the fields from electricity and magnetism!

In this lecture, we shall seek approximate solutions to Maxwell's equations or the wave equations when the frequency is high or the wavelength is short compared to the geometry that the wave interacts with. High-frequency approximate solutions are important in many real-world applications. This is possible when the wavelength is much smaller than the size of the structure. This can occur even in the microwave regime where the wavelength is not that small, but much smaller than the size of the structure. This is the case when microwaves interact with reflector antennas, for instance. It is also the transition from the waves regime to the optics (or ray optics) regime in the solutions of Maxwell's equations. Often times, the term "quasi-optical" is used to describe the solutions in this regime.

In the high-frequency regime, or when we are far away from a source much larger than the Rayleigh distance (see section 30.3.1), the field emanating from a source resembles a spherical wave. Moreover, when the wavelength is much smaller than the radius of curvature of the wavefront, the spherical wave can be approximated by a local plane wave. Thus we can imagine rays to be emanating from a finite source, forming the spherical wave. The spherical wave will ultimately be approximated by plane waves locally at the observation point.[1] This will simplify the solutions

[1]We shall learn later that a ray can be approximated by a bundle of plane waves (or rays) almost parallel to each other.

in many instances. For instance, ray tracing can be used to track how these rays can propagate, bounce, or "ricochet" in a complex environment. In fact, it is now used in the movie industry to give "realism" to simulate the nuances of how light rays bounce around in a room and reflect off objects.

Figure 33.1: Ray-tracing techniques can be used in the movie industry to produce realism in synthetic images (Figure by wikipedia).

33.1 Tangent Plane Approximations

We have learned that reflection and transmission of waves at an infinitely large flat surface can be solved in closed form. The important physics here is that of phase matching. Due to phase matching, we have derived the law of Fresnel reflection, transmission, and Snell's law [63].[2]

When a surface is not flat anymore, there is no closed-form solution. But when a surface is curved such that the radius of curvature is much larger than the wavelength, an approximate solution can be found. This is obtained by using a *local tangent-plane approximation*, which is a good approximation when the frequency is high or the wavelength is short. It is similar in spirit that we can approximate a spherical wave by a local plane wave at the spherical wavefront when the wavelength is short compared to the radius of curvature of the wavefront.

When the wavelength is short, phase matching happens locally, and the Fresnel law of reflection, transmission, and Snell's law are satisfied approximately as shown in Figure 33.2. The tangent plane approximation is the basis for the geometrical optics (GO) approximation [35, 254]. In GO, light waves are replaced by light rays. As mentioned before, a light ray is part of a spherical wave where, locally, the spherical wave can be approximated by a plane wave. The reflection and transmission of these rays at an interface is then estimated using the local tangent plane approximation and local Fresnel reflection and transmission coefficients. This is also the basis for high-frequency solutions where lens, ray optics, or ray tracing can be used. From it, lens technology is derived (see Figure 33.3). [255, 256].[3]

Many real world problems do not have closed-form solutions, and have to be treated with approximate methods. In addition to geometrical optics approximations mentioned above, asymptotic

[2]This law is also known in the Islamic world in AD 984 [253].

[3]Please note that the tangent plane approximation is invalid near a sharp corner or an edge. The solution has to be augmented by an additional diffracted wave coming from the edge or the corner.

methods are also used to find approximate solutions. Asymptotic methods imply finding a solution when there is a large parameter in the problem. In this case, it is usually the frequency. Such high-frequency approximate methods are discussed in [257, 258, 259, 260, 261].

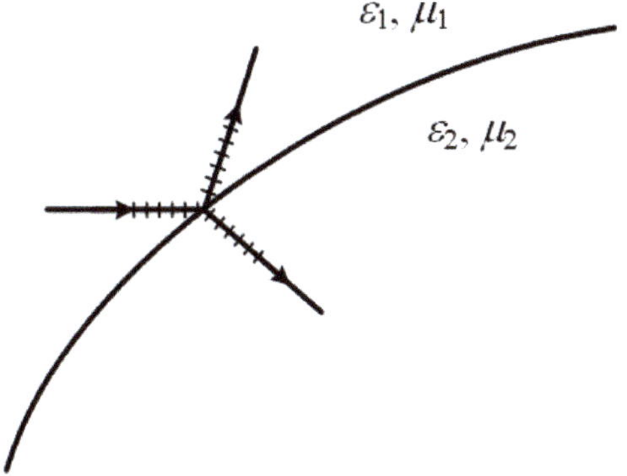

Figure 33.2: In the tangent plane approximation, the surface where reflection and refraction occur is assumed to be locally flat. Thus, phase matching is approximately satisfied, yielding locally the law of reflection, transmission, and Snell's law.

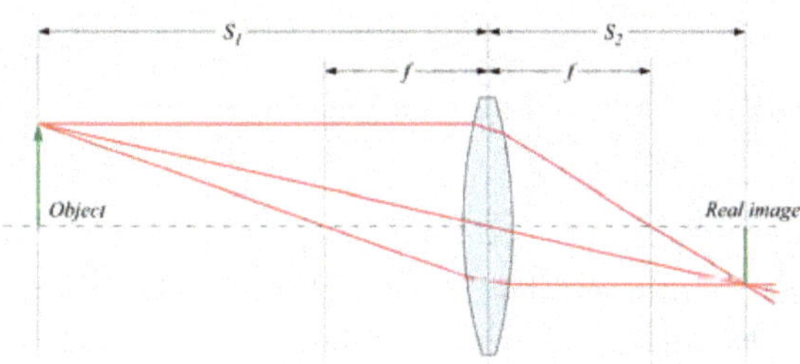

Figure 33.3: Tangent plane approximations can also be made at dielectric interfaces so that Fresnel reflection and transmission coeffients can be used to ascertain the interaction of light rays with a lens. Also, one can use ray tracing to understand the workings of an optical lens (Figure by wikipedia).

33.2 Fermat's Principle

Fermat's principle (1600s) [262, 63] says that a light ray follows the path that takes the shortest time delay between two points.[4] Since time delay is related to the phase shift, and a light ray

[4]This eventually gives rise to the principle of least action, which is a wonderful gift of nature! Nature finds the simplest and most efficient solutions in the real world.

can be locally approximated by a plane wave, this can be restated that a plane wave follows the path that has a minimal phase shift. This principle can be used to derive the law of reflection, transmission, and refraction for light rays. It can be used as the guiding principle for ray tracing as well.

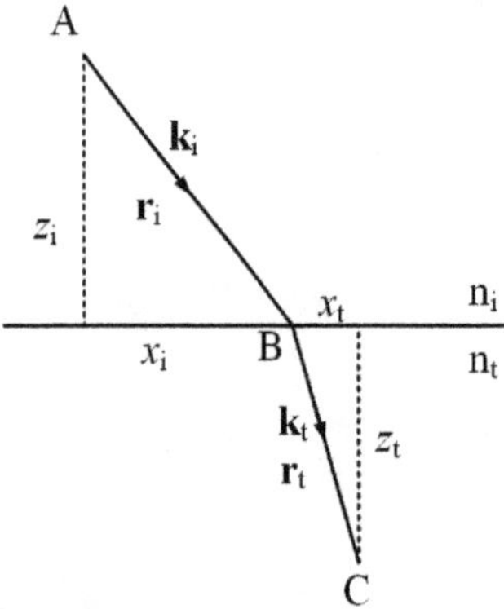

Figure 33.4: In Fermat's principle, a light ray, when propagating from point A to point C, takes the path of least time delay in the time domain, and hence, the least phase shift in the frequency domain.

Assume two points A and C are in two different half spaces as shown in Figure 33.4. Then the phase delay between the two points, per Figure 33.4, can be written as[5]

$$P = \mathbf{k}_i \cdot \mathbf{r}_i + \mathbf{k}_t \cdot \mathbf{r}_t \tag{33.2.1}$$

In the above, $\mathbf{k}_i$ is parallel to $\mathbf{r}_i$, and $\mathbf{k}_t$ is also parallel to $\mathbf{r}_t$. As this is the shortest path with minimum phase shift or time delay, according to Fermat's principle, any other path will be longer, giving rise to more phase shift. In other words, if B were to move slightly to another point, a longer path with more phase shift or time delay will ensue, or that B is the stationary point for the path length or phase shift when the location of B is changed.

Specializing (33.2.1) to a 2D picture, then the phase shift as a function of x_i is stationary. This is shown in Figure 33.4 where we have $x_i + x_t = \text{const}$. Therefore, taking the derivative of (33.2.1) or the phase change with respect to x_i, assuming that $\mathbf{k}_i$ and $\mathbf{k}_t$ do not change as B is moved slightly,[6]

$$\frac{\partial P}{\partial x_i} = 0 = k_{ix} - k_{tx} \tag{33.2.2}$$

[5]As aforementioned, in this course, for wavenumber, we use k and β interchangeably, where k is prevalent in optics and β is used in microwaves.

[6]One can show that as the separations between A, B, and C are large, and if the change in x_i is Δx_i, the changes in $\mathbf{k}_i$ and $\mathbf{k}_t$ are small. The change in phase shift mainly comes from the change in x_i. Alternatively, if one wrote $P = k_i r_i + k_t r_t$ and let $r_i = \sqrt{x_i^2 + z_i^2}$ and $r_t = \sqrt{x_t^2 + z_t^2}$, and took the derivative with respect to x_i, one would get the same answer.

The above yields the law of refraction that $k_{ix} = k_{tx}$, which is just Snell's law; it can also be obtained by phase matching as we have shown earlier. This law was also known in the Islamic world to Ibn Sahl in A.D. 984 [253].

33.2.1 Generalized Snell's Law

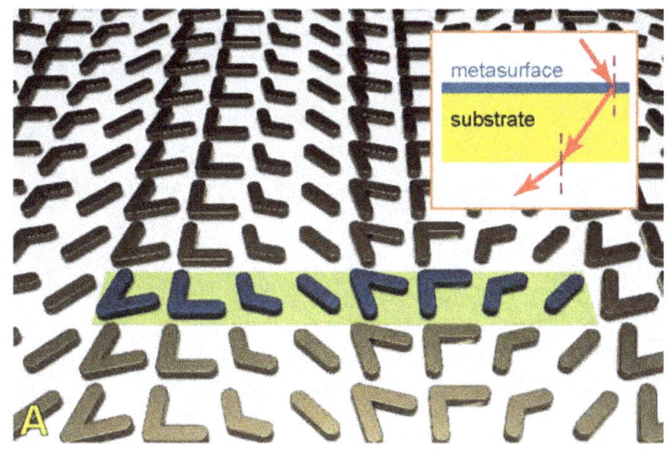

Figure 33.5: A phase screen which is position dependent can be made using nanofabrication and designed with commercial software for solving Maxwell's equations. In such a case, one can derive a generalized Snell's law to describe the diffraction of a wave by such a surface (Figure by Capasso's group [263]).

Metasurfaces are prevalent these days due to advances in nanofabrication and numerical simulation. One of them is shown in Figure 33.5. Such a metasurface can be thought of as a phase screen, providing additional phase shift for the light as it passes through. Moreover, the added phase shift can be controlled to be a function of position because of advances in nanofabrication technology and commercial software for numerical simulation.

To model this phase screen, we add an additional function $\Phi(x, y)$ to (33.2.1), namely that

$$P = \mathbf{k}_i \cdot \mathbf{r}_i + \mathbf{k}_t \cdot \mathbf{r}_t - \Phi(x_i, y_i) \tag{33.2.3}$$

Now applying Fermat's principle that there should be minimal phase delay, and taking the derivative of the above with respect to x_i, one gets

$$\frac{\partial P}{\partial x_i} = k_{ix} - k_{tx} - \frac{\partial \Phi(x_i, y_i)}{\partial x_i} = 0 \tag{33.2.4}$$

The above yields that the generalized Snell's law [263] that

$$k_{ix} - k_{tx} = \frac{\partial \Phi(x_i, y_i)}{\partial x_i} \tag{33.2.5}$$

It implies that the transmitted light can be directed to other angles due to the additional phase screen.[7]

33.3 Gaussian Beam

At this point, we will take a departure to study another high-frequency phenomenon, that of the Gaussian beam. We have seen previously that in a source-free medium, using vector and scalar potential formulation, we arrive at simpler Helmholtz wave equations for **A** and Φ, namely,

$$\nabla^2\mathbf{A} + \omega^2\mu\varepsilon\mathbf{A} = 0 \tag{33.3.1}$$

$$\nabla^2\Phi + \omega^2\mu\varepsilon\Phi = 0 \tag{33.3.2}$$

The above are four scalar equations; and the Lorenz gauge

$$\nabla\cdot\mathbf{A} = -j\omega\mu\varepsilon\Phi \tag{33.3.3}$$

connects **A** and Φ. We can examine the solution of **A** such that

$$\mathbf{A}(\mathbf{r}) = \mathbf{A}_0(\mathbf{r})e^{-j\beta z} \tag{33.3.4}$$

where $\mathbf{A}_0(\mathbf{r})$ is a slowly varying function under the paraxial approximation, while $e^{-j\beta z}$ is rapidly varying in the z direction. (Here, $\beta = \omega\sqrt{\mu\epsilon}$ is the wavenumber.) This is primarily a quasi-plane wave propagating predominantly in the z-direction. We know this to be the case in the far field of a source, but let us assume that this form persists less than the far field, namely, in the Fresnel zone as well. Taking the x component of (33.3.4), we have[8]

$$A_x(\mathbf{r}) = \Psi(\mathbf{r})e^{-j\beta z} \tag{33.3.5}$$

where $\Psi(\mathbf{r}) = \Psi(x,y,z)$ is a scalar slowly varying envelope function of x, y, and z, whereas $e^{-j\beta z}$ is a rapidly varying function of z when β is large or the frequency is high.

33.3.1 Derivation of the Paraxial/Parabolic Wave Equation

Substituting (33.3.5) into (33.3.1), and taking the double z derivative first in the Laplacian operator ∇^2, assuming that $e^{-j\beta z}$ to be fast varying, we arrive at

$$\frac{\partial^2}{\partial z^2}\left[\underbrace{\Psi(x,y,z)}_{\text{slow}}\underbrace{e^{-j\beta z}}_{\text{fast}}\right] = \left[\frac{\partial^2}{\partial z^2}\Psi(x,y,z) - 2j\beta\frac{\partial}{\partial z}\Psi(x,y,z) - \beta^2\Psi(x,y,z)\right]e^{-j\beta z} \tag{33.3.6}$$

Consequently, after substituting the above into the x component of (33.3.1), making use of the definition of ∇^2, we obtain an equation for $\Psi(\mathbf{r})$, the slowly varying envelope as

$$\frac{\partial^2}{\partial x^2}\Psi + \frac{\partial^2}{\partial y^2}\Psi - 2j\beta\frac{\partial}{\partial z}\Psi + \frac{\partial^2}{\partial z^2}\Psi = 0 \tag{33.3.7}$$

[7]Such research is also being pursued by V. Shalaev and A. Boltasseva's group at Purdue U [264].

[8]Also, the wave becomes a transverse wave in the far field, and keeping the transverse component suffices.

where the last term of (33.3.6) containing β^2 on the right-hand side cancels with the term coming from $\omega^2\mu\varepsilon\mathbf{A}$ of (33.3.1).

So far, no approximation has been made in the above equation. Since β is linearly proportional to frequency ω, when $\beta \to \infty$, or in the high-frequency limit,

$$\left|2j\beta\frac{\partial}{\partial z}\Psi\right| \gg \left|\frac{\partial^2}{\partial z^2}\Psi\right| \tag{33.3.8}$$

where we have assumed that Ψ is a slowly varying function of z within the lengthscale of a wavelength, such that $\beta\Psi \gg \partial/\partial z\Psi$, which in turn implies the above. In other words, when $\beta \to \infty$, (33.3.7) can be approximated by

$$\frac{\partial^2\Psi}{\partial x^2} + \frac{\partial^2\Psi}{\partial y^2} - 2j\beta\frac{\partial\Psi}{\partial z} \approx 0 \tag{33.3.9}$$

The above is called the paraxial wave equation. It is also called the parabolic wave equation.[9] It implies that the $\boldsymbol{\beta}$ vector of the wave is approximately parallel to the z-axis, or $\beta_z \cong \beta$ to be much greater than β_x and β_y, and hence, the name.[10]

33.3.2 Finding a Closed-Form Solution

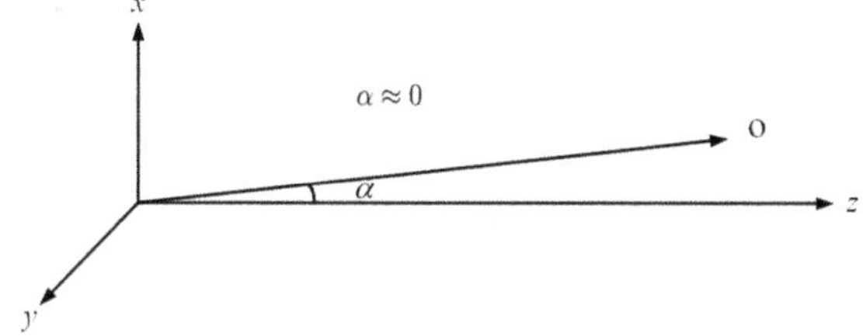

Figure 33.6: This figure shows when the paraxial approximation can be made. The wave physics are different when the field is being observed very close to the z-axis and far from it.

A closed-form solution to the paraxial wave equation can be obtained by a simple mathematical trick.[11] It is known that

$$A_x(\mathbf{r}) = \frac{e^{-j\beta|\mathbf{r}-\mathbf{r}'|}}{4\pi|\mathbf{r}-\mathbf{r}'|} \tag{33.3.10}$$

[9]The paraxial wave equation, the diffusion equation, and Schrödingerś equation are all classified as parabolic equations in mathematical parlance [38, 54, 265, 266].

[10]A succinct derivation of this equation is also by Franz Kartner [267].

[11]Introduced by Georges A. Deschamps of UIUC [268].

is the exact solution to

$$\nabla^2 A_x + \beta^2 A_x = 0 \tag{33.3.11}$$

as long as $\mathbf{r} \neq \mathbf{r}'$. One way to ensure that $\mathbf{r} \neq \mathbf{r}'$ always is to let $\mathbf{r}' = -\hat{z}jb$, a complex number. Then (33.3.10) is always a solution to (33.3.11) for all $\mathbf{r}$, because $|\mathbf{r} - \mathbf{r}'| \neq 0$ always since $\mathbf{r}'$ is complex. Then, we should next make a paraxial approximation to the solution (33.3.10) by assuming that $x^2 + y^2 \ll z^2$. By so doing, it follows that

$$\begin{aligned} |\mathbf{r} - \mathbf{r}'| &= \sqrt{x^2 + y^2 + (z + jb)^2} \\ &= (z + jb) \left[1 + \frac{x^2 + y^2}{(z + jb)^2} \right]^{1/2} \\ &\approx (z + jb) + \frac{x^2 + y^2}{2(z + jb)} + \ldots, \qquad |z + jb| \to \infty \end{aligned} \tag{33.3.12}$$

where the Taylor series has been used in approximating the last term. And then using the above approximation in (33.3.10) yields

$$A_x(\mathbf{r}) \approx \frac{e^{-j\beta(z+jb)}}{4\pi(z + jb)} e^{-j\beta\frac{x^2+y^2}{2(z+jb)}} \approx e^{-j\beta z}\Psi(\mathbf{r}) \tag{33.3.13}$$

Notice that again, the second term in (33.3.12) is ignored in the denominator, but not in the exponent. By comparing the above with (33.3.5), we can identify

$$\Psi(x, y, z) \cong A_0 \frac{jb}{z + jb} e^{-j\beta\frac{x^2+y^2}{2(z+jb)}} \tag{33.3.14}$$

where $A_0 = \frac{\exp(\beta b)}{(4\pi jb)}$ is used to absorb the constants to simplify the expression. By separating the term in the exponent into the real part and the imaginary part, namely,

$$\frac{x^2 + y^2}{2(z + jb)} = \frac{x^2 + y^2}{2} \left(\frac{z}{z^2 + b^2} - j\frac{b}{z^2 + b^2} \right) \tag{33.3.15}$$

and writing the prefactor in terms of amplitude and phase, the result is

$$\frac{jb}{z + jb} = \frac{1}{\sqrt{1 + z^2/b^2}} e^{j\tan^{-1}\left(\frac{z}{b}\right)} \tag{33.3.16}$$

We then have

$$\Psi(x, y, z) \cong \frac{A_0}{\sqrt{1 + z^2/b^2}} e^{j\tan^{-1}\left(\frac{z}{b}\right)} e^{-j\beta\frac{x^2+y^2}{2(z^2+b^2)}z} e^{-b\beta\frac{x^2+y^2}{2(z^2+b^2)}} \tag{33.3.17}$$

The above can be rewritten more suggestively as

$$\Psi(x, y, z) \cong \frac{A_0 e^{j\psi}}{\sqrt{1 + z^2/b^2}} e^{-j\beta\frac{x^2+y^2}{2R(z)}} e^{-\frac{x^2+y^2}{w^2(z)}} \tag{33.3.18}$$

In the above,

$$w^2(z) = \frac{2b}{\beta}\left(1 + \frac{z^2}{b^2}\right), \qquad R(z) = \frac{z^2 + b^2}{z} = z + \frac{b^2}{z}, \qquad \psi(z) = \tan^{-1}\left(\frac{z}{b}\right) \tag{33.3.19}$$

For a fixed z, the parameters $w(z)$, $R(z)$, and $\psi(z)$ are all constants, as shall be explained later, corresponding to the waist, the radius of curvature, and the Gouy phase [269]. Here, $\tan^{-1}\left(\frac{z}{b}\right)$ is a multi-value function, but keeping this function on the branch when z is small such that $\psi \approx \frac{z}{b}$, then $\psi = \frac{\pi}{2}$ when $z \to \infty$.

It will be interesting to compare the constant phase front of the Gaussian beam with that of a spherical wave. For the Gaussian beam, the source location is at $\mathbf{r}' = -\hat{z}jb$, but for a spherical wave, the source location is at $\mathbf{r}' = 0$. Making a paraxial approximation for the spherical wave, assuming that $x^2 + y^2 \ll z^2$ the spherical wave front becomes

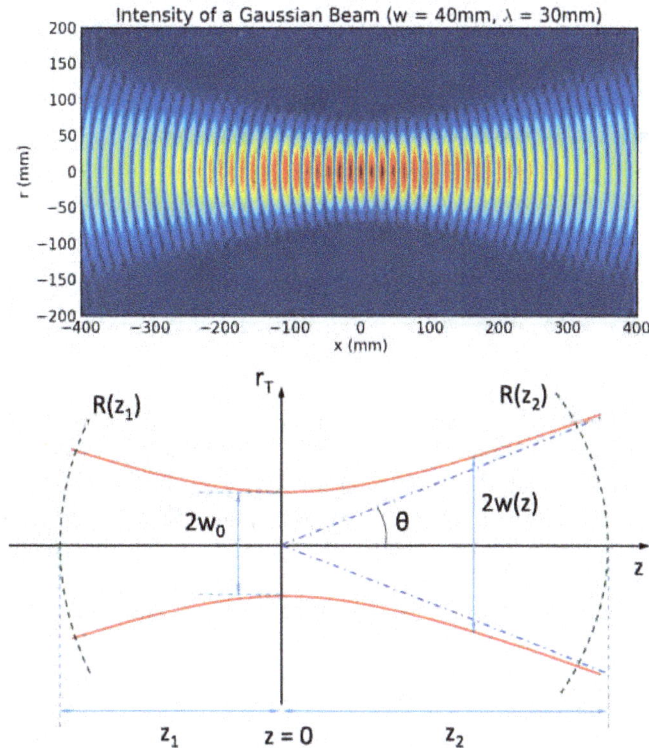

Figure 33.7: The upper figure is a more colorful one (Figure by: https://en.wikipedia.org). Notice that these are homogeneous solutions to the wave equation since the right-hand side is always zero. The lower figure is from Glytsis [148]. Notice that the waist of the Gaussian beam, $w(z)$, and the radius of curvature, $R(z)$ are functions of z. The waist at its narrowest value is $w_0 = w(0)$.

$$e^{-j\beta R} = e^{-j\beta(x^2+y^2+z^2)^{1/2}} = e^{-j\beta z\left(1+\frac{x^2+y^2}{z^2}\right)^{1/2}}$$
$$\approx e^{-j\beta z - j\beta\frac{x^2+y^2}{2z}} \approx e^{-j\beta z - j\beta\frac{x^2+y^2}{2R_0}} \tag{33.3.20}$$

In the last approximation, we assume that $z \approx R_0$.

Now we are ready to give physical interpretations to different factors in (33.3.17). In the above, the beam is Gaussian tapered in the transverse direction, and $w(z)$ is the waist of the Gaussian beam. When radius $\rho = \sqrt{x^2+y^2} = w(z)$ then the value of the field is $1/e$ of its max value when $\rho = 0$. The waist is smallest when $z = 0$. Moreover, $e^{-j\beta\frac{x^2+y^2}{2R}}$ is a spherical wavefront with radius of curvature is $R(z)$ (compare with (33.3.20)). And ψ is the Gouy phase [269] which has a 90° phase shift characteristics of waves that pass through caustics [1, p. 128] in ray optics (see Figure 33.7). Since ψ becomes a constant when z becomes large, the phase change in (33.3.18) is mainly due to $e^{-j\beta\frac{x^2+y^2}{2R(z)}}$.

From (33.3.19), the radius of curvature of the wavefront is $R(z) = \frac{z^2+b^2}{z}$. It becomes large when z is small, implying that the wavefront looks like that of a plane wave. This is also born out by Figure 33.7. This is what we see in a laser pointer: as the wave leaves the aperture of a laser pointer, it is collimated like that of a plane wave, but diverges after a while.

The phase ψ defined in (33.3.19) changes linearly with z for small z, and saturates to a constant for large z. This underscores the fact that $\Psi(\mathbf{r})$ is a slowly varying function, and also, the phase of the entire wave is due to the $\exp(-j\beta z)$ in (33.3.13) which is rapidly varying when β is large or the frequency is high.

A cross section of the electric field due to a Gaussian beam is shown in Figure 33.7. It is to be noted that as the frequency or β increases, the waist of the beam becomes narrower according to (33.3.19). From (33.3.9), the transverse variations of $\Psi(x, y, z)$ due to this narrowing waist becomes large, to balance the third term. Hence, the transverse variations are coupled to the longitudinal variation in the paraxial wave equation. In the next section, we will exploit the transverse variations of the paraxial wave equation to study the richness of possible solutions by exploiting the "homomorphism" of this equation with the Schrödingerś equation.

33.3.3 Other Solutions

In general, the paraxial wave equation in (33.3.9) is of the same form as the Schrödingerś equation, which is of the utmost importance in quantum theory. In recent years, the solution of this equation has made use of spillover knowledge and terms from quantum theory, such as spin angular momentum (SAM) and orbital angular momentum (OAM), even though we are actually still in the classical regime. But it is a partial differential equation that can be solved by the separation of variables just like the Helmholtz wave equation. (Hurrah to the power of separation of variables!)

Therefore, in general, it has solutions of the form[12]

$$\Psi_{nm}(x,y,z) \sim \frac{1}{w(z)} e^{-\frac{(x^2+y^2)}{w^2(z)}} e^{-j\frac{\beta(x^2+y^2)}{2R(z)}} e^{j(m+n+1)\tan^{-1}\left(\frac{z}{b}\right)} H_n\left(\frac{x\sqrt{2}}{w(z)}\right) H_m\left(\frac{y\sqrt{2}}{w(z)}\right) \tag{33.3.21}$$

where $n, m = 0, 1, 2, \ldots$, and $H_n(\xi)$ is a Hermite polynomial of order n. Alternatively, the solutions can also be expressed in terms of Laguerre polynomials, namely,

$$\Psi_{nm}(x,y,z) \sim \frac{1}{w(z)} e^{-j\frac{\beta}{2R(z)}\rho^2} e^{-\frac{\rho^2}{w^2(z)}} e^{j(|n|+2m+1)\tan^{-1}\left(\frac{z}{b}\right)} e^{jn\phi} \left(\frac{\sqrt{2}\rho}{w(z)}\right)^{|n|} L_m^{|n|}\left(\frac{2\rho^2}{w^2(z)}\right) \tag{33.3.22}$$

where $n = 0, \pm 1, \pm 2, \ldots; m = 0, 1, 2, \ldots$, and $L_n^k(\xi)$ is the associated Laguerre polynomial. One can also generate a field due to a multipole source. When the source point is put into the complex space, then a complex Gaussian beam can be generated.

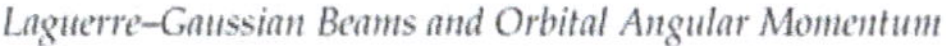

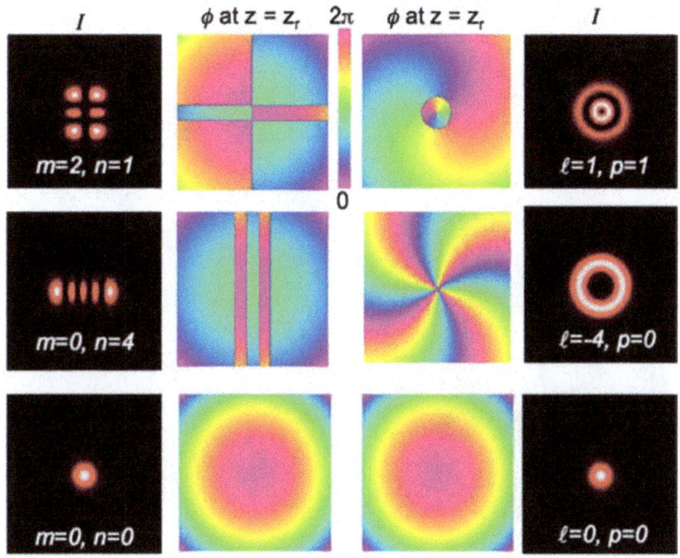

Figure 33.8: Examples of structured light. Structured lights can be used in encoding more information in optical communications (Figure by L. Allen and M. Padgett's chapter in J.L. Andrew's book on structured light [271]).

These Gaussian beams have rekindled recent excitement in the community because, in addition to carrying spin angular momentum as in a plane wave, they can carry orbital angular momentum

[12]See F. Pampaloni and J. Enderlein [270] and E.N. Glytsis [148]. The author also thanks Bo ZHU for pointing out errors in the earlier versions of these equations.

due to the complex transverse field distribution of the beams.[13] They harbor potential for optical communications as well as optical tweezers to manipulate trapped nanoparticles. Figure 33.8 shows some examples of the cross section (xy plane) field plots for some of these beams. They are richly endowed with patterns, implying that they can be used to encode information. These lights are also called structured lights [271].

[13]See D.L. Andrew, *Structured Light and Its Applications*, and articles therein [271].

Exercises for Lecture 33

Problem 33-1:

(i) Verify the statement on footnote 6 in lecture 33 of the lecture book.

(ii) Explain why (33.3.8) is true in the high-frequency and paraxial limit.

(iii) Derive (33.3.18) and the expressions in (33.3.19) and give physical meanings to (33.3.18). Explain why R is the radius of curvature of the wavefront.

Lecture 34

Scattering of Electromagnetic Fields

The scattering of electromagnetic fields is an important and fascinating topic. It especially enriches our understanding of the interaction of light waves with matter. The wavelength of visible light is several hundred nanometers, with atoms and molecules ranging from nanometers onward, and light-matter interaction is richly endowed with interesting physical phenomena! Moreover, the myriad of pictures have been generated by visible light have a great retina impact.

A source radiates a field, and ultimately, in the far field of the source, the field resembles a spherical wave which in turn resembles a plane wave. When a plane wave impinges on an object or a scatterer, the energy carried by the plane wave is deflected to other directions which is the process of scattering. This is akin to the physics of mode conversion at waveguide junctions that we alluded to in section 24.3.3: a pure single mode sprays into many other modes in order to satisfy the boundary condition. In the optical regime, the scattered light allows us to see objects, as well as admire all the beautiful hues and colors of the world. In microwaves, the scatterers cause the loss of energy carried by a plane wave.

A proper understanding of scattering theory reveals many physical phenomena around us. We will begin by studying Rayleigh scattering, which is scattering by small objects (or particles) compared to wavelength. With Rayleigh scattering of a simple sphere, we gain physical insight to many phenomena, such as why the sky is blue and the sunset is red!

34.1 Rayleigh Scattering

Rayleigh scattering is a solution to the scattering of light by small particles. These particles are assumed to be much smaller than the wavelength of light. For example, the size of a water molecule is about 0.25 nm, while the wavelength of blue light is about 450 nm. Since the particle size is much smaller than the wavelength, we can use quasi-static approximation to find a simple solution in the vicinity of the small particle.[1]

This simple scattering solution offers us insight into the nature of light scattering (see Figure 34.1). For instance, it explains why the sunset is so magnificently beautiful and how birds and insects can navigate without the help of a compass. By the same token, it also explains why in ancient times, the Vikings, as a seafaring people, could cross the Atlantic Ocean over to Iceland without the help of a magnetic compass as the Chinese did.

Figure 34.1: The magnificent beauty of nature can be partly explained by Rayleigh scattering [206, 272].

When a ray of light impinges on an object, we model the incident light as a plane electromagnetic wave (see Figure 34.2). The time-varying incident field polarizes the particle, making it into a small time-varying dipole, and it re-radiates like a Hertzian dipole. This is the gist of the scattering process: an incident field induces a current (in this case, a polarization current) on the scatterer. With the induced current, the scatterer re-radiates (or scatters). Without loss of generality, we can assume the electromagnetic wave is polarized in the z direction and propagating in the x direction. We assume that the particle to be a small spherical particle with permittivity ε_s and radius a. Essentially, the particle sees a constant field as the plane wave impinges on it; or in a word, the particle feels a quasi-electrostatic field in the incident field.

[1]Since the above is derived with phasor technique, it is also valid for complex permittivity and permeability as long as the wavelength is long. The power of phasor technique again!

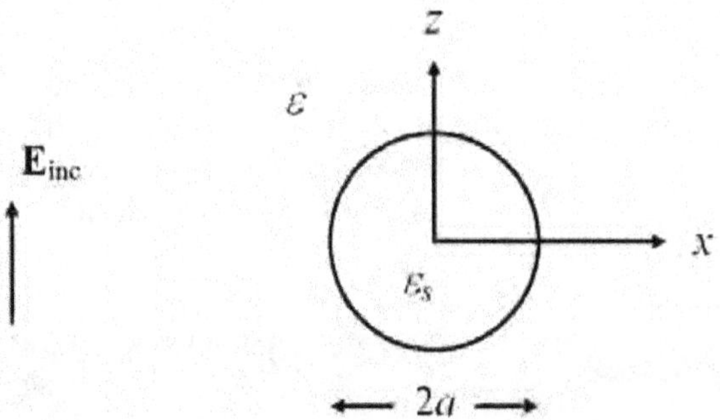

Figure 34.2: Geometry for studying the Rayleigh scattering problem.

34.1.1 Scattering by a Small Spherical Particle

The incident field polarizes the small particle making it look like a small electric dipole. Since the incident field is time harmonic, the small electric dipole will oscillate and radiate like a Hertzian dipole in the far field. First, we will look at the solution in the vicinity of the scatterer, namely, in the near field. Then we will motivate the form of the solution in the far field of the scatterer. (Solving a boundary value problem by looking at the solutions in two different physical regimes and then matching (or patching) the solutions together is known as asymptotic matching or matched asymptotic expansion, a lost art [49].)

A Hertzian dipole can be approximated by a small current source so that

$$\mathbf{J}(\mathbf{r}) = \hat{z}Il\delta(\mathbf{r}) \tag{34.1.1}$$

Without loss of generality, we have assumed the Hertzian dipole to be at the origin. In the above, we let the time-harmonic current $I = dq/dt = j\omega q$. Then

$$Il = j\omega ql = j\omega p \tag{34.1.2}$$

where the dipole moment $p = ql$. As we have seen before, the vector potential $\mathbf{A}$ due to a Hertzian dipole, after substituting (34.1.1), is

$$\begin{aligned}\mathbf{A}(\mathbf{r}) &= \frac{\mu}{4\pi}\iiint_V d\mathbf{r}'\frac{\mathbf{J}(\mathbf{r}')}{|\mathbf{r}-\mathbf{r}'|}e^{-j\beta|\mathbf{r}-\mathbf{r}'|}\\ &= \hat{z}\frac{\mu Il}{4\pi r}e^{-j\beta r} = \hat{z}\frac{j\omega\mu ql}{4\pi r}e^{-j\beta r}\end{aligned} \tag{34.1.3}$$

where we have made use of the sifting property of the delta function in (34.1.1) when it is substituted into the above integral. The above is the exact solution due to a Hertzian dipole. The only approximation is to assume the dipole with current distribution given by (34.1.1), which is a good approximation if the dipole is much smaller than the wavelength.

Near Field

The above gives the vector potential $\mathbf{A}$ due to a Hertzian dipole. Since the dipole is infinitesimally small, the above solution is both valid in the near field as well as in the far field. Since the dipole moment ql is induced by the incident field, we need to relate ql to the amplitude of the incident electric field. To this end, we convert the above vector potential field to the near electric field of a small dipole.

From prior knowledge, we know that the electric field is given by $\mathbf{E} = -j\omega\mathbf{A} - \nabla\Phi$. From dimensional analysis, the scalar potential term dominates over the vector potential term in the near field of the scatterer since $\partial/\partial x \gg 1/c\partial/\partial t$. Hence, we need to derive, for the corresponding scalar potential, the approximate solution.

The scalar potential $\Phi(\mathbf{r})$ is obtained from the Lorenz gauge (see (27.2.23)) that $\nabla \cdot \mathbf{A} = -j\omega\mu\varepsilon\Phi$. Therefore,

$$\Phi(\mathbf{r}) = \frac{-1}{j\omega\mu\varepsilon}\nabla \cdot \mathbf{A} = -\frac{Il}{j\omega\varepsilon 4\pi}\frac{\partial}{\partial z}\frac{1}{r}e^{-j\beta r} \tag{34.1.4}$$

When we are close to the dipole, by assuming that $\beta r \ll 1$, we use a quasi-static approximation about the potential.[2] Then

$$\frac{\partial}{\partial z}\frac{1}{r}e^{-j\beta r} \approx \frac{\partial}{\partial z}\frac{1}{r} = \frac{\partial r}{\partial z}\frac{\partial}{\partial r}\frac{1}{r} = -\frac{z}{r}\frac{1}{r^2} \tag{34.1.5}$$

or after using that $z/r = \cos\theta$, we have

$$\Phi(\mathbf{r}) \approx \frac{ql}{4\pi\varepsilon r^2}\cos\theta \tag{34.1.6}$$

which is the static dipole potential because we are in the near field of the dipole. This dipole induced in the small particle is formed in response to the incident field and its dipole potential is given by the previous expression. In other words, the incident field polarizes the small particle into a small time-oscillating Hertzian dipole which re-radiates.

Next, we calculate the polarizability of a small particle. The polarizability is a measure of the strength of the dipole moment induced by the incident field on a small particle. To find the polarizability, we need only to imagine the particle to be in between two parallel plates which are separated far apart, and with a constant electric field pointing in the z direction. Hence, the inducing (or incident) field can be approximated by a constant local static electric field,

$$\mathbf{E}_{inc} = \hat{z}E_i \tag{34.1.7}$$

This is the field that will polarize the small particle. It can also be an electric field between two parallel plates (as aforementioned). The corresponding electrostatic potential for the inducing field is then

$$\Phi_{inc} = -zE_i \tag{34.1.8}$$

[2]This is the same as ignoring retardation effect.

so that $\mathbf{E}_{inc} \approx \hat{z}E_i$, as $\omega \to 0$. The scattered dipole potential from the spherical particle in its vicinity is quasi-static, and is given by

$$\Phi_{sca} = E_s \frac{a^3}{r^2} \cos\theta \tag{34.1.9}$$

which is the potential due to a static dipole. The electrostatic boundary value problem (BVP) has been previously solved and[3]

$$E_s = \frac{\varepsilon_s - \varepsilon}{\varepsilon_s + 2\varepsilon} E_i \tag{34.1.10}$$

Using (34.1.10) in (34.1.9), we get

$$\Phi_{sca} = \frac{\varepsilon_s - \varepsilon}{\varepsilon_s + 2\varepsilon} E_i \frac{a^3}{r^2} \cos\theta \tag{34.1.11}$$

On comparing with (34.1.6), one can see that the dipole moment induced by the incident field is given by

$$p = ql = 4\pi\varepsilon \frac{\varepsilon_s - \varepsilon}{\varepsilon_s + 2\varepsilon} a^3 E_i = \alpha E_i \tag{34.1.12}$$

where α is the polarizability of the small particle. The above analysis is valid as long as the particle size is much smaller than the wavelength. Hence, the incident field can be a time-harmonic field as well, even in the optical regime.

Far Field

Now, we have learned that a small particle is polarized by the time-harmonic incident field. If the incident field is time harmonic, the small dipole will be time-oscillating and it will radiate like a time-varying Hertzian dipole whose far-field we have previously derived (see section 28.2). In the far field of the Hertzian dipole, we have

$$\mathbf{E} = -j\omega\mathbf{A} - \nabla\Phi = -j\omega\mathbf{A} - \frac{1}{j\omega\mu\varepsilon}\nabla\nabla\cdot\mathbf{A} \tag{34.1.13}$$

But when we are in the far field, $\mathbf{A}$ behaves like a spherical wave which in turn behaves like a local plane wave if one goes far enough. Therefore, $\nabla \to -j\boldsymbol{\beta} = -j\beta\hat{r}$. Using this approximation in (34.1.13), we arrive at

$$\mathbf{E} \cong -j\omega\left(\mathbf{A} - \frac{\boldsymbol{\beta}\boldsymbol{\beta}}{\beta^2}\cdot\mathbf{A}\right) = -j\omega(\mathbf{A} - \hat{r}\hat{r}\cdot\mathbf{A}) = -j\omega(\hat{\theta}A_\theta + \hat{\phi}A_\phi) \tag{34.1.14}$$

where we have used $\hat{r} = \boldsymbol{\beta}/\beta$. This is similar to the far-field result we derived in section 29.1.2. In the above, we have used $\mathbf{A} = \overline{\mathbf{I}}\cdot\mathbf{A}$ and that $\overline{\mathbf{I}} = \hat{r}\hat{r} + \hat{\theta}\hat{\theta} + \hat{\phi}\hat{\phi}$.

[3]It was one of the homework problems. See also section 8.3.7. Kong [35, p. 774] has a different way of deriving this.

34.1.2 Scattering Cross Section

From (34.1.3), upon making use of (34.1.2), noticeably, $A_\phi = 0$ due to symmetry, while

$$A_\theta = -\frac{j\omega\mu ql}{4\pi r} e^{-j\beta r} \sin\theta \tag{34.1.15}$$

Consequently, using (34.1.12) for ql, we have in the far field that[4]

$$E_\theta \cong -j\omega A_\theta = -\frac{\omega^2 \mu ql}{4\pi r} e^{-j\beta r} \sin\theta = -\omega^2 \mu\varepsilon \left(\frac{\varepsilon_s - \varepsilon}{\varepsilon_s + 2\varepsilon}\right) \frac{a^3}{r} E_i e^{-j\beta r} \sin\theta \tag{34.1.16}$$

Using local plane-wave approximation that

$$H_\phi \cong \sqrt{\frac{\varepsilon}{\mu}} E_\theta = \frac{1}{\eta} E_\theta \tag{34.1.17}$$

where $\eta = \sqrt{\mu/\varepsilon}$, the time-averaged Poynting vector is given by $\langle \mathbf{S} \rangle = 1/2\Re e \{\mathbf{E} \times \mathbf{H}^*\} = \frac{1}{2\eta}|E_\theta|^2 \hat{r}$. Therefore, the total scattered power is obtained by integrating the power density over a spherical surface when r tends to infinity. Thus, the total scattered power is

$$P_s = \int d\Omega \langle S_r \rangle = \frac{1}{2\eta} \int_0^\pi r^2 \sin\theta d\theta \int_0^{2\pi} d\phi |E_\theta|^2 \tag{34.1.18}$$

$$= \frac{1}{2\eta} \beta^4 \left| \frac{\varepsilon_s - \varepsilon}{\varepsilon_s + 2\varepsilon} \right|^2 \frac{a^6}{r^2} |E_i|^2 r^2 \left(\int_0^\pi \sin^3\theta d\theta \right) 2\pi \tag{34.1.19}$$

But we can evaluate the integral in closed form, namely,

$$\int_0^\pi \sin^3\theta d\theta = -\int_0^\pi \sin^2\theta d\cos\theta = -\int_0^\pi (1 - \cos^2\theta) d\cos\theta$$

$$= -\int_1^{-1} (1 - x^2) dx = \frac{4}{3} \tag{34.1.20}$$

Therefore,

$$P_s = \frac{4\pi}{3\eta} \left| \frac{\varepsilon_s - \varepsilon}{\varepsilon_s + 2\varepsilon} \right|^2 \beta^4 a^6 |E_i|^2 \tag{34.1.21}$$

In the above, even though we have derived the equation using electrostatic theory, it is also valid for complex permittivity defined in section 7.1.2. One can take the divergence of (7.1.11) to arrive at a Gauss's law for lossy dispersive media, namely, $\nabla \cdot \underset{\sim}{\varepsilon} \mathbf{E} = 0$, which is "homomorphic" to the lossless case where $\underset{\sim}{\varepsilon}$ is the complex permittivity defined in section 8.1. The equations for the lossy

[4]The ω^2 dependence in (34.1.16) of the far field implies that the radiated electric field in the far zone is proportional to the acceleration of the charges on the dipole. This can be traced to that the vector potential $\mathbf{A}$ is proportional to the current, which is proportional to the velocity of the charge. But the electric field is proportional to the time derivative of the vector potential of the charge. Therefore, the total field $\mathbf{E}$ is proportional to the acceleration of the charge.

case in the phasor world is mathematically homomorphic to those for the lossy case. Hurrah again to phasor technique!

The scattering cross section is the effective area of a scatterer such that the total scattered power is proportional to the incident power density times the scattering cross section. As such it is defined as

$$\Sigma_s = \frac{P_s}{\langle S_{\text{inc}} \rangle} = \frac{8\pi a^2}{3} \left| \frac{\varepsilon_s - \varepsilon}{\varepsilon_s + 2\varepsilon} \right|^2 (\beta a)^4 \tag{34.1.22}$$

where we have used the local plane-wave approximation that

$$\langle S_{\text{inc}} \rangle = \frac{1}{2\eta} |E_i|^2 \tag{34.1.23}$$

The above also implies that

$$P_s = \langle S_{\text{inc}} \rangle \cdot \Sigma_s \tag{34.1.24}$$

In other words, the scattering cross section Σ_s is an effective cross-sectional area of the scatterer that will intercept the incident wave power $\langle S_{\text{inc}} \rangle$ to produce the scattered power P_s. This concept is similar to the effective aperture A_e of an antenna. In an antenna, we have

$$P_{\text{received}} = \langle S_{\text{inc}} \rangle \cdot A_e \tag{34.1.25}$$

where A_e is a measure of how much power is received or absorbed by the antenna. A similar idea of absorption cross-section area is also used in the community.

It is seen that the scattering cross section grows as the fourth power of frequency since $\beta = \omega/c$. The radiated field grows as the second power because it is proportional to the acceleration of the charges on the particle. The higher the frequency, the more the scattered power. This mechanism can be used to explain why the sky is blue. It also can be used to explain why a sunset has a brilliant hue of red and orange (see Figure 34.3).

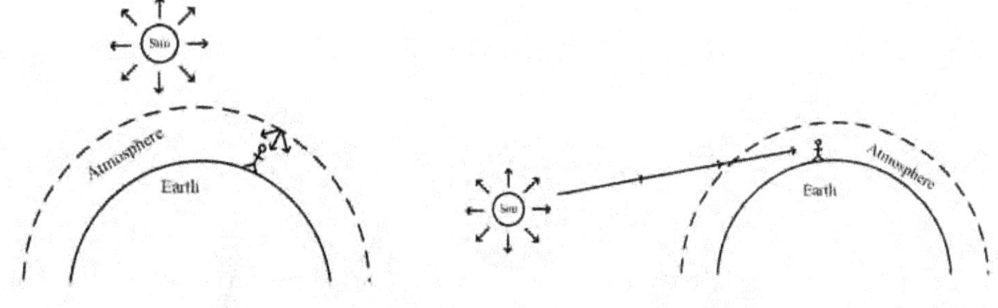

Figure 34.3: (Left) During the daytime, when we look at the sky, we mainly see scattered sunlight. Since high-frequency light is scattered more, the sky appears blue. (Right) At sunset, the sunlight has to go through a thicker atmosphere. Thus, the blue light is scattered away, leaving the red light that reaches our eyes. Therefore, the color of the sunset appears red. (The figures are not drawn to scale.)

The above also explains the brilliant glitter of gold plasmonic nanoparticles as discovered by ancient Roman artisans. For gold, the medium resembles a plasma, and hence, we can have $\varepsilon_s < 0$, and the denominator of (34.1.22) can be very small, giving rise to strongly scattered light (see section 8.3.7).

Furthermore, since the far-field scattered power density of this particle is

$$\langle S \rangle = \frac{1}{2\eta} E_\theta H_\phi^* \sim \sin^2\theta \tag{34.1.26}$$

the scattering power pattern of this small particle is not isotropic. In other words, these dipoles radiate predominantly in the broadside direction but not in their endfire directions. Therefore, insects and sailors can use this to figure out where the sun is even in a cloudy day. In fact, it is like a rainbow: If the sun is rising or setting below the horizon, there will be a bow across the sky where the scattered field is predominantly linearly polarized.[5] It is believed that the Vikings used such a "sunstone" for direction finding to traverse the Atlantic Ocean. A sunstone is shown in Figure 34.4.

Figure 34.4: A sunstone can indicate the polarization of the scattered light. From that, one can deduce where the sun is located, either behind us or in front of us (Figure by: https://en.wikipedia.org).

34.1.3 Small Conductive Particle

The above analysis is for a small dielectric particle. The quasi-static analysis may not be valid for when the conductivity of the particle becomes very large. For instance, for a perfect electric

[5]You can go through a Gedanken experiment to convince yourself of this.

conductor (PEC) immersed in a time-varying electromagnetic field, the magnetic field in the long wavelength limit induces an eddy current in a PEC sphere.[6] Hence, in addition to an electric dipole component, a PEC sphere also has a magnetic dipole component. The scattered field due to a tiny PEC sphere is a linear superposition of an electric and magnetic dipole components. These two dipolar components have electric fields that cancel precisely at certain observation angle. It gives rise to deep null in the bistatic radar cross section (RCS)[7] of a PEC sphere as illustrated by calculation in Figure 34.5.

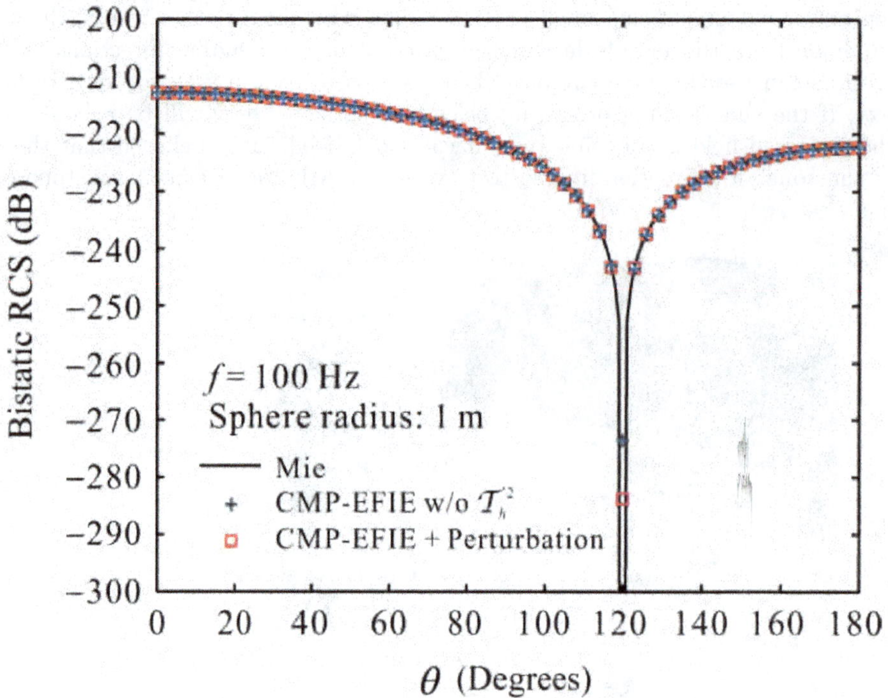

Figure 34.5: RCS (radar cross section) of a small PEC scatterer (Figure by Sheng et al. [273]). There is no true PEC in optics [215], but at RF and microwave, many conductors can be approximated by PEC.

34.2 Mie Scattering

When the size of the scatterer or the sphere becomes larger compared to wavelength λ, quasi-static approximation is insufficient to approximate the solution. Then one has to solve the boundary

[6]Note that there is no PEC at optical frequencies. A metal behaves more like a plasma medium at optical frequencies.

[7]A scattering cross section in the microwave range is called an RCS due to its prevalent use in radar technology.

value problem in its full glory, usually with the full-wave theory or Mie theory [274, 275]. With this theory, the scattering cross section does not grow indefinitely with frequency as in (34.1.22). It has to plateau to a value for increasing frequency. For a sphere of radius a, the scattering cross section becomes πa^2 in the high-frequency limit. This physical feature of this plot is shown in Figure 34.6, and it also explains why the sky is not purple.

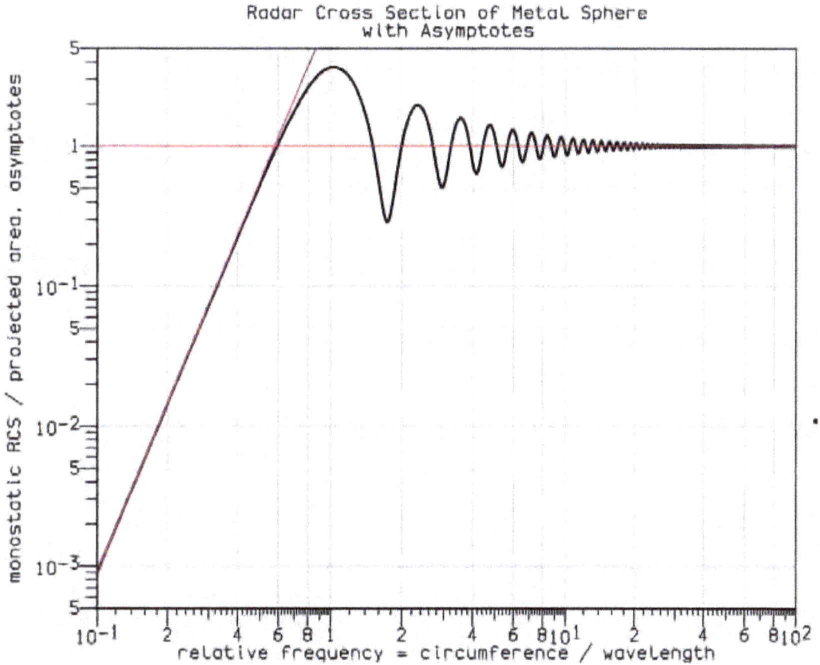

Figure 34.6: Radar cross section (RCS) calculated using Mie scattering theory (Figure by: https://www.researchgate.net) [275].

34.2.1 Optical Theorem

Before we discuss the Mie scattering solution, let us discuss an amazing theorem called the optical theorem. This theorem says that the scattering cross section of a scatterer depends only on the forward-scattering power density of the scatterer. In other words, if a plane wave is incident on a scatterer, the scatterer will scatter the incident wave and power in all directions. But the total power scattered by the object is only dependent on the forward-scattering power density of the object or scatterer. This amazing theorem is called the optical theorem, and a proof of this is given in J.D. Jackson's book [50].

The true physical reason for this is power orthogonality. Two plane waves cannot exchange power with each other unless they are propagating in the same direction, or they share the same $\mathbf{k}$ or

$\boldsymbol{\beta}$ vector. When $\boldsymbol{\beta}$ is both the plane-wave direction of the incident wave as well as the forward-scattered wave, then power can be transferred from the incident plane wave to the forward-scattered wave. This is similar to power orthogonality in a waveguide, where orthogonal modes of a waveguide carry power independently of each other [92, 104].

The scattering pattern of a scatterer for increasing frequency is shown in Figure 34.7. For Rayleigh scattering where the wavelength is long, the scattered power is distributed isotropically (in the azimuthal direction) except for the doughnut shape of the radiation pattern, namely, the $\sin^2\theta$ dependence. As the frequency increases, the power is scattered increasingly in the forward direction. The reason being that for a very short wavelength, the scatterer looks like a disc to the incident wave, casting a shadow in the forward direction. Hence, there has to be scattered field in the forward direction to cancel the incident wave to cast this shadow. This point is often counter-intuitive to students of electromagnetic scattering.

In a nutshell, the optical theorem is intuitively obvious for high-frequency scattering. The amazing part about this theorem is that it is true for all frequencies. But this is a consequence of power orthogonality, as mentioned above.

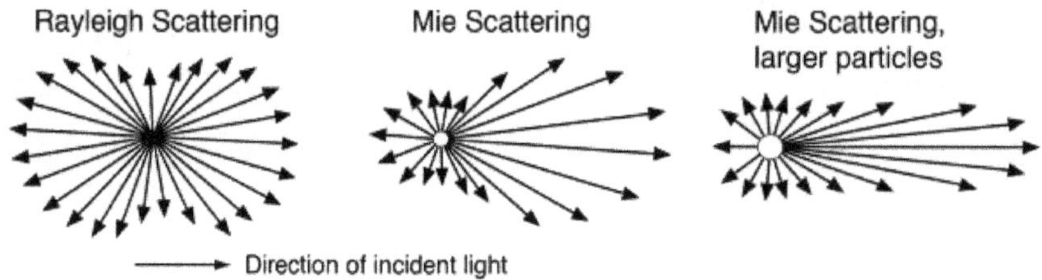

Figure 34.7: A particle scatters increasingly more in the forward direction as the frequency increases (Figure by: hyperphysics.phy-astr.gsu.edu).

34.2.2 Mie Scattering by Spherical Harmonic Expansions

As explained above, as the wavelength becomes shorter, we need to solve the boundary value problem in its full glory without making any approximations like the previous section. This closed-form solution can be found for a sphere scattering by using separation of variables and spherical harmonic expansions that will be discussed in this section.

The Mie scattering solution by a sphere will be discussed later in this lecture.[8] The separation of variables in spherical coordinates is not the only useful for Mie scattering; it is also useful for analyzing spherical cavity problems. So we will present the precursor knowledge so that you can read further into Mie scattering theory next.

[8]It is also treated in J.A. Kong's book [35] and Chapter 3 of W.C. Chew, *Waves and Fields in Inhomogeneous Media* [38], as well as many other textbooks [50, 70, 107, 276].

34.2.3 Separation of Variables in Spherical Coordinates

To this end, we look at the scalar wave equation $(\nabla^2 + \beta^2)\Psi(\mathbf{r}) = 0$ in spherical coordinates. A lookup table can be used to evaluate $\nabla \cdot \nabla$, or divergence of a gradient in spherical coordinates. The Helmholtz wave equation then becomes[9]

$$\left(\frac{1}{r^2}\frac{\partial}{\partial r}r^2\frac{\partial}{\partial r} + \frac{1}{r^2 \sin\theta}\frac{\partial}{\partial\theta}\sin\theta\frac{\partial}{\partial\theta} + \frac{1}{r^2\sin^2\theta}\frac{\partial^2}{\partial\phi^2} + \beta^2\right)\Psi(\mathbf{r}) = 0 \tag{34.2.1}$$

Noting the $\partial^2/\partial\phi^2$ derivative, by using separation of variables technique, we assume $\Psi(\mathbf{r})$ to be of the form

$$\Psi(\mathbf{r}) = F(r,\theta)e^{jm\phi} \tag{34.2.2}$$

This will simplify the $\partial/\partial\phi$ derivative in the partial differential equation since $\frac{\partial^2}{\partial\phi^2}e^{jm\phi} = -m^2e^{jm\phi}$. Then (34.2.1) becomes

$$\left(\frac{1}{r^2}\frac{\partial}{\partial r}r^2\frac{\partial}{\partial r} + \frac{1}{r^2 \sin\theta}\frac{\partial}{\partial\theta}\sin\theta\frac{\partial}{\partial\theta} - \frac{m^2}{r^2\sin^2\theta} + \beta^2\right)F(r,\theta) = 0 \tag{34.2.3}$$

The above can be solved using the separation of variables, by letting that

$$F(r,\theta) = b_n(\beta r)P_n^m(\cos\theta) \tag{34.2.4}$$

where $b_n(\beta r)$ is a spherical Bessel function, and $P_n^m(\cos\theta)$ is the associated Legendre polynomial or function.[10] The associate Legendre function satisfies

$$\left\{\frac{1}{\sin\theta}\frac{d}{d\theta}\sin\theta\frac{d}{d\theta} + \left[n(n+1) - \frac{m^2}{\sin^2\theta}\right]\right\}P_n^m(\cos\theta) = 0 \tag{34.2.5}$$

Note that (34.2.5) is an eigenvalue problem with eigenvalue $n(n+1)$, and $|m| \le |n|$. The value $n(n+1)$ is also known as the separation constant.

Consequently, using (34.2.3), and (34.2.5), we can show that $b_n(\beta r)$ in (34.2.4) satisfies

$$\left[\frac{1}{r^2}\frac{d}{dr}r^2\frac{d}{dr} - \frac{n(n+1)}{r^2} + \beta^2\right]b_n(\beta r) = 0 \tag{34.2.6}$$

The above is the spherical Bessel's equation where $b_n(\beta r)$ is either the spherical Bessel function $j_n(\beta r)$, spherical Neumann function $n_n(\beta r)$, or the spherical Hankel functions $h_n^{(1)}(\beta r)$ and $h_n^{(2)}(\beta r)$. The spherical functions are the close cousins of the cylindrical functions $J_n(x)$, $N_n(x)$, $H_n^{(1)}(x)$, and $H_n^{(2)}(x)$. They are related to the cylindrical functions via [54, 38] [11]

$$b_n(\beta r) = \sqrt{\frac{\pi}{2\beta r}}B_{n+\frac{1}{2}}(\beta r) \tag{34.2.7}$$

---

[9]By a quirk of mathematics, it turns out that the first term on the right-hand side below can be simplified by observing that $\frac{1}{r^2}\frac{\partial}{\partial r}r^2 = \frac{1}{r}\frac{\partial}{\partial r}r$.

[10]Thank goodness these special functions like the Legendre polynomial and the Bessel functions were discovered in the late 1700s before Maxwell's equations were discovered!

[11]By a quirk of nature, the spherical Bessel functions needed for 3D wave equations are in fact simpler than cylindrical Bessel functions needed for 2D wave equations. One can say that 3D is real, but 2D is surreal.

Since (34.2.6) is a second order ordinary differential equation, only two of the four possible solutions are independent similar to the cylindrical Bessel functions'case [1, p. 14], only two of the possible four solutions are linearly independent.

It is customary to define the spherical harmonics as [50, 1]

$$Y_{nm}(\theta,\phi) = \sqrt{\frac{2n+1}{4\pi}\frac{(n-m)!}{(n+m)!}}P_n^m(\cos\theta)e^{jm\phi} \tag{34.2.8}$$

The above is orthonormalized such that

$$Y_{n,-m}(\theta,\phi) = (-1)^m Y_{nm}^*(\theta,\phi) \tag{34.2.9}$$

and that

$$\int_0^{2\pi} d\phi \int_0^{\pi} \sin\theta d\theta Y_{n'm'}^*(\theta,\phi)Y_{nm}(\theta,\phi) = \delta_{n'n}\delta_{m'm} \tag{34.2.10}$$

These functions are also complete (complete in a mathematical sense)[12] like the Fourier series so that

$$\sum_{n=0}^{\infty}\sum_{m=-n}^{n} Y_{nm}^*(\theta',\phi')Y_{nm}(\theta,\phi) = \delta(\phi-\phi')\delta(\cos\theta-\cos\theta') \tag{34.2.11}$$

In general, the solution to the scalar wave equation in spherical coordinates can be expanded into the form

$$\Psi(\mathbf{r}) = \Psi(r,\theta,\phi) = \sum_{n=0}^{\infty}\sum_{m=-n}^{n}[A_{nm}j_n(kr) + B_{nm}n_n(kr)]Y_{nm}(\theta,\phi) \tag{34.2.12}$$

If the solution space includes the origin, and there is no source at the origin, then we can set $B_{nm} = 0$ since $n_n(kr)$ is singular at $r = 0$. In general, the orthonormality of the spherical harmonics expressed in (34.2.11) can be used to find the expansion coefficients A_{nm} and B_{nm} as in finding the coefficients of a Fourier series expansion.

[12] In a nutshell, a set of basis functions is complete in a subspace if any function in the same subspace can be expanded as a sum of these basis functions.

34.3 More on Mie Scattering

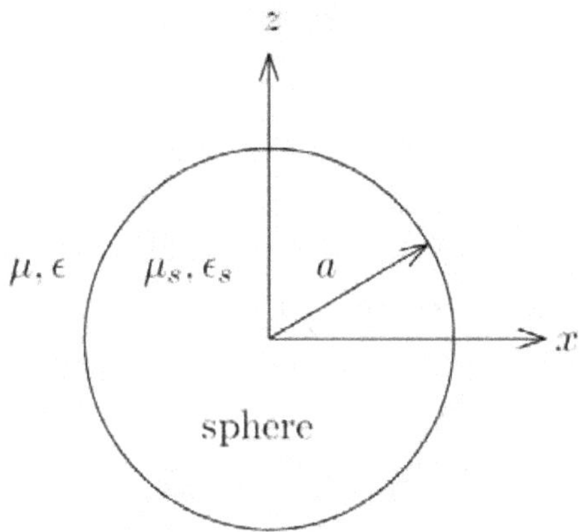

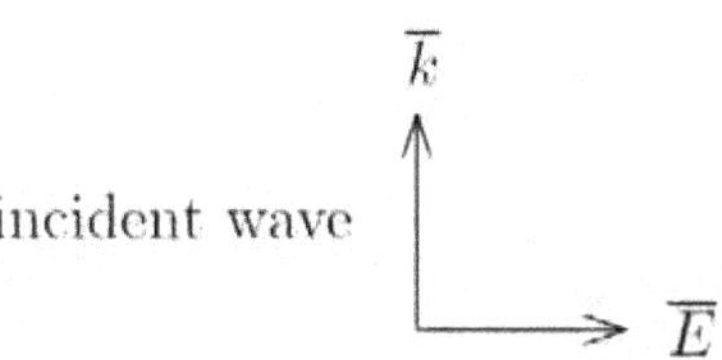

Figure 34.8: Geometry to illustrate the Mie scattering problem. The incident wave is traveling in the z direction so that it is symmetric about the z-axis, which simplifies the analysis. The problem can be solved in closed form using spherical harmonic expansions and Debye potentials.

The solution of electromagnetic scattering by a spherical scatterer can be found exactly. To do this, we need to define the Debye potential π_e and π_m [274]. These are like the pilot potential we defined earlier for hollow waveguides. Unlike a hollow waveguide, the pilot vector is not a constant vector. In spherical coordinates, to characterize the TM to r and TE to r waves, we use pilot vector $\mathbf{r}$. Thus, for TM to r waves, we have

$$\mathbf{H}^{TM} = \nabla \times \mathbf{r}\pi_e \tag{34.3.1}$$

and for TE to r waves, we have

$$\mathbf{E}^{TE} = \nabla \times \mathbf{r}\pi_m. \tag{34.3.2}$$

Using the fact that the above fields are solutions to Helmholtz wave equations, it can be shown that [1]

$$\left(\nabla^2 + k^2\right)\pi_e = 0, \tag{34.3.3}$$

$$\left(\nabla^2 + k^2\right)\pi_m = 0. \tag{34.3.4}$$

The above is incredibly simple even though we have started with a nonconstant pilot vector $\mathbf{r}$.

In general, in a source-free region, we can derive the electromagnetic fields as

$$\mathbf{H} = \nabla \times \mathbf{r}\pi_e + \frac{1}{-j\omega\mu}\nabla \times \nabla \times \mathbf{r}\pi_m, \tag{34.3.5}$$

$$\mathbf{E} = \nabla \times \mathbf{r}\pi_m + \frac{1}{j\omega\epsilon}\nabla \times \nabla \times \mathbf{r}\pi_e. \tag{34.3.6}$$

Furthermore, we can extract the r components of $\mathbf{H}$ and $\mathbf{E}$ in the above to yield [164, Exercise 3.22, Solution Manual]

$$H_r = -\frac{1}{j\omega\mu}\left[\frac{\partial^2}{\partial r^2}r\pi_m + k^2 r\pi_m\right], \tag{34.3.7}$$

$$E_r = \frac{1}{j\omega\epsilon}\left[\frac{\partial^2}{\partial r^2}r\pi_e + k^2 r\pi_e\right]. \tag{34.3.8}$$

Since π_m and π_e are solutions of the scalar wave equations (34.3.3) and (34.3.4), their general solutions are of the form

$$\left\{\begin{matrix} j_n(kr) \\ h_n^{(2)}(kr)\end{matrix}\right\} P_n^m(\cos\theta)\left\{\begin{matrix}\cos m\phi \\ \sin m\phi\end{matrix}\right\} \tag{34.3.9}$$

where the brackets imply "linear superpositions of". In the above, $j_n(x)$ is the spherical Bessel function and $h_n^{(2)}(x)$ is the spherical Hankel function of the second kind. The above is the spherical harmonics representation of the wave solution in spherical coordinates.

Since π_m and π_e are of the form given in (34.3.9), then from the scalar wave equation expressed in spherical coordinates, we can show that (34.3.7) and (34.3.8) simplify to

$$H_r = -\frac{1}{j\omega\mu}\frac{n(n+1)}{r}\pi_m \tag{34.3.10}$$

$$E_r = \frac{1}{j\omega\varepsilon}\frac{n(n+1)}{r}\pi_e \tag{34.3.11}$$

In other words, H_r and E_r uniquely define π_m and π_e

Interface Boundary Conditions

The interface boundary conditions for a spherical surface are that tangential E and H are continuous.

By extracting the transverse to r components of (34.3.5) and (34.3.6), we have

$$\mathbf{H}_s = -\mathbf{r} \times \nabla_s\pi_e + \frac{1}{i\omega\mu}\frac{1}{r}\frac{\partial}{\partial r}r^2\nabla_s\pi_m, \tag{34.3.12}$$

$$\mathbf{E}_s = -\mathbf{r} \times \nabla_s\pi_m - \frac{1}{i\omega\epsilon}\frac{1}{r}\frac{\partial}{\partial r}r^2\nabla_s\pi_e. \tag{34.3.13}$$

One can easily show that if we have only TE to r or only TM to r waves, the boundary conditions can be satisfied still. This implies that the boundary conditions for continuous tangential E and tangential H fields do not couple the TE and TM waves. Or the TE wave solution can be solved independently of the TM wave solution![13]

It can be shown that to ensure continuity of tangential E and H fields, it is necessary to have the continuity of

$$\pi_m, \quad \pi_e \tag{34.3.14}$$

and the continuity of

$$\frac{1}{r\varepsilon}\frac{\partial}{\partial r} r\pi_e, \quad \frac{1}{r\mu}\frac{\partial}{\partial r} r\pi_m \tag{34.3.15}$$

To re-emphasize, the TE and TM boundary conditions can be satisfied independently of each other. We will take advantage of this to simplify the solution to this problem and solve the TE and TM scattering problems independently of each other.

Therefore, to solve the sphere scattering problem of a plane wave, we need first to expand the plane wave in terms of spherical harmonics. Then we need to extract the TE to r or TM to r waves or the Debye potentials.

First, if we have a plane wave polarized in the $\hat{x}$ direction incident on the spherical scatterer, then

$$\mathbf{E} = \hat{x}E_0 e^{-jkz} = \hat{x}E_0 e^{-jkr\cos\theta} \tag{34.3.16}$$

$$\mathbf{H} = \hat{y}\sqrt{\frac{\varepsilon}{\mu}} E_0 e^{-jkr\cos\theta} \tag{34.3.17}$$

Further, we need to use the *wave transformation* that a plane wave can be expressed in terms of spherical harmonics. Namely,

$$e^{-jkr\cos\theta} = \sum_{n=0}^{\infty} j^{-n}(2n+1)j_n(kr)P_n(\cos\theta) \tag{34.3.18}$$

The incident wave is chosen so that it is axially symmetric simplifying the spherical harmonics expansion: only $m = 0$ harmonic is needed (see (34.2.12)). (The above is proved in Harrington [55].) Thus we can write the incident field in (34.3.16) after using $\hat{r}\cdot\hat{x} = \sin\theta\cos\phi$ as

$$E_r^{\text{inc}} = E_0 \sin\theta\cos\phi e^{-jkr\cos\theta} \tag{34.3.19}$$

$$= E_0 \frac{\cos\phi}{jkr}\frac{\partial}{\partial\theta}\left(e^{-jkr\cos\theta}\right) \tag{34.3.20}$$

Thus we expressed the incident electric field, after using the wave transformation, as

$$E_r^{\text{inc}} = \frac{E_0\cos\phi}{jkr}\sum_{n=0}^{\infty} j^{-n}(2n+1)j_n(kr)\frac{\partial}{\partial\theta}P_n(\cos\theta) \tag{34.3.21}$$

[13]It is a mystery of nature that wave problems in odd dimensions are simpler than problems in even dimensions!

Then using $\frac{\partial}{\partial\theta}P_n(\cos\phi) = P_n^1(\cos\theta)$, the above becomes

$$E_r^{\text{inc}} = -j\frac{E_0\cos\phi}{(kr)^2}\sum_{n=1}^{\infty} j^{-n}(2n+1)\hat{J}_n(kr)P_n^1(\cos\theta) \tag{34.3.22}$$

The summation begins with $n = 1$ because $P_0^1(\cos\theta) = 0$. Also, we define $\hat{J}_n(kr) = krj_n(kr)$ which is the normalized spherical Bessel function.

Next, we need to identify the TE and TM waves buried in the incident wave. Comparing the above equations (34.3.10) and (34.3.11), we identify the Debye potentials for the TM and TE polarizations, given the **r** component of the fields. Thus for the incident fields, the Debye potentials are identified as

$$\pi_e^{\text{inc}} = -\frac{E_0\cos\phi}{\omega\mu r}\sum_{n=1}^{\infty} j^{-n}\frac{(2n+1)}{n(n+1)}\hat{J}_n(kr)P_n^1(\cos\theta) \tag{34.3.23}$$

$$\pi_m^{\text{inc}} = \frac{E_0\sin\phi}{kr}\sum_{n=1}^{\infty} j^{-n}\frac{(2n+1)}{n(n+1)}\hat{J}_n(kr)P_n^1(\cos\theta) \tag{34.3.24}$$

Next, we write down the Debye potentials for the TM scattered fields to be

$$\pi_e^s = \frac{E_0\cos\phi}{\omega\mu r}\sum_{n=1}^{\infty} a_n\hat{H}_n^{(2)}(kr)P_n^1(\cos\theta) \tag{34.3.25}$$

$$\pi_m^s = \frac{E_0\sin\phi}{kr}\sum_{n=1}^{\infty} b_n\hat{H}_n^{(2)}(kr)P_n^1(\cos\theta) \tag{34.3.26}$$

Here, $\hat{H}_n^{(2)}(x)$ are normalized spherical Hankel functions of the second kind. These scattered fields have to be outgoing waves when $kr \to \infty$. Hence, spherical Hankel functions of the second kind are chosen with $\exp(j\omega t)$ time dependence.

In addition, we need to write down the solutions inside the spherical scatterer. They have to be regular at the origin or at $r = 0$. Therefore, normalized spherical Bessel functions $\hat{J}_n(x)$ are chosen as the solutions. Consequently,

$$\pi_e^i = \frac{E_0\cos\phi}{\omega\mu_s r}\sum_{n=1}^{\infty} c_n\hat{J}_n(k_s r)P_n^1(\cos\theta) \tag{34.3.27}$$

$$\pi_m^i = \frac{E_0\sin\phi}{k_s r}\sum_{n=1}^{\infty} d_n\hat{J}_n(k_s r)P_n^1(\cos\theta) \tag{34.3.28}$$

As discussed before, the scattering by a spherical scatter can be solved as two independent problems of TE to r and TM to r waves. Thus, after matching boundary conditions that π_e and $\frac{1}{\varepsilon}\frac{\partial}{\partial r}r\pi_e$ (and their dual for the TM case) are continuous at the interface,[14] we can show that

$$a_n = \frac{j^{-n}(2n+1)}{n(n+1)}\,\frac{-\sqrt{\varepsilon_s\mu}\hat{J}_n'(ka)\hat{J}_n(k_s a) + \sqrt{\varepsilon\mu_s}\hat{J}_n(ka)\hat{J}_n'(k_s a)}{\sqrt{\varepsilon_s\mu}\hat{H}_n^{(2)\prime}(ka)\hat{J}_n(k_s a) - \sqrt{\varepsilon\mu_s}\hat{H}_n^{(2)}(ka)\hat{J}_n'(k_s a)} \tag{34.3.29}$$

[14]These boundary conditions also follow from eq. (3.8.1) of [1].

$$b_n = \frac{j^{-n}(2n+1)}{n(n+1)} \frac{-\sqrt{\varepsilon_s\mu}\hat{J}_n(ka)\hat{J}_n'(k_s a) + \sqrt{\varepsilon\mu_s}\hat{J}_n'(ka)\hat{J}_n(k_s a)}{\sqrt{\varepsilon_s\mu}\hat{H}_n^{(2)}(ka)\hat{J}_n'(k_s a) - \sqrt{\varepsilon\mu_s}\hat{H}_n^{(2)'}(ka)\hat{J}_n(k_s a)} \tag{34.3.30}$$

$$c_n = \frac{j^{-n}(2n+1)}{n(n+1)} \frac{-j\sqrt{\varepsilon_s\mu}}{\sqrt{\varepsilon_s\mu}\hat{H}_n^{(2)'}(ka)\hat{J}_n(k_s a) - \sqrt{\varepsilon\mu_s}\hat{H}_n^{(2)}(ka)\hat{J}_n'(k_s a)} \tag{34.3.31}$$

$$d_n = \frac{j^{-n}(2n+1)}{n(n+1)} \frac{j\sqrt{\varepsilon_s\mu}}{\sqrt{\varepsilon_s\mu}\hat{H}_n^{(2)}(ka)\hat{J}_n'(k_s a) - \sqrt{\varepsilon\mu_s}\hat{H}_n^{(2)'}(ka)\hat{J}_n(k_s a)} \tag{34.3.32}$$

The Wronskian of spherical Bessel functions has been used to simplify the above [38, p. 189], [50]; namely,

$$\hat{J}_n(ka)\hat{H}_n^{(2)'}(ka) - \hat{J}_n'(ka)\hat{H}_n^{(2)}(ka) = -j$$

These solutions, complicated though they are, are very useful because they are the exact solution of a scattering by a spherical object in electromagnetics. They are useful for validating numerical solutions of scattering problems.

Exercises for Lecture 34

Problem 34-1:

(i) Explain how the Vikings could have used the physical results of Rayleigh scattering to navigate themselves across the North Atlantic Ocean to arrive at Iceland.

(ii) Verify (34.1.22) and explain why it is also valid for the complex permittivity case.

(iii) Give an intuitive reason as to why the optical theorem makes sense in the high-frequency limit.

(iv) By using the separation of variables, explain how you would solve the Helmholtz wave equation in 3D in spherical coordinates.

$$\left(\nabla^2 + \beta^2\right)\Psi(\mathbf{r}) = 0$$

(v) By using the interface boundary conditions suggested before (34.3.29), derive the constants a_n, b_n, c_n, and d_n and verify if they are correct.

Lecture 35

Spectral Expansions of Source Fields—Sommerfeld Integrals

In previous lectures, we have assumed plane waves in finding closed form solutions. Plane waves are simple waves, and their closed form solutions for reflections off a flat surface or a planarly layered medium can be found easily. But plane waves are figments of mathematical imaginations that are not encountered in the real world.

All sources can be thought of as linear superpositions of point sources. Thus it is important to study the field generated by a point source first, and how its field interacts with the environment. When the source is a point source, it generates a spherical wave. We do not know how to reflect a spherical wave off a planar interface exactly, but we do know how to reflect plane waves off a planar interfaces. But by expanding a spherical wave as a sum of plane waves and evanescent waves using Fourier transform technique, we can solve for the solution of a point source over a layered medium easily in terms of spectral integrals using Fourier transform in space. But these integrals are complicated: because Sommerfeld was the first person to have derived and studied them, these integrals are often called Sommerfeld integrals.

Finally, we shall apply the method of stationary phase to approximate these complicated integrals to elucidate their physics. From this, we can see ray physics and Fermat's principle emerging from the complicated mathematics. It reminds us of a lyric from the musical The Sound of Music—ray, a drop of golden sun! Rays have mesmerized the human mind, and it will be interesting to see if the mathematics behind them are equally enchanting.

By this time, you probably feel inundated by the ocean of knowledge that you are imbibing. But if you can assimilate these concepts, it will be an exhilarating experience as the knowledge will last a lifetime. Also, in the latter part of the course, we will use $e^{-i\omega t}$ time convention as is often used by mathematicians, physicists, and the optics community.

35.1 Spectral Representations of Sources

As mentioned above, a plane wave is a mathematical idealization that does not exist in the real world. In practice, waves are nonplanar in nature as they are generated by finite sources, such as antennas and scatterers. For example, a point source generates a spherical wave, which is nonplanar. Fortunately, these nonplanar waves can be expanded as sum of plane waves. Once this is done, then the study of non-plane-wave reflections from a layered medium becomes routine.

In the following, we shall show how waves resulting from a point source can be expanded in terms of a plane-wave summation (or integral). This topic is found in many textbooks [2, 35, 38, 107, 145, 146, 265, 277].

35.1.1 A Point Source—Fourier Expansion and Contour Integration

There are a number of ways to derive the plane-wave expansion of a point source [1][Lect. 36]. We will illustrate one of the ways. The Fourier expansion in space, or spectral decomposition, or the plane-wave expansion of the field due to a point source could be derived using Fourier transform technique. First, notice that the scalar wave equation for the field due to a point source at the origin is

$$\left(\nabla^2 + k_0^2\right)\phi(x,y,z) = \left[\frac{\partial^2}{\partial x^2} + \frac{\partial^2}{\partial y^2} + \frac{\partial^2}{\partial z^2} + k_0^2\right]\phi(x,y,z) = -\delta(x)\,\delta(y)\,\delta(z). \tag{35.1.1}$$

The above equation could be solved for the field $\phi(x,y,z)$ in the spherical coordinates, yielding the solution given in the previous lecture, namely, Green's function with the source point at the origin, or that[1]

$$\phi(x,y,z) = \phi(r) = \frac{e^{ik_0 r}}{4\pi r}. \tag{35.1.2}$$

The solution is entirely spherically symmetric due to the symmetry and location of the point source.

Next, assuming that the Fourier transform of $\phi(x,y,z)$ exists,[2] we can write using Fourier transform theory that

$$\phi(x,y,z) = \frac{1}{(2\pi)^3}\iiint_{-\infty}^{\infty} dk_x dk_y dk_z\, \tilde{\phi}(k_x,k_y,k_z) e^{ik_x x + ik_y y + ik_z z}. \tag{35.1.3}$$

Then we substitute the above into the left-hand side of (35.1.1), and after exchanging the order of differentiation and integration,[3] one can simplify the Laplacian operator in the Fourier space, or spectral domain, to arrive at

$$\nabla^2 = \frac{\partial^2}{\partial x^2} + \frac{\partial^2}{\partial y^2} + \frac{\partial^2}{\partial z^2} \Rightarrow -k_x^2 - k_y^2 - k_z^2$$

[1]From this point onward, we will adopt the $\exp(-i\omega t)$ time convention to be commensurate with the optics and physics literatures.

[2]The Fourier transform of a function $f(x)$ exists if it is absolutely integrable, namely that $\int_{-\infty}^{\infty}|f(x)|dx$ is finite (see [1]).

[3]Exchanging the order of differentiation and integration is allowed if the integral converges after the exchange.

by Fourier Transforms. Then, together with the Fourier representation of the delta function on the right-hand side of (35.1.1), which is[4]

$$\delta(x)\,\delta(y)\,\delta(z) = \frac{1}{(2\pi)^3}\iiint\limits_{-\infty}^{\infty} dk_x dk_y dk_z\, e^{ik_x x + ik_y y + ik_z z} \tag{35.1.4}$$

we convert (35.1.1) into

$$\iiint\limits_{-\infty}^{\infty} dk_x dk_y dk_z\, [k_0^2 - k_x^2 - k_y^2 - k_z^2]\tilde{\phi}(k_x, k_y, k_z) e^{ik_x x + ik_y y + ik_z z}$$
$$= -\iiint\limits_{-\infty}^{\infty} dk_x dk_y dk_z\, e^{ik_x x + ik_y y + ik_z z}. \tag{35.1.5}$$

where the Fourier transform of the left-hand side of (35.1.1) has been taken, where the second derivatives in x, y, and z are converted to $-k_x^2 - k_y^2 - k_z^2$. Since the above Fouier integral is equal for all x, y, and z, by the uniqueness of Fourier transforms [57, p. 286], we deduce that their integrands are the same. Hence, we arrive at the inverse transform of the above to get

$$\tilde{\phi}(k_x, k_y, k_z) = \frac{-1}{k_0^2 - k_x^2 - k_y^2 - k_z^2}. \tag{35.1.6}$$

Consequently, using this in (35.1.3), we have that $\phi(x, y, z)$ expressed in terms of the Fourier inverse transform–namely,

$$\phi(x, y, z) = \frac{-1}{(2\pi)^3}\iiint\limits_{-\infty}^{\infty} d\mathbf{k}\, \frac{e^{ik_x x + ik_y y + ik_z z}}{k_0^2 - k_x^2 - k_y^2 - k_z^2}. \tag{35.1.7}$$

where $d\mathbf{k} = dk_x dk_y dk_z$. The above expresses the fact the $\phi(x, y, z)$, which is a spherical wave by (35.1.2), is expressed as an integral summation of "plane waves". But these "plane waves" are not physical plane waves in free space since $k_x^2 + k_y^2 + k_z^2 \neq k_0^2$. In other words, these "plane waves" do not satisfy the physical dispersion relation of a plane wave.

[4]We have made use of that $\delta(x) = 1/(2\pi)\int_{-\infty}^{\infty} dk_x \exp(ik_x x)$ three times.

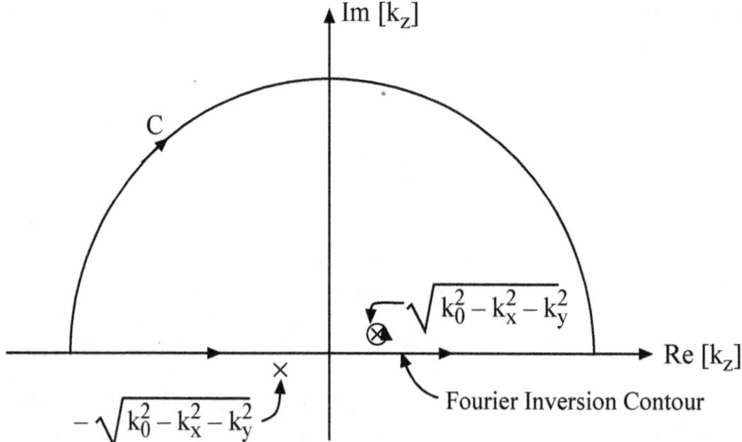

Figure 35.1: By invoking Cauchy's residue theorem, the integration along the real axis is equal to the integration along C plus the residue of the pole at $k_z = (k_0^2 - k_x^2 - k_y^2)^{1/2}$. By invoking Jordan's lemma, the integration around C is zero for most integrals, and thus, we require only the residue evaluation.

Weyl Identity—Plane-Wave Expansion of a Point-Source Field

To make the plane waves in (35.1.7) into physical plane waves, we have to massage it into a different form. We rearrange the triple integrals in (35.1.7) so that the dk_z integral is performed first. In other words,

$$\phi(\mathbf{r}) = \frac{1}{(2\pi)^3} \iint\limits_{-\infty}^{\infty} dk_x dk_y e^{ik_x x + ik_y y} \int_{-\infty}^{\infty} dk_z \frac{e^{ik_z z}}{k_z^2 - (k_0^2 - k_x^2 - k_y^2)} \tag{35.1.8}$$

where we have deliberately rearranged the denominator with k_z being the variable in the inner integral. Then the integrand has poles at $k_z = \pm(k_0^2 - k_x^2 - k_y^2)^{1/2}$.[5] Moreover, for real k_0 and real values of k_x and k_y, these two poles lie on the real axis, rendering the integral in (35.1.7) undefined. However, if a small loss is assumed in k_0 such that $k_0 = k_0' + ik_0''$, then the poles migrate to be off the real axis (see Figure 35.1), and the integrals in (35.1.8) are well-defined. (In actual fact, this is intimately related to the uniqueness principle we have studied before: An infinitesimal loss is needed to guarantee uniqueness in an open space, as shall be explained below (see Lecture 11).

First, the reason is that without loss, $|\phi(\mathbf{r})| \sim O(1/r)$, $r \to \infty$ is not strictly absolutely integrable; hence, its Fourier transform does not exist [57]: The manipulation that leads to (35.1.7) is not strictly correct. Second, the introduction of a small loss also guarantees the Sommerfeld radiation condition needed for the uniqueness of the solution to (35.1.1), and therefore, the equality of (35.1.2) and (35.1.8) [38].

[5]In (35.1.7), the pole is located at $k_x^2 + k_y^2 + k_z^2 = k_0^2$. This equation describes a sphere in $\mathbf{k}$ space, known as the Ewald's sphere [278].

Observe that in (35.1.8), when $z > 0$, the integrand is exponentially small when $\Im m[k_z] \to \infty$. Therefore, by Jordan's lemma [94], the integration for k_z vanishes over the contour C as shown in Figure 35.1. Then, by Cauchy's theorem [94], the integration over the Fourier inversion contour on the real axis is the same as integrating around the pole singularity located at $k_z = (k_0^2 - k_x^2 - k_y^2)^{1/2}$, yielding the residue of the pole (see Figure 35.1). Consequently, after doing the residue evaluation, a triple integral becomes a double integral, and we have

$$\phi(x, y, z) = \frac{i}{2(2\pi)^2} \iint_{-\infty}^{\infty} dk_x dk_y \frac{e^{ik_x x + ik_y y + ik_z' z}}{k_z'}, \quad z > 0, \tag{35.1.9}$$

where $k_z' = (k_0^2 - k_x^2 - k_y^2)^{1/2}$ is the value of k_z at the pole location.

Similarly, for $z < 0$, we can add a contour C in the lower half plane that contributes zero to the integral. Therefore, one can deform the contour to pick up the pole contribution. As such, the integral is equal to the pole contribution at $k_z' = -(k_0^2 - k_x^2 - k_y^2)^{1/2}$ (see Figure 35.1). An expression similar to (35.1.9) can be derived for $z < 0$. Consequently, the result valid for all z can be written as

$$\phi(x, y, z) = \frac{i}{2(2\pi)^2} \iint_{-\infty}^{\infty} dk_x dk_y \frac{e^{ik_x x + ik_y y + ik_z' |z|}}{k_z'}, \quad \text{all } z. \tag{35.1.10}$$

By the uniqueness of the solution to the partial differential equation (35.1.1) satisfying radiation condition at infinity, we can equate (35.1.2) and (35.1.10), yielding the identity

$$\frac{e^{ik_0 r}}{r} = \frac{i}{2\pi} \iint_{-\infty}^{\infty} dk_x dk_y \frac{e^{ik_x x + ik_y y + ik_z |z|}}{k_z}, \tag{35.1.11}$$

where $k_x^2 + k_y^2 + k_z^2 = k_0^2$, or $k_z = (k_0^2 - k_x^2 - k_y^2)^{1/2}$. Now the plane wave in the integrand of the above is a physical plane wave. The above is known as the *Weyl identity* (Hermann Weyl, 1919)[6]. To ensure the radiation condition, we require that $\Im m[k_z] > 0$ and $\Re e[k_z] > 0$ over all values of k_x and k_y in the integration. Furthermore, Equation (35.1.11) could be interpreted as an integral summation of physical plane waves propagating in all directions, including evanescent waves. It is the plane-wave expansion (including the evanescent wave) of a spherical wave.

[6]You will notice that this was derived after Sommerfeld had solved his problem, implying that Sommerfeld should have known this identity before Weyl.

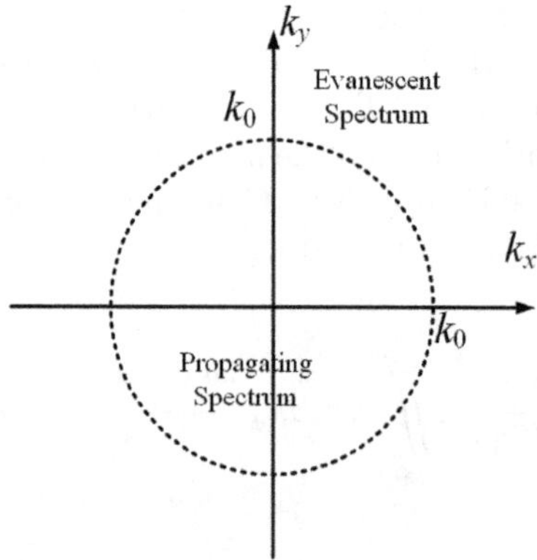

Figure 35.2: The integral in the Weyl identity is done over the entire k_x-k_y plane. The wave is propagating for $\mathbf{k}_\rho = \hat{x}k_x + \hat{y}k_y$ vectors inside the disk, while the wave is evanescent for $\mathbf{k}_\rho$ outside the disk. This is easily seen by the $\exp(ik_z|z|) = \exp(i\sqrt{k_0^2 - k_x^2 - k_y^2}|z|)$ dependence of the integrand in (35.1.11). Therefore, the high spectral frequency components of the Fourier integral that is outside the disk form the evanescent spectrum.

One can also interpret the above as a 2D integral in the Fourier space over the k_x-k_y plane or variables. When $k_x^2 + k_y^2 < k_0^2$, or the spatial spectrum involving k_x and k_y is inside a disk of radius k_0, the waves are propagating waves. But for contributions outside this disk, the waves are evanescent (see Figure 35.2). And the high Fourier (or spectral) components of the Fourier spectrum correspond to evanescent waves. These high spectral components, which are related to the evanescent waves, are important for reconstituting the singularity of the Green's function.[7] Unfortunately, they become exponentially small the further we are from the source point.

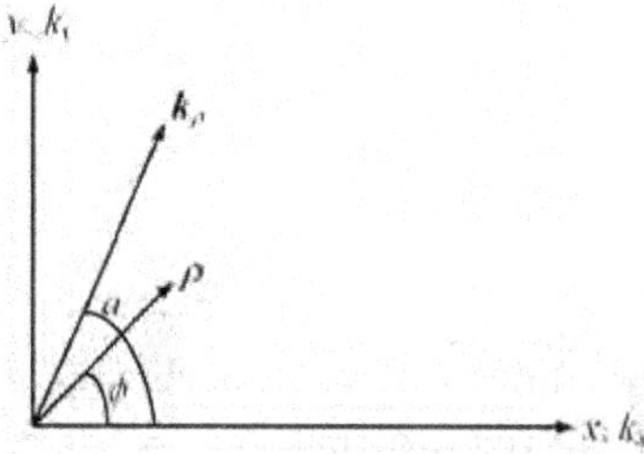

Figure 35.3: The $\mathbf{k}_\rho$ and the $\boldsymbol{\rho}$ vectors on the k_x-k_y plane and the x-y plane. The two planes are superposed in the same figure in this picture. The 2D plane wave is given by $\exp(i\mathbf{k}_\rho \cdot \boldsymbol{\rho})$.

[7]It may be difficult to wrap your head around so many new concepts, and you will have to contemplate and reread them to deeply understand them.

Sommerfeld Identity—A Semi-Infinite Integral

The Weyl identity has a double integral, and hence, is more difficult to integrate numerically. Here, we shall derive the Sommerfeld identity, which has only one semi-infinite integral. First, in (35.1.11), we express the integral in cylindrical coordinates and write $\mathbf{k}_\rho = \hat{x}k_\rho \cos\alpha + \hat{y}k_\rho \sin\alpha$, $\boldsymbol{\rho} = \hat{x}\rho\cos\phi + \hat{y}\rho\sin\phi$ (see Figure 35.3). Then in cylindrical coordinates, $dk_x dk_y = k_\rho dk_\rho\, d\alpha$, and $k_x x + k_y y = \mathbf{k}_\rho \cdot \boldsymbol{\rho} = k_\rho \rho \cos(\alpha - \phi)$, and with the appropriate change of variables, (35.1.11) then becomes

$$\frac{e^{ik_0 r}}{r} = \frac{i}{2\pi} \int_0^\infty k_\rho dk_\rho \int_0^{2\pi} d\alpha \frac{e^{ik_\rho \rho \cos(\alpha-\phi)+ik_z|z|}}{k_z}, \tag{35.1.12}$$

where $k_z = (k_0^2 - k_x^2 - k_y^2)^{1/2} = (k_0^2 - k_\rho^2)^{1/2}$, where in cylindrical coordinates, in the $\mathbf{k}_\rho$-space, or the Fourier space, $k_\rho^2 = k_x^2 + k_y^2$. As such, using the integral identity for the Bessel functions given by[8]

$$J_0(k_\rho \rho) = \frac{1}{2\pi} \int_0^{2\pi} d\alpha\, e^{ik_\rho \rho \cos(\alpha-\phi)}, \tag{35.1.13}$$

We can use the above to replace the $d\alpha$ integral in (35.1.12). Therefore, (35.1.12) becomes

$$\frac{e^{ik_0 r}}{r} = i \int_0^\infty dk_\rho \frac{k_\rho}{k_z} J_0(k_\rho \rho) e^{ik_z|z|}. \tag{35.1.14}$$

The above is also known as the *Sommerfeld identity* (Sommerfeld 1909 [161]; [265, p. 242]). Its physical interpretation is that a spherical wave can now be expanded as an integral summation of conical waves or cylindrical waves in the ρ direction, times a plane wave in the z direction over all wavenumbers k_ρ. This wave is propagating in the $\pm z$ direction when $k_\rho < k_0$, but evanescent in the $\pm z$ direction when $k_\rho > k_0$ as shown in Figure 35.2.

By using the fact that $J_0(k_\rho\rho) = 1/2[H_0^{(1)}(k_\rho\rho) + H_0^{(2)}(k_\rho\rho)]$, and the reflection formula that $H_0^{(1)}(e^{i\pi}x) = -H_0^{(2)}(x)$,[9] then a variation of the above identity can be derived as [38]

$$\frac{e^{ik_0 r}}{r} = \frac{i}{2} \int_{-\infty}^\infty dk_\rho \frac{k_\rho}{k_z} H_0^{(1)}(k_\rho \rho) e^{ik_z|z|}. \tag{35.1.15}$$

[8]See Chew [38][eq. (2.2.15)], or Whitaker and Watson (1927) [279].

[9]The reflection formula allows the analytic continuation of the integrand from the positive k_ρ to the negative k_ρ axis.

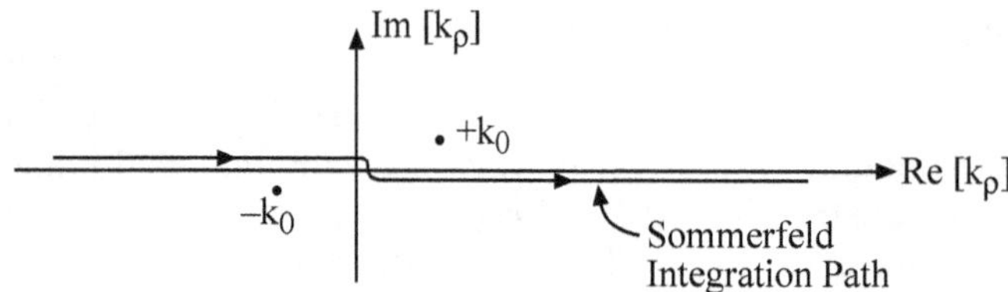

Figure 35.4: This figure illustrates the Sommerfeld integration path. Along this path, the integrand is well defined, as the path does not collide with any of the singularities of the integrand. Due to the existence of branch cuts and branch points, the integration path has to be defined on the correct Riemann sheet (see discussion in the text and also reference [1]).

Since $H_0^{(1)}(z)$ has a logarithmic branch-point singularity at $z = 0$,[10] and $k_z = (k_0^2 - k_\rho^2)^{1/2}$ has algebraic branch-point singularities at $k_\rho = \pm k_0$, the integrand is a multivalue function. For proper bookkeeping of a multivalue function, branch cuts and Riemann sheets need to be defined [94, 1]. Thus, the integral in Equation (35.1.15) is undefined unless we stipulate also the path of integration and the Riemann sheet. Thus, a path of integration adopted by Sommerfeld, which is even good for a lossless medium, is shown in Figure 35.4. Because of the manner in which we have selected the reflection formula for the Hankel functions—that is, $H_0^{(1)}(e^{i\pi}x) = -H_0^{(2)}(x)$–the path of integration should be above the logarithmic branch-point singularity at the origin. With this definition of the Sommerfeld integration, the integral is well defined even when there is no loss—that is, when the branch points $\pm k_0$ are on the real axis.

35.2 A Source on Top of a Layered Medium

Previously, we have studied the propagation of plane electromagnetic waves from a single dielectric interface in section 18.1 as well as through a layered medium in section 20.1. It can be shown that plane waves reflecting from a layered medium can be decomposed into TE-type plane waves, where $E_z = 0$, $H_z \neq 0$, and TM-type plane waves, where $H_z = 0$, $E_z \neq 0$.[11] One also sees how the field due to a point source can be expanded into plane waves in section 35.1.

In view of the above observations, when a point source is on top of a layered medium, it is then best to decompose the field of the point source in terms of plane waves of TE type and TM type. Then, the nonzero component of E_z characterizes TM-to-z waves, while the nonzero component of H_z characterizes TE-to-z waves. Hence, given a field, its TM and TE components can be extracted readily. Furthermore, if these TM and TE components are expanded in terms of plane waves, their propagation in a layered medium can be studied easily.

The problem of a vertical electric dipole on top of a half space was first solved by Sommerfeld (1909) [161] using Hertzian potentials, which are related to the z components of the electromagnetic field. The work is later generalized to layered media, as discussed in the literature. Later, Kong

[10] $H_0^{(1)}(z) \sim \frac{2i}{\pi}\ln(z)$, see Chew [38, p. 14], or Abromawitz or Stegun [115].

[11] Chew, *Waves and Fields in Inhomogeneous Media* [38]; Kong, *Electromagnetic Wave Theory* [35].

(1972) [280] suggested the use of the z components of the electromagnetic field instead of the Hertzian potentials.

35.2.1 Electric Dipole Fields–Spectral Expansion

The representation of a spherical wave in terms of plane waves can be done using the Weyl identity or the Sommerfeld identity. Here, we will use the Sommerfeld identity in anticipation of simpler numerical integration, since only single integrals are involved. The **E** field in a homogeneous medium due to a point current source or a Hertzian dipole directed in the $\hat{\alpha}$ direction, $\mathbf{J} = \hat{\alpha} I\ell\,\delta(\mathbf{r})$, is derivable via the vector potential method or the dyadic Green's function approach. Then, using the dyadic Green's function approach, or the vector/scalar potential approach, the field due to a Hertzian dipole is given by

$$\mathbf{E}(\mathbf{r}) = i\omega\mu \left(\bar{\mathbf{I}} + \frac{\nabla\nabla}{k^2} \right) \cdot \hat{\alpha} I\ell \frac{e^{ikr}}{4\pi r}, \tag{35.2.1}$$

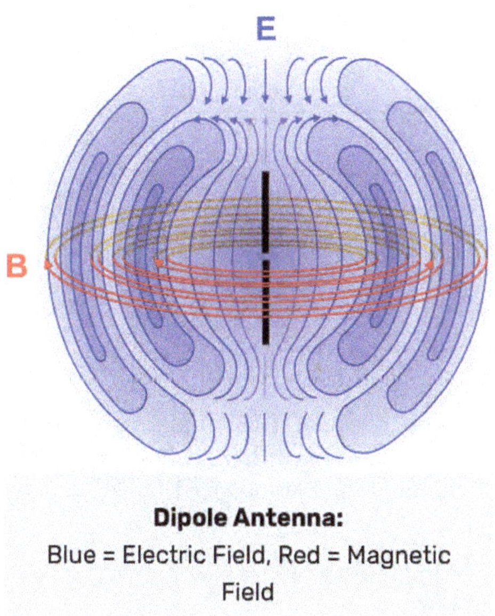

Figure 35.5: Sketch of the time-harmonic EM fields of a VED. The electric field resembles that of Figure 29.1(a) which is the electric field of a radiating dipole. Only part of the magnetic field is shown here (Figure by: Subsonic.ae).

where $I\ell$ is the current moment and $k = \omega\sqrt{\mu\epsilon}$, is the wavenumber of the homogeneous medium. Furthermore, from $\nabla \times \mathbf{E} = i\omega\mu\mathbf{H}$, the magnetic field due to a Hertzian dipole is shown to be given by (after using $\nabla \times \nabla = 0$)

$$\mathbf{H}(\mathbf{r}) = \nabla \times \hat{\alpha} I\ell \frac{e^{ikr}}{4\pi r}. \tag{35.2.2}$$

By picking $\hat{\alpha} = \hat{z}$, this becomes a vertical electric dipole (VED). With the above fields, their TM-to-z and TE-to-z components can be extracted easily in anticipation of their plane-wave expansions for propagation through layered media. A sketch of the EM fields is shown in Figure 35.5. As we are in the near field, the electric field resembles that of a static electric dipole shown previously in Figure 28.4. But when we are away from the source, the field exhibits wave phenomenon like that shown in Figure 29.1(a). However, the magnetic field shown in Figure 35.5 is not true to its form (only magnetic field in one plane is shown). The magnetic field is described by $\mathbf{H} = \hat{\phi} H_\phi$ permeates the space around the dipole.

(a) Vertical Electric Dipole (VED)—Spectral Expansion

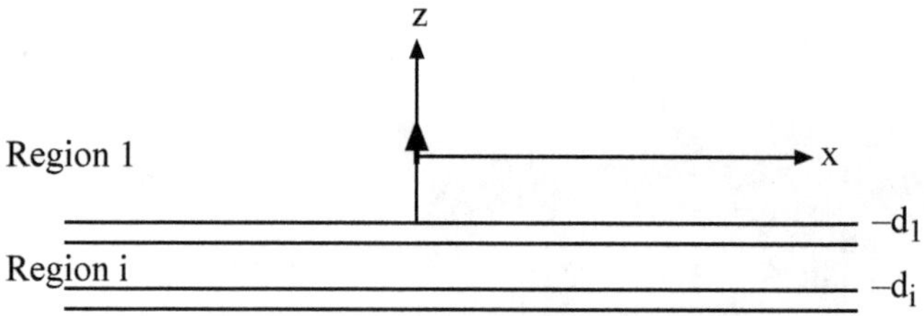

Figure 35.6: A vertical electric dipole over a layered medium.

We start with the vertical electric dipole first because it is simpler as it produces only TM waves. A vertical electric dipole shown in Figure 35.6 has $\hat{\alpha} = \hat{z}$; hence, in anticipation of their plane-wave expansions, the TM-to-z component of the field is characterized by $E_z \neq 0$ or that

$$E_z = \frac{i\omega\mu I\ell}{4\pi k^2} \left(k^2 + \frac{\partial^2}{\partial z^2} \right) \frac{e^{ikr}}{r}, \tag{35.2.3}$$

and the TE component of the field is characterized by (35.2.3) that does not look like plane wave at all, but using the Sommerfeld identity, it can be related to a linear superposition of plane waves.

$$H_z = 0, \tag{35.2.4}$$

in the above, implying the absence of the TE-to-z field.

Next, using the Sommerfeld identity (35.1.14) in the above, and after exchanging the order of integration and differentiation, we have[12]

$$E_z = \frac{-I\ell}{4\pi\omega\epsilon}\int_0^\infty dk_\rho \frac{k_\rho^3}{k_z} J_0(k_\rho\rho)e^{ik_z|z|}, \ |z| \neq 0 \tag{35.2.5}$$

after noting that $k_\rho^2 + k_z^2 = k^2$. Notice that now Equation (35.2.5) expands the z component of the electric field in terms of cylindrical waves in the ρ direction and a plane wave in the z direction. (Cylindrical waves actually are linear superpositions of plane waves. We can work backward from (35.1.15) to (35.1.11) to see this.) As such, the integrand in (35.2.5) in fact consists of a linear superposition of TM-type plane waves. The above is also the *primary field* generated by the source.[13]

Consequently, for a VED on top of a stratified medium as shown, expanding the source field in terms of plane waves, the downgoing plane waves from the point source will be reflected like TM waves with the generalized reflection coefficient $\tilde{R}_{12}^{TM}$. Henceforth, over a stratified medium, the field in region 1 can be written as

$$E_{1z} = \frac{-I\ell}{4\pi\omega\epsilon_1}\int_0^\infty dk_\rho \frac{k_\rho^3}{k_{1z}} J_0(k_\rho\rho)\left[e^{ik_{1z}|z|} + \tilde{R}_{12}^{TM} e^{ik_{1z}z+2ik_{1z}d_1}\right], \tag{35.2.6}$$

where $k_{1z} = (k_1^2 - k_\rho^2)^{\frac{1}{2}}$, and $k_1^2 = \omega^2\mu_1\epsilon_1$, the wavenumber in region 1.

The phase-matching condition dictates that the transverse variation of the field in all the regions must be the same. Consequently, in the i-th region, the solution becomes[14]

$$\epsilon_i E_{iz} = \frac{-I\ell}{4\pi\omega}\int_0^\infty dk_\rho \frac{k_\rho^3}{k_{1z}} J_0(k_\rho\rho)A_i\left[e^{-ik_{iz}z} + \tilde{R}_{i,i+1}^{TM} e^{ik_{iz}z+2ik_{iz}d_i}\right]. \tag{35.2.7}$$

Notice that Equation (35.2.7) is now expressed in terms of $\epsilon_i E_{iz}$ because $\epsilon_i E_{iz}$ reflects and transmits like H_{iy}, the transverse component of the magnetic field or TM waves.[15] Therefore, $\tilde{R}_{i,i+1}^{TM}$ and A_i could be obtained using the methods discussed in Chew [1].

This completes the derivation of the integral representation of the electric field everywhere in the stratified medium. These integrals are known as Sommerfeld integrals. Recently, they have been studied by Michalski and Mosig [281].

The case when the source is embedded in a layered medium can be derived similarly.

[12]By using (35.1.15) in (35.2.3), the $\partial^2/\partial z^2$ operating on $e^{ik_z|z|}$ produces a Dirac delta function singularity. But for simplicity, in (35.2.5), we ignore the delta function since $|z| \neq 0$. Hence, $\partial^2/\partial z^2$ produces a $-k_z^2$, and $k^2 - k_z^2 = k_\rho^2$. Detailed discussion on this can be found in the chapter on the dyadic Green's function in Chew, *Waves and Fields in Inhomogeneous Media* [38].

[13]One can perform a sanity check on the odd and even symmetry of the field's z-component by sketching the fields of a static vertical electric dipole.

[14]It will take quite a bit of work to get this expression, but you just need to know that it can be done, and know where to look for the resources for it.

[15]See Chew [38], p. 46, (2.1.6) and (2.1.7). Or we can gather from (18.1.6) to (18.1.7) that the $\mu_i H_{iz}$ transmits like E_{iy} at a dielectric interface, and by duality, $\epsilon_i E_{iz}$ transmits like H_{iy}.

(b) Horizontal Electric Dipole (HED)—Spectral Expansions

The HED is more complicated. Unlike the VED that excites only the TM-to-z waves, an HED will excite both TE-to-z and TM-to-z waves. For an HED pointing in the x direction, $\hat{\alpha} = \hat{x}$; hence, (35.2.1) and (35.2.2) give the TM-to-z and the TE-to-z components, in anticipation of their plane-wave expansions, as

$$E_z = \frac{iI\ell}{4\pi\omega\epsilon}\frac{\partial^2}{\partial z \partial x}\frac{e^{ikr}}{r}, \tag{35.2.8}$$

$$H_z = -\frac{I\ell}{4\pi}\frac{\partial}{\partial y}\frac{e^{ikr}}{r}. \tag{35.2.9}$$

It is to be noted that the above does not look like TE and TM waves at all. It is only after the expansion of the spherical waves into plane waves using Sommerfeld integrals that each term of the integrands resembles a TE and TM waves. Then, with the Sommerfeld identity (35.1.15), we can expand the above (after exchanging the order of differentiation and integration) as

$$E_z = \pm\frac{iI\ell}{4\pi\omega\epsilon}\cos\phi \int_0^\infty dk_\rho\, k_\rho^2 J_1(k_\rho\rho)e^{ik_z|z|} \tag{35.2.10}$$

$$H_z = i\frac{I\ell}{4\pi}\sin\phi \int_0^\infty dk_\rho \frac{k_\rho^2}{k_z} J_1(k_\rho\rho)e^{ik_z|z|}. \tag{35.2.11}$$

The $\pm$ sign above comes from the ∂_z derivative involving $|z|$. It also indicates that E_z is odd symmetric about $z = 0$, which one can easily verify by sketching the electric field of an HED using physical insight. Now, (35.2.10) represents the wave expansion of the TM-to-z field, while (35.2.11) represents the wave expansion of the TE-to-z field in terms of Sommerfeld integrals which are plane-wave expansions in disguise.[16] Observe that because E_z is odd about $z = 0$ in (35.2.10) because of the $\pm$ sign infront, the downgoing wave has an opposite sign from the upgoing wave. At this point, the above are just the primary fields generated by the source.

On top of a stratified medium, the downgoing wave is reflected accordingly depending on its wave type. Consequently, we have

$$E_{1z} = \frac{iI\ell}{4\pi\omega\epsilon_1}\cos\phi \int_0^\infty dk_\rho\, k_\rho^2 J_1(k_\rho\rho)\left[\pm e^{ik_{1z}|z|} - \tilde{R}_{12}^{TM} e^{ik_{1z}(z+2d_1)}\right], \tag{35.2.12}$$

$$H_{1z} = \frac{iI\ell}{4\pi}\sin\phi \int_0^\infty dk_\rho \frac{k_\rho^2}{k_{1z}} J_1(k_\rho\rho)\left[e^{ik_{1z}|z|} + \tilde{R}_{12}^{TE} e^{ik_{1z}(z+2d_1)}\right]. \tag{35.2.13}$$

Notice that the negative sign in front of $\tilde{R}_{12}^{TM}$ in (35.2.12) follows because the downgoing wave in the primary field has a negative sign as shown in (35.2.10). In the Sommerfeld integrals, $k_{iz} = \sqrt{k_i^2 - k_\rho^2}$ are double-value functions. So the integrands are potentially multivalue functions and care has to

[16] Using (35.1.13), we can convert a Bessel wave back to be a plane wave.

be taken when evaluating these integrals. Proper bookkeeping is done by defining branch cuts and branch points. However, it can be shown that, sometimes, these double-value functions do not make the integrand double value, and this is discussed in [38, p. 112].

35.3 Stationary Phase Method and Fermat's Principle[17]

Sommerfeld integrals are rather complex, and by themselves, they do not offer much physical insight into the field. To elucidate the physics, we can apply the stationary phase method to approximate these integrals when the frequency is high, or when kr is large, or when the observation point is many wavelengths away from the source point. It turns out that this method is intimately related to Fermat's principle described in section 33.2.

Stationary Phase Method (SPM)–A Canonical Case

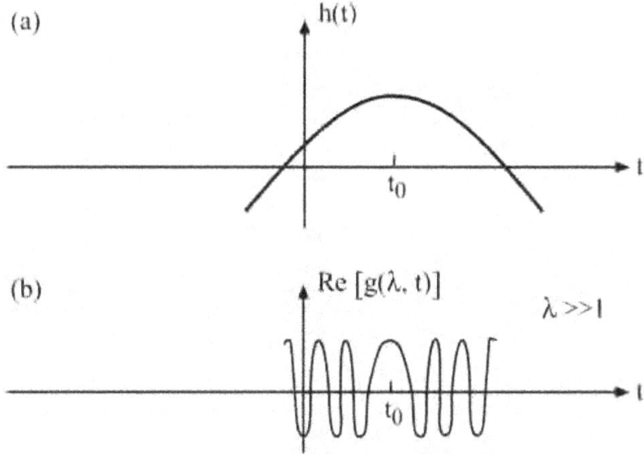

Figure 35.7: We first look at $h(t)$ which is in the exponent of $g(\lambda, t)$ (see (35.3.2)). Here, $h(t)$ is a slowly varying function of t as shown in (a). But it is in the exponent multiplied by λ. When λ is large, the function $g(\lambda, t)$ is rapidly varying where its real part is as shown in (b).

SPM is useful in deriving the leading-order approximation to an integral whose integrand is rapidly oscillating. Such rapid oscillations render the numerical integration of the integral difficult. In this case, an asymptotic approximation is the best way to obtain numerical values for the integral.

As an example, consider a basic, canonical integral of the form

$$I = \int_{-\infty}^{\infty} dt\, f(t)\, g(\lambda, t) \tag{35.3.1}$$

[17]May be skipped on first reading. This section is too complicated for a first course in graduate electromagnetics. But if you have the will, you can learn it.

where $g(\lambda, t)$ can be integrated in closed form, but not $f(t)g(\lambda, t)$. Furthermore, if

$$g(\lambda, t) \sim e^{i\lambda h(t)}, \quad \lambda \to \infty \tag{35.3.2}$$

("$\sim$" means "asymptotic to") then when λ is large, $g(\lambda, t)$ is a rapidly oscillating function of h, and hence, of t, where $h(t)$ may be a function of t as shown in Figure 35.7(a). As a result, the real part of the function $g(\lambda, t)$ will look like that shown in Figure 35.7(b). Also, the imaginary part of $g(\lambda, t)$ has the same feature. If $h(t)$ has a stationary point at $t = t_0$, that is,

$$h'(t_0) = 0 \tag{35.3.3}$$

it varies least slowly as a function of t at $t = t_0$; hence, $g(\lambda, t)$ is least rapidly oscillating at $t = t_0$. Furthermore, we assume that when $\lambda \to \infty$, $f(t)$ is a more slowly varying function compared to $g(\lambda, t)$.

If the integrand is rapidly oscillating, the contribution to the integration is small because of the cancellation of the positive and negative parts of the integrand. However, this cancellation is least at t_0 where the integrand is least rapidly oscillating. Therefore, most of the contribution to the integration will come from the neighborhood of t_0. Consequently, the integral can be approximated as (see Figure 35.7) Chew [282])

$$I \sim f(t_0) \int_{-\infty}^{\infty} dt\, g(\lambda, t), \ \lambda \to \infty \tag{35.3.4}$$

because $f(t)$ is approximately a constant $f(t_0)$ in the neighborhood of t_0, and most of the contribution to the integration comes from this neighborhood, we can take the constant outside the integral and leave the rapidly varying part in the integrand. Moreover, since $g(\lambda, t)$ has a closed-form solution, a simple algebraic approximation is obtained for I. This approximation is asymptotic in the sense that it becomes a better approximation when λ becomes larger as the rapidly varying integrand makes most of the contribution comes from the neighborhood of t_0 (see Figure 35.7).

Application of SPM to Sommerfeld Integrals

Next, we prepare the Sommerfeld integrals for SPM approximation. In order to avoid having to work with special functions like Bessel functions, we convert the Sommerfeld integrals back to spectral integrals in the Cartesian coordinates. (We could have obtained the aforementioned integrals in Cartesian coordinates were we to start with the Weyl identity instead of the Sommerfeld identity.) To do the back conversion, we make use of the identity,

$$\frac{e^{ik_0 r}}{r} = \frac{i}{2\pi} \iint_{-\infty}^{\infty} dk_x dk_y \frac{e^{ik_x x + ik_y y + ik_z |z|}}{k_z} = i \int_0^{\infty} dk_\rho \frac{k_\rho}{k_z} J_0(k_\rho \rho) e^{ik_z |z|}. \tag{35.3.5}$$

We can just focus our attention on the reflected wave term in (35.2.6) and rewrite it in Cartesian coordinates to get

$$E_{1z}^R = \frac{-I\ell}{8\pi^2\omega\epsilon_1} \iint_{-\infty}^{\infty} dk_x dk_y \frac{k_x^2 + k_y^2}{k_{1z}} R_{12}^{TM} e^{ik_x x + ik_y y + ik_{1z}(z+2d_1)}$$
$$= \iint_{-\infty}^{\infty} dk_x dk_y \frac{1}{k_{1z}} F(k_x, k_y) e^{ik_x x + ik_y y + ik_{1z}(z+2d_1)} \tag{35.3.6}$$

where we have put all the complicated terms of the integrand in the function $F(k_x, k_y)$, defined as

$$F(k_x, k_y) = \frac{-I\ell}{8\pi^2\omega\epsilon_1}(k_x^2 + k_y^2) R_{12}^{TM}$$

In the above, $k_x^2 + k_y^2 + k_{1z}^2 = k_1^2$ is the dispersion relation satisfied by the plane wave in region 1. Also, R_{12}^{TM} is dependent on $k_{iz} = \sqrt{k_i^2 - k_x^2 - k_y^2}$ in Cartesian coordinates, where $i = 1, 2$ represent the two media that the dispersion relation can be defined for. For simplicity, we will assume that $d_1 = 0$ to begin. Now the problem reduces to finding the approximation of the following integral:

$$E_{1z}^R = \iint_{-\infty}^{\infty} dk_x dk_y \frac{1}{k_{1z}} F(k_x, k_y) e^{irh(k_x, k_y)} \tag{35.3.7}$$

where

$$rh(k_x, k_y) = r\left(k_x \frac{x}{r} + k_y \frac{y}{r} + k_{1z} \frac{z}{r}\right) \tag{35.3.8}$$

The ratios x/r, y/r, and z/r are deliberately used because they are at most of $O(1)$ when $r \to \infty$. The large parameter here is $\lambda = r$. The above integral is too complicated to see its physics clearly. To elucidate its physics, we want to approximate the above integral when $rh(k_x, k_y)$ is large. This happens when x, y, and z are large compared to the wavelength.

In the above, $e^{irh(k_x,k_y)}$ is a rapidly varying function of k_x and k_y when x, y, and z are large, or r is large compared to the wavelength.[18] In other words, a small change in k_x or k_y will cause a large change in the phase of the integrand, or the integrand will be a rapidly varying function of k_x and k_y. Due to the cancellation of the integral when one integrates a rapidly varying function, most of the contributions to the integral will come from around the stationary point of $h(k_x, k_y)$ or where the function is least slowly varying (see Figure 35.7).[19] (Otherwise, the integrand is rapidly varying away from this point and the integration contributions will destructively cancel with each other, while around the stationary point, they will add constructively.)

The stationary point of $h(k_x, k_y)$ in the k_x and k_y plane is found by setting the derivatives of $h(k_x, k_y)$ with respect to k_x and k_y to zero. By so doing

$$\frac{\partial h}{\partial k_x} = \frac{x}{r} - \frac{k_x}{k_{1z}}\frac{z}{r} = 0, \qquad \frac{\partial h}{\partial k_y} = \frac{y}{r} - \frac{k_y}{k_{1z}}\frac{z}{r} = 0 \tag{35.3.9}$$

[18]The yardstick in wave physics is always wavelength. Large distance from the source point is also synonymous with increasing the frequency or reducing the wavelength.

[19]We have replaced a 1D function $h(t)$ with a 2D function $h(k_x, k_y)$, but the idea is the same.

The above represents two equations from which the two unknowns, k_{xs} and k_{ys}, at the stationary phase point can be solved for. By expressing the above in spherical coordinates, $x = r\sin\theta\cos\phi$, $y = r\sin\theta\sin\phi$, $z = r\cos\theta$, the values of (k_{xs}, k_{ys}), that satisfy the above equations are

$$k_{xs} = k_1\sin\theta\cos\phi, \quad k_{ys} = k_1\sin\theta\sin\phi \tag{35.3.10}$$

with the corresponding $k_{1zs} = k_1\cos\theta$. Here, the wave vector $\mathbf{k}_s = \hat{x}k_{xs} + \hat{y}k_{ys} + \hat{z}k_{zs}$ is parallel to the vector that connects that source and the observation point–namely, $\mathbf{r} = \hat{x}x + \hat{y}y + \hat{z}z$. In this case, $\mathbf{k}_s \cdot \mathbf{r} = k_1 r$.

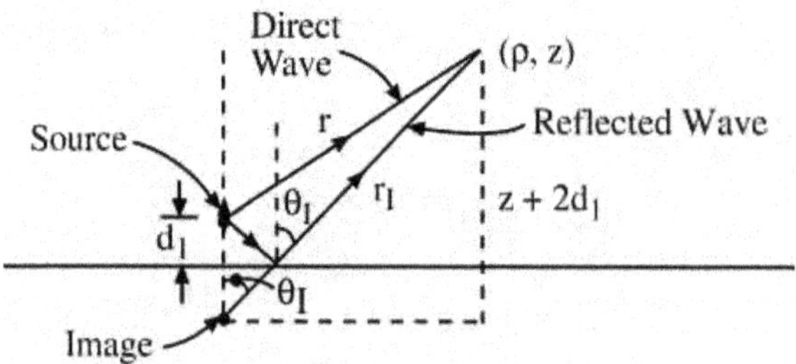

Figure 35.8: At high frequencies, the source point and the observation point are connected by a ray. The ray represents a bundle of plane waves that interfere constructively. This is even true for a bundle of plane waves that reflect off an interface. So ray theory or ray optics prevails here, and the ray bounces off the interface according to the reflection coefficient of a plane wave impinging at the interface with θ_I.

When one integrates on the k_x and k_y plane, the dominant contribution to the integral will come from the point in the vicinity of (k_{xs}, k_{ys}). Assuming that $F(k_x, k_y)$ is slowly varying, we can approximate $F(k_x, k_y)$ to be a constant equal to its value at the stationary phase point (k_{xs}, k_{ys}), and say that[20]

$$E_{1z}^R \simeq F(k_{xs}, k_{ys}) \iint_{-\infty}^{\infty} \frac{1}{k_{1z}} e^{ik_x x + ik_y y + ik_{1z}z} dk_x dk_y = 2\pi F(k_{xs}, k_{ys}) \frac{e^{ik_1 r}}{ir} \tag{35.3.11}$$

In the above, the integral can be performed in closed form using the Weyl identity. The physical meaning of the above is that if we begin with a complicated integral such as that in (35.3.7), which represents an integral summation of plane waves in all directions, it can be approximated by a spherical wave. The above analysis says that only plane waves in the vicinity of the stationary point contribute to the integral, and these plane waves are paraxial. Thus, a spherical wave can be thought of as a summation of a bundle of paraxial plane waves.

When $d_1 \neq 0$, we will have to replace (35.3.8) by

$$r_I h(k_x, k_y) = r_I \left(k_x \frac{x}{r_I} + k_y \frac{y}{r_I} + k_{1z} \frac{z + 2d_1}{r_I} \right) \tag{35.3.12}$$

[20]The rapidly varying function with a stationary phase point is behaving like the sifting property of a delta function.

where $r_I = \sqrt{x^2 + y^2 + (z + 2d_1)^2}$ is the distance from the image point to the observation point (see Figure 35.8). The problem is "homomorphic" to the previous case. By expressing the above in spherical coordinates whose origin is the image point, $x = r_I \sin\theta_I \cos\phi$, $y = r_I \sin\theta_I \sin\phi$, $z + 2d_1 = r_I \cos\theta_I$, the values of (k_{xs}, k_{ys}), that satisfy the above equations are

$$k_{xs} = k_1 \sin\theta_I \cos\phi, \quad k_{ys} = k_1 \sin\theta_I \sin\phi \tag{35.3.13}$$

The subsequent stationary phase approximation yields

$$E_{1z}^R \simeq F(k_{xs}, k_{ys}) \iint_{-\infty}^{\infty} \frac{1}{k_{1z}} e^{ik_x x + ik_y y + ik_{1z}(z+2d_1)} dk_x dk_y = 2\pi F(k_{xs}, k_{ys}) \frac{e^{ik_1 r_I}}{ir_I} \tag{35.3.14}$$

The above expressions have a few important physical interpretations.

(i) Even though a source is emanating plane waves in all directions in accordance with (35.1.11), at the observation point r or r_I far away from the image source point, only one or few plane waves in the vicinity of the stationary phase point are important. This is because they interfere with each other constructively to form a spherical wave that represents the ray connecting the source point to the observation point. Plane waves in other directions interfere with each other destructively, and are not important. That is the reason that the source point and the observation point are connected only by one ray. Since we know that a field point and a source point are connected by a spherical wave, we can think of a ray as the principal component of plane wave at the stationary point that points from the source point to the observation point [283]. Also, a bundle of plane waves (or ray) in the vicinity of the stationary phase point add coherently to forms a spherical wave. These bundles of plane waves are also almost paraxial with respect to each other.

(ii) The function $F(k_x, k_y)$ could be a very complicated function like the reflection coefficient R^{TM}, but only its value at the stationary phase point matters the most. The plane wave component that emanates from the stationary phase point will interfere constructively with other plane wave components in its neighborhood of the stationary phase point to form a spherical wave front. The spherical wave front emanates from an image point source located at $x = 0$, $y = 0$, $z = -d_1$. If we were to make $d_1 \neq 0$, the math remains similar except that now, we replace r with $r_I = \sqrt{x^2 + y^2 + (z + 2d_1)^2}$, the distance from the observation point to the image source point. The reason being that due to the reflecting half-space, the source point now has an image source point as shown in Figure 35.8. Also, we have to be mindful to replace z with $z + 2d_1$ in the above stationary phase analysis. The stationary phase method now extracts a ray (or a bundle of plane waves) that emanates from the source point, bounces off the half-space, and the reflected ray reaches the observer modulated by the reflection coefficient R^{TM}. But the value of the reflection coefficient that matters the most is the angle at which the incident ray impinges on the half-space. Now, the reflected field seems to have emanated from an image source point.

(iii) At the stationary phase point, the ray is formed by the $\mathbf{k}_I$-vector where

$$\mathbf{k}_I = \hat{x} k_1 \sin\theta_I \cos\phi + \hat{y} k_1 \sin\theta_I \sin\phi + \hat{z} k_1 \cos\theta_I$$

This is a $\mathbf{k}_I$ vector that points from the image source point to the observation point. This ray points in the same direction as the position vector of the observation point with respect to the image source point–namely, $\mathbf{r}_I = \hat{x}r_I \sin\theta_I \cos\phi + \hat{y}r_I \sin\theta_I \sin\phi + \hat{z}r_I \cos\theta_I$. In other words, the $\mathbf{k}_I$-vector and the $\mathbf{r}_I$-vector point in the same direction. This is reminiscent of Fermat's principle, because when this happens, the ray propagates with the minimum phase shift and time delay between the image source point and the observation point, which is also the statement of Fermat's principle. When $z \to z + 2d_1$, the ray for the image source is as shown in Figure 35.8 where the ray is minimum phase from the image source to the observation point. The stationary phase point is the minimum phase point that minimizes the "time-delay". Hence, the stationary phase method is intimately related to Fermat's principle.

Exercises for Lecture 35

Problem 35-1: This exercise refers to lecture 35 of the lecture notes.

(i) By taking the residue of a contour integral around a pole, show how you can go from (35.1.7) to (35.1.9) and (35.1.10).

(ii) Show that (35.2.3) can be derived from (35.2.1), and then show that (35.2.5) can be derived assuming that $z \neq 0$. Similarly, from (35.2.1), derive (35.2.8) and (35.2.9), and then show that (35.2.10) and (35.2.11) can be derived.

(iii) (**Optional**) First, for the section on stationary phase, derive (35.3.6) which is the reflected field term in Cartesian coordinates. Assuming that $d_1 \neq 0$, derive the equivalence of (35.3.14) with $r \to r_I = \sqrt{x^2 + y^2 + (z + 2d_1)^2}$ using stationary phase argument, and elucidate the physics expressed in the math in accordance to item (ii) below (35.3.14). Explain why this is related to Fermat's principle.

Lecture 36

Computational Electromagnetics, Numerical Methods

Due to the rapid advancements of digital computers and the blinding speed at which computations can be done, numerical methods to seek solutions of Maxwell's equations have become vastly popular. Massively parallel digital computers now can compute at breakneck speeds of tera, peta, exa-flops throughputs [284], where FLOPS stands for "floating operations per second". They have also spawned terms that we have not previously heard of (see also Figure 36.1).[1]

Name	Unit	Value
kiloFLOPS	kFLOPS	10^3
megaFLOPS	MFLOPS	10^6
gigaFLOPS	GFLOPS	10^9
teraFLOPS	TFLOPS	10^{12}
petaFLOPS	PFLOPS	10^{15}
exaFLOPS	EFLOPS	10^{18}
zettaFLOPS	ZFLOPS	10^{21}
yottaFLOPS	YFLOPS	10^{24}

Figure 36.1: Nomenclature for measuring the speed of modern-day computers. The fastest computer in the world changes from year to year.

We repeat a quote from Freeman Dyson—"Technology is a gift of God. After the gift of life it is

[1]The gaming industry, followed by artificial intelligence needs, have spurred the development of GPU's (graphical processing units). These processors have been used in solving many computational science and engineering problems.

perhaps the greatest of God's gifts. It is the mother of civilizations, of arts, and of sciences." The spur for computer advancement was the Second World War. During that time, men went to war while women stayed back to work as computers, doing laborious numerical computations manually (see Figure 36.2 [285]). The need for faster computers was obvious. (The trajectories of rockets cannot be solved for in closed form; computers are needed.) Unfortunately, in the last half century or so, we have been using a large part of the gift of technology in warfare to destroy God's greatest gift: life!

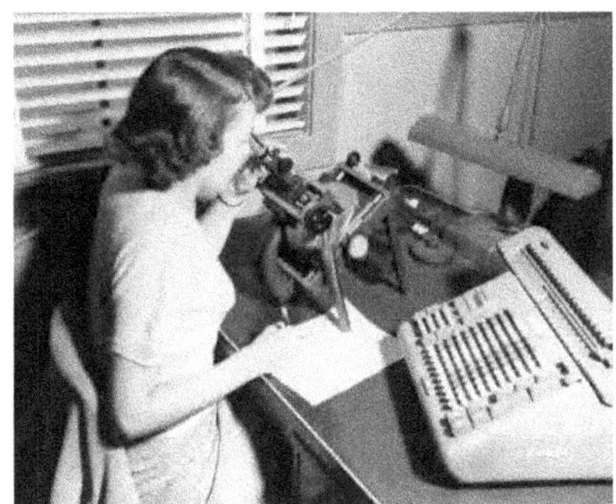

Figure 36.2: A woman working as a "computer" shortly after the second world war (Figure by: https://en.wikipedia.org/wiki/computer [285]).

36.1 Computational Electromagnetics, Numerical Methods

Due to the high fidelity of Maxwell's equations in describing electromagnetic physics in nature, and they have been validated to high accuracy (see section 1.1), oftentimes, a numerical solution obtained by solving Maxwell's equations is more reliable than laboratory experiments. This field of finding numerical solutions to Maxwell's equations is also known as computational electromagnetics (CEM).

The field is a descendent of mathematical modeling as well as a new branch of computational science and engineering (CSE). It began with the quest for closed-form solutions or analytic solutions that could return a number with some simple calculations. These closed-form solutions have greatly benefitted engineering designs. Some examples of these are the Mie scattering solution, the Sommerfeld half-space problems, and many more. As shall be learned from this lecture, many wave scattering problems can be couched in terms of solving a matrix equation. For many highly complex problems, a high-fidelity description of the problem can be gotten using many degrees of freedom, or by increasing the number of unknowns in the pertinent matrix equations. Numerical methods exploit the blinding speed of modern digital computers to perform calculations, and consequently to solve large matrix systems of equations. This manner of solving engineering problems is also termed numerical simulation. It turns out that the design of a modern-day computer chip is done with numerical simulation more than 90 percent of the time.

Computational electromagnetics consists mainly of two classes of numerical solvers: one that solves differential equations directly (the differential equation solvers); and one that solves integral equations (the integral equation solvers). Both these classes of equations are derivable from Maxwell's equations [10].[2]

36.2 Examples of Differential Equations

An example of a differential equation with no closed-form solution but driven by a source is the scalar wave equation:

$$(\nabla^2 + k^2(\mathbf{r}))\,\phi(\mathbf{r}) = Q(\mathbf{r}), \tag{36.2.1}$$

An example of a vector differential equation for a vector electromagnetic field is

$$\nabla \times \overline{\boldsymbol{\mu}}^{-1} \cdot \nabla \times \mathbf{E}(\mathbf{r}) - \omega^2 \overline{\boldsymbol{\varepsilon}}(\mathbf{r}) \cdot \mathbf{E}(\mathbf{r}) = i\omega \mathbf{J}(\mathbf{r}) \tag{36.2.2}$$

These equations are linear equations,[3] but for inhomogeneous media where $k^2(\mathbf{r})$ and $\epsilon(\mathbf{r})$ are functions of position vector $\mathbf{r}$, generally, they do not have closed-form solutions: a numerical solution is the norm. The problems expressed by (36.2.1) and (36.2.2) share one commonality–that is, they can be abstractly written as

$$\mathscr{L} f = g \tag{36.2.3}$$

where $\mathscr{L}$ is the differential operator which is linear,[4] f is the unknown, and g is the driving source. Differential equations, or partial differential equations, as mentioned before in (36.2.1) and (36.2.2), have to be solved with boundary conditions. Otherwise, there is no unique solution to these equations.

In the case of the scalar wave equation (36.2.1), $\mathscr{L} = (\nabla^2 + k^2)$ is a differential operator. In the case of the electromagnetic vector wave equation (36.2.2), $\mathscr{L} = (\nabla \times \overline{\boldsymbol{\mu}}^{-1} \cdot \nabla \times) - \omega^2 \overline{\boldsymbol{\varepsilon}} \cdot$. Furthermore, f will be $\phi(\mathbf{r})$ for the scalar wave equation (36.2.1), while it will be $\mathbf{E}(\mathbf{r})$ in the case of a vector wave equation for an electromagnetic system (36.2.2). The g on the right-hand side can represent Q in (36.2.1) or $i\omega\mathbf{J}(\mathbf{r})$ in (36.2.2).

---

[2] Computations are heavily used in other fields such as computational mechanics, computational fluid dynamics, as well as computational physics.

[3] Nonlinear problems are often approximated as a series of linear problems.

[4] The simple test of linearity is that $\mathscr{L}(a_1 f_1 + a_2 f_2) = a_1 \mathscr{L} f_1 + a_2 \mathscr{L} f_2$. This test is homomorphic to the linearity test in linear algebra.

36.3 Examples of Integral Equations

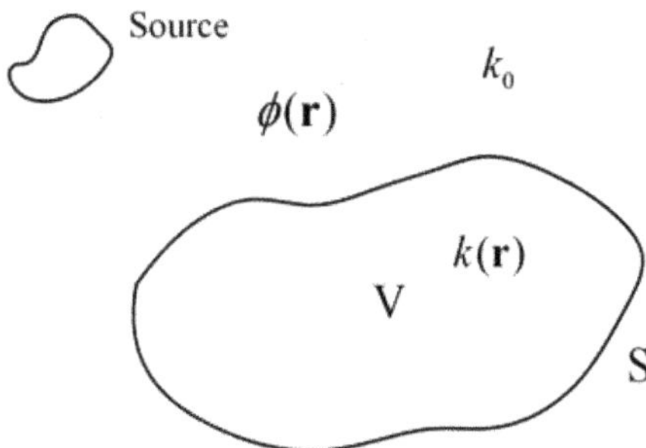

Figure 36.3: Geometry for the derivation of the volume-integral equation for scalar waves. The wavenumber $k(\mathbf{r})$ is assumed to be inhomogeneous, and hence, a function of position $\mathbf{r}$ inside the scatterer, but a constant k_0 outside the scatterer. The dyadic Green's function or vector case is homomorphic to the scalar wave case.

36.3.1 Volume Integral Equation

This course is replete with PDEs, but we have not come across too many integral equations as yet. In integral equations, the unknown is embedded in the integral. The simplest integral equation to derive is the volume integral equation. Hence, we shall first derive the volume integral equation for the scalar wave case.[5] In this case, the pertinent scalar wave equation is

$$[\nabla^2 + k^2(\mathbf{r})]\phi(\mathbf{r}) = Q(\mathbf{r}), \tag{36.3.1}$$

where $k^2(\mathbf{r})$ represents an inhomogeneous medium over a finite domain V, and $k^2 = k_0^2$, which is constant outside V (see Figure 36.3). Next, we define a Green's function satisfying

$$[\nabla^2 + k_0^2]g(\mathbf{r}, \mathbf{r}') = -\delta(\mathbf{r} - \mathbf{r}'), \qquad \forall \mathbf{r}, \mathbf{r}'. \tag{36.3.2}$$

Then, (36.3.1) can be rewritten as

$$[\nabla^2 + k_0^2]\phi(\mathbf{r}) = Q(\mathbf{r}) - [k^2(\mathbf{r}) - k_0^2]\phi(\mathbf{r}). \tag{36.3.3}$$

Note that the right-hand side of (36.3.3) can be considered an equivalent source. Since the Green's function corresponding to the differential operator on the left-hand side of (36.3.3) is known, by the principle of linear superposition, we can write the formal solution to (36.3.3) as

$$\phi(\mathbf{r}) = -\int_{V_s} dV' g(\mathbf{r}, \mathbf{r}')Q(\mathbf{r}') + \int_V dV' g(\mathbf{r}, \mathbf{r}')[k^2(\mathbf{r}') - k_0^2]\phi(\mathbf{r}'). \tag{36.3.4}$$

The first term on the right-hand side is just the field due to the source in the absence of the inhomogeneity or the scatterer, and hence, is the incident field. The second term is a volume

[5]The vector wave case is homomorphic to the scalar wave case.

integral over the space where $k^2(\mathbf{r}') - k_0^2 \neq 0$, or inside the inhomogeneous scatterer. Therefore, (36.3.4) becomes

$$\phi(\mathbf{r}) = \phi_{inc}(\mathbf{r}) + \overbrace{\int\limits_V dV' g(\mathbf{r},\mathbf{r}')[k^2(\mathbf{r}') - k_0^2]\phi(\mathbf{r}')}^{\text{Scattered Field—}\phi_{sca}(\mathbf{r})} . \tag{36.3.5}$$

It is to be noted that the above sources are radiating via the Green's function, and hence they satisfy the radiation condition, since the Green's function satisfies the radiation condition.

In the above equation, if the total field $\phi(\mathbf{r}')$ inside the volume V on the right-hand side is known, then $\phi(\mathbf{r})$ can be calculated everywhere. But $\phi(\mathbf{r})$ is unknown at this point. To solve for $\phi(\mathbf{r})$, an integral equation has to be formulated for $\phi(\mathbf{r})$. To this end, we imposed (36.3.5) for $\mathbf{r}$ in V. Then, $\phi(\mathbf{r})$ on the left-hand side and on the right-hand side are the same unknown defined over the same region V. Consequently, (36.3.5) becomes the desired integral equation after rearrangement as

$$\phi_{inc}(\mathbf{r}) = \phi(\mathbf{r}) - \int\limits_V dV' g(\mathbf{r},\mathbf{r}')[k^2(\mathbf{r}') - k_0^2]\phi(\mathbf{r}'), \quad \mathbf{r} \in V. \tag{36.3.6}$$

In the above, the unknown $\phi(\mathbf{r})$ is defined over a volume V, over which the integration is performed, and hence the name volume integral equation. Alternatively, the above can be rewritten, using shorthand notation, as

$$\phi_{inc}(\mathbf{r}) = \phi(\mathbf{r}) - \mathcal{G}(\mathbf{r},\mathbf{r}')\mathcal{O}(\mathbf{r}')\phi(\mathbf{r}'), \quad \mathbf{r} \in V, \tag{36.3.7}$$

where $\mathcal{G}$ is the integral operator in (36.3.6),[6] and $\mathcal{O}(\mathbf{r}') = [k^2(\mathbf{r}') - k_0^2]$ is the scatterer object function. It is also a *Fredholm integral equation* of the second kind because the unknown is both inside and outside the integral operator. In the above, integration over repeated variable $\mathbf{r}'$ is implied. Nevertheless, it can be written more abstractly as

$$\mathscr{L} f = g \tag{36.3.8}$$

where $\mathscr{L}$ is a linear operator ("homomorphic" to a linear matrix operator), while f represents the unknown function $\phi(\mathbf{r})$ and g is the known function $\phi_{inc}(\mathbf{r})$ (f and g are homomorphic to matrix vectors).[7] In the above

$$\mathscr{L} = \mathcal{I} - \mathcal{G}\mathcal{O}, \qquad f = \phi(\mathbf{r}), \qquad g(\mathbf{r}) = \phi_{inc}(\mathbf{r}) \tag{36.3.9}$$

where $\mathcal{I}$ is the identity operator.

---

[6]Sometimes, this is called the kernel of the integral equation.

[7]As we shall learn, a function should be thought of as an abstract vector.

36.3.2 Surface Integral Equation

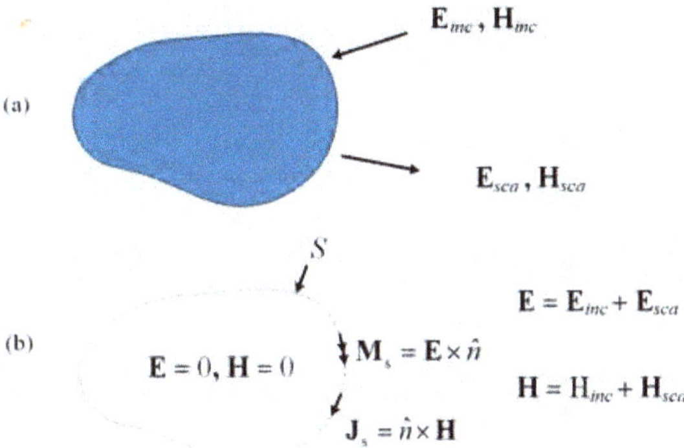

Figure 36.4: Geometry for the derivation of the surface-integral equation for vector electromagnetics waves. (a) The original electromagnetics scattering problem. (b) The equivalent electromagnetics problem by invoking equivalence principle.

The surface integral equation method is rather popular in many applications because it can employ the homogeneous-medium Green's function, which exists in simple closed form,[8] and the unknowns reside on a surface rather than in a volume.[9]

The surface integral equation for a vector electromagnetic field can be derived using the equivalence theorem, also called the Love's equivalence theorem [108]. Given a scattering problem shown in Figure 36.4(a), it can be replaced by an equivalence problem as shown in Figure 36.4(b). One can verify this by performing a Gedanken experiment as we have done for the other equivalence problems discussed in section 13.1.

In this figure, the total fields outside the scatterer are $\mathbf{E} = \mathbf{E}_{inc} + \mathbf{E}_{sca}$ and $\mathbf{H} = \mathbf{H}_{inc} + \mathbf{H}_{sca}$. The impressed equivalence currents are given by $\mathbf{M}_s = \mathbf{E} \times \hat{n}$ and $\mathbf{J}_s = \hat{n} \times \mathbf{H}$. These impressed currents together generate the scattered fields outside the scatterer, while they generate zero field inside the scatterer!

As such, the scattered fields outside the scatterer can be found from the radiation of the impressed currents $\mathbf{M}_s$ and $\mathbf{J}_s$. Notice that these currents are radiating via the free-space Green's function because the scatterer has been removed in this equivalence problem. Also, if the scatterer is a PEC, then the tangential component of the total electric field is zero on the PEC surface.[10] Therefore, $\mathbf{M}_s = 0$ and only $\mathbf{J}_s$ is radiating via the free-space Green's function.

Please remember that this equivalence problem is very different from that of an impressed current on the PEC scatterer as discussed in section 13.2. There, only the magnetic surface current is

[8] Numerical Green's functions have been proposed to enable solutions of inhomogeneous media [286].

[9] This is sometimes called the boundary integral equations method [287, 288].

[10] At microwave frequencies, good metallic conductors can be modeled as PEC as the skin depth is very small in these cases.

radiating in the presence of the PEC, and the Green's function is that of a current source radiating in the presence of the PEC scatterer, and it is not the free-space Green's function.

Now we can write the fields outside the scatterer using (13.4.20)

$$\mathbf{E}_{sca}(\mathbf{r}) = \frac{1}{i\omega\varepsilon}\nabla\times\nabla\times\oint_S dS'\, g(\mathbf{r}-\mathbf{r}')\,\hat{n}'\times\mathbf{H}(\mathbf{r}') = \frac{1}{i\omega\varepsilon}\nabla\times\nabla\times\oint_S dS'\, g(\mathbf{r}-\mathbf{r}')\,\mathbf{J}_s(\mathbf{r}') \tag{36.3.10}$$

In the above, we have swapped $\mathbf{r}'$ and $\mathbf{r}$ compared to (13.4.20). Also, we have kept only the electric current $\mathbf{J}_s(\mathbf{r})$ due to $\hat{n}\times\mathbf{H}(\mathbf{r})$. If we impose the boundary condition that the tangential component of the total electric field is zero, then we arrive at $\hat{n}\times\mathbf{E}_{sca} = -\hat{n}\times\mathbf{E}_{inc}$ and the integral equation

$$-\hat{n}\times\mathbf{E}_{inc}(\mathbf{r}) = \hat{n}\times\frac{1}{i\omega\varepsilon}\nabla\times\nabla\times\oint_S dS'\, g(\mathbf{r}-\mathbf{r}')\,\mathbf{J}_s(\mathbf{r}'). \qquad \mathbf{r}\in S \tag{36.3.11}$$

In the above, $\hat{n}\times\mathbf{E}_{inc}(\mathbf{r})$ is known on the left-hand side on the scatterer's surface, while the right-hand side has embedded in it the unknown surface current $\mathbf{J}_s(\mathbf{r}) = \hat{n}\times\mathbf{H}(\mathbf{r})$ on the surface of the scatterer. Therefore, the above is an integral equation for the unknown surface current $\mathbf{J}_s(\mathbf{r})$. It can be written as a form of $\mathscr{L}f = g$ just like other linear operator equations.

36.4 Function as a Vector

Several linear operator equations have been derived in the previous sections. They are all of the form[11]

$$\mathscr{L}f = g \tag{36.4.1}$$

In the above, f is a functional vector which is the analogue of the vector $\mathbf{f}$ in matrix theory or linear algebra.[12] In linear algebra, the vector $\mathbf{f}$ is of length N in an N-dimensional space. It can be indexed by a set of countable index, say i, and we can describe such a vector in 1D with N numbers such as $f_i, i = 1, \ldots, N$ explicitly. This is shown in Figure 36.5(a).

A function $f(x)$, however, can be thought of as being indexed by x in the 1D case. But the index in this case is a continuum and countably infinite. Thus, it corresponds to a vector of infinite dimension and it lives in an infinite-dimensional space.[13]

To make such functions economical in storage, for instance, in 1D example case, we replace the function $f(x)$ with its sampled values at N locations, such that $f(x_i), \quad i = 1, \ldots, N$. Then the values of the function in between the stored points $f(x_i)$ can be obtained by interpolation.[14] Therefore, a function vector $f(x)$, even though it is infinite-dimensional, can be approximated by

[11]This is "homomorphic" to the matrix equation $\overline{\mathbf{A}}\cdot\mathbf{x} = \mathbf{b}$.

[12]In this course, we have used a boldface letter to denote a 3-vector (namely vector in 3D space) But in the rest of the course, a boldface letter can represent an N-dimensional vector.

[13]When these functions are square integrable, implying finite "energy," these infinite-dimensional spaces are called Hilbert spaces.

[14]This is in fact how special functions like $\sin(x)$, $\cos(x)$, $\exp(x)$, $J_n(x)$, $N_n(x)$, etc. are computed and stored in modern computers. Furthermore, it can be proved that a smooth (or bandlimited) function can be interpolated to exponential accuracy.

a finite-length vector, $\mathbf{f}$.[15] This concept is illustrated in Figure 36.5(b) and (c). This concept can be generalized to a function of 3D space $f(\mathbf{r})$. If $\mathbf{r}$ is sampled over a 3D volume, it can provide an index to a vector $f_i = f(\mathbf{r}_i)$, and thus, $f(\mathbf{r})$ can be thought of as a vector as well.

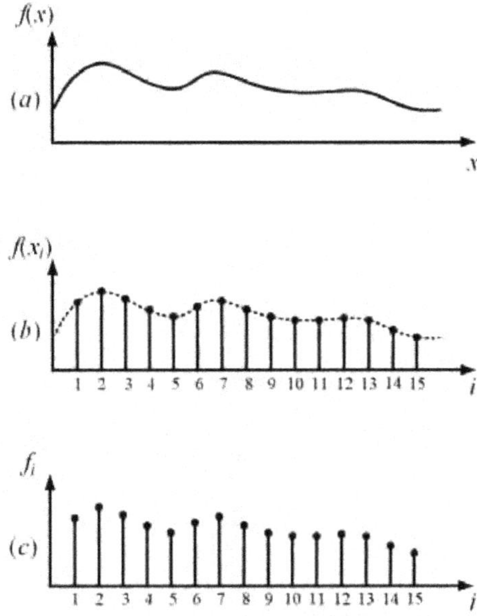

Figure 36.5: A function can be thought of as a vector. (a) A continuum function $f(x)$ plotted as a function of x. (b) Digitally sampled values of the same function. (c) When stored in a computer, it will be stored as an array vector.

36.5 Operator as a Map

36.5.1 Domain and Range Spaces

An operator like $\mathscr{L}$ above can be thought of as a map or a transformation. In this lecture, we will consider linear operators only, and hence, they are like linear matrix operators. A linear operator maps a function f defined in a Hilbert space V to g, another function defined in another Hilbert space W. Mathematically, this is written as

$$\mathscr{L} : V \to W \tag{36.5.1}$$

indicating that $\mathscr{L}$ is a map (or operator) of vectors in the space V to vectors in the space W. Here, V is also called the *domain space* (or domain) of $\mathscr{L}$ while W is the *range space* (or range) of $\mathscr{L}$. We need to familiarize ourselves with this language as research becomes more interdisciplinary.

[15] Especially if it is bandlimited. This is routine in signal processing, where a vector in an infinite dimensional Hilbert space is approximated by a finite-length sample of the vector as in Nyquist's theorem.

36.6 Approximating Operator Equations with Matrix Equations

36.6.1 Subspace Projection Methods

One main task of a numerical method is first to approximate an operator equation $\mathscr{L} f = g$ by a matrix equation $\overline{\mathbf{L}} \cdot \mathbf{f} = \mathbf{g}$. To achieve the above, we first let

$$f \cong \sum_{n=1}^{N} a_n f_n \tag{36.6.1}$$

In the above, $f_n, n = 1, \ldots, N$ are known functions called basis functions (or expansion functions analogous to Fourier harmonics in a Fourier series). Now, a_n's are the new unknowns to be sought. Also the above is an approximation, and the accuracy of the approximation depends very much on the original function f. A set of very popular basis functions are functions that form a piecewise linear interpolation of the function from its nodes. These basis functions are shown in Figure 36.6 in 1D and 2D.

Upon substituting (36.6.1) into (36.4.1), we obtain

$$\sum_{n=1}^{N} a_n \mathscr{L} f_n = g \tag{36.6.2}$$

Then, upon multiplying (36.6.2) by w_m and integrating over the space that $w_m(\mathbf{r})$ is defined, then we have

$$\sum_{n=1}^{N} a_n \langle w_m, \mathscr{L} f_n \rangle = \langle w_m, g \rangle, m = 1, \ldots, N \tag{36.6.3}$$

In the above, the inner product is defined as

$$\langle f_1, f_2 \rangle = \int d\mathbf{r} f_1(\mathbf{r}) f_2(\mathbf{r}) \tag{36.6.4}$$

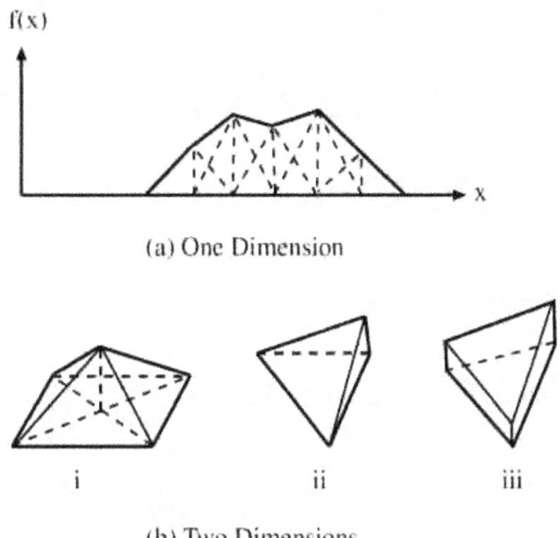

(a) One Dimension

(b) Two Dimensions

Figure 36.6: Examples of basis function in (a) one dimension; and in (b) two dimensions. Each of these functions are defined over a finite domain. Hence, they are also called subdomain basis functions. They can be thought of as interpolatory functions where the values in between the nodes are obtained by interpolation of the nodal values. When generalized to 3D, these basis functions become tetrahedrons, which are harder to draw and visualize.

where the integration is over the support of the functions, or the space over which the functions are defined.[16] For PDEs, these functions are defined over a 2D or 3D coordinate space, while in SIEs, these functions are defined over a surface or a 2D manifold (or a 1D manifold where they become curved lines).[17] In 1D problems, these functions are defined over a 1D coordinate space.

The above (36.6.3) is often referred to as the weak form of the original equation (36.4.1) [290, 291]. The original equation needs to be imposed in an infinite dimensional Hilbert space, while (36.6.3) is only imposed in a smaller, finite dimensional subsspace.

36.6.2 Dual Spaces

The functions $w_m, m = 1, \ldots, N$ in (36.6.3) is known as the weighting functions or testing functions. The testing functions should be chosen so that they can approximate well a function that lives in the range space W of the operator $\mathscr{L}$. Such a set of testing functions lives in the *dual space* of the range space. For example, if f_r lives in the range space of the operator $\mathscr{L}$, the set of function f_d, such that the inner product $\langle f_d, f_r \rangle$ exists, forms the dual space of W. If the inner product $\langle f_d, f_r \rangle$ is of infinite value, then f_d is outside the dual space of W.

[16]This is known as the reaction inner product [55, 38, 289]. As opposed to most math and physics literature, the energy inner product is used [289] where $\langle f_1, f_2 \rangle = \int d\mathbf{r} f_1^*(\mathbf{r}) f_2(\mathbf{r})$.

[17]A 2D manifold is a curved surface where locally, at a given point, it can be approximated by a flat 2D Euclidean space and similarly for 1D manifold. This concept can be generalized to nD manifold.

36.6.3 Matrix and Vector Representations

The above equation (36.6.3) is a matrix equation of the form

$$\overline{\mathbf{L}} \cdot \mathbf{a} = \mathbf{g} \tag{36.6.5}$$

where

$$\begin{gathered} \left[\overline{\mathbf{L}}\right]_{mn} = \langle w_m, \mathscr{L} f_n \rangle \\ [\mathbf{a}]_n = a_n, \qquad [\mathbf{g}]_m = \langle w_m, g \rangle \end{gathered} \tag{36.6.6}$$

What has effectively happened here is that given an operator $\mathscr{L}$ that maps a function that lives in an infinite-dimensional Hilbert space V, to another function that lives in another infinite-dimensional Hilbert space W, via the operator equation $\mathscr{L} f = g$, we have approximated the Hilbert spaces with finite-dimensional spaces (subspaces), and finally, obtain a finite-dimensional matrix equation that is the representation of the original infinite-dimensional operator equation. This is the spirit of the *subspace projection method.*

In the above, $\overline{\mathbf{L}}$ is the matrix representation of the original operator $\mathscr{L}$ in the subspaces, and $\mathbf{a}$ and $\mathbf{g}$ are the vector representations of f and g, respectively, in their respective subspaces. In this case, it is common to call $L_{mn} = \langle w_m, \mathscr{L} f_n \rangle$ the matrix element of the operator $\mathscr{L}$.

When such a method is applied to integral equations, it is usually called the method of moments (MOM). (Surface integral equations are also called boundary integral equations (BIEs) in other fields [288].) When finite discrete basis are used to represent the surface unknowns, it is also called the boundary element method (BEM) [292]. But when this method is applied to solve PDEs, it is called the finite element method (FEM) [293, 210, 294, 295], which is a rather popular method due to its simplicity.

36.6.4 Mesh Generation and Geometry Modeling

To solve for the scattering solution from an arbitrary geometry, we first have to approximate the arbitrary geometry by a mathematical model: this is usually a geometry describable by meshes, which can be mathematically defined as well. In order to approximate a geometry, we need to consider geometries that can be thought of as unions of line segments, triangle patches, or tetrahedrons. A complicated geometry can be formed from the union of three geometry types. This is called the meshing process (also known as tessellation or discretization).

With these mathematically defined geometries, functions can be defined on an arbitrarily shaped line, surface, or volume by a finite sum of basis functions. The basic elements from which more complicated geometries can be built are called simplices.

In 1D, this basic element is a line segment. The union of line segments can be used to approximate arbitrary lines or curves if the curve is segmented fine enough. In 2D, this basic element is a triangle. The union of triangles, with their surface normals pointing in different directions, can be used to model or approximate a surface geometry (see Figure 36.7). In 3D, this basic element is a tetrahedron. The union of tetrahedrons can be use model a 3D volumetric geometry (see Figure 36.8). The smaller these basic elements are, the more accurate the geometry model is. Accuracy of the geometry modeling can be improved by using curvilinear elements as well.

Such meshes are used not only in CEM, but in other fields such as solid mechanics, fluids, and physics. Hence, there are many "solid modeling" commercial software available to generate sophisticated meshes.

Then basis functions, which are defined on these simplices, are used in (36.6.1) and defined to interpolate the field between nodal values in a line segment, triangle, or a tetrahedron.[18]

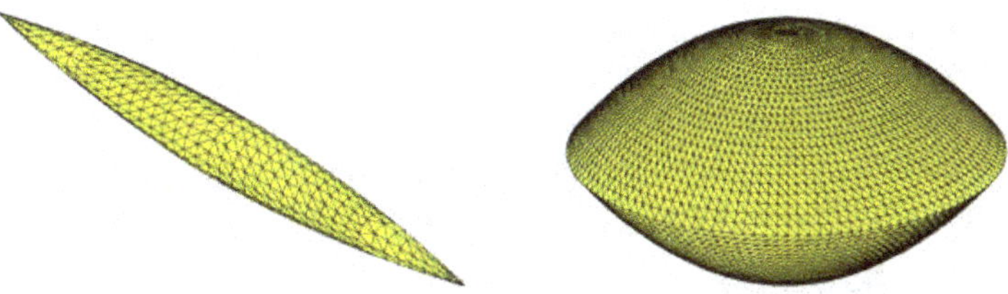

Figure 36.7: An arbitrary surface (also called a 2D manifold) can be meshed by a union of triangles.

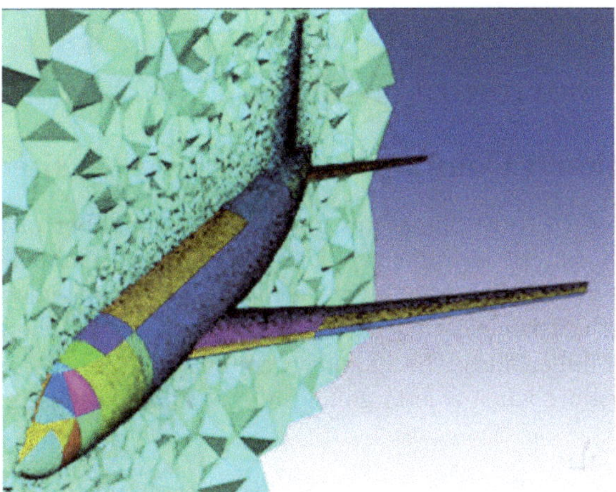

Figure 36.8: A volume region can be meshed by a union of tetrahedra. But the surface of the aircraft is meshed with a union of triangles (Figure by: gmsh.info).

36.6.5 Differential Equation Solvers versus Integral Equation Solvers

As we have shown, the two classes of numerical solvers for Maxwell's equations are differential equation solvers and integral equation solvers. Differential equation solvers are generally easier to implement. As shall be shown in the next lecture, they can also be easily implemented using a

[18]Sometimes, the values of the basis functions are defined on edges–for example, of a triangle or a tetrahedron [296, 297].

finite-difference solver. The unknowns in a differential equation solver are the fields. Since the fields permeate all of space, the unknowns are volumetrically distributed. When the fields are discretized by representing them by their point values in space, they require a large number of unknowns to represent. The plus side is that the matrix system associated with a differential equation solver is usually sparse, requiring less storage and less time to solve.

As has been shown, integral equation solvers are formulated using the Green's functions. In other words, integral equations are derived from Maxwell's equations using a Green's function (or a dyadic Green's function), where the unknowns now are surface unknowns such as surface electric and magnetic currents, or volume uknowns. Therefore, the unknowns are generally smaller in number, living only on the surface of a scatterer (or they occupy a smaller part of space). Hence, they can be approximated by a smaller set of unknowns. Consequently, the matrix systems generally are smaller. Once the currents are found, then the fields they generate can also be computed.[19]

Since the derivation of integral equations requires the use of the Green's functions, they are in general singular when $\mathbf{r} = \mathbf{r}'$, or when the observation point $\mathbf{r}$ and the source point $\mathbf{r}'$ coincide. Care has to be taken to discretize the integral equations in the neighborhood of the singularity. Moreover, a Green's function connects every current source point on the surface of a scatterer with every other source point, yielding a dense matrix system. But fast methods have been developed to solve such dense matrix systems [10].

36.7 Matrix Solution by Matrix-Free Method

36.7.1 Computational Complexity and Curse of Dimensionality

When computers are used to solve real-world problems, the number of unknowns needed are in general humongous. It is easy to see that problems in 1D can be modeled by a small number of unknowns. In CEM, the yardstick is usually wavelength. If n unknowns are used to approximate a function in a wavelength, it is clear that the unknown count is n^d where d is the dimensionality of the problem. In three dimensions, the unknown count scales as n^3 (or $O(n^3)$) (read "of order n^3"), which is terrible scaling when n is large and d is not small. This is also known as the curse of dimensionality [298].

It is seen that the solution of Maxwell's equations can be reduced to solving a matrix equation $\overline{\mathbf{L}} \cdot \mathbf{a} = \mathbf{g}$, where $\overline{\mathbf{L}}$ is a $N \times N$ matrix. The old way of solving a matrix system is to use Gaussian elimination, or Cramer's rule. Unfortunately, the number of arithmetics operations needed to solve the matrix system using these old methods are proportional of N^3 (or of $O(N^3)$). This is again terrible scaling, easily exhausting our computational resources.

The modern way of solving a matrix system is to find one where the CPU scaling scales as $O(N^\alpha)$ where α is made as close to one as possible. For an algorithm, the rate at which the CPU time grows as N is known as the computational complexity of the algorithm. For instance, we say that Gaussian elimination has a computational complexity of $O(N^3)$ which is a terrible complexity. For many problems related to wave physics, the physics can be exploited to expedite the matrix-vector

[19]Volume integral equations are also used, but the reduction in unknown count is not as dramatic.

multiply such that a dense matrix-vector multiply can be performed with as low complexity as $O(N)$ [299].

Another issue that designers of modern day computational algorithms have to grapple with is the issue of memory complexity. For instance, in forming a matrix equation, one has to form (or fill) a matrix system. The matrix and vector elements are generated in accordance to the formulas for their representations as shown in (36.6.5) and (36.6.6). Since there are N^2 elements in a matrix system, filling or forming the matrix system will take $O(N^2)$ time and memory usage. This is a terrible scaling, especially when compounded by the curse of dimensionality. 3D problems especially are notorious for hogging of computer resources in both CPU and memory usage!

36.7.2 Matrix Solutions

As aforementioned, given a matrix equation, there are many ways to seek its solution. The simplest way is to find the inverse of the matrix operator by direct inversions (for example, using Gaussian elimination [201] or Cramer's rule [300]). But on the down side, they have computational complexity[20] of $O(N^3)$, and requiring storage of $O(N^2)$. Due to the poor computational and memory complexity of direct inversion, when N is large, other methods have to be sought. One way is to seek a matrix-free method! In a matrix-free method, the matrix system is never generated, but an algorithm to produce the matrix-vector product $\overline{\mathbf{L}} \cdot \mathbf{a} = \mathbf{c}'$ is needed. The memory needed to store $\mathbf{c}'$ is $O(N)$, which is much less than what is required for storing the matrix $\overline{\mathbf{L}}$. This drastically reduces the memory complexity of the algorithm.

To this end, it is better to convert the solving of a matrix equation into an optimization problem which is matrix-free. These methods can be designed so that a much larger system can be solved with the existing resources of a digital computer. Optimization problems result in finding the stationary point of a functional.[21] First, we will figure out how to find such a functional.

Consider a matrix equation given by

$$\overline{\mathbf{L}} \cdot \mathbf{f} = \mathbf{g} \tag{36.7.1}$$

For simplicity, we consider $\overline{\mathbf{L}}$ as a symmetric matrix.[22] Then the corresponding functional or cost function is

$$I = \mathbf{f}^t \cdot \overline{\mathbf{L}} \cdot \mathbf{f} - 2\mathbf{f}^t \cdot \mathbf{g} \tag{36.7.2}$$

Such a functional is a quadratic functional because it is analogous to $I = Lx^2 - 2xg$, which is quadratic, in its simplest 1D rendition.

To find its optimal value or its stationary point, we take the first variation with respect to $\mathbf{f}$, namely, we let $\mathbf{f} = \mathbf{f}_o + \delta\mathbf{f}$. Upon substituting this into the above, and collecting the leading order and first order terms, we find the first order approximation of the functional I as

$$\delta I = \delta\mathbf{f}^t \cdot \overline{\mathbf{L}} \cdot \mathbf{f}_o + \mathbf{f}_o^t \cdot \overline{\mathbf{L}} \cdot \delta\mathbf{f} - 2\delta\mathbf{f}^t \cdot \mathbf{g} \tag{36.7.3}$$

[20] The scaling of computer time with respect to the number of unknowns (degrees of freedom) is known in computer parlance as computational complexity.

[21] Functional is function of a function [54, 38]. Here, we include a function of a vector to be a functional as well.

[22] Functional for the asymmetric case can be found in Chew, chapter 5 [38].

If $\overline{\mathbf{L}}$ is a symmetric matrix, the first two terms are the same, which is easily verified by taking the transpose of one of them and using that the transpose of a scalar is itself. Then the above just becomes

$$\delta I = 2\delta \mathbf{f}^t \cdot \overline{\mathbf{L}} \cdot \mathbf{f}_o - 2\delta \mathbf{f}^t \cdot \mathbf{g} \tag{36.7.4}$$

For $\mathbf{f}_o$ to be the optimal point or the stationary point, then its first variation has to be zero, or that $\delta I = 0$. Since $\delta \mathbf{f}^t$ above is arbitrary, as a result, we conclude that at the optimal point (or the stationary point), the following equation has to be satisfied:

$$\overline{\mathbf{L}} \cdot \mathbf{f}_o = \mathbf{g} \tag{36.7.5}$$

Hence, the optimal point to the quadratic functional I in (36.7.2) is the solution to (36.7.1) or (36.7.5).

36.7.3 Gradient of a Functional

The above method, when applied to an infinite-dimensional Hilbert space problem, is called the variational method, but the main ideas are similar. The wonderful thing about such a method is that instead of doing direct inversion of a matrix system (which is expensive), one can search for the optimal point or stationary point of the quadratic functional using gradient search or gradient descent methods or some optimization method.

It turns out that the gradient of a quadratic functional can be found quite easily. Also it is cheaper to compute the gradient of a functional than to find the inverse of a matrix operator. To do this, it is better to write out a functional using index (or indicial, or Einstein) notation [301]. In this notation, summations over repeated indices are implied. Then, the functional first variation δI in (36.7.4) becomes

$$\delta I = 2\delta f_j L_{ji} f_i - 2\delta f_j g_j \tag{36.7.6}$$

Also as aforementioned, in this notation, the summation symbol is dropped and summations over repeated indices are implied. In the above, we neglect to distinguish between $\mathbf{f}_o$ and $\mathbf{f}$. It is implied that $\mathbf{f}$ represents the optimal point. In this manner, it is easier to see what a functional derivative is. We can differentiate the above with respect to f_j easily to arrive at

$$\frac{\partial I}{\partial f_j} = 2L_{ij} f_i - 2g_j \tag{36.7.7}$$

Notice that the remaining equation has one index j remaining in index notation, meaning that it is a vector equation. We can reconstitute the above using our more familiar matrix notation that the above is similar to

$$\frac{\delta I}{\delta \mathbf{f}} = \nabla_{\mathbf{f}} I = 2\overline{\mathbf{L}} \cdot \mathbf{f} - 2\mathbf{g} \tag{36.7.8}$$

The left-hand side is a notation for the functional derivative or the gradient of a functional in a multidimensional space which is a vector obviated by indicial notation. And the right-hand side is the expression for calculating this gradient. One needs only to perform a matrix-vector product to find this gradient. The cost of a matrix-vector product is at most $O(N^2)$ for dense matrices, and can be as low as $O(N)$ for sparse matrices. Hence, the computational complexity of finding this gradient is $O(N^2)$ at worst if $\overline{\mathbf{L}}$ is a dense matrix, and as low as $O(N)$ if $\overline{\mathbf{L}}$ is a sparse matrix.[23] Fast methods can be developed to expedite such matrix-vector multiply as documented in [10]. In a gradient search method, such a gradient is calculated repeatedly until the optimal point is found. Such methods are called iterative methods.

If the optimal point can be found in N_{iter} iterations, then the CPU time scales as $N_{\text{iter}}N^{\alpha}$ where $1 < \alpha < 2$. There is a clever gradient search algorithm, called the *conjugate gradient method*, that can find the exact optimal point in $N_{\text{iter}} = N$ in exact arithmetics. But an exact solution is not needed in an optimal solution: an approximate solution suffices. In many gradient search methods, it suffices to obtain an approximate solution where the error is acceptable after N_{iter} where $N_{\text{iter}} \ll N$ is possible. Thus, the total solution time (or solve time) $N_{\text{iter}}N^{\alpha} \ll NN^{\alpha} \ll N^3$ is possible, resulting in great savings in computer time, especially if $\alpha = 1$. This is the case for FEM [210], [295], [302], [303], [304], and fast multipole algorithm [305], [306].[24]

What is more important is that this method does not require the storage of the matrix $\overline{\mathbf{L}}$, but a computer code that produces the vector $\mathbf{g}_o = \overline{\mathbf{L}} \cdot \mathbf{f}$ as an output, with $\mathbf{f}$ as an input. Both $\mathbf{f}$ and $\mathbf{g}_o$ require only $O(N)$ memory storage. Such methods are called matrix-free methods. Even when $\overline{\mathbf{L}}$ is a dense matrix, which is the case if it is the matrix representation of some Green's functions, fast methods now exist to perform the dense matrix-vector product in $O(N \log N)$ operations.[25]

The value I is also called the cost function, and its minimum is sought in finding the solution by gradient search methods. Detailed discussions of these methods are given in [307]. Figure 36.9 shows the contour plot of a cost function in 2D. When the condition number[26] of the matrix $\overline{\mathbf{L}}$ is large (implying that the matrix is ill-conditioned), the contour plot resembles a deep and narrow valley. And hence, the gradient search method will tend to zig-zag along the way as it finds the optimal solution. This implies that convergence is slow for matrices with large condition numbers. It is seen graphically that when the condition number of a matrix system is large, it is deleterious to the convergence of gradient search methods [291, 290].

Figure 36.10 shows a cartoon picture in 2D of the histories of different search paths from a machine-learning example where a cost functional similar to I has to be minimized. Finding the optimal point or the minimum point of a general functional is still a hot topic of research: it is important in artificial intelligence as well as in solving large systems of linear algebraic equations.

---

[23]This is the case for many differential equation solvers such as the finite-element method or the finite-difference method.

[24]It is to be noted that many fast algorithms have computational complexity of $O(N \log N)$, such as FFT and fast multipole algorithm. It is important to note that $N \log N < N^{1+\alpha}$, where $\alpha > 0$ as $N \to \infty$.

[25]Chew et al., *Fast and Efficient Algorithms in CEM* [10].

[26]This is the ratio of the largest eigenvalue of the matrix to its smallest eigenvalue.

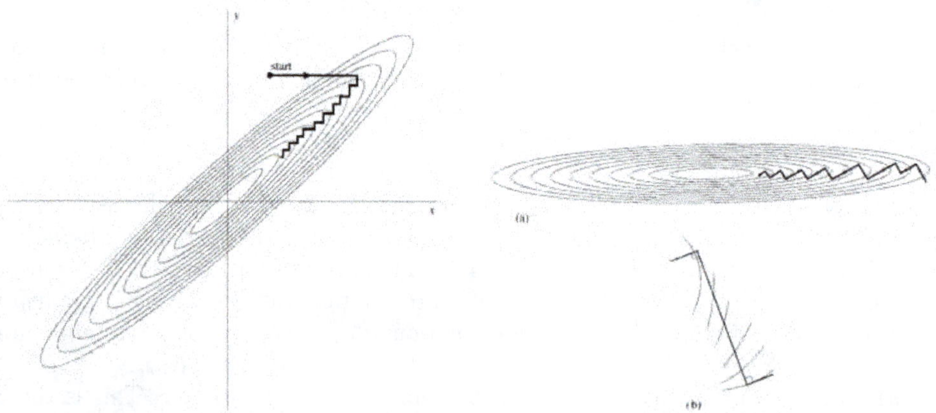

Figure 36.9: Plot of a 2D cost function, $I(x, y)$ for an ill-conditioned system (Figure by Numerical Recipe [307]). A higher-dimensional plot of this cost function will be difficult. The zig-zag search path is the hallmark of an ill-conditioned system. This cost function has only two eigenvalues, one of which is larger than the other one.

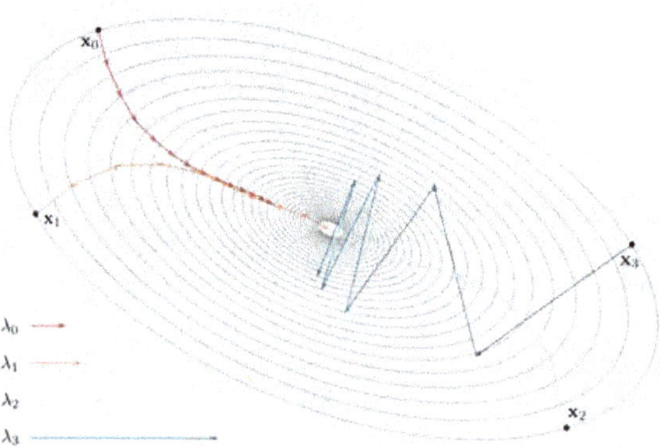

Figure 36.10: Gradient search or gradient descent method is finding an optimal point. As is obvious, the initial value is instrumental in converging into the correct solution (Figure by: Y. Ioannou, https://blog.yani.io/sgd/).

Exercises for Lecture 36

Problem 36-1:

(i) Derive (36.2.2) from Maxwell's equations.

(ii) In the lecture notes, we assume that $\overline{\mathbf{L}}$ is a symmetric matrix. But now, assume that the operator $\overline{\mathbf{L}}$ in (36.7.1) is not symmetric. The optimal solution is obtained by solving two equations. Construct a functional I such that its optimal point is the solution to (36.7.1) and another auxiliary equation. Show that such a functional can be defined as

$$I = \mathbf{w}^t \cdot \overline{\mathbf{L}} \cdot \mathbf{f} - \mathbf{w}^t \cdot \mathbf{g} - \mathbf{g}_a^t \cdot \mathbf{f}$$

By taking the first variation of the above functional with respect to f and w, show that the optimal solutions that minimize the above functional are solutions to the equations

$$\overline{\mathbf{L}} \cdot \mathbf{f} = \mathbf{g}$$

$$\overline{\mathbf{L}}^t \cdot \mathbf{w} = \mathbf{g}_a$$

(iii) Using index notation, for a symmetric system discussed in the lecture notes, show that the gradient of a functional I in the N-dimensional space is given by (36.7.8).

Problem 36-2:

(i) Derive (36.3.5) and explain why the second term on the right-hand side corresponds to the scattered field.

(ii) Explain why the scattered field satisfies the radiation condition.

Lecture 37

Finite-Difference Method, Yee Algorithm

In this lecture, we will introduce one of the simplest methods to solve Maxwell's equations numerically. This is the finite-difference time-domain (FDTD) method first proposed by Yee [308] and popularized by Taflove [309]. Because of its simplicity, a simple Maxwell's equations solver can be coded in one afternoon. Thus almost every physics or electrical engineering laboratory has a home-grown version of the finite-difference time-domain (FDTD) solver. This method is the epitome of that "simplicity rules".[1] Professor Hermann Haus at MIT used to say: find the simplest method to do things. Complicated methods will be forgotten, but the simplest ones will prevail. This is also reminiscent of Einstein's saying, "Everything should be made as simple as possible, but no simpler!"

37.1 Finite-Difference Time-Domain Method

To obtain the transient (or time-domain) solution of the wave equation for a more general, inhomogeneous medium, a numerical method has to be used. The FDTD method, a numerical method, is particularly suitable for solving transient problems. Compounded by rapid growth in computer speed, with its versatility, it has been used with great success in solving many practical problems. This method is based on the simple Yee algorithm [308] and has been vastly popularized by Taflove [309, 310]. If one search using Google Scholar, one will find many papers written on this topic. Among many, here is one highly popular one called MEEP [311] written by the MIT team.

In the finite-difference method, continuous space-time is replaced with a discrete space-time. Thus in the discrete space-time, partial differential equations are replaced with finite-difference equations. These finite-difference equations are readily implemented on a digital computer. Furthermore, an iterative or time-stepping scheme can be implemented without having to store or solve large

[1] "Rule" is used as a verb here.

matrices, resulting in great savings in computer time and memory. In addition, the matrix for the system of equations is never generated, making this a matrix-free method: There is no need to store the matrix system for matrix and memory management as one writes this numerical solver. More recently, the development of parallel processor architectures in computers has also further enhanced the efficiency of the finite-difference time-domain scheme [312].

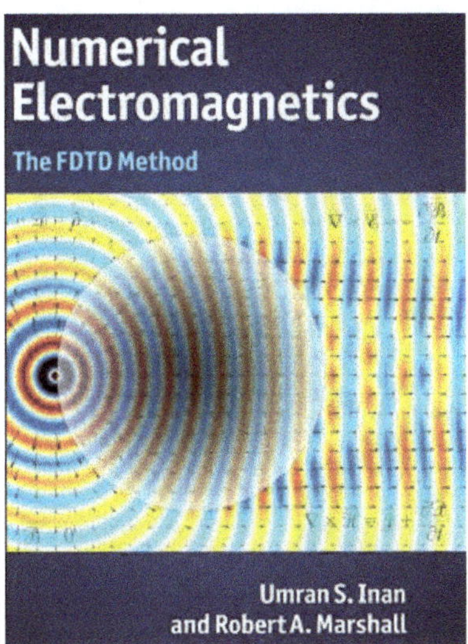

Figure 37.1: The FDTD method is so popular, and so simple that many authors choose to adorn their book covers with the beautiful graphics of FDTD simulations, as shown in the figure here.

The finite-difference method is also described in numerous early works (see, for example, Taflove 1988 [309]; Potter 1973 [313]; Ames 2014 [314]; and Morton 2019 [315].)[2]

37.1.1 The Finite-Difference Approximation

Consider first a scalar wave equation of the form

$$\frac{1}{c^2(\mathbf{r})}\frac{\partial^2}{\partial t^2}\phi(\mathbf{r},t) = \mu(\mathbf{r})\nabla\cdot\mu^{-1}(\mathbf{r})\nabla\phi(\mathbf{r},t). \tag{37.1.1}$$

The above equation appears in scalar acoustic waves or a 2D electromagnetic waves in inhomogeneous media [158, 38]. It clearly has no closed-form solution.

To convert the above into a form that can be solved by a digital computer easily, first, one needs to find finite-difference approximations to the time derivatives. The time derivative can be approximated in many ways. For example, a derivative can be approximated by forward, backward, and

[2]A Google Scholar search will yield a more complete list of recent works in this topic.

central finite-difference formulas [316] (see Figure 37.2).

$$\text{Forward difference:} \quad \frac{\partial\phi(\mathbf{r},t)}{\partial t} \approx \frac{\phi(\mathbf{r},t+\Delta t)-\phi(\mathbf{r},t)}{\Delta t}, \tag{37.1.2}$$

$$\text{Backward difference:} \quad \frac{\partial\phi(\mathbf{r},t)}{\partial t} \approx \frac{\phi(\mathbf{r},t)-\phi(\mathbf{r},t-\Delta t)}{\Delta t}, \tag{37.1.3}$$

$$\text{Central difference:} \quad \frac{\partial\phi(\mathbf{r},t)}{\partial t} \approx \frac{\phi(\mathbf{r},t+\frac{\Delta t}{2})-\phi(\mathbf{r},t-\frac{\Delta t}{2})}{\Delta t}, \tag{37.1.4}$$

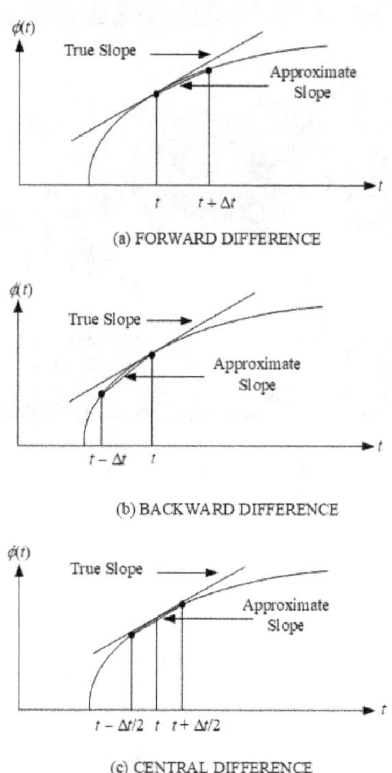

Figure 37.2: Different finite-difference approximations for the time derivative. One can eyeball the above plots and see that the central-difference formula is the best. This can be further confirmed by a Taylor series analysis.

where Δt is a small number. Of the three methods of approximating the time derivative, the central-difference scheme is the best approximation, as is evident from Figure 37.2. The errors in the forward and backward differences are $O(\Delta t)$ (or first-order errors) while the central-difference approximation has an error $O[(\Delta t)^2]$ (or second-order error). This can be easily verified by Taylor-series expanding the right-hand sides of (37.1.2) to (37.1.4).

Consequently, using the central-difference formula twice, we arrive at the approximation for the second derivative as

$$\frac{\partial^2}{\partial t^2}\phi(\mathbf{r},t) \approx \frac{\partial}{\partial t}\left[\frac{\phi(\mathbf{r},t+\frac{\Delta t}{2}) - \phi(\mathbf{r},t-\frac{\Delta t}{2})}{\Delta t}\right] \tag{37.1.5}$$

$$\approx \frac{\phi(\mathbf{r},t+\Delta t) - 2\phi(\mathbf{r},t) + \phi(\mathbf{r},t-\Delta t)}{(\Delta t)^2}. \tag{37.1.6}$$

Next, if the function $\phi(\mathbf{r},t)$ is indexed on discrete time steps on the t-axis, such that for $t = l\Delta t$, then we can index $\phi(\mathbf{r},t) = \phi(\mathbf{r}, l\Delta t) = \phi^l(\mathbf{r})$. Here, l is an integer used to count the time steps. Using this notation, Equation (37.1.6) then becomes

$$\frac{\partial^2}{\partial t^2}\phi(\mathbf{r},t) \approx \frac{\phi^{l+1}(\mathbf{r}) - 2\phi^l(\mathbf{r}) + \phi^{l-1}(\mathbf{r})}{(\Delta t)^2}. \tag{37.1.7}$$

37.1.2 Time Stepping or Time Marching

With this notation and approximations, (37.1.1) can be approximated by a time-stepping (or time-marching) formula, namely,

$$\phi^{l+1}(\mathbf{r}) \cong c^2(\mathbf{r})(\Delta t)^2\mu(\mathbf{r})\nabla\cdot\mu^{-1}(\mathbf{r})\nabla\phi^l(\mathbf{r}) + 2\phi^l(\mathbf{r}) - \phi^{l-1}(\mathbf{r}). \tag{37.1.8}$$

Therefore, given the knowledge of $\phi(\mathbf{r},t)$ at $t = l\Delta t$ or $\phi^l(\mathbf{r})$, and at $t = (l-1)\Delta t$, or $\phi^{l-1}(\mathbf{r})$ for all $\mathbf{r}$, one can deduce $\phi(\mathbf{r},t)$ at $t = (l+1)\Delta t$, or $\phi^{l+1}(\mathbf{r})$ for all $\mathbf{r}$. In other words, given the initial values of $\phi(\mathbf{r},t)$ at, for example, $t = 0$ and $t = \Delta t$, $\phi(\mathbf{r},t)$ can be deduced for all subsequent times, provided that the time-stepping formula above is stable.

At this point, the right-hand side of (37.1.8) involves the space derivatives. There exist a plethora of ways to approximate and calculate the right-hand side of (37.1.8) numerically. Here, we shall illustrate again the use of the finite-difference method to calculate the right-hand side of (37.1.8). Before proceeding further, note that the space derivatives on the right-hand side in Cartesian coordinates are

$$\mu(\mathbf{r})\nabla\cdot\mu^{-1}(\mathbf{r})\nabla\phi(\mathbf{r}) = \mu\frac{\partial}{\partial x}\mu^{-1}\frac{\partial}{\partial x}\phi + \mu\frac{\partial}{\partial y}\mu^{-1}\frac{\partial}{\partial y}\phi + \mu\frac{\partial}{\partial z}\mu^{-1}\frac{\partial}{\partial z}\phi. \tag{37.1.9}$$

Then, one can approximate using central differencing that

$$\frac{\partial}{\partial z}\phi(x,y,z) \approx \frac{1}{\Delta z}\left[\phi\left(x,y,z+\frac{\Delta z}{2}\right) - \phi\left(x,y,z-\frac{\Delta z}{2}\right)\right], \tag{37.1.10}$$

Consequently, using central differencing two times,

$$\begin{aligned}\frac{\partial}{\partial z}\mu^{-1}\frac{\partial}{\partial z}\phi(x,y,z) \approx \frac{1}{(\Delta z)^2}\bigg\{&\mu^{-1}\left(z+\frac{\Delta z}{2}\right)\phi(x,y,z+\Delta z)\\ &-\left[\mu^{-1}\left(z+\frac{\Delta z}{2}\right)+\mu^{-1}\left(z-\frac{\Delta z}{2}\right)\right]\phi(x,y,z)\\ &+\mu^{-1}\left(z-\frac{\Delta z}{2}\right)\phi(x,y,z-\Delta z)\bigg\}.\end{aligned} \tag{37.1.11}$$

Furthermore, after denoting $\phi(x,y,z) = \phi_{m,n,p}$, $\mu(x,y,z) = \mu_{m,n,p}$, on a discretized grid point at $x = m\Delta x$, $y = n\Delta y$, $z = p\Delta z$, we have $(x,y,z) = (m\Delta x, n\Delta y, p\Delta z)$. Then in finite-difference notations,

$$\frac{\partial}{\partial z}\mu^{-1}\frac{\partial}{\partial z}\phi(x,y,z) \approx \frac{1}{(\Delta z)^2}\Big[\mu^{-1}_{m,n,p+\frac{1}{2}}\phi_{m,n,p+1} - \left(\mu^{-1}_{m,n,p+\frac{1}{2}} + \mu^{-1}_{m,n,p-\frac{1}{2}}\right)\phi_{m,n,p} + \mu^{-1}_{m,n,p-\frac{1}{2}}\phi_{m,n,p-1}\Big]. \tag{37.1.12}$$

This cumbersome and laborious-looking equation can be abbreviated if we define a central difference operator as[3]

$$\bar{\partial}_z\phi_{m,n,p} = \frac{1}{\Delta z}\left(\phi_{m,n,p+\frac{1}{2}} - \phi_{m,n,p-\frac{1}{2}}\right) \tag{37.1.13}$$

Then the right-hand side of the (37.1.12) can be written more succinctly as

$$\frac{\partial}{\partial z}\mu^{-1}\frac{\partial}{\partial z}\phi(x,y,z) \approx \bar{\partial}_z\mu^{-1}_{m,n,p}\bar{\partial}_z\phi_{m,n,p} \tag{37.1.14}$$

With similar approximations to the other terms in (37.1.9), (37.1.8) is now compactly written as

$$\phi^{l+1}_{m,n,p} = (\Delta t)^2 c^2_{m,n,p}\mu_{m,n,p}\left[\bar{\partial}_x\mu^{-1}_{m,n,p}\bar{\partial}_x + \bar{\partial}_y\mu^{-1}_{m,n,p}\bar{\partial}_y + \bar{\partial}_z\mu^{-1}_{m,n,p}\bar{\partial}_z\right]\phi^{l}_{m,n,p} + 2\phi^{l}_{m,n,p} - \phi^{l-1}_{m,n,p}. \tag{37.1.15}$$

The above can be readily implemented on a computer for time stepping. The right-hand side of the above is dependent on the present time step l and previous time step $l-1$. These values of the field can be used to predict the left-hand side which is the field value on time step $l+1$, the future time step.

Notice, however, that the use of central differencing results in the evaluation of medium property μ at half grid points. This is inconvenient, as the introduction of material values at half grid points increases computer memory used. Hence, it is customary to store the medium values at the integer grid points for ease of bookkeeping, and to deduce the values at half-grid points using the following approximations

$$\mu_{m+\frac{1}{2},n,p} \simeq \frac{1}{2}(\mu_{m+1,n,p} + \mu_{m,n,p}), \tag{37.1.16}$$

$$\mu_{m+\frac{1}{2},n,p} + \mu_{m-\frac{1}{2},n,p} \simeq 2\mu_{m,n,p}, \tag{37.1.17}$$

and so on. Moreover, if μ is a smooth function of space, it is easy to show that the errors in the above approximations are of second order by Taylor series expansions.

[3]This is in the spirit of [317].

For a homogeneous medium, with $\Delta x = \Delta y = \Delta z = \Delta s$, namely, we assume the space steps to be equal in all directions, (37.1.15) written explicitly then becomes

$$\phi^{l+1}_{m,n,p} = \left(\frac{\Delta t}{\Delta s}\right)^2 c^2 \left[\phi^l_{m+1,n,p} + \phi^l_{m-1,n,p} + \phi^l_{m,n+1,p} + \phi^l_{m,n-1,p} + \phi^l_{m,n,p+1} + \phi^l_{m,n,p-1} - 6\phi^l_{m,n,p}\right] + 2\phi^l_{m,n,p} - \phi^{l-1}_{m,n,p}. \tag{37.1.18}$$

Notice then that with the central-difference approximation, the value of $\phi^{l+1}_{m,n,p}$ is dependent only on $\phi^l_{m,n,p}$, and its nearest neighbors, $\phi^l_{m\pm1,n,p}$, $\phi^l_{m,n\pm1,p}$, $\phi^l_{m,n,p\pm1}$, and $\phi^{l-1}_{m,n,p}$, its value at the previous time step. Moreover, in the finite-difference scheme outlined above, no matrix inversion is required at each time step. Such a scheme is known as an explicit scheme. The use of an explicit scheme is a major advantage of the finite-difference method compared to the finite-element methods. Consequently, in order to update N grid points using (37.1.15) or (37.1.18), $O(N)$ multiplications are required for each time step. In comparison, $O(N^3)$ multiplications are required to invert an $N \times N$ full matrix–for example, using Gaussian elimination. The simplicity and efficiency of these finite-difference algorithms have made them vastly popular.

37.1.3 Stability Analysis

The implementation of the FDTD scheme using time marching does not always lead to a stable scheme. Hence, in order for the solution to converge, the time-stepping scheme must at least be stable. Consequently, it is necessary to find the condition under which this numerical scheme is stable. To do this, one performs the von Neumann stability analysis (von Neumann 1943 [318]) on Equation (37.1.18). We will assume the medium to be homogeneous to simplify the analysis.

As shown in the previous lecture 35, section 35.1, a point source gives rise to a spherical wave that can be expanded as the sum of plane waves in different directions. It also implies that any wave emerging from sources can be expanded as sum of plane waves. This is the spirit of the spectral expansion method in the previous lecture: if a scheme is not stable for a plane wave, it would not be stable for any wave. Consequently, to perform the stability analysis, we assume a propagating plane wave (or a Fourier mode) as a trial solution. To this end, we let

$$\phi(x, y, z, t) = A_0 e^{ik_x x + ik_y y + ik_z z - i\omega t}, \tag{37.1.19}$$

The above is clearly a solution to the scalar wave equation. In discretized form, we let $x = m\Delta x$, $y = n\Delta y$, $z = p\Delta z$, $t = l\Delta t$ and then by letting $\Delta x = \Delta y = \Delta z = \Delta s$, we have a discrete version of the Fourier mode:

$$\phi^l_{m,n,p} = A_0 e^{ik_x m\Delta s + ik_y n\Delta s + ik_z p\Delta s - i\omega l\Delta t}. \tag{37.1.20}$$

It turns out the discrete Fourier modes are eigenfunctions of the finite-difference operators. Using (37.1.20), it is easy to show that for the x space derivative,

$$(\Delta s)^2 \bar{\partial}_x^2 \phi^l_{m,n,p} = \phi^l_{m+1,n,p} - 2\phi^l_{m,n,p} + \phi^l_{m-1,n,p} = 2[\cos(k_x\Delta s) - 1]\phi^l_{m,n,p} = -4\sin^2\left(\frac{k_x\Delta s}{2}\right)\phi^l_{m,n,p}. \tag{37.1.21}$$

The above indicates that the space finite-difference operator acting on the discrete Fourier mode is just a constant times the Fourier mode. This implies that a Fourier mode is an eigenfunction of the finite-difference operator! The space derivatives in y and z directions can be similarly derived.

The second order time derivative in the wave equation can be similarly approximated by a finite-difference equation yielding

$$\frac{\partial^2}{\partial t^2}\phi(\mathbf{r},t)(\Delta t)^2 \approx \phi^{l+1}_{m,n,p} - 2\phi^{l}_{m,n,p} + \phi^{l-1}_{m,n,p} = -4\sin^2\left(\frac{\omega\Delta t}{2}\right)\phi^{l}_{m,n,p} \tag{37.1.22}$$

We need to find the finite-difference approximation of the wave equation, namely that

$$\nabla^2\phi(\mathbf{r},t) - \frac{1}{c^2}\frac{\partial^2}{\partial t^2}\phi(\mathbf{r},t) = 0 \tag{37.1.23}$$

Using (37.1.21) and its equivalence for x, y, z space derivatives, and then (37.1.22) for the time derivatives, it follows that the finite-difference approximation of the above wave equation gives us

$$\begin{aligned} 4\sin^2\left(\frac{\omega\Delta t}{2}\right)\phi^{l}_{m,n,p} &= 4\left(\frac{\Delta t}{\Delta s}\right)^2 c^2\left[\sin^2\left(\frac{k_x\Delta s}{2}\right) + \sin^2\left(\frac{k_y\Delta s}{2}\right)\right. \\ &\qquad \left. + \sin^2\left(\frac{k_z\Delta s}{2}\right)\right]\phi^{l}_{m,n,p} \\ &= 4r^2 s^2\phi^{l}_{m,n,p}, \end{aligned} \tag{37.1.24}$$

where

$$r = \left(\frac{\Delta t}{\Delta s}\right)c, \qquad s^2 = \sin^2\left(\frac{k_x\Delta s}{2}\right) + \sin^2\left(\frac{k_y\Delta s}{2}\right) + \sin^2\left(\frac{k_z\Delta s}{2}\right). \tag{37.1.25}$$

Equation (37.1.24) implies that, for nonzero $\phi^{l}_{m,n,p}$,[4] it is necessary for

$$\sin^2\left(\frac{\omega\Delta t}{2}\right) = r^2 s^2, \tag{37.1.26}$$

The above is the dispersion relation relating ω, the frequency of the Fourier mode, with k_x, k_y, and k_z, the Fourier wavenumbers.

In order for the solution to be stable, ω has to be real when we solve the above equation for ω. In this way, (37.1.19) and (37.1.20) are bounded. Hence, it is necessary that the right-hand side of the above equation is less than 1 so that ω is real in (37.1.26). When ω is real, then the expression in This implies that if

$$r^2 s^2 < 1, \tag{37.1.27}$$

[4]For those who are more mathematically inclined, we are solving an eigenvalue problem in disguise. Remember that a function is a vector, even after it has been discretized.:)

then stability is ensured since a solution with real ω exists. Since from (37.1.25), $s^2 \leq 3$ for all k_x, k_y, and k_z. Also, from (37.1.27), we conclude that

$$r < \frac{1}{s}$$

But we also know that from the definition of s in (37.1.25) that

$$\frac{1}{\sqrt{3}} \leq \frac{1}{s}$$

In other words, the left-hand side of the above is the lower bound for $1/s$. The above two inequalities will be satisfied if the general condition is

$$r < \frac{1}{\sqrt{3}} < \frac{1}{s}, \qquad \text{or} \qquad \Delta t < \frac{\Delta s}{c\sqrt{3}}. \tag{37.1.28}$$

after using that $r = c\Delta t/(\Delta s)$. The above is the general condition for stability.

The above analysis is for three–dimensional problems. It is clear from the above analysis that for an n-dimensional problem where $n = 1, 2, 3$, then the condition for stability is

$$\Delta t < \frac{\Delta s}{c\sqrt{n}}. \tag{37.1.29}$$

One may ponder on the physical meaning of this inequality further, but it is only natural that the time step Δt has to be bounded from above. Otherwise, one arrives at the ludicrous notion that the time step can be arbitrarily large, thus violating causality.

Moreover, if the grid points of the finite-difference scheme are regarded as a simple cubic lattice, then the distance $\Delta s/\sqrt{n}$ is also the distance between the closest lattice planes through the simple cubic lattice. Notice that the time for the wave to travel between these two lattice planes is $\Delta s/(c\sqrt{n})$. Consequently, the stability criterion (37.1.29) implies that the time step Δt has to be less than the shortest travel time for the wave between the lattice planes in order to satisfy causality. In other words, if the wave is time-stepped ahead of the time on the right-hand side of (37.1.29), instability ensues, because of the violation of causality.

The above is also known as the CFL (Courant, Friedrichs, and Lewy 1928 [319]) stability criterion.[5] It could be easily modified for $\Delta x \neq \Delta y \neq \Delta z$ [310]. The above analysis implies that we can pick a larger time step if the space steps are larger. A larger time step will allow one to generate a time-domain response rapidly. However, one cannot arbitrarily make the space step large due to grid-dispersion error, as shall be discussed next.

37.1.4 Grid-Dispersion Error

When a finite-difference scheme is stable, it still may not be accurate to produce good results due to the errors in the finite-difference approximations. Hence, it is useful to ascertain the errors in

[5]It is common to call this the Courant condition in the math community.

terms of the size of the grid and the time step. An easy error to analyze is the *grid-dispersion error.* In a homogeneous, dispersionless medium, all plane waves propagate with the same phase velocity. However, in the finite-difference approximation, all plane waves will not propagate at the same phase velocity due to the grid-dispersion error. As we shall see, the error is dependent on the frequency and the direction of propagation of the waves. We have seen previously that a polychromatic pulse constitute different frequencies. In a dispersive medium, these frequency components propagate with different phase velocities giving rise to pulse distortion. Ways to overcome these errors are suggested in [320, 321].

As a consequence, a pulse in the time domain, which is a linear superposition of plane waves with different frequencies, will be distorted if the dispersion introduced by the finite-difference scheme is intolerable. Therefore, for simplicity, we will analyze the grid-dispersion error in a homogeneous free space medium.

To ascertain the grid-dispersion error, we assume a time-harmonic solution where the Fourier mode has time dependence of the form $A_0 e^{-i\omega l \Delta t}$ in (37.1.20). In this case, the left-hand side of (37.1.24) becomes

$$\left(e^{-i\omega\Delta t} - 2 + e^{+i\omega\Delta t}\right)\phi^l_{m,n,p} = -4\sin^2\left(\frac{\omega\Delta t}{2}\right)\phi^l_{m,n,p}. \tag{37.1.30}$$

Then, from (37.1.24), it follows that

$$\sin\left(\frac{\omega\Delta t}{2}\right) = rs, \tag{37.1.31}$$

where r and $s(k_x, k_y, k_z)$ are given in (37.1.25). Now, (37.1.31) governs the relationship between ω and k_x, k_y, and k_z in the finite-difference scheme, and hence, it is a dispersion relation for the finite-difference approximate solution.

The above gives a rather complicated relationship between the frequency ω and the wavenumbers k_x, k_y, and k_z. This is the result of the finite-difference approximation of the scalar wave equation. As a sanity check, when the space and time discretizations become very small, we should recover the dispersion relation of homogeneous medium or free space.

In contrast, if a medium is homogeneous, it is well known that (37.1.1) has a plane-wave solution of the type given by (37.1.19) where

$$\omega = c\sqrt{k_x^2 + k_y^2 + k_z^2} = c|\mathbf{k}| = ck. \tag{37.1.32}$$

where $\mathbf{k} = \hat{x}k_x + \hat{y}k_y + \hat{z}k_z$ is the direction of propagation of the plane wave. Defining the phase velocity to be $\omega/k = c$, this phase velocity is isotropic, or the same in all directions. Moreover, it is independent of frequency.

But in (37.1.31), because of the definition of s as given by (37.1.25), the dispersion relation between ω and $\mathbf{k}$ is not isotropic (anisotropic). This implies that plane waves propagating in different directions will have different phase velocities. Moreover, unlike (37.1.32), different frequencies will have different phase velocities.

Equation (37.1.31) is the dispersion relation for the approximate solution. It departs from Equation (37.1.32), the exact dispersion relation for free space, due to the finite-difference approximation. This departure gives rise to errors called grid dispersion errors. For example, when c is a constant, (37.1.32) states that the phase velocities of plane waves of different wavelengths and directions are the same. However, this is not true for (37.1.31), as shall be shown.

To elaborate more on the grid dispersion error, we assume that s is small. Then (37.1.31), after using Taylor series expansion, can be written approximately as

$$\frac{\omega \Delta t}{2} = \sin^{-1} rs \cong rs + \frac{r^3 s^3}{6}. \tag{37.1.33}$$

When Δs is small, using the small argument approximation for the sine function, one obtains from (37.1.25) that

$$s \simeq \frac{\Delta s}{2}(k_x^2 + k_y^2 + k_z^2)^{1/2}\left[1 - \frac{\Delta s^2}{24}\left(\frac{k_x^4 + k_y^4 + k_z^4}{k_x^2 + k_y^2 + k_z^2}\right)\right] \tag{37.1.34}$$

Equation (37.1.33), by taking its higher-order Taylor expansion, then becomes

$$\frac{\omega \Delta t}{2} \simeq r\frac{\Delta s}{2}(k_x^2 + k_y^2 + k_z^2)^{1/2}\left[1 - \delta\right] \tag{37.1.35}$$

Or after using the definition of r in (37.1.25), we have

$$\omega \simeq (k_x^2 + k_y^2 + k_z^2)^{1/2}\left[1 - \delta\right] \tag{37.1.36}$$

We further derive that (see [38])

$$\delta = \frac{\Delta s^2}{24}\frac{k_x^4 + k_y^4 + k_z^4}{k_x^2 + k_y^2 + k_z^2} - \frac{r^2 \Delta s^2}{24}(k_x^2 + k_y^2 + k_z^2) \tag{37.1.37}$$

Remember that $r = c\Delta t/\Delta s$, a dimensionless quantity. From the above, (37.1.35) is almost the same as (37.1.32) except for the factor $1 - \delta$. Also, if $\delta = 0$, we retrieve the dispersion relation of the homogeneous free-space medium. So δ is a measure of the departure of the dispersion relation from that of free space due to our finite-difference approximation. More insight can be gotten if we let $(k_x, k_y, k_z) = (k\sin\theta\cos\phi, k\sin\theta\sin\phi, k\cos\theta)$, then the above can be rewritten as

$$\delta = \frac{k^2 \Delta s^2}{24}F(\theta, \phi) - \frac{r^2 k^2 \Delta s^2}{24}G(\theta, \phi) \tag{37.1.38}$$

where $F(\theta, \phi)$ and $G(\theta, \phi)$ are functions of $O(1)$ that depend only on θ, ϕ. Since r^2 is less than one, the dominant term above is the first term even though the two terms are of the same order. It is clear that to reduce the grid-dispersion error, we can make $\frac{k^2 \Delta s^2}{24} \ll 1$. In this case, δ above is small, reducing the grid-dispersion error.

Since $\mathbf{k}$ is inversely proportional to wavelength λ, then δ in the correction to the above equation is proportional to $2\pi\Delta s^2/\lambda^2$. Therefore, to reduce the grid dispersion error, it is necessary for δ to be small or to have

$$\frac{1}{24}\left(\frac{2\pi\Delta s}{\lambda}\right)^2 \ll 1. \tag{37.1.39}$$

Or the space discretization Δs has to be smaller than the simulation wavelength to mitigate the grid-dispersion error. When this is true, using the fact that $r = c\Delta t/\Delta s$, then (37.1.35) becomes

$$\frac{\omega}{c} \approx \sqrt{k_x^2 + k_y^2 + k_z^2}\,. \tag{37.1.40}$$

which is close to the dispersion relation of free space as indicated in (37.1.32). Furthermore, Δt must be chosen so that the CFL stability criterion is met. Therefore, the rule of thumb is to choose about 10 to 20 grid points per wavelength. Also, for a plane wave propagating as $e^{i\mathbf{k}\cdot\mathbf{r}}$, the grid dispersion yields an error $\delta\mathbf{k}$ in the vector $\mathbf{k}$ giving rise to cumulative error $e^{i\delta\mathbf{k}\cdot\mathbf{r}}$. Hence, the larger the distance traveled, the larger is $\mathbf{r}$, and the larger is the cumulative phase error. And hence, the grid size must be smaller in order to arrest such a phase error due to the grid dispersion. The larger the simulation domain, the smaller this per unit length error should be to control this cumulative error!

37.2 The Yee Algorithm

The Yee algorithm (Yee 1966 [308])[6] is a simple algorithm specially designed to solve vector electromagnetic field problems on a rectilinear grid. The finite-difference time-domain (FDTD) method (Taflov 1988) when applied to solving electromagnetics problems usually uses this algorithm. To derive it, Maxwell's equations in the time-domain are first written in Cartesian coordinates:

$$-\frac{\partial B_x}{\partial t} = \frac{\partial E_z}{\partial y} - \frac{\partial E_y}{\partial z}, \tag{37.2.1}$$

$$-\frac{\partial B_y}{\partial t} = \frac{\partial E_x}{\partial z} - \frac{\partial E_z}{\partial x}, \tag{37.2.2}$$

$$-\frac{\partial B_z}{\partial t} = \frac{\partial E_y}{\partial x} - \frac{\partial E_x}{\partial y}, \tag{37.2.3}$$

$$\frac{\partial D_x}{\partial t} = \frac{\partial H_z}{\partial y} - \frac{\partial H_y}{\partial z} - J_x, \tag{37.2.4}$$

$$\frac{\partial D_y}{\partial t} = \frac{\partial H_x}{\partial z} - \frac{\partial H_z}{\partial x} - J_y, \tag{37.2.5}$$

$$\frac{\partial D_z}{\partial t} = \frac{\partial H_y}{\partial x} - \frac{\partial H_x}{\partial y} - J_z. \tag{37.2.6}$$

Before proceeding any further, it is prudent to rewrite the differential equation form of Maxwell's equations in their integral form shown in the beginning of the course. For instance, the first equation above can be rewritten as[7]

$$-\frac{d}{dt}\iint_{\Delta S} B_x dS = \oint_{\Delta C} \mathbf{E}\cdot d\mathbf{l} \tag{37.2.7}$$

[6]Note that this algorithm, together with the method of moments [208] for solving Maxwell's equations, emerge shortly after the advent of IC-based digital computers.

[7]The integral form of Maxwell's equations were preferred in the early days because of the proximity of the concepts to early measurements. The finite integration technique developed by T. Weiland, that uses the integral form, is also of note [322, 323].

where $\Delta S = \Delta x \Delta z$. The approximation of this integral form will be applied to the face that is closest to the observer in Figure 37.3. Hence, one can see that the curl of $\mathbf{E}$ is proportional to the time-derivative of the magnetic flux through the suface enclosed by ΔC, which is ΔS.

One can see this relationship for the other surfaces of the cube in the figure as well: the electric field is curling around the magnetic flux. For the second half of the above equations, one can see that the magnetic fields are curling around the electric flux, but on a staggered grid. These two staggered grids are intertwined with respect to each other. This is the spirit with which the Yee algorithm is written. He was apparently motivated by fluid dynamics when he did the work.

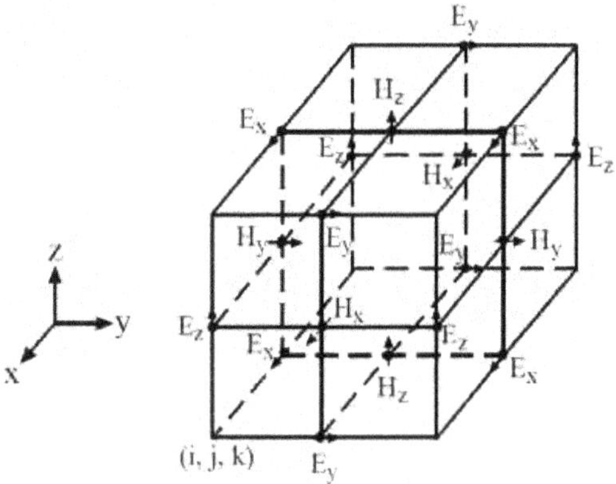

Figure 37.3: The assignment of fields on a grid in the Yee algorithm [308]. This algorithm is vastly popular for electromagnetic simulations [310].

After denoting $f(m\Delta x, n\Delta y, p\Delta z, l\Delta t) = f^l_{m,n,p}$, a more compact notation where the superscript is time and subscripts are space points, and replacing derivatives with central finite-differences in accordance with Figure 37.3, (37.2.1) becomes

$$\frac{1}{\Delta t}\left[B^{l+\frac{1}{2}}_{x,m,n+\frac{1}{2},p+\frac{1}{2}} - B^{l-\frac{1}{2}}_{x,m,n+\frac{1}{2},p+\frac{1}{2}}\right] = \frac{1}{\Delta z}\left[E^l_{y,m,n+\frac{1}{2},p+1} - E^l_{y,m,n+\frac{1}{2},p}\right] - \frac{1}{\Delta y}\left[E^l_{z,m,n+1,p+\frac{1}{2}} - E^l_{z,m,n,p+\frac{1}{2}}\right]. \tag{37.2.8}$$

where the above formula is evaluated at the time step $t = l\Delta t$. Moreover, the above can be repeated for (37.2.2) and (37.2.3). Notice that in Figure 37.3, the electric field is always assigned to the edge center of a cube, whereas the magnetic field is always assigned to the face center of a cube.[8]

In fact, after multiplying (37.2.8) by $\Delta z \Delta y$, (37.2.8) is also the approximation of the integral forms [9] of Maxwell's equations when applied at a face of a cube. By doing so, the left-hand side of

[8]This algorithm is intimately related to differential forms, which has given rise to the area of discrete exterior calculus [324].

[9]This integral form is shown at the beginning of the course.

(37.2.8), by using (37.2.7), becomes

$$(\Delta y\Delta z/\Delta t)\left[B^{l+\frac{1}{2}}_{x,m,n+\frac{1}{2},p+\frac{1}{2}} - B^{l-\frac{1}{2}}_{x,m,n+\frac{1}{2},p+\frac{1}{2}}\right], \tag{37.2.9}$$

which is the time variation of the total flux through an elemental area $\Delta y\Delta z$. Moreover, by summing this flux on the six faces of the cube shown in Figure 37.3, and using the right-hand side of (37.2.8) and its equivalence, it can be shown that the magnetic flux adds up to zero. Hence, $\frac{\partial}{\partial t}\nabla\cdot\mathbf{B} = 0$ condition is satisfied within the numerical approximations of Yee algorithm. The above shows that if the initial value implies that $\nabla\cdot\mathbf{B} = 0$, the algorithm will preserve this condition. So even though we are solving Faraday's law, Gauss's law is also enforced if the cumulative numerical error is kept small. This is important in maintaining the stability of the numerical algorithm [38].

Furthermore, a similar approximation of (37.2.4) leads to

$$\begin{aligned}\frac{1}{\Delta t}\left[D^{l}_{x,m+\frac{1}{2},n,p} - D^{l-1}_{x,m+\frac{1}{2},n,p}\right] &= \frac{1}{\Delta y}\left[H^{l-\frac{1}{2}}_{z,m+\frac{1}{2},n+\frac{1}{2},p} - H^{l-\frac{1}{2}}_{z,m+\frac{1}{2},n-\frac{1}{2},p}\right]\\ &-\frac{1}{\Delta z}\left[H^{l-\frac{1}{2}}_{y,m+\frac{1}{2},n,p+\frac{1}{2}} - H^{l-\frac{1}{2}}_{y,m+\frac{1}{2},n,p-\frac{1}{2}}\right] - J^{l-\frac{1}{2}}_{x,m+\frac{1}{2},n,p}.\end{aligned} \tag{37.2.10}$$

By the same token, similar approximations apply for (37.2.5) and (37.2.6). In addition, the above has an interpretation similar to (37.2.8) if one thinks in terms of a cube that is shifted by half a grid point in each direction to form a staggered grid. Hence, the finite difference approximations of (37.2.4) to (37.2.6) are consistent with the finite difference approximation of $\frac{\partial}{\partial t}\nabla\cdot\mathbf{D} = -\nabla\cdot\mathbf{J}$. This manner of alternatively solving for the **B** and **D** fields in tandem while the fields are placed on a staggered grid is also called the leap-frog scheme.

In the above, $\mathbf{D} = \epsilon\mathbf{E}$ and $\mathbf{B} = \mu\mathbf{H}$. Since the magnetic field and the electric field are assigned on staggered grids, μ and ϵ may have to be assigned on staggered grids. This does not usually lead to serious problems if the grid size is small. Alternatively, (37.1.16) and (37.1.17) can be used to remove this problem and to reduce storage.

By eliminating the **E** or the **H** field from the Yee algorithm, it can be shown that the Yee algorithm is equivalent to finite differencing the vector wave equation directly. Hence, the Yee algorithm is also constrained by the CFL stability criterion [317].

The following figures show some results of FDTD simulations. Because the answers are in the time domain, beautiful animations of the fields are also available online: https://www.remcom.com/xfdtd-3d-em-simulation-software

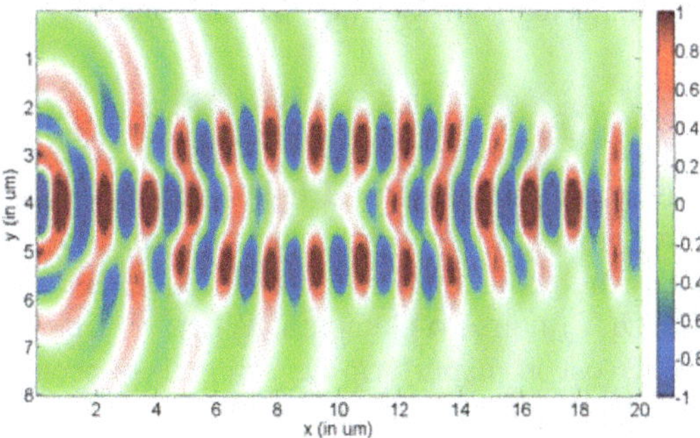

Figure 37.4: The 2D FDTD simulation of complicated optical waveguides. Such simulations can be done from static to optical frequencies (Figure by: https://www.mathworks.com).

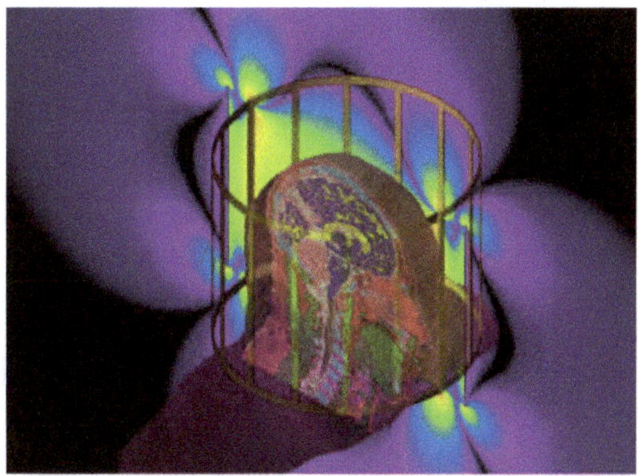

Figure 37.5: FDTD simulation of human head in a squirrel cage of an MRI (magnetic resonance imaging) system. A static magnetic field biases the spins in the human body. Then an RF field is used to tilt the spins, causing them to precess. Their precession gives rise to electromagnetic radiation (also called spin echo) that can be measured by the squirrel cage coils (Figure by REMCOM).

37.2.1 Finite-Difference Frequency-Domain Method

Unlike electrical engineering, in many fields, nonlinear problems are prevalent. But when we have a linear time-invariant problem, it is simpler to use phasor technique to solve the problem in the frequency domain. This is analogous to performing a time Fourier transform of the pertinent linear equations. However, phasor technique cannot be applied when nonlinearity exists. Then the derivative forms of Maxwell's equations have to be solved directly in the time domain [325, 326].

Consequently, one can write (37.2.1) to (37.2.6) in the frequency domain to remove the time derivatives. Then one can apply the finite-difference approximation to the space derivatives using the Yee grid. As a result, in replacement of Maxwell's equations, one arrives at a matrix equation

$$\overline{\mathbf{A}} \cdot \mathbf{x} = \mathbf{b} \tag{37.2.11}$$

where $\mathbf{x}$ is an unknown vector containing $\mathbf{E}$ and $\mathbf{H}$ fields, and $\mathbf{b}$ is a source vector that drives the electromagnetic system containing $\mathbf{J}$.[10] The above matrix-vector product can be effected using the Yee algorithm. Due to the near-neighbor interactions of the fields on the Yee grid, the matrix $\overline{\mathbf{A}}$ is highly sparse and contains $O(N)$ nonzero elements. When an iterative method is used to solve the above equation, the major cost is in performing a matrix-vector product $\overline{\mathbf{A}} \cdot \mathbf{x}$. However, in practice, the matrix $\overline{\mathbf{A}}$ is never generated nor stored, making this a matrix-free method. Moreover, the effect of $\overline{\mathbf{A}} \cdot \mathbf{x}$ can be easily invoked by using the Yee algorithm.

37.3 Absorbing Boundary Conditions

This lecture would not be complete without mentioning absorbing boundary conditions. As a computer has finite memory, space of an infinitely large extent cannot be simulated. Hence, it is important to design absorbing boundary conditions at the walls of the simulation domain or box, so that waves impinging on them are not reflected. This mimics the physics of an infinitely large box.

This is analogous to experiments in microwave engineering. In order to perform experiments in an infinite space, such experiments are usually done in an anechoic chamber[11] An anechoic chamber has its walls padded with absorbing materials or microwave absorbers so as to minimize the reflections off its walls (see Figure 37.6). The electromagnetically quiet environment is important for studying EMC/EMI (electromagnetics compatibility/electromagnetics interference) problems. This is an important problem in many industries that use electromagnetics technologies. Figure 37.7 shows the acoustic equivalent of an anechoic chamber.

Figure 37.6: An anechoic chamber for radio frequency. In such an electromagnetically quiet chamber, interference from other RF equipment is minimized (Figure by Panasonic).

[10] If we are convinced that all linear systems can be mapped to the above equation, then linear Maxwell's equations driven by a source can be mapped to the above equation.

[11] A nonechoing or nonreflecting chamber.

Figure 37.7: An acoustic anechoic chamber. In such a chamber, there is no reflection from the walls of the chamber; even the sound of a baby breastfeeding can be heard clearly (Figure by AGH University, Poland).

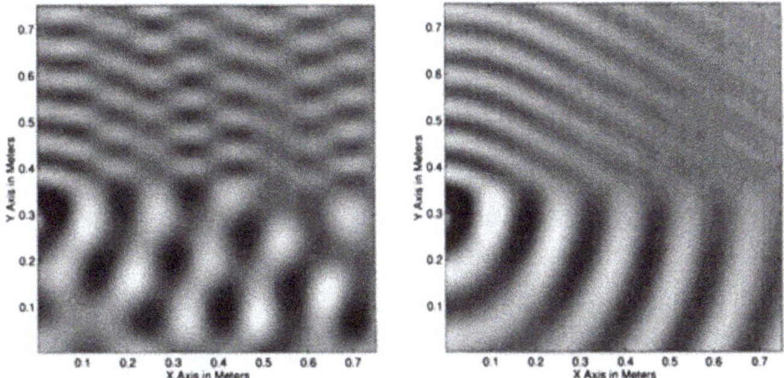

Figure 37.8: Simulation of a source on top of a half-space (left) without stretched coordinates PML ABC; and (right) with stretched coordinats PML ABC [327]. One can see the effect of standing wave patterns due to the reflections from the walls of the simulation domain. The reflected waves give rise to interference patterns that are clearly visible in the left figure. They disappear upon the introduction of stretched coordinates PML ABC (Figure by: https://www.semanticscholar.org/paper/Complex-coordinate-stretching-as-a-generalized-Chew-Jin/ff8da6ca4ad2358016ec169ffa43cb871873a842).

By the same token, in order to numerically simulate an infinitely large box with a finite-size box, absorbing boundary conditions (ABCs) are designed at its walls. The simplest of such ABCs is the impedance boundary condition. (A transmission line terminated with an impedance reflects less than one terminated with an open or a short circuit.) Another simple ABC is to mimic the Sommerfeld radiation condition (much of this is reviewed in [38]).[12]

A recently invented ABC is the perfectly matched layers (PML) [328]. Also, another similar ABC is the stretched coordinates PML [329]. Figure 37.8 shows simulation results with and without stretched coordinates PMLs on the walls of the simulation domain [327].

[12]ABCs are beyond the scope of these lecture notes.

Exercises for Lecture 37

Problem 37-1: This problem refers to lecture 37.

(i) Show that (37.1.14) expands to (37.1.12) after using (37.1.13) and its equivalence in x and y directions. Hence, derive (37.1.18).

(ii) Show that (37.1.28) is indeed true. Give the physical meaning of (37.1.29).

(iii) Derive (37.1.35) and (37.1.37). What is the physical meaning of the extra δ in (37.1.35)?

Lecture 38

Quantum Theory of Electromagnetics

The quantum theory of the world is the culmination of a series of astonishing intellectual exercises. Quantum theory is often termed the intellectual triumph of the twentieth century. One often says that deciphering the laws of nature is like watching two persons play a chess game with rules unbeknownst to us. By watching the moves, conjectures were made about the rules, which were later confirmed by experiments. After much sweat and tears, we finally have pieced together these perplexing rules. But we are grateful that these laws of nature are revealed to us, and upon them, we can build new technologies.

It is important to know that the quantum theory of electromagnetics[1] emerges alongside quantum theory. This new quantum theory of electromagnetics is intimately related to Maxwell's equations as shall be seen. It also inspired quantum field theory [46, 330]. This new theory spawns the possibility for quantum technologies, one of which is quantum computing. Others are quantum communication, quantum cryptography, quantum sensing, and many more.

A recent major advancement in quantum theory is the confirmation of the correctness of quantum interpretation, which cannot be derived but only validated by experiments. This was not done until 1982, and with it, we ushered in the new era of quantum information science. In 2022, the Nobel Prize in physics was given to Alain Aspect [331], John F. Clauser [332], and Anton Zeilinger [333] for their contributions to quantum information science.

[1]This is often called quantum theory of light in many textbooks: Because quantum theory was first manifested in optical frequencies with optical photons. Nowadays, microwave photons are detectable in the laboratory [26], and it will come a day that this quantum theory will manifest itself broadly through the electromagnetic spectrum.

38.1 Historical Background on Quantum Theory

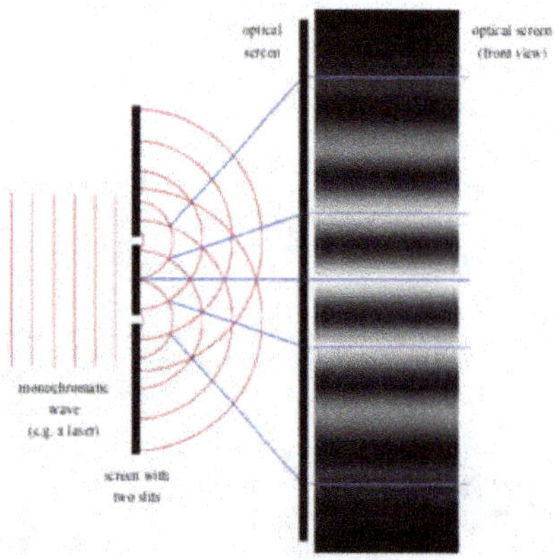

Figure 38.1: A Young's double-slit experiment. Again, the interference pattern reveals the wave nature of light (Figure By: https://www.shmoop.com).

Initially, light was thought to comprise a stream of particles since Grecian times. That light is a wave was demonstrated by the Newton's rings phenomenon [22] in the 18th century (1717) (see Figure 38.2). In 1801, Thomas Young demonstrated the double-slit experiment for light [334] that further confirmed its wave nature (see Figure 38.1). By the beginning of the 20th century, one has to accept that light is both a particle, called a photon, carrying a quantum of energy with a quantum of momentum, as well as a particle endowed with wave-like behavior. This conflicting behavior is called wave-particle duality. We shall outline the historical reason for this development.

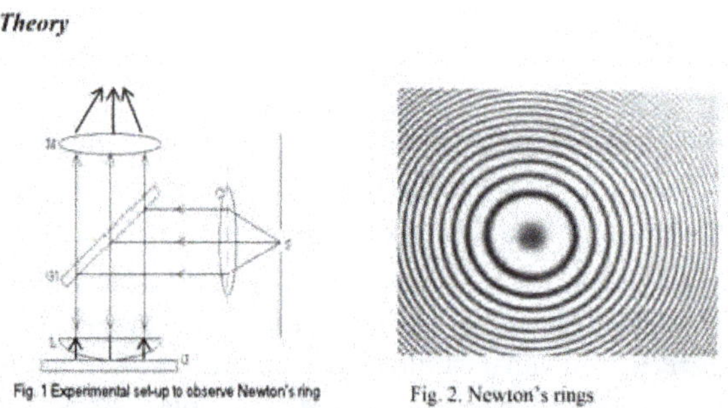

Figure 38.2: A Newton's rings experiment that indicates the wave nature and wave-particle duality of light (Figure By: http://www.iiserpune.ac.in).

As mentioned above, quantum theory is a major intellectual achievement of the 20th century, even though new knowledge is still emerging in it. Several major experimental findings led to the revelation of quantum theory of nature.

In nature, we know that matter is not infinitely divisible. This is vindicated by the atomic theory of John Dalton (1766-1844) [335]. So fluid is not infinitely divisible; as when water is divided into smaller pieces, one will eventually arrive at the water molecule, H_2O, which is the fundamental building block of water.

It turns out that electromagnetic energy is not infinitely divisible either. The electromagnetic radiation out of a heated cavity would have a very different spectrum if electromagnetic energy were infinitely divisible. In order to fit the experimental observation of radiation from a heated electromagnetic cavity, called black-body radiation, Max Planck (1900s) [336] postulated that electromagnetic energy comes in packets or is quantized. Each packet of energy or a quantum of energy E is associated with the frequency of electromagnetic wave, namely

$$E = \hbar\omega = \hbar 2\pi f = hf \tag{38.1.1}$$

where $h = 6.626 \times 10^{-34}$ J·s (Joule-second) is the Planck's constant, and $\hbar = h/(2\pi)$ is the reduced Planck's constant. Since $\hbar$ is very small, this packet of energy is very small unless ω is large. Thus it is no surprise that the quantization of electromagnetic field is first associated with light, a very high frequency electromagnetic radiation. A red-light photon at a wavelength of 700 nm corresponds to an energy of approximately 2 eV $\approx 3 \times 10^{-19}$J $\approx 75\, k_BT$, (where k_BT denotes the thermal energy from thermal law, and k_B is Boltzmann's constant. Hence, k_BT is about 25 meV at room temperature[2]). A microwave photon has approximately 1×10^{-5} eV $\approx 10^{-2}$meV, making it a lot harder to detect compared to an optical photon.

The second experimental evidence that light is quantized is the photo-electric effect [337]. It was found that matter emitted electrons when light was shined on it. First, the component of the light frequency has to correspond to the "resonant" frequency of the atom so that the photon from the light will be absorbed.[3] Second, the number of electrons emitted is proportional to the number of packets of energy $\hbar\omega$ that the light carries. This was a clear indication that light energy traveled in packets or quanta as posited by Einstein in 1905.

This wave-particle duality concept mentioned at the beginning of this section was not new to quantum theory, as electrons were known to behave both like a particle and a wave. The particle nature of an electron was confirmed by the measurement of its charge by Millikan in 1913 in his oil-drop experiment. (The double-slit experiment for electron was done in 1927 by Davison and Germer, indicating that an electron has a wave nature as well [334].) In 1924, de Broglie[4] [90] suggested that there is a wave associated with an electron with momentum p such that

$$p = \hbar k \tag{38.1.2}$$

[2]This is a number that ought to be remembered by semiconductor scientists as the size of the material bandgap with respect to this thermal energy decides if a material is a semiconductor, conductor, or insulator at room temperature.

[3]This is akin to the physics of resonant tunneling in antennas (see Figure 31.4).

[4]de Broglie's great-grandfather fought in the America revolutionary war led by General Lafayette from France. Purdue University is located in Lafayette named after him.

where $k = 2\pi/\lambda$, the electron de Broglie wavenumber. All this knowledge gave hints to the quantum theorists of that era to come up with a new way to describe nature.[5]

Classically, particles like electrons move through space obeying Newton's laws of motion, first established in 1687 [338]. The old way of describing particle motion is known as classical mechanics, but the new way of describing particle motion is known as quantum theory (or quantum mechanics). Quantum theory is very much motivated by a branch of classical mechanics called Hamiltonian theory. We will first use Hamiltonian theory to study a simple pendulum and connect it with classical electromagnetic oscillations. A pendulum[6] in this case consists of two oppositely charged particles connected to each other by a spring. These two particles, can be an electron and a nucleus as in a dipolar molecule. When polarized, they form little dipoles as in a polarization medium. (We conjecture that an electron-positron (e-p) pair in vacuum can be such two particles.) In any case, these two particles can oscillate with respect to each other once set in motion, creating a dipole oscillation very much like a pendulum. A collection of such dipoles in space can be observed macroscopically. Hence, we will study this pendulum both classically and quantumly; and we can focus our attention on one of such pendulums.

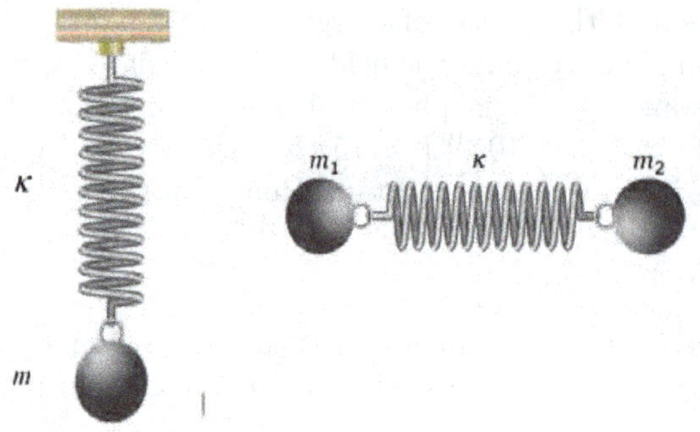

Figure 38.3: A classical pendulum can be used to illustrate classical Hamiltonian theory. The equation of motion of a classical pendulum can be derived using Newton's law or Hamiltonian theory. As such, it is easier to illustrate the quantum theory of a pendulum starting with classical Hamiltonian theory. In the figure, κ and m are spring constant and mass, respectively of the pendulum. The horizontal pendulum on the right can be simplified using the center-of-mass coordinate [339].

We can assume that these dipoles are bosonic particles[7]; they can oscillate cooperatively *en masse*, say, in an electromagnetic cavity. These bosonic oscillations can be observed macroscopically in a cavity describable by classical electromagnetics oscillations. Likewise, an electromagnetic cavity

[5]It is to be noted that both electrons and photons display wave-particle duality, but they are actually quite different quantum particles. An electron has mass whereas a photon is massless.

[6]In this course, we will use the word "pendulum" to replace the polysyllabic word "harmonic oscillator".

[7]Feynman's lectures [76] provide a review of bosons and fermions. Bosons have integer spins while Fermions have half-integer spins.

filled with material medium, the polarization density is collection of dipoles that behave like bosons, and hence, exhibits cooperative bosonic oscillations that can be observed macroscopically. Therefore, we begin by studying a classical pendulum and later connect it to the study of a quantum pendulum.

A Simple Classical Pendulum

As we have seen in the Drude-Lorentz-Sommerfeld model (see lecture 8), for a particle of mass m attached to a spring connected to a wall,[8] where the restoring force is like Hooke's law, the equation of motion of a pendulum by Newton's law is

$$m\frac{d^2q}{dt^2} + \kappa q = 0 \tag{38.1.3}$$

where κ is the spring constant, and we assume that the oscillator is not driven by an external force, but is in natural or free oscillation. Here, q is used to denote the displacement of the pendulum from its quiescent position. The above equation is homomorphic/analogous to a cavity resonance problem. We can see that $q \Leftrightarrow E_0$ where E_0 is the amplitude of the field in the cavity, we can relate the two problems. By letting[9]

$$q = q_0 e^{-i\omega t} \tag{38.1.4}$$

where q_0 is the amplitude of this complex signal, the above (38.1.3) becomes[10]

$$-m\omega^2 q_0 + \kappa q_0 = 0 \tag{38.1.5}$$

Again, a nontrivial solution is possible only at the resonant frequency of the oscillator or that when $\omega = \omega_0$ where

$$\omega_0 = \sqrt{\frac{\kappa}{m}} \tag{38.1.6}$$

This is the natural solution or resonant solution of the system.[11]

---

[8]This particle need not be connected to a wall. It can be connected to another particle, and the two particles are oscillating about their center of mass, which is a fixed point independent of time.

[9]As aforementioned, for this part of the lecture, we will switch to using $\exp(-i\omega t)$ time convention as is commonly used in the optics and physics literatures.

[10]In optics, instead of using phasors as in electrical engineering, it is a custom to use a complex signal with the understanding that the real part of this complex signal is time-harmonic similar to the phasor technique [62]. Whenever a real signal is needed, one adds the "complex conjugate" (c.c.) part. Hence, q_0 in (38.1.4) is not necessarily real-valued. Thus, this does not imply the q here used in optics, and E_0, often used in phasor techniques, are equal to each other. In this case, q_0 is a complex amplitude, while E_0 is real-valued amplitude because of definition of the phasor technique. These two values, q_0 and E_0, are analogous to each other and used by two different communities to denote the amplitude of a signal. What cultural diversities we have even within finding the solutions to Maxwell's equations!

[11]As aforementioned, natural solution, resonant solution, homogeneous solution (used in ODE), and null-space solution (in linear algebra) are similar concepts. They are solutions without a driving term.

38.2 Hamiltonian Theory

Quantum theory is intimately related to Hamiltonian theory. Equation (38.1.3) can be derived by Newton's law, but it is more interesting, as an alternative, to derive it via Hamiltonian theory. This is because Hamiltonian theory motivates quantum theory.

Hamiltonian theory, developed by William Hamilton (1805–1865) [340], is motivated by energy conservation [341]. The Hamiltonian H of a simple pendulum system is given by its total energy, namely that

$$H = T + V \tag{38.2.1}$$

where T is the kinetic energy and V is the potential energy of the system.

For a simple pendulum, the kinetic energy is given by

$$T = \frac{mv^2}{2} = \frac{m^2v^2}{2m} = \frac{p^2}{2m} \tag{38.2.2}$$

where $p = mv$ is the momentum of the particle. The potential energy, assuming that the particle is attached to a spring with spring constant κ and displaced q from equilibrium, is given by

$$V = \frac{1}{2}\kappa q^2 = \frac{1}{2}m\omega_0^2 q^2 \tag{38.2.3}$$

Hence, the Hamiltonian is given by

$$H = T + V = \frac{p^2}{2m} + \frac{1}{2}m\omega_0^2 q^2 \tag{38.2.4}$$

At any instant of time t, we assume that $p(t) = mv(t) = m\frac{d}{dt}q(t)$ is independent of $q(t)$.[12] Thus, they can vary independently of each other but in tandem to keep H a constant. Or $p(t)$ and $q(t)$ time evolve to conserve energy or to keep H, the total energy constant or independent of time.[13] In other words,

$$\frac{d}{dt}H\left(p(t), q(t)\right) = 0 = \frac{dp}{dt}\frac{\partial H}{\partial p} + \frac{dq}{dt}\frac{\partial H}{\partial q} \tag{38.2.5}$$

Therefore, the Hamilton equations of motion for the dynamic variables (or conjugate variables) p and q are derived to be[14]

$$\frac{dp}{dt} = -\frac{\partial H}{\partial q}, \qquad \frac{dq}{dt} = \frac{\partial H}{\partial p} \tag{38.2.6}$$

[12] $p(t)$ and $q(t)$ are termed conjugate dynamic variables.

[13] In math, we have the concept of Cauchy initial value problems where the initial values are the position variable $q(t)$ and its time derivative, $q'(t)$ at $t = 0$ [1, p. 529]. (See also [80, Lec. 2].

[14] Note that the Hamilton equations are determined to within a multiplicative constant, because one has not stipulated the connection between space and time, or we have not calibrated our clock [341].

From (38.2.4), we gather that

$$\frac{\partial H}{\partial q} = m\omega_0^2 q, \qquad \frac{\partial H}{\partial p} = \frac{p}{m} \tag{38.2.7}$$

Applying (38.2.6), we have[15]

$$\frac{dp}{dt} = -m\omega_0^2 q, \qquad \frac{dq}{dt} = \frac{p}{m} \tag{38.2.8}$$

Combining the two equations in (38.2.8) above, we have

$$m\frac{d^2q}{dt^2} = -m\omega_0^2 q = -\kappa q \tag{38.2.9}$$

which is also derivable by Newton's law.

(The above Hamilton equations can be generalized to the case involving multiple particles [342]. Then the dynamic variables of each particle are functions of position i or can be vector in nature meaning there could be three oscillators polarized in the x, y, z directions at one location. This can be incorporated using index notation. For the general case, they are

$$\frac{dp_i}{dt} = -\frac{\partial H}{\partial q_i}, \qquad \frac{dq_i}{dt} = \frac{\partial H}{\partial p_i} \tag{38.2.10}$$

In the above, the index i can represent the three components of the position vector $\mathbf{r}$ as well as the polarization of the vector field. Hence, the general case is no more difficult than the simple case.)

38.3 The Need for a Random Quantum Observable

In classical theory, conjugate dynamic variables, such as position q and momentum p, are deterministic variables. These are conjugate variables since they vary in tandem to maintain energy conservation as in Hamiltonian theory. But in the quantum world, dynamic variables, as revealed by experimental findings, are also random variables. Hence, these dynamic variables, p and q, called observables, are random variables. To fully describe the random properties of these observables, they need to be described by operator-vector pair. As such, these quantum observables, which are random dynamic variables, are represented by operator-vector pair.[16]

The operator is a deterministic quantity. But the vector is related to the probability density function, which entails the random properties of the quantum observable. We shall see how this is achieved via the Schrödinger's equation.

[15] We can also calibrate our clock here so that it agrees with our definition of momentum in the ensuing equation.

[16] This term observable is coined by Dirac, but due to the cavalier way it has been used, it is a source of confusion [343, p. 50]. In this course, we will use the term "observables" for dynamic variables in the quantum world which are also random variables.

38.4 Schrödinger's Equation (1925)

Having seen the Hamiltonian theory used for describing a simple classical pendulum, which is homomorphic to an electromagnetic oscillation, we shall next see the quantum theory description of the same simple pendulum, using it to understand quantum electromagnetics oscillations. We assume that the particles associated with a quantum pendulum to be "massive" quantum particles whose equations of motion are describable by quantum theory.[17]

As aforementioned, in the quantum world, the position and momentum of the "massive" particles that make up the quantum pendulum are random variables, a property of quantum observables. As shall be seen, to incorporate their randomness, Schrödinger's equation is needed. Schrödinger's equation cannot be derived but is postulated: It is the wonderful result of a postulate, inspiration, and a powerful guessing game based on experimental observations [79, 80].[18]

Since Schrödinger's equation is motivated by Hamiltonian theory, we will start with it:[19] Hamiltonian theory for the simple pendulum says that

$$H = \frac{p^2}{2m} + \frac{1}{2}m\omega_0^2 q^2 = E \tag{38.4.1}$$

where E is the total energy of the oscillator, or pendulum. In the above, p and q are dynamic variables: in the classical world are deterministic, but in the quantum world they are random variables as shall be seen.[20]

To this end, Schrödinger proposed his equation, which is a partial differential equation. He was very much motivated by the experimental revelation at his time that the momentum of a quantum particle, which is the electron in this case, is given by $p = \hbar k$ from De Broglie [90], $E = \hbar\omega$ from the black-body radiation law of Planck's [25], and the photoelectric effect.

Therefore, equation (38.4.1) can be rewritten more suggestively as

$$\frac{\hbar^2 k^2}{2m} + \frac{1}{2}m\omega_0^2 q^2 = \hbar\omega \tag{38.4.2}$$

We can associate

$$p = \hbar k \Rightarrow -i\hbar\frac{\partial}{\partial q} \tag{38.4.3}$$

and $-i\omega \Rightarrow \frac{\partial}{\partial t}$ as these operators will become such when operating on the time-harmonic plane wave $e^{ikq-i\omega t}$. To add more texture to the above equation, one lets the above become an operator

[17] Since a photon is massless, we will use this term to imply non-photonic particles, or particles with non-zero mass. According to Einstein theory of special relativity, only a massless particle can move at the speed of light [76].

[18] Rumor has it that he got the inspiration after a retreat in the mountain.

[19] I am indebted to Francis E. Low of MIT for this perspective. I took my first quantum theory course from him in the 1970s.

[20] Here, we will use the word "massive" to indicate a particle with non-zero mass, unlike a photon which has zero mass.

equation that operates on a wave function $\psi(q,t)$ so that[21]

$$-\frac{\hbar^2}{2m}\frac{\partial^2}{\partial q^2}\psi(q,t)+\frac{1}{2}m\omega_0^2q^2\psi(q,t)=i\hbar\frac{\partial}{\partial t}\psi(q,t) \tag{38.4.4}$$

Then defining the following operators[22]

$$\hat{p}=-i\hbar\frac{\partial}{\partial q} \qquad \hat{q}=\hat{I}q \tag{38.4.5}$$

we have

$$\frac{\hat{p}^2}{2m}\psi(q,t)+\frac{1}{2}m\omega_0^2\hat{q}^2\psi(q,t)=i\hbar\frac{\partial}{\partial t}\psi(q,t) \tag{38.4.6}$$

$$\hat{H}\psi(q,t)=i\hbar\frac{\partial}{\partial t}\psi(q,t) \tag{38.4.7}$$

where

$$\hat{H}=\frac{\hat{p}^2}{2m}+\frac{1}{2}m\omega_0^2\hat{q}^2 \tag{38.4.8}$$

The wave function $\psi(q,t)$ above is used to describe the state of the particle, giving it a wave nature.[23] As shall be shown, the above (38.4.7) is more easily solved as an initial value problem.

As aforementioned, we guess the wave function above is of the asymptotic form for high-frequency

$$\psi(q,t)\sim e^{ikq-i\omega t} \tag{38.4.9}$$

By back substitution, we can show that the above satisfies (38.4.4) approximately when k and ω become very large.

Equation (38.4.4) is Schrödinger's equation (or the Schrödinger's wave equation) in one dimension for the quantum version of the simple pendulum.[24] In Schrödinger's equation, we can further posit that the wave function has the general form

$$\psi(q,t)=e^{ikq-i\omega t}A(q,t) \tag{38.4.10}$$

where $A(q,t)$ is a slowly varying function of q and t, compared to $e^{ikq-i\omega t}$ for a wave packet.[25]

---

[21]The following equation can also be derived by math-physics logic in [344], which is substantially longer. Here, we will use Schrödinger's conjecture to shorten the derivation but the end results are the same.

[22]From this point onward, we will use hat or ^ to denote a quantum operator.

[23]The term "state" was later adopted by the control theorist. Thinking of a function as a vector, Schrödinger's equation reminds us of the state-variable approach in control theory where the state of a system is described by a state-variable vector $\mathbf{v}(t)$. The time evolution of the state variable in control theory, the simplest version, is $d\mathbf{v}(t)/dt=\overline{\mathbf{A}}\cdot\mathbf{v}(t)$. This equation is homomorphic to Schrödinger's equation.

[24]1D oscillation can describe many simple harmonic oscillations (pendulum) in the world. The vibration of a dipolar atom or molecule, or even electron-positron pair can be described as such. In this lecture notes, we will use the world "pendulum" to replace the polysyllabic word "harmonic oscillator".

[25]Recall that this is similar in spirit to when we studied high-frequency solutions of Maxwell's equations and paraxial wave approximation in section 33.3.1.

With this wave packet, the $\partial^2/\partial q^2$ can be again approximated by $-k^2$ in the short-wavelength limit, or when $k \to \infty$, as has been done in the paraxial wave approximation (see section 33.3.1). Furthermore, if the wave function is assumed to be quasi-monochromatic, then $i\hbar\partial/\partial_t\psi(q,t) \approx \hbar\omega\psi(q,t)$, we again retrieve the classical equation in (38.4.2) from (38.4.4). Hence, the classical equation (38.4.2) is a short wavelength, quasi-monochromatic approximation of Schrödinger's equation for a wave packet. However, as we shall see, the solutions to Schrödinger's equation are not limited to just waves described by (38.4.10).

Correspondence Principle

Planck postulated that the quantization energy of a photon is $\hbar\omega$. As we shall see, electromagnetic energy is carried by photons: the propagation of an electromagnetic wave (or field) corresponds to a stream of discrete photons each carrying a packet of energy. In the limit when $\hbar \to 0$, a large number of photons is needed to carry this energy, making it look like a classical field. This is the gist of the correspondence principle [344].

Stationary Solution and Eigenvalue Problem

The solution to Schrödinger's equation is easily solved as an initial value problem as shall be shown in the next lecture. To facilitate this, we need to expand the solution in terms of the eigenfunctions or eigensolutions of the equation. In the physics literature, these eigensolutions are also known as the stationary solutions.

The form of Schrödinger's equation presages the use of separation of variables as has been done with many equations in electromagnetics. Eventually, we will solve Schrödinger's equation as an initial value problem. To facilitate this we need to seek the eigensolution of Schrödinger's equation. To this end, we convert (38.4.4) above into an eigenvalue problem, just as in waveguide and cavity problems, using separation of variables. Thus, we let[26]

$$\psi(q,t) = \psi_n(q)e^{-i\omega_n t} \tag{38.4.11}$$

By so doing, (38.4.4) becomes an eigenvalue problem

$$\left[-\frac{\hbar^2}{2m}\frac{d^2}{dq^2} + \frac{1}{2}m\omega_0^2 q^2\right]\psi_n(q) = E_n\psi_n(q) \tag{38.4.12}$$

where

$$E_n = \hbar\omega_n = \left(n + \frac{1}{2}\right)\hbar\omega_0 \tag{38.4.13}$$

is the eigenvalue while $\psi_n(q)$ is the corresponding eigenfunction.[27] We should associate ω_n as a variable associated with the separation of variables, while ω_0 as the frequency of the original classical pendulum in (38.1.6).

[26] Mind you, the following is ω_n, not ω_0. Another way of saying it is that the function $\exp(-i\omega_n t)$ diagonalizes ∂_t operator. This is equivalent to the Fourier series expansion. In general, $\psi(q,t) = \sum_n b_n\psi_n(q)e^{-i\omega_n t}$, but we are looking at just one Fourier mode for simplicity. We can invoke the completeness of Fourier series for such an expansion.

[27] These problems are known as Sturm-Liouville problems. Their eigensolutions form a complete set [94].

The parabolic q^2 potential profile is also known as a quadratic potential well as it provides the restoring force to keep the particle bound to the well (see section 38.2 and (38.2.8). The above equation is analogous to the Schrödinger's equation for a particle trapped by a potential well: in this case, the potential well is a quadratic potential well [80, 79].

The above equation is also similar to the electromagnetic equation for a dielectric slab waveguide, where the second term is a dielectric profile that can trap a waveguide mode. Therefore, the potential well is a trap for the particle both in classical mechanics as expressed in (38.2.4) and also in wave physics as in (38.4.12). It keeps the particle bound to the well classically (see section 38.2 and (38.2.8).

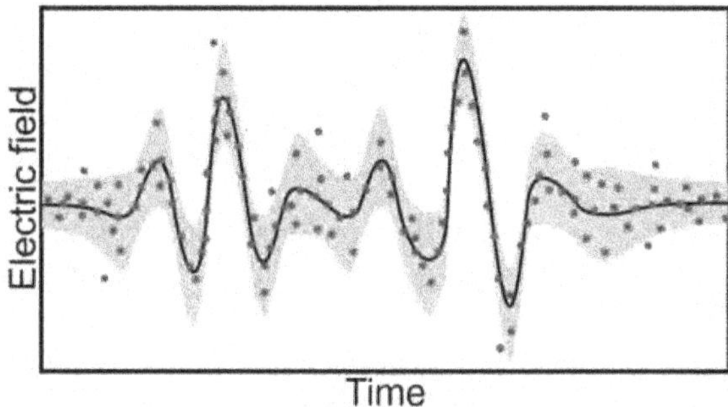

Figure 38.4: Schematic representation of the randomness of the measured electric field. The electric field amplitude, analogous to the amplitude of the quantum pendulum, is a quantum observable, and hence, is a random variable. The electric field amplitude maps to the displacement (position) of the quantum pendulum, which is also a random variable and a quantum observable. (Figure By Kira and Koch [345]).

The above eigenfunction $\psi_n(q)$ in equation (38.4.12) can be solved in closed form in terms of Hermite-Gaussian functions (1864) [346], or that[28]

$$\psi_n(q) = \sqrt{\frac{1}{2^n n!}\sqrt{\frac{m\omega_0}{\pi\hbar}}}\, e^{-\frac{m\omega_0}{2\hbar}q^2} H_n\left(\sqrt{\frac{m\omega_0}{\hbar}}q\right) \tag{38.4.14}$$

where $H_n(y)$ is a Hermite polynomial[29], and the eigenvalues are found in closed form as given in (38.4.13).

Here, the eigenfunction or eigenstate[30] $\psi_n(q)$ is known as the photon number state (or just a number state or Fock state in short) of the solution. The energy levels E_n of these trapped states are exactly spaced one $\hbar\omega_0$ apart, corresponding to the energy of one photon per Planck's law! It

[28]Lucky that we are, all these were figured out by contemporaries of James Clerk Maxwell. We stand on the shoulders of giants!

[29]The mass m is that of the particle that forms the quantum pendulum. It could be an electron, an atom, or some other particles.

[30]These terms, including eigenvectors, are used interchangeably in the literature.

takes exactly the energy of one photon to excite the "atom" from a lower energy state to the next energy state. To excite the "atom" from the ground state to the eigenstate $\psi_n(q)$ will take the energy of n photons, or $n\hbar\omega$.

There were two highlights that were brought about by the wave theory of a particle from the Schrödinger's equation: the prediction of the energy levels of a hydrogen atom and the probabilistic interpretation of quantum theory. When Schrödinger's equation was solved in 3D for a simple hydrogen atom, the analogue of (38.4.13) predicted the spectroscopic lines of the hydrogen atom with resounding success, a feat that had eluded scientists for decades. Schrödinger's theory confirmed the wave-particle duality first postulated by de Broglie.

Moreover, this position $q(t)$, is related to the displacement of the electron from its equilibrium position in a dipolar atom (molecule), an electron-hole pair in a semiconductor [75], and possibly, in the electron-positron (e-p) pair in vacuum. When such a displacement happens, a local dipole occurs which can be modeled by a 1D pendulum. More of such examples can be found in Figure 8.2.[31]

These dipoles, or polarization density are mapped collectively or macroscopically, and proportional to the amplitude $E_x(z,t)$ (for a fixed z) of a traveling field or a cavity field. In other words, the collective dipole oscillation becomes $E_x(z,t)$. Experimental measurements have improved so much that single-photons can be detected. Hence, it is the amplitude of an electromagnetic oscillation that becomes uncertain and fuzzy as shown in Figure 38.4 from Kira and Koch [345]. In this figure, the dots represent the one possible outcome of the measured value at the given time of the observable in the laboratory. Because the observable is a random variable, the measured value is not the same at a given time if the measurement is repeated. The shaded zone indicates the range of possible values of the observable since this is a random variable. The solid line indicates the average (expectation) value of the observable after averaging over many measurements.

38.5 More on Features of Random Quantum Observables

As aforementioned, in the beginning, it was not understood the physical meaning of the wave function in Schrödinger's equation. It was later suggested by Max Born that the wave functions are related to the probabilistic interpretation of quantum measurements. Hence, the wave function is related to the probability density function that describes this random behavior.

Since observables are random variables, they have average values (or expectation values) that can be computed. In the following, we will illustrate the calculation of these expectation values, using math, Dirac, and matrix (or linear algebraic) notations.

Dirac notation is homomorphic to linear algebra notation: a vector is denoted as a "ket," written as $|\psi\rangle$ while the conjugate transpose of a vector in Dirac notation is called a "bra," which is denoted

[31]A photon, as is many quantum particles, is both a wave and a particle. A simple way to think of a photon is that it is a massless particle that is embedded in a quantum pendulum. A series of coupled quantum pendulum in space allows a wave field (with embedded photons) to ride on it. So as the photon particle propagates through space, it exhibits both its particle and wave nature. As we have seen, a photon is a packet of energy that is embedded in a quantum pendulum. A coupled set of quantum pendulums allows a wave field to ride on it; the wave field zips through space at a tremendous velocity, the velocity of light.

as $\langle\psi_n|$. Hence, the inner product between two vectors $|\psi_1\rangle$ and $|\psi_2\rangle$ is denoted as $\langle\psi_1|\psi_2\rangle$ in Dirac notation. These vectors are also called state vectors in physics as they describe the state of a quantum system.[32]

- Now, p and q become random variables or observables, and $|\psi(q,t)|^2$ is the associated probability density function which is related to the wave function. Then it is necessary that

$$\int_{-\infty}^{\infty} dq|\psi(q,t)|^2 = 1 \Leftrightarrow \underbrace{\langle\psi(t)|\psi(t)\rangle}_{\text{Dirac}} = 1 \Leftrightarrow \underbrace{\boldsymbol{\psi}(t)^\dagger \cdot \boldsymbol{\psi}(t)}_{\text{Matrix}} = 1 \tag{38.5.1}$$

  The above is the normalization condition for all state vectors of a quantum system, since $\psi(q,t)$ is related to probability density functions. We have expressed this normalization condition explicitly using math notation, Dirac notation, and then using matrix (or linear algebra) notation.

- The average value or expectation value of the random observable (or variable) q is now given by

$$\int_{-\infty}^{\infty} dqq|\psi(q,t)|^2 = \langle q(t)\rangle = \bar{q}(t) \Leftrightarrow \underbrace{\langle\psi(t)|\hat{q}|\psi(t)\rangle}_{\text{Dirac}} \Leftrightarrow \underbrace{\boldsymbol{\psi}(t)^\dagger \cdot \overline{\mathbf{q}} \cdot \boldsymbol{\psi}(t)}_{\text{Matrix}} \tag{38.5.2}$$

  In the above, the operator $\hat{q} = \hat{I}q$.[33]

- The momentum p is now a random variable, or quantum observable, related to the differential operator representation $\hat{p} = -i\hbar\partial/(\partial q)$. The average or expectation value of this random variable p is now given by

$$\begin{aligned} -i\hbar \int_{-\infty}^{\infty} dq\psi^*(q,t)\frac{\partial}{\partial q}\psi(q,t) &= \langle p(t)\rangle = \bar{p}(t) \\ &\Leftrightarrow \underbrace{\langle\psi(t)|\hat{p}|\psi(t)\rangle}_{\text{Dirac}} \Leftrightarrow \underbrace{\boldsymbol{\psi}(t)^\dagger \cdot \overline{\mathbf{p}} \cdot \boldsymbol{\psi}(t)}_{\text{Matrix}} \end{aligned} \tag{38.5.3}$$

In quantum theory, dynamic variables such as p and q are elevated and represented by operators. These operators, together with the state vector that describe the state of the quantum system, endow these dynamic variables with random properties. These dynamic variables with random properties are also called quantum observables.[34] It is to be noted that these operator representations for the momentum $\hat{p}$ and the position $\hat{q}$ satisfy the commutator relation that [35]

$$[\hat{p},\hat{q}] = \hat{p}\hat{q} - \hat{q}\hat{p} = -i\hbar\hat{I} \tag{38.5.4}$$

[32]There is a one-to-one correspondence of the Dirac notation to the matrix algebra notation. The matrix-vector (or operator-vector) is product $\hat{A}|x\rangle \leftrightarrow \overline{\mathbf{A}} \cdot \mathbf{x}$, $\langle x| \leftrightarrow \mathbf{x}^\dagger$ $\langle x_1|x_2\rangle \leftrightarrow \mathbf{x}_1^\dagger \cdot \mathbf{x}_2$. The preponderance of languages in different communities is like the story of the Tower of Babel. The fact that a function is a vector has been broached by us in section 36.4 in computational electromagnetics.

[33]As is well known in linear algebra, in the above, $\bar{q}$, the average value is measurable in the laboratory which is unique, but the vector $\boldsymbol{\psi}(t)$ (or $|\psi(t)\rangle$) and the operator $\overline{\mathbf{q}}$ (or $\hat{q}$) are not unique.

[34]This term was used by Dirac, but please see [343, p. 50] for clarification.

[35]The symbol $[\hat{p},\hat{q}] = \hat{p}\hat{q} - \hat{q}\hat{p}$ is called a commutator.

It is to be noted that other properties of a random variable, such as higher moments, can be calculated once the probability density function is known [347].

As we have seen above, in the quantum world, the randomness of the observables is embedded in the state vector. Hence, the random properties of a quantum observable need to be described by the associated operator-vector pair. In other words, quantum observables p and q are characterized by quantum operators $\hat{p}$ and $\hat{q}$ and the wave function $\psi(q,t)$.[36] Together with the wave function $\psi(q,t)$, which can be thought of as a vector, the observables p and q are endowed with random properties. For instance, the quantum observables for the position q and momentum p are now random variables in the quantum world. The random properties of an observable are best described by an operator-vector pairs such as

$$p \Leftrightarrow \{\hat{p}, \psi(q,t)\}, \qquad q \Leftrightarrow \{\hat{q}, \psi(q,t)\} \tag{38.5.5}$$

or

$$p \Leftrightarrow \{\hat{p}, |\psi(t)\rangle\}, \qquad q \Leftrightarrow \{\hat{q}, |\psi(t)\rangle\} \tag{38.5.6}$$

In the above, we call operators $\hat{p}$ and $\hat{q}$ the operator representations of the random observables p and q [29]. Their elevation to operators together with the state vector $|\psi(t)\rangle$ is important to endow the observables with random properties in the quantum world. It is important to note that the operators $\hat{p}$ and $\hat{q}$ are deterministic, even though the observables p and q are random variables. (It is noted that definition of observables has been a source of confusion as expressed by [343, p. 50].)

38.6 Beautifying Schrödinger's Equation with Normalized Coordinates

As aforementioned, it is best to solve Schrödinger's equation as an initial value problem. To facilitate this, we need to solve for the eigensolutions of the equation. Rewriting Schrödinger's equation as the eigenequation for the photon number state for the quantum pendulum, we have

$$\hat{H}\psi_n(q) = \left[-\frac{\hbar^2}{2m}\frac{d^2}{dq^2} + \frac{1}{2}m\omega_0^2 q^2\right]\psi_n(q) = E_n\psi_n(q). \tag{38.6.1}$$

where $\hat{H}$ is the Hamiltonian operator defined by the above equation, $\psi_n(q)$ is the eigenfunction also called the photon number state, or Fock state, and E_n is the corresponding eigenvalue. The above can be changed into a dimensionless form first by dividing $\hbar\omega_0$, and then letting $\xi = \sqrt{\frac{m\omega_0}{\hbar}}q$. It is the variable for length, but has been normalized to become a dimensionless variable. The above, expressed in wave function $\Psi_n(\xi)$ where ξ is dimensionless, then becomes

$$\frac{1}{2}\left(-\frac{d^2}{d\xi^2} + \xi^2\right)\Psi_n(\xi) = \frac{E_n}{\hbar\omega_0}\Psi_n(\xi) \tag{38.6.2}$$

[36] In this lecture, we will use "hat" to denote a quantum operator as in many textbooks. These operators, though living in infinite dimensional Hilbert space, are homomorphic to matrix operators in your linear algebra course. Many properties of matrix operators carry over to Hilbert space operators.

where we have used $\Psi_n(\xi)$ to denote the wave function in normalized coordinate. It is to be noted that the above is still a 1D in space wave-like equation with different trapped "modes" with frequency ω_n given in (38.4.13). Yet the frequency is different for different trapped "modes". But this equation is still the quantum equation for a quantum pendulum with fundamental frequency ω_0.

Furthermore, the differential equation on the left-hand side in (38.6.2) looks almost like $A^2 - B^2$, and hence motivates its factorization. To this end, we define two beautiful new differential operators with normalized coordinates

$$\hat{a}^\dagger = \frac{1}{\sqrt{2}}\left(-\frac{d}{d\xi} + \xi\right), \qquad \hat{a} = \frac{1}{\sqrt{2}}\left(\frac{d}{d\xi} + \xi\right) \tag{38.6.3}$$

The first operator above is the creation or raising operator, and the latter operator represents the annihilation or lowering operator.[37] They are also collectively called ladder operators. The reason for their names is obviated later.

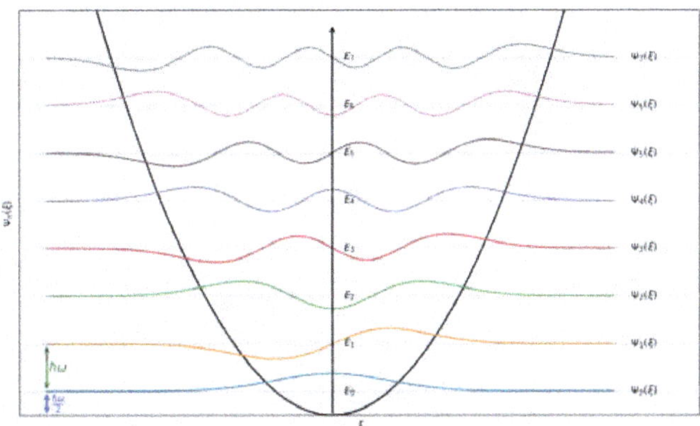

Figure 38.5: Plots of the eigensolutions of the simple quantum pendulum in normalized coordinate ξ. The photon-number states, $\Psi_n(\xi)$, (also called Fock state), are nonclassical states because they do not have a classical analogue. This 1D pendulum describes many dipolar oscillators in nature: the vibration of an atom in a lattice, an atom bound to a molecule, electron-hole pair, an electron sitting in a lattice, or possibly, electron-positron pairs in a vacuum.

From the above, by subtracting and adding the two equations, we arrive at new operators in coordinate basis as

$$\hat{\xi} = \frac{1}{\sqrt{2}}\left(\hat{a}^\dagger + \hat{a}\right) = \xi\hat{I} \tag{38.6.4}$$

$$\hat{\pi} = \frac{i}{\sqrt{2}}\left(\hat{a}^\dagger - \hat{a}\right) = -i\frac{d}{d\xi} \tag{38.6.5}$$

[37]In this course, we assume that quantum operators are "homomorphic" to matrix operators except that the quantum operators can be infinite dimensional.

It is noticed that $\hat{\xi}$ is the position operator, $\hat{\pi}$ is the momentum operator $\hat{p}$ in normalized coordinates. We can use the definition of Hermiticity for Hilbert space operators as expounded in lecture 26 and also in [38, p. 281] to show that these operators are in fact Hermitian. So on first glance, the operators in (38.6.3) are not the conjugate transpose of each other but at a deeper level, after using inner product to define the conjugate transpose, they are.

With the above definitions of the raising and lowering operators, it is easy to show, by straightforward back substitution, that Schrödinger's equation (38.6.2) for a quantum pendulum can be rewritten more compactly as

$$\frac{1}{2}\left(\hat{a}^\dagger\hat{a} + \hat{a}\hat{a}^\dagger\right)\Psi_n(\xi) = \left(\hat{a}^\dagger\hat{a} + \frac{1}{2}\right)\Psi_n(\xi) = \frac{E_n}{\hbar\omega_0}\Psi_n(\xi) \tag{38.6.6}$$

Moreover, by more direct substitution, one can show that the commutator

$$\left[\hat{a}, \hat{a}^\dagger\right] = \hat{a}\hat{a}^\dagger - \hat{a}^\dagger\hat{a} = \hat{I} \tag{38.6.7}$$

where $\hat{I}$ is the identity operator. This fact has been used to obtain the second equality in (38.6.6). From Figure 38.5, it is seen that $\Psi_n(\xi)$ corresponds to a quantum state with n photons embedded in it. In fact, one can define a number operator such that

$$\hat{n} = \hat{a}^\dagger\hat{a} \tag{38.6.8}$$

and

$$\hat{n}\Psi_n(\xi) = n\Psi_n(\xi) \tag{38.6.9}$$

In a word, this number operator reveals the number of photons that have been embedded in the "atom"[38] to be in this quantum state.[39]

It is to be noted that these "modes" are called Fock states or number states. These "modes" have quantum (or quantized) energy level of $(n + \frac{1}{2})\hbar\omega_0$. The amazing thing about them is that they have energy levels which are exactly $\hbar\omega_0$ apart which is equal to the energy of a single photon. Hence, the n-th number state corresponds to n photons embedded in the quantum pendulum (or "atom"). When $n = 0$ this is a curious mode corresponding to zero photons, but there is still an energy level of $\frac{1}{2}\hbar\omega_0$ called the zero point energy. This is the ground state of the quantum pendulum corresponding to no photons, but non-zero field, the vacuum fluctuating field. It has been shown that this fluctuating field is what gives rise to Casimir force [348, 349, 350].

[38]We will use "atom" here to replace the name "quantum pendulum".

[39]Also, particles that satisfy (38.6.7) commutator are known as bosonic particles, or just bosons. Unlike fermionic particles, bosons cooperate with each other forming collective behaviours. A dipole, consisting of a proton and an electron, does not have fractional integer spins, and hence, behaves like a boson. Polarization currents in materials are known to exhibit collective, cooperative behaviour. A photon is another example of bosonic particle that exhibits collective behaviour allowing a collection of them to carry energy in an optical fiber with very little loss. On the other hand, if the minus sign in (38.6.7) is replaced with a positive sign, the commutator becomes that of fermionic particles [75, p. 98]. These kind of particles will exhibit Pauli's exclusion principle, exhibiting non-cooperative behaviours.

It can be further shown by direct substitution that[40]

$$\hat{a}\Psi_n(\xi) = \sqrt{n}\Psi_{n-1}(\xi) \tag{38.6.10}$$

$$\hat{a}^\dagger\Psi_n(\xi) = \sqrt{n+1}\Psi_{n+1}(\xi) \tag{38.6.11}$$

where $\Psi_n(\xi)$ is defined in (38.4.14). It is now clear why $\hat{a}$ is called a lowering operator as it transforms a Fock state associated with n photons to a Fock state associated $n-1$ photons. By the same token, $\hat{a}^\dagger$ is called a raising operator as it transforms a Fock state associated with n photons to one associated with $n+1$ photons. These operators are referred to as ladder operators.

This problem under study here is analogous to that of a trapped particle in a potential well discussed in many quantum theory books. But the trapped particle here is in a quadratic potential, and it takes exactly the energy of one photon, $\hbar\omega_0$, to raise the energy of the particle from a lower level to the next higher level. As aforementioned, this scenario can be used to describe the oscillation of many 1D system found in nature where a particle is trapped in quadratic potential. It is to be noted that the trapped particle is not a photon, but it takes exactly the energy of one photon to change the energy level of the trapped particle (see trapped particle in [79, p. 37]).

The raising (creation) and lowering (annihilation) operators are the fundamental operators of a quantum pendulum. Together, they are also called ladder operators. The root source of these quantum harmonic operators can be traced to the use of quadratic potential in Schrödinger's equation. But they are beautiful operators and have even dominated quantum field theory [330, 75].

In the above, the Fock state or photon number state in dimensionless unit and coordinate basis is

$$\Psi_n(\xi) = \sqrt{\frac{1}{2^n n!\sqrt{\pi}}} e^{-\frac{\xi^2}{2}} H_n(\xi) \tag{38.6.12}$$

where $H_n(\xi)$ is a Hermite polynomial, and the wave function is Gaussian tapered by $e^{-\xi^2/2}$. These Gaussian-tapered Hermite polynomials are orthogonal since they are the eigenfunctions of an ordinary differential equation of the Sturm-Liouville type [94], and they can be orthonormalized, namely, that $\langle n|n'\rangle = \delta_{n,n'}$. The ground state, which corresponds to the lowest energy level or smallest eigenvalue, is given by

$$\Psi_0(\xi) = \sqrt{\frac{1}{\sqrt{\pi}}} e^{-\frac{\xi^2}{2}} \tag{38.6.13}$$

It is easy to show that $\hat{a}\Psi_0(\xi) = 0$. It is the state that corresponds to no photon, but with nonzero mean square. It has energy level of $\frac{1}{2}\hbar\omega$ in (38.4.13), also called the zero point energy of the quantum pendulum.

[40] The followings are written in coordinate basis. It is implied that ladder operators are also in coordinate basis. Given a function $f(x)$ which is a function of space x, it can be related to its Fourier transform as $f(x) = \frac{1}{2\pi}\int_{-\infty}^{\infty} \tilde{f}(k)e^{ikx}dk$. In physicist parlance, x is the coordinate basis while k is the Fourier (or momentum) basis. The Hilbert space has not changed, but the basis used to represent the space has changed (see section 36.6.3) [344].

More on Dirac Notation

As we know that a function is associated with a vector in the earlier part of the course. Dirac notation, developed in 1927, is "homomorphic" to modern linear algebra notation.[41] A vector is expressed as

$$\Psi_n(\xi) \to |\Psi_n\rangle \tag{38.6.14}$$

This vector can be in an infinite dimension vector space often referred to as a Hilbert space. The lowering and raising operators in Dirac notations are now

$$\hat{a}\Psi_n(\xi) = \sqrt{n}\Psi_{n-1}(\xi) \to \hat{a}|\Psi_n\rangle = \sqrt{n}|\Psi_{n-1}\rangle \tag{38.6.15}$$

$$\hat{a}^\dagger\Psi_n(\xi) = \sqrt{n+1}\Psi_{n+1}(\xi) \to \hat{a}^\dagger|\Psi_n\rangle = \sqrt{n+1}|\Psi_{n+1}\rangle \tag{38.6.16}$$

Or with increasing terseness,

$$\hat{a}|n\rangle = \sqrt{n}|n-1\rangle \tag{38.6.17}$$

$$\hat{a}^\dagger|n\rangle = \sqrt{n+1}|n+1\rangle \tag{38.6.18}$$

In the above, the eigenfunction $|\Psi_n\rangle$ is notated by its eigenvalues n in accordance to (38.6.9).

It is to be noted that now, the operator $\hat{a}$ and $\hat{a}^\dagger$ now operate on an abstract vector $|\Psi_n\rangle$ which is an abstract vector in the Hilbert space. These operators previously act on functions, but now, they can be thought of as acting on abstract vectors: they need not be restricted to just differential operators as defined in (38.6.3) but may include abstract operators as well. In the above (38.6.15) and (38.6.16), the operators on the left column are differential operators acting on vectors which are functions. On the right column, they are abstract operators acting on abstract vectors denoted by Dirac notation. We will use context to decide if these operators are differential operators or abstract operators (see further discussion in [1, p. 277]).

Previously, the lowering and raising operators $\hat{a}$ and $\hat{a}^\dagger$ were associated with differential operators. Now we associate these operators with abstract operators that map a Hilbert space vector to another Hilbert space vector. Since these are abstract operators, we need not give their explicit forms, but just their mapping properties of how they map an abstract vector to another abstract vector. In [344], we discussed ways to find the explicit forms of these operators and ways to switch between their representations. Consequently, the Hamiltonian operator can now be expressed concisely and abstractly as[42]

$$\hat{H} = \hbar\omega_0\left(\hat{a}^\dagger\hat{a} + \frac{1}{2}\right) \tag{38.6.19}$$

Then we can write (38.6.6) concisely and abstractly as

$$\hat{H}|\Psi_n\rangle = E_n|\Psi_n\rangle \tag{38.6.20}$$

---

[41] A more in depth discussion of this notation is given in [289].

[42] This is analogous to the abstract linear operator equation $\mathscr{L}f = g$ in the lecture on computational electromagnetics.

The above is Dirac notation for an operator-vector product, where $|\Psi_n\rangle$, called a "ket", is Dirac's way of indicating an abstract vector.[43] Then Schrödinger's equation (38.4.4) can now be written concisely as

$$\hat{H}|\Psi(t)\rangle = i\hbar\partial_t|\Psi(t)\rangle \tag{38.6.21}$$

where we have used Dirac notation $|\Psi(t)\rangle$ to represent a function as a vector which is also called a state vector (see section 36.4). In the above, for energy conservation, $\hat{H}$ is independent of t. Therefore, (38.6.21) can be solved formally to yield

$$|\Psi(t)\rangle = e^{-i\hat{H}t/\hbar}|\Psi(0)\rangle \tag{38.6.22}$$

When $|\Psi(0)\rangle$ is known, then we can write the solution to the above initial value problem (IVP) as in the above. More will be discussed in the next lecture.

38.7 Heisenberg Picture versus Schrödinger Picture

Quantum observables are represented by quantum operators together with wave function or state vectors that endow the observables with random properties [29, 347]. The final measurable quantities of the random observables are their average, variance, and moments, after performing many measurements.

Because the quantum state equation can be solved formally in closed form (see (38.6.22)), we can look at the quantum world in two pictures: The Heisenberg picture or the Schrödinger picture. But if the quantum state of a system $|\Psi(t)\rangle$, is known, the mean value (or expectation value) of the random observable, which is measurable by averaging over many measurements, is given by

$$\bar{O}(t) = \langle\Psi(t)|\hat{O}_s(0)|\Psi(t)\rangle. \tag{38.7.1}$$

In the above, $\bar{O}(t)$ is unique but the vector $\langle\Psi(t)|$, the operator $\hat{O}_s(0)$, and the vector $|\Psi(t)\rangle$ are not unique.

Alternatively, one can rewrite the above using $|\Psi(t)\rangle = e^{-i\hat{H}t/\hbar}|\Psi(0)\rangle$, which is given formally in (38.6.22). Here, $e^{-i\hat{H}t/\hbar}$ is the time-evolution operator which is a unitary operator. Then a new way of expressing the expectation value is

$$\bar{O}(t) = \langle\Psi(0)|e^{i\hat{H}t/\hbar}\hat{O}_s(0)e^{-i\hat{H}t/\hbar}|\Psi(0)\rangle. \tag{38.7.2}$$

In other words, we can identify the complicated expression in the middle, and the above can be rewritten as

$$\bar{O}(t) = \langle\Psi(0)|\hat{O}_h(t)|\Psi(0)\rangle. \tag{38.7.3}$$

where

$$\hat{O}_h(t) = e^{i\hat{H}t/\hbar}\hat{O}_s(0)e^{-i\hat{H}t/\hbar}. \tag{38.7.4}$$

[43]The conjugate transpose of this vector, called a "bra", is written as $\langle\Psi_n|$. The inner product between two vectors is written as $\langle\Psi|\Phi\rangle$. This defines a Hilbert space with defined inner product. Chapter 2 of [289] gives a nice summary of these concepts.

The above is a similarity transform with unitary matrix, or a unitary transform. It illustrates two different ways to look at the quantum world that produce the same expectation value of a quantum operator $\bar{O}(t)$, the measurable quantity in the laboratory.

The first way in (38.7.1), called the Schrödinger picture, keeps the quantum operator $\hat{O}_s(0)$ to be time independent, but retain the quantum state vector $|\Psi(t)\rangle$ as a function of time. In the second way in (38.7.3), called the Heisenberg picture, the operator $\hat{O}_h(t)$ is time dependent, but the state vector $|\Psi(0)\rangle$ is independent of time. Hence, one can derive the equation of motion of an operator in the Heisenberg picture as shall be shown next. Also, note that the operators $\hat{O}_h(t)$ and $\hat{O}_s(0)$ in the Heisenberg picture and the Schrödinger picture, respectively, are related to each other by a unitary transform via the time-evolution operator $e^{-i\hat{H}t/\hbar}$.

38.7.1 Equations of Motion of Operators in Heisenberg Picture

The Heisenberg picture is closer to our classical world, because in this picture, the operator representation of the quantum observable is a function of time. Taking the time derivative of an operator representation in the Heisenberg picture in (38.7.4), one concludes that

$$\frac{d\hat{O}_h}{dt} = \frac{i}{\hbar}\left(\hat{H}\hat{O}_h - \hat{O}_h\hat{H}\right) = \frac{i}{\hbar}\left[\hat{H}, \hat{O}_h\right] \tag{38.7.5}$$

where the commutator $\left[\hat{H}, \hat{O}_h\right] = \hat{H}\hat{O}_h - \hat{O}_h\hat{H}$ is used to abbreviate the above equation. The above is the Heisenberg equation of motion for quantum operator $\hat{O}_h$. One can apply the above to the operators $\hat{p}$, $\hat{q}$, $\hat{a}$ and $\hat{a}^\dagger$ in the Heisenberg picture to arrive at their respective equations of motion. Here, $\hat{p}$ is the operator representation of the momentum operator, which in coordinate basis is given by $\hat{p} = \hbar k = -i\hbar\partial_q$ in the discussion around (38.4.2). Also, $\hat{I}q$ is the operator representation of the position operator $\hat{q}$ in coordinate basis. Using them in the Heisenberg equation of motion above gives rise to

$$\frac{d\hat{p}}{dt} = \frac{i}{\hbar}\left[\hat{H}, \hat{p}\right], \qquad \frac{d\hat{q}}{dt} = \frac{i}{\hbar}\left[\hat{H}, \hat{q}\right] \tag{38.7.6}$$

By repeated use of the commutator $[\hat{p}, \hat{q}] = -i\hbar\hat{I}$, we can show that [341, 351] [352, eq. (160)]

$$\frac{d\hat{p}}{dt} = -\frac{\partial\hat{H}(\hat{p}, \hat{q})}{\partial\hat{q}}, \qquad \frac{d\hat{q}}{dt} = \frac{\partial\hat{H}(\hat{p}, \hat{q})}{\partial\hat{p}} \tag{38.7.7}$$

They are "homomorphic" to the classical Hamilton equations in (38.2.5). Thus we call them the quantum Hamilton equations.[44] [45]

The above shows that the equations of motion in the Heisenberg picture are very similar in form to the classical equations of motion. It also greatly simplifies the derivation of quantum equations

[44] They can also be derived using energy-conservation arguments [341].

[45] The above can be made simpler if we can find an eigenvector that diagonalizes both $\hat{p}$ and $\hat{q}$. Unfortunately, such an eigenvector cannot be found. The deeper meaning is that the uncertainty relationship between $\hat{p}$ and $\hat{q}$ exists in the quantum world.

of motion in the quantum world. Many classical equations of motion are derivable using classical Hamiltonian theory. As we shall see, the use of "homomorphism" greatly simplifies the derivation of quantum equations of motion from quantum Hamiltonian theory [347].

38.8 "Homomorphism" between Classical and Quantum

The above exercise shows that there is homomorphism between a quantum pendulum and a classical pendulum. Classical Maxwell's equations are the consequence of the coupling of a set of classical pendulum which can be expressed in terms of classical Hamiltonian theory as shall be shown below. By the same token, quantum Maxwell's equations are the result of the coupling of a set of quantum pendulum that can be expressed in terms of quantum Hamiltonian theory as shall be shown below as well. Homomorphism allows us to bridge the gap between these two theories easily [353].

38.8.1 Hamiltonian—Classical Picture

In general, the total energy or the Hamiltonian of an electromagnetic system is

$$H = \frac{1}{2}\int_V d\mathbf{r}\left[\varepsilon \mathbf{E}^2(\mathbf{r},t) + \frac{\mathbf{B}^2(\mathbf{r},t)}{\mu}\right]. \tag{38.8.1}$$

It is customary to write this Hamiltonian in terms of scalar and vector potentials. For simplicity, we assume the radiation gauge which is equivalent to setting $\nabla\cdot\mathbf{A} = 0$ with $\Phi = 0$.[46] When $\Phi = 0$, the Coulomb gauge and the Lorenz gauge are the same. Then $\mathbf{B} = \nabla\times\mathbf{A}$ and $\mathbf{E} = -\dot{\mathbf{A}}$, and the classical Hamiltonian from the above for a Maxwellian system becomes

$$H = \frac{1}{2}\int_V d\mathbf{r}\left[\varepsilon \dot{\mathbf{A}}^2(\mathbf{r},t) + \frac{(\nabla\times\mathbf{A}(\mathbf{r},t))^2}{\mu}\right]. \tag{38.8.2}$$

At this juncture, it is better to define another conjugate variable $\mathbf{\Pi} = \varepsilon\dot{\mathbf{A}}$ so that the above Hamiltonian can be rewritten as

$$H = \frac{1}{2}\int_V d\mathbf{r}\left[\frac{\mathbf{\Pi}^2(\mathbf{r},t)}{\varepsilon} + \frac{(\nabla\times\mathbf{A}(\mathbf{r},t))^2}{\mu}\right] \tag{38.8.3}$$

The above Hamiltonian is analogous to (38.2.4). In (38.2.4), the simple scalar conjugate variables p and q vary in tandem to keep the Hamiltonian H a constant. But here, the conjugate variables (or functions) $\mathbf{\Pi}$ and $\mathbf{A}$ vary in tandem to keep H in (38.8.3) a constant. The result is the classical Maxwell's equations [347] as shall be shown below.

[46] This gauge is valid since it implies that there is no charge in the system since $\nabla\cdot\mathbf{E} = 0$ then. This gauge is also discussed in detail in [50] and used in [354] and [141]. This gauge implies that if the incident field has $\Phi = 0$, it remains that way for the subsequent scattered field (see also Exercise 38.4).

38.8.2 Generalization of Hamilton Equations to the Continuum Case

The Hamilton equations of motion in this case, analogous to (38.7.7), are derived in reference [353, eq. (81)][47] Functional derivative need to be used to define these Hamilton equations.

$$\frac{\partial \mathbf{A}(\mathbf{r},t)}{\partial t} = \frac{\delta H}{\delta \mathbf{\Pi}(\mathbf{r},t)}, \qquad \frac{\partial \mathbf{\Pi}(\mathbf{r},t)}{\partial t} = -\frac{\delta H}{\delta \mathbf{A}(\mathbf{r},t)} \tag{38.8.4}$$

They can be derived similar to (81) of [353] except that μ and ε have been set to 1 there. Consequently, using the above, we have

$$\frac{\partial \mathbf{A}(\mathbf{r},t)}{\partial t} = \frac{\mathbf{\Pi}(\mathbf{r},t)}{\varepsilon}, \qquad \frac{\partial \mathbf{\Pi}(\mathbf{r},t)}{\partial t} = -\frac{1}{\mu}\nabla\times\nabla\times\mathbf{A}(\mathbf{r},t) \tag{38.8.5}$$

To arrive at (38.8.5), we have used integration by parts to move the curl operators onto $\mathbf{A}$. Combining the above yields the vector wave equation as the equation of motion for the vector potential, namely

$$\mu\varepsilon\partial_t^2\mathbf{A}(\mathbf{r},t) + \nabla\times\nabla\times\mathbf{A}(\mathbf{r},t) = 0 \tag{38.8.6}$$

After using that $\mathbf{E} = -\dot{\mathbf{A}}$ and that $\mu\mathbf{H} = \nabla\times\mathbf{A}$, Maxwell's equations can be derived by Hamiltonian theory to be

$$\nabla\times\mathbf{H}(\mathbf{r},t) = \varepsilon\partial_t\mathbf{E}(\mathbf{r},t) \tag{38.8.7}$$

$$\nabla\times\mathbf{E}(\mathbf{r},t) = -\mu\partial_t\mathbf{H}(\mathbf{r},t) \tag{38.8.8}$$

The first equation above is obtained from (38.8.6) and using the definition for $\mathbf{E}$ and $\mathbf{H}$. The second equation is obtained by taking the curl of the first equation in (38.8.5).

The above are classical Maxwell's equations for Ampère's law and Faraday's law derived using Hamiltonian theory. The solutions of these equations will maintain the classical Hamiltonian H to be a constant as the electromagnetic fields vary as a function of time. The divergence equations can be derived by taking the divergence of the above equations as shown in section 2.3 before.

38.8.3 Hamiltonian—Quantum Picture

We saw in the pendulum case, to get the equations for the quantum pendulum, we elevate the classical conjugate variables, the momentum p and position q in the classical Hamiltonian, to become quantum operators. For the electromagnetic oscillation, the classical conjugate variables in the classical Hamiltonian are $\mathbf{\Pi}(\mathbf{r},t)$ and $\mathbf{A}(\mathbf{r},t)$ as shown in (38.8.3). They are analogous to p and q in the classical pendulum, and they need to be elevated to become quantum operators to describe quantum Maxwell's equations.

To this end, to get the quantum Hamiltonian, we elevate the conjugate variables to become quantum operators in the quantum world in the Heisenberg picture. The quantum Hamiltonian then, from

[47] Since the derivation is lengthy but detailed in [342], this will be relegated as a homework exercise 38-3.

(38.8.3), becomes[48]

$$\hat{H} = \frac{1}{2}\int\limits_V d\mathbf{r}\left[\varepsilon\hat{\mathbf{E}}^2(\mathbf{r},t) + \frac{\hat{\mathbf{B}}^2(\mathbf{r},t)}{\mu}\right] = \frac{1}{2}\int\limits_V d\mathbf{r}\left[\frac{\hat{\mathbf{\Pi}}^2(\mathbf{r},t)}{\varepsilon} + \frac{1}{\mu}\left(\nabla\times\hat{\mathbf{A}}(\mathbf{r},t)\right)^2\right] \tag{38.8.9}$$

where $\hat{\mathbf{\Pi}}(\mathbf{r},t) = \varepsilon\dot{\hat{\mathbf{A}}}(\mathbf{r},t)$. We also have used the radiation gauge which can be effected by the Coulomb gauge with $\Phi = 0$ [354].

38.8.4 Generalization of Quantum Hamilton Equations to the Continuum Case

The quantum equations of motion can be derived from the above Hamiltonian by generalizing (38.8.4) to the quantum case. Just as the simple single pendulum case where a commutator between the conjugate variables needs to be defined, the commutator for the conjugate variables for the quantum case is [353]

$$[\hat{q}_i(t), \hat{p}_j(t)] = i\hbar\delta_{ij}\hat{I} \tag{38.8.10}$$

In the above commutator, the index i and j can imply position or the polarization of the oscillator.

The commutation relation for the vector potential case is slightly more complicated and can be written compactly as follows. It is

$$\left[\hat{\Pi}_i(\mathbf{r},t), \hat{A}_j(\mathbf{r}',t)\right] = i\hbar\delta(\mathbf{r}-\mathbf{r}')\delta_{ij}\hat{I} \tag{38.8.11}$$

where i and j now represent components of the three-vectors[49] in index notation. The above can be derived from the more general commutator relation given in (38.8.10) where for a vector quantity at location i, it can be thought of three independent oscillators collocated at i. Therefore, this new way of finding equations of motion for the vector case is no more complicated than the scalar case.

Consequently, the above can also be written more succinctly as

$$\left[\hat{\mathbf{\Pi}}(\mathbf{r},t), \hat{\mathbf{A}}(\mathbf{r}',t)\right] = i\hbar\delta(\mathbf{r}-\mathbf{r}')\hat{\bar{\mathbf{I}}} \tag{38.8.12}$$

With the above commutation relation, it is seen that the quantum case is homomorphic to the classical case. It can be shown that by repeated use of the commutator relations that [353][50]

$$\left[\hat{\mathbf{\Pi}}(\mathbf{r}',t), \hat{H}\right] = -i\hbar\frac{\delta\hat{H}}{\delta\hat{\mathbf{A}}(\mathbf{r}',t)} = i\hbar\partial_t\hat{\mathbf{\Pi}}(\mathbf{r}',t) \tag{38.8.13}$$

$$\left[\hat{\mathbf{A}}(\mathbf{r}',t), \hat{H}\right] = i\hbar\frac{\delta\hat{H}}{\delta\hat{\mathbf{\Pi}}(\mathbf{r}',t)} = i\hbar\partial_t\hat{\mathbf{A}}(\mathbf{r}',t) \tag{38.8.14}$$

---

[48] A more elaborate discussion is given in [342].

[49] A three component vector is termed a as three-vector, and a four component one is termed a four-vector in the physics literature [50].

[50] The derivation is lengthy but somewhat straightforward. You are asked to repeat this in the homework exercise 38-3.

where the Heisenberg equation of motion has been used to establish the rightmost equality above. In other words, the commutator acts analogously like a derivative operator with respect to each other just like the lone harmonic oscillator case. The above equations are similar to (38.8.4) of the classical case. Moreover, the Hamilton equations of motion, very similar to (38.8.5), can be obtained, except that conjugate variables are replaced by conjugate operators. As such, we have

$$\frac{\partial \hat{\mathbf{A}}(\mathbf{r},t)}{\partial t} = \frac{\hat{\mathbf{\Pi}}(\mathbf{r},t)}{\varepsilon}, \qquad \frac{\partial \hat{\mathbf{\Pi}}(\mathbf{r},t)}{\partial t} = -\frac{1}{\mu}\nabla \times \nabla \times \hat{\mathbf{A}}(\mathbf{r},t) \tag{38.8.15}$$

The above yields the quantum vector wave equation as the equation of motion, namely

$$\mu\varepsilon\partial_t^2 \hat{\mathbf{A}}(\mathbf{r},t) + \nabla \times \nabla \times \hat{\mathbf{A}}(\mathbf{r},t) = 0 \tag{38.8.16}$$

Quantum Maxwell's equations then can be derived from the above similar to the classical case. A more elaborate derivation is given in [342]. As such, we arrive at quantum Maxwell's equations

$$\nabla \times \hat{\mathbf{H}}(\mathbf{r},t) = \varepsilon\partial_t \hat{\mathbf{E}}(\mathbf{r},t) \tag{38.8.17}$$

$$\nabla \times \hat{\mathbf{E}}(\mathbf{r},t) = -\mu\partial_t \hat{\mathbf{H}}(\mathbf{r},t) \tag{38.8.18}$$

which are quantum Ampère's law and quantum Faraday's law. In the above, $\hat{\mathbf{E}}$ and $\hat{\mathbf{H}}$ are operator representations [29] of the random quantum observables $\mathbf{E}$ and $\mathbf{H}$. For instance, each component of $\mathbf{E}$, namely E_x, E_y, and E_z are now random observables, which are represented by operators $\hat{E}_x$, $\hat{E}_y$, and $\hat{E}_z$. The operators themselves are not random, but together with the state vector $|\Psi\rangle$, they endow the relevant observables with random properties!

The quantum Maxwell's equations above are very similar to the classical Maxwell's equation. But by themselves, they are an incomplete description of a quantum system because they contain field operators that need to act on a quantum state $|\Psi\rangle$. The quantum state $|\Psi\rangle$ time-evolves according to the quantum state equation[51]

$$\hat{H}|\Psi(t)\rangle = i\hbar\frac{\partial|\Psi(t)\rangle}{\partial t} \tag{38.8.19}$$

It is to be noted that we are now in the Schrödinger picture in the above equation. To repeat, it is the field operator, for example $\hat{E}_x$, together with the state vector $|\Psi\rangle$ that endows the quantum field observable, E_x with random properties, such as average and variance [347].

It is to be noted that the operators in (38.8.17) and (38.8.18) are physically meaningful only if they act on a state vector. The state vector is obtained from solving (38.8.19). In other words, the quantum Maxwell's equations can be written, in the Heisenberg picture, as

$$\nabla \times \hat{\mathbf{H}}(\mathbf{r},t)|\Psi(0)\rangle = \varepsilon\partial_t \hat{\mathbf{E}}(\mathbf{r},t)|\Psi(0)\rangle \tag{38.8.20}$$

$$\nabla \times \hat{\mathbf{E}}(\mathbf{r},t)|\Psi(0)\rangle = -\mu\partial_t \hat{\mathbf{H}}(\mathbf{r},t)|\Psi(0)\rangle \tag{38.8.21}$$

The above can be solved using the mode-decomposition picture.

[51] Many authors call this Schrödinger's equation, but it has been applied to all kinds of Hamiltonians including Dirac's Hamiltonian, as well as the electromagnetic Hamiltonian defined in (39.3.1). To avoid this misnomer, we call this the quantum state equation.

38.9 The Tenet of Quantum Interpretations and Measurements

A point is in order regarding the physical interpretations of the above equations (38.8.20) and (38.8.21). The quantum operators are analog of infinite dimensional matrix operators, and the ket vectors are analog of infinite dimensional column vectors. Hence, the above equations are of infinite dimensional column vectors. The ket vector or the state vector represents the infinite many possible linear superpositions of states. The above hence represents infinitely many superposition of solutions to Maxwell's equations that are "Maxwellian".

To connect the above with the real world, a quantum measurement has to be performed. This is analogous to projecting the quantum state onto another quantum state in an inner product operator. It is analogous to a measurement in classical electromagnetics where the measurement of the field generated by a source is obtained by inner product the field thus generated with another source (see lecture 12 on reciprocity in this lecture book.)

However, a tenet of quantum measurement is that the quantum state of a quantum system is unknown until after the measurement. Quantum interpretation says that the quantum state has collapsed to the state found by the measurement. The fact that a linear superposition of states exists, each of which is Maxwellian before a measurement, is the source of quantum parallelism.

There were two prevailing schools of thought on quantum measurements. One school, led by Niels Bohr of Copenhagen, purported that a quantum observable, such as $E_x(\mathbf{r}, t)$ is truly random. It is not known before a quantum measurement. After the measurement, the observable collapses to the value found by the experiment.[52]

However, Einstein, being the *realist* that he was, did not agree with this interpretation. "God does not play dice!" he said. The above interpretation is too "spooky" to be comfortable with him. He posited that a quantum system has already collapsed to the quantum state in a random fashion before the measurement. A measurement only confirms the state to which a quantum system has collapsed to. To explain the randomness of the experimental outcomes, Einstein proposed that there was a hidden random variable involved that decided which state the quantum system was to have collapsed into.

Einstein, being a genius, had his admirers. To prove Einstein right, John Bell, a mathematician, came up with an inequality [355]. One can design experiments in the laboratory to measure certain parameters a and b. If Einstein is correct, then laboratory measurements should confirm $a < b$, while if Niels Bohr is correct, then the measurement outcome should confirm that $a > b$. This famous inequality is known as Bell's inequality.

Fortunately, experiments later designed, most notably by Alain Aspect [331, 356], confirmed that the Copenhagen school of interpretation was correct, to the dismay of John Bell![53] These were important experiments that gave rise to the birth of quantum information science. It means that information can be stored *incognito* in a quantum system before the measurement.

[52] I first learned of this incredulous interpretation from a quantum physics course by Min Chen of MIT. This interpretation is a subject of much debate as it offends the sensibilities of many.

[53] To me philosophically, this is a beautiful outcome as it implies that our Karma is not written on our forehead when we were born. Our futures are in our own hands.

The above linear superposition of states is not possible in our real world, but it is possible in the quantum world. The linear superposition of states can be choreographed to perform tasks in parallel, giving rise to quantum parallelism or quantum advantage.[54]

Figure 38.6: The quantum linear superposition of states in the quantum world can be interpreted as "ghosts-and-angels". Before the measurement, we do not know the quantum state of the quantum system. After the measurement, the quantum state is said to have collapsed to the state that has been measured.

[54]Further reading on these topics related to quantum technology can be found in the lecture notes by Roth [357].

Exercises for Lecture 38

Problem 38-1: This problem refers to lecture 38. If you are busy, you can pick four of seven in the following parts.

(i) Who first postulated that the energy of an electromagnetic field is quantized into packets given by the formula $\hbar\omega$? Find this packet of energy for a 10 GHz microwave photon and an optical photon of 500 nm in wavelength.

(ii) How did Schrödinger arrive at the motivation for Schrödinger's equation (38.4.4)?

(iii) In classical theory, the position and momentum of a particle is given by q and p which are deterministic quantities. But in quantum theory, the position and momentum of a particle are random variables that become uncertain. How are these random properties expressed in quantum theory?

(iv) Show that when the system is in the photon-number state or Fock state, the expectation value of the position operator $\hat{q}$ and the momentum operator $\hat{p}$ are zero. For this, we can use normalized coordinates, $\hat{\xi}$ and $\hat{\pi}$, and use (38.6.4) and (38.6.5) to prove this point.

(v) Explain why two noncommutating operators cannot share a common eigenvector, and how this is related to the uncertainty principle.

(vi) Derive (38.7.5), the equation of motion for a quantum operator.

Problem 38-2:

(i) Derive (38.7.7) by repeated use of the commutator $[\hat{q}, \hat{p}] = i\hbar\hat{I}$. The derivation of these equations was also given in [351] and [352, eq. (160)].

(ii) Derive (38.8.13) and (38.8.14). In order to do so, as before, we follow the procedure outlined in reference [353]. First, we need to show that the commutators, in addition to inducing simple derivatives as eq. (88) and eq. (89), also induce functional derivatives as in eq. (94) and eq. (95) of the said reference. Using this, eq. (101) and eq. (102) of the said reference can be derived, leading to (38.8.13) and (38.8.14) in this lecture.

(iii) Derive (38.8.17). In order to do so, we will follow the procedure outlined in reference [353]. First, we need to show that the Hamilton equations for the discrete case as expounded in eq. (55) of [353] can be generalized to the continuum case as expounded in eq. (70) of the same reference. In a similar manner, show that eq. (81) of the same reference is true. Using this, we can derive (38.8.17) in this lecture.

Problem 38-3: This problem shows that an operator equation is not unique. A new equation can be derived using similarity transform. Convince yourself that the following is true. First, you are given

$$\overline{\mathbf{A}} \cdot \mathbf{x} = \mathbf{b}$$

and

$$\overline{\mathbf{u}}^{\dagger} \cdot \overline{\mathbf{u}} = \overline{\mathbf{I}},$$

The above identity operator, $\overline{\mathbf{I}}$ can be inserted at places where inner products occur, and it is also called the resolution of identity operator. After some manipulation, we obtain a new operator equation that

$$\overline{\mathbf{u}} \cdot \overline{\mathbf{A}} \cdot \overline{\mathbf{u}}^{\dagger} \cdot \overline{\mathbf{u}} \cdot \mathbf{x} = \overline{\mathbf{u}} \cdot \mathbf{b} \Longrightarrow \overline{\mathbf{A}}' \cdot \mathbf{x}' = \mathbf{b}'$$

where

$$\overline{\mathbf{A}}' = \overline{\mathbf{u}} \cdot \overline{\mathbf{A}} \cdot \overline{\mathbf{u}}^{\dagger}$$

$$\mathbf{x}' = \overline{\mathbf{u}} \cdot \mathbf{x},$$

$$\mathbf{b}' = \overline{\mathbf{u}} \cdot \mathbf{b}$$

Problem 38-4: This problem shows that the $\Phi = 0$ is valid under certain condition. Show that for a dispersionless, inhomogeneous medium,

$$\nabla \cdot \varepsilon \nabla \Phi + \omega^2 \varepsilon^2 \mu \Phi = -\varrho_{\text{inc}} \tag{38.9.1}$$

$$-\nabla \times \mu^{-1} \nabla \times \mathbf{A} + \omega^2 \varepsilon \mathbf{A} + \varepsilon \nabla \left(\frac{1}{\mu \varepsilon^2} \nabla \cdot \varepsilon \mathbf{A} \right) = -\mathbf{J}_{\text{inc}}$$

with the use of generalized Lorenz gauge

$$\nabla \cdot \varepsilon \mathbf{A} = i\omega \varepsilon^2 \mu \Phi$$

If $\varrho_{\text{inc}} = 0$ for all ω, then $\Phi = 0$ always. In this case, the above is self consistent only if $\nabla \cdot \mathbf{J}_{\text{inc}} = \mathbf{0}$. For such current, we can use the $\Phi = 0$ gauge, and generalized Lorenz gauge becomes generalized Coulomb gauge, or

$$\nabla \cdot \varepsilon \mathbf{A} = 0$$

Then the above equations can be simplified.

Lecture 39

Quantum Coherent State and More

As mentioned in the previous lecture, the discovery of Schrödinger's wave equation was a rousing success! The coherent state is important as it can involve a large number of photons working coherently with each other. This can portend the easy detectability of this state as many photons are involved: we do not have to design a system for detecting quantum signals at a single-photon level. Therefore, we will study the coherent state in greater details.

At first, we have QED (quantum electrodynamics) which is the study of interaction of microscopic atomic particles with electromagnetics field [358]. Later, we have the field of quantum optics with the observation that the quantization of electromagnetic field is important even when the wavelength of light is much larger than the lattice spacing in matter. There are many textbooks written on the subject such as [359] and many more.

But the wavefunctions in Schrödinger equations have to be connected with classical electromagnetic oscillations. As seen previously, classical electromagnetic oscillations resemble those of a classical pendulum. In the quantum world, the original eigenstates of the quantum pendulum were the photon number states, also called the Fock states.[1] These quantum wave oscillations, the Fock states, do not resemble the classical oscillations of a classical pendulum at all! Their connection to the classical pendulum was tenuous, but required by the correspondence principle: quantum wave phenomena in the quantum pendulum should resemble classical pendulum in the high-energy (high frequency, short wavelength) limit. This connection was finally established by using the coherent state.

[1]Some authors do not distinguish between these two states, the photon number state and the Fock state [79], but some do [360].

After connecting the classical pendulum to the quantum pendulum using the coherent state, we will study the use of quantum electromagnetics/optics[2] for a communication problem as an illustration of the use of this knowledge.

39.1 The Quantum Coherent State

We have seen that the photon number states of a quantum pendulum do not have a classical correspondence as the average or expectation values of the position and momentum of the pendulum in the photon-number state (often just called a number state) are always zero for all time for this state, namely, $\langle n|\hat{a}|n\rangle=0$, and $\langle n|\hat{a}^\dagger|n\rangle=0$. Hence, photon-number states are also known as nonclassical states.[3]

Therefore, we have to seek a time-dependent quantum state that has the classical equivalence of a pendulum. This is offered by the coherent state, which is the contribution of many researchers, most notably, Roy Glauber (1925–2018) [361] in 1963 and George Sudarshan (1931–2018) [362]. Glauber was awarded the Nobel Prize in 2005.

We like to emphasize again that the mode of an electromagnetic field inside a cavity, or a traveling wave oscillation in an open space, is "homomorphic" to the oscillation of a classical pendulum if we gaze at it for a fixed location. Then we can connect the oscillation of a classical electromagnetic mode to that of a quantum electromagnetic mode. In other words, we connect the classical pendulum to the quantum pendulum. The coherent state is a linear superposition of photon number states that makes it look like a localized wave packet. As such, a coherent state can make a quantum pendulum resemble a classical pendulum in the correspondence-principle limit [344].

The above discussion is akin to the fact that a delta pulse $\delta(z-ct)$ is a solution to the wave equation (see section 3.2). The delta pulse can also be expanded in terms of superposition of plane-waves $\exp(-i\omega t + ikz)$ with $k = \omega/c$ (or in Fourier modes). But the plane wave modes are not at all localized, and hence, are not physical, whereas the delta pulse is localized. By the same token, the photon number states are not localized, but a linear superposition of them forms the coherent state that we shall show to correspond to a localized pulse [344].

Another important point to note is that a photon is a very different quantum particle from an electron. A photon is a boson, whereas an electron is a fermion.[4] Bosons like to behave cooperatively with each other, moving in unison through space occupying the same state. But fermions have to obey the Pauli exclusion principle, meaning that they have to be opposite in quantum states (e.g., one with spin-up state, and the other, with spin-down state) before they can pair up with each other. The famous example is the formation of Cooper pairs between two electrons of opposite spins.

[2]Since quantization of electromagnetic field is now detectable down to microwave frequency, we prefer to call this field quantum electromagnetics.

[3]A photon number state to a quantum state is like a Fourier mode to a classical signal. It alone does not have a physical meaning because it has infinite duration. A physical signal has to be of finite duration *de rigueur*. But a summation of Fourier modes can be used to reconstitute any signals of finite duration. A single photon number state does not resemble a physical wave function of a quantum state. But a linear superposition of the photon number state can reconstitute a physical quantum wave function that is physical with a finite duration!

[4]Feynman's lectures [72] give a nice intro to these particles. Bosons are particles with integer spins, while Fermions have half-integer spins.

This happens when the temperature is low enough. As a consequence, Cooper pairs behave like quasi-bosons, allowing them to move cooperatively and seamlessly through a lattice if the thermal vibration of the lattice is low enough. This was first explained by the famous BCS (Bardeen, Cooper, Schrieffer) theory [363] and also explained in textbooks such as [75].

A photon is a wonderful particle: First, it is massless; and second, it zips through space at the speed of light; third, it has been known to traverse galaxies that are billions of light-years away. Photons have been known to propagate cooperatively through an optical fiber with low loss of about 0.25 dB/km [364]. Photons have been used for quantum teleportation[5] and they have been used to confirm the Bell's inequality [331]. Recently, quantum parallelism has been demonstrated by using a boson sampler in the manner of the simple JiuZhang quantum computer [365]. There are recent news on this topic as well [366, 367, 368].

As we have seen in the previous lecture, electromagnetic fields have to be elevated to become quantum field operators to describe the physics of photons correctly. It is on top of this quantum fields that a photon "rides" on.

39.2 Some Details on the Coherent States

As one cannot see the characteristics of a classical pendulum emerging from the photon number states, one needs another way of bridging the quantum world with the classical world as required by the correspondence principle. We need to find a wave-packet description of the quantum pendulum that will bridge this gap. This is the role of the coherent state: It will reveal the correspondence principle, with a classical pendulum emerging from a quantum pendulum when the energy of the pendulum is large.

The derivation of the coherent state is more math than physics. We will briefly introduce it here below, but its derivation can be found in many textbooks such as Gerry and Knight [369, p. 44] and Fox [370, p. 158]. To say this succinctly, the coherent state is the eigenstate of the annihilation operator. If we call this coherent eigenstate vector $|\alpha\rangle$, it satisfies

$$\hat{a}|\alpha\rangle = \alpha|\alpha\rangle \tag{39.2.1}$$

We will keep the above in the Schrödinger picture so that $\hat{a}$ we have introduced previously in the previous lecture is time-independent, but $|\alpha\rangle$ can be time dependent. Here, we use the symbol α as an eigenvalue as well as an index or identifier of the state $|\alpha\rangle$.[6] Since the number state $|n\rangle$ is complete,[7] the coherent state $|\alpha\rangle$ can be expanded in terms of the number state $|n\rangle$.[8] Also, the eigenvalue α in (39.2.1) can be complex since $\hat{a}$ is a non-Hermitian operator.

---

[5]A simple intro to quantum teleportation can be found in [80].

[6]This notation is cryptic and terse, but one can always down-convert it as $|\alpha\rangle \to |f_\alpha\rangle \to f_\alpha(\xi)$ to get a more explicit notation with an intuitive feel. However, using eigenvalue to index a quantum state vector is a potential source of confusion. For instance, when we denote a photon number state by $|n\rangle$, n is not the eigenvalue of the Schrödinger Hamiltonian but the eigenvalue of the number operator $\hat{n}$. One has to be wary of such confusion when parsing these notations which are popular in the physics community.

[7]Because it is the eigenstate of the number operator $\hat{n} = \hat{a}^\dagger\hat{a}$ which is a Hermitian operator.

[8]The math-physics of the problem comes together beautifully: the photon number state (also called the Fock state) is the eigenstate of the number operator $\hat{n} = \hat{a}^\dagger\hat{a}$, but the coherent state is the eigenstate of the annihilation operator $\hat{a}$.

The coherent state can be derived to be [369, 370]

$$|\alpha\rangle = e^{-|\alpha|^2/2} \sum_{n=0}^{\infty} \frac{\alpha^n}{\sqrt{n!}} |n\rangle \tag{39.2.2}$$

where the $e^{-|\alpha|^2/2}$ is a prefactor for normalization. The above can be proved easily by back substitution. A closed-form expression also exists for the above summation [369, p. 50], namely, in coordinate basis [344, other references therein], it is given as

$$\Psi_\alpha(\xi) = \left(\frac{\omega}{\pi\hbar}\right)^{1/4} e^{-|\alpha|^2/2} e^{\xi^2/2} e^{-(\xi-\alpha/\sqrt{2})^2} \tag{39.2.3}$$

(The coherent state can also be expressed as a displaced harmonic oscillator [75], which has a different Hamiltonian from a simple harmonic oscillator.)

Since the coherent state is a linear superposition of the photon number states, an average number of photons can be associated with the coherent state. If the average number of photons embedded in a coherent state is N, then it can be shown that $N = |\alpha|^2$. This is easily shown by finding the expectation of the number operator

$$\hat{n} = \hat{a}^\dagger \hat{a} \tag{39.2.4}$$

with respect to the coherent state. That is

$$N = \langle\alpha|\hat{a}^\dagger \hat{a}|\alpha\rangle = |\alpha|^2 \tag{39.2.5}$$

where we have used $\langle\alpha|a^\dagger = \langle\alpha|\alpha^*$. As can be seen, $|\alpha|^2$ is related to the amplitude of the quantum oscillation: The more photons there are in a coherent state, the larger $|\alpha|^2$ is.

Derivation of the Coherent States[9]

Since the number state $|n\rangle$, which is the eigenstate of the number operator $\hat{n}$, is Hermitian and complete, the coherent state $|\alpha\rangle$ can be expanded in terms of the photon-number state $|n\rangle$. Or that

$$|\alpha\rangle = \sum_{n=0}^{\infty} C_n |n\rangle \tag{39.2.6}$$

When the annihilation operator acts on the above, we have then

$$\hat{a}|\alpha\rangle = \sum_{n=0}^{\infty} C_n \hat{a}|n\rangle = \sum_{n=1}^{\infty} C_n \hat{a}|n\rangle = \sum_{n=1}^{\infty} C_n \sqrt{n}|n-1\rangle = \sum_{n=0}^{\infty} C_{n+1}\sqrt{n+1}|n\rangle \tag{39.2.7}$$

The last equality follows from changing the variable of summation from n to $n+1$. Equating the above with $\alpha|\alpha\rangle$ on the right-hand side of (39.2.1), then

$$\sum_{n=0}^{\infty} C_{n+1}\sqrt{n+1}|n\rangle = \alpha \sum_{n=0}^{\infty} C_n |n\rangle \tag{39.2.8}$$

[9] May be skipped on first reading since this is mainly mathematics. The derivation is also given in Gerry and Knight [369].

By the orthonormality of the number states $|n\rangle$ and the completeness of the set, we arrive at

$$C_{n+1} = \alpha C_n/\sqrt{n+1} \tag{39.2.9}$$

The above can be used recursively to yield

$$C_n = C_{n-1}\alpha/\sqrt{n} = C_{n-2}\alpha^2/\sqrt{n(n-1)} = \ldots = C_0\alpha^n/\sqrt{n!} \tag{39.2.10}$$

Consequently, the coherent state $|\alpha\rangle$ is

$$|\alpha\rangle = C_0 \sum_{n=0}^{\infty} \frac{\alpha^n}{\sqrt{n!}} |n\rangle \tag{39.2.11}$$

But due to the probabilistic interpretation of a quantum state, the state vector $|\alpha\rangle$ is normalized to one, or that[10](see (38.5.1))

$$\langle\alpha|\alpha\rangle = 1 \tag{39.2.12}$$

Then

$$\begin{aligned}\langle\alpha|\alpha\rangle &= C_0^* C_0 \sum_{n,n'}^{\infty} \frac{\alpha^n}{\sqrt{n!}} \frac{\alpha^{n'}}{\sqrt{n'!}} \langle n'|n\rangle \\ &= |C_0|^2 \sum_{n=0}^{\infty} \frac{|\alpha|^{2n}}{n!} = |C_0|^2 e^{|\alpha|^2} = 1\end{aligned} \tag{39.2.13}$$

where we have used the Taylor expansion for $e^{|\alpha|^2}$. Therefore, $C_0 = e^{-|\alpha|^2/2}$ for normalization, or that the expression for the coherent state as given before in (39.2.2), or

$$|\alpha\rangle = e^{-|\alpha|^2/2} \sum_{n=0}^{\infty} \frac{\alpha^n}{\sqrt{n!}} |n\rangle \tag{39.2.14}$$

where $e^{-|\alpha|^2/2}$ is a prefactor for normalization. In the above, to reduce the double summations into a single summation, we have made use of the orthonormality condition that $\langle n'|n\rangle = \delta_{n'n}$, or that the photon-number states are orthonormal.[11] The above is the clear statement that the coherent state $|\alpha\rangle$ is a linear superposition of photon number states.

39.2.1 Time Evolution of the Coherent State

As mentioned before, Schrödinger's equation for the quantum state (also called the state vector) of a quantum particle can be written concisely as (see also (38.6.22))

$$\hat{H}|\Psi(t)\rangle = i\hbar\partial_t|\Psi(t)\rangle \tag{39.2.15}$$

[10]The expression can be written more explicitly, or less tersely, as $\langle\alpha|\alpha\rangle = \langle f_\alpha|f_\alpha\rangle = \int_{-\infty}^{\infty} d\xi f_\alpha^*(\xi) f_\alpha(\xi) = 1$.

[11]The eigenstates of a Hermitian operator can be proven to be complete, but that of a non-Hermitian operator has eigenstates that need not be complete until proven so.

The above not only entails the form of Schrödinger's equation; it is the form of the general quantum state equation. It is clear that $|\Psi(t)\rangle$ is a function of time to make sense of the above equation. Since $\hat{H}$ is time independent for energy conservation, the above can be integrated to yield the formal solution which is[12]

$$|\Psi(t)\rangle = e^{-i\hat{H}t/\hbar}|\Psi(0)\rangle \tag{39.2.16}$$

The above solution is in the Schrödinger picture.

We remind ourselves that the Hamiltonian $\hat{H}$ of a quantum pendulum is of the form $\hbar\omega_0\left(\hat{a}^\dagger\hat{a}+\frac{1}{2}\right) = \hbar\omega_0\left(\hat{n}+\frac{1}{2}\right)$ where

$$\hat{n} = \hat{a}^\dagger\hat{a}$$

is the number operator. Here, $\hat{n}$ is Hermitian, and its eigenvector is $|n\rangle$ with eigenvalue n. One can apply this to the photon number state with $\hat{H}$ being that of the quantum pendulum and $|n\rangle$ being the eigenvector or eigenstate of $\hat{H}$. Then it is quite easy to show that

$$\hat{H}|n\rangle = \hbar\omega_0\left(n+\frac{1}{2}\right)|n\rangle = \hbar\Omega_n|n\rangle \tag{39.2.17}$$

In other words, the eigenvalue of the Hamiltonian is $\hbar\Omega_n = \hbar\omega_0\left(n+\frac{1}{2}\right)$. Then the time evolution of the number state $|n\rangle$ is (see appendix on the meaning of a function of an operator)

$$|\Psi(t)\rangle = e^{-i\hat{H}t/\hbar}|n\rangle = e^{-i\Omega_n t}|n\rangle \tag{39.2.18}$$

where $\Omega_n = \left(n+\frac{1}{2}\right)\omega_0$. The above is in the Schrödinger picture.

Thus, we can time-evolve coherent state, by using (39.2.16) and (39.2.17). In particular, the time dependent coherent state, after using (39.2.2), evolves in time as

$$|\alpha, t\rangle = e^{-i\hat{H}t/\hbar}|\alpha\rangle = e^{-|\alpha|^2/2}\sum_{n=0}^{\infty}\frac{\alpha^n e^{-i\hat{H}t/\hbar}}{\sqrt{n!}}|n\rangle = e^{-|\alpha|^2/2}\sum_{n=0}^{\infty}\frac{\alpha^n e^{-i\Omega_n t}}{\sqrt{n!}}|n\rangle \tag{39.2.19}$$

We express the time dependence explicitly to remind ourselves that we are in the Schrödinger picture still. The state $|\alpha\rangle$ and linear superposition of $|n\rangle$ can be thought of as the initial value of $|\alpha\rangle$ and sum of $|n\rangle$ at $t = 0$. Also, $|n\rangle$ on the right-hand side above are independent of time, or evaluated at $t = 0$. The last equality above is established with the help of (39.2.18). By letting $\Omega_n = \omega_0\left(n+\frac{1}{2}\right)$, the above can be rewritten as[13]

$$|\alpha, t\rangle = e^{-i\omega_0 t/2}e^{-|\alpha|^2/2}\sum_{n=0}^{\infty}\frac{\left(\alpha e^{-i\omega_0 t}\right)^n}{\sqrt{n!}}|n\rangle \tag{39.2.20}$$

By rewriting $|\alpha, t\rangle$ into the above form, it is clearly an eigenvector of the annihilation operator $\hat{a}$ because it is of the same form as in (39.2.2), except for a multiplicative factor $e^{-i\omega_0 t/2}$, and with the replacement of α in (39.2.2) by $\alpha e^{-i\omega_0 t}$. Now the state is clearly time dependent.

[12] The following is a function of an operator acting on a vector. For a review, please read the appendix.

[13] An arbitrary phase is introduced in the quantum state, but it has no role in finding the expectation value of an operator.

We see that the last factor in (39.2.20) is similar to the expression for a coherent state in (39.2.2) except that now, α is replaced by $\alpha e^{-i\omega_0 t}$. Therefore, we can express the above, (39.2.20), more succinctly by replacing α in (39.2.2) with a time-dependent $\tilde{\alpha} = \alpha e^{-i\omega_0 t}$ as

$$|\alpha, t\rangle = e^{-i\omega_0 t/2}|\alpha e^{-i\omega_0 t}\rangle = e^{-i\omega_0 t/2}|\tilde{\alpha}\rangle \tag{39.2.21}$$

The above is clearly an eigenvector of the annihilation operator $\hat{a}$. Hence, we will use the eigenvalue $\tilde{\alpha}$ to index this new eigenvector $|\tilde{\alpha}\rangle$: the difference is that $\tilde{\alpha} = \alpha e^{-i\omega_0 t}$ is a complex number which is also a function of time t. It is to be noted that in the coherent state in (39.2.20), the states time-evolve coherently together in a manner to result in a phase shift $e^{-i\omega_0 t}$ in the eigenvalue, giving rise to a new state with new eigenvalue $\tilde{\alpha}$!

39.2.2 The Correspondence Principle for a Pendulum

From the above, it is seen that the expectation values of ξ and π have to be taken with respect to another state to see the correspondence principle, namely, they should look like the position and momentum of a classical pendulum. This can be found by taking their expectation values with respect to the time-dependent coherent state $|\alpha\rangle$. We begin with the observable for normalized position ξ, which is a random variable. The expectation value of this random variable ξ is derived from its operator representation $\hat{\xi}$ (see (38.6.4))[14]

$$\langle\alpha|\hat{\xi}|\alpha\rangle = \frac{1}{\sqrt{2}}\langle\alpha|\hat{a}^\dagger + \hat{a}|\alpha\rangle \tag{39.2.22}$$

We write the expectation value in the above form, as will be obvious later, to exploit the fact that the coherent state $|\alpha\rangle$ is an eigenstate of the annihilation operator $\hat{a}$. To this end, if we take the complex conjugate transpose of (39.2.1)[15], we have

$$\langle\alpha|\hat{a}^\dagger = \langle\alpha|\alpha^* \tag{39.2.23}$$

since $\langle\alpha|$ is the conjugate transpose of $|\alpha\rangle$, and (39.2.22) becomes

$$\bar{\xi} = \langle\xi\rangle = \langle\alpha|\hat{\xi}|\alpha\rangle = \frac{1}{\sqrt{2}}(\alpha^* + \alpha)\underbrace{\langle\alpha|\alpha\rangle}_{=1} = \sqrt{2}\Re e\,(\alpha) \neq 0 \tag{39.2.24}$$

Here, we are relieved that the expectation value of ξ is not zero. We have to remind ourselves that we have been in the Schrödinger picture all along. Putting in the time-dependence for the coherent state explicitly, we let α become $\tilde{\alpha}(t) = \alpha e^{-i\omega_0 t}$ as in (39.2.21), then letting $\alpha = |\alpha|e^{-i\Psi}$ or $\tilde{\alpha}(t) = |\alpha|e^{-i\Psi - i\omega_0 t}$ yields

$$\bar{\xi}(t) = \langle\xi(t)\rangle = \langle\tilde{\alpha}(t)|\hat{\xi}|\tilde{\alpha}(t)\rangle = \frac{1}{\sqrt{2}}\left[\tilde{\alpha}^*(t) + \tilde{\alpha}(t)\right]\underbrace{\langle\tilde{\alpha}(t)|\tilde{\alpha}(t)\rangle}_{=1} = \sqrt{2}\Re e\,(\tilde{\alpha}(t)) \neq 0 \tag{39.2.25}$$

[14]Griffith explains the concept or operator representation lucidly in his book [29, p. 30].

[15]Here, $\langle\alpha|$ is the conjugate transpose of $|\alpha\rangle$. Dirac notation is homomorphic with matrix algebra notation. In matrix algebra, $(\overline{\mathbf{a}} \cdot \mathbf{x})^\dagger = \mathbf{x}^\dagger \cdot (\overline{\mathbf{a}})^\dagger$.

In both (39.2.24) and (39.2.25), we have made use of that $\langle\alpha|\alpha\rangle = 1$ and $\langle\tilde{\alpha}(t)|\tilde{\alpha}(t)\rangle = 1$, which are the normalization conditions for quantum states. Then, letting $\tilde{\alpha}(t) = \alpha e^{-i\omega_0 t}$ where $\alpha = |\alpha|e^{-i\Psi}$ is also a complex number, we have

$$\bar{\xi}(t) = \langle\xi(t)\rangle = \sqrt{2}|\alpha|\cos(\omega_0 t + \Psi) \tag{39.2.26}$$

In the above, we use $\xi(t)$ to denote the observable which is a random variable. So $\bar{\xi}(t) = \langle\xi(t)\rangle$ refers to the average of the random observable $\xi(t)$ which is also a function of time.

By the same token,

$$\bar{\pi} = \langle\pi\rangle = \langle\alpha|\hat{\pi}|\alpha\rangle = \frac{i}{\sqrt{2}}(\alpha^* - \alpha)\underbrace{\langle\alpha|\alpha\rangle}_{=1} = \sqrt{2}\Im m\,(\alpha) \neq 0 \tag{39.2.27}$$

For the explicit time-dependent case, we let α be replaced by $\tilde{\alpha}(t) = \alpha e^{-i\omega_0 t}$ to arrive at

$$\bar{\pi}(t) = \langle\pi(t)\rangle = -\sqrt{2}|\alpha|\sin(\omega_0 t + \Psi) \tag{39.2.28}$$

In the above, it is seen that an observable is a dynamic variable in the quantum world. It is a random variable which is also a function of time. Hence, it has to be characterize by its operator representation and the state vector indicating the state that it is in. The state vector endows the observable with random properties. Hence,

$$\pi \Leftrightarrow \{\hat{\pi}, |\alpha\rangle\} \tag{39.2.29}$$

$$\xi \Leftrightarrow \left\{\hat{\xi}, |\alpha\rangle\right\} \tag{39.2.30}$$

Now, we see that the expectation values of the normalized coordinate operator $\hat{\xi}$ and momentum operator $\hat{\pi}$ with respect to the coherent state just behave like a classical pendulum. There is however a marked difference: These values, ξ and π, which are quantum observables in the quantum world, are random variables: they have mean values with standard deviations or variances that are nonzero. Thus, they have quantum "fluctuation" or quantum noise, called shot noise, associated with them [370]. Since the quantum pendulum is homomorphic with the oscillation of a quantum electromagnetic mode, the amplitude of a quantum electromagnetic mode will have a mean and a "fluctuation" as well. Now, these are quantum noise associated with a quantum observable. However, this does not imply that these observables are fluctuating with respect to time; they are subject to the probabilistic interpretation of quantum theory! The above assumes the "ghost-angel" interpretation of quantum theory. However, an alternative theory is the parallel or multi-world universe theory, and the path of collapse of the quantum state will be different [371].

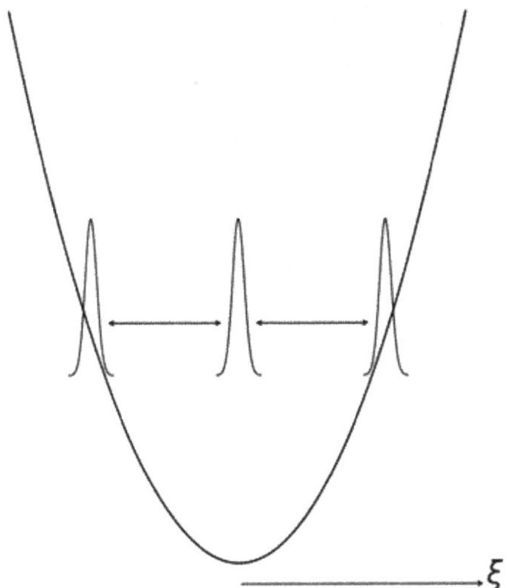

Figure 39.1: The time evolution of the coherent state. It is a wave packet that follows the motion of a classical pendulum or harmonic oscillator albeit with some uncertainty about its position.

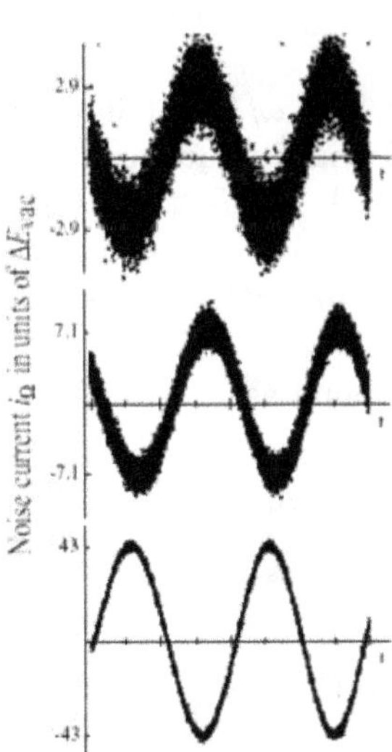

Figure 39.2: The time evolution of the coherent state at the very low average signal level. It is sinusoidal, but with great uncertainty innate in it when the energy of the coherent state is small (Figure By: https://en.wikipedia.org/wiki/Coherent state).

39.2.3 Semiclassical Picture of a Plane Wave

In general, the total energy or the Hamiltonian of a classical electromagnetic system is

$$H = \frac{1}{2}\int_V d\mathbf{r}\left[\varepsilon \mathbf{E}^2(\mathbf{r},t) + \frac{\mathbf{B}^2(\mathbf{r},t)}{\mu}\right]. \tag{39.2.31}$$

As shown in the previous lecture, classical Maxwell's equations can be derived from the above.

For simplicity, we look at the 1D case, such that a plane wave is a solution to the classical Maxwell's equations. Hence, we can assume a 1D traveling wave with periodic boundary condition such that the Hamiltonian above is in fact independent of time. To this end, we let

$$\mathbf{A} = \mathbf{a}_x \mathbf{A}_x$$

In the general case of a plane wave in 1D, we have

$$A_x(z,t) = \sum_l \frac{1}{2} A_l e^{-i\omega_l t + ik_l z} + c.c. \tag{39.2.32}$$

The above is a polychrome plane wave which can be linear superposed to make a localized plane wave traveling in the z direction. For simplicity, we can look at one component of these plane waves even though by itself, it is not physical since it is not localized. Thus, looking at the l-th mode only,

$$A_{x,l}(z,t) = \frac{1}{2} A_l e^{-i\omega_l t + ik_l z} + c.c. \tag{39.2.33}$$

where *c.c.* stands for "complex conjugation", and $k_l = \omega_l / c$.[16] (As we have seen in the lecture on group velocity, a pulse can only be localized if it is multi-frequency or polychrome.) For simplicity, we assume a monochrome plane wave here: we call this its l-th mode. The corresponding E field is

$$\mathbf{E}_l = -\partial_t \mathbf{A}_l = \mathbf{a}_x i\omega_l \frac{1}{2} A_l e^{-i\omega_l t + ik_l z} + c.c. \tag{39.2.34}$$

where $\mathbf{A}_l = \mathbf{a}_x A_{x,l}$. The corresponding B field is found as

$$\mathbf{B}_l = \nabla \times \mathbf{A}_l = \partial_z A_{x,l} = \frac{1}{2} ik_l A_l e^{-i\omega_l t + ik_l z} + c.c. \tag{39.2.35}$$

Consequently, for the l-th mode,

$$\mathbf{E}_l{}^2 = \mathbf{E}_l \cdot \mathbf{E}_l = \frac{1}{2}\omega_l^2 (A_l A_l^*) + \text{OSC} = \frac{1}{2}\omega_l^2 |A_l|^2 + \text{OSC} \tag{39.2.36}$$

where OSC stands for oscillatory terms in t and z. Similarly,

$$\mathbf{B}_l{}^2 = \frac{1}{2} k_l^2 (A_l A_l^*) + \text{OSC} = \frac{1}{2} k_l^2 |A_l|^2 + \text{OSC} \tag{39.2.37}$$

[16] For instance, we can choose the period to be L so that $k_l = 2l\pi/L$.

and $k_l^2 = \omega_l^2 \mu\varepsilon = \omega_l^2/c^2$. With periodic boundary condition, the oscillatory terms will integrate to zero, leaving behind only the constant terms. Thus, integrating over the volume such that $V = \int_V d\mathbf{r} = \mathcal{A}\int_0^L dz$, where $\mathcal{A}$ is the cross-sectional area, and L is the exact integer period of the mode, then we have the Hamiltonian (39.2.31) for the l-th mode becoming (see Exercises)

$$H_l = \frac{V\varepsilon}{2}\omega_l^2\left|\frac{A_l}{2}\right|^2 + \frac{V}{2\mu}k_l^2\left|\frac{A_l}{2}\right|^2 = V\varepsilon\omega_l^2\left|\frac{A_l}{2}\right|^2. \tag{39.2.38}$$

where $V = \mathcal{A}L$, is the volume[17] and letting $k_l^2 = \omega_l^2\mu\epsilon$. In the above, the two terms above have been combined into one term after noticing that $k_l = \omega_l/c$. The above is in fact independent of time, and if the above plane wave for the l-th mode is carrying n_l photons, then the total energy is

$$H_l = n_l\hbar\omega_l \tag{39.2.39}$$

where n_l is the number of photons in the volume V. By equating this with the above, we deduce that

$$\left|\frac{A_l}{2}\right|^2 = \frac{n_l\hbar\omega_l}{V\varepsilon\omega_l^2} \tag{39.2.40}$$

We call this a semiclassical derivation because (39.2.39) is not derived based on quantized electromagnetic field, but motivated by the conjecture that there are n_l photons embedded in the field. In the above, even if we start a polychrome wave, due to the orthogonality of the modes, the total Hamiltonian can be expressed as the sum of individual Hamiltonian for each frequency ω_l. In a word, the Hamiltonian can be considered separately from each other.

39.3 Photon-Carrying Plane Wave–Quantum Case

For this, we can start with the quantum Hamiltonian given in the last lecture, but reproduced here:

$$\hat{H} = \frac{1}{2}\int_V d\mathbf{r}\left[\varepsilon\hat{\mathbf{E}}^2(\mathbf{r},t) + \frac{\hat{\mathbf{B}}^2(\mathbf{r},t)}{\mu}\right] = \frac{1}{2}\int_V d\mathbf{r}\left[\frac{\hat{\mathbf{\Pi}}^2(\mathbf{r},t)}{\varepsilon} + \frac{1}{\mu}\left(\nabla\times\hat{\mathbf{A}}(\mathbf{r},t)\right)^2\right] \tag{39.3.1}$$

Using the vector potential approach, we can easily postulate that a vector-potential plane wave multiplied by a position and time-independent operator is a solution to quantum Maxwell's equations in (38.8.17) after we have found the corresponding electric and magnetic fields.

In actual fact, a physical photon has to ride on a plane wave that is localized. This is only true if the wave carrying the photon is polychrome. That is

$$\hat{\mathbf{A}}_l(z,t) = \mathbf{a}_x\hat{A}_{x,l}(z,t) = \mathbf{a}_x\sum_l \frac{1}{2}\hat{A}_l e^{-i\omega_l t + ik_l z} + h.c. \tag{39.3.2}$$

[17]Here, a sanity check shows that using the $|\mathbf{E}|^2 = \omega^2|\mathbf{A}|$, and the above is equal to $\varepsilon|\mathbf{E}|^2V$, the energy stored in the electric field in the volume V. It includes both the energy stored in the $\mathbf{E}$ and $\mathbf{H}$ fields.

However, for simplicity, we will pick a monochrome plane wave. Thus, focusing on the l-th mode of the plane-wave solution in free space, we let[18]

$$\hat{\mathbf{A}}_l(z,t) = \mathbf{a}_x \hat{A}_{x,l}(z,t) = \mathbf{a}_x \frac{1}{2}\hat{A}_l e^{-i\omega_l t + ik_l z} + h.c. \tag{39.3.3}$$

where "$h.c.$" stands for "Hermitian conjugate" since we have operators now.[19] Moreover, $\hat{A}_l$ is a time independent operator to ensure that the above clearly satisfies quantum Maxwell's equations. To confirm, we find the corresponding E field operator as

$$\hat{\mathbf{E}}_l = -\partial_t \hat{\mathbf{A}}_l = \mathbf{a}_x i\omega_l \frac{1}{2}\hat{A}_l e^{-i\omega_l t + ik_l z} + h.c. \tag{39.3.4}$$

The above is very similar to (39.2.34), the classical field case. The corresponding B field operator is found as[20]

$$\hat{\mathbf{B}}_l = \nabla \times \hat{\mathbf{A}}_l = \mathbf{a}_y \partial_z \hat{A}_{x,l} = \mathbf{a}_y \frac{1}{2} ik_l \hat{A}_l e^{-i\omega_l t + ik_l z} + h.c. \tag{39.3.5}$$

Consequently,

$$\hat{\mathbf{E}}_l^2 = \hat{\mathbf{E}}_l \cdot \hat{\mathbf{E}}_l = \frac{1}{4}\omega_l^2(\hat{A}_l\hat{A}_l^\dagger + \hat{A}_l^\dagger\hat{A}_l) + \text{OSC} \tag{39.3.6}$$

where OSC stands for oscillatory terms. Similarly,

$$\hat{\mathbf{B}}_l^2 = \frac{1}{4}k_l^2(\hat{A}_l\hat{A}_l^\dagger + \hat{A}_l^\dagger\hat{A}_l) + \text{OSC} \tag{39.3.7}$$

where OSC stands for oscillatory terms in t and z, and $k_l^2 = \omega_l^2\mu\varepsilon$. The derivation thus far is similar to the classical case except for now, the operators $\hat{A}_l^\dagger$ and $\hat{A}_l$ are now quantum operators, and they do not commute. Again, as in the classical case, we can make the integration of the OSC term vanish by the appropriate choice of the integration period.

As discussed previously in [372, Fig. 2], a linear time-invariant system can be mode-decomposed into sum of modes, each of which can be quantized separately. Since we need to solve the quantum state equation, the Hamiltonian for the l-th mode in (39.3.1) is then analogous to (39.2.38)

$$\hat{H}_l = \frac{1}{4}V\varepsilon\omega_l^2(\hat{A}_l\hat{A}_l^\dagger + \hat{A}_l^\dagger\hat{A}_l) \tag{39.3.8}$$

where V is integration volume of the Hamiltonian. The subscript l underscores that we are looking at the l-th mode Hamiltonian.[21] Next, if we let[22]

$$\frac{1}{2}\sqrt{V\varepsilon\omega_l^2}\hat{A}_l = \sqrt{\frac{\hbar\omega_l}{2}}\hat{a}_l \tag{39.3.9}$$

[18]We assume that the solutions to (38.8.17) and (38.8.18) are equivalent to the solutions to (38.8.16), which are all termed Maxwellian, even though (38.8.16) is not Maxwell's equations.

[19]By this symmetrization, the above operator is clearly Hermitian. An operator representation of an observable has to be Hermitian so that its expectation value is real.

[20]In this lecture, we will use $\mathbf{a}_y$ etc to represent a unit vector instead $\hat{y}$ as it will conflict with the notation for operators.

[21]The mode Hamiltonians, as in the classical case in (39.2.38), can be considered separately since these modes are orthogonal to each other. The total Hamiltonian of the system is the sum of the Hamiltonian of each individual mode, similar to the mode decomposition approach expressed in [372].

[22]We notice that from (39.3.9), the following is dimensionally correct since the energy density for the electric field is proportional to $\frac{1}{2}\varepsilon\omega_l^2\left(\hat{A}_l\right)^2$ and the right-hand side is proportional to $\hbar\omega_l$ and a_l is dimensionless.

where $\hat{a}_l$ is the annihilation operator for the l-th mode, then the quantum Hamiltonian for the l-th mode becomes

$$\hat{H}_l = \frac{1}{2}\hbar\omega_l \left(\hat{a}_l\hat{a}_l^\dagger + \hat{a}_l^\dagger\hat{a}_l\right) = \hbar\omega_l \left(\hat{a}_l^\dagger\hat{a}_l + \frac{1}{2}\right) \tag{39.3.10}$$

where we have used the commutation relation of the ladder operator to get the last equality. The above is the quantum Hamiltonian of the l-th mode of the electromagnetic oscillator with oscillator frequency ω_l. Therefore, the above is clearly "homomorphic" to the Hamiltonian of the quantum pendulum as expressed previously in (38.6.19) and the ladder operators defined in (38.6.3). Here, we have replaced ω_0 with ω_l, which is the oscillator frequency of this quantum pendulum of the l-th mode. Hence, we denote the ladder operators with subscript l to make this distinction.

Consequently, the quantum vector potential in (39.3.3), assuming that only the l-th mode is propagating for simplicity, can now be expressed in terms of the annihilation operator $\hat{a}_l$ as[23]

$$\hat{A}_{x,l}(z,t) = \sqrt{\frac{\hbar}{2V\varepsilon\omega_l}}\hat{a}_l e^{-i\omega_l t + ik_l z} + h.c. \tag{39.3.11}$$

From (39.3.11), it is customary to write for the l-th mode field[24]

$$\hat{A}_{x,l}(z,t) = \sqrt{\frac{\hbar}{2V\varepsilon\omega_l}} e^{i\theta_l} e^{ik_l z - i\omega_l t}\hat{a}_l + h.c. = \hat{A}_{x,l}^{(+)}(z,t) + \hat{A}_{x,l}^{(-)}(z,t) \tag{39.3.12}$$

$$\hat{A}_{x,l}^{(+)}(z,t) = \sqrt{\frac{\hbar}{2V\varepsilon\omega_l}} e^{i\theta_l} e^{ik_l z - i\omega_l t}\hat{a}_l \tag{39.3.13}$$

$$\hat{A}_{x,l}^{(-)}(z,t) = \sqrt{\frac{\hbar}{2V\varepsilon\omega_l}} e^{-i\theta_l} e^{-ik_l z + i\omega_l t}\hat{a}_l^\dagger \tag{39.3.14}$$

where $\hat{A}_{x,l}^{(+)}(z,t)$ and $\hat{A}_{x,l}^{(-)}(z,t)$ are the positive and negative frequency parts of the signal $\hat{A}_{x,l}(z,t)$. They are also the Hermitian conjugate of each other to ensure that $\hat{A}_{x,l}(z,t)$ is Hermitian with real expectation value so that it is the operator representation of a real observable. The above are operators which need to act on a vector related to the ladder operator. The natural space to act on (called domain space in computational electromagnetics–see the lecture 36 on the definition of range and domain spaces) is a space spanned by a linear superposition of Fock states. The state vector $|\Psi_l\rangle$ can be a linear superposition of Fock states: the coherent state is such a state, or a simple sum of Fock states.

In the quantum world, these oscillators are now converted to quantum oscillators. Moreover, they can carry quantum information if they act on a state vector. Having (39.3.11) acting on a state vector, we have

$$\hat{A}_{x,l}(z,t)|\Psi_l\rangle = \sqrt{\frac{\hbar}{2V\varepsilon\omega_l}}\hat{a}_l e^{-i\omega_l t + ik_l z}|\Psi_l\rangle + \sqrt{\frac{\hbar}{2V\varepsilon\omega_l}}\hat{a}_l^\dagger e^{i\omega_l t - ik_l z}|\Psi_l\rangle \tag{39.3.15}$$

[23]This is nonphysical as only a monochrome wave is propagating.

[24]Here, we have added a benign phase proportional to θ_l. This can be added without bespoiling the fact that the above is still a solution of quantum Maxwell's equations.

In the above, we assume that only l-th mode is propagating with the state $|\Psi_l\rangle$. The quantum state $|\Psi_l\rangle$ is an abstract vector. It has an intrinsic degree of freedom not easily expressed here. We shall call the function associated with $|\Psi_l\rangle$, $\Psi_l(\xi)$ so that the additional intrinsic degree of freedom is ξ. It is $\Psi_l(\xi)$ that is associated with this abstract vector $|\Psi_l\rangle$. The function $\Psi_l(\xi)$ can be expressed as a linear superposition of Fock states or number states.

If this state is in the photon number state $|n_l\rangle$, it can be shown that the expectation of the field in this quantum state is zero, and hence, not detectable. But if we were to square the field, and find its expectation value with respect to number state, it will not be zero, and it is detectable.

39.3.1 Heisenberg Picture versus Schrödinger Picture

The above, in the Heisenberg picture, is an infinitely long vector corresponding to the infinite possible outcomes of the operator-vector product. If the quantum state $|\Psi_l\rangle$ is in the n_l photon number state, it can be thought of as n_l photons riding on the l-th plane wave carrying quantum information. This quantum information is flying (zipping) through space at the speed of light. The above is in Heisenberg picture where the quantum state does not change with time, but the quantum operator is time varying.

We can also time evolve the above in the Schrödinger picture using the coherent state. Here, we time-evolve the quantum system as an initial value problem. Therefore, we set the above field operator and state to $t = 0$. To this end, we have

$$\hat{A}_{x,l}(z, t = 0)|\Psi_l\rangle = \sqrt{\frac{\hbar}{2V\varepsilon\omega_l}}\hat{a}_l e^{ik_l z}|\Psi_l\rangle + \sqrt{\frac{\hbar}{2V\varepsilon\omega_l}}\hat{a}_l^\dagger e^{-ik_l z}|\Psi_l\rangle \tag{39.3.16}$$

Alternatively, we can write the above as

$$\tilde{A}_{x,l}(z, t = 0)|\Psi_l\rangle = \sqrt{\frac{\hbar}{2V\varepsilon\omega_l}}\left(\hat{a}_l e^{ik_l z}|\Psi_l\rangle + \hat{a}_l^\dagger e^{-ik_l z}|\Psi_l\rangle\right) \tag{39.3.17}$$

As before, we can time-evolve the above and take the expectation of the above with respect to the time-dependent coherent state $|\tilde{\alpha}_l(t)\rangle$ where $\tilde{\alpha}_l(t) = \alpha e^{-i\omega_l t} = |\alpha|e^{-i\omega_l t - i\psi_l}$ described in section 39.2.2. Then the average or expectation value of the field becomes

$$\bar{A}_{x,l}(z, t) = \langle\tilde{\alpha}_l(t)|\hat{A}_{x,l}|\tilde{\alpha}_l(t)\rangle = \sqrt{\frac{\hbar}{2V\varepsilon\omega_l}}\langle\tilde{\alpha}_l(t)|\hat{a}_l e^{ik_l z} + \hat{a}_l^\dagger e^{-ik_l z}|\tilde{\alpha}_l(t)\rangle \tag{39.3.18}$$

$$= \sqrt{\frac{\hbar}{2V\varepsilon\omega_l}}\left\{e^{ik_l z}\tilde{\alpha}_l(t)\langle\tilde{\alpha}_l(t)|\tilde{\alpha}_l(t)\rangle + e^{-ik_l z}\tilde{\alpha}_l^*(t)\langle\tilde{\alpha}_l(t)|\tilde{\alpha}_l(t)\rangle\right\} \tag{39.3.19}$$

Consequently,

$$\bar{A}_{x,l}(z, t) = \sqrt{\frac{\hbar}{2V\varepsilon\omega_l}}2\Re e\left(|\alpha_l|e^{ik_l z - i\omega_l t - i\psi_l}\right) \tag{39.3.20}$$

Similar to (39.2.26) and (39.2.28), we can express the above as

$$\bar{A}_{x,l}(z, t) = \sqrt{\frac{\hbar}{2V\varepsilon\omega_l}}2|\alpha_l|\cos(\omega_l t - k_l z + \psi_l) \tag{39.3.21}$$

which is an equation of a traveling wave.

39.4 Wave of Arbitrary Polarization—Quantum Case

The simplest use of quantum electromagnetics is in quantum communications. We will show how this can be done with a simple example, but yet, fully quantum. For simplicity, we will assume a monochrome case, which of course is not physical as it is noncausal, but it will enlighten the underlying physics of this problem.

In general, when both horizontal and vertical polarizations of the plane wave are present, we can write the quantum electromagnetic field for the continuous wave (CW) vector potential as

$$\hat{\mathbf{A}}_l(z,t) = \sum_{s\in\{v,h\}} \sqrt{\frac{\hbar}{2\omega_l\varepsilon V}}\mathbf{e}_s\hat{a}_{l,s}e^{i(k_lz-\omega_lt)} + h.c. \tag{39.4.1}$$

where $\mathbf{e}_h$ and $\mathbf{e}_v$ for horizontal polarization and vertical polarization, respectively. Therefore, there are actually two quantum oscillators working in unison in the above equation: one oscillating horizontally, and the other one oscillating vertically. So the correct Hamiltonian for the l-th mode for this quantum system, accounting for energy storage in both polarizations is,

$$\hat{H}_l = \hat{H}_{l,h} + \hat{H}_{l,v} \tag{39.4.2}$$

With these kinds of oscillators, we will illustrate how it can be used for quantum communications.

Alternatively, we rewrite the above as

$$\hat{\mathbf{A}}_l(z,t) = \sum_{s\in\{v,h\}} \sqrt{\frac{\hbar}{2\omega_l\varepsilon V}}\mathbf{e}_s\hat{a}_{l,s}e^{i(k_lz-\omega_lt)} + h.c. = \hat{\mathbf{A}}_l^{(+)}(z,t) + \hat{\mathbf{A}}_l^{(-)}(z,t) \tag{39.4.3}$$

where $\hat{\mathbf{A}}_l^{(+)}(z,t)$ represents the positive frequency part of $\hat{\mathbf{A}}_l(z,t)$ and $\hat{\mathbf{A}}_l^{(-)}(z,t)$ is its negative frequency part, where

$$\hat{\mathbf{A}}_l^{(+)}(z,t) = \sum_{s\in\{v,h\}} \sqrt{\frac{\hbar}{2\omega_l\varepsilon V}}\mathbf{e}_s\hat{a}_{l,s}e^{i(k_lz-\omega_lt)} \tag{39.4.4}$$

$$\hat{\mathbf{A}}_l^{(-)}(z,t) = \sum_{s\in\{v,h\}} \sqrt{\frac{\hbar}{2\omega_l\varepsilon V}}\mathbf{e}_s\hat{a}_{l,s}^{\dagger}e^{-i(k_lz-\omega_lt)} \tag{39.4.5}$$

where $\hat{\mathbf{A}}_l^{(-)}(z,t) = \left[\hat{\mathbf{A}}_l^{(+)}(z,t)\right]^{\dagger}$, and they are vector quantities. So far, we are still looking at the field in 1D space time, or z, t. For simplicity, we first assume that there is only one photon present in the system even though there are two oscillators. Hence, in this case, the photon is in a quantum linear superposition of two states or polarizations. Or one can think of that the photon can "morph" seamlessly between these two polarizations. The measurement outcome of the photon polarization is randomly found in one of these two polarizations.[25] To facilitate this, the two oscillators have to be in perfect phase coherence.

[25]The random nature of the outcome is more to do with the quantum interpretation of the measurement.

Since there are two independent Hamiltonians, $\hat{H}_{l,h}$ and $\hat{H}_{l,v}$ for the oscillators, each Hamiltonian has its own number state or Fock state. Let us look at the one-photon Fock state more carefully. The outer products of one-photon Fock states are:[26]

$$|\Psi_{\text{one-photon}}\rangle = A_v|1_v\rangle|0_h\rangle + A_h|0_v\rangle|1_h\rangle = A_v|1_v\rangle + A_h|1_h\rangle \tag{39.4.6}$$

where the l-th mode is implied. In the above, the one-photon state is written as a superposition of the horizontally polarized one-photon state and the vertically polarized one. The last symbols are the most succinct ones where the zero-photon states are implied and dropped. When we have two quantum systems working in unison or in coherence, the basis state vectors for the quantum states are the products of the individual basis vectors.

Since the above has a probabilistic interpretation, then $|A_v|^2 + |A_h|^2 = 1$. What is measurable is the expectation of a quantum operator with respect to the quantum state, which is $|\Psi_{\text{one-photon}}\rangle$ in this case.

However, it is well-known that the expectation value of the field operator $\hat{\mathbf{A}}_l(z,t)$ is zero with respect to the Fock states or the photon number states as they are termed nonclassical states. To calculate a non-zero expectation value, we calculate that of the energy, which is proportional to $\hat{\mathbf{A}}_l^{(-)} \cdot \hat{\mathbf{A}}_l^{(+)}$, which is nonzero, because it is proportional to $\hat{a}^\dagger\hat{a}$ which is the number operator. Thus, the expectation value of the above operator with respect to the one-photon state is nonzero. The measurable quantity M_l corresponding to the l-th mode is given by

$$M_l = \langle\Psi_{\text{one-photon}}|\hat{\mathbf{A}}_l^{(-)}(z,t) \cdot \hat{\mathbf{A}}_l^{(+)}(z,t)|\Psi_{\text{one-photon}}\rangle \tag{39.4.7}$$

where the one-photon state is riding on the l-th mode of the system. But this is a random variable because the photon chooses to ride randomly between the two polarizations.[27] The above gives the expectation value or the average of this random variable.

For clarity, after simplification using (39.4.4) and (39.4.5), we define the $\hat{O}$ operator explicitly as

$$\hat{O}_l = \hat{\mathbf{A}}_l^{(-)}(z,t) \cdot \hat{\mathbf{A}}_l^{(+)}(z,t) = \frac{\hbar}{2\omega_l \varepsilon V}\left[\hat{a}_h^\dagger\hat{a}_h + \hat{a}_v^\dagger\hat{a}_v\right] \tag{39.4.8}$$

Though not stipulated, the above ladder operators correspond to the l-th mode of the multi-mode plane wave. In arriving at the above, we have made use of that $\mathbf{e}_v$ and $\mathbf{e}_h$ are orthogonal to each other in the dot product above. Continuing to work on the algebra explicitly, we have

$$M_l = |A_v|^2 C_l\langle 1_v|\hat{a}_v^\dagger\hat{a}_v|1_v\rangle + |A_h|^2 C\langle 1_h|\hat{a}_h^\dagger\hat{a}_h|1_h\rangle = \left(|A_v|^2 + |A_h|^2\right) C_l = C_l \tag{39.4.9}$$

where $C_l = \frac{\hbar}{2\omega_l\varepsilon V}$. In the above, we have made use of that $|A_v|^2 + |A_h|^2 = 1$ and that $\hat{a}_s^\dagger\hat{a}_s|1_s\rangle = |1_s\rangle$ where $s \in \{v, h\}$, and that $\hat{a}_s^\dagger\hat{a}_s$ is the number operator acting on a one-photon state.

The physical interpretation of the above is that the system launched a quantum state $|\Psi_{\text{one-photon}}\rangle$ which is a random superposition of two one-photon states, $|1_v\rangle$ and $|1_h\rangle$. This one-photon state is

[26] These are outer products or tensor products, sometimes denoted as $|1_v\rangle \otimes |0_h\rangle$, but physicists love the more succinct version used here [373, eq. (21.3)] [80, section 12.6.3].

[27] Only ghosts and angels can do that. We may call the random state the "ghost-angel" state.

being operated upon by the $\hat{O}_l$ operator, and the outcome is then projected onto the one-photon state $\langle \Psi_{\text{one-photon}}|$. In quantum theory, the projection entails the concept of quantum measurement where the quantum state collapses into the measured state.

From the mathematics, we see two possible paths of quantum collapse. If the measurement outcome is the state $|1_h\rangle$, then the probability of this outcome is proportional to

$$\langle 1_h|\hat{\mathbf{A}}_l^{(-)}(z,t)\cdot\hat{\mathbf{A}}_l^{(+)}(z,t)|\Psi_{\text{one-photon}}\rangle = \frac{\hbar}{2V\varepsilon\omega_l}|A_h|^2 \tag{39.4.10}$$

If the measurement outcome is the state with $|1_v\rangle$, its value is proportional to

$$\langle 1_v|\hat{\mathbf{A}}_l^{(-)}(z,t)\cdot\hat{\mathbf{A}}_l^{(+)}(z,t)|\Psi_{\text{one-photon}}\rangle = \frac{\hbar}{2V\varepsilon\omega_l}|A_v|^2 \tag{39.4.11}$$

Since $|A_h|^2 + |A_v|^2 = 1$, the probability of detecting a photon is $|A_h|^2$ and $|A_v|^2$ respectively for a horizontally and vertically polarized photon. The above is completely random!

The fact that a photon state remains *incognito* before a measurement can be used for quantum encryption or quantum communication. There is even a no-cloning theorem that guarantees that quantum communication is secure. If Alice and Bob[28] were to exploit the random properties of quantum theory to communicate information buried in a photon, Eve, an eavesdropper, could not steal the photon, alter its properties, and resend while remaining undiscovered [79].

39.5 Polychrome Photons versus Monochrome Photons

In the previous lecture, we have assumed that the photon is associated with a plane wave that is entirely monochrome (CW or time harmonic in other parlances). But a CW signal is unphysical as it has no beginning nor end. Photons in the laboratory have been associated with a flight time, and this is well defined only if the electromagnetic pulse associated with a photon is a localized pulse in time and space. We can think of a photon "riding" on such a wave, and this is only possible and agrees with experimental observation if the photon rides on a wave of finite duration, or a localized pulse. Therefore, the electromagnetic pulse that a photon rides on has to be localized, or of finite duration.

We can also imagine how a photon is emitted from an atom: The atom experiences an electronic transition of its energy level from an upper level to a lower level. Due to energy conservation, a photon has to be emitted that is equal to the energy gap of the transition. A photon can be associated with a packet of electromagnetic energy that is radiated. This pulse of energy has a beginning and an end: hence, it is broadband and causal. Only a photon with a polychrome field can carry this photon away from the atom.

A polychrome field can be expressed as

$$\hat{A}_x(z,t) = \sum_l \sqrt{\frac{\hbar}{2V\varepsilon\omega_l}}\hat{a}_l e^{-i\omega_l t + ik_l z} + h.c. \tag{39.5.1}$$

[28]Two proverbial stars in quantum communications [79].

where l is used to index the frequency of the field or mode. The above can be thought of as a multimodal field where each propagating field is a mode. The corresponding total Hamiltonian for the multimodal field is then

$$\hat{H} = \sum_{l=1}^{\infty} \hat{H}_l \tag{39.5.2}$$

In the above, we can use the index l to include both frequency and polarization. In order to solve the quantum state equation, we need to find the space that (39.5.2) would operate on. The field lives in a Hilbert space that is the outer product of individual field modes of the system, which is a humongous Hilbert space. For simplicity, we will assume only one photon riding on this multimodal field. In general, the state of the system can be written as the outer product of the states of each individual mode, namely,

$$|\Psi_{\text{general}}\rangle = \prod_{l=1}^{\infty} |n_l\rangle \approx \prod_{l=1}^{N} |n_l\rangle \tag{39.5.3}$$

which can be approximated by N modes. In the above, the l index includes both the vectorial nature of the $\mathbf{k}$-th mode and its polarization. If each mode consists of M states due to two polarizations times the Fock number states per state, the size of the Hilbert space is proportional of M^N, which is a humongous Hilbert space.

39.5.1 Single-Photon Case

For simplicity, we will assume a single photon riding on this multi-modal field with N modes, which can be expressed using the same framework as (39.5.3). But the photon can only be riding on one of these many modes: thus most of the plane wave modes are in the ground state with zero photons except for one mode.

This we denote the single photon state as

$$|1\rangle_p = |0\rangle_1 \cdots |1\rangle_p \cdots |0\rangle_N \tag{39.5.4}$$

where the photon in the above now is riding on the p-th mode only even though there are N possible modes that the photon can ride on. The left-hand side is the abbreviation for this single-photon state. Hence, there are N possibilities for single-photon state. Outer product between the state vectors above is implied. A single-photon state can also be written as a linear superposition of many single-photon states each of which rides on a mode of different frequency [374]. It can be expressed as follows:

$$|\Psi^{(1)}\rangle \approx \sum_{p=1}^{N} \tilde{w}_p |1\rangle_p = \sum_{p=1}^{N} \tilde{w}_p \hat{a}_p^\dagger |0\rangle = \hat{\mathbf{a}}^\dagger \cdot \tilde{\mathbf{w}} |0\rangle \tag{39.5.5}$$

Figure 39.3: The ability of a photon to move seamlessly from mode to mode in a resonant cavity reminds me of a monkey being able to easily jump from tree to tree in a tropical jungle! It shows the beauty of nature (Figure By: https://en.wikipedia.org).

where $[\tilde{\mathbf{w}}]_p = \tilde{w}_p$ is the probability amplitude of $|1\rangle_p$ that incorporates the spectral amplitude of the wave packet. The simplicity of the notation $|1\rangle_p$ belies the fact that it is actually the outer product of many photon number state, many of which correspond to zero photon. The above implies that a single photon is "riding" on the multi-modal (or multi-frequency) localized wave packet, with the normalization condition

$$\tilde{\mathbf{w}}^\dagger \cdot \tilde{\mathbf{w}} = 1 \tag{39.5.6}$$

where $\tilde{\mathbf{w}}$ is a vector of length N.

39.5.2 Two-Photon Case

Unentangled Two-Photon Case

For the two-photon case that are not entangled, the state can be written as [375]

$$|\Psi^{(2)}\rangle = |\Psi_A^{(1)}\rangle \otimes |\Psi_B^{(1)}\rangle = \left(\hat{\mathbf{a}}^\dagger \cdot \tilde{\mathbf{w}}_B\right)|0\rangle \otimes \left(\hat{\mathbf{a}}^\dagger \cdot \tilde{\mathbf{w}}_A\right)|0\rangle = \left(\hat{\mathbf{a}}^\dagger \cdot \tilde{\mathbf{w}}_B\right)\left(\hat{\mathbf{a}}^\dagger \cdot \tilde{\mathbf{w}}_A\right)|0\rangle \tag{39.5.7}$$

The last equality means that the outer product between two vectors in the second last equality is implied. These pulses have been used successfully to study the Hong-Ou-Mandel effect [374].

It is seen that when a quantum state is factorized into product of two quantum states of different systems, one can measure one of the quantum states without affecting the other quantum state. A quantum measurement can be thought of the projection of the quantum state onto another state vector. If such a vector lives in Hilbert space A, such a projection or measurement can be mathematically couched as[29]

$$\langle \Phi_A | \Psi^{(2)} \rangle \tag{39.5.8}$$

If the state is factorizable as in (39.5.7), then the resultant measurement gives

$$\langle \Phi_A | \Psi^{(2)} \rangle = \left(\langle \Phi_A | \Psi_A^{(1)} \rangle \right) \otimes | \Psi_B^{(1)} \rangle \tag{39.5.9}$$

One can see that vector in system A can be projected onto $|\Phi_A\rangle$ without affecting the vector in system B.

Entangled Two-Photon Case

For the entangled two-photon case, the quantum state can be written

$$|\Psi^{(2)}\rangle = \sum_i \sum_j \Psi_{i,j} \hat{a}_i^\dagger \hat{a}_j^\dagger |0\rangle \tag{39.5.10}$$

It is easily seen that the above is non-factorizable into a form like the unentangled case. In the above, $\Psi_{i,j}$ has information from system A as well as system B.

In some applications, the index i can be associated with one frequency while the index j can be associated with another frequency (see [375]). Then $\Psi_{i,j}$ can be associated with a joint probability amplitude. These probability amplitudes can be plotted as in Figure 39.4 (see [375]).

To understand the above plots intuitively, it can be shown easily that if the joint probability amplitude is approximately represented by $\delta(\omega_1 + \omega_2)$, it is anti-correlated in frequency but correlated in time, indicating entanglements.

[29]That an inner product is a measurement is similar to the reaction and reciprocity theorem in classical electromagnetics (see the lecture on Reciprocity Theorem).

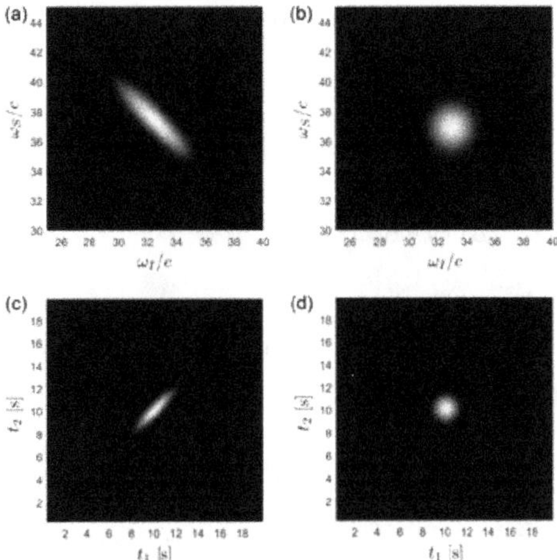

Figure 39.4: Spectral probability amplitudes for (a) two entangled photons and (b) two nonentangled photons. The degree of two-photon coincidence count rate or (c) entangled and (d) nonentangled photon pairs. The entangled photon pairs have both strong temporal correlation and frequency anti-correlation, but not for the non-entangled photon pairs [375].

39.6 Epilogue

In conclusion, the quantum theory of light is a rather complex subject. It cannot be taught in just two lectures, but what we wish is to give you a taste of this perplexing but beautiful theory here. One needs to read and re-read these two lectures to imbibe these concepts as many of these concepts are new to classical electromagnetics. It takes much longer to learn this subject well; after all, it is the by-product of almost a century of intense intellectual exercise. This knowledge is still very much in its infancy in its application to technology. Hopefully, the more we teach this subject, the better we can articulate, understand, and explain the ideas behind this subject.

When James Clerk Maxwell completed the theory of electromagnetics over 150 years ago and wrote a tome on the topic, rumor has it that most people could not read beyond the first 50 pages of his book [19]. But after over a century and a half of regurgitation and articulation, we can now teach the subject to undergraduate students! When Maxwell put his final stroke to the equations named after him, he could never have foreseen that these equations are valid from subatomic length scales to galactic length scales, and from static to gamma ray frequencies. Now, these equations are even valid from the classical to the quantum world as well!

Hopefully, by introducing this frontier of knowledge in electromagnetic field theory in this course, it will pique your interest enough in this subject, that you will take this as a lifelong learning experience. We shall join hands in solving future problems and seeking new understandings of the world.

If technology is the gift of God, may God give us the inspiration to use it wisely!

Exercises for Lecture 39

Problem 39-1:

(i) Show that C_0 in (39.2.13) is as indicated.

(ii) Derive (39.2.26) and (39.2.28) of the text.

(iii) Derive (39.3.8).

(iv) In order to time evolve (39.3.15) in the coherent state, it is best to time-evolve it as an initial value problem. To this end, we set $t = 0$ in (39.3.15), and replace $|\Psi_l\rangle$ with the coherent state.

$$|\alpha_l\rangle = e^{-|\alpha_l|^2/2} \sum_{n=0}^{\infty} \frac{\alpha_l^n}{\sqrt{n!}} |n\rangle$$

Then we time-evolve the above like (39.2.20) of the lecture notes. Derive (39.3.20) and (39.3.21).

(v) Derive (39.4.10) and (39.4.11).

(vi) Show the correspondence principle between classical and quantum cases mentioned after (39.2.39) and (39.3.10).

(vii) Explain why the humongous Hilbert space scales as M^N after (39.5.3).

Appendix A

A.1 Meaning of the Function of an Operator

We will digress here to discuss the meaning of a function of an operator, which occurs in (39.2.16), where the exponential function is a function of the operator $\hat{H}$. This knowledge is used prevalently in control theory, but not so much in other parts of electrical engineering. Since in this course, we will primarily use this knowledge to understand quantum theory, we will restrict ourselves to Hermitian operators or matrices.

A function of an operator is best explained by expanding the pertinent function into a Taylor series assuming convergence; namely, it becomes a polynomial series given as

$$f(\hat{A}) = f(0)\hat{I} + f'(0)\hat{A} + \frac{1}{2!}f''(0)\hat{A}^2 + \ldots + \frac{1}{n!}f^{(n)}(0)\hat{A}^n + \ldots \tag{A.1.1}$$

where $\hat{A}$ is a Hermitian operator. Without loss of generality, we can continue to use matrix algebra notation. The above series has no meaning unless it acts on an eigenvector of the matrix operator $\hat{A}$, where $\hat{A}\mathbf{v} = \lambda\mathbf{v}$. Hence, by applying the above equation (A.1.1) to an eigenvector $\mathbf{v}$ of $\hat{A}$, we have

$$\begin{aligned} f(\hat{A})\mathbf{v} &= f(0)\mathbf{v} + f'(0)\hat{A}\mathbf{v} + \frac{1}{2!}f''(0)\hat{A}^2\mathbf{v} + \ldots + \frac{1}{n!}f^{(n)}(0)\hat{A}^n\mathbf{v} + \ldots \\ &= f(0)\mathbf{v} + f'(0)\lambda\mathbf{v} + \frac{1}{2!}f''(0)\lambda^2\mathbf{v} + \ldots + \frac{1}{n!}f^{(n)}(0)\lambda^n\mathbf{v} + \ldots = f(\lambda)\mathbf{v} \end{aligned} \tag{A.1.2}$$

where, assuming convergence, that

$$f(\lambda) = f(0) + f'(0)\lambda + \frac{1}{2!}f''(0)\lambda^2 + \ldots + \frac{1}{n!}f^{(n)}(0)\lambda^n + \ldots \tag{A.1.3}$$

The last equality follows by re-summing the Taylor series back into a function. Applying this to an exponential function of an operator, we have, when $\mathbf{v}$ is an eigenvector of $\hat{A}$, that

$$e^{\hat{A}}\mathbf{v} = e^{\lambda}\mathbf{v} \tag{A.1.4}$$

The above proof applies to any polynomial function $f(x)$. In a word,

$$f(\hat{A})\mathbf{v} = f(\lambda)\mathbf{v} \tag{A.1.5}$$

For a general vector $\mathbf{V}$, it can be expanded as a sum of the eigenvectors of $\hat{A}$, namely, that

$$\mathbf{V} = \sum_i a_i \mathbf{v}_i \tag{A.1.6}$$

where $\mathbf{v}_i$ is the i-th eigenvector of $\hat{A}$. Then one can proceed to evaluate in the general case of

$$f(\hat{A})\mathbf{V} = \sum_i a_i f(\hat{A})\mathbf{v}_i = \sum_i a_i f(\lambda_i)\mathbf{v}_i \tag{A.1.7}$$

for an arbitrary $\mathbf{V}$. Here, λ_i is the i-th eigenvalue of $\hat{A}$.

A.2 Resolution of Identity Operator

It is prudent to introduce the beautiful operator called the resolution of identity operator in here [344]. Such an operator is useful in a number of manipulations of quantum equations involving operators. It is discussed in Stone and Golbart [376, p. 515], and also in Liu [377, eq. (517)].

Matrix Elements and Matrix Representations

Before proceeding, I would like to introduce the concepts of matrix elements and representations [38, p. 281] (see lecture 36, in this book). Given an operator $\mathscr{L}$, its matrix element, in Dirac notation, is given by

$$\left[\overline{\mathbf{L}}\right]_{mn} = L_{mn} = \langle \Psi_m | \mathscr{L} | \Psi_n \rangle \tag{A.2.1}$$

The matrix $\overline{\mathbf{L}}$ is the matrix representation of the operator $\mathscr{L}$. This is similar to that of computational electromagnetics in section 36.6.3 except that we have use Dirac notation here. Also, in quantum world, the operators are invariably Hermitian.

Assume that an N dimensional linear vector space is spanned by N orthonormal unit vectors

$$\{\mathbf{e}_i, i = 1, \ldots, N\} \tag{A.2.2}$$

Then it is quite clear that in an N dimensional linear vector space, the identity matrix (or operator) can be expanded in terms of the outer products of these unit vectors as

$$\overline{\mathbf{I}} = \sum_{i=1}^{N} \mathbf{e}_i \otimes \mathbf{e}_i = \sum_{i=1}^{N} \mathbf{e}_i \mathbf{e}_i^{\dagger}. \tag{A.2.3}$$

The above can be generalized to an infinite dimensional continuum Hilbert space using the orthonormal eigenvectors of a Hermitian operator. For instance, for the operator $\hat{q}$ such that

$$\hat{q}|q\rangle = q|q\rangle \tag{A.2.4}$$

where $|q\rangle$ is an orthonormalized eigenvector with eigenvalue q, we can use $|q\rangle$ analogous to the unit vector for coordinate space for spanning the Hilbert space. As an analogous to (A.2.3), we can define an identity operator for the case where the eigenvalue q is a continuum as

$$\hat{I} = \int_{-\infty}^{\infty} dq |q\rangle\langle q|, \tag{A.2.5}$$

The same is true to the $\hat{p}$ operator. Namely,

$$\hat{p}|p\rangle = p|p\rangle \tag{A.2.6}$$

Then an identity operator, using $|p\rangle$ orthonormal basis vectors, is

$$\hat{I} = \int_{-\infty}^{\infty} dp|p\rangle\langle p|, \tag{A.2.7}$$

Switching Representations Using Coordinate and Momentum Bases

The identity operator can be inserted in an inner product without changing its value (see 38.9). Therefore, using it, we can switch the representations of the operators in different bases easily.

Using the resolution of the identity operator, we can rewrite

$$\langle q\rangle = \langle\Psi|\hat{q}|\Psi\rangle = \int\int dqdq'\langle\Psi|q'\rangle\langle q'|\hat{q}|q\rangle\langle q|\Psi\rangle = \int\int dqdq'\Psi^*(q')\langle q'|\hat{q}|q\rangle\Psi(q) \tag{A.2.8}$$

where we have defined $\langle q|\Psi\rangle$ as the function $\Psi(q)$ etc.[1] Since $|q\rangle$ is an eigenvector of $\hat{q}$, with eigenvalue q, then $\hat{q}|q\rangle = q|q\rangle$ and

$$\langle q'|\hat{q}|q\rangle = q\langle q'|q\rangle = \delta(q'-q)q \tag{A.2.9}$$

where we have used the orthonormality of the eigenvectors such that $\langle q'|q\rangle = \delta(q'-q)$. By doing so, the above then becomes

$$\langle q\rangle = \int dq\Psi^*(q)q\Psi(q) \tag{A.2.10}$$

The above is the same as what we had before in (38.5.2) proving the correctness of the notation there.

In the above, $\hat{q}$ is an abstract operator, with eigenvector $|q\rangle$ that spans the infinite dimensional, continuum Hilbert space. But $\langle q'|\hat{q}|q\rangle = \delta(q'-q)q$ is the matrix representation of the operator $\hat{q}$ in the space spanned by the basis vector $|q\rangle$.

We can repeat the above exercise for finding $\langle p\rangle$ in the coordinate domain.

$$\langle p\rangle = \langle\Psi|\hat{p}|\Psi\rangle = \int\int dqdq'\langle\Psi|q'\rangle\langle q'|\hat{p}|q\rangle\langle q|\Psi\rangle = \int\int dqdq'\Psi^*(q')\langle q'|\hat{p}|q\rangle\Psi(q) \tag{A.2.11}$$

We could also have calculated the above using the definition of $\hat{p}$ in coordinate domain, which is usually given sloppily as

$$\hat{p} = -i\hbar\frac{d}{dq} \tag{A.2.12}$$

From (A.2.11), we have that

$$\langle p\rangle = \langle\Psi|\hat{p}|\Psi\rangle = \int dq\Psi^*(q)\left(-i\hbar\frac{d}{dq}\right)\Psi(q) \tag{A.2.13}$$

[1] See [344] for an explanation.

Using that, and comparing the two equations, (A.2.11) and (A.2.13), we deduce that[2]

$$\langle q'|\hat{p}|q\rangle = -i\hbar\delta(q'-q)\frac{d}{dq} \tag{A.2.14}$$

It is often said that the operator $\hat{p} = -i\hbar\frac{d}{dq}$. Strictly speaking, this is not rigorous. What is implied is that the diagonal elements of the matrix representation in coordinate basis of the operator $\hat{p}$ is $-i\hbar\frac{d}{dq}$. A matrix representation (or element) of an operator always has two indices, q' and q in this case as in (A.2.14).

A.3 Density Matrix or Operator

A quantum system has to be characterized by a quantum state indicated by the quantum state vector $|\Psi\rangle$. Since the phase of a quantum state is immaterial, an elegant way to represent a quantum state is via the density operator or density matrix which is the outer product of two state vectors. It is defined as

$$\hat{\rho} = |\Psi\rangle\langle\Psi| \tag{A.3.1}$$

It has the same information as the state vector $|\Psi\rangle$, but only relative phases are important, as $\langle\Psi|$ is the conjugate transpose of $|\Psi\rangle$. We shall show that the expectation of the operator $\hat{A}$ in this state $|\Psi\rangle$ is

$$\langle\hat{A}\rangle = \langle\Psi|\hat{A}|\Psi\rangle = \mathrm{tr}\left(\hat{A}\hat{\rho}\right) \tag{A.3.2}$$

Therefore, the density operator $\hat{\rho}$ contains the information on the quantum state $|\Psi\rangle$ of a system [79].

We can calculate the trace of an operator in terms of the sum of the diagonal elements of its matrix representation [80]. Assuming that we have a countable set of orthonormal basis vectors that is complete, we can find the matrix representation of the operator $\hat{A}\hat{\rho}$ so that its trace can be found. The matrix elements or representation of the operator $\hat{A}\hat{\rho}$ is given by

$$M_{mn} = \langle\phi_m|\hat{A}\hat{\rho}|\phi_n\rangle \tag{A.3.3}$$

Hence, the trace of the matrix can be found as

$$\mathrm{tr}\left(\hat{A}\hat{\rho}\right) = \sum_n \langle\phi_n|\hat{A}\hat{\rho}|\phi_n\rangle \tag{A.3.4}$$

We show that the above is basis independent by inserting the identity operator $\hat{I} = \sum_m |\phi_m\rangle\langle\phi_m|$ twice in the inner product on the right-hand side of (A.3.2) to yield[3]

$$\langle\Psi|\hat{A}|\Psi\rangle = \sum_{n,m}\langle\Psi|\phi_n\rangle\langle\phi_n|\hat{A}|\phi_m\rangle\langle\phi_m|\Psi\rangle \tag{A.3.5}$$

[2]Actually, the matrix representation from eq. (34) of [344] is $\langle q'|\hat{p}|q\rangle = -i\hbar\delta'(q'-q)$, but it can be seen to be equivalent to (A.2.14).

[3]Again, we assume that the set of vectors $|\phi_m\rangle$ forms a complete set, since they are eigenvectors of Hermitian operators.

Since the factors in the summand are scalars, we can reorder them to give

$$\langle\Psi|\hat{A}|\Psi\rangle = \sum_{n,m}\langle\phi_m|\Psi\rangle\langle\Psi|\phi_n\rangle\langle\phi_n|\hat{A}|\phi_m\rangle = \sum_{n,m}\rho_{mn}A_{nm} = \text{tr}\left(\overline{\boldsymbol{\rho}}\cdot\overline{\mathbf{A}}\right) \tag{A.3.6}$$

where ρ_{mn} is the matrix representation of $\hat{\rho}$ while A_{nm} is the matrix representation of $\hat{A}$. The above is clearly independent of the orthonormal basis we have chosen. Also, it is quite easy to show that $\text{tr}\left(\overline{\boldsymbol{\rho}}\cdot\overline{\mathbf{A}}\right) = \text{tr}\left(\overline{\mathbf{A}}\cdot\overline{\boldsymbol{\rho}}\right)$. Hence, we can write the expectation value, which is also valid for continuum basis vectors, as (A.3.2).

A.4 Uncertainty Principle—Its Physical Meaning

When two operator representations $\hat{x}$ and $\hat{p}$ of two quantum observables do not commute, it can be expressed by

$$[\hat{x}, \hat{p}] = i\hbar\hat{I} \tag{A.4.1}$$

The above commutator induces a Fourier transform relationship between the state vectors represented in basis p and x [344]. Thus a state vector $|\psi\rangle$ can be represented in the coordinate basis $\psi(x) = \langle x|\psi\rangle$, or the Fourier (or momentum) basis $\psi(p) = \langle p|\psi\rangle$. We use $\tilde{\psi}(p)$ to indicate that it is a different function from $\psi(x)$ even though they are just different representations of the same state vector using different basis in the Hilbert space. Thus,

$$\tilde{\psi}(p) = \int_{-\infty}^{\infty} dx \frac{e^{-ipx/\hbar}}{\sqrt{2\pi\hbar}}\psi(x)$$
$$\psi(x) = \int_{-\infty}^{\infty} dp \frac{e^{ipx/\hbar}}{\sqrt{2\pi\hbar}}\tilde{\psi}(p).$$

In the quantum world, conjugate variables (a classical mechanics concepts) are replaced by conjugate observables, which double as random variables as well (see section 38.5). Observables are characterized by operator-vector pairs: where the operators are matrix (or operator) representations of the observables, and the vectors embed the random properties of the observables. Hence, there are some uncertainties about the values of these observables.

When the operator representations do not commute, it implies that these matrix (or operator) representation cannot share eigenvectors. If they do, then they can be simultaneously diagonalized by the same set of eigenvectors.[4]

When two observables are represented by operators that are non-commuting, then a Fourier transform relationship can be established between these two observables [344].

The state of a quantum system, such as the quantum pendulum, is described by the state vector $|\psi\rangle$ This is an abstract vector in the Hilbert space that describes the state of our quantum system.

[4]A matrix representation can be transformed to a diagonal form using similarity transforms using unitary matrix made by the eigenvectors. If two matrix operators share the same set of eigenvectors, then the matrix representations of these operators can be replaced by diagonal operators making them commuting operators again: this contradicts our assumption to start with. This also means that the eigenfunctions of the two operators are similar to each other if they commute.

This abstract vector $|\psi\rangle$ can be represented in the coordinate basis x or in the momentum basis p. But there is only one abstract vector $|\psi\rangle$ living in this Hilbert space. They look different when we look at this vector in the coordinate basis x or the momentum p (also called Fourier) basis. This means that for a localized pulse basis function in the coordinate basis, it has to be represented by many basis functions in the Fourier basis. A localized pulse in the coordinate basis is spread out in the momentum basis, and vice versa.

The uncertainty principle is usually expressed in a relationship that describes the spread (standard deviation) of a pulse function in the coordinate basis versus the momentum basis. To this end, we will study this principle in the context of the spread of its representation in the coordinate basis versus the momentum basis. We will use a Gaussian function as its Fourier transform can be found easily. Here, we will find the standard deviation of a Gaussian pulse in the coordinate basis versus the momentum basis.[5]

To show the above uncertainty principle relationship, we need to invoke the Fourier transform relationship between Gaussian wave functions in the coordinate basis and those in the momentum basis. First, a Gaussian can be Fourier transformed in closed form, or [116, SU 51]

$$\left(a\sqrt{2}\right)^{-1} e^{-p^2/(4a^2)} \Longleftrightarrow e^{-a^2x^2} \tag{A.4.2}$$

where $\Longleftrightarrow$ implies the Fourier transform pair.

First, we assume the wave function to be Gaussian-tapered in the Fourier domain, or that

$$\tilde{\psi}(p) = Ce^{-\frac{(p-p_0)^2}{4\sigma_p{}^2}} \tag{A.4.3}$$

The probability density function for $\tilde{\psi}(p)$ in (A.4.3) is the magnitude squared of the wave function, which is, after using the same equation that

$$|\tilde{\psi}(p)|^2 \sim e^{-\frac{(p-p_0)^2}{2\sigma_p^2}} \tag{A.4.4}$$

We have chosen the form of (A.4.3) such that a Gaussian probability density function given above will have a standard deviation of σ_p [378].

Next, we need to find the probability density function in the coordinate basis, corresponding to a wave function that is Gaussian-tapered in the Fourier basis. Such a wave function, after using Fourier transform relation given above, and using the shifting property of a Fourier transform, is given by

$$\psi(x) = C\sqrt{2}\sigma_p e^{-(x\sigma_p)^2 + ip_0x} \tag{A.4.5}$$

where σ_p is the standard deviation of the random variable p.[6]

---

[5] It is often said that these two functions are in different spaces in some textbooks, but this is technically wrong as there is only one Hilbert space that we are working with. We can say that these two functions are in different domains. Better still, we say that these two functions are represented by two different basis, using the more precise language we have developed here.

[6] The above, (A.4.3) and (A.4.5) are actually the same vector in the Hilbert space, but represented in different bases.

The above are wave functions in coordinate basis. The probability density function is proportional to magnitude squared of the above which is

$$|\psi(x)|^2 \sim e^{-2x^2\sigma_p^2} = e^{-x^2/(\sigma_p^{-2}/2)} \tag{A.4.6}$$

The standard deviation of a Gaussian probability density function, as aforementiond in (A.4.5), and using the same rule for $|\psi(x)|^2$ here, is $\sigma_x^2 = \sigma_p^{-2}/4$. From the above, we deduce that

$$\sigma_x \sigma_p = \frac{1}{2} \tag{A.4.7}$$

The actual Fourier transform relationship in [344] has $\hbar \neq 1$. Now, letting $\hbar \neq 1$ in the derivation, we have

$$\sigma_x \sigma_p = \hbar \frac{1}{2} \tag{A.4.8}$$

The above relationship is rooted in the Fourier transform relationship between two domains. One commonly encountered in electrical engineering is between time and frequency domains in signal processing. This gives rise to the time-bandwidth product (TBP) relationship in radar signal processing.

A.5 Generalized Uncertainty Principle

In this Appendix, we will follow [91, 29] in the derivation of this generalized Heisenberg uncertainty principle. In quantum theory, observables, which are random variables, are characterized by operator-vector pairs [347]. The operator is termed the operator representation of the observable [29]. The vector (or state vector) contains the random properties of the observable. To connect the operators to real world observables, we take the expectation values of the operators. That is

$$\bar{A} = \langle A \rangle = \langle f|\hat{A}|f\rangle \tag{A.5.1}$$

where A is a scalar random variable of the quantum observable. The quantum observable is characterized by the operator-vector pair, namely, that $\hat{A}$ is the operator representation of this observable, and $|f\rangle$ is a state vector that defines that state the quantum system is in. The expectation value of an operator also gives the statistical mean of the observable expressed as the mean of the eigenvalue of the corresponding operator. This can be proved by expanding the vector $|f\rangle$ in terms of the eigenvector of $\hat{A}$. In general,

$$\langle f|(\hat{A})^n|f\rangle = \langle (A)^n \rangle \tag{A.5.2}$$

$$\langle f|F(\hat{A})|f\rangle = \langle F(A) \rangle \tag{A.5.3}$$

where the left-hand sides are the expectation values (or average) of the operators, while the right-hand sides are the expectation values of their eigenvalues. The above, (A.5.1) to (A.5.3), can be proved by expanding $|f\rangle$ in terms of the eigenvectors of $\hat{A}$.

We can define the deviation from its mean by the operator

$$\Delta\hat{A} = \hat{A} - \bar{A} \tag{A.5.4}$$

On the right-hand side, the first term is an operator while the second term is the mean.[7] The above operator has zero expectation value or zero mean.

We can study the variance of the operator $\left(\Delta\hat{A}\right)^2$ which has nonzero expectation value. Namely,

$$\begin{aligned}\left\langle f|(\Delta\hat{A})^2|f\right\rangle &= \left\langle f|(\hat{A}-\bar{A})^2|f\right\rangle \\ &= \langle (A-\bar{A})^2\rangle \\ &= \langle A^2 - 2\bar{A}A + \bar{A}^2\rangle \\ &= \langle A^2\rangle - 2\bar{A}\langle A\rangle + \bar{A}^2 \\ &= \langle A^2\rangle - \bar{A}^2 = \langle(\Delta A)^2\rangle = \overline{(\Delta A)^2} = \sigma_A^2\end{aligned} \tag{A.5.5}$$

The above is just the definition of variance of random variable A as in statistics. The standard deviation is obtained by taking the square root of the variance to get[8]

$$\sigma_A = \sqrt{\overline{(\Delta A)^2}} \tag{A.5.6}$$

We can derive a similar expression for σ_B, the standard deviation for B.

The generalized uncertainty principle is obtained by using the Schwartz inequality:

$$\left|\int f^* g dx\right|^2 \leq \left(\int |f|^2 dx\right)\left(\int |g|^2 dx\right) \tag{A.5.7}$$

We can rewrite the above using Dirac notation to get

$$|\langle f|g\rangle|^2 \leq \langle f|f\rangle\langle g|g\rangle \tag{A.5.8}$$

The above is the generalization of the cosine inequality we have for 3-vectors[9]

$$|\mathbf{A}\cdot\mathbf{B}|^2 = |\mathbf{A}|^2|\mathbf{B}|^2|\cos\theta|^2 \leq |\mathbf{A}|^2|\mathbf{B}|^2 \tag{A.5.9}$$

It can be generalized to N-vectors or vectors in N dimensional space, and then to vectors in infinite dimensional space if the integrals converge. If we define, for a quantum system in state $|\psi\rangle$,

$$|f\rangle = \left(\hat{A}-\bar{A}\right)|\psi\rangle = \hat{a}|\psi\rangle \tag{A.5.10}$$

$$|g\rangle = \left(\hat{B}-\bar{B}\right)|\psi\rangle = \hat{b}|\psi\rangle \tag{A.5.11}$$

where $\hat{a} = \hat{A} - \bar{A}$ and $\hat{b} = \hat{B} - \bar{B}$. Then

$$\langle f|f\rangle = \langle\psi\left|\hat{a}^2\right|\psi\rangle = \left\langle\psi\left|\left(\hat{A}-\bar{A}\right)^2\right|\psi\right\rangle = \overline{(\Delta A)^2} \tag{A.5.12}$$

$$\langle g|g\rangle = \left\langle\psi\left|\hat{b}^2\right|\psi\right\rangle = \left\langle\psi\left|\left(\hat{B}-\bar{B}\right)^2\right|\psi\right\rangle = \overline{(\Delta B)^2} \tag{A.5.13}$$

[7]To be strictly correct, we should multiply the second term by the identity operator, but this is usually understood.
[8]It is tempting to denote the standard deviation as ΔA but this could be confusing in view of (A.5.4).
[9]The integral for the inner products above can be generalized to 3D space.

Using the inequality in (A.5.8), we have

$$\left(\overline{(\Delta A)^2}\right)\left(\overline{(\Delta B)^2}\right) \geq |\langle\psi|\hat{a}\hat{b}|\psi\rangle|^2 \tag{A.5.14}$$

It can be seen easily that if $[\hat{A}, \hat{B}] = i\hat{C}$, then

$$[\hat{a}, \hat{b}] = i\hat{C} \tag{A.5.15}$$

where $\hat{C}$ is Hermitian. Furthermore,

$$\hat{a}\hat{b} = \frac{\hat{a}\hat{b} + \hat{b}\hat{a}}{2} + \frac{\hat{a}\hat{b} - \hat{b}\hat{a}}{2} = \frac{\hat{a}\hat{b} + \hat{b}\hat{a}}{2} + i\frac{\hat{C}}{2} \tag{A.5.16}$$

Taking the expectation value of the above, we have

$$\langle\hat{a}\hat{b}\rangle = \left\langle\frac{\hat{a}\hat{b} + \hat{b}\hat{a}}{2}\right\rangle + \frac{i}{2}\langle\hat{C}\rangle \tag{A.5.17}$$

In the above, if $\hat{a}$ and $\hat{b}$ are commuting operators, then $\hat{C}$ would be zero.

Since $\hat{a}$ and $\hat{b}$ are Hermitian, the operator in first term on the right-hand side is also Hermitian. Hence, its expectation value is purely real, while the second term is purely imaginary. Therefore, the amplitude squared of the above is

$$|\langle\hat{a}\hat{b}\rangle|^2 = \left|\left\langle\frac{\hat{a}\hat{b} + \hat{b}\hat{a}}{2}\right\rangle\right|^2 + \frac{1}{4}|\langle\hat{C}\rangle|^2 \geq \frac{1}{4}|\langle\hat{C}\rangle|^2 \tag{A.5.18}$$

Using the above in (A.5.14), we have

$$\left(\overline{(\Delta A)^2}\right)\left(\overline{(\Delta B)^2}\right) \geq \frac{1}{4}|\langle\hat{C}\rangle|^2 \tag{A.5.19}$$

The above is the generalized uncertainty principle for two quantum observables A and B which are also random variables. We can take the square root of the above to get

$$\sigma_A\sigma_B \geq \frac{1}{2}|\langle\hat{C}\rangle| \tag{A.5.20}$$

Thus if $\hat{a}$ and $\hat{b}$ commute, then $\hat{C}$ can be zero. Then σ_A or σ_B can be as small as zero.

Appendix B

.1 Some Useful Mathematical Formulas

Useful Vector Identities

$$\mathbf{a}\cdot(\mathbf{b}\times\mathbf{c})=\mathbf{b}\cdot(\mathbf{c}\times\mathbf{a})=\mathbf{c}\cdot(\mathbf{a}\times\mathbf{b}), \tag{.1.1}$$

$$\mathbf{a}\times(\mathbf{b}\times\mathbf{c})=\mathbf{b}(\mathbf{a}\cdot\mathbf{c})-\mathbf{c}(\mathbf{a}\cdot\mathbf{b}), \tag{.1.2}$$

$$\nabla\times\nabla\psi=0, \tag{.1.3}$$

$$\nabla\cdot\nabla\times\mathbf{A}=0, \tag{.1.4}$$

$$\nabla\cdot(\psi\mathbf{A})=\mathbf{A}\cdot\nabla\psi+\psi\nabla\cdot\mathbf{A}, \tag{.1.5}$$

$$\nabla\times(\psi\mathbf{A})=\nabla\psi\times\mathbf{A}+\psi\nabla\times\mathbf{A}, \tag{.1.6}$$

$$\nabla\cdot(\mathbf{A}\times\mathbf{B})=\mathbf{B}\cdot\nabla\times\mathbf{A}-\mathbf{A}\cdot\nabla\times\mathbf{B}, \tag{.1.7}$$

$$\nabla(\mathbf{A}\cdot\mathbf{B})=(\mathbf{A}\cdot\nabla)\mathbf{B}+(\mathbf{B}\cdot\nabla)\mathbf{A}+\mathbf{A}\times\nabla\times\mathbf{B}+\mathbf{B}\times\nabla\times\mathbf{A}, \tag{.1.8}$$

$$\nabla\times(\mathbf{A}\times\mathbf{B})=(\mathbf{B}\cdot\nabla)\mathbf{A}-(\mathbf{A}\cdot\nabla)\mathbf{B}+\mathbf{A}\nabla\cdot\mathbf{B}-\mathbf{B}\nabla\cdot\mathbf{A}, \tag{.1.9}$$

$$\nabla\times\nabla\times\mathbf{A}=\nabla\nabla\cdot\mathbf{A}-\nabla^2\mathbf{A}. \tag{.1.10}$$

In Cartesian coordinates, $\nabla^2\mathbf{A}$ can be decomposed as

$$\nabla^2\mathbf{A}=\hat{x}\nabla^2 A_x+\hat{y}\nabla^2 A_y+\hat{z}\nabla^2 A_z, \tag{.1.11}$$

because ∇^2 commutes with $\hat{x}$, $\hat{y}$, and $\hat{z}$, i.e., $\nabla^2\hat{x}=\hat{x}\nabla^2$ and so on. This is not true in other curvilinear coordinates; hence, this decomposition is not allowed.

Gradient, Divergence, Curl, and Laplacian in Rectangular, Cylindrical, Spherical, and General Orthogonal Curvilinear Coordinate Systems

(a) **Rectangular System;** x, y, z:

$$\nabla\psi = \frac{\partial\psi}{\partial x}\hat{x} + \frac{\partial\psi}{\partial y}\hat{y} + \frac{\partial\psi}{\partial z}\hat{z}, \tag{.1.12}$$

$$\nabla\cdot\mathbf{A} = \frac{\partial A_x}{\partial x} + \frac{\partial A_y}{\partial y} + \frac{\partial A_z}{\partial z}, \tag{.1.13}$$

$$\nabla\times\mathbf{A} = \left(\frac{\partial A_z}{\partial y} - \frac{\partial A_y}{\partial z}\right)\hat{x} + \left(\frac{\partial A_x}{\partial z} - \frac{\partial A_z}{\partial x}\right)\hat{y} + \left(\frac{\partial A_y}{\partial x} - \frac{\partial A_x}{\partial y}\right)\hat{z}, \tag{.1.14}$$

$$\nabla^2\psi = \frac{\partial^2\psi}{\partial x^2} + \frac{\partial^2\psi}{\partial y^2} + \frac{\partial^2\psi}{\partial z^2}. \tag{.1.15}$$

(b) **Cylindrical System;** ρ, ϕ, z:

$$\nabla\psi = \frac{\partial\psi}{\partial\rho}\hat{\rho} + \frac{1}{\rho}\frac{\partial\psi}{\partial\phi}\hat{\phi} + \frac{\partial\psi}{\partial z}\hat{z}, \tag{.1.16}$$

$$\nabla\cdot\mathbf{A} = \frac{1}{\rho}\frac{\partial}{\partial\rho}(\rho A_\rho) + \frac{1}{\rho}\frac{\partial A_\phi}{\partial\phi} + \frac{\partial A_z}{\partial z}, \tag{.1.17}$$

$$\nabla\times\mathbf{A} = \left(\frac{1}{\rho}\frac{\partial A_z}{\partial\phi} - \frac{\partial A_\phi}{\partial z}\right)\hat{\rho} + \left(\frac{\partial A_\rho}{\partial z} - \frac{\partial A_z}{\partial\rho}\right)\hat{\phi} + \frac{1}{\rho}\left(\frac{\partial}{\partial\rho}(\rho A_\phi) - \frac{\partial A_\rho}{\partial\phi}\right)\hat{z}, \tag{.1.18}$$

$$\nabla^2\psi = \frac{1}{\rho}\frac{\partial}{\partial\rho}\left(\rho\frac{\partial\psi}{\partial\rho}\right) + \frac{1}{\rho^2}\frac{\partial^2\psi}{\partial\phi^2} + \frac{\partial^2\psi}{\partial z^2}. \tag{.1.19}$$

(c) **Spherical System;** r, θ, ϕ:

$$\nabla\psi = \frac{\partial\psi}{\partial r}\hat{r} + \frac{1}{r}\frac{\partial\psi}{\partial\theta}\hat{\theta} + \frac{1}{r\sin\theta}\frac{\partial\psi}{\partial\phi}\hat{\phi}, \tag{.1.20}$$

$$\nabla\cdot\mathbf{A} = \frac{1}{r^2}\frac{\partial}{\partial r}(r^2 A_r) + \frac{1}{r\sin\theta}\frac{\partial}{\partial\theta}(\sin\theta A_\theta) + \frac{1}{r\sin\theta}\frac{\partial A_\phi}{\partial\phi}, \tag{.1.21}$$

$$\nabla\times\mathbf{A} = \frac{1}{r\sin\theta}\left[\frac{\partial}{\partial\theta}(\sin\theta A_\phi) - \frac{\partial A_\theta}{\partial\phi}\right]\hat{r} + \frac{1}{r}\left[\frac{1}{\sin\theta}\frac{\partial A_r}{\partial\phi} - \frac{\partial}{\partial r}(rA_\phi)\right]\hat{\theta} + \frac{1}{r}\left[\frac{\partial}{\partial r}(rA_\theta) - \frac{\partial A_r}{\partial\theta}\right]\hat{\phi}, \tag{.1.22}$$

$$\nabla^2\psi = \frac{1}{r^2}\frac{\partial}{\partial r}\left(r^2\frac{\partial\psi}{\partial r}\right) + \frac{1}{r^2\sin\theta}\frac{\partial}{\partial\theta}\left(\sin\theta\frac{\partial\psi}{\partial\theta}\right) + \frac{1}{r^2\sin^2\theta}\frac{\partial^2\psi}{\partial\phi^2}. \tag{.1.23}$$

(d) General Orthogonal Curvilinear Coordinate System; x_1, x_2, x_3:

The metric coefficients (h_1, h_2, h_3) in a general orthogonal curvilinear coordinate system are defined by

$$ds_i = h_i\, dx_i; \quad i = 1 \text{ or } 2, \text{ or } 3, \tag{.1.24}$$

where ds_i denotes a differential length in the direction of dx_i. Moreover, the variable, x_i may not have the dimension of length. One way of finding the metric coefficients is to express the rectangular variables in terms of the variables of that system:

$$x = y(x_1, x_2, x_3), \tag{.1.25}$$

$$y = y(x_1, x_2, x_3), \tag{.1.26}$$

$$z = z(x_1, x_2, x_3). \tag{.1.27}$$

Then

$$ds_i = \left[\left(\frac{\partial x}{\partial x_i}\right)^2 + \left(\frac{\partial y}{\partial x_i}\right)^2 + \left(\frac{\partial z}{\partial x_i}\right)^2\right]^{1/2} dx_i, \quad i = 1, 2, 3. \tag{.1.28}$$

Hence,

$$h_i = \left[\left(\frac{\partial x}{\partial x_i}\right)^2 + \left(\frac{\partial y}{\partial x_i}\right)^2 + \left(\frac{\partial z}{\partial x_i}\right)^2\right]^{1/2}. \tag{.1.29}$$

For instance, in an elliptical coordinate system,

$$x = c \cosh u \cos v, \tag{.1.30}$$

$$y = c \sinh u \sin v. \tag{.1.31}$$

If (x_1, x_2, x_3) represent (u, v, z), then by applying (.1.29), we have

$$h_1 = h_2 = c(\sinh^2 u \cos^2 v + \cosh^2 u \sin^2 v)^{1/2} = c(\cosh^2 u - \cos^2 v)^{1/2}, \tag{.1.32}$$

$$h_3 = 1. \tag{.1.33}$$

In general, for any orthogonal curvilinear coordinate system,

$$\nabla\psi = \sum_{i=1}^{3} \frac{1}{h_i}\frac{\partial \psi}{\partial x_i}\hat{x}_i, \tag{.1.34}$$

$$\nabla \cdot \mathbf{A} = \frac{1}{\Delta}\sum_{i=1}^{3} \frac{\partial}{\partial x_i}\left(\frac{\Delta A_i}{h_i}\right), \quad \Delta = h_1 h_2 h_3, \tag{.1.35}$$

$$\nabla \times \mathbf{A} = \frac{1}{\Delta}\left|\begin{pmatrix} h_1\hat{x}_1 & h_2\hat{x}_2 & h_3\hat{x}_3 \\ \frac{\partial}{\partial x_1} & \frac{\partial}{\partial x_2} & \frac{\partial}{\partial x_3} \\ h_1 A_1 & h_2 A_2 & h_3 A_3, \end{pmatrix}\right|, \tag{.1.36}$$

$$\nabla^2\psi = \frac{1}{\Delta}\sum_{i=1}^{3} \frac{\partial}{\partial x_i}\left(\frac{\Delta}{h_i^2}\frac{\partial \psi}{\partial x_i}\right). \tag{.1.37}$$

Bibliography

[1] W. Chew, *Waves and Fields in Inhomogeneous Media.* Van Nostrand, 1990.

[2] J. A. Kong, *Theory of Electromagnetic Waves.* New York, Wiley-Interscience, 1975.

[3] A. Einstein *et al.*, "On the electrodynamics of moving bodies," *Annalen der Physik*, vol. 17, no. 891, p. 50, 1905.

[4] P. A. M. Dirac, "The quantum theory of the emission and absorption of radiation," *Proceedings of the Royal Society of London, Series A, Containing Papers of a Mathematical and Physical Character*, vol. 114, no. 767, pp. 243–265, 1927.

[5] R. J. Glauber, "Coherent and incoherent states of the radiation field," *Physical Review*, vol. 131, no. 6, p. 2766, 1963.

[6] C.-N. Yang and R. L. Mills, "Conservation of isotopic spin and isotopic gauge invariance," *Physical Review*, vol. 96, no. 1, p. 191, 1954.

[7] G. t'Hooft, *50 years of Yang-Mills theory.* World Scientific, 2005.

[8] C. W. Misner, K. S. Thorne, and J. A. Wheeler, *Gravitation.* Princeton University Press, 2017.

[9] F. Teixeira and W. C. Chew, "Differential forms, metrics, and the reflectionless absorption of electromagnetic waves," *Journal of Electromagnetic Waves and Applications*, vol. 13, no. 5, pp. 665–686, 1999.

[10] W. C. Chew, E. Michielssen, J.-M. Jin, and J. Song, *Fast and Efficient Algorithms in Computational Electromagnetics.* Artech House, Inc., 2001.

[11] A. Volta, "On the electricity excited by the mere contact of conducting substances of different kinds, in a letter from Mr. Alexander Volta, FRS Professor of Natural Philosophy at the University of Pavia, to the Rt. Hon. Sir Joseph Banks, Bart. KBPR S." *Philosophical transactions of the Royal Society of London*, no. 90, pp. 403–431, 1800.

[12] A.-M. Ampère, *Exposé méthodique des phénomènes électro-dynamiques, et des lois de ces phénomènes.* Bachelier, 1823.

[13] ——, *Mémoire sur la théorie mathématique des phénomènes électro-dynamiques uniquement déduite de l'expérience: dans lequel se trouvent réunis les Mémoires que M. Ampère a com-*

muniqués à l'Académie royale des Sciences, dans les séances des 4 et 26 décembre 1820, 10 juin 1822, 22 décembre 1823, 12 septembre et 21 novembre 1825. Bachelier, 1825.

[14] B. Jones and M. Faraday, *The Life and Letters of Faraday.* Cambridge University Press, 2010, vol. 2.

[15] G. Kirchhoff, "Ueber die auflösung der gleichungen, auf welche man bei der untersuchung der linearen vertheilung galvanischer ströme geführt wird," *Annalen der Physik*, vol. 148, no. 12, pp. 497–508, 1847.

[16] L. Weinberg, "Kirchhoff's' third and fourth laws'," *IRE Transactions on Circuit Theory*, vol. 5, no. 1, pp. 8–30, 1958.

[17] T. Standage, *The Victorian Internet: The remarkable story of the telegraph and the nineteenth century's online pioneers.* Phoenix, 1998.

[18] J. C. Maxwell, "A dynamical theory of the electromagnetic field," *Philosophical Transactions of the Royal Society of London*, no. 155, pp. 459–512, 1865.

[19] P. J. Nahin, *Oliver Heaviside: Sage in Solitude.* IEEE Press New York, 1987.

[20] H. Hertz, "On the finite velocity of propagation of electromagnetic actions," *Electric Waves*, vol. 110, 1888.

[21] M. Romer and I. B. Cohen, "Roemer and the first determination of the velocity of light (1676)," *Isis*, vol. 31, no. 2, pp. 327–379, 1940.

[22] Wikipedia, "Newton's Rings," https://en.wikipedia.org/wiki/Newton's_rings.

[23] A. Arons and M. Peppard, "Einstein's proposal of the photon concept–a translation of the Annalen der Physik paper of 1905," *American Journal of Physics*, vol. 33, no. 5, pp. 367–374, 1965.

[24] A. Pais, "Einstein and the quantum theory," *Reviews of Modern Physics*, vol. 51, no. 4, p. 863, 1979.

[25] M. Planck, "On the law of distribution of energy in the normal spectrum," *Annalen der physik*, vol. 4, no. 553, p. 1, 1901.

[26] A. Houck, D. Schuster, J. Gambetta, J. Schreier, B. Johnson, J. Chow, L. Frunzio, J. Majer, M. Devoret, S. Girvin *et al.*, "Generating single microwave photons in a circuit," *Nature*, vol. 449, no. 7160, pp. 328–331, 2007.

[27] Z. Peng, S. De Graaf, J. Tsai, and O. Astafiev, "Tuneable on-demand single-photon source in the microwave range," *Nature Communications*, vol. 7, p. 12588, 2016.

[28] B. D. Gates, Q. Xu, M. Stewart, D. Ryan, C. G. Willson, and G. M. Whitesides, "New approaches to nanofabrication: Molding, printing, and other techniques," *Chemical Reviews*, vol. 105, no. 4, pp. 1171–1196, 2005.

[29] D. J. Griffiths and D. F. Schroeter, *Introduction to Quantum Mechanics.* Cambridge University Press, 2018.

[30] J. S. Bell, "The debate on the significance of his contributions to the foundations of quantum mechanics, Bell's Theorem and the Foundations of Modern Physics (A. van der Merwe, F. Selleri, and G. Tarozzi, eds.)," 1992.

[31] C. Pickover, *Archimedes to Hawking: Laws of science and the great minds behind them.* Oxford University Press, 2008.

[32] R. Resnick, J. Walker, and D. Halliday, *Fundamentals of Physics.* John Wiley, 1988.

[33] J. L. De Lagrange, "Recherches d'arithmétique," *Nouveaux Mémoires de l'Académie de Berlin*, 1773.

[34] S. Ramo, J. R. Whinnery, and T. Duzer van, *Fields and waves in communication electronics, Third Edition.* John Wiley & Sons, Inc., 1995, also 1965, 1984.

[35] J. A. Kong, *Electromagnetic Wave Theory.* EMW Publishing, 2008, also 1985.

[36] H. M. Schey, *Div, Grad, Curl, and All that: An informal text on vector calculus.* WW Norton New York, 2005.

[37] M. N. Sadiku, *Elements of Electromagnetics.* Oxford University Press, 2014.

[38] W. C. Chew, *Waves and Fields in Inhomogeneous Media.* IEEE Press, 1995, also 1990.

[39] A. R. Choudhuri, *The Physics of Fluids and Plasmas: An Introduction for Astrophysicists.* Cambridge University Press, 1998.

[40] V. J. Katz, "The history of Stokes' theorem," *Mathematics Magazine*, vol. 52, no. 3, pp. 146–156, 1979.

[41] J. C. Maxwell, *A Treatise on Electricity and Magnetism.* Dover New York, 1954, first published in 1873, vol. 1 and 2.

[42] A. D. Yaghjian, "Reflections on Maxwell's treatise," *Progress In Electromagnetics Research*, vol. 149, pp. 217–249, 2014.

[43] B. J. Hunt, *The Maxwellians.* Cornell University Press, 2005.

[44] J. C. Maxwell, "Poems of James Clerk Maxwell," https://mypoeticside.com/poets/james-clerk-maxwell-poems.

[45] W. K. Panofsky and M. Phillips, *Classical Electricity and Magnetism.* Courier Corporation, 2005.

[46] T. Lancaster and S. J. Blundell, *Quantum field theory for the gifted amateur.* OUP Oxford, 2014.

[47] J. D. Jackson, *Classical Electrodynamics.* John Wiley & Sons, 1962.

[48] W. C. Chew, "Fields and waves: Lecture notes for ECE 350 at UIUC," *https://engineering.purdue.edu/wcchew/ece350.html*, 1990.

[49] C. M. Bender and S. A. Orszag, *Advanced mathematical methods for scientists and engineers I: Asymptotic methods and perturbation theory.* Springer Science & Business Media, 2013.

[50] J. D. Jackson, *Classical Electrodynamics.* John Wiley & Sons, 1999.

[51] J. M. Crowley, *Fundamentals of Applied Electrostatics.* Krieger Publishing Company, 1986.

[52] C. Balanis, *Advanced Engineering Electromagnetics.* Hoboken, NJ, USA: Wiley, 2012.

[53] P. Moon and D. E. Spencer, *Field theory handbook: including coordinate systems, differential equations and their solutions.* Springer, 2012.

[54] R. Courant and D. Hilbert, *Methods of Mathematical Physics, Volumes 1 and 2.* Interscience Publ., 1962.

[55] R. F. Harrington, *Time-Harmonic Electromagnetic Fields.* McGraw-Hill, 1961.

[56] L. Esaki and R. Tsu, "Superlattice and negative differential conductivity in semiconductors," *IBM Journal of Research and Development,* vol. 14, no. 1, pp. 61–65, 1970.

[57] E. Kudeki and D. C. Munson, *Analog Signals and Systems.* Upper Saddle River, NJ, USA: Pearson Prentice Hall, 2009.

[58] A. V. Oppenheim and R. W. Schafer, *Discrete-Time Signal Processing.* Pearson Education, 2014.

[59] E. C. Jordan and K. G. Balmain, *Electromagnetic Waves and Radiating Systems.* Prentice-Hall, 1968.

[60] G. Agarwal, D. Pattanayak, and E. Wolf, "Electromagnetic fields in spatially dispersive media," *Physical Review B,* vol. 10, no. 4, p. 1447, 1974.

[61] S. L. Chuang, *Physics of Photonic Devices.* John Wiley & Sons, 2012, vol. 80.

[62] B. E. Saleh and M. C. Teich, *Fundamentals of Photonics.* John Wiley & Sons, 2019.

[63] M. Born and E. Wolf, *Principles of Optics: Electromagnetic Theory of Propagation, Interference and Diffraction of Light.* Elsevier, 2013, also 1959 to 1986.

[64] R. W. Boyd, *Nonlinear Optics.* Elsevier, 2003.

[65] Y.-R. Shen, *The Principles of Nonlinear Optics.* New York, Wiley-Interscience, 1984.

[66] N. Bloembergen, *Nonlinear Optics.* World Scientific, 1996.

[67] P. C. Krause, O. Wasynczuk, and S. D. Sudhoff, *Analysis of Electric Machinery.* McGraw-Hill New York, 1986.

[68] A. E. Fitzgerald, C. Kingsley, S. D. Umans, and B. James, *Electric Machinery.* McGraw-Hill New York, 2003, vol. 5.

[69] M. A. Brown and R. C. Semelka, *MRI.: Basic Principles and Applications.* John Wiley & Sons, 2011.

[70] C. A. Balanis, *Advanced Engineering Electromagnetics.* John Wiley & Sons, 1999, also 1989.

[71] Wikipedia, "Lorentz force," https://en.wikipedia.org/wiki/Lorentz_force/, accessed: 2019-09-06.

[72] R. P. Feynman, R. Leighton, and M. Sands, *Feynman Lectures on Physics, Volume III: Quantum Mechanics.* New York, NY, USA: Basic books, 1965.

[73] R. O. Dendy, *Plasma Physics: an introductory course.* Cambridge University Press, 1995.

[74] Wikipedia, "Kennelly-Heaviside Layer," https://en.wikipedia.org/wiki/Kennelly-Heaviside_layer.

[75] H. Haken, *Quantum Field Theory of Solids, an Introduction.* North-Holland, 1976.

[76] R. P. Feynman, R. B. Leighton, and M. Sands, *The Feynman Lectures on Physics, Vols. I, II, & III: The new millennium edition.* Basic Books, 2011, also 1963, 2006, vol. 1,2,3.

[77] Wikipedia, "Spectral line shape," https://en.wikipedia.org/wiki/Spectral_line_shape.

[78] P. Sen and W. C. Chew, "The frequency dependent dielectric and conductivity response of sedimentary rocks," *Journal of Microwave Power*, vol. 18, no. 1, pp. 95–105, 1983.

[79] D. A. Miller, *Quantum Mechanics for Scientists and Engineers.* Cambridge, UK: Cambridge University Press, 2008.

[80] W. C. Chew, "Quantum mechanics made simple: Lecture notes for ECE 487 at UIUC," *http://wcchew.ece.illinois.edu/chew/course/QMAll20161206.pdf*, 2016.

[81] B. G. Streetman and S. Banerjee, *Solid State Electronic Devices.* Prentice Hall New Jersey, 2000, vol. 4.

[82] Smithsonian, "This 1600-year-old goblet shows that the romans were nanotechnology pioneers," https://www.smithsonianmag.com/history/this-1600-year-old-goblet-shows-that-the-romans-were-nanotechnology-pioneers-787224/, accessed: 2019-09-06.

[83] K. G. Budden, *Radio Waves in the Ionosphere.* Cambridge University Press, 2009.

[84] R. Fitzpatrick, *Plasma Physics: an introduction.* CRC Press, 2014.

[85] G. Strang, *Introduction to Linear Algebra.* Wellesley, MA, USA: Wellesley-Cambridge Press, 1993, vol. 3.

[86] K. C. Yeh and C.-H. Liu, "Radio wave scintillations in the ionosphere," *Proceedings of the IEEE*, vol. 70, no. 4, pp. 324–360, 1982.

[87] J. Kraus, *Electromagnetics.* McGraw-Hill, 1984, also 1953, 1973, 1981.

[88] Wikipedia, "Circular polarization," https://en.wikipedia.org/wiki/Circular_polarization.

[89] Q. Zhan, "Cylindrical vector beams: from mathematical concepts to applications," *Advances in Optics and Photonics*, vol. 1, no. 1, pp. 1–57, 2009.

[90] Wikipedia, "Louis de Broglie," https://en.wikipedia.org/wiki/Louis_de_Broglie.

[91] H. Haus, *Electromagnetic Noise and Quantum Optical Measurements*, ser. Advanced Texts in Physics. Springer Berlin Heidelberg, 2000.

[92] W. C. Chew, "Lectures on theory of microwave and optical waveguides, for ECE 531 at UIUC," *https://arxiv.org/pdf/2107.09672*, 2016.

[93] L. Brillouin, *Wave Propagation and Group Velocity.* Academic Press, 1960.

[94] F. B. Hildebrand, *Advanced Calculus for Applications.* Prentice-Hall, 1962.

[95] S. Schelkunoff, "Some equivalence theorems of electromagnetics and their application to radiation problems," *The Bell System Technical Journal*, vol. 15, no. 1, pp. 92–112, 1936.

[96] W. C. Chew, "Lectures on theory of microwave and optical waveguides," *arXiv preprint arXiv:2107.09672*, 2021.

[97] C.-T. Chen, *Linear System Theory and Design.* Oxford University Press, 1998.

[98] C. Kittel and P. McEuen, *Introduction to Solid State Physics.* Wiley New York, 1996, vol. 8.

[99] N. W. Ashcroft and N. D. Mermin, *Solid State Physics.* Cengage Learning, 2022.

[100] S. H. Schot, "Eighty years of Sommerfeld's radiation condition," *Historia mathematica*, vol. 19, no. 4, pp. 385–401, 1992.

[101] D. Hinton, *Analects.* Counterpoint, 2014.

[102] H. Bible, "New Revised Standard Version (NRSV)," *Grand Rapids MI, USA: Zondervan*, 1989.

[103] V. Rumsey, "Reaction concept in electromagnetic theory," *Physical Review*, vol. 94, no. 6, p. 1483, 1954.

[104] R. E. Collin, *Foundations for Microwave Engineering.* John Wiley & Sons, 2007, also 1966.

[105] W. J. Hoefer, "The transmission-line matrix method-theory and applications," *IEEE Transactions on Microwave Theory and Techniques*, vol. 33, no. 10, pp. 882–893, 1985.

[106] A. Ruehli, G. Antonini, and L. Jiang, *Circuit Oriented Electromagnetic Modeling Using the PEEC Techniques.* John Wiley & Sons, 2017.

[107] A. Ishimaru, *Electromagnetic Wave Propagation, Radiation, and Scattering from Fundamentals to Applications.* Wiley Online Library, 2017, also 1991.

[108] A. E. H. Love, "I. the integration of the equations of propagation of electric waves," *Philosophical Transactions of the Royal Society of London, Series A, Containing Papers of a Mathematical or Physical Character*, vol. 197, no. 287–299, pp. 1–45, 1901.

[109] Wikipedia, "Christiaan Huygens," https://en.wikipedia.org/wiki/Christiaan_Huygens.

[110] ——, "George Green (mathematician)," https://en.wikipedia.org/wiki/George_Green_(mathematician).

[111] C.-T. Tai, *Dyadic Green's Functions in Electromagnetic Theory.* PA: International Textbook, Scranton, 1971.

[112] ——, *Dyadic Green Functions in Electromagnetic Theory.* Institute of Electrical & Electronics Engineers (IEEE), 1994.

[113] W. Franz, "Zur formulierung des huygensschen prinzips," *Zeitschrift für Naturforschung A*, vol. 3, no. 8-11, pp. 500–506, 1948.

[114] J. C. Maxwell, *A treatise on electricity and Magnetism.* Oxford: Clarendon Press, 1873, vol. 1.

[115] M. Abramowitz and I. A. Stegun, *Handbook of Mathematical Functions: with formulas, graphs, and mathematical tables.* Courier Corporation, 1965, vol. 55.

[116] I. S. Gradshteyn and I. M. Ryzhik, *Table of Integrals, Series, and Products.* Academic Press, 2014.

[117] W. C. Chew and J. Kong, "Microstrip capacitance for a circular disk through matched asymptotic expansions," *SIAM Journal on Applied Mathematics*, vol. 42, no. 2, pp. 302–317, 1982.

[118] J. D. Jackson, *Classical Electrodynamics, Third Edition.* John Wiley & Sons, 1974.

[119] J. F. Lee, "Finite element methods for modeling passive microwave devices," Ph.D. dissertation, Carnegie Mellon University.

[120] J.-F. Lee, "Analysis of passive microwave devices by using three-dimensional tangential vector finite elements," *International Journal of Numerical Modelling: Electronic Networks, Devices and Fields*, vol. 3, no. 4, pp. 235–246, 1990.

[121] L. Nagel and D. Pederson, "Simulation program with integrated circuit emphasis," in *Midwest Symposium on Circuit Theory*, 1973.

[122] P. H. Smith, "Transmission line calculator," *Electronics*, vol. 12, no. 1, pp. 29–31, 1939.

[123] T. M. Philip and M. J. Gilbert, "High-performance nanoscale topological inductor," in *2017 75th Annual Device Research Conference (DRC).* IEEE, 2017, pp. 1–2.

[124] R. Plonsey and R. E. Collin, *Principles and Applications of Electromagnetic Fields.* McGraw-Hill, 1961.

[125] B. P. Lathi and R. A. Green, *Linear Systems and Signals.* Oxford University Press New York, 2005, vol. 2.

[126] D. M. Pozar, E. J. K. Knapp, and J. B. Mead, "ECE 584 microwave engineering laboratory notebook," http://www.ecs.umass.edu/ece/ece584/ECE584_lab_manual.pdf, 2004.

[127] J. Schutt-Aine, "Experimental-coaxial transmission line measurement using slotted line," http://emlab.uiuc.edu/ece451/ECE451Lab02.pdf.

[128] F. Arute, K. Arya, R. Babbush, D. Bacon, J. C. Bardin, R. Barends, R. Biswas, S. Boixo, F. G. Brandao, D. A. Buell, B. Burkett, Y. Chen, Z. Chen, B. Chiaro, R. Collins, W. Courtney, A. Dunsworth, E. Farhi, B. Foxen, A. Fowler, C. Gidney, and M. G. et al., "Quantum supremacy using a programmable superconducting processor," *Nature*, vol. 574, no. 7779, pp. 505–510, 2019.

[129] Y. Wu, W.-S. Bao, S. Cao, F. Chen, M.-C. Chen, X. Chen, T.-H. Chung, H. Deng, Y. Du, D. Fan *et al.*, "Strong quantum computational advantage using a superconducting quantum processor," *arXiv preprint arXiv:2106.14734*, 2021.

[130] J. P. Gordon, H. J. Zeiger, and C. H. Townes, "The maser—new type of microwave amplifier, frequency standard, and spectrometer," *Physical Review*, vol. 99, no. 4, p. 1264, 1955.

[131] D. R. Smith, J. B. Pendry, and M. C. Wiltshire, "Metamaterials and negative refractive index," *science*, vol. 305, no. 5685, pp. 788–792, 2004.

[132] T. J. Cui, D. R. Smith, and R. Liu, *Metamaterials.* Springer, 2010.

[133] Y. Park, C. Depeursinge, and G. Popescu, "Quantitative phase imaging in biomedicine," *Nature photonics*, vol. 12, no. 10, pp. 578–589, 2018.

[134] Wikipedia, "Automated network analyzer," https://en.wikipedia.org/wiki/Network_analyzer_(electrical).

[135] V. G. Veselago, "Electrodynamics of substances with simultaneously negative," *Usp. Fiz. Nauk*, vol. 92, p. 517, 1967.

[136] J. B. Pendry, "Negative refraction makes a perfect lens," *Physical Review Letters*, vol. 85, no. 18, p. 3966, 2000.

[137] R. E. Collin, *Field Theory of Guided Waves.* McGraw-Hill, 1960.

[138] D. M. Pozar, *Microwave Engineering.* John Wiley & Sons, 2011.

[139] B. Clark, D. F. Allen, D. L. Best, S. D. Bonner, J. Jundt, M. G. Luling, M. O. Ross *et al.*, "Electromagnetic propagation logging while drilling: Theory and experiment," *SPE Formation Evaluation*, vol. 5, no. 03, pp. 263–271, 1990.

[140] B. Vaughn and D. Peroulis, "An updated applied formulation for the Goubau transmission line," *Journal of Applied Physics*, vol. 126, no. 19, p. 194902, 2019.

[141] Q. S. Liu, S. Sun, and W. C. Chew, "A potential-based integral equation method for low-frequency electromagnetic problems," *IEEE Transactions on Antennas and Propagation*, vol. 66, no. 3, pp. 1413–1426, 2018.

[142] Z. Ying, "Opening our eyes to spectacles of the past," https://www.shine.cn/feature/art-culture/1909061557/.

[143] Wikipedia, "Snell's law," https://en.wikipedia.org/wiki/Snell's_law.

[144] ——, "Graphical processing unit," https://en.wikipedia.org/wiki/Graphics_processing_unit.

[145] G. Tyras, *Radiation and Propagation of Electromagnetic Waves.* Academic Press, 1969.

[146] L. Brekhovskikh, *Waves in Layered Media.* Academic Press, 1980, first ed. 1960.

[147] Scholarpedia, "Goos-hanchen effect," http://www.scholarpedia.org/article/Goos-Hanchen_effect.

[148] Glytsis, Elias N., "Introduction to Gaussian Beams." [Online]. Available: http://users.ntua.gr/eglytsis/OptEng/Gaussian_Beams.pdf

[149] K. Kao and G. A. Hockham, "Dielectric-fibre surface waveguides for optical frequencies," in *Proceedings of the Institution of Electrical Engineers*, vol. 113, no. 7. IET, 1966, pp. 1151–1158.

[150] Wikipedia, "Optical fiber," https://en.wikipedia.org/wiki/Optical_fiber.

[151] Atlantic Cable, "1869 indo-european cable," https://atlantic-cable.com/Cables/1869IndoEur/index.htm.

[152] Wikipedia, "Submarine communications cable," https://en.wikipedia.org/wiki/Submarine_communications_cable.

[153] D. Brewster, "On the laws which regulate the polarisation of light by reflexion from transparent bodies," *Philosophical Transactions of the Royal Society of London*, vol. 105, pp. 125–159, 1815.

[154] Wikipedia, "Brewster's angle," https://en.wikipedia.org/wiki/Brewster's_angle.

[155] H. Raether, "Surface plasmons on smooth surfaces," in *Surface Plasmons on Smooth and Rough Surfaces and on Gratings*. Springer, 1988, pp. 4–39.

[156] E. Kretschmann and H. Raether, "Radiative decay of non radiative surface plasmons excited by light," *Zeitschrift für Naturforschung A*, vol. 23, no. 12, pp. 2135–2136, 1968.

[157] Wikipedia, "Homomorphic encryption," https://en.wikipedia.org/wiki/Homomorphic_encryption.

[158] K. Aki and P. G. Richards, *Quantitative Seismology*, 2002.

[159] B. A. Auld, *Acoustic Fields and Waves in Solids*. Ripol Classic, 1973.

[160] Wikipedia, "Surface plasmon," https://en.wikipedia.org/wiki/Surface_plasmon.

[161] A. Sommerfeld, *Über die Ausbreitung der Wellen in der drahtlosen Telegraphie*. Verlag der Königlich Bayerischen Akademie der Wissenschaften, 1909.

[162] A. Baños, *Dipole radiation in the presence of a conducting half-space*. Pergamon Press, 1966.

[163] P. Yeh, A. Yariv, and C.-S. Hong, "Electromagnetic propagation in periodic stratified media. i. general theory," *JOSA*, vol. 67, no. 4, pp. 423–438, 1977.

[164] W. C. Chew, I. Aksun, J. H. Lin, C. C. Lu, G. Otto, Y. M. Wang, R. Wagner, W. H. Weedon, "Solution manual to waves and fields in inhomogeneous media," https://engineering.purdue.edu/wcchew/solution_manual_WFIM.pdf, 1993.

[165] Wikipedia, "Charles K. Kao," https://en.wikipedia.org/wiki/Charles_K._Kao.

[166] R. Knox and P. Toulios, "Integrated circuits for the millimeter through optical frequency range," in *Proc. Symp. Submillimeter Waves*, vol. 20. Brooklyn, NY, 1970, pp. 497–515.

[167] T. Tamir, "Beam and waveguide couplers," in *Integrated Optics.* Springer, 1975, pp. 83–137.

[168] G. B. Hocker and W. K. Burns, "Mode dispersion in diffused channel waveguides by the effective index method," *Applied optics*, vol. 16, no. 1, pp. 113–118, 1977.

[169] J. Buus, "The effective index method and its application to semiconductor lasers," *IEEE Journal of Quantum Electronics*, vol. 18, no. 7, pp. 1083–1089, 1982.

[170] K. Chiang, "Review of numerical and approximate methods for the modal analysis of general optical dielectric waveguides," *Optical and Quantum Electronics*, vol. 26, pp. S113–S134, 1994.

[171] M. Qiu, "Effective index method for heterostructure-slab-waveguide-based two-dimensional photonic crystals," *Applied physics letters*, vol. 81, no. 7, pp. 1163–1165, 2002.

[172] P. Dumon, W. Bogaerts, V. Wiaux, J. Wouters, S. Beckx, J. Van Campenhout, D. Taillaert, B. Luyssaert, P. Bienstman, D. Van Thourhout *et al.*, "Low-loss soi photonic wires and ring resonators fabricated with deep uv lithography," *IEEE Photonics Technology Letters*, vol. 16, no. 5, pp. 1328–1330, 2004.

[173] M. Moharam and T. K. Gaylord, "Diffraction analysis of dielectric surface-relief gratings," *Journal of the Optical Society of America*, vol. 72, no. 10, pp. 1385–1392, 1982.

[174] A. Taflove, S. C. Hagness *et al.*, "Computational electrodynamics: The finite-difference time-domain method," *Norwood, 2nd Edition, MA: Artech House, l995*, 1995.

[175] Shanhui Fan, "Shanhui Fan's Research Group," https://fangroup.stanford.edu/.

[176] H. B. Callen and T. A. Welton, "Irreversibility and generalized noise," *Physical Review*, vol. 83, no. 1, p. 34, 1951.

[177] R. Kubo, "The fluctuation-dissipation theorem," *Reports on Progress in Physics*, vol. 29, no. 1, p. 255, 1966.

[178] C. Lee, S. Lee, and S. Chuang, "Plot of modal field distribution in rectangular and circular waveguides," *IEEE Transactions on Microwave Theory and Techniques*, vol. 33, no. 3, pp. 271–274, 1985.

[179] W. C. Chew, *Waves and Fields in Inhomogeneous Media.* IEEE Press, 1996.

[180] M. Abramowitz and I. A. Stegun, "Handbook of mathematical functions: With formulas, graphs, and mathematical tables," http://people.math.sfu.ca/~cbm/aands/index.htm.

[181] C. A. Balanis and E. Holzman, "Circular waveguides," *Encyclopedia of RF and Microwave Engineering*, 2005.

[182] W. C. Chew, W. Sha, and Q. I. Dai, "Green's dyadic, spectral function, local density of states, and fluctuation dissipation theorem," *arXiv preprint arXiv:1505.01586*, 2015.

[183] J. B. Johnson, "Thermal agitation of electricity in conductors," *Physical Review*, vol. 32, no. 1, p. 97, 1928.

[184] H. Nyquist, "Thermal agitation of electric charge in conductors," *Physical Review*, vol. 32, no. 1, p. 110, 1928.

[185] Wikipedia, "Very Large Array," https://en.wikipedia.org/wiki/Very_Large_Array.

[186] M. Al-Hakkak and Y. Lo, "Circular waveguides with anisotropic walls," *Electronics Letters*, vol. 6, no. 24, pp. 786–789, 1970.

[187] J. Zhu, T. E. Roth, D.-Y. Na, and W. C. Chew, "Generalized helmholtz decomposition for modal analysis of electromagnetic problems in inhomogeneous media," *IEEE Journal on Multiscale and Multiphysics Computational Techniques*, 2023.

[188] *Mathematical Methods for Physicists. Waltham, MA, USA: Elsevier.*

[189] G. L. Matthaei, L. Young, and E. M. T. Jones, *Microwave Filters, Impedance-Matching Networks, and Coupling Structures.* Artech House, 1980.

[190] P. Silvester and P. Benedek, "Microstrip discontinuity capacitances for right-angle bends, t junctions, and crossings," *IEEE Transactions on Microwave Theory and Techniques*, vol. 21, no. 5, pp. 341–346, 1973.

[191] R. Garg and I. Bahl, "Microstrip discontinuities," *International Journal of Electronics Theoretical and Experimental*, vol. 45, no. 1, pp. 81–87, 1978.

[192] G. Matthaei, L. Young, and E. Jones, "Design of microwave filters, impedance-matching networks, and coupling structures." Stanford Research Inst Menlo Park CA, Tech. Rep., 1963.

[193] M. Bozzi, A. Georgiadis, and K. Wu, "Review of substrate-integrated waveguide circuits and antennas," *IET Microwaves, Antennas & Propagation*, vol. 5, no. 8, pp. 909–920, 2011.

[194] J. Liu, D. R. Jackson, and Y. Long, "Substrate integrated waveguide (siw) leaky-wave antenna with transverse slots," *IEEE Transactions on Antennas and Propagation*, vol. 60, no. 1, pp. 20–29, 2011.

[195] P. Smith and E. Turner, "A bistable fabry-perot resonator," *Applied Physics Letters*, vol. 30, no. 6, pp. 280–281, 1977.

[196] A. Yariv, *Optical Electronics.* Saunders College Publ., 1991.

[197] Wikipedia, "Klystron," https://en.wikipedia.org/wiki/Klystron.

[198] ——, "Magnetron," https://en.wikipedia.org/wiki/Cavity_magnetron.

[199] G. Quaranta, G. Basset, O. J. Martin, and B. Gallinet, "Recent advances in resonant waveguide gratings," *Laser and Photonics Reviews*, vol. 12, no. 9, p. 1800017, 2018.

[200] G. Righini, Y. Dumeige, P. Feron, M. Ferrari, G. Nunzi Conti, D. Ristic, and S. Soria, "Whispering gallery mode microresonators: fundamentals and applications," *La Rivista del Nuovo Cimento*, vol. 34, pp. 435–488, 2011.

[201] G. Strang, *Linear Algebra and Its Applications.* Academic Press, 1976.

[202] Y. P. Zhang and D. Liu, "Antenna-on-chip and antenna-in-package solutions to highly integrated millimeter-wave devices for wireless communications," *IEEE Transactions on Antennas and Propagation*, vol. 57, no. 10, pp. 2830–2841, 2009.

[203] Y. Zhang and J. Mao, "An overview of the development of antenna-in-package technology for highly integrated wireless devices," *Proceedings of the IEEE*, vol. 107, no. 11, pp. 2265–2280, 2019.

[204] W. C. Chew, "Vector potential electromagnetics with generalized gauge for inhomogeneous media: formulation," *Progress in Electromagnetics Research*, vol. 149, pp. 69–84, 2014.

[205] W. Chew, "Computational electromagnetics: the physics of smooth versus oscillatory fields," *Philosophical Transactions of the Royal Society of London, Series A: Mathematical, Physical and Engineering Sciences*, vol. 362, no. 1816, pp. 579–602, 2004.

[206] L. Rayleigh, "X. on the electromagnetic theory of light," *The London, Edinburgh, and Dublin Philosophical Magazine and Journal of Science*, vol. 12, no. 73, pp. 81–101, 1881.

[207] Wikipedia, "Guglielmo Marconi," https://en.wikipedia.org/wiki/Guglielmo_Marconi.

[208] R. F. Harrington, "Matrix methods for field problems," *Proceedings of the IEEE*, vol. 55, no. 2, pp. 136–149, 1967.

[209] C. A. Balanis, *Antenna Theory: Analysis and Design.* John Wiley & Sons, 2016.

[210] J.-M. Jin, *The finite element method in electromagnetics.* John Wiley & Sons, 2015.

[211] S. A. Schelkunoff and H. T. Friis, *Antennas: theory and practice.* Wiley New York, 1952, vol. 639.

[212] H. G. Schantz, "A brief history of UWB antennas," *IEEE Aerospace and Electronic Systems Magazine*, vol. 19, no. 4, pp. 22–26, 2004.

[213] E. Kudeki, "Fields and Waves," http://remote2.ece.illinois.edu/~erhan/FieldsWaves/ECE350lectures.html.

[214] Wikipedia, "Antenna Aperture," https://en.wikipedia.org/wiki/Antenna_aperture.

[215] L. Novotny and B. Hecht, *Principles of nano-optics.* Cambridge university press, 2012.

[216] L. Josefsson and P. Persson, *Conformal Array Antenna Theory and Design.* John Wiley & Sons, 2006, vol. 29.

[217] R. J. Mailloux, *Phased Array Antenna Handbook.* Artech House, 2017.

[218] J. G. Proakis, *Digital Signal Processing: Principles Algorithms and Applications.* Pearson Education India, 2001.

[219] R. L. Haupt, *Antenna arrays: a computational approach.* John Wiley & Sons, 2010.

[220] D. Rutledge and M. Muha, "Imaging antenna arrays," *IEEE Transactions on Antennas and Propagation*, vol. 30, no. 4, pp. 535–540, 1982.

[221] D. K. Cheng, "Optimization techniques for antenna arrays," *Proceedings of the IEEE*, vol. 59, no. 12, pp. 1664–1674, 1971.

[222] R. W. P. King, G. S. Smith, M. Owens, and T. Wu, "Antennas in matter: fundamentals, theory, and applications," *NASA STI/Recon Technical Report A*, vol. 81, 1981.

[223] H. Yagi and S. Uda, "Projector of the sharpest beam of electric waves," *Proceedings of the Imperial Academy*, vol. 2, no. 2, pp. 49–52, 1926.

[224] Wikipedia, "Dipole antenna," https://en.wikipedia.org/wiki/Dipole_antenna.

[225] ——, "Twin-Lead," https://en.wikipedia.org/wiki/Twin-lead.

[226] D. Dregely, R. Taubert, J. Dorfmüller, R. Vogelgesang, K. Kern, and H. Giessen, "3D optical Yagi–Uda nanoantenna array," *Nature communications*, vol. 2, no. 1, pp. 1–7, 2011.

[227] Antenna-theory.com, "Slot Antenna," http://www.antenna-theory.com/antennas/aperture/slot.php.

[228] Y. Lo, D. Solomon, and W. Richards, "Theory and experiment on microstrip antennas," *IEEE Transactions on Antennas and Propagation*, vol. 27, no. 2, pp. 137–145, 1979.

[229] A. D. Olver and P. J. Clarricoats, *Microwave Horns and Feeds.* IET, 1994, vol. 39.

[230] B. Thomas, "Design of corrugated conical horns," *IEEE Transactions on Antennas and Propagation*, vol. 26, no. 2, pp. 367–372, 1978.

[231] P. J. B. Clarricoats and A. D. Olver, *Corrugated Horns for Microwave Antennas.* IET, 1984, no. 18.

[232] Y. T. Lo and S. Lee, *Antenna Handbook, Volume III: Applications.* Springer Science & Business Media, 2012.

[233] P. Gibson, "The Vivaldi aerial," in *1979 9th European Microwave Conference.* IEEE, 1979, pp. 101–105.

[234] Wikipedia, "Cassegrain Antenna," https://en.wikipedia.org/wiki/Cassegrain_antenna.

[235] ——, "Cassegrain Reflector," https://en.wikipedia.org/wiki/Cassegrain_reflector.

[236] W. A. Imbriale, S. S. Gao, and L. Boccia, *Space Antenna Handbook.* John Wiley & Sons, 2012.

[237] J. A. Encinar, "Design of two-layer printed reflectarrays using patches of variable size," *IEEE Transactions on Antennas and Propagation*, vol. 49, no. 10, pp. 1403–1410, 2001.

[238] D.-C. Chang and M.-C. Huang, "Microstrip reflectarray antenna with offset feed," *Electronics Letters*, vol. 28, no. 16, pp. 1489–1491, 1992.

[239] G. Minatti, M. Faenzi, E. Martini, F. Caminita, P. De Vita, D. González-Ovejero, M. Sabbadini, and S. Maci, "Modulated metasurface antennas for space: synthesis, analysis and realizations," *IEEE Transactions on Antennas and Propagation*, vol. 63, no. 4, pp. 1288–1300, 2014.

[240] X. Gao, X. Han, W.-P. Cao, H. O. Li, H. F. Ma, and T. J. Cui, "Ultrawideband and high-efficiency linear polarization converter based on double v-shaped metasurface," *IEEE Transactions on Antennas and Propagation*, vol. 63, no. 8, pp. 3522–3530, 2015.

[241] D. De Schweinitz and T. L. Frey Jr., "Artificial dielectric lens antenna," Nov. 13 2001, US Patent 6,317,092.

[242] K.-L. Wong, "Planar antennas for wireless communications," *Microwave Journal*, vol. 46, no. 10, pp. 144–145, 2003.

[243] H. Nakano, M. Yamazaki, and J. Yamauchi, "Electromagnetically coupled curl antenna," *Electronics Letters*, vol. 33, no. 12, pp. 1003–1004, 1997.

[244] K. Lee, K. Luk, K.-F. Tong, S. Shum, T. Huynh, and R. Lee, "Experimental and simulation studies of the coaxially fed U-slot rectangular patch antenna," *IEE Proceedings-Microwaves, Antennas and Propagation*, vol. 144, no. 5, pp. 354–358, 1997.

[245] K. Luk, C. Mak, Y. Chow, and K. Lee, "Broadband microstrip patch antenna," *Electronics Letters*, vol. 34, no. 15, pp. 1442–1443, 1998.

[246] M. Bolic, D. Simplot-Ryl, and I. Stojmenovic, *RFID Systems: Research Trends and Challenges.* John Wiley & Sons, 2010.

[247] Z. N. Chen, X. Qing, and H. L. Chung, "A universal UHF RFID reader antenna," *IEEE transactions on microwave theory and techniques*, vol. 57, no. 5, pp. 1275–1282, 2009.

[248] D. M. Dobkin, S. M. Weigand, and N. Iyer, "Segmented magnetic antennas for near-field UHF RFID," *Microwave Journal*, vol. 50, no. 6, p. 96, 2007.

[249] Wikipedia, "Faraday cage," https://en.wikipedia.org/wiki/Faraday_cage.

[250] J. A. Stratton, *Electromagnetic Theory.* McGraw-Hill Book Company, Inc., 1941.

[251] W. Meissner and R. Ochsenfeld, "Ein neuer effekt bei eintritt der supraleitfähigkeit," *Naturwissenschaften*, vol. 21, no. 44, pp. 787–788, 1933.

[252] D. Sievenpiper, L. Zhang, R. F. Broas, N. G. Alexopolous, and E. Yablonovitch, "High-impedance electromagnetic surfaces with a forbidden frequency band," *IEEE Transactions on Microwave Theory and techniques*, vol. 47, no. 11, pp. 2059–2074, 1999.

[253] Wikipedia, "Snell's law," https://en.wikipedia.org/wiki/Snell's_law.

[254] H. Lamb, "On sommerfeld's diffraction problem; and on reflection by a parabolic mirror," *Proceedings of the London Mathematical Society*, vol. 2, no. 1, pp. 190–203, 1907.

[255] W. J. Smith, *Modern Optical Engineering.* McGraw-Hill New York, 1966, vol. 3.

[256] D. C. O'Shea, T. J. Suleski, A. D. Kathman, and D. W. Prather, *Diffractive Optics: Design, Fabrication, and Test.* Bellingham, WA, USA: SPIE Press, 2004, vol. 62.

[257] J. B. Keller and H. B. Keller, "Determination of reflected and transmitted fields by geometrical optics," *JOSA*, vol. 40, no. 1, pp. 48–52, 1950.

[258] G. A. Deschamps, "Ray techniques in electromagnetics," *Proceedings of the IEEE*, vol. 60, no. 9, pp. 1022–1035, 1972.

[259] R. G. Kouyoumjian and P. H. Pathak, "A uniform geometrical theory of diffraction for an edge in a perfectly conducting surface," *Proceedings of the IEEE*, vol. 62, no. 11, pp. 1448–1461, 1974.

[260] R. Kouyoumjian, "The geometrical theory of diffraction and its application," in *Numerical and Asymptotic Techniques in Electromagnetics.* Springer, 1975, pp. 165–215.

[261] S.-W. Lee and G. Deschamps, "A uniform asymptotic theory of electromagnetic diffraction by a curved wedge," *IEEE Transactions on Antennas and Propagation*, vol. 24, no. 1, pp. 25–34, 1976.

[262] Wikipedia, "Fermat's principle," https://en.wikipedia.org/wiki/Fermat's_principle.

[263] N. Yu, P. Genevet, M. A. Kats, F. Aieta, J.-P. Tetienne, F. Capasso, and Z. Gaburro, "Light propagation with phase discontinuities: generalized laws of reflection and refraction," *Science*, vol. 334, no. 6054, pp. 333–337, 2011.

[264] X. Ni, N. K. Emani, A. V. Kildishev, A. Boltasseva, and V. M. Shalaev, "Broadband light bending with plasmonic nanoantennas," *Science*, vol. 335, no. 6067, pp. 427–427, 2012.

[265] A. Sommerfeld, *Partial Differential Equations in Physics.* Academic Press, 1949, vol. 1.

[266] R. Haberman, *Elementary Applied Partial Differential Equations.* Englewood Cliffs, NJ, USA: Prentice Hall, 1983, vol. 987.

[267] F. Kärtner, "6.974 fundamentals of photonics: Quantum electronics. spring 2006. massachusetts institute of technology: Mit opencourseware."

[268] G. A. Deschamps, "Gaussian beam as a bundle of complex rays," *Electronics Letters*, vol. 7, no. 23, pp. 684–685, 1971.

[269] L. G. Gouy, *Sur une propriété nouvelle des ondes lumineuses.* Gauthier-Villars, 1890.

[270] J. Enderlein and F. Pampaloni, "Unified operator approach for deriving Hermite–Gaussian and Laguerre–Gaussian laser modes," *JOSA A*, vol. 21, no. 8, pp. 1553–1558, 2004.

[271] D. L. Andrews, *Structured Light and Its Applications: An Introduction to Phase-Structured Beams and Nanoscale Optical Forces.* Academic Press, 2011.

[272] J. W. Strutt, "Xv. on the light from the sky, its polarization and colour," *The London, Edinburgh, and Dublin Philosophical Magazine and Journal of Science*, vol. 41, no. 271, pp. 107–120, 1871.

[273] S. Sun, Y. G. Liu, W. C. Chew, and Z. Ma, "Calderón multiplicative preconditioned efie with perturbation method," *IEEE Transactions on Antennas and Propagation*, vol. 61, no. 1, pp. 247–255, 2012.

[274] G. Mie, "Beiträge zur optik trüber medien, speziell kolloidaler metallösungen," *Annalen der physik*, vol. 330, no. 3, pp. 377–445, 1908.

[275] Wikipedia, "Mie scattering," https://en.wikipedia.org/wiki/Mie_scattering.

[276] K. Sarabandi, *Foundations of Applied Electromagnetics.* Michigan Publishing, 2023.

[277] L. B. Felsen and N. Marcuvitz, *Radiation and Scattering of Waves.* John Wiley & Sons, 1994, also 1973, vol. 31.

[278] P. P. Ewald, "Die berechnung optischer und elektrostatischer gitterpotentiale," *Annalen der Physik*, vol. 369, no. 3, pp. 253–287, 1921.

[279] E. Whitaker and G. Watson, *A Course of Modern Analysis.* Cambridge Mathematical Library, 1927.

[280] J. Kong, "Electromagnetic fields due to dipole antennas over stratified anisotropic media," *Geophysics*, vol. 37, no. 6, pp. 985–996, 1972.

[281] K. A. Michalski and J. R. Mosig, "Multilayered media green's functions in integral equation formulations," *IEEE Transactions on Antennas and Propagation*, vol. 45, no. 3, pp. 508–519, 1997.

[282] W. Chew, "A quick way to approximate a Sommerfeld-Weyl-type integral (antenna far-field radiation)," *IEEE Transactions on Antennas and Propagation*, vol. 36, no. 11, pp. 1654–1657, 1988.

[283] J. E. Greivenkamp, *Field guide to geometrical optics.* SPIE press Bellingham, Washington, 2004, vol. 1.

[284] Wikipedia, "FLOPS," https://en.wikipedia.org/wiki/FLOPS.

[285] ——, "Computer," https://en.wikipedia.org/wiki/computer.

[286] H. Gan, "Numerical Green's function in surface integral equation method and hydrodynamic model for solar cell analysis," Ph.D. dissertation, University of Illinois, 2019.

[287] W. C. H. McLean, *Strongly Elliptic Systems and Boundary Integral Equations.* Cambridge University Press, 2000.

[288] G. C. Hsiao and W. L. Wendland, *Boundary Integral Equations.* Springer, 2008.

[289] W. C. Chew, M. S. Tong, and B. Hu, "Integral equation methods for electromagnetic and elastic waves," *Synthesis Lectures on Computational Electromagnetics*, vol. 3, no. 1, pp. 1–241, 2008.

[290] R. Wait, A. R. Mitchell, and A. R. Mitchell, *Finite Element Analysis and Applications.* John Wiley & Sons, 1985.

[291] O. Axelsson and V. A. Barker, *Finite Element Solution of Boundary Value Problems: Theory and Computation.* SIAM, 2001.

[292] P. K. Banerjee and R. Butterfield, *Boundary Element Methods in Engineering Science.* McGraw-Hill London, 1981, vol. 17.

[293] O. C. Zienkiewicz, R. L. Taylor, P. Nithiarasu, and J. Zhu, *The finite Element Method.* McGraw-Hill London, 1977, vol. 3.

[294] J.-F. Lee, R. Lee, and A. Cangellaris, "Time-domain finite-element methods," *IEEE Transactions on Antennas and Propagation*, vol. 45, no. 3, pp. 430–442, 1997.

[295] J. L. Volakis, A. Chatterjee, and L. C. Kempel, *Finite Element Method Electromagnetics: Antennas, Microwave Circuits, and Scattering Applications.* John Wiley & Sons, 1998, vol. 6.

[296] J.-C. Nédélec, "Mixed finite elements in r 3," *Numerische Mathematik*, vol. 35, pp. 315–341, 1980.

[297] S. Rao, D. Wilton, and A. Glisson, "Electromagnetic scattering by surfaces of arbitrary shape," *IEEE Transactions on Antennas and Propagation*, vol. 30, no. 3, pp. 409–418, 1982.

[298] M. Köppen, "The curse of dimensionality," in *5th online world conference on soft computing in industrial applications (WSC5)*, vol. 1, 2000, pp. 4–8.

[299] W. Chai and D. Jiao, "An h-matrix-based integral-equation solver of reduced complexity and controlled accuracy for solving electrodynamic problems," *IEEE transactions on antennas and propagation*, vol. 57, no. 10, pp. 3147–3159, 2009.

[300] Cramer and Gabriel, *Introduction a l'analyse des lignes courbes algebriques par Gabriel Cramer...* chez les freres Cramer & Cl. Philibert, 1750.

[301] J. A. Schouten, *Tensor Analysis for Physicists.* Courier Corporation, 1989.

[302] A. C. Polycarpou, "Introduction to the finite element method in electromagnetics," *Synthesis Lectures on Computational Electromagnetics*, vol. 1, no. 1, pp. 1–126, 2005.

[303] J. P. A. Bastos and N. Sadowski, *Electromagnetic Modeling by Finite Element Methods.* CRC Press, 2003.

[304] Ö. Özgün and M. Kuzuoğlu, *MATLAB-based Finite Element Programming in Electromagnetic Modeling.* CRC Press, 2018.

[305] R. Coifman, V. Rokhlin, and S. Wandzura, "The fast multipole method for the wave equation: A pedestrian prescription," *IEEE Antennas and Propagation Magazine*, vol. 35, no. 3, pp. 7–12, 1993.

[306] J. Song, C.-C. Lu, and W. C. Chew, "Multilevel fast multipole algorithm for electromagnetic scattering by large complex objects," *IEEE Transactions on Antennas and Propagation*, vol. 45, no. 10, pp. 1488–1493, 1997.

[307] W. H. Press, S. A. Teukolsky, W. T. Vetterling, and B. P. Flannery, *Numerical Recipes, Third Edition: The Art of Scientific Computing.* Cambridge University Press, 2007.

[308] K. Yee, "Numerical solution of initial boundary value problems involving maxwell's equations in isotropic media," *IEEE Transactions on Antennas and Propagation*, vol. 14, no. 3, pp. 302–307, 1966.

[309] A. Taflove, "Review of the formulation and applications of the finite-difference time-domain method for numerical modeling of electromagnetic wave interactions with arbitrary structures," *Wave Motion*, vol. 10, no. 6, pp. 547–582, 1988.

[310] A. Taflove, S. C. Hagness, and M. Piket-May, "Computational electromagnetics: the finite-difference time-domain method," *The Electrical Engineering Handbook*, vol. 3, no. 629-670, p. 15, 2005.

[311] A. F. Oskooi, D. Roundy, M. Ibanescu, P. Bermel, J. D. Joannopoulos, and S. G. Johnson, "Meep: A flexible free-software package for electromagnetic simulations by the fdtd method," *Computer Physics Communications*, vol. 181, no. 3, pp. 687–702, 2010.

[312] W. Yu, R. Mittra, T. Su, Y. Liu, and X. Yang, *Parallel Finite-Difference Time-Domain Method.* Artech House Norwood, 2006.

[313] D. Potter, "Computational physics," 1973.

[314] W. F. Ames, *Numerical Methods for Partial Differential Equations.* Academic press, 2014, also 1977.

[315] K. W. Morton, *Revival: Numerical Solution Of Convection-Diffusion Problems (1996).* CRC Press, 2019.

[316] F. B. Hildebrand, *Introduction to Numerical Analysis.* Courier Corporation, 1987.

[317] W. C. Chew, "Electromagnetic theory on a lattice," *Journal of Applied Physics*, vol. 75, no. 10, pp. 4843–4850, 1994.

[318] J. v. Neumann, *Mathematische Grundlagen der Quantenmechanik, Berlın.* Springer, New York, Dover Publications, 1943.

[319] R. Courant, K. Friedrichs, and H. Lewy, "Über die partiellen differenzengleichungen der mathematischen physik," *Mathematische Annalen*, vol. 100, no. 1, pp. 32–74, 1928.

[320] M. Okoniewski, E. Okoniewska, and M. A. Stuchly, "Three-dimensional subgridding algorithm for fdtd," *IEEE Transactions on antennas and propagation*, vol. 45, no. 3, pp. 422–429, 1997.

[321] W. H. Weedon and C. M. Rappaport, "A general method for fdtd modeling of wave propagation in arbitrary frequency-dispersive media," *IEEE Transactions on Antennas and Propagation*, vol. 45, no. 3, pp. 401–410, 1997.

[322] T. Weiland, "A discretization model for the solution of maxwell's equations for six-component fields," *Archiv Elektronik und Uebertragungstechnik*, vol. 31, pp. 116–120, 1977.

[323] M. Clemens and T. Weiland, "Discrete electromagnetism with the finite integration technique," *Progress in Electromagnetics Research*, vol. 32, pp. 65–87, 2001.

[324] M. Desbrun, A. N. Hirani, M. Leok, and J. E. Marsden, "Discrete exterior calculus," *arXiv preprint math/0508341*, 2005.

[325] R. M. Joseph and A. Taflove, "Fdtd maxwell's equations models for nonlinear electrodynamics and optics," *IEEE Transactions on Antennas and Propagation*, vol. 45, no. 3, pp. 364–374, 1997.

[326] T. Weiland, "On the numerical solution of maxwell's equations and applications in the field of accelerator physics," *Part. Accel.*, vol. 15, no. DESY-84-006, pp. 245–292, 1984.

[327] W. C. Chew, J. Jin, and E. Michielssen, "Complex coordinate system as a generalized absorbing boundary condition," in *IEEE Antennas and Propagation Society International Symposium 1997. Digest*, vol. 3. IEEE, 1997, pp. 2060–2063.

[328] J.-P. Berenger, "A perfectly matched layer for the absorption of electromagnetic waves," *Journal of Computational Physics*, vol. 114, no. 2, pp. 185–200, 1994.

[329] W. C. Chew and W. H. Weedon, "A 3D perfectly matched medium from modified maxwell's equations with stretched coordinates," *Microwave and optical technology letters*, vol. 7, no. 13, pp. 599–604, 1994.

[330] A. Zee, *Quantum Field Theory in a Nutshell.* Princeton University Press, 2010, vol. 7.

[331] A. Aspect, P. Grangier, and G. Roger, "Experimental realization of Einstein-Podolsky-Rosen-Bohm Gedankenexperiment: a new violation of Bell's inequalities," *Physical Review Letters*, vol. 49, no. 2, p. 91, 1982.

[332] J. F. Clauser and A. Shimony, "Bell's theorem. experimental tests and implications," *Reports on Progress in Physics*, vol. 41, no. 12, p. 1881, 1978.

[333] D. M. Greenberger, M. A. Horne, A. Shimony, and A. Zeilinger, "Bell's theorem without inequalities," *American Journal of Physics*, vol. 58, no. 12, pp. 1131–1143, 1990.

[334] Wikipedia, "Double-slit experiment," https://en.wikipedia.org/wiki/Double-slit_experiment.

[335] ——, "John Dalton," https://en.wikipedia.org/wiki/John_Dalton.

[336] ——, "Max Planck," https://en.wikipedia.org/wiki/Max_Planck.

[337] ——, "Photoelectric effect," https://en.wikipedia.org/wiki/Photoelectric_effect.

[338] ——, "Newton's laws of motion," https://en.wikipedia.org/wiki/Newton's_laws_of_motion.

[339] H. Goldstein, C. Poole, and J. Safko, *Classical Mechanics.* College Park, MD, USA: American Association of Physics Teachers (AAPT), 2002.

[340] Wikipedia, "William Rowan Hamilton," https://en.wikipedia.org/wiki/William_Rowan_Hamilton.

[341] W. C. Chew, A. Y. Liu, C. Salazar-Lazaro, D. Na, and W. E. I. Sha, "Hamilton equation, commutator, and energy conservation," *Quantum Report*, pp. 295–303, Dec. 2019.

[342] W. C. Chew, A. Y. Liu, C. Salazar-Lazaro, and W. E. I. Sha, "Quantum electromagnetics: A new look—Part I and Part II," *J. Multiscale and Multiphys. Comput. Techn.*, vol. 1, pp. 73–97, 2016.

[343] L. E. Ballentine, *Quantum mechanics: a modern development.* World Scientific Publishing Company, 2014.

[344] C. J. Ryu, E. Kudeki, D.-Y. Na, T. E. Roth, and W. C. Chew, "Fourier transform, dirac commutator, energy conservation, and correspondence principle for electrical engineers," *IEEE Journal on Multiscale and Multiphysics Computational Techniques*, vol. 7, pp. 69–83, 2022.

[345] M. Kira and S. W. Koch, *Semiconductor Quantum Optics.* Cambridge University Press, 2011.

[346] Wikipedia, "Gaussian beam," https://en.wikipedia.org/wiki/Gaussian_beam.

[347] W. C. Chew, D.-Y. Na, P. Bermel, T. E. Roth, C. J. Ryu, and E. Kudeki, "Quantum maxwell's equations made simple: Employing scalar and vector potential formulation," *IEEE Antennas and Propagation Magazine*, vol. 63, no. 1, pp. 14–26, 2020.

[348] F. S. S. Rosa, D. A. R. Dalvit, and P. W. Milonni, "Electromagnetic energy, absorption, and casimir forces. ii. inhomogeneous dielectric media," *Phys. Rev. A*, vol. 84, p. 053813, Nov 2011.

[349] T. G. Philbin, "Casimir effect from macroscopic quantum electrodynamics," *New J. Phys.*, vol. 13, no. 6, p. 063026, jun 2011.

[350] T. Xia, P. R. Atkins, W. Sha, and W. C. Chew, "Casimir force: Vacuum fluctuation, zero-point energy, and computational electromagnetics," *IEEE Antennas and Propagation Magazine*, vol. 63, no. 5, pp. 14–24, 2021.

[351] W. H. Louisell, *Quantum Statistical Properties of Radiation.* Wiley New York, 1973, vol. 7.

[352] K. Gottfried and T.-M. Yan, *Quantum Mechanics: Fundamentals.* Boca Raton, FL, USA: CRC Press, 2018.

[353] W. C. Chew, A. Y. Liu, C. Salazar-Lazaro, and W. E. I. Sha, "Quantum electromagnetics: A new look—part i," *IEEE Journal on Multiscale and Multiphysics Computational Techniques*, vol. 1, pp. 73–84, 2016.

[354] J. Zhu, C. J. Ryu, D.-Y. Na, and W. C. Chew, "The radiation gauge: When is it valid?" *arXiv preprint arXiv:2407.08764*, 2024.

[355] J. S. Bell, "On the Einstein Podolsky Rosen paradox," *Physics Physique Fizika*, vol. 1, no. 3, p. 195, 1964.

[356] A. Aspect, J. Dalibard, and G. Roger, "Experimental test of Bell's inequalities using time-varying analyzers," *Physical Review Letters*, vol. 49, no. 25, p. 1804, 1982.

[357] T. E. Roth, "Fundamentals of quantum technology," 2025. [Online]. Available: https://nanohub.org/resources/38552

[358] Wikipedia, "Quantum Electrodynamics," https://en.wikipedia.org/wiki/Quantum_electrodynamics.

[359] M. O. Scully and M. S. Zubairy, "Quantum optics," 1999.

[360] L. Mandel and E. Wolf, *Optical Coherence and Quantum Optics.* Cambridge, UK: Cambridge University Press, 1995.

[361] Wikipedia, "Roy J. Glauber," https://en.wikipedia.org/wiki/Roy_J._Glauber.

[362] ——, "E.C. George Sudarshan," https://en.wikipedia.org/wiki/E._C._George_Sudarshan.

[363] J. Bardeen, L. N. Cooper, and J. R. Schrieffer, "Theory of superconductivity," *Physical Review*, vol. 108, no. 5, p. 1175, 1957.

[364] Techopedia, "Optical fiber loss," https://www.fiberoptics4sale.com/blogs/archive-posts/95049798-calculating-fiber-loss-and-distance-estimates.

[365] H.-S. Zhong, Y.-H. Deng, J. Qin, H. Wang, M.-C. Chen, L.-C. Peng, Y.-H. Luo, D. Wu, S.-Q. Gong, H. Su, Y. Hu, P. Hu, X.-Y. Yang, W.-J. Zhang, H. Li, Y. Li, X. Jiang, L. Gan, G. Yang, L. You, Z. Wang, L. Li, N.-L. Liu, J. J. Renema, C.-Y. Lu, and J.-W. Pan, "Phase-programmable gaussian boson sampling using stimulated squeezed light," *Phys. Rev. Lett.*, vol. 127, p. 180502, Oct 2021. [Online]. Available: https://link.aps.org/doi/10.1103/PhysRevLett.127.180502

[366] Scientific American, "Chinese quantum teleportation," https://www.scientificamerican.com/article/china-reaches-new-milestone-in-space-based-quantum-communications/.

[367] Tiernan Ray, "Why Googles quantum breakthrough is truly remarkable and what happens next?" [Online]. Available: https://www.zdnet.com/article/why-googles-quantum-breakthrough-is-truly-remarkable-and-what-happens-next/

[368] Gong Zhe, "Quantum supremacy race heats up China's new test blasts past Google." [Online]. Available: https://news.cgtn.com/news/2024-12-18/Quantum-supremacy-race-heats-up-China-s-new-test-blasts-past-Google-1zqZGatID0Q/index.html

[369] C. Gerry and P. Knight, *Introductory Quantum Optics.* Cambridge, UK: Cambridge University Press, 2004.

[370] M. Fox, *Quantum Optics: an introduction.* OUP Oxford, 2006, vol. 15.

[371] H. Everett and J. Wheeler, "DeWitt, Bryce; Graham, R. Neill (eds.). the many-worlds interpretation of quantum mechanics," *Princeton Series in Physics. Princeton, NJ: Princeton University Press. pv ISBN 0-691-08131-X*, 1973.

[372] W. C. Chew, A. Y. Liu, C. Salazar-Lazaro, and W. E. I. Sha, "Quantum electromagnetics: A new look—part ii," *IEEE Journal on Multiscale and Multiphysics Computational Techniques*, vol. 1, pp. 85–97, 2016.

[373] E. Merzbacher, *Quantum Mechanics.* John Wiley & Sons, 1998.

[374] D.-Y. Na, J. Zhu, W. C. Chew, and F. L. Teixeira, "Quantum information preserving computational electromagnetics," *Physical Review A*, vol. 102, no. 1, p. 013711, 2020.

[375] D.-Y. Na, J. Zhu, and W. C. Chew, "Diagonalization of the hamiltonian for finite-sized dispersive media: Canonical quantization with numerical mode decomposition," *Physical Review A*, vol. 103, no. 6, p. 063707, 2021.

[376] M. Stone and P. Goldbart, *Mathematics for physics: a guided tour for graduate students.* Cambridge University Press, 2009.

[377] A. Liu, "A systematic computational study of cavity and waveguide quantum electrodynamics," Ph.D. dissertation, University of Illinois at Urbana-Champaign, 2019.

[378] Wikipedia, "Normal Distribution," https://en.wikipedia.org/wiki/Normal_distribution.

Index

About the Author

Weng Cho Chew received all his degrees from MIT. His interests include wave physics, computational electromagnetics, and quantum electromagnetics. He joined Schlumberger-Doll Research Center in 1981 and then the University of Illinois in 1985. There, he was director of the Electromagnetics Lab, founder professor, Y. T. Lo chair professor, and Fisher distinguished professor. From 2007 to 2011, he served as the dean of engineering at Hong Kong University. In 2017, he joined Purdue University as a distinguished professor. He has coauthored three books, many lecture notes, over 450 journal papers, and more than 600 conference papers. He received two IEEE Technical Field Awards in Graduate Teaching and Electromagnetics. Moreover, he received the IEEE AP-S CT Tai Distinguished Educator Award, ACES Computational Electromagnetics Award, and SP-WLA Pioneer Award. He is a member of the National Academy of Engineering, and in 2018, he was president of the IEEE AP-S.

www.ingramcontent.com/pod-product-compliance
Lightning Source LLC
Chambersburg PA
CBHW081211220326
41598CB00037B/6751
9781626710962